# Advances in Intelligent Systems and Computing

Volume 908

**Series editor**

Janusz Kacprzyk, Systems Research Institute, Polish Academy of Sciences, Warsaw, Poland
e-mail: kacprzyk@ibspan.waw.pl

The series "Advances in Intelligent Systems and Computing" contains publications on theory, applications, and design methods of Intelligent Systems and Intelligent Computing. Virtually all disciplines such as engineering, natural sciences, computer and information science, ICT, economics, business, e-commerce, environment, healthcare, life science are covered. The list of topics spans all the areas of modern intelligent systems and computing such as: computational intelligence, soft computing including neural networks, fuzzy systems, evolutionary computing and the fusion of these paradigms, social intelligence, ambient intelligence, computational neuroscience, artificial life, virtual worlds and society, cognitive science and systems, Perception and Vision, DNA and immune based systems, self-organizing and adaptive systems, e-Learning and teaching, human-centered and human-centric computing, recommender systems, intelligent control, robotics and mechatronics including human-machine teaming, knowledge-based paradigms, learning paradigms, machine ethics, intelligent data analysis, knowledge management, intelligent agents, intelligent decision making and support, intelligent network security, trust management, interactive entertainment, Web intelligence and multimedia.

The publications within "Advances in Intelligent Systems and Computing" are primarily proceedings of important conferences, symposia and congresses. They cover significant recent developments in the field, both of a foundational and applicable character. An important characteristic feature of the series is the short publication time and world-wide distribution. This permits a rapid and broad dissemination of research results.

More information about this series at http://www.springer.com/series/11156

Svetlana Ashmarina · Anabela Mesquita ·
Marek Vochozka
Editors

# Digital Transformation of the Economy: Challenges, Trends and New Opportunities

Springer

*Editors*
Svetlana Ashmarina
Department of Applied Management
Samara State University of Economics
Samara, Russia

Anabela Mesquita
School of Accounting and Administration
Polytechnic Institute of Porto
São Mamede de Infesta, Portugal

Marek Vochozka
Institute of Technology and Business
České Budějovice, Czech Republic

ISSN 2194-5357 ISSN 2194-5365 (electronic)
Advances in Intelligent Systems and Computing
ISBN 978-3-030-11366-7 ISBN 978-3-030-11367-4 (eBook)
https://doi.org/10.1007/978-3-030-11367-4

Library of Congress Control Number: 2018966835

This Springer imprint is published by the registered company Springer Nature Switzerland AG
The registered company address is: Gewerbestrasse 11, 6330 Cham, Switzerland

# Contents

**Digital Economy in the Socio-Economic Development  of Enterprises**

# Russia and the World in the Digital Epoch

# Digital Economy Institutional Traps: A Regional-Sectoral Approach

T. E. Stepanova and R. K. Polyakov[✉]

Kaliningrad State Technical University, Kaliningrad, Russia
tatyana.stepanova@klgtu.ru, polyakov_rk@mail.ru

**Abstract.** The article presents the results of the authors' research the subject of which are issues of the region fisheries industry development in a digital economy. The paper presents the main problems of entrepreneurship in the region, due to its exclave position. Those problems were the main deterrent on its way to a competitive position among the Baltic Sea countries.

The article is devoted to the analysis of the institutional space of the Kaliningrad region in the fisheries industry from the standpoint of the institutional traps theory. The study allowed the authors to identify existing institutional traps in the industry. There are institutional traps associated with the development subsidy model, with the format of the region as a special economic zone (SEZ), with the lack of small enterprises.

The authors of the article studied the issues related to the growing pressure of the digital economy and determined the regulating directions of the fisheries industry institutional traps in the Kaliningrad region.

The study offers the author's view on a significant impact of the institutional traps in the fisheries sector of the region in the context of the global transformation of the country and the increasing pressure of universal digitalization.

**Keywords:** Digital economy · Digitalization · Fisheries industry · Innovation · Institutional traps · Technology

## 1 Introduction

The theory of institutional traps is one of the most priority areas of scientific research, which allows developing practical recommendations for the transformation of institutional space at various levels of management, including digital economy. Let us turn to the study of the institutional traps of a specific region - the Kaliningrad region, which has more than once played the role of a pilot region for Russia, where advanced models of economic development were introduced and tested such as the "Yantar" SEZ, and free customs zone. This is the first in Russia special economic zone which brought various innovations (Kaliningrad citizens were the first who benefited) and then "exported" them to other regions of the country. However, in the last five to seven years, Kaliningrad's status of the innovator has been lost in a large extent, which immediately affected its socio-economic situation. Almost all exclave territories have a limited domestic consumer market, which is the main problem of the region entrepreneurship development. The problem can be solved by expanding markets for

S. Ashmarina et al. (Eds.): *Digital Transformation of the Economy: Challenges, Trends and New Opportunities*, pp. 3–19, 2019.
https://doi.org/10.1007/978-3-030-11367-4_1

goods either for export or to other regions of the country in transit through neighboring states. As a result, there is an increase in the cost of goods due to an increase in transportation costs in terms of customs clearance and cargo escort, transit fees. In addition, we can see a decline in the competitiveness of goods and products of the exclave, a slowdown in the turnover of the working capital invested in them, an increase in the payback period of the invested capital.

Consequently, the most acute problems of entrepreneurship in the region, due to its exclave position, become a brake on the development of the regional economy. Among the factors that negatively affect the economic development of the Kaliningrad region, it is necessary to highlight the reduction in industrial production and the number of enterprises, especially in the small business sector, the low level of solvency of the population and enterprises, the underdevelopment of the general infrastructure of entrepreneurship as a whole, and transport and information infrastructure.

Based on the most of the indicators of innovation impact on the economy of the Kaliningrad region we can say that the exclave is seriously lagging behind the leading innovative economies, including those of the Baltic region. The region has not become a leader in the European market, because its high-tech products are characterized by insufficient competitiveness. In addition, the risks of falling into a "loop of stagnation" increase with a multiple lag from European countries in most of the considered indicators.

## 2   Materials and Methods

Liberalization of foreign economic activity, as follows from the experience of the exclave region, sharply stimulates the development of entrepreneurship, increasing its investment attractiveness due to the possibility of profit transfer. Attracting and securing capital in the regional economy can be facilitated by a multi-level business stimulation system, where first-level incentives encompass enterprises whose activities include export-import operations on priority goods for the region, second-level firms that carry out only part of their trade operations (warehousing, packing, partial processing of goods, transshipment of goods), the third level - local producers.

The adoption of the Federal Law "On the Special Economic Zone in the Kaliningrad Region" [1] significantly worsened the working conditions of newly created small enterprises. The law did not solve the existing problems with import substitution, but reinforced the "shadow" sector of the regional economy: it prolonged the special economic zone effect of customs privileges for ten years for the old participants of the market. The main factor in increasing socio-economic stability could be the cultivation of efficient small business in local conditions. However, government reckoned on large investors, hoping to strengthen the influence of domestic capital as an alternative to Western capital. Such an approach could be justified only if a local industrial park of a certain specialization was formed in Kaliningrad. In practice, in the conditions of the Kaliningrad region, the Law created a conflict situation with unequal rules.

In addition to the above factors, other factors also have a significant deterrent effect in the Kaliningrad region [2]:

- the presence of a high level of criminalization of the business sphere, the "shadow" sector of the economy, whose indicators are significantly higher than the permissible parameters;
- significant tax burden on small businesses, insufficient differentiation of taxation of various categories of entrepreneurs are one of the main factors for the growth of the "shadow" sector of the economy;
- unresolved problems associated in practice with the elimination of unjustified administrative barriers, and, above all, by the regulatory authorities;
- high values of investment risk indicators, the current structure of investment, not stimulating the development of production;
- increasing manifestations of unfair competition in relation to small businesses in commodity markets (imposing contracts on unfavorable terms, price dumping, creating artificial barriers to entering the market, etc.);
- the predominance of an inadequate level of professional training in the business environment, the existence of discrimination in labor relations between employers and employees in this segment of economic activity.

It should be noted that the problems of small business are determined by state policy, as well as the confrontation of small business with administrative barriers.

The latter are not just an obstacle to the development of especially small business and entrepreneurship, but a state problem: when confronted with administrative barriers, small enterprises go into a "shadow" economy in an effort to reduce transaction costs (overhead) and bureaucratic risks [3].

The reasons are objective: the interception of initiatives by other Russian SEZs, the abolition of customs privileges, the ruble fall, etc. Nevertheless it does not make the situation easier. In fact, the subsidy model of development was imposed on the region - the most conservative format, which can implemented only in a Russian hinterland. We believe that by all immanent signs this is a real institutional trap (subsidy model of development). Given this, this region is not capable of becoming a guide for the European exclave of Russia, which is one-step away from industrial centers of world importance.

The subsidy model was originally conceived as a transitional one. It should be replaced by an investment model when the region develops and earns itself, using the benefits of an exclave position and compensating for its shortcomings. This is not only about the new Law on Territory - in any case, it will be a compromise between the needs of the region and the possibilities of Moscow. We can do a lot for the development of the region itself, relying on our own resources.

The advanced model of the Kaliningrad region in the era of the digital economy - what should it be at the new historical stage? What specialization will ensure the sustainability of our economy and what is the role of small and medium businesses in it? Can it only become one of the pillars of the development of the domestic market, if the SME (small and medium enterprise) sector is freed from administrative pressure? Are there any world analogues of the Kaliningrad exclave - and how have the local modernization practices provided these territories with an economic breakthrough? Is Kaliningrad capable of becoming a "European window of Russia"? All these questions are remained open.

## 3   Results

Today, some regions are growing faster than others are, and at times, this is contrary to economic theory. In the current digital whirlwind, the economic climate for the regions is becoming unpredictable, and the tools that policymakers choose are not always the ways to stimulate new and sustainable growth in the region. Therefore, these features are central to this work. The authors of the article investigated the regulation of the institutional traps of the fishing industry of the Kaliningrad region in the conditions of digitalization of the economic space.

The study by Antonelli [4], the OECD "Regions Matter" [5] suggest that regions grow in very different ways, and the simple concentration of resources in the region does not lead to long-term growth. The authors of the latter report [5] state that there is no unique model of sustainable growth. Concentration of economic activity does not necessarily lead to higher levels of productivity or higher growth rates. Growth opportunities exist in all types of regions throughout the territory and depend on how well the region is able to mobilize its assets in order to take full advantage of its potential growth.

The speed with which technological changes occur is constantly increasing [6]. This is confirmed by a number of studies in recent years. Therefore, in their work, Wei and Liu [7] analyze technological changes and evaluate the role they play in economic growth. In another paper, the author Perera-Tallo [8], using the developed growth model, answers the question of how technological changes increase the share of revenues of reproducible factors at the expense of non-reproducible.

Technological congruence understanding is an important factor in economic growth, both at the firm level and at the aggregate level [4]. Using this knowledge, one can increase the competitiveness of regions and significantly stimulate technological progress [9].

The accumulation of technological knowledge creates an increasing return [10], while a region with such a technological knowledge base has more chances to activate learning processes that will increase the ability to create new technological knowledge compared to the regions without it [11]. Consequently, the stock of knowledge accumulated in the region increases its future inventiveness/innovative ability [9].

The rapid development of breakthrough technologies poses new challenges to global leaders in digitalization, those who are already involved in the digital transformation processes, and those who are only at the beginning. The risks of "subversive" impact, naturally associated with new technologies, make decision-making processes at the state level more complex and multifaceted. The government has to ensure an increasingly complex balance between protecting the basic interests of the country and its subjects, on the one hand, and using new technologies to ensure the country's competitiveness and accelerate economic growth, on the other.

Despite the fact that Russia has developed a clear vision and strategy for digital transformation and set ambitious goals, much work needs to be done to prepare detailed action plans and roadmaps to implement this strategy. The results of the assessment of Russia's readiness for the digital economy are presented in Fig. 1.

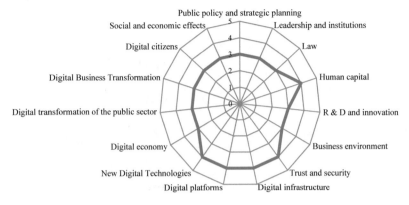

**Fig. 1.** Results of an assessment of Russia's readiness for the digital economy [12] (Source: Analysis of the current level of development of the digital economy in the Russian Federation. World Bank, Institute of the Information Society, October 2017)

The more detailed Russia's assessments in the framework of digital development are presented in the World Bank's 2016 "World Development Report: digital dividends review" [13]. This report examined the socioeconomic effects of digital transformation, that is, digital dividends, and the conditions for obtaining them.

However, it is already clear what is necessary for the country to make significant efforts to optimize the management of this process, in order to identify the best tools for monitoring and evaluating ensuring the effectiveness of Russia's development in the digital space.

It is also extremely important to develop a mechanism for the active participation of all major stakeholders in the transition to a digital economy and constant coordination at the federal, regional and municipal levels.

In recent years, Russia has invested in the development of broadband Internet access and has built a fairly powerful and developed digital infrastructure, including a competitive telecommunications market. This infrastructure has allowed the creation of large domestic and localized digital platforms.

An assessment of Russia's readiness for the digital economy in terms of the digital infrastructure is on the Fig. 2.

In recent years, Russia has achieved some success in the development of digital government, with technological changes affecting the relationship between the customer and the seller. There are tools online sales that allow you to meet the basic needs of customers: speed, convenience and simplicity.

Next, we will conduct an economic assessment of the development of the Kaliningrad region on the subject of fisheries industry institutional traps of the region in the digital economy.

A general description of enterprises and organizations of the Russian Federation is presented below [14] (Fig. 3).

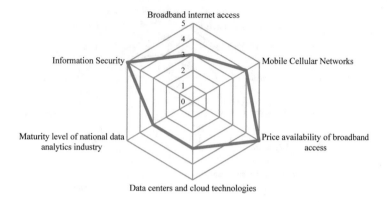

**Fig. 2.** Assessment of Russia's readiness for the digital economy: Digital infrastructure [12] (Source: Analysis of the current level of development of the digital economy in the Russian Federation. World Bank, Institute of the Information Society, October 2017)

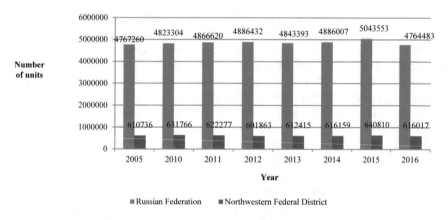

**Fig. 3.** The number of enterprises and organizations in Russia (by the end of the year) (Source: compiled by the authors according to Rosstat) [14]

During the analyzed period, the number of enterprises and organizations in the country decreased by 2,777 units or 0.1% (Fig. 4). In the Northwestern Federal District, the opposite trend is observed: their number increased by 5281 units or 0.9%. Moreover, the proportion of enterprises of the exclave in the district grew steadily (from 7.6 to 9.0%) (Fig. 4).

The general trend in the number of enterprises in the Kaliningrad region indicates their significant growth. Thus, their number increased in 2016 compared to the same indicator in 2005 by 8957 units or 19.3% (Fig. 5).

However, at the beginning of 2017, the number of enterprises in the fisheries sector in the region was only 157, which is less than a percent of the total number of enterprises in the region (Fig. 5).

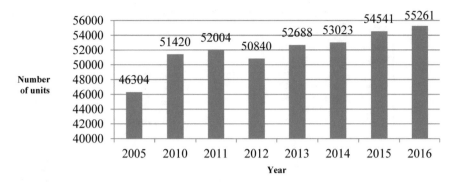

**Fig. 4.** The number of enterprises and organizations in the Kaliningrad region (at the end of the year) (Source: Compiled by the authors according to Rosstat) [14]

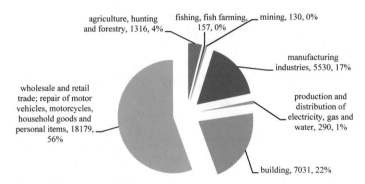

**Fig. 5.** Distribution of the number of enterprises and organizations by types of economic activity in 2016 (at the end of the year) in the Kaliningrad region (Source: compiled by the authors according to Rosstat) [14]

The share of enterprises in the fisheries sector of the North-West Federal District in the total number of enterprises in this sector of Russia in 2016 was 13.3%, and the Kaliningrad region - 1.9%. The share of enterprises of the exclave industry in the district is also insignificant - 14.4% (Table 1).

Another institutional trap for the fishery in the Kaliningrad region is the shortage of small enterprises in the industry. This is eloquently shown by the following data (Tables 2 and 3) [15].

Over the past five years, the dynamics of the number of enterprises in the fisheries industry in the region shows a reverse trend in the development of the total number of organizations in the region. There was a reduction of 14 economic entities or 8.2%, i.e. there is an outflow of capital from the industry.

The data in Table 3 confirm the previously made conclusion with respect to small enterprises in the industry: over the past five years, their number has decreased by 12 units or 8.7%. As shown in Fig. 6, they make up only 0.03% of the total number of small enterprises of the exclave.

**Table 1.** Distribution of the number of enterprises and organizations by type of economic activities in Russia in 2016

| Name | Enterprises in total | By type of economic activities among them | | | | | | |
|---|---|---|---|---|---|---|---|---|
| | | Rural household, hunting and forestry | Fishing, fish farming | Mining | Manufacturing industries | Production and distribution of electricity, gas and water | Building | Wholesale and retail trade; repair of motor vehicles, motorcycles, household goods and personal items |
| Russian Federation | 4764483 | 133140 | 8187 | 18187 | 38717 | 31846 | 49786 | 1585030 |
| Northwestern federal district | 616017 | 14846 | 1088 | 1684 | 51091 | 3318 | 68286 | 205538 |
| Kaliningrad region | 55261 | 1316 | 157 | 130 | 5530 | 290 | 7031 | 18179 |

Source: compiled by the authors according to Rosstat [15]

**Table 2.** Distribution of economic entities by type of economic activity in the Kaliningrad region

| Number of organizations (end of year) | Year | | | | |
|---|---|---|---|---|---|
| | 2012 | 2013 | 2014 | 2015 | 2016 |
| Total: | 50845 | 52695 | 53029 | 54549 | 55261 |
| Including: | | | | | |
| Agriculture, hunting and forestry | 3230 | 3131 | 1295 | 1293 | 1316 |
| Fishing, fish farming | 171 | 173 | 174 | 170 | 157 |
| Mining | 100 | 103 | 101 | 114 | 130 |
| Manufacturing industries | 5137 | 5230 | 5361 | 5446 | 5530 |
| Production and distribution of electricity, gas and water | 268 | 268 | 278 | 285 | 290 |
| Building | 5613 | 6037 | 6480 | 6852 | 7031 |
| Wholesale and retail trade; repair of motor vehicles, motorcycles, household goods and personal items | 15862 | 16538 | 17329 | 17923 | 18179 |

Source: compiled by the authors according to Rosstat [15]

The following Table 4 presents the main economic indicators of the activities of small enterprises in the Kaliningrad region.

Despite the fact that the Kaliningrad region occupies a leading position in comparison with the all-Russian indicators, the general situation in the development of small business in the region and the city is rather controversial.

The risks associated with small businesses in the Kaliningrad region are specific: the domestic consumer market is organic in capacity due to the exclave position of the region, its remoteness from the rest of the country does not allow entrepreneurs to redistribute goods to their markets in response to market changes. On the other hand,

**Table 3.** Number of small enterprises (including micro enterprises) by type of the economic activity in the Kaliningrad region (units)

| Indicator | Year | | | |
|---|---|---|---|---|
| | 2012 | 2013 | 2014 | 2016 |
| Total | 24340 | 21538 | 21779 | 28308 |
| Including: | | | | |
| Agriculture, hunting and forestry | 522 | 367 | 378 | 436 |
| Fishing, fish farming | 93 | 59 | 47 | 81 |
| Mining | 47 | 30 | 38 | 64 |
| Manufacturing industries | 2883 | 2156 | 2345 | 2900 |
| Production and distribution of electricity, gas and water | 109 | 81 | 74 | 99 |
| Building | 2357 | 2650 | 2783 | 3671 |
| Wholesale and retail trade; repair of motor vehicles, motorcycles, household goods and personal items | 10699 | 8319 | 8179 | 10359 |
| Hotels and restaurants | 805 | 641 | 717 | 816 |
| Transportation and communication | 2316 | 2173 | 2150 | 2960 |
| Real estate operations, rental and provision of services | 3600 | 4089 | 4028 | 5514 |
| Health and social services | 275 | 298 | 278 | 423 |
| Education | 53 | 61 | 52 | 64 |
| Provision of other community, social and personal services | 395 | 423 | 474 | 616 |

Source: compiled by the authors according to Rosstat [15]

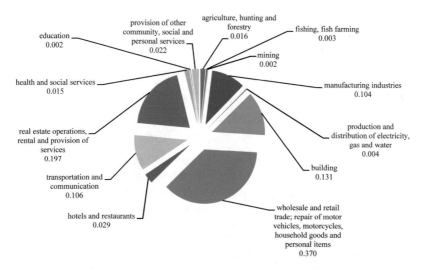

**Fig. 6.** Distribution of small enterprises (including micro enterprises) by types economic activity in the Kaliningrad region, in %. (Source: compiled by the authors according to Rosstat) [15]

access to the markets of the European Union countries is hampered by existing barriers in the form of stringent quality requirements, certification and labeling of goods, and customs restrictions.

**Table 4.** Main economic indicators of Kaliningrad region small enterprises for the period from 2012 to 2016

| Год | Total number of enterprises | Including microenterprises |
|---|---|---|
| Number of enterprises (by the end of the year), units | | |
| 2012 | 24340 | 22173 |
| 2013 | 21538 | 19437 |
| 2014 | 21779 | 19145 |
| 2015 | 27014 | 24904 |
| 2016 | 28308 | 26105 |
| Average number of employees, thousand people | | |
| 2012 | 88,5 | 27,9 |
| 2013 | 94,6 | 32,6 |
| 2014 | 120,3 | 49,7 |
| 2016 | 128,5 | 64,1 |
| The average number of employees (without external part-timers), thousand people | | |
| 2012 | 83 | 25,5 |
| 2013 | 90 | 30,5 |
| 2014 | 115,5 | 48,2 |
| 2016 | 123 | 61,1 |
| Turnover of enterprises, mln. rub. | | |
| 2012 | 192195,8 | 92608,7 |
| 2013 | 201583,1 | 90542,4 |
| 2014 | 256277,1 | 87650,9 |
| 2016 | 363557,5 | 197811,5 |

Source: compiled by the authors according to Rosstat [15]

Consequently, the state and development of small business in the region is directly dependent on the economic situation in it. Economic ties with other regions of the country and foreign countries are burdened with high transport and other expenses.

The weak export orientation of small businesses in the Kaliningrad region has become a threat during the liberalization of Russian customs legislation related to Russia's accession to the WTO.

At the same time, the small business of the Kaliningrad region continues to operate in international and regional markets. Accents are made on the growth of the competitive goods production, the search and attraction of new investments, the search for business partners and markets, the provision of modern equipment and the introduction of innovative technologies.

At the beginning of December 2017, the governor of the region A. Alikhanov noted: "Amendments to the law on the Kaliningrad SEZ improve conditions for business in the region, reduce the fiscal burden, and also enable the introduction of electronic visas for foreigners" [16].

The Federal Law "On Amendments to Certain Legislative Acts of the Russian Federation on the Socio-Economic Development of the Kaliningrad Region" was signed by President Vladimir Putin on December 5. Earlier, on November 28, the head of state approved amendments to the Tax Code, which also concern residents of the SEZ in the region.

He noted that many entrepreneurs were waiting for the adoption of amendments and postponed investment in new projects, until conditions improved. "In the full sense of the word, we can say that it (the law) is long-awaited for all of us. We worked on it for quite a long time, almost for two years" - said Governor Alikhanov [16], adding that not all proposals from the region were included in the final document, but work on improving the law could be continued. According to him, the Kaliningrad SEZ has always been a leader in innovations for business and amendments to the law can return this title to it.

Amendments to the law on the Kaliningrad SEZ suggest, in particular, the minimum amount of investment 10 million rubles required to obtain the status of a SEZ resident for projects in the field of health and pharmaceuticals and in 1 million rubles - for the IT sphere; as well as exemption of residents from VAT on export goods, and for goods imported from the territory of the Kaliningrad region - deferment of VAT for

**Table 5.** Investments in the fixed capital of small enterprises by type of economic activities (in actual prices; mln. rubles)

| Indicator | Year | | | |
|---|---|---|---|---|
| | 2012 | 2013 | 2014 | 2016 |
| Total | 2384,6 | 4186,1 | 3774 | 9084,9 |
| Including: | | | | |
| Agriculture, hunting and forestry | 607,6 | 1835 | 949,9 | 1487,7 |
| Fishing, fish farming | – | – | – | 18,6 |
| Mining | 29,4 | 21,9 | 79,9 | 3 |
| Manufacturing industries | 997,1 | 1060,2 | 956,9 | 1358,2 |
| Production and distribution of electricity, gas and water | – | 296,5 | 156,7 | 130 |
| Building | 276,3 | 99,2 | 101,3 | 2712,8 |
| Wholesale and retail trade, repair of motor vehicles, motorcycles, household goods and personal items | 136,8 | 126,4 | 154,9 | 650,6 |
| Hotels and restaurants | 0,2 | 30 | 15,8 | 34 |
| Transportation and communication | 35,9 | 303,9 | 458,2 | 321,9 |
| Real estate operations, rental and provision of services | 140,5 | 247,8 | 719,2 | 1904 |
| Health and social services | 10 | 14,7 | 23 | 56,7 |
| Provision of other community, social and personal services | 2,3 | 0,1 | 0,3 | – |

Source: compiled by the authors according to Rosstat [15]

180 days, compensation for railroad transportation expenses as well as zero profit tax will be available for residents throughout the first six years of work in the SEZ.

Electronic visas are introduced for up to eight days, when issued, residents are exempted from consular fees. For projects in the SEZ, the timeline for conducting a state environmental review is reduced from three months to 45 days, and for state examination of project documentation and engineering survey results from 60 to 45 calendar days.

One of the most important points in the regional government consider a reduction in the rate of social contributions for residents from 30 to 7%, as well as the abolition of the utilization fee for agricultural and construction equipment imported into the region. The validity of the Kaliningrad SEZ is extended until 2045, while it was proposed to establish a period until 2095. The proposal to lift the ban on commission trade in goods from local agricultural producers, which were expected in the region, was not included in the law.

Amendments to the law "On the Special Economic Zone in the Kaliningrad Region" and the Tax Code turn the region into the most profitable territory for doing business in the country from the costs standpoint. As a result - investments from South Korea and Germany. One of the institutional traps in this branch of the region is the investment trap (Table 5).

It should be noted that in the given data on investments in the fixed capital of small enterprises there is no sector at all. The investment to fisheries industry small enterprises began only in 2016.

Further from Fig. 7 it is clear that the use of information and communication technologies in organizations is growing rapidly. Organizations are actively investing in the development of the digital capabilities of their organizations.

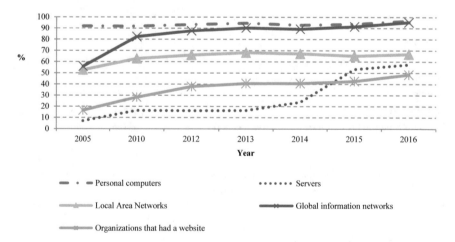

**Fig. 7.** The use of information and communication technologies in organizations. (Source: compiled by the authors according to Rosstat) [14]

However, this is not enough for the breakthrough opportunities of the region. Studies show that budget allocations for the development of fisheries and fish farming in the region do not provide for articles on increasing the rate of digitization of the regional industry.

The volume of budget allocations of the state program: the total funding of the state program at the expense of the regional budget is 567,750.56 * thousand rubles. (excluding subventions from the federal budget to the regional budget for the implementation of powers in the field of organization, regulation and protection of aquatic biological resources in the amount of 686 thousand rubles), including on an annual basis (see. Fig. 8) [17].

Let us consider the expected results of the program:

- comprehensive modernization of seven Kaliningrad region fisheries industry enterprises;
- private investment in the regional sector amount to more than 2 billion rubles;
- the expected increase in the efficiency of use of the raw material base up to 95%;
- production of fish products by enterprises of the region should reach 410 thousand tons per year.

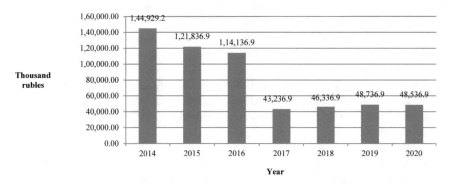

**Fig. 8.** The amount of state program funding at the expense of the regional budget for 2014–2020. (Source: compiled by the authors according to the data of the state program of the Kaliningrad region "Development of the fisheries complex") [17]

Over the past three years, the modernization of the fishing fleet has begun at the expense of budget support within the regional state program for the development of the fisheries sector of the exclave. Eight small fishing trawlers were re-equipped: modern fish pumps and cooled fish tanks were installed. That will allow, according to experts, to increase labor productivity twice as much and to improve the quality of fish entering processing significantly. A large-scale project is being implemented by a group of companies-investors of the Unified Marine Complex in the city of Svetly, the Kaliningrad region. A small fishing trawler was rebuilt at the local ship repair site: the hull was rebuilt, the first fishing tanks were installed, and field and electrical equipment, power plants, navigation and communications equipment were completely replaced. As a result, cargo capacity will increase to 120 tons. Raw materials of higher quality will go for processing. Fishing trawlers comparable in class are not built anywhere in the Russian Federation. Investments exceeded eighty million rubles. According to the Kaliningrad region Agency for Fishery, 30% of the costs are subsidized to fishers from the regional budget: for the modernization of vessels and the payment of loans interest. Since 2013, about 20 million rubles have been allocated for these purposes.

Currently, in connection with the introduction of amendments to the law of the Russian Federation "On the Special Economic Zone (SEZ) in the Kaliningrad Region", the investment situation in the region may improve. It is expected that deferred investment will come in 2018 to the economy of the region. Summing up all the above mentioned, it can be argued that the shortcomings of the institutional environment are the main causes of the unstable situation in the Kaliningrad economy.

## 4  Discussion

Why the Kaliningrad "tiger" did not jump?

The main reason is as follows. The Kaliningrad Special Economic Zone was unable to adapt to changes in economic conditions. Among the Russian regions, simultaneously with the creation of local SEZs in the Kaluga, Lipetsk regions, Tatarstan and other territories that provide the same privileges on property, land and profit taxes, competition for investors is intensifying. In addition, competing special economic zones as additional preferences began to provide, for example, transport tax rebates or free connection to utilities. Nothing of the kind was provided for residents of the Kaliningrad SEZ.

Later on, in the Far East and in Siberia, special areas of special socio-economic development (ASSED) were organized with facilitated payments to social funds, which became a powerful incentive for the creation of highly skilled jobs and the development of the digital economy. A special economic zone has appeared in Crimea with an unprecedented amount of both tax and non-tax benefits.

What else did seriously reduce the attractiveness of the Kaliningrad SEZ? There was a significant depreciation of the ruble exchange rate, as a result, the rise in prices for imports of raw materials, components and equipment. The process of modernization of regional enterprises began to slip, the competitiveness of products in the markets of the Eurasian Economic Union decreased, taking into account transport costs and the loss of customs preferences. All this allows us to consider its Kaliningrad SEZ format as an institutional trap for the region.

The need to develop and transition to a new economic model in the digital era, focused on the development of the production of high-tech goods and specialized services, including those intended for export, was raised in the economic literature more than a decade ago (for example, in the monograph edited by G.Z. Bunatyan "Kaliningrad region: new challenges, new chances" [18]).

According to the national innovation report [19], it is clear that the current state of the Russian innovation ecosystem is in an unsatisfactory state. Weaknesses can be traced to a whole group of areas, such as research and development, commercialization, innovation, infrastructure and industry maturity, in state institutions and business values.

The results of cluster analysis (Fig. 9) clearly show that countries with a similar level of development of the innovation system have approximately the same development parameters. Therefore, when improving the legislation and management system of the Kaliningrad region, it is necessary to pay attention to the best practices that the group of countries "Technological Leaders" have developed.

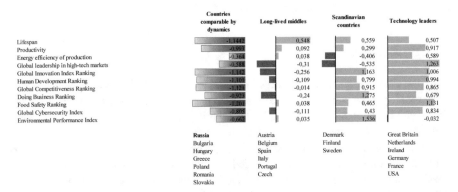

**Fig. 9.** Results of cluster analysis: countries with a similar level of development of the innovation system (Source: compiled by the authors according to National Innovation Report in Russia - 2017 [19])

Experts believe that it is necessary to take a closer look at the international experience of the development of exclaves. This is an example of institutional transplantation, import institutions. There are examples of successful exclaves (Macau, Hong Kong, British Gibraltar) that can be used for Kaliningrad.

It is worth paying attention to the West Berlin experience. The geopolitical position of West Berlin in the second half of the twentieth century was very similar: not very friendly geographical environment. Systemic transit restrictions, remoteness from the economically developed centers of the state, and, consequently, the loss of competitiveness of local goods are characteristic of both regions. The economic situation significantly worsened in both cases: the lack of funds for the "reanimation" of the economy, the lack of production links with other regions of the country, the urgent need for qualified personnel, and high political risks reduced the investment attractiveness. West Berlin, like the Kaliningrad region today, reacted more sharply to the slightest changes on its borders than others. At the same time, the city's economy coped: a well-built system of economic incentives allowed West Berlin to not only survive, but also to develop progressively. In the future, it was the West Berlin experience that was used to reform the economy of East Germany. We have a clearly visible analogy with the Kaliningrad region, which experience in creating a special economic zone has become a model for other regions of Russia.

In the domestic economic literature there are concerns that the wide preferences of the Kaliningrad region may lead to its excessive autonomization, but this position is quite controversial. In the work of E.Y. Vinokurov "The Theory of Enclaves" [20] it is substantiated that the ethnocultural community of the parent state and exclaves make such a development illusory. In regard to these features, the Kaliningrad region can really become a special region for testing and improving innovative technologies, creating high-tech industries in the growing conditions of the digital economy.

We believe that Russian economics has not fully studied the world experience; therefore, it is advisable to investigate it systematically and use legislation to develop similar exclaves in practice. In the future, the Kaliningrad experience can be successfully used in other regions of the country.

## 5  Conclusions

Thus, the current expansion of access to digital technology brings countries a wealth of choice and greater convenience. By enhancing integration processes, increasing efficiency and introducing innovations in the fisheries industry, the region is able to avoid problems associated with probable institutional traps.

By implementing the steps identified by the government of the Russian Federation in practice within the framework of the federal target program "Digital Economy of the Russian Federation" [21], the regions have a chance to receive dividends and create stable conditions for the flourishing of competitive digital economies.

At the same time, the Kaliningrad region should take into account that in the current digital whirlwind, the economic climate for the region becomes unpredictable, and tools that choose politicians are not always ways to stimulate new and sustainable growth in the region. Under these conditions, digital pressure will only increase and the exclave needs to identify and regulate institutional traps, including in relation to the fisheries sector of the Kaliningrad region.

Institutional traps studied on the example of the fisheries industry in the region, under the conditions of the process of digitization of the economy, are systemic and, as analysis showed, they are not visible in key regional bills. In this regard, the key focus on the development of the exclave region should be a policy to stimulate innovation and the development of innovative ecosystems, including in the fisheries industry.

Due to these features, the Kaliningrad region can really become a special region for testing and improving innovative technologies, creating high-tech industries under conditions of digital pressure.

We believe that Russian economics has not fully studied world experience; therefore, it is advisable to systematically investigate and use in practice legislation to develop similar exclaves. In the future, the Kaliningrad experience can be successfully used in other regions of the country.

## References

1. Executive Order of the Government of the Russian Federation of 04.05.2005 N 536-p "On the draft federal law" On the Special Economic Zone in the Kaliningrad Region
2. Information on the socio-economic development of the Kaliningrad region https://gov39.ru/economy/situation. Accessed 16 Oct 2018

3. Lesnykh VV (2007) Theoretical and methodological foundations of the institutional transformation of the defense industry. Ekaterinburg, p. 12
4. Antonelli C (2016) Technological congruence and the economic complexity of technological change. Struct Chang. Econ. Dyn. North-Holland 38:15–24
5. OECD (2009) Regions Matter: Economic recovery, innovation and sustainable growth. Publications, Paris
6. Katz R (2017) Social and Economic Impact. ITU, GSR-17 Discussion paper
7. Wei T, Liu Y (2018) Estimation of resource-specific technological change. Technol Forecast Soc Change, North-Holland
8. Perera-Tallo F (2017) Growing income inequality due to biased technological change. J Macroecon North-Holland, T 52:23–38
9. Apa R et al (2018) Resource to the European regions. Res Policy North-Holland, 47 (9):1700–1712
10. Grossman GM, Helpman E (1991) Trade, knowledge spillovers, and growth. Eur Econ Rev T 35(2–3):517–526
11. Arthur WB (2007) The structure of the invention. Res Policy 36(2):274–287
12. World Bank (2018) Competition in the digital age: Strategic challenges for the Russian Federation. September, Washington, DC
13. World Bank. World Development Report (2016) Digital Dividends. Overview. World Bank, Washington, DC
14. Rosstat (2018) Regions of Russia. Socio-economic indicators. Stat Sat Moscow
15. Kaliningrad region in numbers (2017) Statistical compilation in 2 tons./ Territorial body of the Federal State Statistics Service of the Kaliningrad region (Kaliningradstat). Kaliningrad, T. 2
16. Alikhanov: amendments to the law on the Kaliningrad SEZ will improve conditions for business in the region. TASS Electron Dan. http://tass.ru/ekonomika/4789404. Accessed 06 Oct 2018
17. Kaliningradskaya Pravda (2014) Resolution of the Government of the Kaliningrad region of 04.02.2014 No 40 (as amended on July 25, 2016) On the State program of the Kaliningrad region "Development of the fisheries industry". (insert paper "Official Gazette of the Government of the Kaliningrad Region"), No 23, 11.02.2014
18. Bunatyan GZ (2005) Kaliningrad region: new challenges, new chances. Mechanisms for the development of a unique region: monograph. Foundation "ARER". Kaliningrad
19. National report on innovations in Russia-2017. JSC "PBK". http://www.rvc.ru/upload/iblock/c64/RVK_innovation_2017.pdf. Accessed 16 Oct 2018
20. Vinokurov EY (2007) Theory of enclaves. Terra Baltic, Kaliningrad
21. Meeting of the legislation of the Russian Federation (2017) Order of the Government of the Russian Federation of 28 July 2017 N 1632-p "On approval of the program Digital economy of the Russian Federation", 07 Aug 2017, N 32, Article 5138

# Countering the Financing of Terrorism in the Conditions of Digital Economy

Anastasia Andrianova(✉) ⓘ

Moscow Academy of the Investigative Committee of the Russian Federation,
Moscow, Russia
an.andrianova@mail.ru

**Abstract.** Terrorist activity is a particularly dangerous form of criminal activity that poses a threat to all mankind, along with environmental disasters and wars. The problems of countering the financing of terrorism are relevant in the context of the fight against crime for Russia and for the whole world. The financing of terrorism has long roots, but now the methods of financing are largely developed through the global Internet and the development of digital economy technologies. The continuous improvement of the Internet's capabilities not only opens up additional prospects for the development of the world community, but is also accompanied by the emergence of a number of new global threats.

Traditional forms of financing terrorism are transformed using modern digital technologies - cryptocurrency and virtual currencies, crowdfunding, fundraising, social networks and Telegram channels are actively used by terrorist organizations. In a digital economy, terrorist acts can have an impact on government operations and finances, the cost of securities and the investment climate, significantly reduce tax revenues, help increase the share of military spending in the state budget, and have a devastating effect on the state and society as a whole.

The development of effective solutions to the problem of countering the financing of terrorism is one of the priority areas of scientific research. The article presents the results of an assessment of the main existing methods of financing terrorist activities and elaborated proposals for countering their proliferation and reducing the risks of activating new modern sources of enrichment of criminal communities.

**Keywords:** Anti-money laundering · Bitcoin · Crowdfunding · Cryptocurrency · Digital economy · Hawala · Shadow economy · Terrorism financing

## 1 Introduction

Crime is one of the most destabilizing factors of development, including financial and economic activity of the state. The ability to legalize criminal proceeds and fund criminal activities motivates criminals to continue it [27]. Unlike wars, terrorist activity does not necessarily have a long-term and deep impact on a country's economy. In countries with a large number of deaths, a cumulative effect may be observed in demographics. Developing countries are more vulnerable and subject to negative effects than developed countries with a rich diversified economy [4]. However, even in

S. Ashmarina et al. (Eds.): *Digital Transformation of the Economy: Challenges, Trends and New Opportunities*, pp. 20–31, 2019.
https://doi.org/10.1007/978-3-030-11367-4_2

developed countries, the economic damage can be enormous, especially in the tourism industry or passenger traffic. The reality of the terrorist threat influences electoral processes, radicalization of society and state policy. Decisions made by the authorities of states subjected to a terrorist attack can lead to a tightening of the migration regime, the introduction of additional border controls, a reduction in commodity circulation, and even military actions. For example, the abolition of simplified border controls in the European Union, as a consequence of terrorist attacks in Europe, could lead to a reduction in annual turnover of 220 billion euros [31] and to the European Union's annual GDP decline by 100 billion euros [29].

The fight against extremism and its extreme form of terrorism has a global dimension. This leads to the emergence and strengthening of armed conflicts throughout the world. Monetary and material resources are a source of prosperity for terrorist organizations. Currently, terrorist organizations have resources comparable to he budgets of some states. Financial support for terrorist formations in individual states is provided both by international terrorist organizations and by forces of other states due to global political, economic and ethnic competition. Often, a developed shadow economy becomes a reliable source of enrichment for terrorist organizations. The shadow sector of the economy weakens the resource base of states, leads to a tax deficit, rising inflation and loss of control, destroys the foundations of the state financial system, and hinders the economic, technical, and scientific development of the country.

The shadow economy is also a threat to investment and innovation, which today are becoming the leading factors of economic security. Important components of the shadow economy are the proceeds from criminal activity, corruption, cashing and tax evasion, unofficial banking [18]. The influx of money from criminal activity leads to disruption of the market economy, criminalization of production and trade, increased inflationary pressure, has a devastating effect on the state and society, poses a threat to the economy and the social sphere, and contributes to the further financing of crime, extremism and terrorism [15]. The trade in drugs, weapons and people generates huge financial flows that allow arming real armies, hiring the best specialists and even waging wars with entire states.

The purpose of this study is to conduct a critical analysis of the main existing methods of financing terrorist activities, both traditional and generated by virtual technologies, as well as the development of an effective behavior algorithm to counter it, taking into account modern realities.

## 2 Materials and Methods

Sources of information for the study are materials of the global terrorist database, FATF recommendations, the work of Russian and foreign scientists in the field of economics, digital economy and digital technologies, the fight against terrorism and extremism, financial control.

An integrated approach to the study of various fields of activity of individual states and the world community as a whole, as well as their mutual influence, predetermined the need to use at the first stage of research methods of analysis with synthesis, synthesis and systematization, as well as an assessment method. This approach made it possible to determine the main range of problems on the research topic for further analysis and development of an effective behavior algorithm for countering the financing of terrorist activities, which constitutes the second stage of research activities. At the third stage, the obtained results are processed and analyzed, and goals for further research are formulated.

Such a sequence of research activities allowed to identify the key problems of the relationship between the development of digital technologies and criminal activity, current trends and to determine the range of unresolved issues of interest to research in a given topic.

# 3   Results

According to the consolidated analytical report Global Terrorism Index for 2017 of the Researchers of the Institute of Economics and Peace (IEP) based on the Global Terrorism Database (GTD) data, only the direct damage to the global economy from terrorism amounted to 84 billion US dollars in 2016. This damage does not include indirect losses, long-term negative consequences for the economy. For example, after the 9/11 attacks, the indirect damage only to the US economy was estimated by experts to $190 billion with $65 billion direct damage [12, p.80].

In the modern world, the boundaries between legal money, money dishonest in the hands of businessmen, drug dealers and terrorists are blurred. One of the sources of financing of terrorism is the system of shadow taxation of large legal entrepreneurs in favor of terrorist organizations under the pretext of zakat (Islamic charity) created in the beginning of the 2010s. In this case, participants in the legal economy are forced to pay the Islamists, under the threat of physical violence, which provides sufficiently large financial resources to terrorist and extremist organizations.

With the help of zakat, the terrorist organization IS (Islamic State - terrorist organization, banned in Russia) collected millions of dollars from businesses and people monthly in Iraq and Syria with "taxes", annually approved and spent billions of dollars in budgets [17].

The terrorists use the Black Market Peso Exchange (BMPE), a money laundering system specifically designed to launder the proceeds of drug sales from Colombian rebels and cartels whose members are involved in terrorist attacks [16].

Also in the articles of IS revenues are the sale of crude oil and petroleum products, products of mining and processing enterprises, chemical plants, cement plants, human trafficking, ransom requirements for abducted people, smuggling of cultural property and extortion. Remittances from their foreign supporters and philanthropists, misled, make up a significant portion of the income of terrorists.

Some remittances are remittances through the international banking system, through individuals and organizations affiliated to terrorists, some coming from the traditional hawala system and such innovative digital economy technologies as e-currencies and cryptocurrencies [8].

In countries with a war-torn economy, such as Afghanistan, Iraq, Somalia, Syria, and Yemen, where Al Qaeda and the IS have their hawala strongholds, only a small proportion of the population has bank accounts and even large transactions are made in cash. Together with unfavorable economic conditions and political instability, this makes the modern banking system give way - instead of using Western banks in developed countries, immigrants rely on hawala to send remittances home [10].

Hawala - the traditional system of money transfer of Islamic countries, is an integral part of the global shadow economy. The most common hawala in India and Pakistan, where there are several thousand hawala shops, issuing transfers according to various estimates from tens of billions to a trillion dollars a year. Hawala branches operate almost throughout Asia and Europe, in many countries in Africa, North and South America, and even in Australia. The hawala system also interacts with Islamic banks and Asian commodity exchanges.

Hawala in most states is illegal, does not pay taxes and is out of any regulation. Transfer fee is from 1 to 2%.

In most cases, hawala brokers operate behind the scenes, do not provide advertising and, without attracting attention, can work in the market, in a pharmacy, gardeners or janitors. Usually, the sender of the transfer finds a hawala broker in his ethnic environment or diaspora, gives him money and indicates the address and name of the payee. The hawala broker transmits the recipient's name via the Internet, SMS or by phone to another broker at the destination, adding a code (usually the bill number). The sender reports the code via the Internet or telephone to the recipient, and the recipient receives the money from a local broker within 24 h. The money itself does not actually move, so it is very difficult to track the fact of the transfer. Even if the hawala transaction was discovered by the state authorities, clients and brokers can explain it by purchasing a car or expensive clothes in Europe or helping relatives in Syria and protecting themselves by compiling relevant documents [25].

To equalize payments between brokers, a netting system is used. Clearing centers located in London or Dubai are usually used to offset international settlements. The rest of the debt is covered by bank transfers, smuggling of precious metals and stones, couriers with money. There are various analogues of hawala in the world: fei-chien, hui kuan, xa-vilaad, fei kwan, etc. They operate in different countries, may not intersect with hawala and are also characterized by low levels of control, lack of identification and documentation, cheapness and great popularity among migrants [2].

In modern conditions there are opportunities for the active use of virtual technologies in a negative way to ensure the functioning of criminal activity.

Today, terrorists have already mastered the most advanced technologies of fundraising on Internet platforms for fundraising, crowdfunding and crowdsourcing. There are cases of using the names of these charitable organizations, on whose behalf funds were raised to finance terrorists.

Various online communication channels, such as: social networks, encrypted Telegram channels, instant messengers, chat rooms and websites on the Darknet (the shadow segment of the Internet) allow terrorists to attract like-minded people to expand support, cash flow and the entire flow of resources the world [26].

Often, individuals and real non-profit organizations claiming such fundraising purposes, such as organizing training for the poor or building religious buildings, actually transferred the collected funds or part of them to organizations in contact with terrorist groups. There are also open calls on the Internet, especially on Telegram channels and Darknet, to raise funds for prohibited and terrorist activities. Funds are usually collected anonymously through electronic money and cryptocurrency transfers [7].

Raising funds through the Internet using various payment systems allows terrorist groups to confuse traces so that even experts cannot track such payments.

The president of the International Counter-Terrorism Association, Joseph Linder, said: "Even the special services cannot track all the channels. Because otherwise, all work on the Internet will stop, and all Internet payments will be blocked. All private contacts cannot be traced" [32].

The development of innovative technologies has allowed members of terrorist organizations to expand their operations using virtual platforms such as social networks, online games, mobile applications and even peripheral devices - VR headsets, contactless payment devices NFC. These platforms are used both for collecting and laundering money, and for recruiting via online communication. Terrorists successfully use these innovative methods, primarily because of the lack of a mechanism for controlling and legislatively regulating the unlawful use of these technologies, or because of the irrelevance of current legislation [21].

The freedom granted to a virtual (gaming) currency also presents difficulties with its control, contributing to the implementation of such crimes as money laundering and the financing of terrorism in a virtual environment [3].

Private companies, including social networks, airlines and banks, are at the forefront of the fight against terrorism and other security threats, as they are required by law to pay attention to suspicious transactions. This analytical work is the beginning of the chain in which transaction data is analyzed, collected, communicated, transmitted, and ultimately becomes the basis for blocking and prosecution [5].

Digital economy and innovative technologies are actively developing the market of electronic payment systems and digital currencies. Currently, he has already reached 400 trillion dollars, it is expected to increase by 2020 another 11.2%, which will be about 433,1 trillion dollars. With the increase in volume and availability, digital currencies have become the focus of attention of organized criminal groups engaged in the laundering of criminal proceeds and the financing of terrorism [13].

Digital currencies are a sophisticated financial tool based on network distributed computing and encryption algorithms. This tool is poorly amenable to oversight and pressure from states. Because of this, for example, in Russia, the legislator for a long time did not recognize the legitimacy of operations with cryptocurrencies. A cautious policy of phased introduction of these technologies into the legal field is being implemented in different countries [22].

Despite the technological complexity of cryptocurrency, almost any Internet user can learn how to use it. Cryptocurrency was widely used on Darknet long before everyone's attention. Currently, dozens of different cryptocurrency are actively used.

The most famous and popular cryptocurrency is Bitcoin. Bitcoin is a payment system distributed between network members that uses bitcoin cryptomonettes to account for transactions. Bitcoin was the first decentralized convertible currency and the first cryptocurrency. To ensure the functioning and protection of the system, cryptographic methods and distributed databases are used, the identical copies of which are kept by all users of the system.

Bitcoins are calculated units in the form of a unique chain of numeric and alphabetic characters. Bitcoins are traded with a high degree of anonymity, they can be exchanged for US dollars, euros and other fiat or virtual currencies. Anyone can download a free open source software application on the Internet for sending, receiving and storing bitcoins and controlling transactions. Users can also get Bitcoin addresses that function as accounts on the sites of the bitcoin exchange service providers or on the online wallet service sites.

Bitcoin address accounts do not contain any customer identification information. A distributed system does not have a single server or service provider, so the Bitcoin protocol does not require and does not ensure the identity of the participants and does not provide any control over the legality of operations and the possibility of their cancellation [9].

According to the research of the Recorded Future group conducted by Darknet among hackers, Bitcoin is the most common payment method among the cryptocurrencies in the world, but most Darknet users prefer other cryptocurrencies. Most of the users of Darknet prefer for illegal activities Monero, positioned as the most anonymous and protected. Then follows the popularity of Dash - an easy-to-use, low-cost payment platform. In addition to low confidentiality when buying goods in the dark web, an important reason for reducing the popularity of Bitcoin is the high cost and slowness of transactions.

It should be noted that cryptocurrencies and, in particular, Bitcoin, were used by terrorists, such as IS, in a small amount, because: first, the main senders were European migrants who did not have access to the latest achievements of the digital economy and therefore sent money through the system hawala; and secondly, oil revenues and extortion from enterprises and businessmen in the territory under control brought much more money.

However, in the future this situation will change - for the younger generations, who are targeted by the propagandists of terrorism, it will be easier and safer to transfer money in two clicks of the mouse than to contact hawala brokers.

Cryptocurrencies are used to finance extremism and terrorism and open up an affordable and relatively reliable way to store and launder criminal money. They can be used in all corners of the world without fear of centralized supervision. Terrorists use digital currencies to circumvent state control, since cash flows in the banking system are closely monitored by law enforcement and financial authorities [28].

One of the first cases of the use of bitcoin by terrorists was recorded on social networks, which is promoted by a jihadist group from the Gaza Strip, which raised a small amount of funds [23].

Despite the use of encryption algorithms and anonymous wallets, the absolute elusiveness of transactions is a myth. For example, the implementation of the blockchain technology in the Bitcoin protocol, on which bitcoin transactions are organized, is a continuous sequence of operations. Each of the transactions leaves an indelible mark in the overall chain, which allows you to analyze and identify all cash flows. The difficulty is only a large amount of data and the identification of the real sender and receiver of funds. Criminal methods are constantly being improved: various mediators, cryptobirds and coin-mixers are used to mix (mix) the "clean" and "dirty" cash flows. Already in other criminal areas, such as drug trafficking, digital technologies are being actively used that exclude tracing the seller's and buyer's contacts: cryptocurrency, Darknet sites, role distribution and communication using encrypted instant messengers and Telegram channels, satellite caches. Terrorist organizations are actively adopting the experience of criminal communities, luring highly qualified specialists and mastering advanced technologies. Using online banking, cryptocurrency and crowdfunding makes it easy to collect and transfer enough money to create the conditions for the commission of terrorist acts [12, p.86].

The four most dangerous terrorist organizations (IS, Al-Qaeda, Boko Haram and the Taliban - banned in Russia) have various ways of financing, including remittances, donations, human trafficking, taxation and extortion. IS - the largest, richest and deadly terrorist organization. IS financing reached a maximum level of $2 billion in 2015, with half of these funds being smuggled oil. IS produced up to 75,000 barrels per day, generating revenues of $1,3 million per day. The main sources of financing are the use of the occupied regions, therefore, with the loss of control over the territory, there was a decrease in income. Researchers estimate that as a result of the military defeat in Syria and Iraq, IS revenues decreased from $81 million per month in 2015, to $56 million in 2016 and an estimated $16 million per month in 2017 [12, p.83].

A study of 40 terrorist cells that planned or committed attacks in Western Europe from 1994 to 2013 showed that they were mostly financed independently from the revenues from the legal activities of their members. Only 5% relied entirely on external support from international terrorist organizations, such as Al-Qaida or IS. This means that most attacks are self-funded and do not require external support.

If the cost of preparing and conducting terrorist acts of 9/11 in the United States is estimated at about $500,000, then the cost of conducting the latest attacks in Europe has significantly decreased with the transition to a simpler attack. Most attacks in Europe cost less than $10,000. For example, the attacks on pedestrians in Nice in 2016 were inexpensive and were based on easily accessible cars. This was one of the reasons for the further use of vehicles as weapons.

## 4   Discussion

The national legislation of the Russian Federation on countering the legalization (laundering) of proceeds from crime and terrorist financing (AML/CFT), like the law of most states, is based on international standards developed by the Financial Action Task Force (on Money Laundering) (FATF). The FATF, which is the main international institution in this field, regularly analyzes financial processes related to the trafficking of criminal money and the financing of terrorism and issues relevant recommendations for the whole world. The main instrument of the FATF in the implementation of its mandate are 40 recommendations in the field of AML, which are audited on average once every 5 years, as well as 9 special recommendations in the field of countering the financing of terrorism, which were developed after the events of 9/11 due to the increasing the threat of international terrorism [7].

In accordance with UN Security Council Resolution No 1617 (2005), 40 + 9 of the FATF Recommendations are mandatory international standards for implementation by UN member states. I agree with the opinion of the leading role of the UN and the

Security Council in countering the financing of terrorism, in particular, the exchange of information, but I consider its methods rather political [16].

The FATF cooperates with the International Monetary Fund (IMF), the World Bank, and the United Nations Office on Drugs and Crime. At the state level, the FATF implements its recommendations through the national Financial Intelligence Units (FIUs), which are responsible for collecting and analyzing financial information within the country. Also created are FATF-type Regional Groups (RGTF), which pursue similar goals within the framework of the association with the FATF within the borders of their regions, for example, the Eurasian Group on Combating Money Laundering and Financing of Terrorism (EAG), which includes Russia and China [30].

In Russia, the main financial intelligence unit is the Federal Service Rosfinmonitoring, which on its behalf is a member of the FATF, the associated RGTF (EAG, APG, MONEYVAL) and the informal association « Egmont ».

Some researchers pay special attention to the fact that in the banking sector, in the course of customer identification, it is necessary to obtain information on the purpose of the transaction, the sources of money and property with which the transactions are carried out and the beneficiaries of the customer. If there is any information about involvement in extremist activities or terrorism, restrictions are applied to the client, transactions are blocked, information is transferred to financial and special services [24].

This practice reflects the implementation of the "know your customer" principle, based on the FATF risk-based approach, aimed at obtaining detailed up-to-date information about the client and its counterparties and preventing illegal actions.

The principles and recommendations of the FATF are constantly updated and supplemented by practice and specific instructions that allow States to adjust their policies in a timely manner. Having analyzed the recommendations of the FATF, I believe that the organization conducts internationally analytical, informational and legal work, and its results in practice help prevent the financing of terrorism and the legalization of criminal money.

I agree with the authors, who believe that at the moment the effectiveness of the FATF methods in the field of AML/CFT cannot be fully realized in countering the financing of terrorism. I will add that today's terrorist organizations use a variety of different schemes and tactics than cash-out, tax evacuating, or laundering of criminal and corruption proceeds. For example, during the implementation of the FATF recommendations within the framework of the AML/CFT system on 115-FZ in Russia, control over cash withdrawal and tax evasion was tightened. This type of control is really aimed at increasing tax collection and countering cash withdrawal, and often creates unjustified difficulties for businesses that use cash payments (for example, buying scrap metal from the public) [11, 19].

My view is supported by a study that demonstrates the lack of a link between the violation of AML/CFT requirements and the level of extremism for Russia [1].

I commend the legislative novelty - the US Justice Against Sponsors Of Terrorism Act (JASTA) law that victims of terror and their relatives can sue the country or organization sponsoring the terrorists. Despite obvious international legal conflicts, financial and reputational risks, taking into account close economic ties and globalization, they can force potential terrorist sponsors to abandon their plans [6].

I believe that the FATF policy and recommendations on preventing the use of hawala, digital currencies and various innovative technologies for financing terrorism are not sufficiently developed. This thesis is supported by hawala convenience and AML/CFT circumvention studies [25].

I agree with those researchers who claim that Russia has achieved significant success in AML/ CFT, and today the Russian legislation in this area generally conforms to international standards [15]. However, the historical development of the legal regulation of the financial sector in Russia has given rise to a complex implementation mechanism based on a plurality of interacting entities (Rosfinmonitoring, Bank of Russia, Federal Tax Service, Roskomnadzor, Ministry of Finance of Russia, Federal Security Services, Ministry of Internal Affairs, etc.) with a corresponding regulatory and legal array. There are also problems of synchronization of anti-corruption legislation and AML/CFT, some conflicts of national and international norms and practices. The proliferation of financial and regulatory technologies will exacerbate these problems and contradictions. I believe that these topical issues can be effectively resolved only through research, modeling, developing a concept, strategy and regulatory tactics, and a scientific approach in general, in which Khabrieva T.Ya. agrees with me [15].

I agree with the opinion that the most reasonable solution in the field of digital currency is moderate regulation, which will give the development of the industry, but will not allow the use of cryptocurrency in illegal operations [33].

I suppose that in the near future, along with the development of smart contract technology, methods of embezzlement and fraud involving them will develop, and the stolen funds may be used to finance terrorism [14].

Research into modeling the Bitcoin network as a social network and using the detection of community anomalies to detect related accounts is worth exploring. The proposed approaches include an innovative method for detecting Bitcoin accounts associated with money mixing, and demonstrates high efficiency on real data. This mechanism will allow tracing the chain of transactions and allocate a pool of records with one recipient, despite the mixing of cash flows, which can lead to its identification and blocking [20].

# 5   Conclusions

An analysis of terrorist activity shows that funding is secondary for the commission of single terrorist acts—supporters and tactics are needed. The main tasks of terrorist organizations are recruiting and ideological treatment of their supporters, simple and deadly solutions, for example - the use of trucks to run into the crowd. In such cases, serious external financing is not required, either self-financing or donations, hawala, crowdfunding, income from legal or illegal activities are used.

In regions with unstable political and economic conditions and conflict zones, terrorist organizations have the opportunity to unleash large-scale hostilities and seize resources. In the future, qualified mercenaries are recruited for resources, weapons and military equipment are bought, networks of recruiters are organized, training camps are being built to train supporters. In such conditions, the main sources of funds in the occupied territory (zakat, taxes, trade in petroleum products and other natural resources, human trafficking, looting, trafficking in artifacts and smuggling). It is also possible to finance organizations and states sympathizing with terrorists, a combination of new and traditional methods of financing (crowdfunding, cryptocurrency, hawala).

Today, the loss of most of the controlled territory deprives the IS of a significant part of revenues from the sale of petroleum products, human trafficking and "taxes" from the local population. I note that these circumstances increase the value of financing methods that are not tied to terrain, such as hawala, fundraising, fundraising, electronic and cryptocurrency payments for terrorists. Actively developing information technologies in all sectors create new opportunities for terrorists to attract supporters and funds. The legislator and the supervisory authorities do not have time to regulate and prevent such actions, focusing on already accomplished facts. It is necessary to anticipate and counteract new tactics of terrorists, analyze, model and preemptively regulate these areas in the context of countering the financing of terrorism. Realistically, an effective policy of countering terrorism in modern conditions can only be based on the latest technology, a scientific approach and international cooperation.

# References

1. Aidaralieva AA, Krylov GO (2017) Investigation of money laundering and extremism indicators relationship. In: 5th International conference on future internet of things and cloud workshops (FiCloudW), 25–30. https://doi.org/10.1109/ficloudw.2017.80
2. Arzamastseva EK, Nekhaychuk YS (2017) Financing of terrorism through the hawala system. In: The collection: European research: innovation in science XXIV international scientific and practical conference, 70–72
3. Atli D (2016) Cybercrimes via virtual currencies in international business. Cybersecurity breaches and issues surrounding online threat protection, pp 121–143. https://doi.org/10.4018/978-1-5225-1941-6.ch006
4. Cevik S, Ricco J (2018) Shock and awe? Fiscal consequences of terrorism Received. Springer-Verlag GmbH Germany, part of Springer Nature. https://doi.org/10.1007/s00181-018-1543-3
5. de Goede M (2018) The chain of security. Rev Int Stud, 24–42. https://doi.org/10.1017/s0260210517000353
6. Fahmy W (2017) Some legal aspects of the justice against sponsors of terrorism act. BRICS Law J 4(1):40–57. https://doi.org/10.21684/2412-2343-2017-4-1-40-57
7. FATF (2015) Emerging terrorist financing risks, FATF, Paris. www.fatf-gafi.org/publications/methodsandtrends/documents/emerging-terrorist-financing-risks.html. Accessed 20 Mar 2018
8. FATF (2015) Financing of the terrorist organisation Islamic state in Iraq and the Levant (ISIL), FATF. www.fatf-gafi.org/topics/methodsandtrends/documents/financing-of-terrorist-organisation-isil.html. Accessed 20 Mar 2018
9. FATF (2015) Guidance for a risk-based approach to virtual currencies convertible virtual currency exchangers. http://www.fatf-gafi.org/media/fatf/documents/reports/Guidance-RBA-Virtual-Currencies.pdf
10. Frolova YN (2017) Counteraction to the financing of terrorism as a key element in ensuring economic and national security. In the collection: Safety of a person and society as a problem of social sciences and humanities Materials of the IV international scientific conference, 30–33

11. Gilmour N, Hicks T, Dilloway S (2017) Examining the practical viability of internationally recognised standards in preventing the movement of money for the purposes of terrorism: A crime script approach. J Financ Crime 24(2):260–276. https://doi.org/10.1108/JFC-04-2016-0027
12. GLOBAL terrorism index (2017). http://visionofhumanity.org/app/uploads/2017/11/Global-Terrorism-Index-2017.pdf. Accessed 2 May 2018
13. Guryanov KV, Shatilo YS (2018) The development of electronic payment systems and the legal framework to counter the legalization of proceeds from crime and the financing of terrorism 1(3):23–37
14. Juels A, Kosba A, Shi E (2016) The ring of gyges: investigating the future of criminal smart contracts. In: 23rd ACM Conference on computer and communications security. Vienna, Austria, pp 283–295. https://doi.org/10.1145/2976749.2978362
15. Khabrieva TY (2018) Counteraction to the legalization (Laundering) of proceeds from crime and the financing of terrorism in the context of the digitization of economy: strategic objectives and legal solutions. Russ J Criminol 12(4):459–467. https://doi.org/10.17150/2500-4255
16. Manapov AT (2018) International legal framework for combating the financing of terrorism. Global Science and Innovations 2018 Materials of the International Scientific Conference. Eurasian Center for Innovation Development DARA. 352–355
17. Naumov YG, Latov YV (2017) The role of organized crime in financing terrorism and extremism. Russ J Leg Stud 3(12):230–234
18. O strateghii natsionalnoy bezopasnosti Rossiyskoy Federatsii do 2020 goda: Ukaz Prezidenta RF ot 12 maya 2009 № 537-FZ (red. ot 01.07.2014) [On the strategy of national security of the Russian Federation up to 2020: Order of the RF President dated May 12, 2009 No. 537-ФЗ (version of 01.07.2014)] // "Rossiyskaya gazeta" [the "Russian Gazette"], No. 88. – Art.59. [In Russian]]
19. Pol RF (2018) Uncomfortable truths? ML = BS and AML = BS2. J Financ Crime 25(2):294–308. https://doi.org/10.1108/JFC-08-2017-0071
20. Prado-Romero MA, Doerr C, Gago-Alonso A (2018) Discovering bitcoin mixing using anomaly detection. Progress in Pattern Recognition, Image Analysis, Computer Vision, and Applications, Publisher: Springer, Cham. 534–541, https://doi.org/10.1007/978-3-319-75193-1_64
21. Ramos P, Funderburk P, Gebelein J (2018) Social media and online gaming: a masquerading funding source. Int J Cyber Warf Terror 8(1):25–42
22. Serebrennikova AV, Lebedev MV (2017) Economic aspects of countering the financing of terrorism in the russian federation. Int J Econ Educ 3(3):30–42
23. Simanovich LN (2018) Financing of terrorism through transactions. In: the collection: expert of the year 2018 collection of articles of the III international scientific research contest, pp 72–76
24. Tarasova NV, Belovitsky KB, Lotareva KV (2018) The main methods of countering the legalization (laundering) of illegally obtained proceeds and the financing of terrorism. Collection of scientific papers Actual problems of ensuring economic security and countering the shadow economy, pp 219–224
25. Teichmann FMJ (2018) Financing terrorism through hawala banking in Switzerland. J Financ Crime 25(2):287–293. doi:https://doi.org/10.1108/JFC-06-2017-0056
26. Tierney M (2018) Terrorist financing: an examination of terrorism financing via the internet. Int J Cyber Warf Terror (IJCWT) 8(1):1–11. https://www.researchgate.net/publication/322729273_TerroristFinancing_An_Examination_of_Terrorism_Financing_via_the_Internet. Accessed 2 May 2018. https://doi.org/10.4018/ijcwt.2018010101

27. Zhakenova AK (2018) Features of the system to counter the laundering of criminal proceeds and the financing of terrorism. News of St. Petersburg State University of Economics 3 (111):131–134

28. Aryanova T (2018) Bitcoin and crime: How criminals use cryptocurrencies. Electronic Journal iHODL. (Electronic resource). https://ru.ihodl.com/analytics/2018-07-06/kak-silno-prestupniki-lyubyat-kriptovalyuty/. Accessed 2 May 2018

29. Aussilloux V, Le Hir B (2016) The economic consequences of abandoning the Schengen agreements. France Strategy (Electronic resource). http://www.strategie.gouv.fr/publications/consequences-economiques-dun-abandon-accords-de-schengen. Accessed 2 May 2018

30. Eurasian Group on Combating Money Laundering and Financing of Terrorism. (Electronic resource). https://eurasiangroup.org/ru/fatf. Accessed 2 May 2018

31. Felbermayr G, Gröschl J, Steinwachs T (2016) Trade costs of border controls in the Schengen area (Electronic resource). https://voxeu.org/article/trade-costs-border-controls-schengen-area. Accessed 2 May 2018

32. Kashevarova A and Kazakov I (2015) Rosfinmonitoring: 3.5 thousand Russians sent money to terrorists. The newspaper "Izvestia" (Electronic resource). https://iz.ru/news/596157. Accessed 2 May 2018

33. Sergeeva E (2018) The cryptocurrency turned out to be unbearable for terrorists. Electronic journal Cryptonews (Electronic resource). https://ru.crypto-news.io/news/kriptovalyuty-neprigodny-dlja-terroristov.html. Accessed 2 May 2018

# Robots Liability or Liability for Products?

Sergey Petrovich Bortnikov[✉]

Samara State University of Economics, Samara, Russia
serg-bortnikov@yandex.ru

**Abstract.** The article studies the relationship of liability arising during production, acquisition and use of robots and software. It is necessary to create the minimum comfortable conditions from the economic, legal and administrative points of view to develop the robots market. This is first and foremost the creation of an environment open to innovation, which requires legal transparency. The issue of the need to adjust the legal regulation especially in the field of civil and business legislation is to be considered to create conditions for the robots implementation. The study analyzes the existing institutions and concepts of the long-term legal regulation. The issue of the means of communication between a human and machine is also to be considered including the impact on the legal principles of regulation of relations of liability. The definition of liable parties implies the possibility of giving the robot a special legal personality, for example, giving it the status of an electronic person. The current European legislation on liability in tort and damage is studied against the rules of Directive 85/374/EEC. Various mechanisms to ensure and protect the interests of consumers and users of electronic devices are considered. It is assumed the obligation of robots liability insurance, as it has been already implemented for vehicles, the creation of special funds, the establishment of a liability limit, etc. for the practical use of liability mechanisms.

**Keywords:** Autonomous systems and IoT-Devices · Data economic · Fault-based liability · Liability for products · Rules on Robotics · Tort Law

## 1 Introduction

It is difficult to overestimate historically the importance of the mutual influence of the development of engineering, scientific and technological progress on society and its institutions. Currently, the technology development is in rapid progress. There are new types of mechanisms; robots are increasingly being introduced in production, in household use, as well as in the public sphere. The rapid development of robotics in recent years, as well as the plans of leading manufacturers to get its use to a new stage, have intensified international competition in this field and have caused the need to regulate the technology field that belongs to the future according to many analysts. The world's major economies have recently put forward programs for the development of robotics and its penetration into all spheres of society. The leader was Japan. The Headquarters for Japan's Economic Revitalization, which operates under the Prime Minister, developed the Japan's Robot Strategy - Vision, Strategy, Action Plan, in

© Springer Nature Switzerland AG 2020
S. Ashmarina et al. (Eds.): *Digital Transformation of the Economy: Challenges,*
*Trends and New Opportunities*, pp. 32–41, 2019.
https://doi.org/10.1007/978-3-030-11367-4_3

2015. The strategy was developed until 2020 and until 2025 on a number of aspects. Its basic object is to introduce robots into all spheres of life in Japanese society. The National Robot Initiative was adopted at the beginning of 2015 in the USA.

On the other hand, the progress requires the corresponding rules. The appearance of vehicles, railway service and other mechanisms has caused the development of direct liability regimes in many European jurisdictions. As to the U.S., the locus classicus is described by Morton J. Horwitz [12], as to Germany – Olaf von Gadow [13].

For example, in 2015, there were more than 43,000 industrial accidents in Russia. About 20,000 people die in traffic accidents each year, and more than 12 million people with reduced capabilities and the population is aging rapidly.

Unlike the processes of the beginning of the 20th century, robots and mechanisms are no longer controlled by people; they operate by a specific algorithm and machine code. This raises the question about liability for the mechanisms, as well as its distribution among manufacturers, suppliers of components, owners, custodians and operators of such devices. Many robots, programs and mechanisms have already been sold to date. They are mainly controlled by the machine code that is determined in the sense that it does not allow autonomous solutions for the machine or even machine learning. Another problem is that the machine operation is determined not so much by the actions of its operator as by interaction with the service consumer who becomes an operator and consumer.

## 2  Materials and Methods

The problem under study certainly needs to determine the legal basis for determining liability or its use for the Normative Foundations, new technologies. The purpose of liability is often considered only in terms of the regulation of compensation for victims. But it is important to determine the risk carrier; the problem is not the distribution of compensation to parties to a conflict.

Nowadays it is proposed to protect companies from liability for damage they may cause, for example, in order to promote innovation. It seems that the liability risk and deprivation of the right of the developer or manufacturer to make mistakes will deprive them of appropriate incentives to innovate. Such approach seems to be problematic [17].

Release of certain parties from liability for the harm they actually caused may result in an overabundance of such actions. In addition, such immunity discourages and cancels incentives to take precautions. Such an approach imposes costs of compensation for damage and harm on society at the expense of victims by themselves or the society. New technologies can and should find their own way into the world; they do not need immunity or protection from liability.

When developing the liability concept and pattern for autonomous systems the task of the legislator is to ensure maximum benefits for society with costs minimization for the victim related to compensation for harm and damages. This objective requires the identification and consideration of various facts and circumstances that affect the cost and the nature of expenses, as well as the necessary preventive measures to prevent accidents. These two factors were together placed under the rubric of "primary accident costs" in the classic work by Guido Calabresi [8]. Since the loss for the injured party is an additional burden that is often intolerable for individuals, the costs for different risks (so-called "secondary accident costs" [8]), the types of which also need to be determined, should be provided. The administrative costs for work with the liability system should also not be ignored. The liability rules should not be based on elements that are difficult to implement and expensive, as well as preventing the pre-trial settlement of disputes or the necessary proceedings in a court or other authority with jurisdiction to settle a dispute with responsible parties or their insurers ("tertiary accident costs" [8]).

Preference in settlement should be given to those persons who take maximum precautions against harm, i.e. to develop safety measures that are least costly than other safety measures available to other persons, and less than the costs for compensation for harm they help to avoid. All persons involved in potentially hazardous activities must bear the risk of compensation. The legislator should regulate the internalization in case of potential hazardous activities.

## 3    Results

### 3.1    Creation of One Level of Interaction Between Counterparties as a Condition for Legal Communication

A robot machine makes certain demands in collaboration with a consumer or any third party. From a technical perspective it is a process algorithmization, of course. In economic terms, it is required to simplify any business process as far as possible. Programmers can create very difficult algorithms, but a number of errors and exceptions are growing exponentially. All attempts to correct them lead to a complexity increase; therefore, there are new errors… No, every algorithm shall be utterly binary: flipped for it – a ticket fell out.

The second requirement for robot algorithmization is to program a human client. Who implements a program? – Yes, a 32-bit application in adaptation of WinXP in an iron box (cash dispenser). But a bank's client also does. It implements the same program. It is rather created not so much for a computer as for a client. Indeed, any application can be caused to do much more! It can dance squatting and emulate humor in communication. But, there will be problems concerning a consumer therefore. It will therefore cause decades of different reactions to be processed… It is much easier to program a person.

Who is more restricted by the frameworks of an intentionally simple algorithm in your "communication" with a cash dispenser – a robot machine or a consumer? Indeed, a consumer is unambiguously more restricted. A robot machine requires a PIN code to be entered several times per a session. It is a consumer who is programed, whose actions are programed.

An operator the human is still better than a robot machine. Telling the truth, it is not that it is exponentially or substantially better. An operator can meet the challengers which are not met in general by a robot machine at all. But it is so thus far.

And in case of simple operation performance a robot machine may be even more effective. Why? – A consumer understands that it speaks with a robot machine, submits it, loses its own emotional sensitivity (then it is useless) and shortens its speech.

A person stays in tune with a robot machine, speaks as a robot machine does – plain and simply.

It is the same "cash dispenser effect". A robot-answering machine programs a person. In 95% of cases robot-answering machine efficiency will be more than human-operator efficiency. Due to a fact the robot machine is indeed more effective (releases the line in less time) and can program a client, makes a primitive robot of it too (and it also releases the line in less time). Then effective managers will file a claim against human operators related to their low efficiency. And it will be quite logical:) In fact, an operator will make 20 sales after 100 long conversations and a robot machine will make 80 sales after 100 short conversations.

In 2018 during VIII St.-Petersburg international legal forum the panel, on which Megafon presented a judicial robot machine, was created. Moreover, not only did it present but also let it attend a real trial as a litigator. A person was an opponent of a robot machine. A judicial robot machine is a chart-bot operating on the basis of the neuronet. The neuronet is able to detect words in information which is loaded up to it, and it is able to analyze how often certain words come together. The more often it takes place the bigger proportion of combinations, which the neuronet assigns to an appropriate thread. And, replying to the question, the neuronet can provide texts based on these combinations. Moreover, the neuronet is constantly trained to provide texts – the more one speaks with it and loads up any information to it the "cleverer" bot becomes. It is reflected in the fact that its responses correspond to the topic in increasing frequency. Both a robot machine and a human were litigators in the trial. During debates, i.e. after the bot listened to what a human said, the bot did what everybody had been waiting it to do for so long time – it demonstrated skills of saying legal arguments!

Judges had ten criteria, according to which they marked from 0 to 10. A human scored 243 points, a robot machine – 178. This is not that bad for a first time and for a robot, which had been trained for just two months.

The conflict of a human with a robot machine acquires its specifics, however, at present it shall be considered in terms of existing legal forms. Introduction of robot machines into the system of justice, transfer of corporate lawyers', brokers', consultants' and others' functions to them, however, require additional legal regulation. For example, the argument relates to the fact that robot-advisors lack human perception. A human adviser can offer personalized investment guidance, and encourage investors to save more, diversify, and engage in less speculative trading. The SEC and FINRA have flagged lack of human judgment and oversight as a potential issue for automated investment tools. Financial professionals can ask the client questions to gather supplementary information and develop a nuanced understanding of the client's needs. By contrast, client-facing digital advice tools rely on a discrete set of questions to develop a customer profile.

If the automated investment tool does not allow you to interact with an actual person, consider that you may lose the value that human judgment and oversight, or more personalized service, may add to the process.

And critics of robot-advisors—from certified financial planners, to Massachusetts Secretary of the Commonwealth William Galvin, to Rutgers Law Professor Arthur Laby—have implied that human connection and judgment are essential elements of the investment adviser fiduciary duty [18]. They posit that only humans can connect with clients on a personal enough level to fully understand a client's financial situation. Robot-advisors are likely to miss the subtleties of a client's situation that arise in conversation. Wealth managers understand the value of knowing their clients - on a human and personal level. By having a deeper understanding of their clients, wealth managers can advise them based on [knowledge acquired] through human interaction, not merely by relying on the basic data that's entered into a [robot-advisor question-naire] [10].

## 3.2  National Tort Law and Legal Background

In the European Union tort law is an area in member states legal systems. Each EU member state governs its own liability system, and there are various differences among these systems, however, they share common principles [2]. These principles formed fields, which due to the efforts of comparative law researchers made it possible to determine a "common base" of European civil law. A clear example is Book VI of the Draft Common Frame of Reference known as the "Principles of European Tort Law" [1], prepared by the European Group on Tort Law. Without going into any details, it can be argued that the general rule of liability is a part of all member states' legal systems, and it is still a highlight for all principles, reiterating the common base of European private law. Thus, if a human fails to conduct due diligence, and this neg-ligence impairs any other human, or if an offender does such damage intentionally, this person shall compensate for the victim's damage. The principle mentioned implies reparation of damages to the main human interests, i.e. life, health, bodily integration, freedom of movement and private ownership; in some legal systems the list of interests protected also includes merely economic interests and human dignity.

Common principles of liability also clearly refer to the humans engaged in robot and IoT-device production and use. There are no immunities provided for them. However, one should be skeptical that the Directive 85/374/EEC specifies that at least 18 member states have difficulty in regulating the service provider tort rules [2]. Of course, there are some shortcomings, but differences between legal regulation of damages caused to software and service consumers in various European legal systems are insubstantial. It is clear that the common tort rules are applied both to service providers in the framework of commercial relations (business) and to consumers.

## 3.3  The Products Liability Directive

European Union Law does not quite have a lack of tort liability rules. The Product Liability Directive 85/374/EEC is an exception. It establishes an approach to deter-mination of damages on the basis of harm to the products which in accordance with Article 2 of the Directive (Art. 2 of the Directive) are determined as "movables". A demand for compensation for damages according to the Directive does not require the manufacturer's fault to be determined, as stipulated by Russian Federal Act # 2300-1 "On consumer rights protection". The Directive highlights that, according to its rules, "liability is without fault". However, one fact of damages themselves is not enough for liability to occur. It is rather required the product to be defective and such a defect to cause damages. The term "defect" is defined in Article 6 of the Directive for the purpose of launched product safety presumption (Art. 6 (1 (c) Directive). It is not quite clear what a clause related to application of the rules mentioned to special cases means [16].

A school of international comparative legal science postulates that the mode of manufacturer's product liability, provided for by the Directive, is coherent with fault-based liability, at least in case of liability for defects or nonoperational state of products [3, 7, 15]. And even in case of manufacturing defects, the Directive stipulates no direct liability, as, for example, the French doctrine "responsabilite de fait de choses" suggests [5], but it rather stipulates a "light" version of liability for imprudence.

According to the Directive, liability for damages caused by any defective product is not unlimited. Article 9 implies reparation of damages to life and health, as well as damages to property, provided that it is not a defect of the product itself and that it was intended for personal use. Even in such a case a 500-euro limit is established. Some EU member states consider such a limit as a discount, applied to all demands for compensation for material damages, while all other states allow the victim to file a claim for full compensation, but always provided that there is an excess of the limit mentioned. Liability is not applied if the relevant circumstances are not included in the list of violations, determined in Article 9. Merely economic interests as well as harm to human dignity are not regulated by the Directive.

### 3.4 Liability of Service Providers

All attempts to add any other legal instruments, related to service provider liability, to the Directive, were unsuccessful. Commission, Proposal for a Council Directive on the liability of suppliers of services, COM (90) 482 final, has not been introduced. According to European law the Directive insulation may seems to be tragic. However, it would be wrong to come to a conclusion that service providers are discharged from liability. Legal systems of member states constantly provide for faultless liability of all above mentioned persons, including service providers.

## 4 Discussion

### 4.1 Creation of the Habitat

On that basis, the thesis about the untimely and inexpedient regulation of robotics in Russia seems quite disputable. In 2015, almost 69,000 industrial robots were sold in China. Today, Korea is a world leader by a significant margin in terms of robotics density in the industry: 531 robots per 10,000 workers. Korea managed to overtake even Japan, because the index is 305 there, while the worldwide average level is 69. According to the latest data, the index of robotics density in the industry is only two robots per 10,000 workers in Russia. It is necessary to create the minimum comfortable conditions from the economic, legal and administrative points of view to develop the market. This is first and foremost the creation of an environment open to innovation, which requires legal transparency and clarity.

With this respect, the comfort of the legal system is essential. For example, the Japanese development concept clearly indicates the areas in which the adjustment of legal regulation is required to create conditions for the introduction of robots (medicine, insurance, transport and road traffic, etc.). In the USA, the Federal Automated Vehicles Policy, one of the first comprehensive acts regulating the use of highly automated machines, was published in September 2016. However, such machines are not yet common in the vast majority of States.

At the same time, the urgency of the regulatory issue often raises questions. Firstly, robots are practically not used in many countries; there is no special demand for them. That is, there is nothing to regulate. Secondly, the legal system will be able to cope with the problems of robotics in its current form. That is, regulation is not necessary. Thirdly, there are more important problems now, especially in the economy. That is, the regulation is out of time.

Regulation does not necessarily come after the market. Moreover, the regulation is needed to set standards. This can be shown by the example of a country that until the middle of the XX century did not belong to the giants of the world industry.

In order to increase the rate of robotics, Korea adopted the law "On Creation and Distribution of Smart Robots" at the end of 2008. This law was perhaps the first law of its kind in the world.

All of that allows looking at the problem of adoption of laws on robots the other way. The issue of legal regulation of what can already be found on streets, from drones to cars with self-driving capabilities, is on the agenda. Creation, discussion and adoption of the concept of state regulation of robotics are even more important. Of course, the general business climate, economic realities and social conditions in each country have their own peculiarities. That is why the experience of robotics regulation is different everywhere. The example of South Korea shows that this is good practices. It will be impossible neither to win, nor even to take part in the robotics race without the active role of the state (in whatever form it may be).

## 4.2   Civil Law Rules on Robotics and European Parliament Recommendations

It was determined that the interaction between people with robots needs a legislative authorization, therefore the European Parliament, Resolution dated February 16, 2017, with recommendations to the Commission on Civil Law Rules on Robotics, P8_TA-PROV(2017)0051, has defined the civil liability for damage caused by robots as the "major problem". The European Parliament brings this issue to the EU level, where should be legal certainty with regard to citizens, consumers and companies.

There is a choice between two different approaches: "risk management" and "direct liability".

Direct liability requires the availability of the following three elements: (a) harm, (b) "malicious" robot operation, and (c) a causal connection between harm to the consumer and the robot operation. Availability of intent (or its machine version) is not yet discussed. Of course, there are a lot of technical issues: from the definition of the "malicious" robot operation and its identity to technical failures to the features of the program and the algorithm set by the manufacturer. The behavioral design of the robot can still be generally considered as the liability of the manufacturer, which is currently quite successfully used for almost one hundred years (for example, Article 1079 of the Civil Code of the Russian Federation).

The "risk management" approach is considered as an alternative to direct liability; the liability is imposed on a person who acted negligently, or rather on a person who should minimize the risks of negative impact. The question remains unresolved as to what circumstances should be considered sufficient to impose liability on the user or the manufacturer: the person who created the risk or the person who did not minimize it, having such an opportunity.

Another aspect of the robots liability is the determination or establishment of their special legal status, i.e. their recognition as electronic people or entities (by the example of a legal personality). Such an electronic entity would be liable for any damage caused by the autonomous robot's behavior. Of course, this is the most innovative, interesting and stimulating idea within the parliamentary resolution.

Finally, it is assumed the obligation of robots liability insurance, as it has been already implemented for vehicles. But should it be mandatory or voluntary? In addition, the compulsory insurance mechanism could be implemented through a special fund that could incur some of the damage not covered by the civil liability insurance. Similar solutions already exist in the field of motors.

### 4.3   Building a European Data Economy

In terms of the raised problem, the Communication of the Commission to the European Parliament, the Council, the European Economic and Social Committee and the Committee of the Regions "Building a European Data Economy", 10.1.2017 COM (2017) 9 final was adopted. It reflects the results of discussions on the liability for the use of IoT-Devices, because they are deemed to be "a priority for the development of the digital economy". The existing structure of Council Directive 85/374/EEC dated July 25, 1985, on the approximation of the laws, regulations, and administrative provisions of the Member States concerning liability for defective products is considered insufficient to be used with robots: for example, the classification of autonomous systems does not allow you to define them as goods or, rather, as services. In addition, there is a question of insurance, which can be both voluntary and mandatory.

### 4.4   Responsible Parties

Persons involved in the creation and use of autonomous systems and IoT-Devices can be divided into two different categories: manufacturers and users. The manufacturer's group includes all persons, usually entrepreneurs, who facilitate the development, design and production of autonomous systems, including software developers and programmers. Another group includes everyone who interacts with an autonomous system or an IoT-Device after it has been set to work, i.e. owners, custodians and operators of such devices. The membership of the two groups is not closed or limited. This classification allows determining the possibility of sharing the burden of risk coverage between individual entities or their groups. It stands to reason that one of these means should be an agreement. Standard supply agreements, contracting and other agreements allow redistributing risks among suppliers of goods to the consumer for a long time, including costs for recall of goods, insurance, packing and wrapping materials, as well as costs caused by defective components, etc. [14].

The same can happen in a group of users and consumers, between owners and holders, owners and employees. Let's take a car, for example. The car owner is obliged to bear the costs of civil liability insurance in accordance with the applicable European directives and the current Russian legislation on Federal Act "On Compulsory Insurance of the Civil Liability of Vehicle Owners" No. 40-FA dated April 25, 2002.

If the car is rented to someone else, the cost of such insurance may be assigned to the renter driver as part of the rental price or without it. As long as the liability is placed on a particular entity of the group, the redistribution of costs for compensation of harm can be left to the parties and the freedom of agreement between them.

## 5  Conclusions

Thus, the current situation is characterized by a fragmentary nature of European and national law. Liability of business, launching the product to be sold, and liability of raw material providers are similarly regulated by the Directive mentioned (the Product Liability Directive). If in turnover there is not a traditional "product", but rather licensed rights or services, the Product Liability Directive is not applied and, therefore, no similar European liability system is also applied. In addition, liability of owners, holders or managers of an appropriate product is regulated by the national legislation if there is no EU regulation at all.

The reported research was funded by Russian Foundation for Basic Research and the government of the region of the Russian Federation (Samara Region), grant № 18-411-630011.

## References

1. Von Bar C, Clive E et al (2009) Principles, definitions and model rules of European private law. Munich, Germany
2. Von Bar C (2000) The common European law of torts. Clarendon Press, Oxford
3. Owen DG (201) Products liability law (3rd ed) West Academic Publishing
4. Busnelli FD, Comandé G, Cousy H (2005) Principles of European tort law. Springer, Vienna. https://doi.org/10.1007/3-211-27751-X
5. Terre F, Simler Ph, Lequette Y (2013) Droit civil - Les obligations (11th ed). Dalloz
6. Schwartz GT (1981) Tort Law and the economy in nineteenth-century america: a reinterpretation. https://digitalcommons.law.yale.edu/ylj/vol90/iss8/1. Accessed 24 Oct 2018
7. Bruggemeier G (2015) Tort Law of the European Union. Wolters Kluwer, Germany
8. Calabresi G (1970) The costs of accidents: a legal and economic analysis. Yale University Press

9. Kotz H (1991) Ist die Produkthaftung eine vom Verschulden unabhängige Haftung? In: Pfister B (ed) Festschrift für Werner Lorenz zum siebzigsten Geburtstag. Tübingen: J.C.B. Mohr, German
10. Bernatz K (2015) Preserving human judgment in the age of machines. Westlaw J https://www.firstamtrust.com/trust/news?page=3. Accessed 24 Oct 2018
11. Li M (2018) Are robots goods fiduciaries? Regulating robo-advisors under the investment advisers Act of 1940. Columbia law review. https://columbialawreview.org/content/are-robots-good-fiduciaries-regulating-robo-advisors-under-the-investment-advisers-act-of-1940-2/. Accessed 24 Oct 2018
12. Horwitz MJ (1979) The Transformation of American Law, 1780–1860. Oxford UP
13. Olaf von Gadow (2002) Die Zahmung des Automobilsdurch die Gefahrdungshaftung. Duncker & Humblot, Berlin
14. Ben-Shahar O, White JJ (2006) Boilerplate and economic power in auto manufacturing contracts. 104 Michigan Law Review. http://home.uchicago.edu/omri/publications.html. Accessed 24 Oct 2018
15. Whittaker S (1985) The EEC Directive on product liability. Yearbook of European Law. https://doi.org/10.1093/yel/5.1.233
16. Whittaker S (2005) Liability for products. Oxford UP
17. Shavell S (2004) Foundations of economic analysis of law. President and Fellows of Harvard College, USA
18. Bernard TS (2016) The pros and cons of using a robot as an investment adviser. N.Y. Times. http://www.nytimes.com/2016/04/30/your-money/the-pros-and-cons-of-using-a-robot-as-an-investment-adviser.html. Accessed 24 Oct 2018
19. Wagner G (2012) Custodian's liability. In: Basedow J, Hopt KJ, Zimmermann R, Stier A (ed) The max planck encyclopedia of European private law. Oxford UP
20. Wagner G (2018) Robot liability. https://ssrn.com/abstract=3198764 or http://dx.doi.org/10.2139/ssrn.3198764
21. Schubert W (2000) Das Das Gesetz über den Verkehr mit Kraftfahrzeugen vom 3. 5. 1909. https://doi.org/10.7767/zrgga.2000.117.1.238

# The Power Grid Complex of Russia: From Informatization to the Strategy of Digital Network Development

O. V. Danilova[(⊠)] and I. Yu. Belayeva

Financial University under the Government of the Russian Federation,
Moscow, Russia
danilovaov@yandex.ru, belayeva@mail.ru

**Abstract.** The relevance of the study is due to the need for technological improvement of the Russian energy industry, the development of digital technologies in the network complex. The key problems in the functioning of the Russian electric power industry remain the reliability, safety, quality and availability of electricity, the impact on the environment. In this regard, this article is aimed at disclosing the results of the reform and identifying prospects for the development of the electric power industry. The leading approaches to the study of this problem are formalized, research and system approaches that allow to comprehensively consider solving the problem of transition of electric power industry to digital technologies as the only possible way to improve the quality and reliability of power supply, solve financial and economic problems of the grid complex without increasing tariffs and additional burden on consumers. The article presents empirical data on the successful development and regulatory support of electrical grid regulation. of the country's economy, disclosed, it is revealed that in terms of the volume of introduction of new technologies in the banking and financial sector Russia is among the world leaders, which is facilitated by the relative youth of the industry (by world standards), which makes it relatively easy to implement high-tech solutions, "stepping over some steps of traditional development. The materials of the article are of practical value for workers of electric grids and specialists in the electric power industry, regulators of the electric power system when justifying strategies for investment and tariff policy.

**Keywords:** Digitalization · Electricity reform · Economic system ·
Smart grids state support · Tariff setting

## 1 Introduction

The urgency of the transition of the Russian electric power industry to the implementation of the network digitalization strategy is associated with the need to create an effective innovative energy sector of the economy, ensuring energy security and sustainable growth of the national industry and the quality of life of the population, strengthening the country's foreign economic position. The growth of industrial production, which began in Russia in 2017 and continues in 2018, significantly increased

© Springer Nature Switzerland AG 2020
S. Ashmarina et al. (Eds.): *Digital Transformation of the Economy: Challenges,
Trends and New Opportunities*, pp. 42–53, 2019.
https://doi.org/10.1007/978-3-030-11367-4_4

the requirements for energy information security, flexibility and reliability of the entire electric grid complex [3].

The development and regulation of the country's electric grid industry is carried out in accordance with the set of measures for reforming the electric power industry and creating a competitive electricity (capacity) turnover market determined in 1992. During the years of reform, more than a hundred and fifty laws on restructuring, privatization and streamlining the rules for operating energy companies in a market environment have been adopted, and considerable experience has been gained in working with energy consumers and services of energy companies. In 2009, the Energy Strategy of Russia for the period until 2030 was adopted, approved by the decree of the Government of the Russian Federation dated November 13, 2009. No. 1715-p [15], General Layout of Electric Power Industry Objects until 2035, State Program «Energy Efficiency and Energy Development» [12]. The goals of reducing tariffs for consumers, set during the liberalization of the electricity market, have not been resolved.

Relevant to the study remains the determination of ways to solve the problem of creating a reliable and complete system for accounting of consumed energy resources. Practice shows that this requires:

- develop an effective mechanism of competition of producers for contracts with consumers;
- to reduce the network component in the final price for electricity, since a significant network component causes a steady increase in tariffs, a factor hampering the development of the entire national economy.
- improve industry technical and economic indicators. At present, almost all indicators have changed for the worse: specific fuel consumption for power generation increased, generation capacity and power utilities decreased, installed power transmission losses increased, the number of production personnel increased significantly (accordingly, labor costs) the cost of construction of energy facilities has increased not only in comparison with the pre-reform period, but also in comparison with analogues of the construction of energy Projects by leading foreign construction companies.
- to bring to the world level the indicators of the share of the network component in the price of electricity for consumers (currently, in Russia this figure on average in the country reaches 50%). In the regions, the situation is even worse: the cost of electricity for consumers due to the network component in the prices of the wholesale market grows from 1.5 to 3 times. Given that the cost of gas, which employs more than 60% of thermal power plants in Russia, is below the world level, this price structure indicates an extremely low efficiency of the entire electric industry. The uncontrolled growth in the number of territorial grid companies, regardless of the volume of services rendered, the component of the "boiler" tariff based on the provision of the required gross revenues, also contributed to the growth of electricity tariffs.

The failure of these tasks led to an increase in the costs of all economic agents to pay for electricity. These issues were considered by Aleshina [2], Aleshina [1], Lyubimova [9], Spiridonov [12], Danilova [5]. According to Russian experts, the amount of financial resources diverted from the financial turnover of enterprises in the real sector as a result of the growth in electricity tariffs was at least 550 billion rubles a year, including only due to cross-subsidization of the population and similar groups of consumers by almost 300 billion rubles Kutovoy [8], Chebanov [4], Karamyan [4], Solovyova [4].

Currently, the total maximum power of consumers with a maximum power of at least 670 kW connected to the power grids of distribution subsidiaries of PJSC ROSSETI is 87 GW, and is used by consumers at about 44%. Such inefficient use of capacity occurs against the background of a chronic lack of investment in the electric grid complex, significant physical and technological deterioration of electrical networks. The average technical level of the installed equipment in distribution electrical networks in a number of parameters corresponds to equipment that was operated in developed countries 25–30 years ago. In fact, 50% of the distribution electrical networks have developed their regulatory deadline, and 75 - two regulatory deadlines. The total deterioration of distribution electric grids reached 70%, of main electric grids - about 50%, which is significantly higher than the similar indicators in other countries with a similar territory, where the depreciation rate is 27–44%. In the investment program of PJSC FGC UES for the period 2016–2020. 25% of the funds will be allocated for the implementation of technological connection of consumers, 29% - for the development of electrical networks and 46% - for the modernization of fixed assets. The investment programs of power grid facilities of subsidiaries of PJSC FGC UES provide for the modernization (renovation) of power grid facilities for 2017–2026. in the amount of 495.7 billion rubles, and for 2021–2026. −329.5 billion rubles. The main source of funding for these programs should be their own funds (depreciation and profit) - 64%, attracted funds - 15%, payment for technological connection - 9%, budget financing, funds for additional issue of shares - 2%, other sources - 10% (Research Institute of the Russian Academy of Economics 2016). Construction and maintenance of excess capacity requires appropriate operating and investment costs, to which network organizations direct funds intended for the modernization and renovation of electrical networks.

Taking into account the above, the purpose of the study is to develop proposals for the formation of a coordinated strategy for the development of the electric power industry. Special attention is paid to the need to solve the problem of introducing digital technologies into the electric grid infrastructure, the development of intelligent control systems, the organization of automated data centers, and the creation of intelligent metering systems for electric energy.

The study pursued the following tasks:

(1) to develop proposals for the formation of a well-coordinated strategy for the development of the electric power industry;

(2) to argue the need to implement the strategy for the development of the electric power industry based on the introduction of digital technologies in the electric grid infrastructure;

(3) to establish opportunities for the development of intelligent control systems, the organization of automated data centers, the development of systems for intelligent metering of electric energy;
(4) to determine the criteria for assessing the possibilities of using systems for the smart metering of electric energy.

## 2   Materials and Methods

The theoretical and methodological basis of this research is the concept of an actively adaptive network (smart network) as the basis of the intellectual power system of Russia, scientific concepts of the theory of organization and management, as well as studies of the work of Russian and foreign scientists related to the problems of increasing the energy efficiency of electrical networks, creating control systems smart grids that use information and communication systems to gather information about energy production and energy pg consumption and allow you to automatically increase efficiency, reliability, economic benefits, as well as the sustainability of the production and distribution of electricity.

The methodological basis of the study is a system analysis used to create digital substations, improve energy efficiency and reduce losses, apply digital design and improve power quality, remote control and security.

To achieve the goal and solve the problems identified in the study, the methods of generalization and logical analysis, statistical methods, methods of observation and comparison, empirical description, control and decision theory, a structural approach to system research were used.

The study of the ratio of useful power consumption and installed capacity of power substations was carried out in the whole country based on information obtained on the official website of the Ministry of Energy of Russia. The problem of the lack of comparable growth in electricity consumption among newly affiliated consumers is typical for practically all regions of the country. Therefore, the study of the dynamics of reserved but unclaimed network capacities was carried out on the basis of statistical information on the availability of grid infrastructure of the grid companies of PJSC IDGC of the North-West and VologdaEnergo, PJSC IDGC of Center - Kurskenergo. To study the reserves of network capacity for consumers with Pmax over 3 MW, the subjects of the Russian Federation, which have over fifty of the largest electricity consumers, were included in the statistical sample.

## 3   Results

### 3.1   Descriptive Analysis

A serious problem in reforming the Russian electric power industry since 2005 has been the availability of technological connection of consumers to electric networks in the regions. In order to solve the problem in 2015 at the legislative level, a number of significant changes aimed at simplifying the procedures of technological connection were introduced: the number of administrative approval procedures and the connection time (up to 90 days) decreased, the cost of accession for preferential categories of consumers decreased, 50% of the investment component was excluded from the fee for technical connection. According to the Federal Law on modification of Article 23.2 of the Federal Law "On the Electric Power Industry" the payment for construction of objects of the power grid economy performed within technical accession of the power accepting devices with the maximum power to 150 kW is not levied. Construction of distribution networks from existing power grid facilities to the applicant's site for connection is made entirely at the expense of the network company. As a result, according to the Ministry of energy of Russia, if 5 years ago, according to the criterion of accessibility of the electric grid infrastructure in the world Bank's DoingBusiness rating, Russia took almost the last 184 place, in 2017 Russia entered the top ten countries in this indicator Lyubimova [9].

Along with the positive result of simplification of the order of technical connection, the problem of optimization of network capacity reserves has sharply escalated. Every year the power company receives more than 500 thousand applications for new connection to electric networks. In accordance with the acts of technological connection in 2009–2016, the increase in the maximum capacity was 65 GW. Power generation by power plants of the Unified energy system of Russia in 2016 amounted to 59 576.3 million kWh, including thermal power plants – 59324.0 million kWh., nuclear power plants – 252.3 million kWh.

However, in practice, the connection of new consumers was not accompanied by a proportional increase in power consumption. Over the years of reforms in the electricity sector in the whole country, the useful power consumption has not increased by 1 kW-h. at the same time, the installed capacity of power plants has increased so much that more than 30 million kW of generation have been unclaimed (there are over the necessary reserves).

Table 1 presents data on reserve capacities in the regions of the Russian Federation (the sample includes only entities with more than fifty largest consumers of electricity).

The dynamics of real energy consumption was twice lower than the increase in capacity and the average for the Russian Federation is 58%, which indicates a low utilization of network capacity introduced during this period.

In 2015, the non-performance of obligations on the part of applicants for the supply of capacities amounted to more than 10 GWh in the whole country, which is 38% of the total amount of reserved but unclaimed network capacity. The connection of new consumers in the regions is not accompanied by a comparable increase in power consumption, which indicates that the newly joined consumers do not use the requested value of the maximum power of 100%. For example, according to PJSC «Interregional distribution grid company of the North – West» and «Vologdaenergo», in 2012–2016 in this region the average load of newly introduced power centers with voltage of 35 kV and above did not exceed 45%. This problem is typical for almost all regions of Russia.

**Table 1.** Reserves of network capacity in consumers with Pmax over 3 MW, for 2016

| Subject Russian Federation | Number of consumers | Electricities thousand kWh | P Fact. mW | Pmax. mW | Share of provision, % |
|---|---|---|---|---|---|
| Tyumen region | 80 | 36226163 | 4389,3 | 6406,7 | 31 |
| Belgorod region | 61 | 7627078 | 990,5 | 1485,6 | 33 |
| Kaluga region | 62 | 982772 | 229,9 | 440,7 | 48 |
| Vladimir region | 70 | 1940008 | 297,0 | 571,7 | 48 |
| Omsk region | 68 | 3238991 | 616,9 | 1256,9 | 51 |
| Moscow region | 396 | 3143956 | 2035,4 | 4228,4 | 52 |
| Perm region | 61 | 8003835 | 1635,5 | 3489,1 | 53 |
| Moscow | 258 | 3819295 | 848,9 | 1834,8 | 54 |
| Samara region | 80 | 6424396 | 876,3 | 1942,9 | 55 |
| Sverdlovsk region | 181 | 12282534 | 2026,0 | 4550,3 | 55 |
| Kirov region | 50 | 2132445 | 319,7 | 750,3 | 57 |
| Krasnodar region | 122 | 2235663 | 473,8 | 1238,2 | 62 |
| Tula region | 63 | 1755980 | 350,8 | 924,7 | 62 |
| Ryazan region | 55 | 1847718 | 263,5 | 740,1 | 64 |
| Kemerovo region | 68 | 1837353 | 408,3 | 1203,9 | 66 |
| Chelyabinsk region | 113 | 9003518 | 1643,3 | 4922,9 | 67 |
| Voronezh region | 66 | 2502657 | 421,4 | 1302,1 | 68 |
| Altai territory | 58 | 981017 | 181,2 | 589,3 | 69 |
| Saint-Petersburg | 158 | 2767104 | 324,2 | 1054,6 | 69 |
| Orenburg region | 57 | 4537990 | 631,5 | 2091,1 | 70 |
| Nizhny Novgorod region | 115 | 5120870 | 748,7 | 2776,9 | 73 |
| Volgograd region | 103 | 3738274 | 626,3 | 2427,1 | 74 |
| Chuvash Republic | 50 | 1495703 | 244,1 | 957,2 | 74 |
| Rostov region | 121 | 2035914 | 445,3 | 1923,8 | 77 |
| Tver region | 83 | 1199107 | 195,4 | 954,1 | 80 |
| Krasnodar region | 96 | 1773101 | 323,5 | 1698,8 | 81 |
| The remaining 38 subjects | 1053 | 63643209 | 21546,7 | 51762,2 | 53 |
| Grand total | 3748 | 192296651 | 29893,6 | 71856,8 | 58% |

Source: compiled by the authors on the basis of these sources: [7, 11, 15, 16]

## 3.2 Control Mechanism

The implementation of the state program to encourage consumers to join the electric grid has led to an overestimation and often irresponsible increase in the demand for new connections. According to the tariff setting in the electric power industry established by the Russian tariff system, the cost of services in transfer of electric energy does not depend on size of the maximum power declared within implementation of technology accession to the network organization, and is defined only by the actual volumes of consumption. Networked organizations bear the additional burden and financial costs of building and maintaining facilities that are not actually used. The increase in costs for the design, construction and reconstruction of power grid facilities and the operation of unloaded power grids leads to an increase in tariffs for consumers. Unreasonable forecasts lead to the construction of additional generating capacities that are operated underutilized and do not provide economic efficiency of investment.

As a result, the investment component accounted for in tariffs for services is spent by network organizations not for its intended purpose - not for the modernization of existing electrical networks, but for activities on technological connection. Redundancy of excessive demand leads to increased capital investment, increased maintenance costs and maintenance of electrical networks in constant readiness to provide consumers with the full amount of declared maximum capacity. For example, in 2016 the volume of investment funds spent by the branch of PJSC « Interregional distribution grid company of Center» - «Kurskenergo» for technological connection amounted to 64% of the targeted investment of resources, and in 2017, the volume of these funds amounted to 58%. In 2014–2016 the amount of funds a branch of the PJSC «Interregional distribution grid company of the South» - «Astrakhanenergo» for the accession of the new regional consumers - the construction and reconstruction of substations and power lines, accounted for more than 60% of the investment program of the branch.

For the purpose of rational use by consumers of the maximum power the Ministry of energy of the Russian Federation developed the mechanism of economic responsibility of consumers for use of incomplete volume of the declared capacities. The implementation of the proposed mechanism should ensure fair payment by consumers of all network capacity, including the reserved one Kutovoy [8]. Payment of the reserved maximum power shall be made by the consumers having the maximum power of the power accepting devices not less than 670 kW. Tariffs for transmission services will be set for consumers with hourly electricity metering. Compensation of the corresponding expenses to owners of the reserved capacities is provided at observance of the following conditions:

- the monthly actual capacity of the consumer in the preceding calendar year shall not exceed the maximum capacity by more than 60%;
- actual capacity for the current calendar year shall also not exceed 60% of maximum capacity;
- the amount of the maximum capacity to be paid for shall be determined as the difference between the maximum capacity and the actual capacity and the rate of payment, which shall gradually increase from 0.05 to 0.5 over a period of three years;
- the fee for the reserved maximum capacity shall be determined by the rate for the maintenance of electric networks of the two-exhibition rate for services in transfer of the electric power differentially on voltage levels;
- consumers from block stations for which the external network serves as a backup power source must pay 50% of the actual own power consumption, thereby compensating network organizations for the maintenance of networks.

Implementation of the fair pricing mechanism should lead to the rejection of the use of excess maximum capacity and the fair distribution of its payment among consumers. The introduction of the charge for the reserved maximum capacity includes a mechanism in a single step, according to which deductions from revenue in tariff regulation should be applied to network organizations in case of insufficient load of power centers.

For individual network organizations fee for the unused maximum power can lead to a decrease in the necessary gross revenue to 10%. This is a fairly high figure. Therefore, currently the decision a "phased transition" to paying major consumers of underutilized reserve capacity (discussing the period of 5 years). Over the years, it is planned to carry out a full-scale modernization of the Russian power grid, replace the equipment with a fundamentally new, innovative, development and introduction of digital technologies, an intelligent electricity metering system.

## 3.3 Management Tool

Replacement of technically and morally outdated and already economically inefficient infrastructure of the unified power system is an objective necessity for further development not only of the electric industry, but also of the entire national economy. Modern Supervisory control systems using existing Automated control system of technological process and SCADA TRACE MODE (High-tech Russian software system for process automation) cannot flexibly respond to emerging situational changes, which leads to failures and fan outages of network segments with appropriate financial, technical, material and other consequences Spiridonov [12]. In recent years, specific measures for the transition of energy to digital technologies have been regularly reviewed at the government level. The implementation of a national project to develop an intelligent energy system in Russia will reduce infrastructure costs and create conditions for expanding investments not only in the energy sector, but also in other industries (Fig. 1).

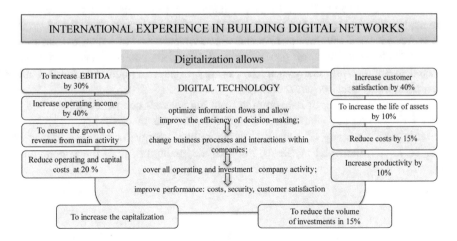

**Fig. 1.** International experience in building digital networks (Source: compiled by the authors)

A fairly new direction of energy modernization is the implementation of the national technology initiative (EnergyNet) project aimed at the introduction of Smart Grids technology with modern control and management devices in distribution networks. EnergyNet-the concept is quite extensive, uniting several large blocks in the field of electric power infrastructure. In addition to creating smart distribution networks, the EnergyNet project includes intelligent distribution power (solar panels, wind turbines, heat pumps, drives) and digital consumer services (flexible power consumption, transportation, etc.).

Smart Grid includes a wide range of technological processes, modern information and communication technologies, innovative equipment and applications designed for the transmission of electricity from the manufacturer to the consumer. The intellectual network represents set of the consumers of software and hardware connected to generating sources and electro-installations and information and analytical and control systems providing reliable and qualitative transfer of electric energy from sources to the receiver at the necessary time and in the necessary quantity. New principles, technology transfer and process control allow to combine at the technological level of consumers and producers of electricity in a single intelligent automated system.

The main objectives of the introduction of such a system is:

- ensuring high quality and reliability of power supply through the introduction of remote monitoring of equipment operation;
- the ability of the power system to self-recovery as soon as possible after emergency shutdowns;
- high resistance to physical and informational attacks;
- integration of all types of energy generation and storage, use of communication and information technologies;
- active involvement of electricity consumers in network management.

A key characteristic of the smart grid lies in its ability automatically to prevent (reduce) the continuous supply of electricity, to solve the problem of quality management of power supply and control of accidents. It should be emphasized that the main safety requirement of Smart Grids technology is to prevent the risk of cascading failures. In order for the "smart" network to work, it is necessary to form the so - called power clusters – a single information and technological space of individual territories, which include enterprises of generation and transportation of energy, companies engaged in engineering, energy services, energy engineering and instrument-making enterprises, educational organizations.

Intelligent processing of data coming from the network components, based on the Smart Grid technology platform, will optimize the use of electricity, improve the reliability and efficiency of energy systems, reduce energy losses, reduce resource costs, solve environmental problems, improve the quality of life of the population.

# 4  Discussion

According to SAP's expert estimates, the potential for GDP growth associated with the development of the digital power industry will be 200 billion rubles, of which 100 billion rubles will be in the power grids. As a result of the digital transformation of the network business, the profit growth of energy companies will be 4.3% of current indicators [6].

In this study, the assessment of the efficiency of digitalization of power grids was carried out on the basis of studying the positions of the expert community and the practice of digitization of the management system, collection, processing and analysis of energy consumption data in the constituent entities of the Russian Federation. Most experts in this field agree that digital energy should be beneficial both for consumers and for all market participants and should not lead to an increase in the cost of electricity. Based on the study, the position of the authors of the article is that digitalization activities should be financed not by electricity tariffs, but by the efficient operation of qualified market participants. As recommendations, the authors of the study propose:

1. To create conditions for increasing the economic efficiency of digitalization through the implementation of cheap domestic solutions and technologies. Import substitution will allow to stimulate the intensive growth of domestic production
2. Not to allow restriction of social functions of network organizations within the framework of their obligations on technological connection of consumers to the infrastructure.
3. Solve the problem of cross-subsidization.
4. To optimize all processes related to the production and consumption of electricity. To carry out measures to improve the tariff menu for consumers in terms of deepening differentiation by zones and time of day, which will encourage consumers to more actively implement digitalization programs.

These measures allow redistributing load schedules, freeing up and withdrawing unnecessary inefficient capacities, reducing construction and putting new unused capacities into operation.

The main risks associated with the introduction of an intelligent power grid are associated, firstly, with a high cyber danger, which is explained by the complex architecture of information and communication networks. Therefore, in the conceptual design of the Smart Grid, considerable attention should be paid to ensuring cyber security, including confidentiality, integrity and completeness of all information systems. Secondly, security risks are associated with the prevention of cascading failures. Other problems hindering the widespread adoption of Smart Grid technology both in Russia and around the world include:

- various consumer requirements for the quality of electricity consumed;
- Lack of reliable energy storage devices;
- high cost of technology - the implementation and operation of the Smart Grid requires significant financial investments;

- there are no standards and regulations;
- weak motivation of the management of generating companies, since the introduction of technology can cause a significant reduction in electricity consumption and, consequently, a decrease in revenues.

# 5 Conclusions

As a result of the provision of benefits to certain categories of electricity consumers in the regions of Russia, the demand for connection to electric networks has significantly increased. However, the increase in the number of new consumers did not result in a comparable increase in power consumption. The increase in the demand for electricity connections was not due to the activation of market mechanisms, but only to the measures of state support for the creation of new electrical facilities. Network organizations spend money not only on the creation of the necessary network infrastructure to consumers, but also on the maintenance of already built electrical networks, to maintain their readiness to issue to consumers the full amount of the declared maximum capacity, defined in the documents on technological connection. The source of funding for excess capacity is investment and operating funds that network organizations divert from the modernization and renovation of electrical networks. The solution to this problem may be the introduction of economic responsibility of consumers for reserving maximum capacity.

A promising direction for solving the accumulated problems may be the introduction of an intelligent electricity metering system. Intelligent accounting will completely change the stereotypes of technological connection, as without digital authorization and binding to the metering device, the consumer will not be able to use the services of the network. Installation of electricity meters will allow consumers to access hourly consumption schedules, optimize the cost of electricity through the use of different tariffs. State bodies, in turn, will have the opportunity to monitor the reliability and quality of services, reliability of electricity balances for tariff regulation.

# References

1. Aleshina EV (2014) Problems and prospects of development of electric grid business Russian Railways holding. Bull Samara State Univ Econ 3(113):36–40
2. Aleshina EV (2016) Modern practice and problems of reforming infrastructure holdings in Russia. J Manag Sci Modern World 2(1):218–220
3. Belayeva I (2014) Socially responsible activities of the state and business proceedings of the International multidisciplinary scientific conferences on social sciences and arts (SGEM 2014), Albena, Bulgaria, 357–363
4. Chebanov KA, Karamyan O Yu, Solovyova Zh (2015) The result of the power industry reform in Russia. Technological development of the Russian fuel and energy complex under the influence of economic sanctions. J Modern Probl Sci Educ 5:16–18

5. Danilova O (2014) Substainable development of territories of presence big business proceedings of the International multidisciplinary scientific conferences on social sciences and arts (SGEM 2014), Albena, Bulgaria: 373–380
6. Expert-Ural. It's time to digitize the network. http://www.acexpert.ru/. Accessed 20 April 2018
7. Information about the availability of the network infrastructure of the branch PJSC «Interregional distribution grid company of the North – West» and «Vologdaenergo» for technological connection. http://www.invest35.ru/assets/files/docs/1_6____.pdf Accessed 20 October 2018
8. Kutovoy GP (2015) Formation of forms and methods of state regulation of electric power industry during reforms of economic relations and privatization. Analytical review, Annex to the magazine «Energetik» 12:13–22
9. Lyubimova NG (2016) Long-term supply of reliable and affordable power supply to consumers. Bull Univ 10:76–79
10. Report on the functioning of the EES of Russia. http://www.so-ups.ru/
11. ROSSETI. http://www.invest35.ru/assets/files/docs/1_6____.pdf. Accessed 22 Mai 2018
12. Spiridonov VV (2014) Intellectual technologies in electric power systems proceedings of the International conference "Applied research and technology", Moscow, MIT «VTU», 51–53
13. State Program "Energy efficiency and energy development" approved by the Government of the Russian Federation 29.05.2015 (№3384p-P9). http://i.cons-systems.ru/u/8c/0a55a8bcc311e4aca5fbaefc9347ae. Accessed 22 Mai 2018
14. Spiridonov VV (2014) Intellectual technologies in electric power systems proceedings of the International conference "Applied research and technology", Moscow, MIT «VTU», 51–53
15. The main provisions of the concept of the national project of the intellectual energy system of Russia (2016) The Ministry of energy of Russia. Research Institute of the Russian Academy of Economics. https://minenergo.gov.ru/. Accessed 14 Mai 2018
16. The strategy of the electric grid complex, approved by the Order of the government of the Russian Federation № 511-R (2013) http://www.rosseti.ru. Accessed 22 Mai 2018

# Features of Information Support of Export Marketing in the Conditions of Digitalization of the Global Economy

S. V. Grankina, N. A. Kryuchkova, and Ya. G. Sayamova[✉]

Samara State University of Economics, Samara, Russia
{svetav_grankina, popova_yana}@mail.ru,
kryuchkova_n.a@bk.ru

**Abstract.** The issue of export of goods by Russian enterprises requires a preliminary assessment of the potential entity in a foreign market. One of the fundamental elements of the decision-making process is to assess the conditions to enter the market of a particular country—define a market niche. According to experts, the algorithm for export marketing does not have significant differences from marketing in the national market. The Internet has been used in conducting marketing research, especially when studying foreign markets in recent years. At the same time, the study of the foreign market is complicated by various kinds of barriers - linguistic, informational, specifics of customs authorities, as well as representatives of the enterprise on the territory of the state to conduct field research. In this case, the research algorithms used and their elements may not always be applicable for a particular country, product or service, and a number of them require mandatory adaptation to factors and trends inherent in the object of study. The markets for goods are characterized by the range, and high dynamics in supply of new models by manufacturers. As a result, there is a complex of problems in case of remote research that can reduce the results achieved. The authors proposed methods and practical recommendations for information retrieval upon mentioned groups of products in the global information space in economy digitalization using various Internet technology tools.

**Keywords:** Digitalization · Information · Information space ·
Internet technology · International market · Global information space ·
Problems of information support for research · Web-scarping

## 1 Introduction

In the ideal situation, the study of a new market should be based on desk and field research. However, with the right approach, the primary information retrieved on the basis of desk research allows drawing preliminary conclusions and necessitating for field research in case of positive results obtained through open information sources available in economy digitalization.

1. If we consider the development of survey methods in research, then the current key stage is the active introduction of the Internet and mobile technologies [22, 23]. New trends in marketing research are a shift in favor of the global information space [1].

© Springer Nature Switzerland AG 2020
S. Ashmarina et al. (Eds.): *Digital Transformation of the Economy: Challenges,*
*Trends and New Opportunities*, pp. 54–65, 2019.
https://doi.org/10.1007/978-3-030-11367-4_5

Information technology tools in the Internet environment allow enterprises to reduce time and financial costs in studying their potential capabilities in foreign markets, and it is increasingly relevant in terms of cost optimization [2, 5].

2. An indispensable condition for the potential competitive advantage of a Russian enterprise in a foreign market is reliable market information, which will make it possible to formulate a strategy for entering a new market.

The purpose of the study is to identify the problems of information support for export marketing for the commodity market, which is characterized by the range and high dynamics in supply of new models by manufacturers, and to suggest possible solutions for them.

To achieve the purpose, the following tasks were set and solved:

1. Determine the optimal application of Internet technology tools in the study of foreign markets for goods and services;
2. Evaluate the effectiveness of email requests for primary information on foreign markets for goods and services, with a wide and deep range and high dynamics of development;
3. Determine the list of problems in the foreign market for goods and services;
4. Provide recommendations on primary information retrieval in conducting export marketing using Internet technology tools.

## 2   Materials and Methods

The study was conducted using theoretical and empirical methods. Theoretical research methods (analysis and synthesis, induction and deduction) made it possible to interpret the collected facts, to identify patterns in the algorithm in studying the foreign market. As for practical methods, the authors used observation, survey, comparison, computational methods. The graphical method made it possible to visually demonstrate the totality of the problems of information retrieval in the global information space in conducting export marketing, as well as the consistency and variability of primary information retrieval using Internet technology tools. The combination of these methods allowed us to ensure the validity and consistency of the results.

## 3   Results

The representativity of information in conducting marketing research is one of the significant problems faced by enterprises in the study of domestic and foreign markets. A scientific and methodological approach, including a set of principles, methods, tools, determines general aspects and specificity of research at the international level. However, it should be borne in mind that, apart from the general problems of representativity for any market and any product or service, there are also difficulties, the nature of which associated with the specifics of the subject of the study [12]. In the context of a general economic downturn, a complex of sanctions and an aggravation of political relations between countries in the world community, the digital channel for information retrieval remains the most accessible and optimal in terms of primary costs. In turn, a remote study of a foreign market is accompanied by a set of problems and factors that adversely affect the reliability of the final data.

Figure 1 shows the problems accompanying necessary information retrieval in conducting export marketing for product markets with a wide range and high development dynamics.

Information retrieval should begin with the definition of a potential group of countries for possible foreign trade operations. Macroeconomic parameters for the development of these countries are available in open sources - public and private statistical services, which record the dynamics of the main economic indicators of the country's development. An important indicator necessary for analysis is the size of exports and imports of the country of potential presence. An example of a source of such information is the Information Portal of the Central Statistical Office of Poland.

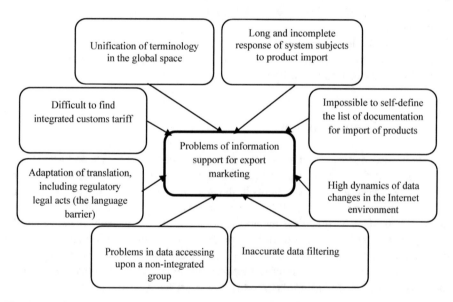

**Fig. 1.** Problems of information support for export marketing (*Source* compiled by the authors)

Also, the structure of export-import can be presented on the basis of such information resources as The Observatory of Economic Complexity [16], Trade Map [20], and others, with the indicators of export and import efficiency, as well as the main market players [3]. These information resources have convenient filters that allow you to make a sample both by country and by product.

Traditions and customs, consumer culture and consumer behavior of the population of the country of potential presence require compulsory study. The source of this information can be websites of manufacturers of products, analysis of which allows receiving information about consumer preferences, as well as sites of large retail chains that give feedback to their customers, which is also a source of valuable information about consumer tastes and their satisfaction with products of the enterprise.

Assessment of the enterprise – exporter's capacities in the foreign market requires analyzing the activities of potential competitors – studying their production capacity, range, production and sales volumes, the number of enterprises and their employees. Many enterprises publish reports on their activities on websites. However, not all parameters of their functioning are publicly available. It is difficult to study the fact that a number of manufacturers and trade enterprises belong to international groups of companies, and statistical reporting is available for companies as a whole without revealing the share of enterprises operating in a particular country.

For example, a company operating on the Polish market under the KOŁO brand became part of the global Geberitt Group. Quantitative data on production in this group are freely available and information on the share of enterprises located in one country is also available, which allowed obtaining primary information on production under the KOŁO brand in Poland [19].

With this in mind, it is possible to model the data obtained by identifying the share of each enterprise in the group of companies, based on the general data for the group.

At the same time, when researching international markets, in order to retrieve information on the volume of production by local producers in a particular country, it is necessary to send an e-mail to request the necessary data for an international study in addition to searching for open data. So, from 10 sent letters (among them Ravak, Sapho, Laufen and others) there was only one response. Contact details were taken from the official pages of manufacturing companies. At the same time, even 1 received response did not fully contain the data required for evaluation.

The language barrier can be solved by using the "Page Translation" service; however, a literal translation does not always correctly reflect special terms, which can complicate the analysis of information. In particular, the analysis of the range by models can be difficult due to bad translation of model names. As a result, a number of product items may not be taken into account, or duplication of the same information, presented in different verbal form, may occur. The solution to this problem is to attract reliable information from a native speaker of the country whose market is to be studied, or to carefully compare the range presented in the trade enterprise with the manufacturer's range, with key quantitative and qualitative features of products.

One of the problems of information retrieval in conducting marketing research on the Internet at the international level is the following: it is difficult to select reliable sources and sources containing unified information, and it is difficult to convert data collected at various sites into a single form.

For example, the level of development of Internet navigation sites of foreign manufacturers and intermediaries in the countries of the European Union showed that: firstly, many sites are not equipped with an automatic commodity item counter, which slows down the data retrieval process (piece counting) and increases errors when independently calculating data for evaluating the structure of supply in the context of retail sales channels; secondly, filtering positions for data retrieval for a particular product category sometimes results in a single-item recalculation, because the poor quality of the filter setting leads to the so-called re-grading. This aspect requires additional study in terms of site configuration and improving the quality of information retrieval during filtering. Accordingly, when retrieving information, the researcher should be guided by the need to re-check the data of the built-in counter. This leads to additional labor costs and an increase in the time for obtaining research results.

Information retrieval for export marketing includes the need to analyze the cost of delivery and estimate the cost of importing goods to the country of destination. The study of the specifics of customs clearance and fees in the EU countries should be started by studying the EU Council Directives, the Council Regulations on the Tariff and Statistical Nomenclature and the General Customs Tariff [18]. The Tariff code system has been developed for imported products and the customs duty for groups of goods in EU countries is basically the same, but the value added tax varies, for example, for sanitary ware from 19 to 23%, and local taxes or import restrictions are still possible [6]. To understand the size of costs when importing products into a foreign state, it is necessary to calculate the integrated customs tariff. The search for this portal is difficult due to the lack of links both on official legal websites of the EU and on websites of the customs administration of the countries. Requests by e-mail to advisory and information centers of the customs administration of foreign countries help in the study of this resource, but are characterized by a long waiting time for a response. The exception is the Republic of Latvia. The response time of customs clearance centers was 1–2 days, but the probability of getting a response is 40%. The Czech Republic showed the highest probability of a response among the EU countries - 100%, but the duration of waiting for a response to a request ranged from 8 to 13 days. Figure 2 clearly reflects the conversion of responses to email –queries, in terms of the time factor and completeness of the response.

According to the results, the queries can be classified into 4 groups:
Group I - badly informative (no response or the wrong email);
Group II - poorly informative (requiring information by telephone);
Group III - well-informative (a link to the Internet resource for calculating the integrated customs tariff);
Group IV - well informative (including the results of the calculation of rates and customs fees according to the Tariff code).

The peculiarity of information retrieval from authorities is characterized by inten-sified activities of the researcher, due to the lack of efficiency in updating information, a request via e-mail must be sent to all the addresses provided on official resources. Often requests do not reach because the addressee is not available.

Recently, enterprises of the market of transport and logistics services working for export have become very cautious about external demands. If earlier a letter received by e-mail from a free e-mail address was considered normal, then now it is increasingly necessary to prove the existence of the declared organization. Inquiries by e-mail will be more successful if you call representatives of the transport and logistics company beforehand and provide the following information: the name of the represented enterprise, its location and field of activity, contacts for communication. The proba-bility of obtaining results without prior telephone communication is 5–7%. Among the factors that reduce the representativity of information on foreign markets, it is neces-sary to note the high dynamics of updating data on the range (dynamically there is a reduction or increase in the number of positions).

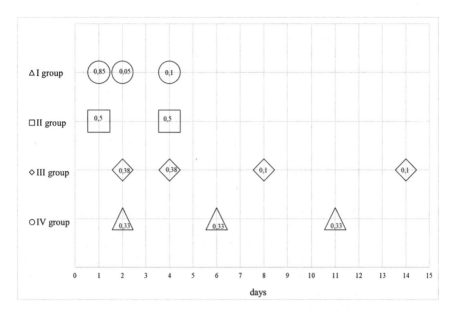

**Fig. 2.** Dynamics of conversion of email requests to customs authorities (*Source* compiled by the authors)

For example, a study of the assortment in online stores in Latvia, Poland, Germany and several other countries of the European Union showed that in a situation where the work with the site was carried out in stages over several days, it was necessary to correct the total number of positions each time and carry out already collected data for individual product categories so that the current data correspond to the site data. The range of change was from 0.3% to 5%. This feature is typical for markets for goods and services with a wide and deep range and high development dynamics.

At the present stage, a specific trend in the development of foreign markets is the active distribution of no name products in individual countries. This complicates the analysis of information about the proposal in the context of several signs. This aspect should be taken into account when analyzing the structure of supply by countries, the structure of supply by brands. The country of origin is not always indicated on sites in sales channels and there is practically no information about a particular enterprise.

The unification of terminology in the global space is one of significant problems of international market research. This is due to both objective and subjective factors. It should be noted that individual data retrieval upon the structure of supply in e-sales channels is complicated by the fact that some large companies do not provide a catalog of goods and, as a result, complicate the assessment of aggregate supply in the country's market. For example, the international networks "Metro", "BN Kurši" and a number of others place only an offer of goods for a promotion, and not the entire range of products represented in the trading network.

Based on the problems described above that arise in necessary information retrieval for export marketing, the authors suggested using Internet technology tools in the system for collecting information on foreign markets (Fig. 3.).

As part of the methodology, principles of information retrieval in the global information space are formulated using a set of Internet technology tools:

1. Parallel use of Internet technology tools;
2. A large number of requests using postal services;
3. Minimization of the list of questions of interest in the e-mail request and their maximum specification;
4. Optimization of the verbal request structure.

The initial stage of research determines a set of initial conditions for information retrieval. That is, primary information retrieval in the global network is carried out in accordance with the tasks set on the basis of the general methodology of international marketing research, while it is advisable to plan and organize several search processes in parallel, since modern Internet technologies provide access to a set of information retrieving and processing tools. In particular, the method of export marketing in the framework of information retrieval can be based on various search services, and such important tools as web-scarping and specialized software products (information platforms tailored to distinctive features of goods or services). The launch of information retrieval processes at once using the above three tools ensures high speed of information gathering and its completeness taking into account the scale of the global information space.

It should also be noted that in most cases specialized software products will require certain costs. However, when deciding whether to expand the market at the international level, these costs are justified. In turn, web-scarping can be paid and free.

Work with web sites that are relevant to a request can be completed at the initial stage either by retrieving the necessary and complete information or partially limited, or by the lack of necessary data. Taking into account the variability of results, it is necessary to use such Internet technology tools as web forums and mail services in the system of information retrieval for export marketing of goods with a wide and deep assortment and high dynamics of development. Their application will be more effective when using the principles proposed by the authors and data on the possible conversion of responses to email requests.

It is recommended to carry out the second stage of information processing by complementing the data from the first stage of retrieving, processing and evaluating information with information that can be retrieved on the basis of direct contacts (using skype, telephony and ICQ chat), as well as indirect methods.

The transition from stage to stage occurs consistently and from the perspective of the researcher means that information retrieved from global digital sources is insufficient and there are problems compiling the necessary representative data, especially those that characterize production volumes in quantitative terms (these data are especially important for the exporting company), production and marketing structure by brands and countries.

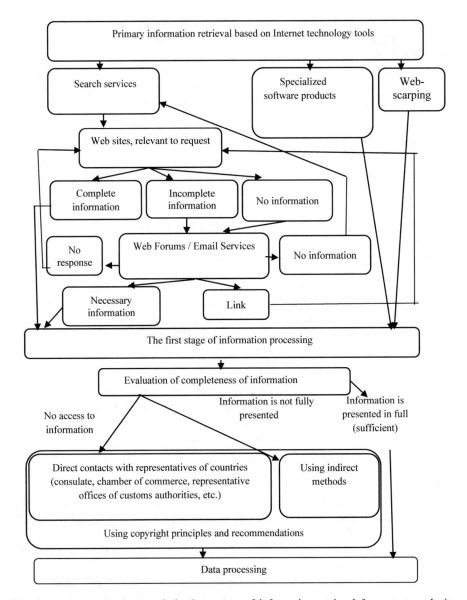

**Fig. 3.** Internet technology tools in the system of information retrieval for export marketing (*Source* compiled by the authors)

It should be noted that the duration of each stage can be significant, since the expectation of a backlash within the framework of organizing direct contacts with representatives of the country of study, according to the authors, can take on average up to two weeks.

In addition, the possibility of response or conversion of contacts is quite low, which means that it is necessary to ensure a wide coverage of possible sources of information, having previously formed a contact list with a set of contact data.

Since the problems arising from conducting research for export marketing can affect the representativity of the reporting data and the duration of work, the authors suggest the following recommendations for solving them and minimizing the risks they cause:

- In order to improve the accuracy of translation and to retrieve reliable information about the activities of subjects in the foreign market, as well as legislation, it is possible to involve a specialist in technical translation or a native speaker of a country of potential presence, using the page translation service with careful comparison of the range presented in the commercial enterprise with the manu-facturer's one. The key parameters should be quantitative and qualitative features of the product;
- For the unification of terminology, it is necessary to study qualitative features of the product, its technical features, and in identifying their identity, to establish fixed concepts that can be used in further research. The same principle should be followed when identifying homogeneous processes and fixing their definitions in a report;
- In order to complete the response at the initial stage, it is required to form a request and send it to the civil service consulting centers responsible for contacts with potential partners, to employees representing the producers of goods and retail networks in the countries of possible presence.

If there is inefficient interaction via the Internet, it is necessary to establish personal contacts with representatives of producers of goods, retail networks and transport and logistics companies by phone. When forming an Internet request, the authorities need to minimize the list of questions of interest and maximize their specificity.

Provided that the market of a particular country has its own specifics, the terms used in the global space relating to the assortment of the product group to be studied should be clarified.

Information retrieval about the assortment of a specific sales channel from the site should be carried out in a limited single time interval (most optimally, within 1 day). If the information on the assortment on the site of the intermediary is of a truncated nature, then field studies are needed directly at outlets or by e-mail request in order to increase the representativity of information.

To minimize the risk of errors in filtering data using the built-in capabilities of the site, you should perform a visual check of the conformity of counted positions based on filtering. If a discrepancy is found, then make an adjustment on the basis of a piece-wise recalculation of positions in a separate product group.

To access data on a non-integrated product group, it is possible to request such information to the statistics authorities of the countries whose market research is being carried out. The recommendations given by the authors allow enterprises, in the context of their limited capabilities and thanks to economy digitalization, to obtain primary information about the markets of countries with a potential presence in order to make an informed decision on field research.

## 4   Discussion

The essence of the concept of "marketing research", its theoretical and practical significance was determined in foreign studies of the first quarter of the XX century. In 1921, P. Byte in his work "Market Analysis" (one of the first works devoted to market research) described the main types of marketing research [14]. A. Nelson, one of the founders of international marketing research, introduced the concept of "market segment" into scientific circulation in 1923 [22]. In turn, F. Kotler [8] proposed a classification of marketing research. The need and importance of marketing research contributed to the emergence of the research association ESOMAR in 1948, which unites companies and organizations conducting marketing research around the world. This association has developed the ESOMAR code, which defines the basic concepts and presents the principles of conducting marketing research, including international ones [17].

The problems of the representativity of information in conducting marketing research over the past decade are solved due to information and analytical systems. This tool allows you to effectively and independently solve the problems of processing and analyzing materials from open sources. One of the first developments of the market in Russia is the information retrieval system "Marketing", as well as information and analytical systems: "Medialogia", "Artefact", Public.ru and others. The main drawbacks of the systems under consideration when studying foreign markets is the absence of the "Study of sectoral markets" add-on, mainly Russian-language sources and a monthly subscription [10]. Without a constant need to study export markets, the feasibility of acquiring is very low. Among the publications on decision-making support systems, M. Lutovac [9] made a valuable contribution to the description of such systems and the specifics of international research. The features of international marketing research are studied in detail by scientists and practitioners from various countries. However, at the present stage much attention has been paid to the development of remote research in the global space. The dynamics of changes in information flows and the amount of diverse data increase every year, but their closeness also increases, and access to certain areas is complicated [4, 13, 15, 21]. It should be noted that distinctive features of specific goods or services that determine the specialization of the exporting company and, accordingly, the market [7, 11] have a significant impact on research. Identification of these aspects, taking into account the specifics of the object-sale allows you to complement the capabilities of researchers and increase the speed of information retrieval.

## 5   Conclusions

The desire of Russian manufacturers to optimize the budget for export marketing leads to a shift in preferences in favor of remote research based on a variety of information sources, as well as direct contacts using global information systems that have been widely developed in economy digitalization. At the planning stage of information retrieval on foreign markets, the identified problems of the representativity of information should be taken into account and their impact on the results should be minimized using the proposed author's methodology, including various Internet tools, principles and recommendations. This will allow ensuring a high level of reliability of the data obtained, and objectively evaluating the necessary time, labor and financial resources even in the framework of the initial remote study.

# References

1. Agafonova A, Pogorelova E, Yakhneeva I (2018) Marketing-mix evolution in e-commerce. Econ Entrepreneurship 7(96):709–714
2. Agic E, Cinjarevic M, Kurtovic E, Cicic M (2016) Strategic marketing patterns and performance implications. Eur J Market 12:2216–2248
3. Bank Danych Makroekonomicznych (BDM). https://bdm.stat.gov.pl/. Accessed 3 Mai 2018
4. Evtodieva TE, Chernova DV, Voitkevich NI, Khramtsova ER, Gorgodze TE (2017) Transformation of logistics organizational forms under the conditions of modern economy. Russia and the European Union. Development and Perspectives Cep. "Contributions to Economics" Cham, Switzerland
5. Frösén J, Tikkanen H (2016) Development and impact of strategic marketing – a longitudinal study in a nordic country from 2008 to 2014. Eur J Market 12:2269–2294
6. Information system of the integrated customs tariff in the Republic of Poland. https://ext-isztar4.mf.gov.pl/taryfa_celna/Browser?lang=PL&cssfile=tarbro&date=20180518/. Accessed 25 Mai 2018
7. Kipnis E, Emontspool J, Broderick AJ (2012) Living diversity: developing a typology of consumer cultural orientations in culturally diverse marketplaces: consequences for consumption. Adv Consum Res 40:427–435
8. Kotler Ph (1967) Marketing management: analysis, planning and control. PrenticeHall, Englewood Cliffs, NJ
9. Lutovach Mitar et al (2016) Identification of information relevant for international marketing. Int Res J 7(49):43–48. https://doi.org/10.18454/IRJ.2016.49.185
10. Plesovskikh AV, Kovalenko NI (2014) Marketing research as an informational occasion for advancement in the external environment. Serv Market 3(39):166–181
11. Popadynets N, Shults S, Barna M (2017) Differences in consumer buying behaviour in consumer markets of the EU member states and Ukraine. Економічний Часопис-XXI 7–8:26–30
12. Preece M (2015) Managing information and knowledge in service industries. Adv Bus Market Purch 22:3–154
13. Rana S, Sharma SK (2015) Advances in international marketing. A literature review, classification, and simple meta-analysis on the conceptual domain of international marketing 1990–2012(25):189–222
14. Safarova IM (2014) Evolution of development of methods of research of the trade markets. Econ Manag New Chall Prospect 6:29–31
15. Steenkamp JB, Baumgartner H (1998) Assessing measurement invariance in cross-national consumer research. J Consum Res 25:78–90
16. The Observatory of Economic Complexity. https://atlas.media.mit.edu/ru/. Accessed 25 Mai 2018
17. The official website ESOMAR https://www.esomar.org/. Accessed 23 Mai 2018
18. The official website European Commission. http://ec.europa.eu/taxation_customs/. Accessed 30 Mai 2018
19. The official website of the brand KOLO. https://www.kolo.com.pl/. Accessed 28 Mai 2018

20. Trade statistics for international business development. https://www.trademap.org/Index. aspx. Accessed 20 Mai 2018
21. Voitkevich N, Chernova D, Sosunova L, Astaf'eva N (2018) Evaluation of the effectiveness of marketing solutions in the channels of distribution of goods and services. Econ Sci 159:22–25
22. Williams JEM (2006) Export marketing information-gathering and processing in small and medium-sized enterprises. Market Intell Plan 24(5):477–492
23. Young RB, Javalgi RG (2007) International marketing research: a global project management perspective. Bus Horiz 50(113–22):1

# Human Capital Evaluation in the Digital Economy

Tatiana Anatolyevna Korneeva(✉), Olga Nikolaevna Potasheva,
Tatiana Evgenyevna Tatarovskaya,
and Galina Aleksandrovna Shatunova

Samara State University of Economics, Samara, Russia
korneeva2004@bk.ru, {olgakuzmina0212,
tatarovskaya.tatyana}@gmail.com, shatunova.g@ya.ru

**Abstract.** The purpose of the study is to identify the importance of human capital in the digital economy. The authors substantiate the relevance of knowledge as an element of human capital to ensure its effective functioning. The impact of widespread digitalization in all socio-economic processes on the development of this type of capital was observed. The importance of knowledge led to the emphasis in the study of attention to the identification of problems of the modern education system through the prism of the difficulties faced by graduates of educational institutions. According to the authors, the assessment of human capital should be expressed in assessing the level of qualifications of both graduates and employees of economic entities. In the framework of this study, a survey of business entities in the Samara region is conducted to identify the graduates' employment problems (the human capital formation problem at the beginning of the individual career path in the digital economy) and to determine the level of employees' financial literacy (the problem of human capital growth in the digital economy). In order to form comprehensive view of human capital development prospects the scoring the results method was applied for these surveys.

**Keywords:** Digital economy · Digitalization · Education system
Human capital · Labor market · Professional standard · Qualification assessment

## 1 Introduction

The current stage of economic development is characterized by its continuous filling with innovations, new knowledge, information systems to solve problems of any complexity, including the global ones. The penetration of computer technologies in all spheres of social and economic processes has formed a new understanding of the economy as a social relation set [8]. In recent years, researchers from all over the world use the concept of "digital economy" to describe the observed phenomena in modern conditions.

In the Russian Federation, the task of developing the information society has been set at the state level. Implementation of the strategy is planned at the period from 2017 to 2030. As part of this strategy, a special place is occupied by the program "Digital

S. Ashmarina et al. (Eds.): *Digital Transformation of the Economy: Challenges, Trends and New Opportunities*, pp. 66–78, 2019.
https://doi.org/10.1007/978-3-030-11367-4_6

economy of the Russian Federation", the implementation of which should be carried out at several levels: strategic, operational and tactical. Elements of the tactical level are: the legal framework, personnel, infrastructure, information security and technology. Thus, human resources represent one of the key factors in the Russian Federation economy development in the context of global digitalization. Consequently, assessment of personnel constituent capacity at entities of the economy today is very relevant.

Human capital as an intellectual capital element at economic entity is a set of knowledge, skills and creative abilities of the company staff. Assessment of human capital within a business entity is a certification of personnel to determine the level of qualification, the necessary competencies availability and compliance with the position. Federal law No. 238-FL dated 03.07.2016 "On independent assessment of qualifications" became an innovation in this area, giving a special status to professional standards. These standards contain a description of the qualifications necessary for the employee to work a particular profession.

The system of qualifications independent assessment was developed to solve the problem of assessing the human capital at economic entity. The proposed certification procedures, a detailed list of work functions and other aspects were to unify the evaluation procedure and allow determining the employees' potential level at the stage of entry into the digital economy. However, this innovation has created the preconditions for the emergence of new, no less urgent problems and challenges. In particular, business entities that have tested employee assessment programs have concluded that the assessment of knowledge, skills and abilities should be carried out for the first time before, and not when the employee is already employed in the organization. The state and society should pay attention to the system of education, which forms the basic level of knowledge, skills and the individual abilities. It is advisable to assess the educational institutions graduates' qualifications in order for these institutions to identify what needs to be improved in the educational process to form a competitive and competent specialist adapted to the digital economy conditions.

In this situation, employers receive many benefits. Students of educational institutions will be interested in the period of study to master various competencies, as this will depend on their competitiveness in the labor market. And in the event that the certification at the end of the educational institution qualification level would be unacceptable for a career in the planned direction, students will be motivated to improve their skills in training programs of additional education.

Besides, nowadays there is an urgent task of improving the students' financial literacy and educational institutions graduates. Considering financial literacy as a basis for the formation of knowledge, skills and the individual abilities in the digital economy, it is possible to ensure the functioning human capital effectiveness in the economic entity.

## 2 Materials and Methods

In this study, a survey of employers as participants of the qualifications independent assessment system was conducted. The survey was attended by 120 representatives of business entities at the Samara region. The study was conducted in two directions:

- overview of the graduates educational institutions problem employment in the digital economy conditions;
- improving economic entities employees financial literacy in the digital economy.

The composition of the questions included in the questionnaire was formed on the basis of the National agency for the development of qualifications recommendations, whose tasks include "the development of the workforce quality, the education results evaluation, training and work experience" [12].

The survey, devoted to the problems of graduates' employment, was conducted in relation to the professions of the financial market, important for the development of the digital economy in Russia: accountant, specialist in work with overdue debts, specialist in remote banking, specialist in credit brokerage.

The survey, devoted to the problems of graduates' employment of, was conducted in relation to the professions of the financial market, important for the development of the digital economy in Russia: accountant, specialist in work with overdue debts, specialist in remote banking, specialist in credit brokerage:

- short-answer question;
- question with detailed answer;
- question with a choice of one answer from the list.

The list of questions in the questionnaire was compiled in such a way in order to provide a correlation between the data characterizing the Respondent and his answers (Table 1).

The questionnaire was carried out according to the following algorithm:

1. Preparatory stage: setting goals and objectives of the survey, formulation of questions, preparation of a form for respondent's answers using information technology, respondents' list compilation who will be invited to participate in the survey.
2. Operational stage – collection of the respondents answers with the use of information technology – spreadsheet Google Forms.
3. The resulting stage is the processing, analysis and interpretation of the responses received from the respondents.

According to the results of the survey, the primary processing of the results was carried out in Google Forms, the secondary – in the program for working with Microsoft Excel spreadsheets. The further method of work with the survey results was to form their graphical representation, which will demonstrate the distribution of answers to various questions.

In the conducted research expediency of point estimates method application as a kind of an ordinal scaling method was revealed. In the questions that provided answers, each of the answers was given a score, where the highest score characterized the positive picture in the framework of the problem under consideration, and the lowest score – negative. The highest score of the scale is limited by the answers number to the question. Therefore, the higher the index of integrated assessment, the more favorable the situation in the surveyed economic entity or in the relationship with graduates of educational institutions. Realization of the scores method was implemented in the program to spreadsheet Microsoft Excel.

**Table 1.** Employers survey structure as participants of the independent qualifications assessment system

| Point № | Survey questions «Review of graduates employment problems» | Survey questions «Improving financial literacy of employees» |
|---|---|---|
| 1 | Title of enterprise | Title of enterprise |
| 2 | Size of enterprise | Size of enterprise |
| 3 | The average number of employees | The average number of employees |
| 3 | Has your company hired graduates in financial market professions over the past 5 years? | Economic activity |
| 4 | What problems did your company face in the process of graduates employment? | Have your employees ever received financial literacy training? |
| 5 | Do you think there are specific difficulties in hiring in the following professions: accountant, specialist in work with overdue debts, specialist in remote banking services, and specialist in credit brokerage? | Do your employees use credit instruments? |
| 6 | If you answered question № 5 "Yes", then specify the types of difficulties (in brief) | Do your employees use bank cards for purposes other than ATM cash withdrawals? |
| 7 | What are the characteristics of the applicant are decisive in deciding whether to hire him? | Do your employees keep records of family income and expenses? |
| 8 | What actions of the educational system could neutralize the problems in the graduates' employment? | What is the attitude of your employees to creating their own business? |

Source: compiled by the authors

# 3 Results

This study includes a survey of employers in the Samara region. Below are the results of the survey "review of the graduates' employment problems" on some issues, the answer to which implied the choice of a specific option.

In order to identify the shortcomings inherent in the modern education system, the issue of identifying the difficulties faced by companies in the graduates' employment process (Fig. 1).

The most common problem is the low level or complete lack of professionalism and qualifications needed in a specific position. Together with the low level of business qualities (discipline, initiative, etc.), this problem indicates that modern educational programs are focused on the fact that graduates have mastered a certain amount of knowledge, but they are poorly correlated with future practice. This proves the selected answers high proportion, reflecting the psychological graduates' difficulties: high expectations in matters of wages and difficulties in professional adaptation.

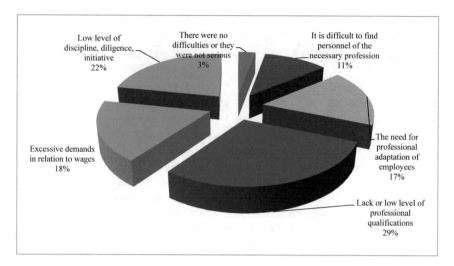

**Fig. 1.** The results of the answer to the question «What problems did your company face in the process of graduates employment?» (Source: compiled by the authors)

In addition, of particular interest is the set of applicants' characteristics that are decisive in employment (Fig. 2). Almost a third of employers (27%) welcome the applicant work experience presence, even a graduate of the institution. At the same time, if a graduate is a full-time student, then building a career at the educational process expense will not be an unambiguously acceptable solution for him. Internship during the training period by the employer is not taken into account in any way.

Another important point is the attention of employers to the personal qualities of the applicant (discipline, initiative, etc.). With a high level of training, erudition and a certain back-ground candidate for the position, almost a quarter of employers (22%) are ready to make a decision on hiring an applicant without work experience and higher education.

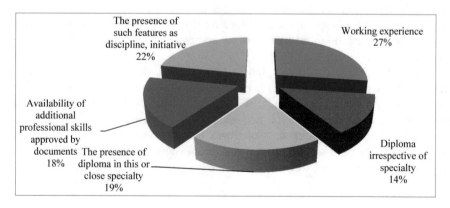

**Fig. 2.** The results of the answer to the question «Do you think there are specific difficulties in hiring in the following professions?» (Source: compiled by the authors)

The purpose of the survey "improving financial literacy of employees" was to identify the position of the economic entities management in relation to the importance of training employees in the basics of financial literacy, as well as to determine the degree of information ownership about how employees are adapted to the digital economy.

As a result of the economic entities heads survey at the Samara region, it was found that almost half of the respondents (47%) did not train employees in the basics of financial literacy (Fig. 3). A third of the respondents (32%) have not previously conducted training, but plans to conduct it in the near future. At the same time, as most respondents noted, training plans are focused only on employees holding senior or key positions. Line staff is not included in these plans.

To assess the degree of employees' adaptation to the conditions in the digital economy, the heads of business entities were asked about the purposes of using Bank cards by employees (Fig. 4). Since the survey was conducted using Google Forms tabular processor, managers who agreed to participate in the study had the opportunity to clarify this information from employees.

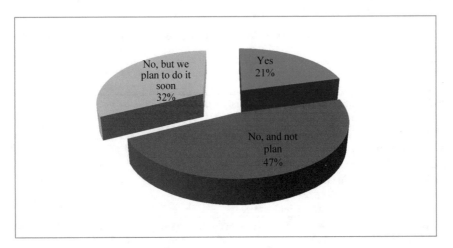

**Fig. 3.** The results of the answer to the question «Have your employees ever received financial literacy training?» (Source: compiled by the authors)

At the same time, a third of respondents (33%) said that they do not have information. More than half of the respondents said that their employees use Bank cards not only to withdraw cash. At the same time, 45% said that there are few such employees. In the comments to the questionnaire, they noted that they are mostly people under 40 years old.

The Microsoft Excel tools spreadsheets made it possible to make a point assessment based on the questions results. The distribution of points on the questionnaire was carried out by an expert method. The method of scoring, which was applied to the results of the surveys, is presented in Tables 2 and 3.

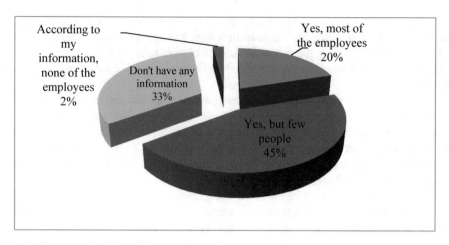

**Fig. 4.** The results of the answer to the question «Do your employees use bank cards for purposes other than ATM cash withdrawals?» (Source: compiled by the authors)

**Table 2.** The results of the surveys scoring method «Review of graduates employment problems»

| Survey questions | Number of questions | Minimum score | Maximum score |
|---|---|---|---|
| Has your company hired graduates in financial market professions over the past 5 years? | 2 | 1 | 2 |
| What problems did your company face in the process of graduates employment? | 6 | 1 | 6 |
| Do you think there are specific difficulties in hiring in the following professions: accountant, specialist in work with overdue debts, specialist in remote banking services, and specialist in credit brokerage? | Open type question | Short answer - 1 | Detailed answer - 2 |
| If you answered question № 5 "Yes", then specify the types of difficulties (in brief) | Open type question | No answer - 0 | There is an answer -1 |
| What are the characteristics of the applicant are decisive in deciding whether to hire him? | 5 | 1 | 5 |
| What actions of the educational system could neutralize the problems in the graduates' employment? | Open type question | Short answer - 1 | Detailed answer - 2 |
| Total (min/max score) | | 5 | 18 |

Source: compiled by the authors

To evaluate the results analyzed by the numerical score method, the method of ordinal calibration was applied. Consequently, the results were grouped into the following groups:

**Table 3.** The results of the surveys scoring method «Review of graduates employment problems»

| Survey questions | Number of questions | Minimum score | Maximum score |
|---|---|---|---|
| Have your employees ever received financial literacy training? | 3 | 1 | 3 |
| Do your employees use credit instruments? | 3 | 1 | 3 |
| Do your employees use bank cards for purposes other than ATM cash withdrawals? | 4 | 1 | 4 |
| Do your employees keep records of family income and expenses? | 3 | 1 | 3 |
| What is the attitude of your employees to creating their own business? | 3 | 1 | 3 |
| Total (min/max score) | | 5 | 16 |

Source: compiled by the authors

(1) according to the survey "review of graduates employment problems":

- 5–9 points-according to respondents, the level of graduates employment problems is high, urgent measures are required;
- 10–14 points-there are problems in the field of graduates employment, but their impact is insignificant;
- 15–18 points-the situation in the field of graduates' employment is considered by respondents as positive.

(2) according to the survey "improving financial literacy of employees":

- 5–8 points – the economic entities employees financial literacy level is low;
- 9–12 points – the economic entities employees financial literacy level is at the average level;
- 13–16 points – the economic entities employees' financial literacy level is at a high level.

According to the results of the study, a comprehensive assessment of the survey "Review of graduates employment problems" was 12 points, according to the survey "Improving financial literacy of employees" - 9 points.

In the comments and open-ended questions, respondents noted low levels of computer literacy and poor knowledge of information technology on the part of both graduates and staff. Despite the active use of social networks, neither the first nor the second object of research has the knowledge, skills and abilities to use the tools of digitalization in the professional sphere.

Therefore, for the positions considered in the survey, an analysis of the relevant professional standards was carried out to identify the requirements for candidates for these positions. The prerequisites for their formation are created, in turn, by the requirements for the processes taking place in the business environment and its participants from the digital economy.

1. Position "Accountant":
   (a) required skills:

      - prepare (execute) primary accounting documents, including electronic documents;
      - use computer programs for accounting, information and legal reference systems, office equipment.

   (b) knowledge required:

      - fundamentals of computer science and engineering.

Thus in the professional standard "Accountant" for the position "Chief accountant" it is specified that the candidate for this position has to know:

- procedure of information exchange via telecommunication channels;
- modern technologies of automated information processing.

Position «specialist on work with overdue debts»:

(a) required skills:

   - find contact details of the borrower in open sources and specialized databases;
   - work with large amounts of data.

(b) knowledge required:

   - information technologies in the professional sphere.
     In the context of employment actions also highlight the formation of dedicated software notification templates on the availability of debt.

2. Position "Specialist in remote banking services":

Among the job functions separately allocate the function of a data entry client in a system of remote banking services.

(a) knowledge required:

   - working with systems of remote banking services;
   - specialized software for remote banking;
   - methods of collection, processing and analysis of information using modern means of communication, hardware and computer technology;
   - trends in the development of automated banking systems and technologies.

(b) required skills:

   - use of remote banking systems;
   - carry out the communications necessary to connect to remote banking systems;
   - use a database on the remote banking.

In addition, the specialist must understand the peculiarities of business processes in the field of remote banking; know the techniques of drawing up technical specifications to improve the efficiency of such systems.

3. Position "Specialist in credit brokerage":

Work activities in the field of information technology include the search and analysis of information from open sources that affect the value and liquidity of collateral, and maintaining a database in the information system.

(a)  required skills:

- use modern technical means of financial information search and analysis;
- use reference databases containing relevant information;
- use universal and specialized software required for data collection and analysis;
- possess basic computer skills;
- work in automated systems of professional activity information support;
- use credit calculators.

(b)  required knowledge:

- devices and programs used for information activities;
- technical means of data collection and processing;
- methods of collection, processing and analysis of information using modern communication means, hardware and computer technology;
- modern information technologies, reference and information systems in the field of law, accounting.

Thus, a sample study of the requirements for the three positions showed that in modern conditions of the applicant is required to have a high degree of various tools knowledge in the information technology field to ensure its competitiveness in the labor market.

## 4  Discussion

Among other components of the intellectual capital at an economic entity, human capital seems to be the most effective, as it is directly related to a person [10]. The essence of a person is reflected in the fact that due to its innovative component, the need for change, the ability to think creatively and take into account past experience, the successful development of the business entity in which it occupies a certain position, allowing to realize its potential [5].

In the context of the digital economy, it is important to understand the human capital development importance, as widespread information and the introduction of various electronic technologies form new requirements for a person as a participant in these processes [7]. The tools of the digital economy have replaced human labor in a number of areas, and the task facing man as a specialist has not been simplified: his work has become more intellectual, requiring more accumulated knowledge, skills and abilities than the previously existing traditional system [6].

In classical economic theory, for several centuries, attempts have been made to correlate the economy development and the ability of man to work. W. Petty, A. Smith, L. Jacob, W. Farr, and other researchers have concluded that the welfare of the people and the state as a whole depends directly on the size of the population and the quality of its labor potential [3]. The concept of "human capital" was first introduced in the twentieth century by T. Schultz. D. Becker, his follower, justified the need to stimulate the development of the education system as the basis for the formation of highly effective human capital. In addition, G. Becker has conducted significant research in the field of assessing the economic education efficiency to determine how the level of education and the time spent on it have affected the income of an individual and society as a whole [9].

John Tomer for devoted more than 20 years to the study of human capital and its impact on the increase of labor productivity, financial-economic activity of the economic entity, as well as the study of its evaluation and possible inclusion in commercial organizations intangible assets [2].

Aganbegyan's special attention is paid to the study of human capital in the Russian Federation. The scientist notes that Russia is slightly behind the Western countries in assessing the share of the economically active population potential economic effect in the country's GDP [1].

H. Hayakawa, Y. Venieris consider human capital through the prism of the modern economic development problem of – inequality in the distribution of income and wealth. The researchers suggest adopting a strategic choice that will be implemented through the accumulation of human capital as an endogenous response to technology innovation [4].

This study is based on the views of the twentieth century scientists, who were the first to actualize the issue of the education level importance in the human capital formation. In addition, the position formed at the Russian Federation President level includes the need to develop a set of measures to ensure the qualification assessment systems interaction and accreditation of educational institutions educational programs:

- professional standards are developed for independent assessment of qualifications;
- for professional and public accreditation of educational programs, Federal state educational standards for basic and additional educational programs are developed.

The above directions are implemented taking into account the results of the labor market monitoring, which identifies the employers needs in the level of applicants' qualifications for positions and in the their education level [11].

Methods of interaction between the education system and the labor market in the task framework of ensuring the human capital growth and improving its efficiency, implemented in accordance with the current regulatory legislative base in the Russian Federation, are presented in Table 4.

In February 2018, the international scientific conference "Human capital in the format of digital economy" was held in the Russian Federation. The participants of the conference as an "epigraph" to the planned excursions led the words of the famous Russian scientist S. P. Kapitsa that our era will come in history not as an era in which there were significant studies of the atom or space, but as an era of "global demographic transition". Thus, man as the highest stage of all living creatures development, is a key participant in the development of all socio-economic processes in the world, the creation of computers and the spread of information technology have exacerbated the crisis of the individual, his personality and potential. The digital economy has posed new challenges to humanity: the computer is displacing people from many areas. And only a high professionalism level, a set of individual qualities, competence and other features allow a person to maintain its relevance in the era of high technology.

**Table 4.** Methods of interaction between the education system and the labor market to ensure the human capital growth

| Point № | Education system | Labor market |
|---|---|---|
| 1 | State final certification as way to assess the graduates knowledge on the basis of training in an educational institution | Professional standards system formation for defining the requirements to applicants for various positions |
| 2 | Professional certification of graduates on the training in an educational institution results | Independent qualification assessment (both graduates and lecturers of educational institutions) |

Source: compiled by the authors

International corporations have concluded that human capital may be evaluated as a half of the company assets value. The process of continuous training will help to form a new generation of professionals. Orientation of educational programs to the requirements of the digital economy and employers are necessary to increase the chances of graduates to become competitive in the labor market and realize their potential during their life.

## 5   Conclusions

The study confirmed the relationship of human capital with the education system, which, in turn, should be improved in order to meet the labor market requirements and the conditions in the digital economy. Employers note the significant difficulties of graduates in finding employment after graduation. The low level of qualification, competence, as well as personal qualities such as discipline and initiative, create certain barriers for applicants.

The requirements of the digital economy are quite serious and complex. They include not only financial, but also the individual computer literacy. Without these aspects, it is impossible to form an effectively functioning human capital and the full realization of human potential. The qualifications assessment system development of both already working professionals and educational institutions graduates will make it possible to introduce innovations in the education system so that the product of its functioning is competitive in the digital economy.

## References

1. Aganbegyan AG (2017) Investments in fixed assets and human capital: two interconnected drivers of socioeconomic growth. Stud Russ Econ Dev 28:361. https://doi.org/10.1134/S1075700717040025
2. Altman MJ (2015) Tomer J (2017) Integrating human capital with human development: the path to a more productive and humane economy. Palgrave Adv Behav Econ Palgrave Macmillan, Lond 19:247. https://doi.org/10.1007/s10818-016-9237-4

3. Fitzsimons P (2015) Human Capital Theory and Education. In: Peters M (ed) Encyclopedia of educational philosophy and theory. Springer, Singapore
4. Hayakawa H, Venieris YP (2018) Duality in human capital accumulation and inequality in income distribution. Eurasian Econ Rev. https://doi.org/10.1007/s40822-018-0110-8
5. Lucas RE Jr (2015) Human capital and growth. Am Econ Rev 105(5):85–88. https://doi.org/10.1257/aer.p20151065
6. Mane F, Miravet DJ (2016) Using the job requirements approach and matched employer-employee data to investigate the content of individuals' human capital. Labour Market Res 49:133. https://doi.org/10.1007/s12651-016-0203-3
7. Roy IJ (2018) Role of human resource practices in absorptive capacity and R&D cooperation. Evol Econ 28:885. https://doi.org/10.1007/s00191-018-0573-5
8. Stewart L (2015) The job of human capital: what occupational data reveal about skill sets, economic growth and regional competitiveness. (Electronic Thesis or Dissertation). https://etd.ohiolink.edu/. Accessed 10 Mai 2018
9. Teixeira PN (2014) Gary Becker's early work on human capital – collaborations and distinctiveness. IZA J Labor Econ 3:12. https://doi.org/10.1186/s40172-014-0012-2
10. Zysman J, Kenney M (2017) Intelligent tools and digital platforms: implications for work and employment. Intereconomics 52:329. https://doi.org/10.1007/s10272-017-0699-y
11. International scientific conference "Human capital in the format of digital economy". http://rosnou.ru/pub/diec/index.html. Accessed 10 Mai 2018
12. «National qualifications development Agency: goals and objectives» https://nark.ru/about/mission.php. Accessed 10 Mai 2018

# The Problem of Energy Saving and Its Solution in the Conditions of Formation of a New Model of Economic Development

Olga Evgenevna Malikh[1], Maria Evgenevna Konovalova[2(✉)],
Olga Yurevna Kuzmina[2], and Alexander Michailovich Michailov[2]

[1] Ufa State Oil Technical University, Ufa, Russia
kafedra-et@mail.ru
[2] Samara State University of Economics, Samara, Russia
{mkonoval,2427994}@mail.ru, pisakina83@yandex.ru

**Abstract.** The relevance of the analyzed issue is caused by the need of an integrated assessment of mineral raw material base for determination of efficiency of stereoscopic subsurface management. The purpose of the article is the analysis of prospects for economic development of subsoil use in the conditions of active use of digital technologies. The inevitability of saving natural resources tendencies in economy in the short term, and also the increasing world competition in this field of activity predetermines the need of further usage of new methods and technologies of prospecting, mining and processing of mineral raw material resources to the finished product in the subsurface management. The leading approach in the study of this issue is the systemic approach allowing justifying effective development of linked industries within innovative subsurface management. The possibility of transition of the Republic of Bashkortostan to low-carbon economy and green growth is assessed on the basis of the data on the potential of exhaustible resources, criteria of economic growth, and also characteristics of the used equipment and digital technologies. The set of offers, promoting increase in resilience of socio-economic and ecological systems of the region, due to the diversification of economy and sustainable governance of natural resources, is proved by the authors. Materials of the paper may be useful in generation of a complex socio-economic development strategy of the region, in particular the section, devoted to the development of innovative subsurface management in the conditions of formation of a new model of the digital economy.

**Keywords:** Deterioration of equipment · Digital technology · Potential of exhaustible resources · Subsoil use · Urban development

## 1 Introduction

Implementation of the climate policy faces various difficulties. This is about incongruous approaches to the usage of exhaustible resources in developed and developing countries [17]. It is also important to consider changes in consumer welfare as a result of impact on socio-economic system of the climate policy, and to understand the

© Springer Nature Switzerland AG 2020
S. Ashmarina et al. (Eds.): *Digital Transformation of the Economy: Challenges,*
*Trends and New Opportunities*, pp. 79–87, 2019.
https://doi.org/10.1007/978-3-030-11367-4_7

balance between benefit and cost if there are significant external factors [15]. Studies of the economy of climate change, including theoretical conclusions and empirical results, should be offered to the government. But it is difficult to predict the consequences of control, because of uncertainty regarding time frames and rigidity of actions of the policy, aimed at elimination of possible climate change [4]. Institutional investors may be reluctant to join climate projects, because they are expensive, and time of implementation is about 30–50 years. Some authors believe that capital investment projects, which are carried out at the regional municipal government levels, can be one of the solutions to this problem. The project becomes less expensive, and there is a possibility for an investor to control it [6, 7, 13].

Consequently, the assessment of resource base of certain regions becomes important for implementation of the climate policy. In this regard the purpose of the research is to evaluate the potential of exhaustible resources of the Republic of Bashkortostan and the possibility of its transition to low-carbon economy and "green growth". The Republic of Bashkortostan is chosen as the object of the study as it is a well-resourced region, which is among ten largest constituents of the Russian Federation.

## 2    Materials and Methods

In the course of research the following methods were used: theoretical (systemic, structural and functional, neopositivist, rational), methods of mathematical statistics and tabular display of results.

## 3    Results

Fuel and energy raw materials in the Republic of Bashkortostan are presented by hydrocarbon and brown coal. According to data of Department of Subsoil Use of the Volga Federal District (Bashnedr), as of 1.01.2018 there are 204 oil-and-gas fields in the territory of the Republic. Bashkortostan possesses considerable reserves of iron and manganese ore, copper and sulphide, gold-copper-zinc, and gold-sulphidic ore. In the Republic of Bashkortostan there are almost all kinds of widespread subsoil assets, found in Russia – 13 types: sand-gravel mix and mason sand, brick and tile raw materials and building stone, gypsum and anhydrite, agrochemical raw material, peat and other. There are almost 13 thousand rivers with total length over 57 thousand km. The volume of average annually renewable total stocks of surface water, formed in the territory of the Republic, accounts for 25, 5 $km^3$. Taking into account water from the neighboring areas and the Republic of Tatarstan, the volume of water resources increases up to 35 $km^3$. In general, the Republic of Bashkortostan is less provided with water resources than the Russian Federation: in Bashkortostan it is 8750 $m^3$ of water per capita per year, or 24 $m^3$ per day, and in Russia it is 29380 $m^3$ per year, or 80 $m^3$ per day; in the Perm region this indicator is twice higher.

The existing pollution of water objects of the Republic is connected, first of all, with inefficient work or lack of pollution control facilities. The main reasons of its inadequate performance, as well as in previous years, are use of outdated technologies and worn-out state of the basic production assets.

The objects of accumulated ecological deprivation of previous years, which are in the territory of the Republic, because of its former industrialization, lead to the fact that population health is at risk. Nowadays the priorities are deactivation and subsequent involvement of these objects in economic growth of the region.

**Table 1.** Gross regional product of the Volga Federal District, bn. rub.

| № | Territorial subject of the Volga Federal District | 2015 | 2016 | 2017 |
|---|---|---|---|---|
| 1 | The Republic of Tatarstan | 1437,0 | 1547,1 | 1671,4 |
| 2 | The Republic of Bashkortostan | 1149,4 | 1266,9 | 1248,8 |
| 3 | The Samara Region | 937,4 | 1040,7 | 1152,0 |
| 4 | The Nizhni Novgorod Region | 842,2 | 925,8 | 1018,4 |
| 5 | The Perm Krai | 860,3 | 893,4 | 967,9 |
| 6 | The Orenburg Region | 628,6 | 709,5 | 731,3 |
| 7 | The Saratov Region | 478,3 | 528,6 | 562,3 |
| 8 | The Udmurt Republic | 372,8 | 404,8 | 442,0 |
| 9 | The Penza Region | 239,9 | 270,8 | 297,7 |
| 10 | The Ulyanovsk Region | 240,5 | 260,3 | 279,0 |
| 11 | The Kirov Region | 208,5 | 224,7 | 250,3 |
| 12 | The Chuvash Republic | 217,8 | 224,4 | 235,1 |
| 13 | The Republic of Mordovia | 119,9 | 149,3 | 170,9 |
| 14 | The Mari El Republic | 117,2 | 124,4 | 144,1 |

Source: compiled by the authors on the base of data of the Federal State Statistics Service.

However, the dynamics of economic growth of the Republic of Bashkortostan is unstable (Table 1). The available resources are used non-optimal. We observe that GRP of the region is 25% lower, than in the Republic of Tatarstan. Herein, in average the population of the Republic of Bashkortostan is 350 thousand people more. This fact is also confirmed by data on energy consumption of GRP and degree of depreciation of fixed assets (Tables 2 and 3). According to power consumption, the Republic of Bashkortostan lags behind the Republic of Tatarstan by 48,6%, and according to the degree of depreciation – by 21%. The same data indicate considerable emissions of greenhouse gases as outdated equipment and unworkable technologies are used there.

It goes without saying that there is connection between high energy consumption of GRP (the 10th place in the district) and low GRP per capita (the 6th place in the district). From the point of view of social effects of the climate policy, the Republic of Bashkortostan can be characterized as follows. In five largest cities of the region live 37, 61% of the population and the biggest number of stationary and mobile sources of

**Table 2.** Energy consumption of GRP (kg of oil equivalent by 10 thousand rubles) in territorial subjects of the Volga Federal District

| № | Territorial subject of the Volga Federal District | 2015 | 2016 | 2017 |
|---|---|---|---|---|
| 1 | The Republic of Tatarstan | 184,12 | 153,25 | 143,77 |
| 2 | The Penza Region | 177,02 | 153,32 | 153,03 |
| 3 | The Mari El Republic | 211,15 | 196,86 | 158,85 |
| 4 | The Udmurt Republic | 168,39 | 187,39 | 165,72 |
| 5 | The Chuvash Republic | 194,08 | 180,25 | 173,22 |
| 6 | The Ulyanovsk Region | 209,58 | 190,82 | 175,51 |
| 7 | The Saratov Region | 243,92 | 214,92 | 193,95 |
| 8 | The Nizhni Novgorod Region | 246,97 | 234,01 | 203,37 |
| 9 | The Kirov Region | 246,22 | 224,43 | 205,16 |
| 10 | The Republic of Bashkortostan | 232,18 | 227,35 | 213,69 |
| 11 | The Perm Krai | 327,72 | 307,93 | 217,50 |
| 12 | The Samara Region | 278,36 | 248,64 | 225,28 |
| 13 | The Republic of Mordovia | 264,18 | 233,76 | 227,96 |
| 14 | The Orenburg Region | 367,54 | 263,44 | 260,21 |

Source: compiled by the authors on the base of d data of the Federal State Statistics Service.

**Table 3.** Degree of depreciation of fixed assets of the Volga Federal District, %

| № | Territorial subject of the Volga Federal District | 2015 | 2016 | 2017 |
|---|---|---|---|---|
| 1 | The Republic of Tatarstan | 43,7 | 43,4 | 44,2 |
| 2 | The Ulyanovsk Region | 46,6 | 46,9 | 48,0 |
| 3 | The Nizhni Novgorod Region | 50,2 | 49,7 | 48,7 |
| 4 | The Penza Region | 53,2 | 51,3 | 49,7 |
| 5 | The Kirov Region | 51,3 | 51,9 | 51,0 |
| 6 | The Republic of Bashkortostan | 52,1 | 52,2 | 53,3 |
| 7 | The Samara Region | 53,7 | 53,5 | 53,4 |
| 8 | The Saratov Region | 54,0 | 53,5 | 53,8 |
| 9 | The Chuvash Republic | 54,6 | 53,5 | 56,0 |
| 10 | The Republic of Mordovia | 57,3 | 56,4 | 56,9 |
| 11 | The Orenburg Region | 56,9 | 55,9 | 58,1 |
| 12 | The Mari El Republic | 60,6 | 58,7 | 60,2 |
| 13 | The Perm Krai | 59,6 | 60,2 | 60,3 |
| 14 | The Udmurt Republic | 61,0 | 62,3 | 62,0 |

Source: compiled by the authors on the base of data of the Federal State Statistics Service.

emissions is concentrated there. According to the total index of quality of the urban environment, all cities, except the capital –Ufa – are included only in the second hundred rating of the cities of Russia. Therefore, the index of natural and ecological situation amounts 0, 7. And this is worse than in many other cities of the country.

The index of dynamics of population size fluctuates between 168, 39 (Neftekamsk) and 211, 15 (Sterlitamak), in the capital (Ufa) – 184, 12. Low quality of the environment, medical and demographic problems define low dynamics of population size. This reflects the region's insufficient attractiveness for life [8, 9].

The above-mentioned problems confirm the pressing necessity to pay more attention to the questions of the climate policy, defining good scope of work both for the state, and for business.

However, the government of the region interprets the climate policy in a strict sense, in fact replacing it with environmental actions. In accordance with the state program "Ecology and natural resources of the Republic of Bashkortostan", 2,5 billion rubles were implemented in 2017. The water resources utilization system of the Republic of Bashkortostan (381,4 million rubles), inexhaustible natural management (54,7 million rubles), production and consumer waste management system (783,45 million rubles), ecological safety (1088,1 million rubles) became the program objects. The state program realization required 161, 45 million rubles. The state program doesn't distinguish by ambitiousness of purposes and does not define objectives aimed at reduction of greenhouse gases.

The existing structure of an industrial complex in the Republic caused not only character and rates of economic growth, but also high man-made load on the environment and a number of environmental problems. Emissions of pollutants from stationary and mobile sources, increasing level of man-made danger as a consequence of aging of basic production assets, generation and accumulation of production and consumer waste, lack of effective management systems of environmental quality are essential factors limiting rates of socio-economic development of the economic complex of the Republic.

Consequently, we observe that there are problems and possibilities of formation of the climate policy:

– existence of nature-oriented infrastructure and scientific-and-technological potential will allow solving modern problems of the climate policy;
– extensive woodlands 6,3 million hectares (44,1% of the region's territory) promote emission abatement;
– there is experience of volume reduction of accumulated ecological deprivation;
– application of experience of using renewable energy sources. Renewable sources are widespread and pure. But unlike traditional energy sources, they function regularly. It is necessary to develop technologies (to use available) for energy accumulation. The Republic of Bashkortostan has enough sunny and windy days during the year to star realization of such projects
– manufacture of environmentally clean production at the regional level is small-scale, and it can't enter even the markets of neighboring regions.

Lack of evaluation criteria of ecologically clean production, certification authority, and business entities, which could occupy this niche of economy, are deterrents. In the course of work performance on emission and greenhouse gases inventory, it is necessary to consider existing distribution of emission volumes in Russia: 81–83% - energy production; 6–7% - production sector; 5–6% - agricultural industry and 2–3% - waste handling. As for the Republic of Bashkortostan, it accounts (according to our calculations): 71–74% - energy production, 11–14% - production sector, 13% accounts for agricultural industry and waste. The main emissions from stationary sources fall at enterprises of fuel and energy complex (oil refining, petrochemical, oil extracting and electric power industries) - about 70%. Moreover, in the Republic of Bashkortostan emissions from industrial enterprises do not practically decrease for the last six years. So, in 2008 emissions of pollutants in the atmosphere accounted for 417,4 thousand tons, and in 2015 they reached 448,9 thousand tons.

It indicates that existing production capabilities in the Republic of Bashkortostan should be modernized. The regional government should create a system of incentives for transition to the best available technologies in chemistry, petrochemistry and bio-chemistry, technology, which are based on energy - and resource-saving processes.

## 4   Discussion

Strategic tendencies of innovative development of Russia's regions, set out in frame-work documents, serve as the basis for development of this sort of programs in sub-surface management. They were noted in the Energy Strategy of Russia for the period until 2030, Development strategies of geological branch until 2030 and also in the Long-term government program of studying subsoil assets and rehabilitation of mineral resources based on the balance of production and consumption of mineral raw materials until 2030, the General Development Plan for the Russian Gas Sector until 2030. On the basis of the listed regulatory legal acts and their development, regional projects on prospecting and field development of concrete regions and territories, for example the Republic of Bashkortostan, should be developed. Many experts believe [10] that regarding innovation, working legislation shows either absence, or paucity of stan-dards, aimed at the latest achievements in the field of science, technology and engi-neering decisions. Lack of direct standards, promoting implementation of innovations in the course of geology study and subsurface use, is observed [5].

Efficiency of innovative development of subsurface management depends on the fact how the activity of elements of innovative system is accomplished (government, business, science, universities), on their cooperation, on the level of compliance of its institutional assistance. In this case, the government should act as a partner, who owns considerable resources, and as a facilitator of development of national innovative business. Creating conditions for innovative activity, the government cannot influence on subsoil users directly, forcing them into its implementation. The positive result can be received only by means of economic encouragement, government support or strengthening economic responsibility [3]. The institute of public private partnership, which is mainly used in projects with high risk level and high share of innovative and investment expenditures, begins to play an important role. Some forms of PPP like contractual joint ventures, production sharing agreements, government contracts, rent of subsoil areas are developing in Russia, and such forms as concessions and brand new companies, which are widespread in the world, unfortunately, aren't used.

Organizational novation, for example, formation of development of innovation clusters of subsurface management and mineral raw material centers of economic growth, is an important component of institutional assistance of innovative development

[12, 16]. Formal conditions for implementation of such institutional forms are already prepared. Clusters should supplement each other, cooperate in activity, and be in competition with each other. Only in this case, it will be possible to gain advantages of cluster approach for development of regional subsurface management: conservative system of dissemination of new technologies, knowledge, and output, which is based on its own scientific base; cost minimization on innovations; existence of flexible business organization, which will allow forming of innovative growing points of economy; creation of conditions for development of small business.

The innovative mode of development of subsurface management involves not only basing on new scientific knowledge and technologies, but also obligatory occurrence of personnel, prepared for innovative activity. Development of the initial stage of innovative process demands essential redeployment of organizational structure of management and the ideology of executive staff, renewal of network of regional vocational rehabilitation centers on the basis of sectoral higher educational institutions, constant system of stage-by-stage retraining of personnel for innovative activity [1, 14].

Only creating favorable conditions for development of innovative activity and acceleration of technological development can provide formation of innovative subsurface management. As previously noted, the process of innovative development of economy in Russia only starts to progress, and the aspects of systemic approach to innovative development of subsurface management are not fully explored. There are only studies of legal, organizational and economic elements. There is no uniform system of formation and functioning of innovative subsurface management. It is necessary to consider foreign experience, where innovative activity stopped being linear functional, notably the process consisting of several successive steps: state support of scientific-research and technological analysis and their commercialization. In many countries, these stages are built in the hierarchical system, where government institution holds the key place. Its leading role does not reduce the importance of participation of science and business institutions, realizing direct and indirect participation in innovative subsurface management in the form of theoretical and methodological, technical and technological innovations by means of the market-based mechanism.

## 5 Conclusions

Consequently, the Republic of Bashkortostan has opportunities for economic development in the context of implementation of the Paris climate agreement of 2015. On the basis of data on the potential of exhaustible resources, indicators of economic growth, and also characteristics of the used equipment and technologies we can observe that transition of the region to low-carbon economy and "green growth" requires institutional changes. The offers, promoting increase in resilience of socio-economic and ecological systems of the Republic of Bashkortostan are based on diversification of economy and sustainable governance of natural resources. Transition from a narrow ecological interpretation to broad understanding of the climate policy is necessary. Primary goals can be:

1. Updating of the region's environmental regulations by reference to new requirements of new climate policy.
2. Decisions in the field of economic policy should be made with consideration for the changing approaches of functioning of the world commodities and financial markets. Adjustment of investment policy – energy saving – decrease in emissions will be required. Another important tendency should be work with business community regarding clarification of expenditures and interests, introduction of contracts for increase in effectiveness.
3. Development of adsorptive capacity of woods, increasing $CO^2$ accretion.
4. Creation of "green" development plans for the cities will allow solving two problems – to make cities of the region comfortable for living and to promote decrease in emissions of greenhouse gases [3, 11]. There should be a detailed economic model of the city with the forecast of growth trends taking into account a broad set of economic and sociological variables and with a target for decrease in emissions of greenhouse gases by a certain percentage from the current level. Reequipment of residential, industrial buildings – embedding of small electrical generating stations, zoning of the urban environment, careful specification of the urban development master plan, contribution of private and municipal sectors. Investments of municipal sector – 5% within 2 decades are necessary for the changes in the expenditure cross-section (from GRP) on economy maintenance. Support of old infrastructure is more expensive, than investment into new one.
5. Development of the system of environmental education, including background of the "green economy" and practice-oriented programs.

# References

1. Bogatyreva MR (2014) Transformation of the labour migration management system. World Appl Sci J 30(11):1556–1558
2. Deliktas E, Günal GG (2016) Economic growth and input use efficiency in low, upper - middle and high incomed countries (1991–2011): a data envelopment analysis procedia. Econ Financ 38:308–317
3. Egorova M, Pluzhnic M, Glik P (2015) Global trends of "Green" Economy development as a factor for improvement of economic and social prosperity procedia. Soc Behav Sci 166:194–198
4. Goulder LH, Pizer WA (2005) The Economics of climate change. New palgrave dictionary of economics, 2nd edn, Macmillan Publishing, Ltd
5. Henckens MLCM, Driessen PPJ, Ryngaert C, Worrell E (2016) The set-up of an international agreement on the conservation and sustainable use of geologically scarce mineral resources original research article. Resour Policy 49:92–101
6. Hui Z, Danxiang A (2011) Environment, energy and sustainable economic growth. Proced Eng 21:513–519
7. Kamal A, Shamseldin M (2016) Considering coexistence with nature in the environmental assessment of buildings. HBRC J 13:2362–2373
8. Malikh OE, Hurmatullina AF, Konovalova ME, Kuzmina OY, Titova NB (2016) Integral assessment of the social and economic development of megacities in Russia. Iejme – Math Educ 11(7):2455–2469

9. Malikh OE, Polyanskaya IK, Konovalova ME, Kuzmina OY, Tarasyuk OV, Osipova IV (2016) Implementation of the State Economic Policy in the Field of Education. Iejme – Math Educ 11(8):3104–3113
10. Orlov VP (2010) Geological-prospecting branch in the context of modernization of the economy. Mineral resources of Russia. Econ Manag 2:3–6
11. Pearson RG (2016) Reasons to conserve nature. Trends Ecol Evol 31:366–371
12. Petrov OV (2010) Strategic tendencies of innovative use of mineral and raw materials potential of Russian subsoil assets. Mineral resources of Russia. Econ Manag 3:37–41
13. Rodionov II, Smirnov AL (2016) Equator Principles as a competitive factor of credit institutions on the global financial market. Vestnik of Vladimir State University named after Alexander and Nikolay Stoletov. Ser Econ Sci 1(7):120–123
14. Sandifer PA, Sutton-Grier AE, Ward BP (2015) Exploring connections among nature, biodiversity, ecosystem services, and human health and well-being: Opportunities to enhance health and biodiversity conservation. Ecosyst Serv 12:1–15
15. Sunstein CR, Reisch LA (2013) Automatically green: behavioral economics and environmental protection. Harv Environ Law Rev 38(1)
16. Tatarkin AI, Petrov OV, Mikhailov BK (2009) Richness of subsoil assets of the Russian Federation: state and tendencies of innovative use. Vestn Russ Acad Sci 9:771–780
17. Van der Ploeg R (2010) Natural Resources: Curse or Blessing? CESifo Working Paper Series 3125

# Regulation of Tax Havens in the Age of Globalization and Digitalization

Svetlana Nikolaevna Revina, Pavel Alexandrovich Paulov,
and Anna Viktorovna Sidorova(✉)

Samara State University of Economics, Samara, Russia
29.revina@mail.ru, paulovpavel@ya.ru, an.sido@bk.ru

**Abstract.** The relevance of the study is due to the growth of the offshore sector in the global economy and the concern of the international community emanating from this potential threats to globalization and digitalization of international business. A vigorous growth of the offshore business has not left unmoved the governments of the developed countries, wherefrom the capital began to "out-migrate" to the countries with more favourable conditions. In this regard, this authors analyze the problems of regulation of offshore centers in the era of digitalization of international business. The leading method in the study of this problem is the dialectical method that allows to comprehensively consider the problems of regulation of offshore centers, to analyze offshore financial centers as a new object of developing international economic law in the era of digitalization. The contribution covers issues of offshore financial centers as a new object of developing international economic law in the era of digital economy; the role of international organizations in the formation of the international legal framework of offshore financial centers and in the process of monitoring changes in the national legislation of States that are offshore financial centers. Findings of this research can be used in the practice of the legislative authorities of the RF and international organizations. Theoretical conclusions of the research can be used in research and development institutes, in higher education institutions for teaching certain modules of disciplines: "Constitutional Law", "Financial Law", "Tax Law".

**Keywords:** Capital · Digitalization · Financial assets · Financial Crimes · Law · Political pressure · Regional economy · Tax havens

## 1 Introduction

In recent years, owing to general globalization of the international business, both the participation of the offshore sector in the global economy and the world community's concern with the potential threats to the global financial system have gone up. Activities of tax havens are under scrutiny of the international organizations such as OECD, FATF (Financial Action Task Force on Money Laundering), as well as the UN, EU, Forum of Financial Stability, etc. Over the last years, the International Monetary Fund's and the World Bank's interest in regulation of offshore companies has grown up [9, 15].

S. Ashmarina et al. (Eds.): *Digital Transformation of the Economy: Challenges, Trends and New Opportunities*, pp. 88–95, 2019.
https://doi.org/10.1007/978-3-030-11367-4_8

Aims and objectives of the research. This research is aimed at identification of international legal problems to regulate functioning of the offshore financial centres and their scientific characteristics. To achieve this goal, the following tasks have been solved: analysis of the offshore financial centres as a new object of the developing international economic law; identification of the role of the international organizations in formation of the international legal framework for activity of the offshore financial centres, as well as while monitoring changes in the national law of the countries being the offshore financial centres.

International regulation of the offshore activity is a new phenomenon in the global economy [6, 7]. In practice, it is carried out within two areas: fight against unfair tax competition and counteraction to financing the criminal activity. Regulation of the offshore business focused on prevention of unfair tax competition is a consequence of the global competition between the industrially developed countries, the countries possessing rich resources and the developing countries that do not have such resources. In its essence, it is a tool of political pressure and therefore causes opposition from the offshore countries. At the international level, OECD works in this area, focusing its activity on monitoring both the offshore jurisdictions and preferential tax territories of the industrially developed countries, including the USA. At the national level, the anti-offshore regulation in this area consists in framing the law, which is intended to reduce losses from application of the offshore method [8].

## 2   Materials and Methods

In the course of the research, the authors have applied both the general methods of scientific knowledge (dialectical method) and the particular methods of theoretical analysis (historical, comparative and factorial), as well as the comprehensive and criteria-based approaches, problem-chronological principle of research, methods of differentiation, classification and generalization.

## 3   Results

Some "unpleasant" aspects that repel a significant number of business structures have appeared in the offshore business recently. For example, the US and EU financial control authorities closely monitor all the transactions involving offshore capital. In addition, in some countries, the special law has been enacted to reduce tax losses from such firms. The offshore companies are also treated with utmost mistrust by the governmental and banking structures, thereby causing difficulties with lending.

It is safe to say that offshore financial centres are a new subject of evolving international economic law.

With regard to the regulation of offshore centres, it was fully supported by most countries, which had voluntarily agreed to make the necessary changes to national legislation. However, the analysis of methods of struggle of the leading States with "tax havens" testifies to a certain limitation and declarative nature of the decisions.

## 4 Discussion

The London International Business Conference "Offshore-2000" made significant impact on the course of development of the world economy. The conference partici-pants expressed concern as for the rapid expansion of tax havens, international financial frauds and considered the ways to restrict them. The conference was attended by the lawmakers from majority of tax havens in the world, representatives of leading banking circles, law firms, of the EU and European Commission for Financial Crimes.

Recently, there has been rather a steady trend towards tightening of the tax and banking laws in the UK island territories (under pressure from the industrially devel-oped countries) by the metropolitan countries [13]. Financial frauds in tax havens are capable of destabilizing the global financial system [11]. It is expected to oppose them by restricting a special tax regime in those jurisdictions, which contributes to main-taining the full anonymity of the depositor, as well as by introducing a single global system of relations between the banks and the investment companies. The common communication software will make it possible to automatically detect the suspicious transactions, to lock them and to bring the required information to the investigating authorities [4, 7].

One of the adverse manifestations of globalization is a dramatic intensification of criminalization of the national economies and international business relations in the area of laundering the illegally gained capitals. According to various estimates, the amount of dirty gains as legalized in the world annually goes up to enormous figures. The dramatic growth of those alarming trends stimulated the world community's opposition thereto, which fact was demonstrated, in particular, by creation of the specialized international organizations.

For example, the Financial Action Task Force on Money Laundering (FATF) was established at the G-7 Meeting in Paris in July 1989, with the aim to conduct com-prehensive analysis of findings of the prompt actions to prevent using the banking system and financial institutions for money laundering; to develop the typical (refer-ence) legislative instruments and preventive measures in that area, as well as to enhance coordination of the relevant joint efforts of the member-states.

FATF coordinates its work with the UN, takes direct part in counteracting legal-ization of dirty money in recent years. Thus, in June 1998, at the 20th Special Session of the UN General Assembly dedicated to fighting against crime in the financial area and counteracting the drug trafficking worldwide, the Action Plan against Money Laundering was approved. In this regard, in December 1998, the document: United Nations Global Programme against Money Laundering, Proceeds of Crime and Financing of Terrorism (GPML) was issued in the form of a special leaflet, where the detailed legal definitions and characteristics of many crimes in the financial area were presented [15]. It is outlined in the preamble that termination of money laundering shall be based upon the conceptual ground, which is common for all the countries in the world.

The UN experts consider identification of the money laundering transactions through analysis of the stage of growth of criminal capitals in the credit and financial area to be the key point, suggesting that it should focus on tracking the stage of integrating the dirty money from the national into the global financial system. That's how the issue was raised at the UN Summit on Fighting against the Organized Transnational Crime, which took place in December 2000 and which was attended by delegations from more than 140 countries. The Summit adopted the UN Convention against Transnational Organized Crime. One of its goals is to settle the issues of seizure of the property gained from criminal actions and to abolish the bank secrecy, should the financial contribution or financial transaction involve laundering of dirty capitals.

As well, some regional entities founded similarly to FATF carry out, within their competences and at the international level, the fight against money laundering. They are, in particular, the Asia-Pacific Group, Caribbean Group, Group of the Southern and Eastern African Countries and the Committee of Experts of the Council of the European Community on evaluation of measures to counteract money laundering. The Money Laundering Panel of the Inter-American Drug Abuse Control Commission monitors introduction of the anti-money-laundering action plan adopted at the Meeting of Ministers of the Western Hemisphere in Buenos Aires in 1995. Interpol, being a powerful and authoritative international entity that makes significant efforts to combat money laundering, investigating various international economic crimes, as well as collecting and analyzing information about the frauds taking place in various countries, holds a special place in the system of the relevant international organizations. The so-called financial intelligence services (i.e., the national anti-money laundering authorities) cooperate within the Egmont Group, founded in 1995.

The following should be mentioned among other influential international organizations of this profile:

- Commonwealth Commercial Crime Unit (CCCU), which brings together 49 countries providing it with the necessary information of international nature about economic crimes;
- Offshore Group of Banking Supervisors, which arranges cooperation in fighting against money laundering in international offshore centres and in free economic zones;
- Bern Club, which gives membership to representatives of the law enforcement agencies of some Western European countries. The main focus of work is to ensure efficient exchange of information among the members;
- International Organization of Securities Commissions (IOSC), which brings together the representatives of the securities market controlling authorities of 136 members from more than 70 countries;
- International Maritime Bureau (IMB), which prevents economic crimes in the sea transport, first of all smuggling, including the currency one;
- Business Security Services of the International Chamber of Commerce, which fights against various types of offences in respect of international trading and bank accounts;
- International Association of Investigators to combat forgery of credit documents;
- International Association of Professional Security Officers of Banks.

Activities of the above organizations have ensured the formation of a fairly large regulatory base governing international cooperation regarding the issues considered. Indeed, the tax haven is the main filter in money circulation, where they launder the dirty money proceeded from human trafficking, sale and distribution of drugs and smuggling of weapons [6, 10].

Actually, the large-scale fighting against money laundering was commenced only after the acts of terrorism of September 11, 2001, when the US Government made a real decision to abridge the finance support of the international terrorism. As late as on October 25, 2001, the US Treasury commenced the large-scale Green Quest Operation. During the four months, the Americans coped to seize the cash in excess of USD 10 mln and the securities for USD 4 300 000. At the same time, the American diplomacy started active pressure on the countries and territories that are known to have offshore traditions and non-transparent banking system.

Following the Top-20 Summit in April 2009, OECD published the list of offshores comprising of three sections. The first section includes the jurisdictions, which have made, in the opinion of OECD, significant achievements in implementation of the developed international standards of tax cooperation, being: Argentina, Australia, American Virgin Islands (USA), Barbados, UK, Hungary, Germany, Greece, Guernsey, Denmark, Jersey, Iceland, Ireland, Spain, Italy, Canada, Cyprus, Korea, Malta, Mauritius, Mexico, Netherlands, New Zealand, Norway, Isle of Man, UAE, Poland, Portugal, Russian Federation, Seychelles, Slovakian Republic, Turkey, Finland, France, Czech Republic, Sweden, South Africa and Japan.

The second section contains the jurisdictions, which have assumed the obligations to adopt the prescribed standards, but have not performed them yet or performed them not to the fullest extent, being: Andorra, Anguilla, Antigua and Barbuda, Aruba, Bahamas, Bahrain, Belize, Bermudas, British Virgin Islands, Cayman Islands, Cook Islands, Dominican Republic, Gibraltar, Grenada, Liberia, Lichtenstein, Marshall Islands, Monaco, Montserrat, Nauru, Netherlands Antilles, Niue, Panama, Saint Kitts and Nevis, Saint Lucia, Saint Vincent and Grenadines, Samoa, San Marino, Turks and Caicos and Vanuatu.

The third section contains the list of the countries, which are out of line with the standards and which will be, most likely, sanctioned in a particular way: Costa-Rica, Labuan, Philippines and Uruguay.

Austria, Belgium, Brunei, Chile, Guatemala, Luxembourg, Singapore and Switzerland are singled out into a separate subgroup: "other financial centres".

In 2007–2010, the UK agreements for avoidance of double taxation came into force with Saudi Arabia (2009), Slovenia (2008), Moldova (2008), Switzerland and New Zealand (2008), Netherlands, France, Libya, Mexico and Cayman Islands. The UK accelerated its actions on the Tax Information Exchange Agreement with up to 20 tax havens and offshore jurisdictions, including Australia, Austria, Belgium, British Virgin Islands, Cayman Islands, Croatia, Ethiopia, Germany, Hungary, Israel, Luxembourg, New Zealand, Oman, Qatar, Spain, Switzerland, Thailand, and especially with three offshore territories: Anguilla, Gibraltar and Turks and Caicos Islands, in respect to adoption of the international tax standard of transparency and information exchange for tax purposes as established by OECD. The TIEA agreement between the UK and Bermudas came into effect from 2008, with the Isles of Man, Guernsey and Jersey from 2009 [2].

In 2009, the owners of offshore companies in a number of prestigious European tax havens were already given a chance to experience the effect of many recommendations of the international financial organizations. So, in the Isle of Man and also in Ireland, the provisions abolishing non-resident status were adopted. As a result, Ireland completely ceased to be functioning as a tax haven. Control over nominee directors was intensified and licensing of the companies providing the trust, secretarial and registration services was introduced in the Channel Islands and in the Isle of Man. Commitment to the OECD standards, as well as discourse about the necessity to implement the financial information standards are not subject to confidentiality and suggest that those jurisdictions can be used only for the reputable and reliable businesses.

One should single out a special set of arrangements (analytical and computer intelligence) among the areas of information and technical counteraction to contemporary crime, carrying out its activity directly through use of tax havens. The analytical intelligence's essence can be defined as surveying the information environment, aimed at obtainment of new knowledge about the surveyed object or phenomenon subject to analytical processing of existing intelligence information and statements about the well-known facts. The analytical intelligence involves examination of the covert surveillance materials, criminal intelligence analysis, reports of unofficial collaborators, intercepted data from various communication channels, as well as analysis of messages, publications and speeches in mass media, statistical data, information contained in the public and private automated data banks and information networks.

The anti-offshore regulation aimed at prevention of laundering of capitals and financing of terrorism meets a full support from majority of the offshore jurisdictions, which have, in a voluntary manner, agreed to make the required amendments into the national law. The body coordinating the intrastate efforts in that area is FATF, which monitors the schemes of illegal capital legalization, initiating the respective counteractive measures, as well as forming the specialized units of the US Federal Services. It is essential to ensure, at the national level, transparency of business-making as a whole and of the financial settlements in the first line. The major consequences of the rigid governmental and interstate regulation of the offshore business on a global scale include: reduction of the level of banking secrecy; expansion of the authorities' ability to access information as for the actual owners of offshore tools; restrictions on use of the bearer shares; expansion of the international exchange of information as regards the taxation; significant limitation of the offshore banking business; changes in the offshore business structure: departure of some jurisdictions from there, significant change of others and arrival of the new ones.

At the same time, analysis of the methods that the leading states use to fight against tax havens indicates to a certain limitation and declarative nature of the decisions made [3].

The G-20 policy will by no means seriously affect profitability of the offshore business, because it will eliminate only one of the many advantages that the offshore jurisdictions have: confidentiality of financial information. Alongside, an unimpeded access to the data may be provided solely to the competent governmental authorities upon their request. Such advantages of tax havens as low tax rates, easier currency control and option of the capital free flow will further continue to invite business, since in the short run no single secret offshore will be available, meanwhile no haven will lose its competitiveness. This fact is emphasized by Switzerland, which controls more than one third of the offshore flows and agrees to fully accept the uniform tax standards, provided that such steps are taken by all and sundry tax havens.

For the developed countries, especially the United States, disclosure of information will promote a better controlling of the excess liquidity flow within the monetary policy, filling the budget with resources without using the printing presses, as well as fighting against global terrorism and money laundering through arresting accounts of particular legal entities and individuals in offshore banks of any jurisdiction [1]. On the other hand, the non-democratic countries' governments can use exclusivity of the governmental authorities for obtaining confidential information as a way to combat individual oligarchic groups by conducting selective hunting for offshore business transactions and political competitors [8]. Thus, the new on-demand transparency scheme will let the governments of many countries turn a blind eye to some offshore schemes and publicly expose others. Confidentiality will further continue protecting the information of offshore companies from outside observers [14].

Presently, there is a rapid adaptation of the offshore jurisdictions' policy to the conditions of tightening control conditions [5].

The countries ceasing their activities in the area of offshore business include those that could not adapt to the new, more stringent international regulation: Nauru, Turks and Caicos, Anguilla, Vanuatu, Western Samoa, Cook Islands and St. Kitts and Nevis; where the governments decide to abandon the offshore business in the countries because they have already restructured their economies: British Virgin Islands, Guernsey and Jersey, Isle of Man, Gibraltar and Madeira [12]; and those leaving the offshore business due to natural disasters, political collapses, etc.: Montserrat, Netherlands Antilles, Grenada, Niue and Palau Republic.

The countries that continue their activity in the offshore sector of economy include: without transformation as a result of political support: Panama, Belize; having transformed their legislation and offshore sector in accordance with the requirements of the international organizations, or having found a new niche in the offshore business: Cyprus, Antigua and Barbuda, Bahamas, Bermuda, Cayman Islands, Saint Vincent and Grenadines, Liechtenstein, Barbados, Mauritius, Seychelles, Labuan and Bahrain [9].

There are countries that enter the offshore business once again: Montenegro, Hungary.

## 5 Conclusions

Under modern conditions, the success factors for offshore jurisdictions include: stability of the domestic political and economic situation; political flexibility, willingness to compromise; cooperation with the international organizations (OECD and FATF); reforms of the national law to achieve greater transparency in the offshore sector; use of the proceeds from registration of offshore companies for development of other sectors of economy.

It can be assumed that those jurisdictions that have already reformed the national law will be actively developing in future. The countries that preserve the law on registration of offshore companies unaltered have an unstable competitive advantage (political ties) and their position in the market may further deteriorate.

# References

1. Andreevskaya TS, Fedorova SV (2018) Tax havens for IT business. In the collection: problems of management of production and innovation systems. Materials of articles of the 3rd regional scientific and practical conference with international participation, p 5
2. Chugunov VI, Nacharkin VV, Zakharov AV (2016) Offshore business in Russia and abroad: problems and ways to solve them. Innov Sci 6:283–288
3. Egnatosyan KS, Bocharova OF (2018) Taxation in tax havens. Alley Sci 3(6):22
4. Goryunova AA, Rybnikova AS, Shatovich MA (2017) Money concealment in tax havens. In the collection: methods, mechanisms and factors of international competitiveness of the national economic systems. Collection of articles of the international scientific and practical conference. In 2 parts, p 64
5. Heifetz BA (2017) Deoffshorization of economy: world experience and Russian specificity. Econ Issues 7:37
6. Rollins J, Wyler LS (2013) Terrorism and transnational crime: foreign policy issues for congress. Congressional research service. https://fas.org/sgp/crs/terror/R41004.pdf
7. Cypher JM (2014) The process of economic development, 4th edn. Taylor Francis Group, London and New York, pp 81–120
8. Kontaurov AG, Dimitrieva EN (2017) Tax havens and features of their jurisdiction. Approbation 2(53):166
9. Leading corporate portal of offshore industry. Electronic resource. https://offshorewealth.info/. Accessed 10 April 2018
10. Michael M, Brewer J (2013) Convergence: illicit networks and national security in the age of globalization. National Defence University Press, Washington DC
11. Money laundering regulations (2017) A public consultation issued by HM treasury. Comments from ACCA. https://www.accaglobal.com/gb/en/technical-activities/technical-resources-search/2017/april/money-laundering-regulations-2017.html/. Accessed 10 April 2018
12. Nechaev A, Antipina O, Matveeva M, Prokopeva A (2015) Offshore schemes as an effective tax planning tool of enterprises' innovative activities. Econ Her XXI 7–8(2):40–43
13. Taxation and investment in United Kingdom (2015) Reach, relevance and reliability. https://www2.deloitte.com/content/dam/Deloitte/cn/Documents/international-business-support/deloitte-cn-ibs-uk-tax-invest-en-2015.pdf. Accessed 10 April 2018
14. The world of offshore financial centres and the role of the EU in it (2017) International Business, pp 15–17
15. United Nations global programme against money laundering (2011) Proceeds of crime and financing of terrorism (GPML), Independent evaluation unit. New York

# Socio-ethical Problems of the Digital Economy: Challenges and Risks

A. V. Guryanova[1]([✉]), I. V. Smotrova[1], A. E. Makhovikov[1],
and A. S. Koychubaev[2]

[1] Samara State University of Economics, Samara, Russia
annaguryanov@yandex.ru, i-smotrova@mail.ru,
shentala_sseu@inbox.ru
[2] Shakarim University, Semei, Kazakhstan
koychubayev_as@mail.ru

**Abstract.** The article discusses the concept and the main conceptions of the digital economy. The authors offer a new approach to digitalization interpreted as a wide-spread social phenomenon. The research idea was to study not only the traditional economic and technological components of digitalization, but to accent its socio-ethical aspects. Digitalization changes much economic structure, traditional markets and besides causes radical transformations in the system of social relations. At the stage of the digital economy and digital society fundamental social and ethical problems are formed. These problems are vitally important for modern civilization. Therefore, the article discusses their risks and perspectives. Socio-ethical problems must be necessarily considered when forecasting the prospects for the digital economy development. In the case of their positive solution the digital economy will develop successfully and the digital transformation of the human being will lead to its positive changes.

**Keywords:** Digital economy · Digital revolution · Digital society ·
Digitalization · Economy · Human being · Socio-ethical problems

## 1 Introduction

Not very long ago (a few decades only) the world has entered the digital age and the digital revolution has started [8]. From the beginning of the 20th century information and communication technologies have developed rapidly, changing traditional ways of human life [4], causing formation of new industries and professions. Such changes touched an economic sphere too [2, 3]. Development of economic relations in conditions of competitive environment between different countries and manufacturers, importance of innovative technologies using and a need of lowing the cost of products and services have created a new phenomenon, which was called «digital economy». It's clear, that in this term the determining impact of the Internet and mobile communications on the sphere of modern economic relations is accented.

The modern digital economy is a real economy of innovations [14]. Its development depends on an effective use of the new technologies. The number of Internet users is growing exponentially in such an economy, information and computer technologies

© Springer Nature Switzerland AG 2020
S. Ashmarina et al. (Eds.): *Digital Transformation of the Economy: Challenges,*
*Trends and New Opportunities*, pp. 96–102, 2019.
https://doi.org/10.1007/978-3-030-11367-4_9

change much the way of human being. Radical transformations concerning the capabilities of consumers, the structure of industries, the role of the state system take place in the global economic system. Traditional economic postulates united with the digital technologies manifest themselves in a new way, acquire a new content. Under an influence of scientific, technical and economic progress significant changes in the seemingly canonic rules of the market economy and doing business start immediately [5]. Now we can see the new manifestations of traditional economic principles and laws. There are also many other problems, including social and ethical, connected with the development of the digital economy.

## 2  Materials and Methods

The present research is realized with a help of the following methods:

- Information method is used to search the important facts about the digital economy and its development;
- Descriptive method is used to show the special characters of the new, socially oriented model of the digital economy;
- Classification and typology methods are used to identify and organize the various socio-ethical problems of the digital economy;
- Analysis and synthesis methods are used to specify and generalize the problem field of the digital ethics;
- Dialectical method is used to consider the dynamics and relations of the main concepts of the research (digitalization, economics, ethics, society and human);
- System method is used to identify various aspects of digitalization in their interdependence as the elements of the united economic system;
- Structural and functional methods are used to find out the features of organization and functioning of the digital economy;
- Modeling method is used to create a certain vision of the great changes of the future digital civilization;
- Prognostic method is used to consider the perspectives of development of the digital economy, which are directly connected with solution of its socio-ethical problems.

## 3  Results

### 3.1  Digital Economy as a Social Phenomenon

#### 3.1.1  Concept of the Digital Economy

The term of the «digital economy» is very popular today. It is spread all over the world and actively used in the everyday life, by common people, scientists, journalists, politicians and businessmen. However, there is no single definition of this concept. So, it offers a wide variety of interpretations. For example, «the digital economy is a set of economic relations concerning production, distribution, exchange and consumption of goods and services of a techno-digital form of existence» [14]. Or «digital (electronic) economy is an economy characterized by the maximal satisfaction of needs of all its participants by the way of using information, including personal» [5]. And even such a one: «…it is an economy that exists in a hybrid world», which means in its tern «the result of connection of the real and virtual worlds» [5].

To understand what the digital economy is, it's necessary first of all to give a definition of the «normal» («not digital») economy. It is an economic activity of the society and also a set of relations taking place in the system of production, distribution, exchange and consumption. Strictly speaking using of the computers, of the Internet and mobile phones can already be considered as a form of «consumption» [3]. Therefore, the digital economy can be defined as a part of economic relations that are mediated by the Internet, cellular network and ICT. Digital economy is a special economic activity basing on the digital computer technologies and using the new methods of generation, processing, storing and transmission of the data.

### 3.1.2    Conceptions of the Digital Economy

One of the first conceptions of the digital economy was created by Nicholas Negroponte [6] – an American computer scientist from the University of Massachusetts. With a help of this term he explained the advantages of the «new» economy in comparison with its traditional version. The «old» economy has been overcome for a reason of the intensive development of information and communication technologies. The «new» digital economy doesn't deal with the movement and processing of atoms which was typical for the matter of physical substances but with the movement and processing of bits that make up the matter of program codes. According to Negroponte, the advantages of the digital economy considered as a «new» type of economic relations, are the same: an absence of physical weight of the product, replaced by information volume; the lower costs of resources for the production of electronic goods; significantly smaller area occupied by the products (usually electronic media); the quick global movement of products by the Internet [6].

Another famous conception of the «digital economy» belongs to Don Tapscott [10] – a famous Canadian economist and a well-known specialist in the field of digital technologies. In 1995 Don Tapscott published his book «The Digital Economy: Promise and Peril in The Age of Networked Intelligence». In this book he gave an exact definition of the «digital economy»: it is based on the use of information computer technology. The main features of this «new society» are the same: focus on knowledge, digital form of objects' representation, virtualization of production, innovative character, integration, convergence, dynamism, globalization, transformation of the relations between the manufacturer and the consumer, exclusion of intermediaries etc. According to Tapscott, exclusion of intermediaries that played an important role in the pre-electronic economy is inevitable. Because in the digital society the manufacturers are able to interact themselves with the potential customers, making their own websites where they advertise the goods they produce. In this case no estate firms, travel agencies and other intermediary organizations are needed.

It's significant that most of the hypotheses presented by Tapscott in 1995 have already been realized in practice. The scientist paid special attention to this fact in the last edition of his book related to its twentieth anniversary [11]. For example, constantly updated multimedia news web-portals have become a daily reality of the modern Internet. The problems of security are also very serious nowadays. That's why all sites containing the private information are protected in one or another way.

### 3.1.3 Social Essence of the Digital Economy

As we have already mentioned this concept and the basic conceptions of the digital economy appeared in the late twentieth century. At that time, they were primarily associated with an intensive development of information and communication technologies [2]. It's obvious that development of the Internet and mobile communications are the basic technologies of the digital economy. But the digital economy is not only a new economic system. Digitalization of the modern economy includes the changes of economic structure and traditional markets, but also the transformation of social relations due to the increasing influence on them of the digital technologies. In other words, when we analyze the phenomenon of the digital economy, we must necessarily pay attention to its social essence.

This tendency is already traced in a number of normative and analytical documents. For example, the experts of the World Bank define the digital economy as a system of economic, social and cultural relations based on the use of the digital information and communication technologies [13]. A similar interpretation accenting the social essence of the digital economy is presented in «The Program of digital economy development in the Russian Federation until 2035». It's given the following definition here: «Digital (electronic) economy is a set of social relations that are formed by using electronic technologies, electronic infrastructure and services, technologies for analyzing Big Data and forecasting in order to optimize production, distribution, exchange, consumption and increase the level of socio-economic development of the country» [12].

### 3.2 Socio-ethical Aspects of the Digital Economy

### 3.2.1 Digital Economy as a Result of the Digital Revolution

The global digital changes taking place in the modern world are often called «the point of non-return» [1] or the «fourth industrial revolution» [8]. It was announced by Klaus Schwab [8] – a famous economist, founder and Executive Director of the World economic forum in Davos. Revolution, wherever it happens (in political, social, cultural, scientific, religious spheres), always involves a quick, radical change. Klaus Schwab writes in his book «The Fourth industrial revolution» that the first of them took place in 1760–1840s. It was caused by an invention of the steam machine. The second industrial revolution dates from the end of XIX – beginning of the XX century. It was determined by the spread of an incandescent lamp, electricity and conveyor manufacturing. The third industrial revolution happened in 1960-ies of the XX century. It was caused by the spread of semiconductors. The fourth industrial revolution falls immediately on our days.

Today the industry all over the world is going through the real revolutionary changes. The scale, volume and complexity of this fourth industrial revolution have no analogues in human history. The three previous revolutions were driven solely by an emergence of new technologies. The fourth differs much from them because of the speed of technology diffusion and the global character of its dissemination and application. The fourth industrial revolution is fundamentally changing the traditional way of human life. We are witnessing significant technological innovations in a wide range of areas – artificial intelligence, robotization, 3D printing, nano- and biotechnology, etc. At the same time, the fourth industrial revolution with its innovative technologies causes many problems, especially in the social and ethical fields.

### 3.2.2    Ethical Risks and Challenges of the Digital Revolution

The «digital revolution» generates ethical challenges and risks in the following aspects:

– Problems of interaction into the system «human/machine» , changing ethical norms and values in the professional activity and in the sphere of «professional/consumer of services» relations;
– Social and ethical problems between the people when they use different kinds of «technical intermediaries» , including problems of interpersonal relations and professional ethics;
– Ethical problems affecting changes in the labor market, reduction and even disappearance of a number of professions in a short time period;
– Absence of equity in access to the achievements of the digital revolution; the appearance of new forms of discrimination or the revival of the old ones. These trends, without paying due attention to them, may become grounds for risks of the social inequality increasing and the growth of social tension;
– Problems of responsibility of the autonomous intelligent technologies which are known as «machines' morality». In this context the most important problem is a problem of responsibility distribution between the creator, the user and the machine. It's also necessary to set ethical boundaries of the possible using of «moral machines» and the trust level to them;
– Changes in moral norms of human behavior caused by the widespread and accessible character of information about the activities of social and political structures, institutions and commercial companies [9];
– Changes in the boundaries of confidentiality in the professional and private spheres, up to disappearance of the privacy itself;
– Social problems of unethical using of the achievements of the digital revolution for commercial or political manipulations realized by individuals or social groups;
– Risk of the privacy violation and the creation of total control systems;
– Danger of automatic systems out of control, such as computer errors that can cause technogenic catastrophes.

### 3.2.3    Ethical Problems of the Education System

Special attention must be paid to an introduction of digital technologies in the sphere of education. Modern colleges and universities offer different online courses; online teachers develop the new methods of learning materials; educational institutions are regularly equipped with the newest models of computers and other technologies. So, the fields of education, science, research, culture and media are the main areas for introduction of the new digital achievements. They are themselves the most important factors causing further development of the digital technologies. Today everyone can use the great opportunities of education system in the areas of training, professional development, continuous professional education [7], development and participation in the economic and social life.

However, the education system needs some more changes to equip people with skills and knowledge corresponding to the demands of the digital environment and knowledge society. Education is also responsible for literacy degree in the media. Therefore, it's necessary to provide greater use of the digital media in continuous education process. A new digital learning strategy must be created. It will be used systematically to introduce and expand the resources of digital media for a high-quality education. The questions of human adaptation to the challenges of the digital economy are very important today. They concern the continuous skill development and the formation of the new skills actual for digitalization. In this context an active policy on the labor market, a continuous form of education and a more flexible education system must be achieved.

## 4  Discussion

The perspectives for the digital economy development get different reviews today. They range much from really optimistic to tragically pessimistic. In our opinion to form a correct view on the essence of the digital economy and to realize its proper strategic planning it's necessary to analyze this phenomenon objectively. Digitalization (it doesn't matter how to treat it – positively or negatively) is a reality of nowadays: humanity have already entered the era of global digital changes. The new digital technologies contain a great potential for development, but they also include great risks. In order to overcome them it's necessary to take in view the human factor, to solve the socio-ethical problems arising in the process of digitalization [2].

The question about the consequences (positive or negative) of the digital economy development is related closely to the human nature: are the humans ready for the changes that are taking place? Human nature changes slowly, but the world transforms quickly under the progress in information technologies and telecommunications [4]. In the nearest future all main spheres of human life – economics and management, science and security – will get new forms and content. The human himself will become the different one. This will inevitably transform the system of social relations and ethical principles [9]. All these tendencies must be analyzed specially. They need assessment, planning and effective regulation. In this case consideration of socio-ethical problems of the digital economy is very actual nowadays.

## 5  Conclusions

Digital economy is a driver of growth and a tool for qualitative changes of the modern society. Digital economy is a special economic activity based on the digital computer technologies, using the new methods of generation, processing, storing and transmission of the data. It is also a wide system of economic, social and cultural relations based on the use of the digital information and communication technologies. Digital economy is a result of the fourth industrial revolution with its significant technological innovations. It is fundamentally changing a world around us and a traditional way of human existence. At the same time, the fourth industrial revolution causes many problems, especially in the social and ethical fields.

Socio-ethical problems of the digital economy generate vital challenges and risks in the way of human activity. They transform moral norms and values of the modern society, cause radical changes in education system. Therefore, the social and ethical aspects of the digital economy must be a special subject of theoretical consideration and the basis of the modern forecasting. If they are positively resolved, the digital economy will progress. And the modern humans will acquire the necessary qualities they need for adaptation to the digital economy.

# References

1. Digital Russia: a new reality (2017) McKinsey & Company. http://www.tadviser.ru/images/c/c2/Digital-Russia-report.pdf
2. Guryanova A, Astafeva N, Filatova N, Khafiyatullina E, Guryanov N (2019) Philosophical problems of information and communication technology in the process of modern socio-economic development. In: Advances in intelligent systems and computing, vol 726, Springer, Cham, pp 1033–1040. https://doi.org/10.1007/978-3-319-90835-9_115
3. Guryanova A, Guryanov N, Frolov V, Tokmakov M, Belozerova O (2017) Main categories of economics as an object of philosophical analysis. In: Contributions of economics. Russia and the European union: development and perspectives, Springer International Publishing AG, Cham, Switzerland, pp 221–228. https://doi.org/10.1007/978-3-319-55257-6_30
4. Guryanova A, Khafiyatullina E, Kolibanov A, Makhovikov A, Frolov V (2018) Philosophical view on human existence in the world of technic and information. In: Advances in intelligent systems and computing, vol 622, Springer, Cham, pp 97–104. https://doi.org/10.1007/978-3-319-75383-6_13
5. Keshelava AV (ed) (2017) Introduction to digital economy. Vniigeosystems
6. Negroponte N (1995) Being digital. Knopf, NY
7. Pecherskaya E, Averina L, Kochetckova N, Chupina V, Akimova O (2016) Methodology of project managers' competency formation in CPE. IJME – Math Educ 11(8):3066–3075
8. Schwab K (2017) The fourth industrial revolution. Exmo, Moscow
9. Shestakov A, Noskov E, Tikhonov V, Astafeva N (2017) Economic behavior and the issue of rationality. In: Contributions of economics. Russia and the European union: development and perspectives Springer International Publishing AG, Cham, Switzerland, pp 327–332. https://doi.org/10.1007/978-3-319-55257-6_43
10. Tapscott D (1995) The digital economy: promise and peril in the age of networked intelligence. McGraw-Hill, NY
11. Tapscott D (2014) The digital economy, Anniversary edn. Rethinking Promise and Peril in the Age of Networked Intelligence, McGraw-Hill, NY
12. The Program of digital economy development in the Russian Federation until 2035 (2017). http://innclub.info/wp-content/uploads/2017/05/strategy.pdf
13. World Bank. World development report (2016) «Digital dividends». Review, Washington, DC. https://doi.org/10.1596/978-1-4648-0671-1.a https://openknowledge.worldbank.org/bitstream/handle/10986/23347/21067
14. Zubarev AE (2017) The digital economy as a form of manifestation of the new economy. Bull Pac Natl Unv 4(47):177–184

# The Transition to a Digital Society in the People's Republic of China (Development and Implementation of the Social Credit Score System)

E. A. Timofeeva[1,2(✉)]

[1] Samara State University of Economics, Samara, Russia
anna0474@mail.ru
[2] Research Institute of the Federal Penitentiary Service of Russia,
Moscow, Russia

**Abstract.** The relevance of the research topic is determined by the most important role of the processes of digitalization of the modern society of the People's Republic of China (hereinafter China), affecting the development of its digital economy and national security. The purpose of this article is to analyze and identify the features of the implementation of the social Credit Score system throughout China. In the process of writing the article used General scientific and private scientific research methods. The leading method of research of this problem is the comparative analysis allowing to reveal features of the implementation of the SCS system by assignment to each citizen of an individual rating. Based on the analysis, the main economic companies involved in the development of pilot algorithms of large-scale digital data processing system are identified, the positive aspects of the system implementation for the development of the country's economy are revealed, the possible negative consequences in the field of human rights violations are revealed. The materials of the article are of practical value for scientists specializing in the study of social problems, problems of the digital economy and economic growth of Asian States, as well as specialists in the field of national security.

**Keywords:** Digital society · Digital economy · Digitalization ·
Rating · Economy · National security

## 1 Introduction

The article is relevant due to the fact that for the first time it presents the analysis and synthesis of data on the implementation process in China by digitization of a new system called Social Credit Score (hereinafter - SCS). The process of digitalization, which is a transition from analog to digital transmission of information, is effectively implemented in the country's economy and contributes to the improvement of various levels of national security. China as one of the largest countries in the world, with a population of over 1.5 billion people successfully ensured that the operation of the information filter of the Internet by the Great Firewall.

© Springer Nature Switzerland AG 2020
S. Ashmarina et al. (Eds.): *Digital Transformation of the Economy: Challenges,*
*Trends and New Opportunities,* pp. 103–110, 2019.
https://doi.org/10.1007/978-3-030-11367-4_10

In modern China, a new type of society is being actively formed-the digital society. A digital society is a society governed by the use of information and communication technologies based on the application of microelectronics, local and global computer networks that collect, process, generate and distribute information through systems of global telecommunication networks. Digital society, in its essence, is a network information society.

It is China that is the first country to introduce a system of social credit of trust by 2020, the registration of which will be mandatory for every citizen and legal entity (each company). The system aims to evaluate and award a social rating to all Chinese citizens on a variety of parameters. For the first time this information appeared in the document of the State Council of the PRC "Project plan for the establishment of a social credibility", published on June 14, 2014, the State Council of China has updated the document on 25 September 2016 under the title "Mechanisms of prevention and punishment of people who are prone to breach of trust". According to the amendment, "if a person has violated trust in one area, restrictions are imposed on him in all" [1].

The main purpose of the introduction of the system "building a harmonious socialist society." The main value of such a society is honesty. It should manifest itself in everything from online behavior to honoring parents. Confucius, whose teaching had a profound influence on Chinese civilization, emphasized honesty as one of the most important qualities of personality. In his life there is a parable: when he was an official, someone offered him a bribe and assured that no one would know. The sage replied, " I know, you know, Heaven knows. Who doesn't?". The Chinese government, like the sky, has a huge amount of information about the life of its citizens in economic and private life [2, 322–325].

The Chinese authorities are promoting the system as a means of measuring and improving the "trust" in society and the economy, which will help to develop a culture of "sincerity". The Chinese government believes that trust will flourish in society through the system. The system will strengthen sincerity in the interaction of government agencies, Commerce and society, as well as enhance legal reliability [3].

In the result of the study will be solved the following objectives: summarizes the economic companies are developing algorithms for the SCS system and have access to the array of information about Chinese citizens; defines the rating categories of citizens (the range from 350 to 950 points); the identified possible positive and negative outcomes associated with the implementation of the system [4].

Each Chinese citizen will be assigned a rating by the system. The rating will be publicly available for every citizen. It is believed that the rating will develop honesty in citizens. Today, China has already started the voluntary implementation of the SCS rating. A number of pilot projects are in full swing in the country. So, in several tens of cities, separate elements of the system are already started. From our point of view, the study of the experience of implementing such a system, the study of its impact on the economy and security of the country [5] will be of interest to Russian researchers.

# 2  Materials and Methods

## 2.1  Method of Research

In the process of research General scientific (analysis, synthesis, comparison, generalization), private scientific (historical and legal, etc.) and sociological observations, methods of empirical, social analysis, secondary analysis of sociological data) methods of data collection and processing were used.

These methods made it possible to comprehensively study the problem of digitalization of society in China and the features of the introduction of a new system of social rating of citizens as a factor affecting the quality of their lives in the field of economy, education, social and healthcare, etc.

## 2.2  Base of Research

The base of the research was scientific research, publications of Russian and foreign economists, sociologists and lawyers studying various aspects of digitalization of society, problems of digital economy of national security problems.

## 2.3  Investigation Phase

The study of the problem was conducted in two stages:

At the first stage: the analysis of the existing scientific literature on the subject of research, as well as legislation in this area; highlighted the problem, purpose, and methods of research.

At the second stage: the conclusions obtained during the analysis of scientific literature and legislation were formulated, the article was prepared.

# 3  Results

3.1. The government of the country is working with a number of economic companies to develop separate algorithms necessary for the operation of such a large-scale data processing system.

Such companies as China Rapid Finance (partner of the networking giant Tencent) and Sesame Credit (the subsidiary of Ant Financial Services Group (AFSG), partner of Alibaba) are working on this project. The aforementioned China Rapid Finance and Sesame Credit have access to a significant amount of personal data. The first company does this through the chat messaging app (currently 850 million active users), and the second—through its own payment system Alipay, used for shopping online, in restaurants, cinemas, to pay for training, money transfers. Ant Financial is engaged in lending to small and medium-sized businesses and insurance. In addition, Sesame Credit collaborates with other data collection platforms. It's a car-sharing service called Didi Chuxing, which was Uber's main competitor in China until it bought out its Chinese division in 2016. With SCS cooperates Baihe, which is the largest Chinese

Dating service. The above services supply Social Credit Score with a significant amount of data about users and calculate the rating on their basis.

3.2. The rating of citizens is calculated as follows. For the absence of violations of the law, useful social activities, timely payment of loans to each citizen will earn points. Tencent implements the SCS rating in the QQ chat application, where the citizen rating ranges from 350 to 950 and is divided into 5 main subcategories:

- material condition (income, credit history of the citizen, payment of bills (electricity, telephone, etc.));
- security (personal data, confirmation of personal information);
- social connections (education, friends);
- law-abiding (diligence, positive feedback about the government, " the ability of the user to fulfill obligations under the contract »);
- consumer behavior (purchases, preferences of goods and services).

The system systematizes information about citizens taking into account their financial situation, creditworthiness, and purchases. For example, if a person often buys diapers, the system will assume that he is most likely a family, reliable. People whose account balance is constantly in the black will be considered responsible.

Starting with 600 points, a citizen is eligible for an $800 unsecured loan for online purchases. The citizen with a rating of 650 points-can rent a car, too, without collateral, are entitled to a quick check-in at hotels and VIP-registration at Beijing International airport. If his rating is more than 666 points, he can take a loan of 50 thousand yuan (433 thousand rubles) in Ant Financial Services. Starting from 700 points, Express processing of travel permits to Singapore without accompanying documents (for example, an invitation from an employer) is available. With a rating of 750 points, you have accelerated Schengen visa. Alibaba, of course, itself does not give a visa, but many contractors of the Internet giant take into account indicators Sesame Credit-with the same trips help tour operators.

The rating goes far beyond a simple assessment of the creditworthiness of citizens. Determining the value of an individual as a citizen of the country, SCS rating will affect his ability to get the desired job, mortgage, as well as in what school his children will be able to study. Rating a person in Social Credit Score can affect the chances to go on a date or get married, because the higher it is, the more noticeable your profile becomes in the application for Dating Baihe.

3.3. Investigating the problem of the attitude of society to the implementation of the SCS system, we note the following. In October 2015, the BBC conducted a survey of respondents in Beijing to get their opinion on the rating system being implemented. The majority of citizens evaluated it from the positive side. Only a few of the respondents noted that a low rating could harm them in the future.

Millions of Chinese citizens have voluntarily signed up for a test version of the state surveillance system. The high rating became a symbol of high status just a few months after the launch. Some sociologists point to a possible fear of repression.

Economists point to the attractiveness for citizens of the awards and "special privileges" they receive, proving their "reliability" in the SCS.

3.4. Analysts note a number of positive aspects expected by the government from the introduction of the system. In order to increase the "honest attitude and level of trust of the entire population", the State Council of China plans to "improve the overall economic competitiveness of the country." This system can affect the reduction of public debt and encourage citizens to timely payment of their debts. Some believe that such changes will give a positive result: people will be motivated to learn to be responsible for themselves and their habits for the sake of good points in the ranking and the status of the so-called "trustworthy citizen".

The introduction of the system can have a positive impact on the improvement of public services. The SCS rating will provide Chinese citizens with much-needed access to a range of financial services. In "the Wall Street Journal", the government notes that the rating "will allow trustworthy citizens to go where they want, and will not allow the discredited to take a step." Financial institutions and banks are already developing credit assessment systems. This data helps them not to give loans to those who will not be able to pay them. This system will contribute to the security of society through the interaction of entrepreneurs and law enforcement agencies. The Chairman of the Board of Alibaba suggested using the Internet data of companies to identify criminals. "It's normal for a person to separately buy a pressure cooker, a timer, and even steel bearings or gunpowder, but it's not normal for him to buy it all at the same time."

The Chinese government believes that the introduction of a civil rating system is also necessary due to the lack of an extensive public credit system in the country. Some Chinese people do not have credit points because they do not have their own home, car or credit card. The country's Central Bank has financial data for 800 million people, but only 320 million have a credit history. According to the Ministry of trade of China annually due to the lack of full credit information about the citizens of the country's economy is losing more than 600 billion yuan.

# 4 Discussion

The problem of using information systems and technologies in digital society as a means of improving the welfare and quality of life of citizens is actively studied by domestic and foreign scientists.

So, the design of management information systems paid attention to in academic research as in G.R. Gromov [8], A. Zaichenko [1], M. Castells [7], S. Prokopenkov [9], K. Shannon [10], I.O. Senkiv [9] etc. Researches of various economic, legal and social aspects of carrying out reforms in Chinese society are reflected in works of the Russian scientists: L.D. Boni [11], E.P. Pivovarova [12], E.A. Timofeeva [5] and others. However, studies of the problem of transition to a digital society in China and the analysis of the features of the introduction of the Social Credit Score system have not been carried out by domestic scientists before.

## 5  Conclusions

Opponents of system note that it can rigidly limit citizens to violate their right for private life. Skeptics consider that China takes a step to the totalitarian police state controlling the citizens violating their right to confidentiality.

With the help of SCS, the Chinese government will be able to promote its vision of "socially acceptable behavior" in the emerging digital society and at the same time will be able to control various aspects of citizens' lives. For offenses of varying severity, points will be deducted. For example, points are deducted for crossing the road in the wrong place, not visiting elderly parents, exceeding the limit on the number of children in the family, etc. (today, couples are allowed to have 2 children, but previously it was possible to have no more than 1 child in the family (policy "one family—one child" 1979).

The Internet will work for citizens with a low rating more slowly, they won't be able to visit some restaurants, clubs and other institutions. He will forbid to go abroad freely. The rating will influence an opportunity to lease something, to obtain the credit and even social privileges. Citizens with low rating won't be able to get a job to certain positions, for example, in public authorities, media or law where reliability, certainly, is the obligatory criterion. Low rating, you won't be able to get an education in private educational institutions or to give there the children.

Conclusions about the reliability of the citizen proceeding from the analysis of his purchases are considered by the system but can be wrong. The basic SCS system can create an inexact and incomplete image of the citizen. So, if the person plays video games for 10 h a day, then the algorithm can declare him the unemployed, without understanding the reasons for his actions. However, perhaps, he works as the engineer and just tests these games. But in this situation automatically he, most likely, will be noted as the unemployed. Actually he, perhaps, just performed the work.

As a result, the SCS system not just studies the behavior of people, it begins to define it, sometimes "forces" to refuse the purchases and actions undesirable to the state.

For 2018 Sesame Credit doesn't punish directly "unreliable" citizens. However, the system is arranged so that unreliable persons won't be able to rent the car, to take the credit and even to find work. Information on citizens gathers from different sources. For example, for replenishment of the violations of citizens of Sesame Credit given about stories the list of pupils who wrote off at state exams that those were responsible then for the offense in the future has requested the Chinese Bureau of education.

The modern digital world passes to the SCS system algorithms defining the quality of life of people. However, it is necessary that system considered all nuances and subtleties of life of people. For example, one person cannot pay bills because he was hospitalized, and another intentionally evades from payments. The SCS system in February 2017 has begun the actions for restriction of trips of citizens with the low social rating. The Supreme national Court of China has reported that 6,15 million citizens have received the ban on a departure from the country within the next 4 years because of public offenses. The ban has been recognized as the first step on the way of introduction of a system of the credit of social trust. 1,65 million more citizens who were included in the blacklist won't be able to use the railway system of the country.

In the conclusion, generalizing the aforesaid, it is possible to draw a conclusion on the formation at the realization of the Social Credit Score system of the whole list of so-called sanctions, restrictions by whom citizens with the low rating will be exposed:

- the ban on work in state institutions;
- the ban on the occupation of senior positions in the pharmaceutical and food industry;
- especially careful examination at customs;
- refusal in social security;
- refusal in tickets of an avian and railway transport (or in a berth in night trains);
- refusal in places at luxury restaurants and hotels;
- the ban on training of children in expensive private schools.

According to forecasts of experts, in the near future in China, there will be black markets of reputation in which roundabout methods of improvement of the rating will be on sale.

Perhaps the Chinese model of digital control over society, which many consider totalitarian, will eventually be established in democratic countries. The governments of several countries around the world already not the first year engaged in the monitoring and evaluation of the activities of their citizens by analyzing the information passing through the network and the country's financial system. For example, us National Security Agency is an official digital eye, which, without hiding, monitors the actions of us citizens. In 2015 US Department of transportation security has proposed to include in the preflight inspection verification of data from social networks, information about the location of the person and his purchase history. The decision was not made because of harsh criticism, but this does not mean that it was completely abandoned. Digital society and the digital economy increasingly offer algorithms that can determine whether a citizen is a threat or not and help to assess the level of its reliability.

The problem of the development of China's digital society and the implementation of the digital system Social Credit Score in the prism of the development of economic and social relations, improvement of security of the country, of course, require further reflection, learning, and research.

# References

1. Zaichenko A (2017) Digital society is a super intellectual society. "News." http://www.gazetaprotestant.ru/2017/08/cifrovoe-obshhestvo-eto-sverx-intellektualnoe-obshhestvo/. Accessed 11 Aug 2017
2. Perelomov LC (1998) Confucius: "LUN Yu". Moscow: publishing company "Eastern literature" RAS, p 588. ISBN 5-02-018024-6
3. Tarasov V (2017) The Chinese "social rating system" will determine the value of people. Behind the great wall of China. 20 Feb 2018. https://inosmi.ru/social/20180220/241513806.html, "Social credit system" of China will assess how valuable you are as a person. https://futurism.com/china-social-credit-system-rate-human-value circulation date, 2 Dec 2017

4. The Global Competitiveness Report 2001–2002 (2002). World economic forum. Oxford University Press, New York, p 587

5. Timofeeva EA, Vilkova AV (2017) Activities of penitentiary institutions and training in the penitentiary system of the People's Republic of China (on the example of the Shanghai region): analytical materials – Samara. Samara law Institute of the FSIN of Russia, p 83. ISBN 978-5-91612-153-7

6. Golubev SS, Chebotarev SS (2018) Normative-legal and methodological measures for implementation of the strategy of economic security of the Russian Federation. Jurisprudence and practice: Bulletin of the Nizhny Novgorod Russian Interior Ministry Academy. No 1 (41), pp 127–132. https://doi.org/10.24411/2078-5356-2018-00018

7. Kastel's M (2001) The internet galaxy: reflections on the internet, business, and society. Oxford University Press Inc., New York. C.249

8. Gromov GR (1982) National information resource: problems of commercial operation. Puschino, NCBI USSR Academy of Sciences, p 120

9. Prokopenkov SV, Senkiv IO (2016) Issues of regional informatization development. Economics and business, № 12-2 (77-2)

10. Shannon C (1963) Computing devices and automata. In the book: works on information theory and Cybernetics, Moscow, pp 162–179

11. Boni LD (2013) Russian Sinology. (Chinese President XI Jinping's meeting with Russian sinologists). Problems of the Far East" No. 4, 3–7

12. Pivovarova EP (2016) Lines of continuity and novelty in economic policy of five generations of the leadership of the People's Republic of China. Economic strategy. No. 6, 64–75

# Target Indicators and Directions for the Development of the Digital Economy in Russia

N. N. Belanova[1]([✉]), A. D. Kornilova[1], and A. V. Sultanova[2]

[1] Samara State University of Economics, Samara, Russia
{bnn371, adkornilova}@yandex.ru
[2] Samara State Technical University, Samara, Russia
sultanovaav@mail.ru

**Abstract.** Information and communication and digital technologies have been rapidly developing in recent years. They are the basis for innovative development, productivity growth, competitiveness and a prerequisite for improving people's living standards changing their quality. The authors identified the main directions and indicators for the development of the digital economy. To determine the level of development of the digital economy in Russia, a comparative analysis of international indices was carried out, reflecting the development of information and telecommunications, computer and digital technologies. It was noted that as information-computer and telecommunication technologies develop, an increasing number of citizens master digital competencies, but the level of development of the digital economy in Russia is lower than in developed countries. The rapid development of information and telecommunication technologies, the creation of new institutions of the digital economy require appropriate macroeconomic conditions and a regulatory mechanism encouraging the development of ICTs and eliminating obstacles in the course of development. The authors studied the mechanism of state planning, based on a comparison of planned and actual growth indicators, and evaluated the performance of the state program.

**Keywords:** Digital economy · Government program ·
Information and communication technologies · World development indicators

## 1 Introduction

The development of digital technologies has a significant impact on society, economic structure, education. A large amount of research has been devoted to this issue in recent years. So, the influence of world megatrends, high-tech digital economy on various aspects of life [18], on the education system [3], on corporate social responsibility [17], on financial markets [11], on quality of life and digital health [7] are considered in many scientific works. A significant body of research is devoted to changes in labor productivity, employment of workers and labor relations in the era of digitalization [5, 6, 15]. However, in most works, both domestic and foreign authors, the focus is on the development of information and communication infrastructure and the market for

© Springer Nature Switzerland AG 2020
S. Ashmarina et al. (Eds.): *Digital Transformation of the Economy: Challenges,*
*Trends and New Opportunities*, pp. 111–118, 2019.
https://doi.org/10.1007/978-3-030-11367-4_11

scientific and technical information, while the problems of planning and managing the development of the digital economy are not sufficiently covered.

The aim of the study is to identify the priority directions for the development of the digital economy, to analyze the feasibility of planned indicators of the implemented state policy.

Based on purposes of the study, the following tasks were set:

- Determine the planned directions and goals for the development of the digital economy;
- Carry out international comparisons based on indices reflecting the development of information and telecommunications, computer and digital technologies; determine the place of Russia in global trends;
- Study indicators of information society development; determine the performance of the state program in the field of information and ICT.

## 2    Materials and Methods

The methodological basis of the contribution is a systematic approach that allows us to consider the digital economy as a holistic object that includes many elements. The following research methods were used: formal-logical (deduction, induction, justification, argumentation), abstract logical, empirical (observation and experimentation). As a statistical toolkit, the authors used the methods of grouping, averages, as well as graphical and tabular methods presenting the results of the study.

## 3    Results

The core of the digital economy is the digital goods and services sector. World statistics indicates a steady growth of the digital economy, an increase in world trade in products of the digital economy. The expenses of enterprises for research related to digital technologies are increasing, digital technology is developing and becoming more accessible, which, in the end, makes it possible to predict the increasing coverage and development of digital technologies in the world.

We will conduct international comparisons and describe the level of development of digital and information technology in Russia compared to other countries. To do this, we analyze the dynamics of a number of indices.

The first of these is the E-Government Development Index (EGDI). It is prepared every two years by the UN Department of Economic and Social Affairs (UN DESA). The index consists of three components characterizing the state of the ICT infrastructure, human capital and online public services.

According to Table 1, Russia has made a significant breakthrough in development, rising in the ranking from 60th place (2008) to 32nd place (2018). Especially successful was the period from 2010 to 2012. In 2018, Russia returns the lost position, and its index value is estimated as very high.

**Table 1.** Russia in EGDI rating

| Indicators | 2008 | 2010 | 2012 | 2014 | 2016 | 2018 |
|---|---|---|---|---|---|---|
| EGDI value | 0,5120 | 0,5136 | 0,7395 | 0,7296 | ... | 0,7969 |
| Place in rating | 60 | 59 | 27 | 27 | 35 | 32 |

Source: compiled by the authors based on data from the Ministry
of Digital Development, Communications and Mass
Communication of the Russian Federation and the UN Department
of Economic and Social Affairs (UN DESA)

Consider the ICT Development Index. It is measured annually by the International Telecommunication Union, a specialized unit of the United Nations. The index consists of three components reflecting the access of the population to ICT, the use of ICT in the country and practical skills in the use of ICT by the population of 190 countries of the world (Table 2).

**Table 2.** Russia in the ICT development index rating

| Indicators | 2008 | 2012 | 2014 | 2016 | 2017 |
|---|---|---|---|---|---|
| IDI value | 4,42 | 6,19 | 6,7 | 6,91 | 7,07 |
| Place in rating | 49 | 40 | 42 | 43 | 45 |

Source: compiled by the authors based on data
from the Ministry of Digital Development,
Communications and Mass Communication of the
Russian Federation and the International
Telecommunication Union (ITU)

According to the ICT Development Index in 2008–2017, Russia occupies 40–50 places in the overall rating. Despite a slight decrease in positions that began in 2012, there has been a steady increase in the ICT Development Index (from 6.19 in 2012 to 7.07 out of 10 in 2017). When calculating the index, the geographical features of the country, the length of the territory, the density and the pattern of population dispersion are not taken into account. These factors have a restraining development in ICT in Russia.

Consider the Networked Readiness Index (NRI), which is calculated by the international organization World Economic Forum in conjunction with the International Business School INSEAD.

**Table 3.** Russia in the NRI rating

| Indicators | 2008 | 2010 | 2012 | 2014 | 2016 |
|---|---|---|---|---|---|
| NRI value | 3,77 | 3,69 | 4,02 | 4,3 | 4,5 |
| Place in rating | 80 | 77 | 56 | 50 | 41 |

Source: compiled by the authors based on data
from the World Economic Forum website.

The Index reflects the networked readiness of countries to the widespread use of ICT for socio-economic development. It includes three parameters: the availability of conditions for the development of ICT and the willingness of citizens, business and government agencies to use ICT and the level of ICT use in public, commercial and government sectors. The data of Table 3 show a positive dynamics of the index for the studied period.

Consider the International Digital Economy and Society Index (I-DESI). It is calculated for non-EU countries (similar to the DESI index of EU countries). The main components of the index are communications, human capital, the use of the Internet and the introduction of digital technologies in business and digital services to the public (Fig. 1).

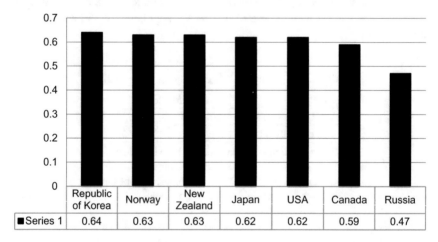

**Fig. 1.** International Digital Economy and Society Index (I-DESI). Source: compiled by the authors based on indicators of the Digital Economy 2018: statistical collection, p. 24)

According to the International Digital Economy and Society Index, Russia lags behind the EU countries, the Republic of Korea, New Zealand, Japan, Norway, the USA, and Canada. Its index value in 2016 was 0.47 (with an average for the EU countries of 0.54).

Russia must occupy a competitive position in the international market, which under conditions of rapid development of ICT and innovative technologies requires the accelerated development of the digital economy and information technology. During the periods of catching up development, an active role of the state is necessary for a breakthrough leap related to the planning and financing of the process.

In 2017, in order to manage the development of the digital economy, the Government of the Russian Federation approved the State Program "Digital Economy of the Russian Federation". It defines goals and objectives within the framework of 5 basic directions of development for the period up to 2024. These include regulations, personnel and education, research competencies and technical reserves, information infrastructure and information security. Consider the goals and priorities in each area:

1. Policy management: comprehensive legislative regulation of relations in connection with the development of the digital economy; adopt measures aimed at stimulating economic activities related to modern technologies, data retrieval (digital economy).
2. Personnel and education: create key conditions for training in the digital economy, improve the education system and the labor market; develop a system of motivation for mastering necessary competencies and the participation of personnel in the development of the digital economy in Russia.
3. Formation of research competencies and technological groundwork: develop a support system for search and applied research in the field of the digital economy (research infrastructure of digital platforms), ensuring technological independence in each of the areas of cross-cutting digital technologies that are globally competitive, and national security.
4. Information infrastructure: develop communication networks, Russian data processing centers, introduce digital data processing platforms to meet the needs of government, business and citizens; develop an effective system for collecting, processing, storing and providing users with spatial data.
5. Information security: protect an individual, society and the state from internal and external information threats, which ensures the realization of constitutional rights and freedoms of a person worthy quality and standard of living of citizens, sovereignty and sustainable socio-economic development of the Russian Federation in the digital economy.

This program is closely related to the Strategy for the Development of the Information Society in the Russian Federation for 2017–2030 and the State Program of the Russian Federation "Information Society (2011–2020)". Since the State Program "Digital Economy" is being implemented relatively recently, the feasibility of the State Program "Information Society" based on a study of planned and actual indicators was analyzed (Table 4).

**Table 4.** Indicators of the state program "Information Society"

| Indicators of the program | 2015 | | 2016 | | 2017 | |
|---|---|---|---|---|---|---|
| | Plan | Result | Plan | Result | Plan | Result |
| Share of citizens using state and municipal services in electronic form, % | 40 | 39,6 | … | … | 60,0 | 64,3 |
| Number of subscribers of mobile broadband access to the information and telecommunication network "Internet" per 100 population, units | 75 | 68,1 | 74 | 71,1 | – | 79,9 |
| Number of high-performance jobs by type of economic activity "communication", thousand units | 381,5 | 233 | 401,5 | 248 | 290,2 | … |
| Share of the population that does not use the information and telecommunication network "Internet" for security reasons, in the total population, % | 7 | 0,4 | 5 | 0,5 | … | … |
| Share of households with access to the Internet information and telecommunication network in the total number of households, % | 69,9 | 68,4 | 72,8 | 70,3 | 83 | 76,3 |
| Place of the Russian Federation in the international IDI ranking, units | 20 | … | 10 | 43 | 42 | 45 |

Source: compiled by the authors based on the data of the state program "Information Society"

The table presents the main development indicators of the State Program "Information Society". Most of the planned development indicators have been achieved, some have been exceeded, and some have not been fulfilled. In 2015, 17 out of 46 indicators were reached, in 2016–27 out of 45, in 2017–56 out of 81. Let us evaluate the performance of the State Program in 2017.

We use the formula to calculate the performance:

$$R = \frac{I}{F} \tag{1}$$

I - the integral evaluation of program performance,

F - the level of financial support of the program

The integral evaluation of program performance is calculated by the formula:

$$I = \frac{1}{N}\sum_{n=1}^{N}\frac{Xnf}{Xnp} \tag{2}$$

N – the number of indicators;

$Xnf$ – the actual value of the indicator,
$Xnp$ – the planned value of the indicator

For 2017, the value of this indicator was 1.14, that is, on average, according to development indicators the over-fulfillment of planned values is 14%.

The level of financial support of the program is calculated by the formula:

$$F = \frac{Ff}{Fp} \tag{3}$$

$Ff$ – actual costs to implement the state program;

$Fp$ – planned costs to implement the state program.

In 2017, the value of this indicator was 1,037, it means that 3.7% more were budget allocations for the program implementation than it was planned.

The integral evaluation of the program performance in 2017 was 1.10. Thus, this program implementation was rather high.

## 4  Discussion

The development of the digital economy makes new demands on the communications system, services and information systems. Transaction costs decrease, digital data becomes a new asset. In the works of B. Pashin and A. Kotiranta [12, 13], macroeconomic aspects, general trends in the development of the digital economy are studied. In the context of rapid growth of ICT and innovative technologies, there is an urgent need to ensure the competitive position of Russian enterprises in the international market [4, 16], which requires the necessary conditions for the accelerated development of digital and information technologies in the domestic market. Revolutionary changes in the field of IT-technologies transform traditional rules, forms of regulation and public policy, including government spending [1, 10, 14]. With all the variety of trends in information society development, identified by researchers, the latter are united in the opinion that humanity has entered a new stage in the development of civilization, when information and digitalization play a decisive role in all spheres of human activity. New reality needs to be managed, planned for development, and financially supported in priority areas. However, the use of financial resources requires evaluating their performance. Methods of evaluating the performance of state programs, including the use of public finances are contained in the work of N. Tulyakova [19]. However, the author, when calculating indicators, suggests comparing not only planned and actual indicators, but also data for the previous period. This is not always possible, since target indicators are developed for each year and do not always coincide over different periods of time. Accounting for only indicators that are comparable in time would limit the parameters of the study. Therefore, the authors used a methodology based on the ratio of planned and actual indicators. In the future, the contribution can be deepened by assigning various criteria (weights) to the indicators in terms of their significance. This will allow result in evaluation at a new level: based on the ranking of indicators and determining the importance of their achievement for the economy.

## 5 Conclusions

Russia must occupy a competitive position in the international market, which under conditions of rapid development of ICT requires the accelerated development of the digital economy and information technology. In periods of catch-up development, an active role of the state is necessary for a breakthrough leap associated with the planning process. Priority areas for the development of the digital economy in Russia are: regulation, personnel and education, research competencies and technical reserves, information infrastructure and information security. According to them, the state has developed a list of target indicators for the development in economy digitalization. An important point of programming is the feasibility and performance of the program. The contribution evaluates the feasibility of the State Program "Information Society" from the point of view of the feasibility of planned indicators and performance on the basis of comparing the intermediate results of the Program with the costs of it.

# References

1. Ales E, Curzi Y, Fabbri T, Senatori I, Solinas G (2018) Working in digital and smart organizations: legal, economic and organizational perspectives on the digitalization of labour relations, 3 edn. New York
2. Ahmad N, Schreyer P (2016) Are GDP and productivity measures up to the challenges of the digital economy? Int Prod Monit 30:4–27
3. Ashmarina SN, Kandrashina EA, Izmailov AM (2017) World trends, trends and their influence on the education system in Russia. Trends Manag 2:55–64
4. Babkin AV, Chistyakova OV (2017) Digital economy and its impact on the competitiveness of business structures. Russian Entrepreneurship 24:76–84. https://doi.org/10.18334/rp.18.24.38670
5. Feldstein MS (2017) Underestimating the real growth of GDP, Personal income and productivity. J Econ Perspect 31(2):145–164
6. Gorelov ON, Korableva ON (2017) Performance problems in the context of the formation of a knowledge-intensive digital economy. Russian Business 19:2749–2758. https://doi.org/10.18334/rp.18.19.38343
7. Fernandes S, Lucas J, Madeira MJ, Cruchinho A, Honório ID (2018) Circular and collaborative economies as a propulsion of environmental sustainability in the new fashion business models. Lect Notes Electr Eng 505:925–932
8. Gupta S, Keen M, Shah A, Verdier G (2017) Digital revolutions in public finance, international monetary fund. http://www.elibrary.imf.org. Accessed 28 Apr 2017
9. Han D (2018) Proprietary control in cyberspace: three moments of copyright growth in China Media. Cult Soc 40(7):1055–1069
10. Kanbur R (2017) The digital revolution and targeting public expenditure for poverty reduction. In: Digital Revolutions 9th edn. https://static1.squarespace.com. Accessed 18 Dec 2017
11. Konovalova ME, Kuzmina OJ (2018) Transformation of financial institutions in the conditions of the emergence of the digital economy. Bull SSEU 6(164):9–13
12. Kotiranta A, Kosk, H, Pajarinen M, Rouvinen P, Ylhäinen I (2017) Digitalization changes the world – are new statistics needed to support economic policy? Prime Minister's Office, February 2017
13. Pashin B (2016) Digital economy: features and development trends. Sci Innov 7:12–18
14. Runciman D (2015) Digital politics: Why progressives need to shape rather than merely exploit the digital economy. Juncture 22(1):11–16
15. Seppänen L, Hasu M, Käpykangas S, Poutanen S (2017) On-Demand work in platform economy: implications for sustainable development advances in intelligent systems and computing 825:803–811
16. Streltsov AV, Eroshevsky CA (2014) Investment support for sustainable economic development of industrial enterprises. Econ Manag 4:11–18
17. Studenikin NV (2017) Digital technologies and new opportunities for CSR in Russia in the context of a green economy. Public Private Partnership 4:28–36. https://doi.org/10.18334/ppp.4.4.38648
18. Sutherland W, Jarrahi MH (2016) The sharing economy and digital platforms: a review and research agenda. Int J Inf Manag 43:328–341
19. Tulyakova NV (2017) Evaluation of the performance of state programs: problems and prospects. Financ Control 4:36–40

# Reforming the Institutional Environment as a Priority for Creating the Digital Economy in Russia

K. N. Ermolaev, Yu. V. Matveev, O. V. Trubetskaya$^{(\boxtimes)}$, I. A. Lunin, and A. V. Snarskaya

Samara State University of Economics, Samara, Russia
ermolaevkn@yandex.ru, matveev@gmail.com,
olgatrub@gmail.com, Luninia0576@gmail.com

**Abstract.** The urgency of the contribution is due to the fact that the theory of institutional change has not been fully reflected in the Russian economic literature, while it forms the main prerequisites for economic growth and development, and provides the necessary conditions for the digital economy in Russia. The authors represent their point of view concerning institutional change contributing to economy digitalization in Russia. The leading methods of research are the methods of mathematical statistics, which allow a comprehensive quantitative evaluation of institutional change for the period 2010–2017 and evaluate the Future ready of Russian society for digitalization. The authors' calculations showed that the reform of the institutional environment is heterogeneous, the density of the institutional environment is low in the digital sector, there is a discrepancy between change in formal and informal institutions, and this indicates the complexity of the transition to the digital economy and requires targeted incremental institutional change. The materials of the contribution are of practical value for the development of management decisions to improve the institutional system of Russian economy.

**Keywords:** Digital economy · Digitalization · Institutional change ·
Institutional system

## 1 Introduction

The process of institutional change means the change of formal and informal rules, norms and coercion that constitute the institutional environment of society [15]. Institutional change is discrete and incremental, spontaneous and targeted. Discrete change is the change in formal rules that occurs as a result of conquest and revolution. Incremental change means that participants in the act of exchange voluntarily renegotiate their contractual relationships in order to gain some potential gain from the transaction [21].

Spontaneous change is such institutional change that is carried out, arises and spreads, without anyone's preliminary intention and plan. Targeted institutional change, on the contrary, arises and spreads in greater or lesser accordance with some consciously designed plan. On the basis of the origin of institutional change, in addition

© Springer Nature Switzerland AG 2020
S. Ashmarina et al. (Eds.): *Digital Transformation of the Economy: Challenges,*
*Trends and New Opportunities*, pp. 119–128, 2019.
https://doi.org/10.1007/978-3-030-11367-4_12

to the two types mentioned, a mixed type can also be distinguished, when the new rule appears unplanned, and its distribution is carried out quite consciously and purposefully.

A number of scientists propose to single out four directions in the theory of institutional change: neoclassical economics, sociological, historical school and evolutionary institutional economics [3, 7, 12, 19].

In Russia, as the reference point for the development of the digital economy can be considered the Message of the President of the Russian Federation to the Federal Assembly on 12/01/2016, in which he instructed to propose system approaches to building up personnel, intellectual and technological capabilities of the Russian Federation in the digital economy [16]. The main objectives of the Digital Economy Program, developed by the Expert Council under the Government of the Russian Federation for the digital economy on January 23, 2017, include the involvement of citizens and business entities working in the digital space; the development of the infrastructure that ensures the interaction of subjects in the digital space; the formation of sustainable digital ecosystems for economic entities; the reduction in costs of economic entities and citizens when interacting with the state and among themselves; the enhancement of the competitive advantage of the economy, business entities and citizens through digital transformations in all spheres of society.

The program assumes that the digital economy will form at three levels:

- Markets and sectors of the economy, where the direct interaction of specific subjects take place (suppliers and consumers of goods, works and services);
- Platforms and technologies, where competencies are formed for the development of markets and industries;
- The environment that creates the conditions for the development of platforms, technologies, the effective interaction of market entities and economic spheres and covers regulations, information infrastructure, personnel and information security.

The purpose of the contribution is a quantitative evaluation of institutional change based on the analysis of the legislative practice of modern Russia, which characterizes the vector of institutional change, as well as an evaluation of the Future ready of Russia for the digital economy.

In the process of research, it is proposed to solve the following tasks: using the data from the Lawstream.ru database, as well as using the authors' own calculations, evaluate the density of the institutional environment and its stability; evaluate the Future ready of the institutional environment for economy digitalization.

## 2   Materials and Methods

The following methods were used: theoretical (analysis, synthesis, synthesis, method of analogy); empirical study of regulatory documentation; and methods of mathematical statistics and graphical presentation of results.

The use of theoretical methods of cognition allowed identifying the features of institutional change that can be quantified. The evaluations of the density of the institutional environment and the vector of institutional development have become possible through the use of the empirical comparison method. The methods of mathematical statistics made it possible to form groups of statistical data with evaluated quantitative parameters of the model of institutional change.

## 3 Results

According to experts, Russia has noticeably advanced in many areas of digital development. There was an increase in domestic spending on research and development in organizations of information and computer technology (ICT) from 1.3% to 3.6% in 2016, the number of broadband Internet subscribers increased from 12.2% in 2011 to 21% in 2017. Also, the share of organizations using the Internet increased from 56.7% in 2010 to 81.8% in 2017, the share of organizations placing orders on the Internet increased during the period under review by 6.6%, and the share of organizations receiving orders for manufactured goods on the network increased by 2.4%.

At the same time, the share of gross value added of the sector to GDP decreased from 3.4% in 2010 to 2.9% in 2016 and it is 2–3 times less than in world major economies, which indicates a lag in the digital development of Russian economy. Further development of the digital economy will require a transformation of formal and informal rules and restrictions, a system of incentives for economic agents, that is, institutional change.

To evaluate the existing and ongoing institutional change in Russian economy, the authors took Lawstream.ru database, which contains data on federal laws formed by synthesizing data from the official website of the State Duma of the Federal Assembly of the Russian Federation and information provided in the Consultant Plus legal reference system. The authors supplemented this database with calculations for 2017, and also separately identified the information and communication technology sector, which includes data related to information, information media, communication, and information security of legal entities and individuals to evaluate the Future ready of the institutional environment for the digital economy.

A quantitative evaluation of institutional change in the database is carried out on the basis of change in the sphere of federal laws regulating all types of activities in Russia. This approach is supported by research of the World Bank, which puts legislation and institutions in the economic sphere in the first place when classifying institutional change.

To evaluate institutional change, such features of the institutional environment are used: density- the number of necessary laws and their structure, as well as the role of the main legislative groups in the implementation of institutional change. Despite the fact that the Lawstream base methodology has a number of limitations (federal laws are only part of the institutional environment, adopted laws may not be implemented and thus do not reflect real institutional change, there is a time lag between the adoption of the law and its implementation in practice), However, this base allows quantifying institutional change and evaluating the country's Future ready for the development of the digital economy.

In order to evaluate the density of the institutional environment, the authors consider data on the number of laws, their structure and areas of regulation.

The peaks in the number of adopted laws were in 2011 and 2014 and amounted to 431 and 560 laws, respectively, in 2015–2016. The level of adopted laws is equalized (their number is 383–388). In 2017 there was a reduction in the number of adopted laws to 253. At the same time, the peaks and minima of adopted federal laws correlate with the dates of authoritative powers of deputies of the State Duma of the Federal Assembly of the Russian Federation. Thus, the peak of the adoption of federal laws in 2011 was at the end of the work of the 5th convocation, and the activity decay was associated with the beginning of the work of the 7th convocation in 2016–2017. The growth of legislative activity of deputies by the end of their term of office may be related to their desire to finish what has accumulated over the years of parliamentary activity, since new members of parliament may complicate or even stop working in the promotion of a specific legislative initiative. Trends in the field of federal legislation make it possible to evaluate the growth of the density of the institutional environment in which laws play a significant role.

The distribution of adopted federal laws by the spheres of regulation (see Table 1) shows that the majority of laws were adopted in the fields of economics and finance (21.2% on average), codes (24.1%) and other laws (38 on average 6%). In the field of information and computer technologies, the number of laws adopted was minimal (0.65% on average), which indicates a low density of the institutional environment in the field of information technologies, and also that the growth of the digital sector does not coincide with change in relevant institutions.

**Table 1.** Distribution of adopted federal laws by regulation (by the date of introduction, in %)

| Years | Economy and finances | Politics and state structure | Codes | ICT | Other laws |
|---|---|---|---|---|---|
| 2010 | 31 | 19,7 | 14,4 | 0,2 | 34,7 |
| 2011 | 23,8 | 15,2 | 16,3 | 0,1 | 44,7 |
| 2012 | 20,9 | 16,1 | 20 | 0,1 | 43 |
| 2013 | 22,2 | 15,3 | 20,2 | 0,1 | 42,2 |
| 2014 | 9,8 | 15,8 | 25,6 | 1,5 | 47,3 |
| 2015 | 13,3 | 14 | 20,4 | 0,2 | 52,1 |
| 2016 | 26,5 | 18 | 17 | 1 | 38,5 |
| 2017 | 22,1 | 9,8 | 59,6 | 2 | 6,5 |
| In average | 21,2 | 15,5 | 24,1 | 0,65 | 38,6 |

Source: compiled by the authors based on Lawstream.ru data

In general, it can be concluded that there is some stabilization of the institutional environment, since the focus is more often on improving legislation than on adopting new legislative forms. On the other hand, a high proportion of amendments can be characterized by the fact that laws are not being adopted deliberately and that they have to be constantly amended.

As for the digital sphere, a low proportion of new laws adopted in the overall small number of adopted laws in this area rather indicate the underdevelopment of the institutional environment.

Let us evaluate the quantitative parameters of the model of institutional change in Russia taking into account federal legislation by the group structure and parameters of transaction costs. We will evaluate the group structure by social actors with the right of legislative initiative. These include the President, the Government, deputies of the Federal Assembly, regional authority bodies and courts. We will evaluate the role of each of them in shaping the institutional environment through the initiation of relevant laws (see Fig. 1). It reflects the number of all laws, including change and amendments proposed by the initiators in each year.

The Russian Government more often than others initiated to introduce or amend federal laws for 2010–2017.The Deputies of the Federal Assembly are lagging far behind, for which legislative activity is one of the main ones. If in 2010 they initiated 145 laws, then in 2017 - only 80. In the third place is the President of the Russian Federation, who submitted 348 laws to parliament for 2010–2017. The initiative of regional legislative bodies in 2017 returned to the level of 2017 and amounted to 36 laws. Courts rarely initiated the adoption of federal laws - an average of 4 per year.

It is possible to evaluate the "innovation" activity of legislators in the field of creating new institutions indirectly by the number of new laws in the total number of legislative initiatives. The adoption of a new law illustrates the changing "rules of the game" as new norms emerge, becoming objects of legislative regulation and forming a new future routine. As can be seen from Table 2, the President of the Russian Federation most often initiated the adoption of new laws, and the share of new legislative initiatives averages 59.1% of all adopted laws for 8 years.

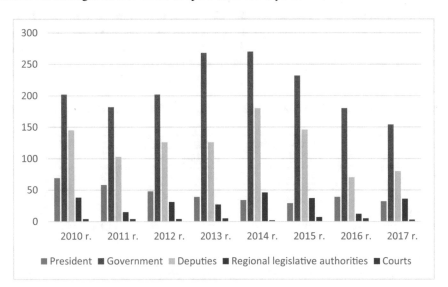

**Fig. 1.** Distribution of adopted federal laws by the initiators for 2010–2017 (by the date of introduction). *Source* compiled by the authors based on data from Lawstream.ru

The share of new laws among the initiatives of the Government and regional legislative authorities is also relatively high - on average, 30.3% and 26.5%, respectively. The deputies of the State Duma are busy with their main work on the development and improvement of legislation - most initiatives are clarifying and amending existing laws, innovation activity is less characteristic for them - the share of new laws introduced by deputies of the Federal Assembly is only an average of 6, 01%. The courts initiated new initiatives in 2015 and 2017. In the rest of the period, amendments were made to already existing legislation.

**Table 2.** The proportion of new federal laws adopted by the initiators for 2010–2017 (by the date of introduction, in %).

| Years | President | Government | Deputies | Regional legislative authorities | Courts |
|---|---|---|---|---|---|
| 2010 | 47 | 41,1 | 4,8 | 42,1 | 0 |
| 2011 | 64 | 35,7 | 5,8 | 26,7 | 0 |
| 2012 | 56 | 31,6 | 4,7 | 25,8 | 0 |
| 2013 | 46 | 20,5 | 3,1 | 18,5 | 0 |
| 2014 | 67,6 | 22,5 | 9,4 | 10,8 | 0 |
| 2015 | 72,4 | 29,7 | 6,8 | 18,9 | 14,3 |
| 2016 | 53,8 | 22,7 | 7,1 | 33 | 0 |
| 2017 | 65,6 | 15,5 | 1,3 | 13,9 | 33,3 |
| In average | 59,1 | 30,3 | 6,01 | 26,5 | 5,9 |

*Source* compiled by the authors based on Lawstream.ru data

If we consider the structure of laws for 2010–2017, then most of the laws in the economic sphere were initiated by Deputies and the Government of the Russian Federation (in average of 23% and 27%, respectively), the President and regional authorities paid very little attention to this area; 4.8% for the President of the Russian Federation and 10% for the regional authorities (Table 3).

**Table 3.** The initiators of laws on the areas of regulation for 2010–2017 (on average for the period, %)

| Initiator | Economy and finances | Politics and state structure | Codes | IKT | Other laws |
|---|---|---|---|---|---|
| President | 4,8 | 22 | 13 | 0 | 5 |
| Government | 37 | 11 | 18 | 47 | 48 |
| Deputies | 33 | 20,5 | 23 | 53 | 35 |
| Regional legislative authorities | 19 | 36 | 19,5 | 0 | 9 |
| Courts | 6,2 | 10,5 | 26,5 | 0 | 3 |

*Source* compiled by the authors based on Lawstream.ru data

Most decisions in the field of politics and state structure are made by the Deputies, the President of the Russian Federation and regional authorities. The Courts for the period under review were mainly engaged in changing the codes (53% on average), with the exception of 2017, when the leadership on legislative initiatives was taken by the Government of the Russian Federation.

In the sector of information and communication technologies, the legislative initiative belongs to the Government of the Russian Federation - 47% and to the Deputies of the State Duma - 53%. It is noteworthy that the change in legislation was carried out after the initiatives of the President of the Russian Federation, in the annual presidential State of the Nation Address.

## 4  Discussion

Initially, the theory of institutional change was formed taking into account the principle of methodological individualism and the limited rationality of an individual's behavior, later a postulate about non-zero transaction costs was added to it.

D. North, considering the causes of change in institutions concluded that institutional change occurs because it becomes profitable for economic agents or groups when certain costs to implement this change occur, as they allow individuals to make additional profits [14]. In the theory of institutional change of Rattern and Khiami the transformation in institutions is associated with the existence of disequilibrium in markets [18]. The reason for change is costs and profits generated by the institutional structure existing in the society. Shifts in prices trigger change in production technology, and institutions act as a constraint maximizing profits.

Chen in his works drew attention to such an important element of institutional change as the structure of property rights, which sets restrictions on the possibilities of competition and exchange [2]. Institutional change means change in property rights. At the same time, an important factor in changing institutions is the cost of new rules and the cost of convincing people to move to new rules.

Tang proposed to distinguish two approaches: harmony and conflict [20]. The harmonic approach contains three key assumptions: the presence of common interests among economic agents; cooperative behavior of subjects; most of the social outcomes are produced by agents to improve their private wealth, but at the same time as a result of cooperation, collective wealth is often improved.

V. Elsner associated the theory of institutional change with the theory of games in its evolutionary-institutional interpretation, which allows for a more in-depth analysis of institutions, identifying the value basis in game theory, instrumental-ceremonial asymmetry [6].

For Russian science, the issue of institutional change is relatively new. However, today we can already note the significant contribution that V. Tambovtsev, V. Volchik, A. Shastitko, S. Kirdin and others [8, 21, 22] made to the study of the theory of institute change. Their studies present the features of institutions, the institutional environment, the theory and methodology of analysis, the features of institutional change in Russian economy, as well as the regional aspect of institutional development [8, 9].

V. Volchik analyzed institutional change using the evolutionary approach, which allows economists to view economic laws as adaptively complex systems [22].

The term "digital economy" first appeared in the works of N. Negroponte in 1995, thanks to the intensive development of information and communication technologies that form the fourth technological mode [13].

L. Anderson and I. Wladawski-Berger considered digital economy as an economy of unlimited opportunities for some economic agents and a source of ruin for others [1].

In studies of the international monetary fund, the digital economy is defined as all activities that use digitized data [4]. A. Goldfarb and K. Tucker believe that the emergence of digital technologies in society does not require a new economic theory, but suggests research on how digital technologies affect costs and limit the economic activities of individuals [5]. A number of economists in their studies dealt with negative effects of the digital economy: spam, online crime, etc. [10, 17].

R. Meshcheryakov studied the digital economy from two perspectives: first, it is based on the development of digital technologies, primarily the area of services and electronic goods. Secondly, the digital economy can be studied as an economic production using digital technologies [11]. T. Yudina considered the digital economy as a new socio-cultural-economic reality in the new world, a "smart" reality. Economy digitalization can be understood in a narrow and broad sense. In a narrow sense it is a change in production or economic relations, a change of their subject-object orientation. In this case, there are relations called "machine-machine", and the person is no longer acting as a subject. In a broad sense, the economy of numbers should be understood as the creation of information-digital platforms and operators at different levels of the economy, allowing solving various economic problems: the development of science, education, government regulation and planning [23].

## 5   Conclusions

In the course of the study, the following results were obtained on quantitative parameters of institutional change and the Future ready for transition to the digital economy:

- The number of adopted federal laws for the period 2010–2017 increased, which indicates an increase in the density of the institutional environment, thus the norms and rules prevailing in society are increasingly acquiring a formal character;
- The adoption of federal laws has recessions and peaks correlated with the dates of authoritative powers of the Deputies of the State Duma of the Federal Assembly of the Russian Federation.

The reform of the institutional environment is heterogeneous: most of the adopted laws relate to the economic sphere and the field of codes. In the digital sector, the number of adopted laws is minimal, which indicates a low density of the institutional environment, as well as the fact that formal change does not keep pace with informal transformations of norms and rules.

The general vector of institutional change indicates an increase in the stability of the institutional environment, which is reflected in a decrease in the share of new laws as compared with the adoption of amendments to the already existing federal legislation. At the same time, we can identify areas where legislation is more often amended: these are codes. In the context of the main trends, there is the ICT sector, here an increase in the share of new laws is observed, which means underdeveloped institutions in this area.

The analysis of the role of legislators in conducting institutional change for the period 2010–2017 shows that the most active initiator of laws was the Government of the Russian Federation. The Deputies of the State Duma have lost their activity and are noticeably inferior to the Government in the number of laws. The activity of regional legislative authorities has also increased.

The role of the President of the Russian Federation in the implementation of institutional change is increasing, as evidenced by the share of the President's initiatives to adopt new laws. The results suggest that institutional change forms a more efficient economic model, but at the same time, it is still difficult to talk about economy digitalization, since formal institutional change is very slow, it is adapted to the existing informal institutions and norms that have been established in this area. Therefore, the further development of the digital economy will require targeted incremental institutional change.

# References

1. Anderson L, Wladawsky-Berger I (2016) The 4 things it takes to succeed in the digital economy. Harvard Business Review. March 24. https://hbr.org/2016/03/. (The 4 Things It Takes to Succeed in the Digital)
2. Cheung SNS (1974) A theory of price control. J Law Econ 17(1)
3. Campbell John L (2004) Institutional change and globalization. Princeton University Press, Princeton, NJ
4. Ducharme LM (2017) Measuring the digital economy. International monetary fund
5. Goldfarb A, Taker C (2017) Digital economics. National Bureau of Economic Research, Cambridge, August, p 91
6. Elsner V (2017) Again about the institutionalist theory of institutional change: institutional dichotomy in a more formal presentation. J Inst Stud 9(2):6–17
7. Christopher Kingston, Cabellero Gonzalo (2009) Comparing theories of institutional change. J Inst Econ 5(2):151–180
8. Kirdina SG, Kirilyuk IL, Tolmacheva IV, Rubinstein AA (2010) The Russian model of institutional change: an experience of empirical-statistical research. Econ Issues 11:97–114
9. Matveev YV, Trubetskaya OV, Lunin IA, Matveev KY (2018) Institutional aspect of the Russian regional economic development. Probl Perspect Manag 16(1):381–391
10. Moore T, Clayton Anderson R (2009) The economics of online crime. J Econ Perspect 23 (3):3–20
11. Mescheryakov R, Savchuk M (2011) Approaches to the implementation of ERP-systems in large enterprises. Bus Inform 2(16):63–67
12. Victor Nee (2001) Introduction. In: Brinton Mary C, Nee Victor (eds) The new institutionalism in sociology. Stanford University Press, Stanford, CA
13. Negroponte N (1995) Being digital/N. Negroponte.-NY:Knopf, p 256
14. North DC, Thomas RP (1970) An economic theory of the growth of the Western World. Econ Hist Rev 23(1):3
15. North DC (1993) Five propositions about institutional change. Econ WPA. Economic History: 9309001
16. Message of the President of the Russian Federation to the Federal Assembly of December 1, 2016. http://www.consultant.ru/document/cons_doc_LAW_207978
17. Rao J, Reiley D (2012) The economics of spam. J Econ Perspect 26(3):87–100

18. Ruttan VW, Hayami Y (1984) Toward a theory of induced institutional innovation. J Dev Stud 20(4)
19. Scott Richard W (2008) Institutions and organizations, 3 edn. Beverly Hills, C.A. Sage
20. Tang S (2011) A general theory of institutional change. London Routledge
21. Tambovtsev V (2006) Theories of institutional change: Manual study. M.: INFRA-M, p 153
22. Volchik V (2012) An evolutionary approach to the analysis of institutional change. TERRA Economicus 10(4)
23. Yudina TN, Tushkanov I Digital economy as a result of the industrial-technological revolution (Theoretical and Practical Aspects). http://reosh.ru/t-n-yudina-i-m-tushkanov-cifrovaya-economika-kak-resuktat-proyshlenno-texnologicheskoj-revolyucii-teoreticheskie-i-prakticheskie-aspekti.htm

# Adapting of International Practices of Using Business-Intelligence to the Economic Analysis in Russia

S. Mitrovic[(✉)]

University of Novi Sad, Novi Sad, Republic of Serbia
Mitrovic.Stanislav@hotmail.com

**Abstract.** The present-day pace in development of the modern information technologies in Russia greatly exceeds the speed at which the methodological, recommendatory, standardization and regulatory/reference framework is being developed for the governing documents that are in force in our country and are actually used by economic entities. In many cases the present-day methodical tools of business intelligence used in the domestic environment lag behind the evolution of the information tools or turn out to be insufficiently adapted to the peculiarities of the economy.

This study is aimed at identifying the basics for integrating and adapting the world-wide experience of using business intelligence solutions (Business Intelligence, BI) for the business entities' economic performance analysis in order to optimize this domain in the Russian environment for various purposes, including elaboration of an information quality improvement program and development of corporate-wide business intelligence systems.

Based on the analysis of foreign experience, the author substantiates the national companies' capability of elaborating the data control methodology and ensuring data transparency. The author concludes the wider methodological directions in using the business intelligence, which may be extrapolated from the international practice to the national companies' business. The author developed the data handling algorithm in the course of the economic performance analysis in companies by means of the BI-technologies. This algorithm implies successive conversion of such data into information, information - into understanding, understanding - into knowledge, and general knowledge - into the goal-oriented applied knowledge that facilitates decision-making.

**Keywords:** Business intelligence · Data management · Digitalization · Economic analysis · Information quality

## 1 Introduction

Business process management is the foundation and the basics for implementing the organizational procedures, information systems, financial and management accounting processes [19]. The latest analytical research carried out by Forrester Research (USA) shows that business intelligence technology continues to be at the top of the list of the business information solutions that foreign companies are planning to buy or

© Springer Nature Switzerland AG 2020
S. Ashmarina et al. (Eds.): *Digital Transformation of the Economy: Challenges, Trends and New Opportunities*, pp. 129–139, 2019.
https://doi.org/10.1007/978-3-030-11367-4_13

upgrade in 2018 [11]. The data collected during the inquiry of the heads of the largest European enterprises that was carried out by experts from the Merill Lynch financial conglomerate (USA) in 2016–2017 show similar trends. They confirm that the introduction and the more profound and detailed integration of BI technologies into business analysis is still one of the three main priorities of contemporary organizations (previously, this trend ranked only fifth) [15].

For the purposes of this study, the term business intelligence (Business Intelligence, BI) refers to a system for information support of organizations, which combines the technology (tools) and the concept (methodology and algorithm) mainly aimed at researching and analysing large amounts of data in order to identify general trends in economic processes, understand the situation and develop the position that enables the company to take effective management decisions in the future.

Inquiries and a series of studies carried out in September–November 2017 by experts from the foreign research magazine Information Week.com [8], which studies data management and usage of the advanced information tools for the companies' business, showed that financial analysis, business activity monitoring and forecasting will be most popular with foreign companies for integrating business intelligence technologies and the corresponding solutions and analytical tools in 2018.

According to the forecast data obtained by the same periodical [8], business intelligence specialists and IT management will still remain the most active users of the BI solutions and analytical tools in the short term, as in the previous year. However, the interest of analysts and data processing specialists in these technologies and tools is expected to grow in 2018.

The integration of the business intelligence technology with the Big Data technology is an important aspect of using the BI technology for economic analysis. The consulting company Forrester briefly defines this phenomenon as follows: "Big Data combine techniques and technologies that extract sense of data at the extreme limits of practicality" [20]. There are four attributes that are essential to defining the Big Data: volume, variety, velocity and veracity [6].

In Big Data environments, it is very important to analyse, decide, and act quickly and frequently. The Big Data may make a huge contribution to science and all aspects of our society, but extracting information from the Big Data is a challenging task [17].

Unlike the Russian market, the foreign market of modern information technologies usage for the analytical processing of economic information has a long history and is developing more rapidly [14]. However, the study of business practices showed that despite the difficulties, the Russian companies have succeeded significantly in assimilation of the BI in recent years. They demonstrate the growing interest and activity of specialists at various levels in using these technologies and tools for the economic analysis and survey of its outputs. The methods they apply to implementing the BI in the economic analysis, with the information method gaining prevalence and the key importance, become more effective and mature. The extent of development of the Russian market for introducing the BI into economic performance analysis of companies still differs from that of the foreign market in terms of quality, scope and methodological level of project implementation in this field, despite the significant number of innovative software solutions created by Russian developers.

That being said, it is logical to conclude that it is possible and necessary to adapt foreign experience of introducing BI-technologies into the economic analysis of domestic business entities. Companies need a reliable tool that will allow them to independently identify the problems that need to be solved and to substantiate the measures aimed at enabling their sustainable development [4].

## 2 Materials and Methods

This study is aimed at identifying the basics for integrating and adapting the world-wide experience of using business intelligence solutions (Business Intelligence, BI) in the business entities' economic performance analysis in order to optimize this domain in the Russian environment for various purposes, including elaboration of a program for improving the information quality, identifying the general trends in the economic processes, understanding the situation and developing the position that will enable the company to take effective management decisions in the future.

The basics of the information theory and business intelligence; the theory of economic and business analysis; the concept of information and economic security standardization and management; and the present-day fundamental scientific research, theoretical concepts, methodological and practical works of the national and foreign authors on implementation of business intelligence, management, introduction and development of the related systems, solutions and technologies in order to improve the companies' economic efficiency lie in the theoretical and methodological foundation of this research.

A wide range of statistical and analytical materials from reviews of the international specialized, standardizing and research organizations; regulatory and conceptual acts, standards and statistical data; publications in specialized journals that reveal the expert opinions of international organizations and specialists from domestic and international companies on using business intelligence solutions for improving the economic and business analysis constitute the reference basis of the study.

## 3 Results

The author suggests that the present-day economic processes in Russia set a new vector for developing the economic analysis methodology in the information environment towards the economic performance analysis of business entities acting in the market economy.

To date, the Russian economic science has developed a relatively integral and scientifically grounded concept of organizing the economic performance analysis in the information and communications environment. This concept is based on generalizing the several decades' experience of designing and applying the modern information technologies and approaches to the economic analysis aided by such technologies and approaches to be organized in companies. It also relies on the development stages of the information technologies, information tools and methodology of the economic analysis in the Russian and foreign practice. The surging progress in the information technologies development and the vast introduction of the interactive, cloud-based technologies require the further continuous development of the related theoretical and methodological concepts.

The Business Intelligence makes it possible to collect, store, manage, distribute, analyse and provide information for working out the vision of the problem that enables making the best decision. This research resulted in an algorithm for handling data, when the BI technologies are applied for the companies' economic performance analysis. This algorithm implies successive conversion of such data into information, information - into understanding, understanding - into knowledge, and general knowledge - into the goal-oriented applied knowledge that facilitates decision-making.

The continuous improvement and development of the business intelligence solutions in the economic analysis make it possible to integrate the strategy for development of the contemporary companies, which was derived from the study results, with the key processes and operational business objectives. They also make it possible to effectively communicate the management's vision to employees and expediently control their performance and contributions to achieving the business objectives in the course of the decision-making process. The foregoing actualizes the need to consider and subsequently develop (in terms of the methodology) the usage of the Business Intelligence systems as an integral instrument for managing various processes in the present-day business.

## 4   Discussion

Let us consider the main directions for updating the methodology of integrating business intelligence solutions into business analysis of economic entities operating in the Russian environment. We will rely on the foreign experience and advanced methodological trends in addition to the perspective trends in development of the methodology for introducing information technologies into the economic analysis of organizations, which have been identified as fundamental.

### 4.1   BI as a Tool for Improving the Quality of Information in Economic Analysis

**Information quality improvement** is one of the main directions. It is the basis for identifying a significant number of problems in the integration projects and, along with that, the basis for their effective implementation.

It should be noted that several years ago, as may be seen from our observations, almost no Russian company was ready to admit the poor quality of its data and solve this problem on a large scale, that is, at the level of the entire company. This aspect has been rarely taken into account in terms of methodological development. However, a large number of Russian businesses have come to realize that the quality of information should be the starting point for improving the quality of economic performance analysis.

At present the methodology of the BI integration into economic analysis, which has been developed by foreign companies, includes the **information quality improvement**

**program** as an integral part of the integration projects and their respective development. It is a highly important component that has been scarcely used in the Russian practice. It is a long-term project that is complex from the methodological point of view. The recommended ways to implement this program step by step may be derived from the foreign practices and used in the Russian conditions. Many foreign organizations have come to the conclusion that the regulatory and reference information management is an effective basis for improving the quality of all information used in the economic performance analysis of companies.

The three key components mentioned below shall be pointed out in any information quality improvement program on the general methodological level: systematic measurement as a starting point, quality improvement as a continuous process and information quality assessment.

We analysed and presented the elements of the information quality improvement program as separate components of the methodology. Let's start the analysis with the **regulatory and reference information management**.

The methodology for managing the regulatory and reference information and its practical implementation processes have been developed recently, and this activity has shown a significant growth. For many foreign companies the regulatory and reference information management is the starting point for solving the information quality problems. This practice may also be adopted by Russian organizations that almost fully lack the methodological basis for handling this element and using its economic potential. Experts estimate that the rather disconnected attempts to create systems for maintaining the regulatory and reference information that have been undertaken in Russia show that most organizations face multiple problems, when trying to develop and implement a system of that kind on a corporate level. The difficulties are caused by the lack of knowledge of the methodology and lead to significant material, labour and time expenditures [13]. Although the reference source dates back to 2009, we have observed no significant changes in the situation with the methodology of handling the regulatory and reference information and using it as a source for the computer-aided economic analysis that may be seen in the Russian practice.

Let us proceed to studying the **data management** aspect from the methodological point of view. We should note that data management has gained particular importance in the economy of business entities and enterprises after the global financial crisis.

Data management is "a business function that provides a strategic direction for information improvement programs, establishes standards and processes, and contributes to achieving the goals of information improvement programs" [7]. According to the analytical agency Experian, a high percentage of sales decisions will be taken based on the available customer information. However, errors that may be found in about a quarter of all customer data negatively affect the economic analysis quality and results [7]. Data management is also complicated by the rapidly growing volumes of information and changes in the use of communication channels.

The proper organization and distribution of authorities is one of the important aspects in ensuring the effective data management. The international practices show that the authorized persons should include IT-specialists, as well as the top management representatives, business units represented by department management, whose tasks include carrying out the economic analysis, and the key employees of these departments, since they possess the higher scope of the information on the data ownership and management in the company. Foreign companies often engage a third party consultant who ensures the required impartiality and guarantees that the methodology will be met and the company will not lose sight of the more general picture when identifying and achieving the goals of information quality improvement.

Besides, a position of a Chief Data Officer (CDO), who is responsible for working with data and ensuring the higher-quality business performance analysis, is gaining importance in the present-day international practice. The current research results show that 82% of foreign companies plan to employ data specialists in the next five years [9]. However, this position remains rare in Russia at the top and medium management level and at the lower level of employment.

## 4.2 Developing Corporate-Wide Business Intelligence Systems for the Economic Performance Analysis

Development of **corporate-wide BI systems** is a tool for obtaining an integral business picture; it should be mentioned as the next line for development of the methodology. It is advisable to adapt it to the Russian conditions. Such systems have already been created in many foreign companies. Such cases still remain scarce and the methodology in this area is still poorly developed in the Russian practice. However, there is an urgent need in such effort.

Recently, both Russian and foreign companies have attempted to practically use the methodology for introduction of corporate BI systems. The study by Knightsbridge Solutions [12] that covered more than 500 business representatives and information technology specialists showed that such attempts were more successful abroad: approximately one fifth of the respondents achieved tangible results in creating and implementing such systems. However, it is impossible to formulate the standard recommendations that could be applicable to each particular company. The companies that attempt to improve the data management efficiency should first analyse the ways to introduce or extend the existing components to the level of a corporate BI system and its structure that reflects the inputs of business units, as well as the IT experts. The effective development and implementation of the methodology requires financial and organizational support from the company management, a continuous and systematic approach and a long-term strategy. Attempts to immediately create a corporate BI system in any company, even in the largest and most experienced in operating BI systems in the field of economic analysis, are doomed to failure.

As we consider the methodological reserves for improving the quality of implementing the BI projects for economic analysis based on international experience, we cannot help mentioning such an important aspect as **ensuring compliance with legal requirements and standards**. Today foreign organizations practice a comprehensive approach to meeting the relevant legal requirements. As the legislative requirements differ in various industries and the standards have certain implementation peculiarities, compliance may not be achieved through implementing any generic solution, but requires setting out the special methodological tools.

Development of a **data control** methodology also plays an important role in fulfilling the legal requirements. It allows companies to analyse the business information collected during longer periods. However, expanding the analysis period (for example, over the previous five years) makes the task much more complex. The possibility to carry out the "historical" data analysis (analysis of the organization's business history) contributes much to fulfilment of the existing legal requirements and the contemporary company's stable development in the highly competitive market environment and inflationary economy.

From the methodology point of view, the concept of data control is closely related to the need to ensure the **corporate data transparency** in the Russian environment. The methods of data control are introduced and improved in order to ensure data transparency, which means that the data source and data alteration algorithm is always traceable.

Foreign companies are using the Big Data management technologies, including the extraction, transformation, and loading (ETL) applications, more often at the present stage. Let us note that the Big Data "allow the extraction of complex data at a high level of abstraction" [3]. This technology helps to maintain data transparency during the economic analysis by means of the business intelligence tools.

Nowadays IT specialists (mainly the IT specialists in the domestic practice) are the most active users of the Big Data tools. However, business users (such as management, accountants and analysts) also gain access to these tools over time. Nowadays the Big Data tools in the BI solutions are seen to evolve to a certain extent towards their simplification for business users. This trend should not be ignored by Russian organizations. That implies development of an appropriate methodology at both the generic and corporate levels.

It is also necessary to draw the attention of the Russian companies that implement the BI into their economic analysis to active development of the methodology in the so-called **Actionable analytics (Actionable BI)**. As the modern companies face the growing need for corporate efficiency management, companies often notice that there is no clear connection between the overall strategy and the tactical solutions, including daily processes. Then the organization naturally raises the question of how to use a large amount of the existing information for the business performance analysis and how to use the obtained results in order to improve this connection and enhance the efficiency.

The analysis carried out by the author shows that the currently established methodological approach to computerization of economic performance analysis in companies, including BI-aided analysis, may be characterized as mature, but requiring further qualitative development. The functionality of business intelligence solutions used in the economic performance analysis in companies has provided opportunities to develop and implement business solutions with various functions:

- business planning;
- investment projects attractiveness analysis and management, short-term and long-term decision making. The short-term period involves the creation and usage of the information in the real time mode within the periods ranging from several minutes to three months (a quarter)' [18];
- procurement planning;

- customer relations management (including the "social networks development, which leads to a larger increase in users and digital content" [5], as well as the CRM (customer relations management) systems development "with a recognized crucial role of the information technology in their implementation" [2]);
- large-scale logistic and production systems management;
- conditioning the productivity increase, since "the information technologies have been considered for decades as an important lever for increasing productivity at the level of both an individual enterprise and overall economy" [21];
- financial state assessment, which "is an obligatory element of all methods that are used to determine the investment attractiveness of a business entity" [16].

At the same time, the studies revealed that in the foreign practice BI is also required for the analytical and mandatory reporting, including standardized reporting; it is primarily needed to establish non-obvious patterns that may have a great impact on the business outcome. For example, in the last decade substantial progress has been made in the field of communication technologies. That led to a surge in the intelligent transport systems development and usage. Driving styles identification using data from vehicle sensors is an interesting research challenge and an important real requirement in the automotive industry. A good representation of the traffic characteristics may be extremely valuable for protection against theft, for car insurance, autonomous driving and many other application scenarios [10]. Another example is the problem of quantifying how certain words in a text (in advertising) may positively or negatively affect a certain numerical signal. Those words may lead to meaningful solutions for such important applications as e-commerce. The problem is to identify the exact keywords that affect the price of the property [1].

Taking into account the well-grounded potential of tools for information and analytical support of the economic analysis, the author based the research on the domestic and foreign theory and practice surveys. Based on the practical implementation and testing of the surveys results, the author developed an algorithm for implementing BI solutions with data for the informational and analytical support of the economic analysis. The algorithm relies on the Big Data processing and the principles of business intelligence systems. The structure of this algorithm includes five consecutive components (steps), which were detailed in the research and which provide reliable, objective and measurable results in developing the companies' economic analysis:

1. Preparing a BI solution

This stage should identify the key elements of the companies' economic performance, such as the business type and volume, industry, business results (products and services), etc. Then particular factors that influence the companies' business results and that should be included into the model for forecasting and scenario-based economic analysis are to be identified. Defining the goal and developing the technical solution (writing the program code) is also an important task at this stage.

## 2. Data collection and storage

First, the periodic (operational, accounting, statistical) reporting data should be used. They present the overall information about the company's business expressed in the summarized and generalized indicators. They make possible the effective ongoing control and management in the company. Data on manufacturing and service provision, progress of raw materials and ordinary materials supply, fulfillment of supply contracts, etc. should be studied during the economic analysis and ongoing reports processing. The interrelated indicators that characterize company's cash circulation in the monetary and physical terms should be extracted and analyzed when processing the accounting data. Such indicators may be found in primary documents, consolidated accounting ledgers and reporting forms. The technical and economic data of the business entity, such as the details of the new equipment commissioning, the equipment state, productive-capacity balance, etc. should be studied during the economic analysis of statistical reports.

## 3. Data integration and analysis, forecasting

The methodical tools for extracting and transforming data, that is, bringing them to the required format, processing the data in accordance with certain rules, combining with other data, etc. should be used at the stage of integrating the data into the BI solutions. Besides, the data should be loaded and recorded into a repository or other database for subsequent processing by the intermediate or end users for solving various problems of economic analysis.

## 4. Data presentation and development of management decisions

The so-called dashboards and scorecards based on the key performance indicators (KPIs) analysis are one of the main ways to present the results of data analysis by using the BI solutions and tools. They provide additional means for visualization of the final analytical data by representing them as scales and indicators. They make it possible to monitor the current values of the selected indicators, compare the actual and the planned (target) indicators, display the dynamics of their changes over time, compare them with the critical (minimum/maximum) values and thus identify the potential business threats in the interactive, rather than the delayed mode.

## 5. BI-solutions model check and improvement

The proposed algorithm is based on the assumption that development of the economic analysis in companies based on survey, processing and integration of large amounts of data for the unbiased and reliable identification of the general trends in economic processes, understanding the situation and developing the company's position for the future effective decisions is the main task of implementing the BI solutions with data for the informational and analytical support of the economic analysis.

So the necessity to obtain reliable, objective and measurable results from the economic analysis and forecasts to be used for the future management decisions is the main focus of attention.

## 5  Conclusions

To conclude the analysis, it is necessary to state that the majority of the areas analyzed above are methodically interrelated. They have two main components: **ensuring the information quality and creating the corporate business intelligence system**. These components reflect the broadest methodological areas that can be extrapolated from foreign practice to the businesses of the national companies.

The scientific novelty of this research is that it deals with development of a number of theoretical, methodological and practical issues of using the business intelligence systems in the economic analysis of companies. These issues are based on the foreign experience analysis and the possibility to adapt it in order to optimize it in the Russian context with due account of the following areas: information quality improvement and elaborating an information quality improvement program, working out the methodology for monitoring the data and ensuring their transparency, determining the optimal algorithm for handling such data, when the BI solutions are used for the economic performance analysis of the Russian organizations.

The main findings of the study indicate that we observe some difficulties in introduction and using the business intelligence for economic analysis both in our country and abroad. However, the BI methods and technologies are progressing rapidly. In this respect it becomes easier for the Russian companies to obtain methodological support based on the foreign developments (among other things), as needed for the projects on integrating business intelligence into the economic performance analysis of companies.

## References

1. Abdallah S (2018) An intelligent system for identifying influential words in real-estate classifieds. J Intell Syst 27(2):183–194. https://doi.org/10.1515/jisys-2016-0100
2. Agapitou C, Bersimis S, Georgakellos D (2017) Appraisal of CRM implementation as business strategy option in times of recession: the role of perceived value and benefits. Int J Bus Sci Appl Manag 12(2):8–31
3. Arel I, Rose DC, Karnowski TP (2010) Deep machine learning—a new frontier in artificial intelligence research [research frontier]. IEEE Computat Intell Mag 5(4):13–18. https://doi.org/10.1109/MCI.2010.938364
4. Barilenko VI (2014) Business analysis as a tool for sustainable development of business entities. Account Anal Audit 1:25–31
5. Boyd DM, Ellison NB (2007) Social network sites: definition, history, and scholarship. J Comput Med Commun 13(1):210–230. https://doi.org/10.1111/j.1083-6101.2007.00393.x
6. Chen H, Chiang RHL, Storey VC (2012) Business intelligence and analytics: from big data to big impact. MIS Q 36(4):1165–1188

7. Data Management (31.03.2016) Accessed 27.10.2016 from Tadviser.ru http://www.tadviser.ru/index.php/Статья:управление_данными_(Data_management)
8. Data Management: [Analysis 2017] (2017) Accessed 19.01.2018, from InformationWeek: https://informationweek.com/data-management.asp
9. Davenport T, Petl DJ (2012) Data Specialist: the most sought-after profession of the 21st century. Harward Bus Rev 11:58–65
10. Ezzini S, Berrada I, Ghogho M (2018) Who is behind the wheel? Driver identification and fingerprinting. J Big Data 5:9. https://doi.org/10.1186/s40537-018-0118-7
11. Forrester Predictions 2018: a year of reckoning: [predictions guide] (2017). Forrester.com. Accessed 15.01.2018, from Forrester Research: https://go.forrester.com/research/predictions/
12. Information Management 2017–2020: [Knightsbridge Solutions analytical report] (2017). Knightsbridge Solutions, Chicago
13. Konoval DG (2009) Creating a unified regulatory information management system: approaches, technologies, stages and results. Gas Ind 8:26–29
14. Mitrović S (2017) The specifics of integrating business intelligence and big data technologies into economic analysis processes. Bus Inform 4(42):40–46. https://doi.org/10.17323/1998-0663.2017.4.40.46
15. Research & Insights [2016–2017] (2017) Retrieved 23.12.2017, from Merill Lynch: Bank of America Corporation: https://www.ml.com/financial-research-and-insights/all.html
16. Sheremet AD (2017) Comprehensive analysis and evaluation of financial and non-financial indicators showing sustainable development of companies. Audit 5:6–9
17. Sohangir S, Wang D, Pomeranets A, Khoshgoftaar TM (2018) Big Data: Deep Learning for financial sentiment analysis. J Big Data 5:3. https://doi.org/10.1186/s40537-017-0111-6
18. Suits VP (2012) The issues of management accounting organization and technology. The Moscow University Bulletin. Series 6: Economics, 3:94–102
19. Suits VP, Bayev AB (2018) The process-based approach to management accounting and analysis data generation. Audit Financ Anal 1:415–420
20. Yuhanna N, Leganza G, Lee J (15.06.2017). The Forrester Wave™: Big Data Warehouse, Q2 2017: [report]. Accessed 11.07.2017 from Forrester Research: https://www.forrester.com/report/The+Forrester+Wave+Big+Data+Warehouse+Q2+2017/-/E-RES136478
21. Zimin KV, Markin AV, Skripkin KG (2012) The impact of information technology on the performance of a Russian enterprise: methodology of empirical research. Bus Inform 1:40–48

# Human Transformation Under an Influence of the Digital Economy Development

A. V. Guryanova[(⊠)], S. V. Krasnov, and V. A. Frolov

Samara State University of Economics, Samara, Russia
{annaguryanov, frolov5070}@yandex. ru,
tacit63@rambler. ru

**Abstract.** The article examines the process of human transformation in conditions of the digital economy. The main philosophical aspect of this process is the radical change of the human consciousness. Transformations of the economic consciousness cause radical changes of its paradigm. So, a new paradigm of the economic thinking forming in the digital age is analyzed in the article. Attention is paid to its prehistory too. Its place and role in the field of the modern company management technology are also considered. The forecasts of the further development of these processes are formulated. The authors specially accent the humanly oriented points of digitalization such as the changes in the character of management in the digital company, the system of the specialists' training, the areas of their professional activity and education. New requirements to the leadership in the digital economy are considered. Problems of dialectical co-operation between the humans, the state and the market are also discussed.

**Keywords:** Digital economy · Digitalization · Economic consciousness · Economic thinking · Economy · Human · Management · Market · Paradigm · State

## 1 Introduction

In the XX century the process of digitalization has begun in economically developed regions such as Europe, North America and East Asia. Digital economy is a type of economic activity where a primary factor of economic, social and humanitarian development is connected with production and using of scientific, technical and other kinds of digital information [3].

On an example of the highly developed countries we can see an obvious fact: the economic behavior assumes not only the special organization forms of the state governing [16]. One of the most important components of this process is human transformation [5] which is, first of all, the change of human consciousness. Human consciousness is the major element of the economic subject [4]. In fact, transformation of the economic thinking is a transformation of the human himself in conditions of the

S. Ashmarina et al. (Eds.): *Digital Transformation of the Economy: Challenges, Trends and New Opportunities*, pp. 140–149, 2019.
https://doi.org/10.1007/978-3-030-11367-4_14

modern digital economy. But human consciousness, as it was proved by T. Kuhn [9], is always based on a certain paradigm. «Paradigm» is a basic conceptual scheme that sets the model of problem statement, its following decision and the methods of research activity dominating during the certain historical period in the certain scientific association. This article is devoted to study the content of this paradigm and its transformation in conditions of the modern digital economy.

## 2   Materials and Methods

Methodological basis of the research is formed by the classical works of philosophers and sociologists related to the analyzed topic. These are the works of Nicolay Berdyaev, Max Weber, Karl Jaspers and Thomas Kuhn.

Methods of philosophical, dialectical and categorical analysis are both used in the article.

### 2.1   Method of Philosophical Analysis

Method of philosophical analysis let the authors focus themselves on synthesizing and generalizing consideration of the research theme. Using the method of philosophical analysis helped them to study the basic theoretical concepts and scenarios of human transformation in conditions of the digital economy. The system of values and ideals accepted by people existing in a digital world and using intensively different digital technologies is also considered.

### 2.2   Method of Dialectical Analysis

As it is known, dialectical method always played an important role in the field of philosophical analysis. It is the most general, universal method of cognitive activity. It makes possible vision of the research object in its historical evolution, changeability and dynamics. Using of dialectical method gave the authors a chance to describe the wide changes taking place in the modern society, economy, in human's existence and thinking under the growing influence of the digital economy.

### 2.3   Method of Categorical Analysis

We are sure in modern conditions it's necessary to expand the sphere of traditional philosophical concepts by inclusion in its field the new terms describing an impact of information and digital technology on the society and the human being [4]. For example, it can be such concepts as «digitalization», «digital society», «digital economy», «digital thinking», etc. Moreover, they are often used in our everyday life. So, they need integration in the field of the special philosophical analysis.

## 3   Results

### 3.1   Motive Powers of Digital Society and Economy

#### 3.1.1   Economic Thinking as a Cause of Human Consciousness and Behavior

The motive power of the modern society development is not any more the material production but the digital one. Certainly, the material products still exist but they have become greatly dependent on the digital technologies [3]. This naturally leads to the rising cost of the products of research and development, design and marketing. In comparison with an industrial society where everything was directed by production and consumption of goods, the digital society (or the «society knowledge») produces and consumes mainly intelligence. So, the human must be much more creative in this new type of the social reality where knowledge is the greatest value.

According to the famous Russian philosopher N. Berdyaev [1], human creativity develops within the limits of those representations of the world which are dominant at the present moment. The modern world is a world information and communication technology. After the works of N. Berdyaev, K. Jaspers and M. Heidegger it's obvious that the digital society has mostly a technical character. Moreover, technique is a special sphere of spirituality as it is based on a human desire to transform actively nature by spirit, on domination over it. This activity also determines the modern economic thinking which is inevitably changing under an influence of the digital economy. Successful economic development is connected with the model of effective economic behavior [16]. The last one is formed within the limits of human transformation in conditions of the digital society.

#### 3.1.2   Historical Transformations of Economic Consciousness

An example of economically developed countries such as the North America, Europe, South-East Asia and Russia shows us that the economic behavior assumes not only the special organization forms of government. It assumes also a special type of institute functioning in people management. A famous theorist of the European capitalism M. Weber named it as a «capitalist spirit» [20]. Weber himself belonged to an uncertain in conceptual terms philosophical school of German idealism, so its terminology must be corrected a little bit. If we clarify the philosophical category of the «capitalist spirit» we can assume that Weber properly meant the special factor of economic development that can be called «economic thinking» in the spirit of our modernity. Economic thinking is the major component of the economic subject. Transformation of the economic thinking is a transformation of the human himself in conditions of the digital economy.

Strictly speaking, the economic thinking is not limited only by the sphere of economy. It creates a special vision of the world and a specific way of human life. It applies to all the areas of social being – social relations, policy, religion, art, etc. Thus, the principles of practical action in economic thinking determine much the behavior of subject of this thinking in economic sphere. K. Jaspers wrote about components of economic thinking: «In origin of the modern technical world inseparably linked among themselves natural sciences, spirit of invention and labor organization. These three factors together possess rationality. Each of these factors has the sources and is connected therefore with a number of problems independent of another factors» [6]. Thus, the economic thinking of the previous industrial society was based on the rational approach to life developed by the ethics of Protestantism.

### 3.1.3  Mechanism of Economic Thinking's Transformation

The economic thinking represents specific historical formation. In other words, it exists within the limits of a certain historical epoch. Each new type of economy refuses the old economic thinking because it is not enough actual. This eliminates possibility of the old types of economic thinking functioning. Therefore, economic development represents periodical changes of the types of economic thinking. The certain type of economic thinking usually dominates during this or that historical period. But in the case when one historical epoch replaces another, collective and individual consciousness start interactions, interventions and competitions. In such a way various types of economic thinking struggle for existence. But the old type of economic thinking doesn't finally disappear, while at least one individual guides by it in his way of live.

In general, economic thinking is a complex, integrative phenomenon. But there is a main point in its organization that determines the typical for this or that model of economic thinking relations between the world and the human. Every time it sets a specific economic model of construction and solution the problems (first of all, of course, economic problems) and focuses attention on one or another aspect of life the human is facing. Such a basic point is called «paradigm» in the modern philosophy of science.

## 3.2  A New Paradigm of Economic Thinking

### 3.2.1  Concept of Paradigm, Its Structure and Functions

At first, the term of «paradigm» was entered in philosophy of science by T. Kuhn [9]. Now it's really widespread in the philosophical and scientific literature. A paradigm, according to Kuhn, is an initial conceptual scheme that gives its followers a model of problem statement and decision. It also provides dominate methods of the research typical for the certain historical period and the certain scientific community. Finally, this set of methods and acceptances united by this or that scientific or philosophical community becomes its general scientific or philosophical ideology. At the same time the other communities can be united by the other ideology and, accordingly, they realize another paradigm use.

These methods and acceptances are also known as a special «conceptual world» that provides only the standard theories. Its participants – the members of the scientific community – realize classical experiments and use traditional methods. Both scientists relevant to the certain scientific community usually accept an official paradigm and expand its possibilities. They try to clear their theory of the contradicting facts and statements, to explain them with an aim to delete every contradiction, threatening the stability of the paradigm. But sooner or later these efforts of the scientists lead to appearance of insoluble theoretical problems or experimental anomalies that prove together inexactness or insufficiency of the official paradigm. These difficulties cause the crisis which means the change of the paradigms and the start of the «scientific revolution». In sum all these seemly separate revolutions have radically changed the old paradigm replaces by the new.

In the XX century scientific revolutions took place for several times according to the different areas of scientific knowledge. For example, occurrence of psychoanalysis, the relativity theory, the quantum mechanics, etc. In their sum all these seemly separated revolutions have led to radical changes in the scientific thinking.

### 3.2.2 Traditional Paradigm of Economic Thinking: The Way of Forming

Traditional paradigm of economic thinking was formed within the limits of the European rationalism. It was based on the several philosophic conceptions of the XVII-XVIII centuries such as J. Locke's philosophy and I. Bentham's utilitarian ethics. They both believed that human consciousness, as well as its body, is determined by the mechanic laws. So, the human activity in the economic sphere is a chain of events, summing causes and consequents. We can probably control and manage it with a help of specially developed technologies. Such an approach is known under the name of «technocracy». Technocracy is a paradigm of economic thinking of an industrial society. Its followers are sure that the human economic activity can be regulated by the principles of scientific and technical rationality and an economic gain.

The type of economic thinking based on this paradigm includes the following components:

1. The work is a main employment of human life.
2. Labor must be rationally organized.
3. The main goal of the work is receiving the salary and its growth.
4. The profession statement is a form of human existence.
5. Desire for profit is an end in itself.
6. Desire for the expanded production.
7. The human is only a factor of production process.
8. Orientation on an exclusive production of goods.
9. The principles of western economy have universal character.
10. Honesty is a best guarantee of the successful business.

Already in the second half of the XX century this approach has become an object of critics of pragmatism, phenomenology and Russian religious philosophy. So, it was declaimed as ineffective. This happened under an influence of the high technologies' development. This process involves the role of personal responsibility and the initiative of the certain worker. The modern digital technology makes new requirements to the humans and forms a basis of the new paradigm of economic thinking connected with the processes of informatization and digitalization.

### 3.2.3 A New Paradigm of Economic Thinking and Management

A new paradigm of economic thinking includes the following components:

1. The work turns into a form of creative activity, at least, for the greater part of the society members.
2. Self-realization becomes a basic stimulus of working.

3. Domination of material benefits producing is just overcome; development of human skills and capacities has become an aim of human life.
4. Education, leisure spheres, etc. have become the real sectors of economy.
5. Management is not guided only by the universal scheme «working with people»; it is mainly focused on the national (including ethical) traditions of human relations.
6. Technologies of human management are replaced by the other forms of relations basing on human individuality.
7. Specific national models of economic development combining principles of market economy with national traditions of managing are formed.
8. Intellectual property becomes a major type of property.

A new paradigm of economic thinking forms a new «ideology of management». Modern economists think it's one of the major factors providing the growth of companies. And they are absolutely right because the ideas of this ideology set the forms of representation and explanation of the purposes, principles and tasks of management, its values, rates, rules, stereotypes of behavior and relations within the business team. Attitudes, preferences and other elements of social psychology are formed with their help. Nowadays there are many examples of companies guided by the highly developed ideology of management. In future such companies will be the most successful.

To create an effective model of modern economic management it's necessary to use the better ideas and methods of social projecting – a new scientific tendency which has developed intensively in the last years and has already well-proved itself.

## 3.3    New Type of Management as a Result of Human Transformation

### 3.3.1    Management Efficiency of Economy and the Human Factor

It's obvious that management efficiency depends much on the working personnel and its professional qualification. The last one depends on the system of personnel training and retraining. In conditions of the digital economy, the workers must be necessarily guided in the following areas: digital technologies, strategic planning, legal bases, technic of commercial operations, financial and accounting skills. They also must be competent in foreign languages to work with the foreign partners. Working in conditions of the digital economy, using the modern digital technic and technologies is impossible without a high-level qualification of the personnel [12, 13].

That's why the problem of training and retraining of the workers is very actual today. The level of their intellectual skills and abilities should make them highly-qualified specialists. So, modern enterprises do not spare money (it's usually from 5 to 10% of receiving salary) for training and retraining their personnel. As the foreign practice shows, this pays off in full in a short time. One of the main strategic tasks of every enterprise is an exit to the world market. For integration with the world market the high professionalism and competence of the participants of the foreign trade activities is inevitably needed.

And one more thing about the management efficiency of economy: its control system should provide a high quality of the produced goods. As it is known, quality is connected with the possibility of goods to satisfy the human requirements. It is necessary to develop business in such a way that the consumer appears as an absolute authority for the manufacturer, and his desire is the universal law. This assumes a really good work of the manufacturer. As a result, the volume of the external control is lowed, but the level of the internal control – the so called «self-checking» – rises on the contrary. The most important thing in the quality management is not the strict control procedures, but the really perfect way of working.

### 3.3.2  New Requirements to the Leadership in the Digital Economy

Requirements for the leaders of the enterprises are greatly increased in conditions of the digital economy. In his famous book «The Digital Economy: Promise and Peril in the Age of Networked Intelligence» D. Tapscott [18] discusses a problem for gaining leadership in the network community. He points that in this type of the society traditional vertically-hierarchical decision-making structures are replaced by the collective intelligence of the whole team working in the Network. In modern conditions such a joint intelligence is better than the individual one belonging to the greatest leader. According to Tapscott, the factors providing the primacy in the digital economy are the same:

1. constant focus on transformation;
2. ability to learn quickly and constantly;
3. ability to work in the team;
4. straightening of the production processes with a help of the digital networks;
5. desire of the company's top leaders to take personal part in the occurring transformations [18].

All these requirements concern not only the common humans, the members of the network community but the leaders, first of all. A considerable quantity of various tests, courses of advanced training and skills development are applied in different spheres of professional activity [14]. The increased requirements of the digital economy cause an importance of the business qualities' assessment, of the capacities and creative activity development.

### 3.3.3  Dialectic of State and Market in the Digital Economy

Development of the control system in economy is impossible without dialectical interaction of two components: the state and the market as regulator of economic processes. Three problems follow from here:

1. Organization of the state institutes;
2. Forming of the market;
3. Creation of the interaction mechanism between the state institutes and the market.

These three problems should be solved both at the strategic and tactical levels. In the first case they must be analyzed deeply, on a distant perspective, in the second – conditions of the certain historical reality must be taken in view.

As for the first problem, the state institutes protect the rule of law, democracy, requirements of the market system and the new economic relations. In general, the state should carry out the reasonable economic policy including credit and financial mechanisms, taxes, establishment of priorities and standards, various inspections, etc. Concerning the second problem, there is no clearness in the ways of the market development methods. In fact, the market is created by the state in the name of the state structures, and also by businessmen. The ideal model of their interaction is the same: the state provides initial production and financial services for realization of the big projects in digital sphere, while the private sector realizes commercial exploitation of the received results that is frequently rather profitable.

Thus, the state and the businessmen co-operate closely in this area. The state tries to create the most favorable conditions for organization of the market of digital goods. The private sector realizes functional weight of production, purchase and sale of these goods. It should be recognized that the described model is quite effective. And it allows in general to solve the third problem – forming of the adequate mechanism of inter-actions between the state institutions and the market.

## 4 Discussion

Various aspects of human transformation in a digital society are actively discussed in the modern scientific literature. However, most of these studies are devoted to the economic aspects of the digital society development [11, 17, 18, 21]. An analysis of the problems of human transformation in conditions of the digital society from the philosophical point of view is presented in the number of modern research works [5, 7, 8]. There is also a number of works on the psychological aspects of human existence in a digital world [2, 10, 15, 19]. It should be specially noted that in the modern philo-sophical literature the problems of the human consciousness and its transformation in conditions of the digital economy are the most debatable and undeveloped. But they are very important, especially, taking in view their impact on the methods of economic management, technologies of the leadership and development of the new forms of education.

## 5 Conclusions

A new paradigm of economic thinking is forming just now. It basically concerns all humanitarian sciences and especially philosophy. The article presents the results of philosophical research of the process of human transformation in conditions of the digital society.

The authors develop an idea that human transformation is first of all the change of his consciousness and thinking. In the field of economics, these are economic con-sciousness and economic thinking as the major components of the economic subject. Thus, transformation of the economic consciousness and thinking in conditions of the modern digital economy can be interpreted as a transformation of the modern human himself.

It's proved in the article that economic thinking is not limited only by the sphere of economics. It creates a special worldview and a specific way of human life. It applies to all the areas of social being – social relations, state, management, religion, etc. So, the principles of practical actions approved in the sphere of economic thinking determine much the human behavior in the economic sphere.

Human thinking is always based on a certain paradigm. The modern economic paradigm is formed under an influence of the digital technologies' development. The modern digital technology makes new requirements to the humans and especially to the leadership. Digital humans must be always able to transform themselves, to learn, to work in the team, to use the digital networks. These factors provide the primacy in the digital economy. The role of personal responsibility and the initiative of the certain participants of the economic relations are valued too.

# References

1. Berdyaev NA (2018) The meaning of the creative act. AST, Moscow
2. Ershova RV (ed) (2016) Digital society as a cultural and historical context of human development. Kolomna State Socio-Humanitarian University, Kolomna
3. Guryanova A, Astafeva N, Filatova N, Khafiyatullina E, Guryanov N (2019) Philosophical problems of information and communication technology in the process of modern socio-economic development. In: Advances in intelligent systems and computing. vol 726. Springer, Cham, pp 1033–1040. https://doi.org/10.1007/978-3-319-90835-9_115
4. Guryanova A, Guryanov N, Frolov V, Tokmakov M, Belozerova O (2017) Main categories of economics as an object of philosophical analysis. In: contributions of economics, Russia and the European Union: development and perspectives. Springer International Publishing AG, Cham, Switzerland, 221–228. https://doi.org/10.1007/978-3-319-55257-6_30
5. Guryanova A, Khafiyatullina E, Kolibanov A, Makhovikov A, Frolov V (2018) Philosophical view on human existence in the world of technic and information. In: Advances in intelligent systems and computing. Springer, Cham 622, pp 97–104. https://doi.org/10.1007/978-3-319-75383-6_13
6. Jaspers K (1953) The origin and the goal of history. Yale University Press, New Haven, CT
7. Keshelava AV (ed.) (2017) Introduction to digital economy. Vniigeosystems
8. Kolmakov VYu (2009) Human of digital civilization. Litera Print, Krasnoyarsk
9. Kuhn T (2012) The structure of scientific revolutions. University of Chicago Press, Chicago
10. Manafy M, Gautschi H (2011) Dancing with digital natives: staying in step with the generation that's transforming the way business is done. CyberAge Books, Medford, NJ
11. Negroponte N (1995) Being digital. Knopf, NY
12. Pecherskaya E, Averina L, Kamaletdinov Yu, Tretyakova N, Magomadova T (2016) Assessment of critical success factors transformation in ERP projects. IJME Math Educ 11 (7):2608–2625
13. Pecherskaya E, Averina L, Kochetckova N, Chupina V, Akimova O (2016) Methodology of project managers' competency formation in CPE. IJME Math Educ 11(8):3066–3075
14. Pecherskaya E, Averina L, Kozhevnikova S (2018) ERP Implementation challenge: case-study of the Russian Federation. Astra Salvensis. Special issue: 411–423
15. Pepperell R (2003) The Posthuman condition: consciousness beyond the brain. Intellect Books, Portland, Oregon
16. Shestakov A, Noskov E, Tikhonov V, Astafeva N (2017) Economic behavior and the issue of rationality. In: Contributions of economics, Russia and the European Union: development and perspectives. Cham, Switzerland, Springer International Publishing AG, pp 327–332. https://doi.org/10.1007/978-3-319-55257-6_43
17. Schwab K (2017) The fourth industrial revolution. Exmo, Moscow
18. Tapscott D (2014) The digital economy anniversary edition: rethinking promise and peril in the age of networked intelligence. McGraw-Hill, NY

19. Turkle S (2010) Alone together. Basic Books, New York
20. Weber M (2001) The protestant ethic and the spirit of capitalism. Routledge
21. Zubarev AE (2017) The digital economy as a form of manifestation of the new economy. Bull Pac Natl Univ 4(47):177–184

# Targeting as an Instrument of State Financial Policy in the Digital Economy

M. B. Tershukova[✉], O. G. Savinov, N. G. Savinova,
and L. N. Milova

Samara State University of Economics, Samara, Russia
{tershukova.marina, savinovog, larisamilova2009}
@yandex.ru, savnad@bk.ru

**Abstract.** The contribution considers the impact of targeting in the monetary and public sector on individual macroeconomic indicators of Russian economy. The authors analyzed the use of such financial instruments as targeting in the management of the most important financial and monetary indicators based on targets of economic and monetary policies, the statistics of the International Monetary Fund, the Bank of Russia, and the Treasury of Russia.

**Keywords:** Currency targeting · Digital economy · Inflation targeting ·
Instruments of monetary policy · Monetary targeting ·
Targeting budget balances on a single treasury account

## 1   Introduction

The strengthening of exogenous factors in the development of Russian economy makes it necessary to adjust targets of state economic policy and regulate macroeconomic indicators in the financial sector. Targeting macroeconomic indicators for the development of a stable national economy determines the relevance of domestic price stability and national currency stability. The expansion of the range of tasks for inflation targeting, evaluation of monetary, non-monetary and monetary-non-monetary factors of inflation determines the use of digital technologies.

Targeting macroeconomic indicators includes a set of measures and instruments of monetary and financial policy that restrain inflation, provide employment and certain rates of economic growth. The close relationship between monetary and financial policies is due to the susceptibility of national economies to crises and the need to smooth out economic fluctuations. In this regard, the role of targeting as an instrument of financial policy for the enforcement of counter-cyclical regulation is being rethought.

Within the digital economy in Russia, the main characteristics of monetary policy pursued by the Central Bank of the Russian Federation are: informational openness, "transparency" of the regulator's activities and its accountability to the public. That is why the Central Bank of the Russian Federation publishes the main targets and instruments of monetary policy being pursued on its Internet site. These include: the rate of inflation in the country, the key interest rate of the Bank of Russia, the minimum

S. Ashmarina et al. (Eds.): *Digital Transformation of the Economy: Challenges, Trends and New Opportunities*, pp. 150–162, 2019.
https://doi.org/10.1007/978-3-030-11367-4_15

reserve requirements for banks, the conditions for granting refinancing loans, and deposit operations with credit institutions.

As part of public accountability for the results of its monetary policy, the Bank of Russia regularly publishes the following data on the Internet site: a monetary policy report; materials of the Board meeting of the Central Bank of the Russian Federation on a key rate; information about the volume of gold and foreign exchange reserves of the state and others. All this is aimed at increasing the predictability and transparency of monetary policy for the general public, and reducing inflation expectations.

At the level of monetary policy, the role of targeting monetary indicators by central banks is increasing. The essence of targeting is forecasting macroeconomic indicators, choosing the instruments of their regulation that are adequate to the set targets.

Targeting is a fairly new phenomenon in the practice of the Bank of Russia. Financial stability and the balanced development of the state's economy largely depend on the correct choice of targets, mode and instruments of monetary policy. However, macroeconomic and institutional conditions for conducting monetary policy are not controlled by the Bank of Russia. Therefore, the establishment of stable relationships between the targets of economic policy, the main directions and decisions of the Bank of Russia in the field of monetary regulation is necessary to ensure stable GDP growth, employment and curbing inflation at a certain level.

The targets of the Russian Development Strategy for the period 2018–2024 are GDP growth and increase in household consumption to 4% due to increased investment in the economy and increased labor productivity. The contribution of the Bank of Russia to the creation of predictable economic conditions and the strengthening of confidence in the policies being pursued is the steady consolidation of growth rates of consumer prices and the rate of inflation near 4%. Thus, in the context of targeting, the increase in household consumption equals the rate of inflation.

Stable national currency and price stability create a favorable environment for the implementation of long-term investment programs. The growth of investment in the economy in order to diversify it is extremely important for maintaining financial stability of society. In this regard, the financial market in Russia should be more focused on lending to the real sector of the economy.

With regard to the global economic and financial instability, central banks are changing the targets of monetary policy, experimenting with its various instruments and modes [20]. The role of monetary policy is being rethought, approaches to targeting macroeconomic indicators are being improved, and measures of crisis management are being developed. At the same time, the efforts of monetary authorities can be directed at regulating the rate of inflation, maintaining the national currency, and setting targets for the money supply. The number of countries moving to inflation targeting is expanding. A set of measures is being developed to establish and control the rate of inflation in the country, to maintain the price stability [18]. These problems are extremely significant for Russian economy. It seems appropriate to study inflation targeting policy, which depends both on the choice of the mode and instruments of monetary policy, and on its macroeconomic and institutional conditions.

The purpose of the contribution is to study the domestic and foreign experience of targeting as an important instrument of state financial policy.

In accordance with the purpose, the following tasks were defined:

1. To analyze domestic and foreign experience in the application of targeting financial indicators in the monetary and credit sphere, as well as in public finance in Russia in the period of economy digitalization.
2. To identify the advantages and disadvantages of this instrument of the country's monetary policy in terms of inflation targeting, as well as financial policy.
3. To recommend ways for improving targeting as the mode of monetary policy and targeting balances on a single account of the Federal Treasury in Russia.

## 2   Materials and Methods

The authors used general scientific methods of cognition (synthesis, analysis, analogy), as well as such methods of empirical cognition as classification, observation, measurement, description, comparison, and generalization.

The results of the contribution were supported by such special methods as: a method for identifying targeting modes of monetary policy in various countries, a system-functional analysis of macroeconomic indicators in the digital economy.

As an information base, publications by domestic and foreign scientists in the field of research on targeting were used as an important instrument for conducting the state's financial policy.

## 3   Results

To ensure financial stability and conduct independent financial policy in the context of globalization of the capital market, central banks of various countries choose adequate modes and targeting instruments for the economy [19].

Thus, the majority of developed countries that have made the transition to targeting inflation use the mode of the floating rate. The US Federal Reserve and the European Central Bank adhere to consistently low inflation, without officially identifying themselves as targeting inflation.

In developing countries, the monetary authorities, when targeting inflation, prefer to manage the exchange rate in order to reduce its volatility.

Some countries target inflation under the free floating rate. These countries include Russia, Mexico and other developing countries. The Czech Republic has chosen a mode of the stabilized rate and adheres to low inflation (Table 1).

**Table 1.** Modes of the exchange rate in countries with inflation targeting

| Exchange rate (number of countries) | Modes of inflation targeting | | | | | |
| --- | --- | --- | --- | --- | --- | --- |
| | 1990–1999 | | 2000–2013 | | 2014–2018 | |
| | Developed countries | Developing countries | Developed countries | Developing countries | Developed countries | Developing countries |
| Floating rate (28) | 2 | 3 | 1 | 18 | 0 | 4 |
| Free floating rate (11) | 4 | 2 | 3 | 0 | 0 | 2 |
| Mode of the stabilized rate (1) | 1 | – | – | – | – | – |

Source: compiled by the authors based on data from Annual Report on Exchange rate arrangements and exchange rate restrictions. International Monetary Fund, 2017, imf.org

Since the beginning of the 90s of the last century, countries began to switch to inflation targeting. Currently, the mode of inflation targeting is used by central banks in forty countries. Central banks of many countries set quantitative benchmarks for the rate of inflation in the form of a range, a "corridor" of values - Israel, Poland, Hungary, Romania and India. Central banks of such countries as the United Kingdom, Sweden, Norway, and Russia are forecasting inflation for a specific value. The period of forecasting the rate of inflation is mainly medium-term, not more than 3 years.

Inflation and economic activity are influenced by central banks using standard instruments of monetary policy - interest rates. Mainly, these are interest rates of central banks, as well as other instruments of permanent action: refinancing of banks; open market operations, including repos; deposit operations.

In recent years, the Bank of Russia has been intensifying its interest rate policy - every month the issue of another change in the key rate, which has been an indicator of monetary policy since October 2013, is being considered at the Board meeting of the Central Bank of the Russian Federation. This takes into account external and internal factors such as the rate of inflation in the country, high risks of economic recession, the situation in the banking sector, etc.

Changing the key rate, the Central Bank of the Russian Federation affects money market rates, then, through them, the interest rates of banks on credit and deposit operations. In the end, it has an indirect impact on the volume of investments and savings in the economy, on economic activity and inflation [12].

Using instruments such as bank refinancing, repurchase transactions, deposit operations, the Central Bank has an indirect impact on the monetary sphere when regulating total banking liquidity. Under conditions of its deficit, the following instruments are actively used: refinancing of banks, repo operations for the purchase of securities from banks. On the contrary, in case of excess banking liquidity, the Bank of Russia uses the instruments of "tying it up", attracting bank funds and, mainly, deposit operations (Fig. 1).

The Central Bank regularly publishes information on cases of aggregate banking liquidity, the amount of its deficit (surplus) on the Internet site.

Inflation targeting in 2014–2018 contributed to price stabilization and lower inflation. At the end of 2017, inflation in Russia fell to a record 2.5%, with a target of 4%. This was partly due to a decrease in consumer demand and a significant debt load of the population [11]. To ensure a predictable rate of inflation, it is necessary to make wider use of digital technologies (blogging, Big Data) to regulate the imbalances in the economy.

Macroeconomic features and institutional conditions for conducting monetary policy in Russia had a significant impact on the decline in GDP growth rates in 2015–2016. For 2017, GDP grew by 1.5% and was below inflation by 1%.

The reduction in the key rate was much slower, which worsened the availability of credit resources and did not contribute to the growth of economic activity. Interest rates on loans to non-financial institutions turned out to be higher than the profitability of individual sectors of the economy.

In the transition to inflation targeting under conditions of removing currency restrictions and full openness of the currency and financial market, the Bank of Russia

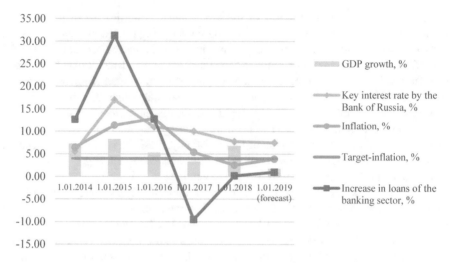

**Fig. 1.** Dynamics of GDP growth, loans of the banking sector of the Russian Federation in terms of inflation targeting and changes in the key rate of the Bank of Russia for 2014–2018, %. Source: compiled by the authors based on data from the Bank of Russia, cbr.ru

faced problems of high volatility in foreign currency and speculative operations. The destabilization of the ruble exchange rate influenced the growth of inflation.

The reduction of the key rate by the Bank of Russia did not allow restoring the growth rates of lending to the real sector of the economy. The measures taken in the framework of monetary policy have weakly ensured the investment process with the innovation component (Fig. 2).

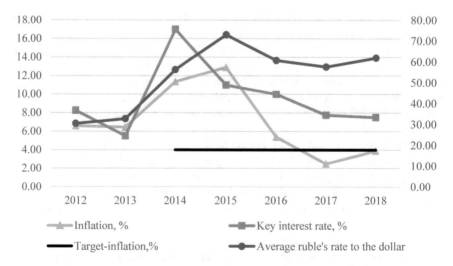

**Fig. 2.** Dynamics of inflation in the Russian Federation, the key rate of the Bank of Russia and the average ruble's rate to the dollar for 2012–2018 (at the beginning of the year). Source: compiled by the authors based on data from the Bank of Russia, cbr.ru

The banking sector has accumulated excess liquidity, which is not used for economic growth [14]. In the field of public finance, the policy of targeting demand balances on a single account of the Russian Treasury and the development of management instruments for free demand balances is based on the use of advanced world experience and is pursued in coordination with the Bank of Russia, taking into account liquidity of the banking sector (Fig. 3).

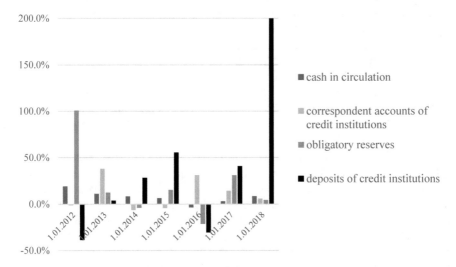

**Fig. 3.** Dynamics of growth rates of monetary base indicators for 2014–2018. Source: compiled by the authors based on data from the Bank of Russia, cbr.ru

Cash Management is widely used in foreign practice. Traditionally, it is understood as a system of banking technologies for managing financial flows of a corporate client [15]. Accordingly, its main users are commercial banks. However, today the Cash Management system, as an innovative method, is actively used by the Treasury of Russia.

The most important activity of Cash Management in the Federal Treasury system is to consolidate funds on STA and hereon to improve management mechanisms for free balances of budget funds. The operations consolidating funds on STA are comparable to a banking product such as a cash-pooling system - consolidation of liquidity. It is a system for managing several accounts opened in a bank, the balance of which is concentrated on one account (cash concentration). Cash Management technologies are used in public finance in Germany, France and other countries [3].

Under the conditions of the financial crisis in 2009 and 2010, the Russian Treasury took measures to concentrate budget fund balances on a single budget account received by budget recipients from their extra budget activities and funds held at their temporary disposal. The concentration of treasury fund balances has become a significant internal reserve for increasing liquidity of a single federal budget account, which can be estimated at about 100 billion rubles of daily minimum balance of funds, of which 80 billion rubles - extra-budget balances, 20 billion rubles - balances at the temporary disposal of federal budget beneficiaries [10]. Thus, the organizational and economic prerequisites for the formation of a mechanism managing liquidity of a single treasury account were created. It should be noted that in this period there was a relative excess of liquidity in the public sector, while there was a shortage of liquidity in the banking and real sector. It was necessary to introduce a mechanism for non-inflation credit financing of the real sector of the economy.

One of the first methods to manage funds on a single treasury account was the method of depositing free funds of the federal budget in national currency with banks. By Decree No. 227 of March 29, 2008, the Government of the Russian Federation approved the procedure for depositing funds from the federal budget with banks. On the basis of this resolution, the Ministry of Finance of the Russian Federation, the Federal Treasury and the Central Bank of Russia have developed and enforced a detailed mechanism for depositing funds of the federal budget with banks.

Analyzing the experience of the Russian Treasury in depositing budget funds with banks, it is necessary to consider this direction of work in the management of STA liquidity from the standpoint of its participants.

For the Federal Treasury, such work is an opportunity to make funds work and to get additional revenues. The results of depositing budget funds with banks indicate the efficiency of this method of cash management on STA. The total amount of revenues from depositing federal budget funds with banks was the following: in 2011 - 19.62 billion rubles, in 2012 - 22.52 billion rubles, in 2013 - 30.55 billion rubles; in 2014 - 40 billion rubles, in 2015 - 68.15 billion rubles, in 2016 - 46.91 billion rubles, in 2017 - 37.7 billion rubles [8]. For commercial banks, such an instrument is a source of additional funds, a liquidity reserve, an alternative to resources of the interbank lending market, and resources attracted from the Central Bank.

During periods of liquidity shortage of the banking sector, federal budget funds deposited with banks are a significant factor in their financial stability. On the basis of their attraction, banks decide mainly the issues of short-term liquidity. If the rates for attracting budget funds are set lower than for attracting from other sources, then federal budget funds can become a full-fledged source for lending operations, especially in case of increased periods for the allocation of budget funds.

The funds of the federal budget were deposited with banks for a period from 7 to 182 days in 2017. In 2017, 13,343.0 billion rubles were offered for the deposit with banks, which is 25.6% more than in 2016. Demand from credit institutions amounted to 4,482.1 billion rubles, which is 60.6% less than in 2016. On the terms of bank deposit agreements, the Federal Treasury deposited funds of the federal budget with banks in the amount of 4,012.4 billion rubles in 2017, which is 47.9% less than in 2016. The decrease in demand from credit institutions for budget funds was due to the presence of a structural liquidity surplus in 2017 according to the Bank of Russia [6].

In order to approximate the interest rates to market indicators when depositing federal budget funds with banks, floating rate for deposits with terms of more than 30 days was applied. The indicative weighted rate of the short-term financial market (one-day ruble loans) RUONIA was used as the base floating rate.

At present, a new mechanism is being created for managing balances on STA in terms of buying and selling foreign currency and concluding transactions in organized currency swap trades. It should be noted that the demand of banks for currency refinancing in 2017 increased mainly through "currency swap" transactions. On average, the volume of such a transaction for the provision of currency liquidity to the banking system was $576.7 million in 2017. The rate on the ruble part of the transaction was 6.75–9%, and on the currency part, 2.2–2.9%. According to analysts, a significant increase in demand for currency swap transactions occurred in December 2017. During this period, the average size of the transaction increased 2.3 times compared to December 2016 and reached $ 1.4 billion. Scientists and economists note that the growth in demand for currency swap transactions is associated with an increase in banks' demand for currency liquidity to the end of the year in terms of the termination of currency repos [1].

One of the key features of the money market in 2017 was the structural surplus of liquidity of the banking sector. In this regard, the Bank of Russia took measures to reduce the instruments for providing liquidity to banks and expanding approaches to its absorption. In addition, in 2017 the Bank of Russia introduced a mechanism for emergency liquidity provision (MELP), aimed at bank recovery. Now banks that will experience temporary liquidity problems can contact the Bank of Russia and receive funds for up to 90 days at a fixed rate corresponding to the key rate increased by 1.75 percentage points. The decision to provide funds to the bank within the framework of MELP is made by the regulator, taking into account its financial sustainability, as well as its systemic significance.

Periods of excess liquidity of the banking sector are accompanied by the actions of the Central Bank to stimulate the search for alternative lines of investment, which is characterized by a decrease in interest rates on the Bank of Russia, deposit auctions and a decrease in the need for funds from the federal budget. Under these conditions the Federal Treasury has to: first, take counter-cyclical measures, increasing the supply of resources and reducing rates faster than alternative rates change - the key rate of the Central Bank of the Russian Federation and the MIACR rate; secondly, it develop alternative methods for allocating budget balances to financial instruments [2].

Despite the fact that there were no cases of non-return of funds deposited with banks, the Federal Treasury hedges possible risks and uses collateral financial instruments in its work. In particular, since 2013, the Treasury of Russia has been using repos in the financial market as a method of managing balances on a single treasury account. Such transactions are more attractive for commercial banks, since they can be used to instantly convert securities into cash. This allows concentrating financial resources in other areas in the short term, and after conducting the second part of the transaction - returning securities.

The parties to a repo agreement are the Federal Treasury and credit institutions that meet a number of requirements established by the RF Government Decree No. 777 of 04.09.2013 "On the Procedure for Managing Balances of Funds on a Single Federal Budget Account for the Purchase (Sale) of Securities under Repurchase Agreements" and concluded general agreements on the purchase (sale) of securities under repurchase agreements with the Federal Treasury body. The term of the general agreement is one year. The objects of such transactions are only bonds of federal loans.

One of the problems of conducting repo transactions involving the Federal Treasury was the absence of federal loan bonds in the assets of credit institutions. Since 2015, as a result of additional capitalization through the Deposit Insurance Agency, commercial banks received additional collateral in the form of federal loan securities for 1 trillion rubles, which made it possible to actively develop a mechanism for allocating temporarily available budget funds based on repos.

It should be noted that the Bank of Russia conducts similar deals with securities from the Lombard list, among which there are both government papers and blue chips, which have a high reliability rating. Of course, if the deposit with repos by the Treasury is extended to other securities, this will increase the volume of these operations and, accordingly, the yield on them.

In 2017, the basket of securities was expanded and included securities for which coupon payments are accounted for during the term of the repo agreement. The system of accounting for repo operations has been improved. To secure the purchase (sale) of securities under repurchase agreements in terms of recording rights to securities, storing them, making settlements under repurchase agreements, the Federal Treasury attracts the National Settlement Depository CJSC, a non-bank credit institution.

In 2017, operations were with different due date: daily repos in the overnight mode, for up to 92 days. The amount of federal budget funds under repo agreements was 108,071.0 billion rubles in 2017, which is 96.8% more than in 2016, the demand from credit institutions amounted to 45,916.6 billion rubles, which is 4.7% more than in 2016 [6].

Thus, even under the conditions of the structural surplus of liquidity of the banking sector, the relevance of raising funds from the federal budget remains high.

The development of the functions of the Treasury of Russia when depositing budget funds with banks and when conducting repo transactions has a significant effect on increasing the efficiency of both the budget and the banking system of the country. Due to the inflow of budget funds into the banking sector, the need of credit institutions in refinancing the Bank of Russia is reduced. At the same time, budget funds are mainly deposited with large banks, which have begun to upsurge liquidity in the money market. Under these conditions, there is a decrease in spreads of one-day ruble interbank rates and repos to the key rate of the Bank of Russia.

Thus, the Treasury of Russia, using various methods and instruments to regulate the volume of funds from the federal budget on STA, also influences the conditions of state monetary policy.

The introduction of the mechanism for budgets of the subjects of the Russian Federation and local budgets is an essential reserve for expanding the capabilities of the Treasury of Russia managing funds on a single federal budget account.

The implementation of this mechanism allowed the subjects of the Russian Federation and municipalities to prevent temporary cash gaps, and taking into account the preferential interest rate (0.1%), to reduce the level of debt burden on regional and local budgets by replacing bank loans with budget loans. The term of loans in 2017 ranged from 10 to 50 days. In November 2017, amendments were made to the budget code of the Russian Federation in terms of extending the period for granting budget loans to 90 days. The total amount of federal budget funds allocated for the provision of budget loans to the subjects of the Russian Federation and municipalities amounted to 839.9 billion rubles. Revenues paid to the federal budget for this instrument amounted to 105.6 million rubles in 2017 [8].

The Treasury of Russia regularly presents information on the results of targeting balances of the federal budget on the official website. In addition, under the conditions of transition to the state integrated information system (SIIS) for managing public finances "Electronic budget" is formed in the subsystem - cash management, in the module - cash planning in order to forecast the balance on the accounts of the Federal Treasury, required to make payments.

## 4  Discussion

With regard to the monetary sphere, targeting is presented as the mode of state monetary policy, developed and conducted by central banks of various countries. Targeting, as the mode of monetary policy, means that central banks set targets for various monetary indicators over the forecast period. As the world experience of monetary and credit regulation of the economy shows, the list of such indicators includes: money supply (monetary targeting), exchange rate (currency targeting), and inflation (inflation targeting).

In domestic practice, targeting of the money supply by the M2 aggregate was used by the Central Bank of the Russian Federation in the early 1990s. In subsequent years up to the end of 2014, the Central Bank of the Russian Federation used the exchange rate indicator as a benchmark, setting the range of changes in the ruble value of the two-currency basket, the so-called "currency corridor", for the forecasted short-term period.

In recent years, monetary policy in our country has been pursued in the targeting mode of such an important indicator of the monetary and credit sphere, as the level of inflation. Let us dwell on the essence of this mode, its conditions, the instruments used by the Central Bank of the Russian Federation to conduct this mode.

Fouejieu, Ebeke, Moiseev consider inflation targeting as the mode of monetary policy based on "choosing a certain inflation value as a target and using the central bank's operating instruments, most often interest rates, to achieve an inflation target for which the authorities are responsible" [17, 7].

M. Malkina, disclosing the content of inflation targeting as the mode of monetary policy, notes that "the target that guides the central bank is the well-known and well-understood published rate of inflation, often its CPI or its equivalent ..." [5].

L. Tolstolesova also considers inflation targeting as the mode of monetary policy when central banks of countries set inflation targets, the achievement of which becomes mandatory for monetary regulators in the medium term [13].

Some authors, such as R. Ibarra, D. Trupkin, F. Kartaev, associate the content of this mode of monetary policy with its conditions, without which inflation targeting is impossible. These conditions include: public announcement of the medium-term inflation target by central banks; recognition of price stability as a priority target of monetary policy; accountability of the regulator to society for the commitments made to achieve the inflation target; constant public announcement on the decisions taken by the monetary authority [4, 16].

Other authors rightly consider these conditions as characteristics of inflation targeting, since they reflect the content of this mode and its elements.

Monetary policy of the Central Bank of the Russian Federation emphasizes that with inflation targeting "it has been established that the main target of the central bank is to ensure price stability. Within the framework of this mode, a quantitative inflation target is formulated and announced, the achievement of which is the responsibility of the central bank ..." [9].

Thus, inflation targeting, in our opinion, is the target of reducing inflation as the main and priority one over other targets of monetary policy, defining the rate of inflation for the forecast period as a specific value or range of values by central banks.

As regards the conditions and prerequisites for conducting the mode of monetary policy under consideration, the majority of authors point out the following:

- Stable, diversified economy;
- Independence of the central bank' activity when choosing the instruments of monetary regulation;
- National floating rate;
- Development of the financial market and the national banking system and others.

The term "targeting" for the first time as applied to public finance was used in the Order of the Ministry of Finance of Russia dated August 29, 2013 No. 227 "On Approval of the Concept for Reforming the System of Budget Payments for the Period up to 2017" in order to develop management instruments for free demand balances on a single treasury account, opened with the Bank of Russia.

Targeting demand balances on STA means managing the size of the daily balance on a single treasury account and ensuring its liquidity. Targeting allows planning the amount of excess cash and using it to generate additional yield by depositing free balances of budget funds with financial instruments.

## 5  Conclusions

Thus, targeting used in the monetary and credit spheres, as well as in public finances, is an instrument that influences certain parameters of monetary policy (interest rates, total banking liquidity, inflation), the functioning of the financial market (proposed resources, terms of provision, differentiation of financial instruments in the form of collateral - collateral and unsecured), certain parameters of financial policy (reducing the debt burden on budgets of the subjects of the Russian Federation, municipalities).

Additional instruments used in targeting balances of federal budget funds on STA include:

- Purchase and sale of foreign currency and conclusion of contracts, which are derivative financial instruments, the subject of which is foreign currency, in organized trading;
- Deposit of funds of the federal budget with banks on demand;
- Purchase (sale) of securities under repo agreements with a fixed and floating rate.

To improve the efficiency of monetary policy, it is more promising to stabilize cyclical changes in GDP, to target GDP that is less dependent on external shocks. It is necessary to build a mechanism for managing market expectations, as confidence in the future entails capital inflows and investment growth.

# References

1. Bozhechkova A, Kiyutsevskaya A, Knobel A, Trunin P (2018) Russian Economy in 2017. Trends and Prospects. Section 2. Monetary and Budgetary Spheres: 37–45. https://www.iep.ru/files/text/trends/2017/02.pdf
2. Zhantimirov PA (2018) Managing balances in a single federal budget account under liquidity surplus: theory and Practice of Social Development 6. https://doi.org/10.24158/tipor.2018.6
3. Kovaleva TM (2011) Financial and credit relations in the context of globalization: monograph. In: Kovaleva TM (ed) Publishing House of Samara State University of Economics, Samara p 22–42
4. Kartaev FS (2017) Is inflation targeting useful for economic growth? Econ Issues 2:63–65
5. Malkina MYu (2016) Inflationary processes and monetary regulation in Russia and abroad. M.: INFRA-M, pp 297–300
6. Materials of the extended board of the Ministry of Finance of the Russian Federation. Execution of the federal budget and budgets of the budget system of the Russian Federation for 2017 pp 109–117. https://www.minfin.ru/common/upload/library/2018/03/main/Ipolnenie_federalnogo_budzheta.pdf
7. Moiseev SR (2011) Monetary policy: theory and practice. M.: Moscow Academy of Finance and Industry, pp 232–234
8. The official website of the Treasury of Russia www.roskazna.ru. Accessed 3 May 2018
9. The main directions of the unified state monetary policy of the Russian Federation for 2019 and the period 2020 and 2021. Official site of the Central Bank of the Russian Federation, pp 96–100. http://www.cbr.ru/Content/Document/File/48125/on_2019(2020-2021).pdf
10. Prokofiev SE (2010) On the main tasks of the Federal Treasury for 2010 to improve the efficiency of the cash management of the federal budget financial resources. Speech by the Deputy Head of the Federal Treasury at the extended meeting of the Russian Treasury Board. www.roskazna.ru/upload/iblock/kollegiya-materialy/doc/10032010pr1.doc. Accessed 3 May 2018
11. Savinov OG, Savinova NG (2016) Adaptation of bank lending to modern economic conditions. Bulletin of Samara State University of Economics, pp 88–91. http://elibrary.ru/item.asp?id=26584149. Accessed 3 May 2018
12. Tershukova MB (2018) Independence of the Central Bank of the Russian Federation as a condition for inflation targeting. Interuniversity collection of scientific papers: Problems of improving the management of industrial enterprises, 347–351
13. Tolstolesova LA (2015) Inflation targeting: advantages and limitations of use. Int Res J 11 (42) part 1:116–118
14. Eskindarov MA, Maslennikov VV, Abramova MA, Lavrushin OI, Goncharenko LI, Solyannikova SP, Morkovkin DE, Abdikeev NM (2017) Strategy of STA 2018–2024: claim, myths and reality (position of experts of Financial University). Finance: Theory and Practice 21(3):6–24. https://doi.org/10.26794/2587-5671-2017-21-3-6-24

15. Gadot G, Dalmia (2014) Liquidity management: best practices for banks. Improving Cash Forecasting to Enhance Liquidity Management, Information week: Bank systems & Technology: October 06. Reference date is 04 June 2018. https://www.banktech.com/compliance/liquidity-management-best-practices-for-banks/a/d-id/1297050d41d.html. Accessed 3 May 2018
16. Ibarra R, Trupkin DR (2016) Reexamining the relationship between inflation and growth: do institutions matter in developing countries? Econ Model 52(Part B):332–351
17. Fouejieu A, Ebeke (2015) Inflation targeting and exchange rate modes in emerging markets. IMF Working Paper, p 228
18. Mavroeidis S, Plagborg-Møller M, Stock JH (2014) Empirical evidence on inflation expectations in the new Keynesian Phillips curve. J Econ Literat 52:124–188
19. Demertzis M, Marcellino M, Viegi N (2009) Anchors for inflation expectations. European University Institute Economics Working Paper: 10. https://ideas.repec.org/p/dnb/dnbwpp/229.html. Accessed 12 May 2018
20. Rey H (2015) Dilemma not trilemma: The global financial cycle and monetary policy independence. NBER Working Paper: 21162 (May 2015). https://www.nber.org/papers/w21162. Accessed 12 May 2018

# Digital Economy – Information Era: Retrospective Analysis

V. S. Grodskiy$^{(\boxtimes)}$ and G. R. Khasaev

Samara State University of Economics, Samara, Russia
omega2017@bk.ru, gr.khas@mail.ru

**Abstract.** The economy of the post-industrial society is interpreted as the unity of information technologies and global production institutions. The method-ological and instrumental study of the economy, which is an open non-equilibrium self-organizing system, is shown. One of the newest trends in science - economic kinetics, or meta-economics, exploring the laws of world evolution are considered. The contribution presents a critical analysis of existing approaches to the selection of historical types of business, demonstrates a new model of eco-nomic formations and inter-formational transitions, substantiates the introduction of wagesability in the digital economy, which ensures the priority development of human capital.

**Keywords:** Capital profitability · Digital economy · Digitalization ·
Economic formation · Meta-economics · Production technology ·
Sociology of production · Wagesability

## 1 Introduction

The expression "digital economy" is a metaphor that characterizes a certain quality of a post-industrial society, or more precisely, a system of information technologies and institutions adequate to them. Information technology is a new reality on a global scale. As for institutions, scientists as a whole and especially social scientists will have to work hard to create new rules and institutions necessary for the full functioning of the existing and expected information technologies. Economic theory (economics), which has succeeded in the study of the modern market economy, should play a significant role in this matter. But it is still unable to adequately describe both the historical past and the long-term future of social production.

American economist Paul Samuelson (1915–2009), the author of the famous textbook called "Economics: Introduction to Analysis" (1948), noted that from the point of view of the "mainstream economic science" [1], "an analytical economic theory" [1], there are six stages of its development: (1) the study of the static equi-librium of the economy; (2) the achievement of comparative economic statics; (3) the theory of maximizing behavior of the economic entity; (4) the introduction of the

S. Ashmarina et al. (Eds.): *Digital Transformation of the Economy: Challenges,*
*Trends and New Opportunities*, pp. 163–179, 2019.
https://doi.org/10.1007/978-3-030-11367-4_16

"principle of conformity" of comparative statics and dynamics into the scientific circulation; (5) the theory of economic dynamics; (6) the theory of "comparative economic dynamics" [1]. According to the author, at the sixth stage of the theory development, "it should include" ... all five subjects considered above, but it should also cover a wider "territory" [1]. This stage, which can be conventionally called "economic kinetics", is not "economic history", but the science of the laws and dimensions of the development of human civilization. It is the science of interconnected non-equilibrium processes of self-organization in society, using the synergy toolkit. It is a relatively new economic synergetic that will allow us to satisfactorily describe the trend of economic development and those institutional innovations that are necessary for our current economic practice.

This contribution does not attempt to comprehensively characterize the "digital economy" and institutions of the information community, but attempts to retrospectively analyze the historical development of social production and create a new model of its formation, which, in our opinion, will facilitate the solution of the necessary tasks: institutional transformations.

## 2 Materials and Methods

The authors used the following standard and innovative methods of scientific research: (1) the formal logic on the formation of concepts, judgments, conclusions, definitions and cause-effect relationships of economic phenomena; (2) dialectics, in the form of the laws of the unity and struggle of opposites, the transition of quantity into quality and back, the negation of negation; (3) mathematics and a graphic method to illustrate the course and results of research; (4) synergetic methods that allow describing non-equilibrium processes of self-organization in the economy; (5) forecasting necessary for the strategy development of transformations in economic theory and practice.

For the first time the authors used the "historical-logical" tool to represent a new, generalized development model of world economic thought, which allows revealing its cross-cutting directions and the essence of the modern "dispute between two Cambridge's", to show the possibilities of its completion. In addition, the authors widely uses the method of comparative (comparative) analysis, which makes it possible to critically approach existing concepts of economic theory and design new models in it.

## 3 Results

The ideas embodied in the formation model of Marx, currently attract attention of economists, but need some development and improvement, supplementing with institutional elements. It is also necessary to revise the system of generalized economic coordinates and indicators introduced by Marx.

The just Marxian thesis that each historical level of development of economic activity has special, own production relations and does not exclude, on the contrary, implies the need to develop universal initial concepts and categories, the use of which makes it possible to trace the natural-social continuum, the continuity of the transition from the "economy nature" to the "nature of the economy" was designated by Adam Smith (1723–1790).

The fundamental characteristics of the economy correspond to its most stable, deep structure. The morphology (structure) of the economy in a generalized form was also considered by the theorists of the period of classical political economy - Francois Quesnay (1694–1774), James Lauderdale (1759–1839) and David Ricardo (1772–1823). The greatest contribution to solving the issue of generalized structural economic indicators was made by Marx. Although Marx considered the most important definition of the type of society, relations between its classes, property and the organization of labor, he did not exclude the possibility of studying the "production process in general", in its form, "in which it is characteristic of all social structures, that is, the process of production outside of its historical character, a universal human process" [2].

The analysis of the "production process in general" indicates its duality, its existence as a labor and cooperation of a person at the same time. "Production … as a natural (labor) … [and] social relation, public in the sense that it refers to the cooperation of many individuals, it does not matter under what conditions, in what way and for what goals" [2]. "Labor is, first of all, a process between man and nature, a process in which a man mediates, regulates and controls the exchange of substances between himself and nature …" [2]. But in production, "people enter relationships not only with nature. They cannot produce without joint activities and they do it for the mutual exchange of their activities" [2].

Relationships of communication can be called "relations of cooperation" in a general, essential sense. Historically, these relations may be far from cooperation and even oppose it, for example, they take the form of rivalry in market conditions. Historical, concrete modifications of production communication cannot completely suppress the original and permanent cooperativity, interdependence and assistance of producers in it. It is the general nature of production in contrast to the individuality of labor that is taken into account by the category of "cooperation relationship". The combination of "labor" and "collaboration" can be called "human life activity".

Production relations cannot exist and develop on their own. In their pure form, they are always mediated and materialized. "Political economy deals with relations between people … But these relations are always connected with things and they are manifested as things" [2]. Therefore, the interaction of people with the surrounding nature (production relations of labor) is materialized in the means of labor, and the interaction of people among themselves (production relations of cooperation) is materialized in the means of communication, in social institutions, that is, in the "economic mechanism" in the broad sense of the concept, covering not only economic, but also all other institutions of society. If the instrumentalization of labor mediates the interaction of a person with an external object, then the institutionalization of cooperation organizes, regulates and ensures sustainability of human life activity.

The means of labor and cooperation are the "means of production" in the broad sense of this concept, which characterize the level of development of all production relations. People are usually declared "carriers of production relations", but in reality the means of production are the material substratum of relations. Man is the subject of production relations.

The analysis of the duality of production relations and the means of their realization inevitably affects the concepts of "productive forces", "basis", "superstructure" and "socio-economic formation", which Marx considered in detail.

First of all, we note that with the expansion of the concept of "basis" with the inclusion of production relations of labor, there is a "superstructure" concept, since relations of labor have their own "superstructure", their material form by means of labor. We call these tools of labor and the labor process as "production technology" and consider this concept as the first universal, generalized coordinate of meta-economics.

The most concentrated approach of Marx is expressed in the concept of "basis" as a combination of relatively stable production relations of society, relations representing its structure. However, the "over-building" of society, as we noted above, does not contain "basic" institutions. It turns out that the institutional arrangement of production relations, according to Marx, is outside the economy and society, and the corresponding concepts are outside the subject of economic theory. In reality, the economic superstructure exists in the form of an "economic mechanism" as a material substrate, means of implementing industrial relations of cooperation developed in the course of economic practice, that is, a system of informal customs, principles and norms, rules that regulate and ensure sustainability of economic life of society. The institutional mechanism organizes the existing system of production cooperation relations.

Since material production is the main sphere of human life, the real relationship of interdependence that develops between people in the course of their activity underlies other political, legal, religious, moral and other social relations. The Russian Marxist Vladimir Lenin (1870–1924), in our opinion, rightly pointed out that "by structuring society, we separate production relations from different areas of social relations as basic, initial, defining all other relations" [3]. Therefore, the economic part of the "superstructure" is its main component, which determines all other social orders and formations. This means that, for example, "the political system … is only an official expression of civil society" [2] that politics as a "concentrated expression of the economy" [3] "always reflects, records the requirements of economic relations" [2]. The economic essence can be found in other social institutions. One can even say that non-economic social institutions are a kind of infrastructure of the economic mechanism. This infrastructure completes and fully implements the structure of society. That is, the provision on the primacy of economy means that non-economic institutional formations are derived from economic relations and are necessary for their existence and reproduction. At the same time, we can call all existing institutions of society as an "extended economic mechanism."

From the momentary complex system of modern social institutions, in the historical aspect, its various elements acted as an economic mechanism, the sequence of which was the following: institutions of kinship, political, religious, market, and in the future ethical institutions will appear. That is, the means of industrial cooperation relations were alternately taboos and customs, the state, the church, the establishment, and in the future self-organization would be used. This circumstance makes it possible to consider the entire complex of social relations and institutions as a single economic mechanism. In such a broad form, we call it the "sociology of production". This concept as the second generalized coordinate of economic theory is opposed to the coordinate "production technology". "Social relations are also produced by people, like canvas, flax, etc." [2], as well as institutions of society. Therefore, sociology of production and production technology are integrated systems and closely interrelated entities.

Thus, the economic structure of society can be represented with the help of the dualism "technology-sociology of production", which can replace the triad "productive forces – basis – superstructure" of society and become a more productive tool for understanding the economy. If the triad of Marx was traditionally considered to be the subject of so-called "historical materialism", then the proposed generalized coordinates should, in our opinion, be included in the subject of economic theory in its broadest understanding, which we called meta-economics.

Above, we gave a qualitative description of the technology and sociology of production, which is universal, generalized, pervasive history of social production, coordinates of meta-economics. But these coordinates require their quantitative measures, which would allow carrying out a comparative analysis of historical forms of the economy, to study the laws of society development. In this aspect, economics cannot differ from natural sciences, which have had generalized indicators. The development of general historical economic indicators is a very complex task, but necessary for the current stage of its development, characterized by the transition from the stage of analytical differentiation and accumulation of knowledge to the stage of its integration and synthesis.

To introduce the desired indicators for coordinates "production technology" and "sociology of production", we will use the methodology for assessing the levels of development of production relations of Marx, who used such indicators as "organic composition of capital" and "rate of surplus value". To give these indicators a general historical character, we express the structure of the total product of society in the following form: $Q = C + V + M$, where $C$ is the natural rent, which contains the bulk of material costs of production, $V$ is the cost of labor in monetary terms, and $M$ is the cumulative profit in the form of the so-called "cash flow", which includes depreciation deductions and net profit. We call the $C/V$ ratio "the level of development of production technology" ($P$), and the $M/V$ ratio—"the level of development of sociology of production" ($S$). Both figures historically increase in general, the first - as the replacement of labor with institutionalized ones, and the second - due to the increase in the number of capital goods in the economy, their depreciation and net profit per unit of labor.

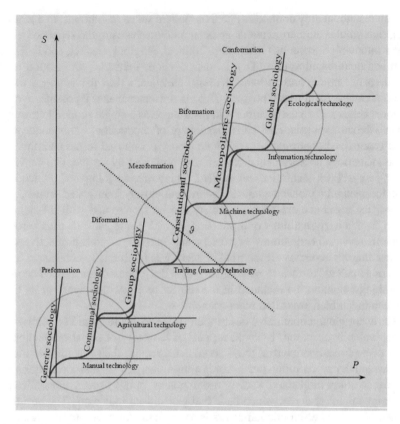

**Fig. 1.** The historical development of social production in the coordinates "Production technology (*P*) - Sociology of production (*S*)" (Source: compiled by the authors)

Above, we have shown that labor is a production relation. However, the relationship of communication is a combined "labor – cooperation", "that is, a multiply productive force arising due to the division of labor caused by the joint activity of various individuals ..." [2]. Profit is an emergent (systemic) effect of the division and cooperation of labor. In other words, it can be said that production technology creates the value of a product, and sociology of production redistributes it in such a way that a profit of cooperation is formed. Technological economy is realized only through "political economy", since "combined activity means organization" [2].

As a consequence of the exchange between nature and society, there is a growth and improvement of the means of labor, which are becoming more and more adapted to man. And they are an integral part of his workforce. But the development of the "technological body" of production is not the only result of strengthening the connection between society and nature. Man, "influencing the external nature and changing it, at the same time changes his own nature. He develops the dormant forces in it and subordinates these forces to his own power" [2]. The same can be said about the "sociological body" of man: the development of the institutionalized cooperation is impossible without the development of personality. Thus, in the general historical aspect as a whole, the harmony of the evolution of objective and subjective components of productive forces is realized.

Proportional-disproportionate super-wave development of human civilization in the coordinates of "$P - S$", presented in Fig. 1 corresponds to Smith's theory of the natural-social continuum (continuity) and Marx's discrete general development theory, since the model reflects the possibility of rapid technological changes only with a certain stability of sociology of production and vice versa. In addition, the model synthesizes the concept of society development with impulse from Marxian technology with the concept of society development with impulse from sociology of production of Max Weber (1864–1920).

For a relatively developed economy, the alternate causal relationship between technologies and sociologies of production is obvious, while for the stage of the emergence of society and economic activity, the issue of first causes becomes important, fundamental. The solution of this issue influences the periodicity of economic development, the economic unity of the history of society, to which many theoretical economists have sought, the explanation of many complex historical phenomena of the economic life of society, as well as the long-term forecast of changes in production.

"History can be viewed from two sides. It can be divided into the history of nature and the history of people. However, they are connected with each other" [2]. In anthropology, as a rule, the "human consciousness" is taken for the fundamental factor of antropogenesis. However, homo sapiens are derived from homo faber: "The first historical activity of … individuals, due to which they differ from animals, is not that they begin to think, but that they begin to produce the means of life they need" [2]; "Consciousness … from the very beginning is a social product and remains with it, as long as people exist at all" [2]. The well-known position of the German historian and philosopher Friedrich Engels (1820–1895) that "labor created man" cannot be considered accurate, since the productivity of "labor" from "cooperation", the productivity of primitive production technology from its sociology, techno genesis from socio-genesis was shown. The "intellectual" interpretation of anthropogenesis, like the labor theory of the origin of mankind, does not solve the problem. There are still some questions: "Why does an individual become a rational being and why does an individual begin to work, produce material goods?" Therefore, only "ecological" concept of the emergence of society deserves attention. The economic literature presents the cooperation and instrumentation of the ancient man' activity. In reality, anthropogenesis requires some explanation from economic theory. And it also has an economic essence, which, in our opinion, can be cognized through meta-economics.

First, anthropogenesis can be explained only from the standpoint of the natural-social continuum, and, first of all, it should be divided into two parts – socio and technogenesis. The nature of the interaction between the natural environment and proto-society is described by Henri Le Chatelier (1850–1936) using the universal "Le Chatelier principle", which brings the system out of balance causing processes in it that tend to weaken this effect. The sharp deterioration of the habitat associated with climatic cooling due to the next geological glacial period on the planet about a million years ago, led to such an internal restructuring in proto-society, which largely compensated for the external catastrophic adverse environmental impact. The change in the internal structure of the system at the same time could be reduced only to a closer contact of individuals, to their cooperation. Rupture of subjects with a familiar object—with a useful nature for them—meant the establishment of additional inter-subject relationships. This cooperation of individuals resulted in the primitive "sociology of vital activity" in the form of a set of specific rules, norms, taboos, traditions, language, and the mythological consciousness of people. Thus, it is sociogenesis that is primary in history. Figure 1 shows the sociotechnological cycle with the initial deviation in the direction of the "sociology of production", $S$. It is important to bear in mind that such a beginning of history determines the entire subsequent cyclical evolution of society and the economy. The trend equation of economic development $S = P$ in the second approximation of our model to reality, taking into account the frequency of society motion, can be written in the following form: $S = P + \varepsilon \cdot sin\sigma$, where $\varepsilon$ is an indicator of the amplitude of fluctuations, $\sigma$ – is a frequency indicator of fluctuations.

Secondly, since the cooperation of individuals always has a reverse side - the specialization of their activities, the demographic expansion of the primary cooperative meant the deepening of the specialization of individuals, and this, in turn, allowed them to use external tools for a more successful fulfillment of certain roles in the cooperation. This significant event of technogenesis came only after many millennia after socio-genesis. The first external technological reaction of society takes place at the stage of technogenesis in the form of a weapon transformation of the natural habitat.

Thirdly, the original diachronic sociotechnogenesis is preserved throughout the subsequent development of the economy and is expressed in the alternation of sig-nificant historical technologies and sociology of production. In fact, the whole history of society is a staged and cyclical process of restoring the natural balance through economic activity. Adaptation super-waves spread from the primitive "oecumene" to the future developed highly-organized "noosphere". Since anthropogenesis is associ-ated with a significant rarity of life benefits, the economy is a system of survival and the development of society, the economic theory solves the problem of overcoming this rarity, ensuring the growth of human welfare.

And fourthly, from the point of view of the primitive-social continuum, the occurrence of people occurred at a certain level of their natural herd cooperation, the initial phase of the development of society is somewhat beyond the limits of the coordinate system. The corresponding equation for the development of civilization $S = f(P)$ in the third approximation of the model will look like this: $S = P + \varepsilon \cdot sin(k \cdot \sigma)$, where $k$ is the proportionality coefficient characterizing the initial phase of the mega-cycle. The horizontal sections of the model, which follow the ascending on the abscissa "$P$", correspond to the historical sequence of significant production technologies - manual, agricultural, product, machine, informational and ecological; the areas that characterize the advancement on the ordinate "$S$" and represent, respectively, the clan, communal, caste, constitutional, monopolistic and global sociology of production.

The change of technologies and sociologies of social production takes place in such a way that each of them is not destroyed, and in a transformed form is included in the economic mechanism of a more developed society. That is, there is a process of "removing" the simple in the complex as cyclical deployment of the initial syncretic state of society takes place. Transitions to higher levels of technology and sociology of production also mean achieving higher levels of production efficiency.

Specific historical forms of business management are the unity of significant technologies and sociologies of production. In the global wave process, it is possible to distinguish "Preformation", "Difformation", "Mesoformation", "Biopformation" and "Conformation", indicated in Fig. 1 by circles, covering specific pairs of "technology-sociology". Each socio-economic form is institutionally defined, has its own distinctive sociology of production, which forms its basic technology, but this sociology itself produces a new technology. Thus, all six historical technologies are links of formations and are located in transitional zones (in Fig. 1, these zones are located in the overlap of adjacent "formations" circles).

"Preformation" refers to the first historical cycle of production, the cycle of its formation, in which the asymmetry of anthropogenesis is manifested and the whole "program" of the further development of society (the phenomenon of preformism most studied in the developmental biology of a living being) is contained. "Preformation" includes two production technologies—manual and agrarian—and two sociologies—generic and community-based.

"Diffraction" means duplication of the second stage of society, its formation on the basis of extensive and intensive agrotechnologies implemented in the history of different countries simultaneously (see Fig. 1 for the first bifurcation, that is, the ramification of the development pathway of society). For example, Ancient Rome as an empire developed by territorial expansion and an increase in production mainly due to the seizure of land and slavery, while neighboring tribes could develop only intensively. Such effective farming and subsistence farming are established in "Diformation" everywhere and institutionalized in the form of "estate sociology". "Diformation" includes "slavery", "feudalism" and the so-called "Asian mode of production" - the beginning of agrotechnology in Mesopotamia (Sumerians), Egypt of the 4th millennium, Egead (West Asia Minor, the Indus river valleys, the Yellow River millennium BC) and other regional modifications of early owner-owning society, since for all the twenty centuries of history covered by this cycle, subsistence farming is typical. The end of the Roman political empire was the beginning of the Christian-Catholic church empire, since the weakening of the state required the strengthening of the church to continue the development of the natural economy. "Diformation" seems to be a single one, which absorbs the Greek-Latin and medieval world.

"Mesoformation" represents an early simple market economy and occupies an intermediate, middle position in history. Civil society is formed in "Mesoformation" as the social basis for the further progress of civilization. Its "commodity technology" means a process of social production that continues beyond individual specialized farms that do not create final products or the entire set of necessary goods and therefore turn out to be interconnected by exchange market relations. The emergence of commodity exchange technology and the formation of "commercial society" in history occurred much earlier, than is commonly believed, from about the end of the XIII century. The name "constitutional sociology" denotes a system of institutes for a market economy that ensures individualism, the rights of an individual and private property, the freedom of entrepreneurship and competition. At the heart of this system of institutions were relations established on the "Protestant ethics" of reformation, which freed people from church prohibitions on the "monetary technology of interest" and proclaimed hard work, rationality and thrift.

The concept of "Biformation" means a forked formation, which is formed on the basis of machine technology. Corporate monopoly sociology in the West, which has shaped and reinforced sectoral control by the largest producers over the past 150 years, is supplemented in Central Asia with the centralist monopolistic sociology of countries with a planned economy (see the second vertical bifurcation [4] in the historical rice cycle Fig. 2). In this regard, the French sociologist Raymond Aron (1905–1983) wrote: "I don't wonder if there is a contradiction between socialism and capitalism, I consider capitalism and socialism as two types of the same industrial society" [5]. And it is possible to agree with this characteristic, adding only that, despite the differences in the institutionalization of sectoral and national economic monopolies, both varieties can be designated by the concept of "social economy". The structure of "Biformation" is such that there is the initial technology common to the market and planned economies, and also a common way for them to further develop production along with its informatization. The transformation of scientific information into the main means of labor and cooperation causes the "removal" of the laws of value and commodity-money relations in general in the global information network, since the exchange of information, unlike the usual commodity exchange, does not lead to the alienation of this universal good. It can be said that information is the only inexhaustible, unrelenting resource.

The fifth, largely not manifested, but predicted, cycle of social production falls on the completion phase of human adaptation to the surrounding nature due to the global development of information and then ecological technologies. "Ecological technology" is the mass distribution of environmentally friendly and waste-free industries that provide high quality habitat for people, and the term "global sociology" or "megasociology" characterizes unified, highly mobile and quickly-restructured institutions of the world community, which correspond to decentralized and demonopolized, adapted to each other economies of the world.

In the literature, there are many names for the new formation. In addition to "communism" of Marx, there are other names of the future society, given by economists and sociologists of different directions and countries: "superindustrial" by American philosopher Alvin Toffler (1928–2016) [6], "post-industrial society" by American sociologist Daniel Bell (1919–2011) [7], "advanced industrial" by German philosopher Herbert Marcuse (1898–1979) [8], "new industrial" by American economist John Galbraith (1908–2006) [9], "post-bourgeois" by American historian George Lichtheim (1912–1973) [10], "post-capitalist" by German philosopher Ralph Dahrendorf (1929–2009), "posteconomic" by Russian economist Vladislav Inozemtsev (born 1968) [11], "post-civilized" by American economist Kenneth Boulding (1910–1993) [12], "active" by American sociologist Emit Etzioni (1929) [13],"technotronic" by American politician Zbigniew Brzezinski (1928–2017) [14], "informational" by Japanese anthropologist Tadao Umesao (1920–2010) and deployed by his compatriot, philosopher Yoneji Masuda (1905–1995) [15].

Names with the prefix "post-" and including "post-industrial society" we consider not appropriate in view of their non-specificity, the name "information society" is the most adequate to the object studied, but it is more suitable for characterizing society itself, and technology applied in economics. Taking into account the unification in the names of formations used by us, as well as the above-mentioned natural-social adaptation essence of the fifth formation, we call it the "Conformation".

Technological duality of antiquity and sociological duality of the XX century are interconnected. Countries that underwent a slave-owning civilization in their development created a mechanism of coercion to work, institutes of public authority and democratic traditions, and then in a monopolized industrial economy they passed the state-centralized system of planned economic management. Conversely, countries and regions that did not go through slavery later became so-called "socialist" and, under certain other circumstances, inevitably develop in this totalitarian form of industrial society. So, contrary to the theory of Marx, there is no need to transit to the economy of a general state monopoly for all countries. At the same time, such transitions can be considered natural, and not just "experiments of leaders" or manifestations of other subjective factors of history.

## 4 Discussion

Formational analysis of society shows that the trend of the profitability cycle, as well as the entire history of social production, is the absolute constant of nature in its broadest sense. It includes society, the so-called "golden proportion" of nature, which operates in all its specific formations of chemical, biological, social origin and establishes their balance and harmony. Most simply, the "golden proportion" is determined by geometrically dividing an ordinary straight line segment into two parts. If its length is denoted per unit, its smaller part is $x$, and the largest one is $1 - x$, then the harmony of the segment will be observed when it is cut in proportion: the ratio of the small part to the large one is the ratio of the latter to the whole segment: $x/(1 - x) = (1 - x)/1$. We find $x$ in the form of a quadratic equation $x = 1-2x + x^2$, the solution of which gives us the numerical value of the "golden proportion": $x = 0.382$. We denote this universal harmony constant as $G_g$. In relation to society, $G_g$ characterizes the harmony of production profitability ($G_p$) and net profit in general in any mode of production: $G = (A - B)/B$, where $A$ and $B$ are specific indicators of results and costs, respectively.

Figure 2 presents the general historical and perspective dynamics of indicator $G$. The trajectory is wave damped in nature and reflects the gradual compensation of the ecological catastrophe—a sharp cooling in the immediate "ice age" that led to the appearance of human community in the biota of the Earth. In the long term, its production activity comes to stabilization of a broken "golden proportion" during the final formation of a coherent system of "nature - society". Separate trajectory cycles characterize the historical sequence of economic criteria - rentability, profitability, wagesability and ecoligibility. The design of these four criteria is the same and is such that all three elements of the cost of produced product, that is, $C$, $V$ and $M$, become the numerator of total production profitability $G$ as residual target result and they are the denominator of fraction costs. So, since the "golden proportion" began to decline, then the criterion of profitability of economic activity ($G_p$) introduced natural rent $C$ ($G_P = C/(V + M)$), which allowed people not only to successfully overcome the adverse effects of ecological disaster, but also improve their well-being. Growth $C$ also made it possible to reproduce land as the main factor of production and develop agriculture. In the conditions of the industrial economy, the first historical criterion of production was already ineffective and was replaced by profitability ($G_p$, $G_p = M/(C + V)$). And in this design, the priority of profit was imposed by the task of growth and development of capital, that is, by machine technology itself. The *Gp* dynamics is cyclic, but the industrial cycle (in Fig. 2, this is the second wave highlighted in blue and red) was less in amplitude than the agrarian cycle and thus significantly brought the society closer to the absolute harmony of indicator $G_g$.

With the completed formation of industrial capital and the transition to a co-rational market economy, wagesability *V* becomes the residual income, and profit *M* must be attributed to costs. The corresponding criterion of wagesability ($G_3$, $G_3 = V/(C + M)$) – is not the result of a pure theory, it is predetermined by the whole course of historical development of social production and is adequate to modern information technology, the task of development "human capital", education and innovation economy. It is important to note that the new criterion is intended to achieve not only economic, but also social goals, since for the first time in history it is directly related to meet person's complex needs and requirements. In essence, this is about "real socialism", and not about some "formal" socialism that existed, where people were still the tool for centralized industrialization of economies. The economic calculation that was used in the planned economy did not basically differ from the usual market commercial calculation. Here and there, profit has become paramount. It is primarily interested in the owner of capital goods, whether it is a private entrepreneur or a centralized state apparatus. The orientation on wagesability should sharply increase its overall efficiency and directly the welfare of workers.

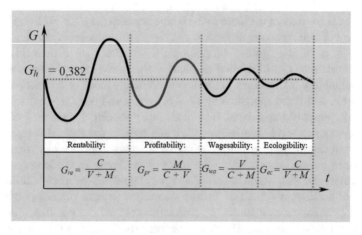

**Fig. 2.** The general historical process of economy harmonization by changing production criteria (Source: compiled by the authors)

The objective need for a full transition to the criterion of wagesability in modern conditions is indicated by the global decline in production profitability (see its second downward phase in Fig. 2) and in the modern market it already exists an alteration of the criterion of production profitability by expanding the so-called system of participation in the property and profits of corporations. Wagesability as a criterion and universal tool for solving social problems does not depend on modern forms of ownership, which under the conditions of information technology is subject to dissipation (dispersion) in society. Therefore, a new criterion was born in the period of the crisis of the planned economy in our country due to economic practice of the so-called "team contract". It was practically implemented in the 70s and 80s in the construction by Nikolay Travkin, in health care by surgeon Svyatoslav Fedorov and other industries and showed an explosive positive effect, but then it was subjected to tough attacks precisely from the bureaucratic system and was curtailed in the course of further profitable market reforms in the country. It seems that the introduction of wagesability, even in the conditions of preservation of national resources would bring much more benefit to society than the protracted transition to the classical market by privatizing state property.

The transition to a new criterion of production is associated with the so-called financial and economic crisis, in the conditions of which the world community has been living since 2008. Experts, politicians and many ordinary people consider it as an ordinary cyclical phenomenon, which is regularly observed in the market economy. Some experts note the seriousness of the situation and compare the current crisis with the Great Depression that swept the Western world in the 1920s–1930s. Despite the fact that since the emergence of negative processes in the financial sector, and then in real sectors of economies, a lot of time has passed and there have been different summits, conferences and even various systems of measures to overcome the crisis and to prevent its repetition. But we still have no surprising answers to questions about the causes, nature, duration of the crisis, as well as the nature of the global economy in the post-crisis perspective. There is a widespread version of mistakes made by financiers, the so-called "over-lending" and artificial "overheating" of stock markets. However, the explanation of the crisis by purely market, opportunistic causes cannot be considered sufficient. It is known that the financial sphere first of all reacts to socio-political events that are taking place, but it is also quickly stabilized with appropriate measures by economic entities and the state. This time, after the "financial bubbles" began to burst, the credit system did not stabilize. It indicates the presence of sustainable capital over-accumulation in relation to decreasing growth rates of production and its level of profitability in industrialized countries. And this phenomenon turns out to be completely unlike what happened during the Great Depression of the market: then, as you know, the problem was in the insufficient investment activity of the business.

Each of the three specified historical criteria provided the preferential growth and development of the corresponding factor of production, although it made the economy homogeneous. It also introduced resource disharmony into it, which is eliminated only with the complete attenuation of the non-equilibrium process of inter-formational transitions. Figure 3 shows the proportions of the historical joint dynamics of the agrarian $(N)$, industrial $(K)$ and information $(L)$ factors of production. The growth line of industrial capital goods is associated with the dynamics (dash line) of financial capital $(K_{fi})$, which, although it has its own market, all the time adjusts to the volume of capital goods. In the conditions of modern stabilization of the world production, there is a separation from it by inertia, that is, in the mode of monetary "pyramid", financial "fictitious" capital. Restoration of the financial sector will occur if it starts to invest in growth not in the industry, but in the service sector of the "human capital", which is still relatively undeveloped. But the growth of investment in the sphere of education and the formation of modern labor in the conditions of preserving private property and the market require using wagesability that allows you to create a powerful foundation for real socialism.

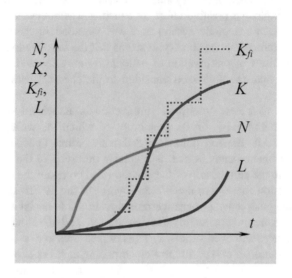

**Fig. 3.** Current crisis of capital profitability and the move to a new production criterion (Source: developed by the authors)

In the long-term historical perspective, after solving the basic social problems of modernity due to wagesability and overcoming the inevitable problems of nature management, society will have to put into operation another, fourth, production criterion - ecologibility $(G_e, G_e = C/(V + M))$, which outwardly coincides with the historically initial criterion of rentability by its design, but at a new level of economic culture of society will allow forming "the profit of nature" $C$ and thereby definitively stabilize the overall index of human civilization efficiency $G$, adjusting it to the "golden proportion" $G_g$.

## 5   Conclusions

Not only information technology should correspond to the digital economy, but also global sociology, which is a universal system of social institutions that defines the effect of wagesability. Wagesability does not depend on the form of ownership of the material production factors, since the priority development of human capital turns rentability and profitability into social costs. In accordance with this fundamental conclusion, the discussion of planned housekeeping, the height of which was in the 20–30s of the last century, when the world still had two management systems - market and planned and there were not only ideological, but also scientific disputes between their supporters and opponents. The controversy took place everywhere, and in the Western economics too. From the past we know about the "theorem on the impossibility of socialism" by Mises, which arose under the conditions of the real functioning of centralist systems. And we know about the idea of the "intellectual delusion of socialists" by Hayek, which can be reduced to three postulates, informational properties: (1) any compulsory organization of society, including a planned economy, does not provide reliable information necessary for the coordination of production activities - carrying out economic calculation; (2) the central planning authority cannot technically collect, retrieve and use the entire array of socio-economic information; (3) the planned information is static in nature, whereas the economy is a dynamic system.

Some Western economists, observing the initial successes of centralist organized economies, tried to challenge these postulates of the "Austrians" and oppose them with their theoretical arguments in defense of socialism. These were, first of all, English economists Henry Dickinson, Frank Durbin and Polish economist Oscar Lange in collaboration with M. Breit, who introduced into the analysis the so-called "parametric prices", similar those that are used by counterparties of markets as "price beneficiaries" in the conditions of their perfect competition. In economic theory, there was even a section called planometry, and Lange, obsessed with the desire to realize his idea of calculating market prices without a market, after a long successful creative stay in England and the United States and being strongly influenced by the ideas of Western economists such as Joseph Schumpeter, Paul Souisi and Vasily Leontie, returned to Poland. In 1955 he headed the State Economic Council to introduce a real policy-making entity.

However, attempts to theoretically substantiate centralized planning undertaken by the abovementioned and other economists, representatives of the Optimal Functioning System of the Economy in the USSR and similar scientific schools in other countries, were unsuccessful. Crisis events in the socialist countries of the late XX century, which led to the restoration of markets in them, allowed many economists, both those who deny the possibility of economic calculation, and those who support the global plan experiment of the twentieth century, to talk about the final fiasco of socialism. So Huerta de Soto without sufficient grounds and even references to the studies of the Italian economist, non-rationalist Piero Sraffa (1898–1983) (substantiating the reality of macroeconomic "production price" by Marx ($Q = B + G \cdot K$, where $Q$ is the cumulative output, $B$ is the cumulative production cost, $G$ is the average production profitability ("interest on capital"), $K$ is the cumulative mass of capital goods), called the creation of market socialism as "squaring the circle" [16], that is, a fundamentally unsolvable task. American economist Robert Heilbroner (1919–2005) wrote that Mises was right that socialism was a great tragedy. But, in our opinion, it is too early to close both theoretical and practical disputes about the plan and the market. Moreover, the issue of directive planning is becoming highly relevant on a global scale.

It is important to understand that the Austrian school is not the only one in modern economic theory in general, and the theory of socialism in particular. There are different concepts of socialism that are not to be verified by the three Mises-Hayek postulates on the impossibility of economic calculation presented above. An equilibrium market and non-market economic theory, rejected by the Austrian school, can develop the former and create new concepts of planometrics on its own principles.

Analysis and proposals for the development of basic ideas of economic theory [17] show, in contrast to the tenets of "Austrian anti-socialism", that: (1) the plan does not require primary comprehensive microeconomic information, but only one macro-indicator ("national dividend", $G$), optimally combining extensive opportunities for the development of the economy; (2) this indicator does not express the diversity of interests of the entire mass of economic entities, but, on the contrary, their uniform and universal interests, employment and growth of well-being; (3) this indicator does not reflect the statics, but the dynamics of the economy.

How can these new theoretical concepts be used constructively?

Unfortunately, Lange, while developing his own economic accounting model, which only imitated market balances, considered that state producers of goods should maximize profits as in the case of commercial calculations. Sraffa, going further in his creative work to Lange, also could not completely get rid of market ideas when creating the macro-model of the calculated "price of production" and suggested equalization in all sectors of capital profitability. Both representations, in our opinion, close the way for the development of economic accounting. A centralized plan does not need max-imization and uniformity of production profitability, but it needs macroeconomic optimization of profitability according to the criteria of employment and innovation, with subsequent differentiation of the indicator by industry. In the "country-single company" all its production units can be emitters of such "parametric prices", in which different expenses are based on inter-sectoral balances and different profits in the amount equal to the optimal national dividend. Such a system of economic calculation does not contradict the market, does not force it out of the life of society, meets the criterion of growth of human capital and can cover the whole complex of social relations.

The observance of the principle of classical political economy about the non-intervention of the state in private business in the market development (the principle of laissez-faire), supposedly guaranteeing the well-being of the whole society, in fact, in various formations led and leads to the distribution of created values in favor of individual factors of production. So, if a long-term historical increase in rentability increased land wealth, and then profitability ensured rapid growth of capital goods, then with wagesability, conditions would be created for the free and valuable devel-opment of the worker. Therefore, the sectoral investment list of the national optimal dividend turns it into an element of costs, and all savings in the production of goods allow being attributed to wages as a financial source of human development.

When economic reform was being prepared in the USSR at the beginning of the 1990s, it was possible to radically change the nature of the nationalized and centralized economic development through the universal introduction of enterprises' rentability and productivity, which had been locally and spontaneously applied and given excellent results. However, then the circumstances were such that the "shock therapy" methods were used in the reform, which doomed the country to unfair privatization of state property, hyperinflation and its subsequent existence with a corrupt "pseudo-market". But such a stupid economic situation should serve as a stimulus for the speedy introduction of institutional innovations in Russia. In this case, the "digitization" of the economy is of great importance.

# References

1. Aron R (1993) Imaginary marxism. Progress, Moscow
2. Bell D (1973) The coming of post-industrial society. A venture of social forecasting. Basic Books, New York
3. Boulding KE (1965) The meaning of the twentieth century. Harper & Row, New York
4. Brzezinsky Z (1970) Between two ages: America's root in the technotronic era. Viking Press, New York
5. Etzioni A (1968) The active society: a theory of societal and political processes. Free Press, New York
6. Galbraith JK (1968) The new industrial state. Houghton Mifflin, Boston
7. Grodsky VS (2016) The development of the basic ideas of economic theory. IN-FRA-M, Moscow
8. Huerta de Soto X (2007) Austrian school of economics: the market and entrepreneurial creativity. Socium, Chelyabinsk
9. Inozemtsev VL (1996) Essays on the history of the economic social formation. Taurus Alpha, Century, Moscow
10. Kulpin ES (1996) Bifurcation West-East. Science, Moscow
11. Lenin VI (1958–1965) Complete works. vol 55. Politizdat, Moscow
12. Lichtgeim G (1961) Maxism: an historical and critical study. Harper Collins, New York
13. Marcuse H (1965) Tolerance: freedom of speech. In: a critique of pure tolerance. Beacon Press, Boston
14. Marx K, Engels F (1955–1982) Works. vol 50. Politizdat, Moscow
15. Masuda Y (1981) The information society as post-industrial society. Free Press, Washington
16. Samuelson PE (2002) Foundations of economic analysis. Economic School, St. Petersburg
17. Toffler A (1970) Future shock. Bantam Books, New York

# Digital Transformation of Economy:
# Trends and Prospects

# Analysis of Entrepreneurial Activity and Digital Technologies in Business

A. V. Shafigullina[1]([✉]), R. M. Akhmetshin[1], O. V. Martynova[1],
L. V. Vorontsova[2], and E. S. Sergienko[3]

[1] Kazan Federal University, Kazan, Russia
{anna88-19, renakhmet, olgavl982}@mail.ru
[2] Kazan Innovative University Named After V.G. Timiryasov, Kazan, Russia
vorontsova@ieml.ru
[3] Southern Federal University, Krasnodar, Russia
admurzin@sfedu.ru

**Abstract.** In the paper "Analysis of entrepreneurial activity and digital technologies in business" we consider the typologies of economies proposed by the Global Entrepreneurship Monitor (GEM) research project. According to this typology, countries are subdivided into groups with a factor-driven economy, an efficiency-driven economy and an innovation-driven economy. We also consider indicators of entrepreneurial activity in various countries: the level of activity of emerging entrepreneurs; activity level of new business owners; total early-stage entrepreneurial activity (TEA); the level of activity of established entrepreneurs; the level of exit from the business. We also consider the reasons for going out of business, the key ones being: business unprofitability, and this indicator is the main one both in the Russian Federation and in countries with an innovation-driven economy and a factor-driven economy. Along with these reasons, we can also add the following: the possibility of another employment as an employee, various personal reasons, pre-planned exit from business, lack of access to financing business, some element of chance, the possibility of selling the business, and the presence of bureaucracy and tax burden. The article analyzes the state of entrepreneurship in general by a number of indicators, including the ratio of early entrepreneurs (emerging entrepreneurs and owners of new businesses) and established entrepreneurs. The authors consider the impact of digital technologies on business, the role and importance of digital technologies in the process of supporting small and medium-sized businesses.

**Keywords:** Artificial intelligence technology · BigData · Blockchain ·
Digital technology · Digitalization · Entrepreneurial activity ·
Entrepreneurship · Index of entrepreneurial activity

## 1 Introduction

In the modern economy of the 21st century, the phrase "digitalization of the economy" is becoming popular, including the development of new digital technologies. These include BigData, blockchain, artificial intelligence technology. The application of these

© Springer Nature Switzerland AG 2020
S. Ashmarina et al. (Eds.): *Digital Transformation of the Economy: Challenges,
Trends and New Opportunities*, pp. 183–188, 2019.
https://doi.org/10.1007/978-3-030-11367-4_17

technologies creates the possibility for flexible regulation and effective support for small and medium-sized businesses. Nowadays, it is important to consider the relationship of the concepts of small and medium-sized businesses with the concepts of entrepreneurial activity. Entrepreneurship and entrepreneurial activity - let's consider these concepts in more detail. There are many definitions of the concept of "entrepreneurship", the most well-known and most popular are the definitions given by such scientists as J. B. Say, A. Smith, P. Drucker, J. Schumpeter and many others. In the most general form, entrepreneurship can be defined as a specific form of activity of an individual who takes the responsibility in decision-making, organizes and initiates various business processes, develops one or another form of business [1]. According to the Civil Code of the Russian Federation (Part One) No. 51-FZ of November 30, 1994 (ed. of December 29, 2017) "…entrepreneurship is an independent activity carried out at your own risk, the purpose of which is to systematically obtain profits from the use of property, the sale of goods, the performance of work or the provision of services by persons registered in this capacity in the manner prescribed by law". Based on the most frequently used definitions of entrepreneurship, we conclude that entrepreneurial activity forms the basis of entrepreneurship. This indicator is the subject of various studies, one of which is the Global Entrepreneurship Monitor (GEM) project.

## 2   Materials and Methods

To describe entrepreneurial activity the analysis method was used, which identified the following main points of the study: it is recommended to use in different countries the typology of economies proposed by the Global Competitiveness Report. According to this typology, countries are subdivided into countries with a factor-driven economy, an efficiency-driven economy, and an innovation-driven economy [5]. Let's consider this typology in more detail. The key characteristics of countries with a factor-driven economy are the following – firms within a country compete based on prices, use their factor endowments and, as a rule, unskilled labour and natural resources. Countries with an efficiency-driven economy use more-efficient production to increase productivity, competitiveness of organizations is achieved by attracting employees with higher education, as well as the ability to benefit from using efficient technologies. The characteristics of countries with an innovation-driven economy are as follows: an economy must produce innovative products by using the most sophisticated production processes; in this type of economy, a firm is able to sustain if it competes based on innovations.

## 3   Results

Based on the Global Entrepreneurship Monitoring (GEM) data for 2016, a table of entrepreneurial activity in countries by stages of economic development was created (Table 1). Each country participating in the Global Enterprise Monitoring has a unique set of socio-economic pillars that affect the level of entrepreneurial activity.

**Table 1.** Entrepreneurial activity in countries by stages of economic development, % (2016)

| | Factor-driven economy | Efficiency-driven average | Efficiency-driven average |
|---|---|---|---|
| Average value | | | |
| The level of activity of emerging entrepreneurs | 9,98 | 8,06 | 5,54 |
| Activity level of new business owners | 7,3 | 6,54 | 3,69 |
| Total early-stage entrepreneurial activity (TEA) | 16,8 | 14,29 | 9,07 |
| The level of activity of established entrepreneurs | 11,18 | 8,65 | 6,68 |
| The level of exit from the business | 12,35 | 9,7 | 8,1 |

*(Source: compiled by the authors)

In the Russian Federation, from 2012 to 2016, the indicators of early entrepreneurial activity were approximately at the same level and amounted to 4.9% in 2012 and 3.2% in 2016. The highest rate of early entrepreneurial activity was recorded in 2013 and was 5.8%. Analyzing these indicators, we can conclude that it is just as many, in relative terms, Russians of working age are the owners of a new business, or emerging entrepreneurs.

However, these relatively high values of the level of early entrepreneurial activity can not be indicative of the prosperity and development of entrepreneurship in certain countries. The fall of the TEA index may indicate an increase in employment opportunities as a sign of economic development and the development of institutions. This can be confirmed by the fact that the average TEA value in countries with a factor-driven economy is 16.8%, in countries with an efficiency-driven economy - 14.29%, and with an innovation-driven economy - 9.07%.

It should be noted that even having a positive idea of entrepreneurship and showing entrepreneurial intentions, this, unfortunately, is not a guarantee of the opening and successful development of business. Various factors can play a role here, for example, "red tape" is a factor of a negative order, representing an unfavorable administrative burden. Among the positive factors open market, access to various factors and technical assistance, cultural values in entrepreneurial behavior can be noted [2].

The state of entrepreneurship as a whole can be analyzed by a number of indicators, which include the ratio of early and established entrepreneurs. Let's consider these definitions in more detail:

- early-stage entrepreneurs - this type includes emerging entrepreneurs and owners of new businesses. Emerging entrepreneurs are considered to be those who took active steps to create a business during the year, or they own a business for less than 3 months, but no wages or other types of incentives have been received. The owners of a new business include those who manage a newly created organization for more than 3 months, but less than 3.5 years and receive income during the activity.

– established businessmen or owners of an established business are those who own
and operate a business and receive dividends and income for more than 3.5 years
from this type of business activity.

Emerging entrepreneurs and owners of new businesses make up a group of early-
stage entrepreneurs. As a calculated indicator a dynamic index - total early-stage
entrepreneurial activity (TEA Index) - is used.

Let's return to the ratio of early entrepreneurs and established entrepreneurs. In the
Russian Federation, this ratio was as follows: the index of established entrepreneurs
was always lower than the index of early entrepreneurial activity. Another negative
point in the overall analysis of entrepreneurial activity is the high level of exit of
existing entrepreneurs from business.

According to the Global Entrepreneurship Monitor (GEM) study, there could be the
following reasons for going out of business: the most common is business unprofitability (almost 33%), almost 8% of respondents closed their business due to the
inability to access finance. 10% of respondents considered that having a job offer as an
employee would be more efficient and attractive comparing to running their own
business. The second most popular reason to stop the business is a personal motive,
accounting about 30% of responses. Only 6% of respondents in Russia consider the
possibility of selling their business as a reason to exit from business, this situation was
observed in previous years [5].

The presence of bureaucratic obstacles and a strong tax burden are the reasons for
going out of business for 15% of respondents. In other countries one of the options for
going out of business among others was the option "retirement", but this answer was
not chosen by any of the respondents in the Russian Federation, that may be a sign of
some shortcomings in the Russian pension system [1].

A rather low share of established entrepreneurs in the total volume of entrepreneurs
indicates a negative conditions for the development of businesses. In countries with
high-developed economy the share of established entrepreneurs is approximately equal
to the share of the early-stage entrepreneurs, and in some countries it exceeds this
indicator (Germany, Greece, Italy, etc.).

In countries with a factor-driven economy, the opposite situation is observed: the
share of the early-stage entrepreneurs is much higher than the share of the established
ones. On average, in these countries the index of early entrepreneurial activity exceeds
the level of activity of established entrepreneurs by 5.62%, in efficiency-driven
countries by 5.64%, and innovation-driven countries by 2.39%. In the aggregate, the
low level of early entrepreneurial activity and the low share of established entrepreneurs in the total number of entrepreneurs negatively characterize the state of
entrepreneurship in Russia.

When analyzing the GEM data, it can be traced that the mining sector in Russia is
11% including agriculture, forestry and fishing. Since this indicator is higher in our
country than in the average for countries belonging to an efficiency-driven economy
and an innovation-driven economy, this is one of the reasons for including Russia in
the number of countries with a factor-driven economy. It is interesting to note that in
the consumer sector, new entrepreneurs dominate (almost 50%), and this suggests that

this sector will continue to attract the attention of new entrepreneurs and will receive its further development.

It should also be noted that the development of the digital economy is now becoming an increasingly important rationale for global economic growth of the market and transformational economy of the Russian Federation. In the future, under appropriate conditions, it can lead to an accelerated economic development of the domestic economy and an increase in labor productivity in all sectors of the economy [4].

## 4   Discussion

In the paper "Analysis of entrepreneurial activity and digital technologies in business", we considered a typology of entrepreneurial activity based on data from the Global Entrepreneurship Monitoring (GEM) Research Agency in conjunction with expert opinions from rb.ru (technology and business analytics). We can conclude that digital technologies are applied in various types of economies and affect different types of entrepreneurs (both emerging and established). The influence of advanced digital technologies will give the greatest effect in the following areas:

- facilitation of administrative procedures for launching a new business, minimizing risks;
- optimization of regulation of small and medium businesses;
- managing the development of small and medium-sized businesses: focusing on targeted, most promising and important areas for the state, attracting funding, providing effective feedback from entrepreneurs to the state and investors.

## 5   Conclusions

Today, the world is standing on the cusp of the fourth stage of the digital revolution. Further development of infrastructure, reducing the cost of processing, storing and transmitting data lead humanity to the threshold of a new, most ambitious stage of the digital revolution. The influence of digital technologies on different industries differ significantly, that could result from the specific forms of interaction between various structures, the influence of external macroeconomic factors or the investment attractiveness of the industry, but also from the owners' concerns about changes in their traditional activities. Although the term "uberization" is often used as a synonym for digital threat to any traditional industry, there is no doubt that all industries and players will sooner or later be forced to go through a digital transformation.

In the modern world, the changes have already affected the B2C-industries (media, retail, banking and insurance services). Here they are caused by fierce competition for two very valuable resources - time and the consumer's wallet. However, experts believe that a fundamentally new mechanism for influencing business processes in the B2B sector, in the context of the current digital revolution in business, will cause a more global restructuring of economic relations in general. It is in B2B that digitalization opportunities are not limited to owning limited consumer resources, but allow infinitely approaching new heights of efficiency and productivity [3]. And, while business structures in the B2B sector are only observing the tools of global digitalization, domestic companies, in order to increase the competitiveness of Russian business structures, in our opinion, should begin to implement and independently increase the digitization of their business activities. As a follow-up to the topic, several key areas of the digital economy in the Russian Federation can be identified: cyber security, cross-border cooperation, creation of common IT platforms, digitalization of public services, B2B marketplaces.

# References

1. Akhmetshin RM, Shafigullina AV (2015) Government regulation of small and medium entrepreneurship under the influence of value-time benchmarks. Mediterr J Soc Sci 6(1):151–154. https://doi.org/10.5901/mjss.2015.v6n1s3p151
2. Antonchenko NG, Kalenskaya NV (2014) Developing a methodology for assessing the efficacy of managerial decisions in entrepreneurial establishments. Life Sci J 11(7):365–369
3. Stough Roger (2009) Knowledge spillovers, entrepreneurship and economic development. Peter Nijkamp Ann Reg Sci 43:835. https://doi.org/10.1007/s00168-009-0301-z
4. Strauss T (2001) Growth and government: is there a difference between developed and developing countries? T. Econ Gov (2001) 2:135. https://doi.org/10.1007/pl00011023
5. Verkhovskaya OR, Dorokhina MV, Sergeeva OV (2014) Global monitoring of entrepreneurship. Russia 2013. National report. St. Petersburg, 64 p

# Risk Identification in the Sphere of Quality Under the Conditions of Digital Economy Development

T. A. Korneeva, I. A. Svetkina, E. S. Morozova,
and A. S. Zotova$^{(\boxtimes)}$ (ID)

Samara State University of Economics, Samara, Russia
korneeva2004@bk.ru, svetkinairina@yandex.ru,
dhrsseu@gmail.com, azotova2012@gmail.com

**Abstract.** The essential character of the publication is based on the fact that management of an economic entity in the context of the digital economy development and continuous transformation of technologies and management methods requires the development of new approaches to quality management. It should cover the entire set of organizational forms of coordination of actions in the field of quality, implementation and management of these activities in the enterprise, as well as maintaining relationships with the external environment. One of such approaches is risk-based quality management performed at all stages of the product life cycle. The process of identifying risks in the field of quality is a characteristic feature of an economic entity activity, especially in a digital economy, when any product with specified quality parameters is available to a consumer using digital technologies. During the process of identifying risks associated with quality, a number of organizational and methodological issues are expected to be addressed, this article is devoted to a discussion of the main ones.

**Keywords:** Digitalization · Digital economy · Economic security · Risk management policy · Risk identification · Risk · Quality

## 1 Introduction

The Program "Digital Economy of the Russian Federation", approved by the decree of the Government of the Russian Federation dated July 28, 2017 No. 1632-p, states that the Russian Federation occupies the 41st place in rankings at the point of readiness for the digital economy with a significant margin from the dozen of the leading countries, and 38th place on the economic and innovative results of digital technology usage. Russia's lag in the development of the digital economy from the leading countries is explained by the "gaps in the regulatory framework for the digital economy and the insufficiently favorable environment for doing business and innovation, and the low level of digital application of business structures" [1].

On the other hand, the results of the international survey of Executive Perspectives on Top Risks 2018 show [2] that at present the risks of transition to the digital economy are of particular concern of companies management. Risks associated with economic

© Springer Nature Switzerland AG 2020
S. Ashmarina et al. (Eds.): *Digital Transformation of the Economy: Challenges, Trends and New Opportunities*, pp. 189–199, 2019.
https://doi.org/10.1007/978-3-030-11367-4_18

conditions and regulatory control are being superseded by risks associated with the introduction of digital technologies, since digital technologies require changes in business models and will determine the ability of organizations to maintain their competitiveness. The success of introducing digital innovations largely depends on the proper risk management of the transition process, the development of new, effective approaches that take into account the full range impact of competitiveness risks [3].

Risks in a modern digital economy are an objective reality inherent by entrepreneurship itself, however, a single approach to understanding the economic nature of the category under consideration has not developed in the theory and practice of entrepreneurship, which is explained by the complex nature of risk, resulting in various definitions that explain the essence of risk indirectly, in some cases, through the consequences of the event (loss, threat, damage, etc.), in others, through the degree of possibility (probability) of the occurrence of the event in definite conditions.

Quality is a complex, multidimensional and at the same time universal category, therefore in the scientific-methodical and practical literature there is no single approach to its definition. A number of authors consider various aspects of quality (philosophical, social, technical, economic, legal) noting that from an economic point of view it characterizes the consumer value of products [4, 5].

The ISO 9000: 2005 standard defines the concept of quality as the degree to which a set of its own characteristics (distinctive properties of a product, process, system) fulfills the requirements (needs or expectations of the consumer). Quality is a complex concept characterizing the effectiveness of all aspects of activity: strategy development, organization of production, accounting, etc. In the sphere of quality, which has a direct impact on production and economic relations, the formation of conceptual foundations is difficult, but they have more value if they are able to provide positive practical impact. Thus, quality as a combination of consumer properties of products is a specific control object and has significant features.

Risks in the area of product (work, service) quality are risks primarily for the consumer. The manufacturer is responsible to the consumer for the effectiveness and safety of products entering the market. The only effective way to ensure product safety is to implement an effective quality risk management system as an element of producer responsibility to society. The consumer needs to have guarantees of efficiency and safety of products. Safety does not mean complete absence of danger. A safe condition is confidence in knowing what "dangerous" events can occur and what effect they will have on the quality of work, the quality of products and, as a result, on the consumer. Product safety is the development of effective procedures designed to ensure this safety.

Quality risk management is a systematic process for the overall implementation of monitoring, informing and reviewing product quality risks from its development to its use by the consumer. Risk management is the only way to solve a problem after identifying important risks. The process of product quality management consists of several stages: the collection and processing of information; monitoring the facility, identifying deviations from budgets and assessing their significance for determining the nature of regulation; preparation and decision making; the organization of their execution, i.e. development of ways, methods and means of influence on the controlled object.

The most important tasks of modern quality management practices are the development and implementation of management decisions aimed at achieving quality indicators of products and the entire enterprise. The management decision is a targeted impact on a management object based on specially collected information. The process of management decision making is an adequate response to the problem situation that has developed during the operation of the facility. Consequently, the economic entity needs to have certain tools that regulate the accumulation and further use of data relating to the process of product quality management. The use of innovative technologies, new communication channels, telecommunications will change the risk management system, including those associated with quality.

The process of product quality management unites the financial, technical, intellectual and organizational potential of the economic entity and the most general concepts are coordinated as part of the integration of the management system functions. Solving the problem of organizing an effective quality risk management system is becoming a priority in the development of the digital economy.

## 2 Materials and Methods

The study is based on the analysis of corporate integrated financial statements of a number of Russian companies. The methods of empirical observation were used to obtain the missing data.

Using the methods of comparison and chronological analysis, groups of risks that have a significant impact on the company's strategy are identified and systematized. Analysis of literature sources allowed to model the resulting operational structure for risk management in the organization.

The synthesis of information was carried out by means of formalization and graphical methods of data presentation. The construction of a matrix model of functions and responsibilities distribution in the risk management system allows to assess the performance of functions in the risk management system. In the course of the study, methods of analysis of literary sources and scientific reasoning, abstraction, and graphic modeling were used.

The following methods were also used: analysis, synthesis, comparison, method of chronological analysis, building models, etc.

## 3 Results

It was revealed that the success of introducing digital innovations into the system of economic entities managing largely depends on proper risk management in the development of the digital economy, on the development of new effective approaches that allow to take into account the whole range of quality and competitiveness risks. To solve this problem, it is necessary to create a system for identifying, analyzing and managing risks in the field of quality, which is unique (individual) for each economic entity [4].

To identify risks in the field of quality, an economic entity analyzes business processes separately for each business area, on the basis of which a separate business process model is built.

The process of ensuring the quality of products should be understood as all sub-processes for the development and implementation of necessary preventive and corrective measures: to meet the requirements established by the consumer for the product quality system; to eliminate identified or prevent potential non-compliance of products with the proposed or implied requirements for them. The complex of periodic and one-time sub processes is formed, aimed at ensuring stable quality characteristics in accordance with standards, certificates and other requirements within an acceptable level for a given product group.

The diversity of sub processes to ensure product quality, their correspondence to the production processes predetermines the need for a scientifically based study of sub processes to ensure the interpretability of information that managers must perceive in order to make management decisions.

The division of the product quality assurance process into sub processes is necessary for internal analytical purposes in order to establish the functional responsibility of the units. Based on the content of each sub process, it is necessary to group them so that you can clearly determine the cost capacity of each factor of the production process and its impact on the cost of production.

Therefore, the volume and composition of the output information about the sub processes to ensure product quality cannot be identical for all levels of management: the higher the level, the more generalized the indicators of the output information.

Therefore, management personnel who control product quality risks should:

- to participate in the system of information integration;
- to carry out inventory (accounting) information;
- to carry out the identification and evaluation of information systematically;
- to send the information report to the consumer timely;
- to formulate and specify measures for obtaining and providing information;
- to assign the information coordinator within the unit;
- to follow the general line of the industrial enterprise strategy;
- to ensure the high quality of information events (relevance, completeness, accuracy);
- to ensure high quality of information reporting (relevance, completeness, accuracy, efficiency, reliability, accuracy, objectivity, balance);
- to provide high quality accounting (inventory) information.

To obtain the effective management decision, information on risks in the area of product quality can be processed by different methods (Table 1), which implies the creation of an integrated system for generating and processing data while maintaining sufficiently autonomous functional subsystems, each of which performs a specific target management function, solving its tasks using specific methods of information and telecommunication technologies. The time of getting the information should be taken into account in the process of product quality management.

**Table 1.** Methods of risks data processing in the field of goods quality

| Methods | Characteristic |
|---|---|
| Supervision | Allows to get general vision of existing economic phenomenon, to set differences between the information objects, to sort out the information, to choose necessary information units |
| Measurement | Gives the economic phenomenon some quality assessment |
| Registration | Can be performed in a definite system, simplifies memorizing and studying of observed phenomena |
| Calculation | The information units are united to find the necessary result |
| Comparison | Comparing the incoming quantity with the controlled one it can be defined that these data are greater or less than controlled quantities |

Source: compiled by the authors

Information about risks in the field of product quality represents all the features of an industrial enterprise, fully discloses the economic, technical and technological features of production processes and production capabilities, established internal relations of production, resources used, financial capabilities, ensures the creation of an effective system of control and management of the enterprise.

Addressing issues of identifying risks at the enterprise level enables business structures to expand their understanding of the company's risk profile and value creation model, as well as consideration how these issues affect shareholders and society.

Institutional investors are interested in obtaining information on how companies in a changing business environment solve quality problems to achieve long-term, sustainable growth [5].

Companies take different approaches to defining a list of risks that should be disclosed. Some companies focus on risks that affect the performance of the statement of their financial position. Others indicate risks that could have a negative effect on profits in the short term, while others focus more on external factors causing fluctuations in share prices. Most often, companies disclose information about risks associated with ordinary business activities.

And this disclosing is generally limited to their general description and the indication of measures that can be taken in this regard by the company. However, this approach does not reveal the important risks affecting the factors of business value creation. When choosing risks for reporting to interested users, it is necessary to proceed from whether they help in assessing the intrinsic value of a business, i.e. his ability to generate profit in the long run.

Today, the largest Russian companies have groups of specialists in monitoring, controlling and managing risks, including risks in the field of quality. In corporate integrated reporting, companies disclose key risks and opportunities specific for definite economic entity, including those opportunities and risks that relate to the organization's impact on certain capitals. Risk disclosures in integrated corporate reporting can be made on the following key issues (Fig. 1).

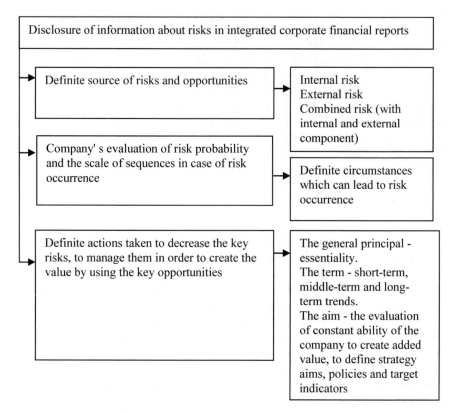

**Fig. 1.** Disclosure of information about risks in integrated corporate financial reports (Source: compiled by the authors)

The study of the corporate integrated reporting of a number of major Russian companies has been performed and as a result the groups of risks have been identified and systematized as well as the measures taken to manage these risks (Table 2).

Sustainable development (or integrated reporting) reports disclose risk factors, but the core corporate reporting does not indicate the valuation of these risks disclosed in the company's sustainable development report [6]. Thus, it can be concluded that there is a mismatch between the risks that are considered "significant" in the sustainability report and the risks disclosed in public financial statements. The complexity of assessing risks in terms of money does not allow the proper allocation of resources, especially in case of long-term risks, with uncertain consequences that arise over an unknown period of time. This is also explained by the fact that in many companies the risks in the field of quality are managed and disclosed by a group of specialists on sustainable development and are considered as separate or less significant than the usual strategic, operational or financial risks.

Most companies that disclose their reports, including those on sustainable development, have risk management systems and processes for managing risks. The

**Table 2.** Disclosure of information about risks in the integrated financial reports of JSC GAZPROM, JSC Sibur

| Types of risks | The description of risks | Measures for risk management |
|---|---|---|
| Risks, linked with staff resources | – Inability to attract and retain qualified staff with right skills and experience in the right quantity<br>– The growing shortage of qualified technical specialists<br>– Increasing competition in the labor market in Russia and abroad<br>– Shortcomings in high-quality professional technical training<br>– The level of social and cultural infrastructure of cities and regions in which the enterprises of the studied companies are located | 1. Competitive rewards, including wages and reward for the result, as well as a package of social benefits<br>2. Programs for the formation of personnel reserve, training and development of personnel, aimed at ensuring its needs for qualified personnel at the current moment and in the future<br>3. Recruitment procedures and measures aimed at personnel development and reducing staff turnover<br>4. Social investment program, aimed at improving the quality of life of employees and their families |
| Risks, linked with industrial safety and labour safety | Risks of employees safety and production activities safety:<br>– due to breakdowns<br>– or equipment failure<br>– natural disasters<br>– terrorist attacks<br>– actions or inaction of staff. The results of risk realization can be: injury or death of people, bringing the company to prosecution process, damage to corporate business reputation | 1. Ensuring safe working conditions for employees<br>2. Constant monitoring and control of dangerous situations and security threats<br>3. Ensuring compliance with industrial safety<br>4. Accidents and incidents are investigated and identifying the reasons the actions are taken to prevent such incidents in the future |
| Environmental risks | Risk of damaging the environment or its pollution may lead to civil liability to prosecution and the need for remedial work. The growth of environmental risks is associated with the expansion of industrial activity and the growth of company assets | 1. Ensuring compliance with environmental standards<br>2. Monitoring changes in environmental legislation in countries of operation<br>3. Programs to minimize the negative impact on the environment, nature conservation and the conservation of biological diversity<br>4. Environmental Monitoring of the situations using mobile labs<br>5. Measures to reduce emissions and waste |

Source: compiled by the authors on the materials of the financial reports [8, 9]

resulting operational structure for managing risk in an organization is, as a rule, consistent with the managerial and legal structure. The example of the organizational risk management structure could be the following (Fig. 2).

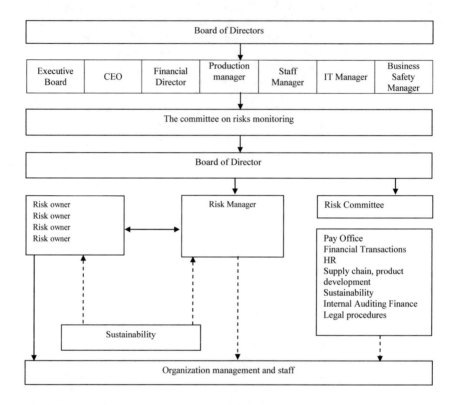

**Fig. 2.** Risk management structure (Source: compiled by the authors on the basis of COSO [10])

Quality risk management is carried out at every level of the company. Based on the proposed structure, risk and sustainability specialists can identify areas of interaction with departments concerning the activities related to quality risks, and on this basis make informed managerial decisions.

Effective quality risk management requires the monitoring of such system implementation. To assess the performance of functions in the risk management system, it is recommended to use the following matrix, which allows to evaluate the distribution of functions performed and responsibilities in the risk management system (Table 3).

This matrix allows to define the roles and responsibilities for incorporating the experience of sustainable development into the quality risk management system in the company framework of ensuring economic security.

**Table 3.** The matrix of function and responsibility distribution in risk management system

| | Component of risk management system | Board of directors and the Committee on risks monitoring | Director or Executive Board | Risk management director | Risk owners | Sustainable development manager |
|---|---|---|---|---|---|---|
| Administration | Creation of management system and the culture of risk management | Set the basics and principles for management and culture | Introduction of principles, management and culture set by the committee | Implementation of structured approach to risk management and culture | Understanding and use of risk management for specific risks | Understanding and use of risk model for risk management of specific risks |
| | Creation of risk management process | Creation of concept of risk recognition, assessment, management and monitoring | Creation and assistance to risk management system. Responsible director appointment | Creation and assistance to transparent risk management process during the whole LLC | Assistance in risk management process creation and execution of various roles in the process of risk management | Understand and use risk management system for quality risks integration into company sustainability management |
| Management process | Business context and strategy study | Accepting the information about significant changes in external and internal environment | Setting the business strategy, aims and risk inclination | Assistance to business context and strategy study | Understanding of internal and external changes in business context and recognition of changes that can lead to risks | Understanding correlation between internal and external changes in business context and risks in business model; Giving information to risk owners and company authorities |
| | To identify risks that will feel impact of business strategy and goals | Knowledge of key risks affecting strategy | Identify and disclosure significant risks that affect business strategy | Contribute to the process of risks identification and business impact | Determination of factors which contribute to risk identification and risk awareness | Support risk owners with tools and knowledge to identify and understand risks |
| | Assessment and determination of the likelihood degree of identified risks | Awareness of the likelihood and importance of key risks and opportunities | Assessment and ranking of key risks and opportunities | Use risk assessment tools | Business risk assessment and its impact on strategy | Supporting risk owners with tools for quantifying and prioritizing risks |
| | Design and implementation of response measures for prior risks | Aware of risk response measures; approval of critical risks identified by management may be required | Resource Allocation for Priority Risk Management | Assist in developing risk response measures for each risk | Developing appropriate response measures to eliminate risk | Support risk owners in developing innovative responses to priority risks |
| | Review of risk indicators and revision as necessary | Consultation on risk status and risk management process | ERM activity monitoring and provision of risks recognition within the company's risk appetite | Developing a consolidated presentation of risk monitoring indicators | Development of indicators for risks monitoring and business context, when risk goes beyond acceptable levels | Assistance to risk owners in developing risk measurement indicators and defining report aspects for internal and external interested users |
| | Risk communication and reporting | Consulting on the activities and processes of external risk management | Informing about the activities and processes of internal and external risk management | Development of internal and external communications on risk management activities and processes | Providing information for internal and external reporting on risk management activities and processes | Information support for risk owners on risk-related aspects of risk management activities and processes |

Source: compiled by the authors on the basis of COSO [10]

## 4   Discussion

The identification of quality risks in a risk management system is crucial for overcoming difficulties that an economic entity may face in managing these risks (for example, problems of quantitative risk assessment). The presented approach can be used to improve the transparency and accountability of quality risk management.

In each economic entity, it is necessary to develop measures aimed at integrating quality risk management into the corporate risk management system. A company's strategy and related objectives must be associated with each stage of the risk management system process.

Product quality risks can be difficult to assess and prioritize. This problem can be aggravated by the fact that the company usually has limited knowledge about quality risks, a tendency to focus on short-term risks, not paying due attention to the risks that may arise in the long term, difficulties in quantifying these risks. Even when a quality risk can be quantified, a company cannot always determine its priority due to a bias towards risks that are known or better understood.

## 5   Conclusions

The process of identifying risks in the field of quality is an important factor in business development, which gives a company great opportunities to improve the efficiency of its activities in the digital economy. Quality risk management helps companies in all stages of the value chain - from reducing costs to gaining competitive advantages in the market. The task of managing risk in the field of quality is becoming increasingly important for companies taking into account the trend of growing expectations from shareholders, consumers and other interested parties. If they are not given due attention, such specific risks can lead to serious consequences for business. But at the same time, certain risks - provided that they are properly identified, evaluated, taken under control and communicated to interested parties - can help the organization increase its effectiveness.

It is important to understand the strategic and operational plans of the business to manage risks effectively. Risks or opportunities associated with product quality should not be identified, assessed or eliminated in isolation from the company's strategic direction or risk appetite. In addition, a risk manager, not reacting to the risks being studied, may lose sight of the significant risks.

The use of the considered approaches to the organization of the risk management process in the field of quality will provide long-term benefits for the company in the conditions of the development of the digital economy.

## References

1. State program "Digital economy of the Russian Federation", approved by the order of the Government of the Russian Federation, dated July 28, 2017, № 1632-p
2. 2018 Top Risks Report: Executive Perspectives on Top Risks for 2018. https://erm.ncsu.edu/library/article/2018-top-risks-report-executive-perspectives-on-top-risks-for-2018. Accessed 14 Apr 2018

3. Risk management in transition to digital economy: focus on the client. http://rass.moscow/wpcontent/uploads/2013/04/СОКОЛОВ_КУЗНЕЦОВ.pdf
4. Garvin DA (1984) What does product quality really mean? Sloan Manag Rev 26:25–43
5. Fields P, Hague D, Koby G, Lommel A, Melby A (2014) What is quality? A management discipline and the translation industry get acquainted. Revista Tradumàtica: tecnologies de la traducció, vol 12. ISSN 1578-7559
6. Korneeva T, Svetkina I, Naumova O, Noskov V (2017) Organizational aspects of economic safety support for retailing sphere. Bull Samara State Univ Econ 9(155):69–79
7. The report of global risks in 2018: issue XIII, The World Economic Forum. https://www.weforum.org/reports/the-global-risks-report-2018. Accessed 14 Apr 2018
8. Who searches - finds. How the company's financial director can manage risks in the sphere of sustainable development and supply long-term benefits for it. https://www2.deloitte.com/content/dam/Deloitte/ru/Documents/risk/russian/gra-sustainability-reporting-ru.pdf
9. Korneeva T, Potasheva O (2016) Developing the methods for the analysis of costs associated with quality. Actual Probl Econ 178(4):294–303
10. GAZPROM annual report. http://www.gazprom-neft.ru/annual-reports/2016/GPN_SR16_RUS_s.pdf
11. Sibur annual report. https://www.sibur.ru/upload/iblock/f93/f93dabf1494add5835a1e9ef3963ae75.pdf
12. Applying enterprise risk management to environmental, social and governance-related risks (COSO). https://www.coso.org/esg/Pages/default.aspx. Accessed 14 Apr 2018

# Innovation Clusters in the Digital Economy

Ruslan Polyakov$^{(\boxtimes)}$ and Tatjana Stepanova

Kaliningrad State Technical University, Kaliningrad, Russia
polyakov_rk@mail. ru, tatyana. stepanova@klgtu. ru

**Abstract.** The article presents the results of the authors' research, the subject of which are the issues of building innovative clusters in the digital economy. The first stage of research work, which is dedicated to identifying the factors affecting the performance of digital economy innovative clusters is described in the paper. Authors revealed that the new institutional environment introduces significant changes in investments, innovations and political priorities of countries, and their changing nature reinforces the risk of humanity's dependence on digital technologies.

The dynamics of key indicators of the new digital economy shows that we suffer the times of cardinal transformations, and with the development of electronics and digital communication channels, the world entered the trajectory of accelerated exponential growth because digital flows have become responsible for GDP growth and determine the future landscape of the countries. We become aware of the fact that the government should assume predictable terms and the formation of digital trust in the new conditions; they will also play a decisive role in creating a reliable development of the digital environment and in forming a responsible attitude of users to key institutions and organizations. This issue is of great interest to all participants in the digital economy, and given the best practices for the development of high-tech companies in the world, the global experience of their cluster development is of particular research interest. The authors believe that this study offers a valuable perspective on how innovation clusters can influence the digital transformation of countries in the face of new growth.

**Keywords:** Digital economy · Digitalization · Innovation clusters ·
Internet · Technology

## 1 Introduction

Goods demand and supply new conditions, rapid changes in innovations and transformation of the institutional environment set the resulting framework today, which fixes the state of the economic environment as a digital evolution. Even today, the new institutional environment introduces significant changes in investments, innovations and political priorities of countries, and the changing nature of risks in this framework shows that the constant dependence of humanity on digital technologies emerge to the prominence in society.

The widespread introduction of processes of automation, artificial intelligence, the industrial Internet and the Internet of things, robotics, unmanned systems and new

S. Ashmarina et al. (Eds.): *Digital Transformation of the Economy: Challenges,*
*Trends and New Opportunities*, pp. 200–215, 2019.
https://doi.org/10.1007/978-3-030-11367-4_19

business models for the consumption of goods and services dramatically change the traditional way of modern life. It also require more effective measures from states for stimulation and development of innovative companies.

Modern reality with a rapidly changing environment reveals the inefficiency and inflexibility of vertically integrated firms in comparison with more innovative and advanced cluster structures. Their effectiveness as a new paradigm of socio-economic and spatial development was first tested in Silicon Valley, Cambridge Science Park and Sophia Antipolis, and then adopted by leading cities in Europe, North America, Australia and Asia, such as Austin, Barcelona, Boston, Delft, Manchester, Melbourne, Singapore, Toronto [1, 2].

New digital economies are pushing states and the innovation ecosystem itself [3] to find more breakthrough approaches to building a cluster. Relying on digital infrastructure, green technologies and infrastructure solutions, you can more efficiently revitalize the environment and increase the global competitiveness of the cluster [4].

Building innovation clusters in the digital economy provides firms with three main groups of benefits: achieving high production rates and levels of innovation, and provokes the creation of new firms in the industry, which only enhances the competitiveness of the cluster.

The cluster allows all its members to achieve high performance, due to the synergistic effect of the concentration of a larger number of different firms in the same territory, moreover, it has exactly the same advantages as hierarchical structures, but without loss of flexibility in decision making and business management by each individual company [5].

Being a part of a cluster, the firm gets unlimited access to the necessary materials for the production of finished products, technologies and information; cooperation is organized with special institutions of government and civil society, which provide companies with the opportunity to implement improvements in the production process. A synergistic effect is achieved as a result of cluster activity as a single mechanism, and not of separate elements.

One of the most important conditions for the successful development of innovation activity is the availability of an appropriate innovation infrastructure, which is a system of economic entities that are not directly involved in innovation activity, but provide general conditions for its effective implementation [6].

The main elements of the innovation infrastructure are innovative business incubators, innovation centers, including technology transfer centers, science and technology parks and technopolises.

In addition, the innovation infrastructure includes financial and credit institutions that ensure the accumulation of investment resources and their distribution among the subjects of innovation activity (banks, stock exchanges, innovation funds, including venture funds), as well as insurance companies that ensure the reduction of losses in the implementation of risky innovation projects.

In recent years, in connection with the intensive development of information technologies, information networks have become important elements of the innovation infrastructure, bringing together various information centers focused on the provision of information services in the field of innovation.

The government plays special role in the innovation infrastructure, which is on the one hand is subjects of innovation activity, and on the other hand, has a significant regulatory impact on this activity. The innovation infrastructure, due to the specifics of the innovation activity, is closely connected with the production and market infrastructure [7].

Cluster innovation activity is a major factor for the socio-economic development of territories. As part of the regional innovation development strategies, mechanisms have been laid for creating complexes of innovation clusters, which should be the center for combining production, investors and consumers of innovative products and services.

## 2 Materials and Methods

Today, the countries of the world live in times of cardinal transformations, the main of which is the digital economy. With the advent of the first industrial revolution and the beginning of scientific and technological progress, the world irrevocably plunged into the era of the global integration process of "globalization". With the development of electronics and digital communication channels we entered the trajectory of accelerated exponential growth due to the fact that digital flows became responsible for GDP growth and determine the future landscape of the economy in countries [8].

The emergence of the Internet has become the greatest technological breakthrough, and today billions of people use it every day to spread information, trade, share ideas, keep in touch with family, friends and colleagues. Thanks to the global Internet penetration of almost 50%, the global digital economy has become a huge opportunity, and now all this power depends on the bandwidth and speed of connecting new customers [9].

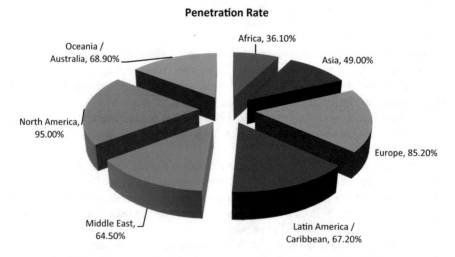

**Fig. 1.** World internet usage and population statistics, June 30, 2018 (Source: compiled by the authors of article based on the data of Internet World Stats. Miniwatts Marketing Group. https://www.internetworldstats.com/stats.htm)

Based on the research of Miniwatts Marketing Group [10], we can say that more than 55.1% of the 7.6 billion people living on the planet today use the Internet. About half of these people are regular users. More detailed statistics on the regions are presented in Fig. 1. The degree of Internet penetration (the ratio of the number of Internet users to the total population) by regions is very uneven - from 36.1% in Africa to 95.0% in North America.

Russia ranks first place in Europe and sixth in the world by the number of Internet users, and over the past three years, the number of smartphones has doubled and now 60% of the population are the owners. This is more than in Brazil, India and the countries of Eastern Europe [10].

The number of users of state and municipal services portals doubled in 2016 alone and reached 40 million people [11].

The degree of penetration of the Internet in Russia is 76.1% [12], and the share of organizations that used information and communication technologies in 2017 was 88.9% (Rosstat) [13].

The dynamics of key indicators in the new digital economy shows that the achievement of a certain growth by the government of the country is truly a competitive advantage in the global digital arena. It is also clear that the impulse that the state sets with its predictable behavior in the framework of the "Strategy for the Development of the Information Society in the Russian Federation for 2017–2030" [14] creates trust and plays a crucial role in digital development.

In July 2017, the Digital Economy program [15] was approved in the Russian Federation. It includes several areas: information infrastructure, the formation of research competencies and technological reserves and information security. It is designed to provide support for technologies such as big data, quantum computers, new production methods, artificial intelligence. Achieving the planned characteristics of the digital economy of the Russian Federation is achieved by achieving the following indicators by 2024:

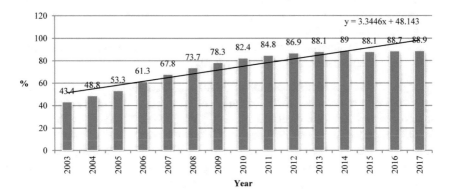

**Fig. 2.** Share of organizations that used information and communication technologies in Russia (Source: compiled by the authors of article based on the data of Federal State Statistics Service, 2017)

Regarding the ecosystem of the digital economy:

- successful operation of at least 10 leading companies (ecosystem operators) competitive in global markets;
- successful operation of at least 10 sectoral (industrial) digital platforms for the economy main subject areas (including digital health care, digital education and smart city);
- successful operation of at least 500 small and medium enterprises in the field of creating digital technologies and platforms and providing digital services.

Regarding personnel and education:

- the number of graduates of educational institutions of higher education in training areas related to information and telecommunication technologies - 120 thousand people per year;
- the number of graduates of higher and secondary vocational education with competences in the field of information technology at the average world level - 800 thousand people per year;
- the share of population with digital skills is 40%;
- Regarding research competencies building up and technological groundwork:
- the number of implemented projects in the field of digital economy (at least 100 million rubles) - 30 units;
- the number of Russian organizations participating in the implementation of large projects ($3 million) in the priority areas of international scientific and technical cooperation in the field of the digital economy - 10;

Regarding information infrastructure:

- the share of households with broadband access to the Internet (100 Mb/s) in the total number of households – 97%;
- sustainable coverage of 5G and above in all large cities (1 million people and more);
- Regarding information security:
- the share of entities using the standards of safe information interaction of state and public institutions – 75%;
- the share of the internal network traffic of the Russian segment of the Internet network routed through foreign servers is 5%.

Budget allocations in the amount of 3,040.4 million rubles were set to finance priority activities in the areas of "Information Infrastructure", "Formation of Research Competences and Technological Resources", "Information Security" of the program "Digital Economy of the Russian Federation" from the reserve fund of the Government of Russia according to the order under data of March 29, 2018 №528-p [16].

Creating predictable conditions and forming digital trust, the government assumes a decisive role in the reliability of the digital environment in the country, forms a responsible attitude of users to key institutions and organizations. This issue is of great interest to all participants in the digital economy, and given the best practices for the development of high-tech companies in the world, the global experience of their cluster development is of particular research interest.

## 3   Results

Over the past eight years, Russia has managed to significantly improve its position in the leading international rankings characterizing the conditions for economic growth (Global Competitiveness Index[1] и Doing Business[2]) and its quality (Global Innovation Index)[3] (see Fig. 3).

The greatest breakthrough occurred both in the long and in the short term in terms of creating favorable business conditions. Thus, in the Doing Business rating, Russia rose by 85 positions: from 124th place in 2011 to 35th place in 2017. At the same time, growth in the last year alone (2016–2017) amounted to +16 rating values.

Today, according to the World Bank, the success of government actions to ensure the business environment for all types of companies in Russia is comparable to such countries as the Netherlands (32nd place), Switzerland (33rd), Japan (34th), Slovenia (37th), Poland and Slovakia (39th). At the same time, there is still a lag behind the majority of EU countries (Great Britain - 7th place; Denmark - 3rd; Sweden - 10th) and OECD (New Zealand - 1st; Norway - 8th; USA - 6th (Fig. 3).

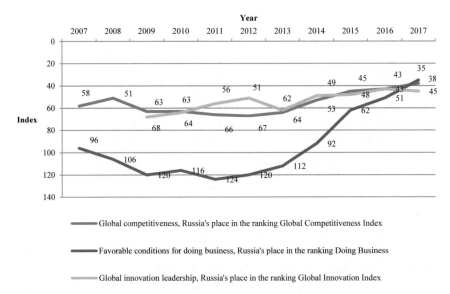

**Fig. 3.** Russia dynamics in the ratings of the Global Competitiveness Index, Doing Business and Global Innovation Index (Source: compiled by the authors of article based on the data of The Global Competitiveness Index Historical Dataset. 2007–2017 (Suggested citation: Cornell University, INSEAD, and WIPO. 2011–2017; The World Bank Group) [17–26].

---

[1]  http://www3.weforum.org/docs/GCR2017-2018/GCI_Dataset_2007-2017.xlsx.

[2]  http://russian.doingbusiness.org/ru/data.

[3]  http://www.wipo.int/publications/ru/series/index.jsp?id=129.

Despite the outlined growth in the authoritative international rankings (Global Competitiveness Index, Doing Business, Global Innovation Index), the share of articles by Russian scientists in international scientific journals remains low. Against the background of the positive internal Russian dynamics of patenting (Fig. 2), there is an insufficient level of international patent activity. Table 1 shows the dynamics of filed applications and patents received in the Russian Federation.

**Table 1.** Dynamics of filed applications and received patents in the Russian Federation

| Indicator | Year | | | | | | | |
|---|---|---|---|---|---|---|---|---|
| | 2010 | 2011 | 2012 | 2013 | 2014 | 2015 | 2016 | 2017 |
| **Patent applications filed:** | | | | | | | | |
| For inventions - total | 42500 | 41414 | 44211 | 44914 | 40308 | 45517 | 41587 | 36454 |
| For utility models - total | 12262 | 13241 | 14069 | 14358 | 13952 | 11906 | 11112 | 10643 |
| For industrial designs - total | 3997 | 4197 | 4640 | 4994 | 5184 | 4929 | 5464 | 6487 |
| **Patents issued:** | | | | | | | | |
| For inventions - total | 30322 | 29999 | 32880 | 31638 | 33950 | 34706 | 33536 | 34254 |
| For utility models - total | 10581 | 11079 | 11671 | 12653 | 13080 | 9008 | 8875 | 8774 |
| For industrial designs - total | 3566 | 3489 | 3381 | 3461 | 3742 | 5459 | 4455 | 5339 |
| **Number of valid patents - total** | **259698** | **236729** | **254891** | **272641** | **292048** | **305119** | **314615** | **326624** |

Source: compiled by the authors of article based on the data of Federal State Statistics Service, http://www.gks.ru/free_doc/new_site/business/nauka/innov6.xls

Russia domestic dynamics of patenting (Fig. 4).

According to Clarivate Analytics[4], in the Web of Science, Russian scientists published 50,401 papers in 2016. According to the Web of Science, the share of Russian publications from the world at the moment is just over 2.48%. Regarding the Scopus database of Elsevier[5], Russian scientists published 66,775 papers in 2016, including 46,901 articles, 14,321 conference materials and 2,114 reviews. In 2015, the total volume of publications was 64,925 works. The share of publications by Russian scientists increased by 0.17% and in 2016 was 2.48%.

The presented indicators were achieved primarily by launching the project "Improving the competitiveness of the leading universities of the Russian Federation

---

[4] https://clarivate.com/.

[5] https://www.elsevier.com/.

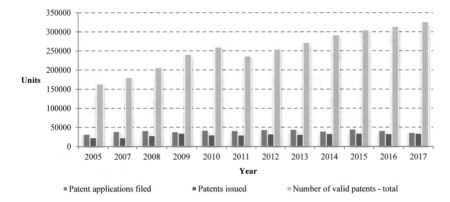

**Fig. 4.** Dynamics of inventive activity in Russia (Source:compiled by the authors of article based on the data of Federal State Statistics Service, 2017)

among the world's leading scientific and educational centers (5–100)" (Project 5–100) and the priority national project "Education" [27] (Fig. 5).

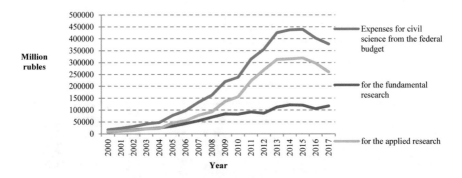

**Fig. 5.** Dynamics of expenses on civil science from the federal budget (Source:compiled by the authors of article based on the data of Federal State Statistics Service, 2017)

However, under the conditions of sanctions pressure, the innovative activity of organizations was significantly reduced, which confirms the data below in the diagram (Figs. 5 and 6).

At the same time, under the conditions of sanctions pressure, Russian companies continue to make technological innovations (Fig. 7).

Studies show that technology exports in the scale of the Russian economy remain insignificant in scope, and its structure is dominant by non-protectable results of intellectual activity and engineering services. For exports of high-tech products, Russia is comparable with foreign countries, but it happens at the expense of a limited number of niche product groups. Identified problems in research and development and

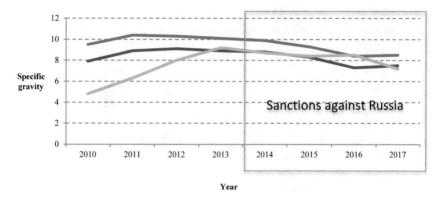

Innovative activity of organizations (the proportion of organizations that carried out technological, organizational, marketing innovations in the reporting year, in the total number of surveyed organizations)

The share of organizations implementing technological innovations in the reporting year, in the total number of surveyed organizations

The proportion of innovative goods, works, services in the total volume of goods shipped, work performed, services

**Fig. 6.** Russian companies innovation activity dynamics (Source:compiled by the authors of article based on the data of Federal State Statistics Service, 2017)

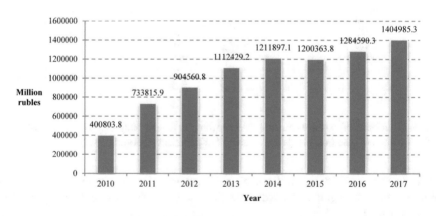

**Fig. 7.** The companies' costs dynamics implemented in technological innovation (Source: compiled by the authors of article based on the data of Federal State Statistics Service, 2017)

increasing sanctions pressure can erase the results achieved by the levels of innovative development of the country.

The chosen course on the digital economy is able to revive the Russian economy; however, it seems to us possible to solve these problems due to the accumulated modern experience in the framework of cluster development.

## 4  Discussion

According to the Global Innovation Index (GII) of 2017 [18], it was revealed that innovation activities tend to be geographically concentrated in specific clusters. This phenomenon allows us to make the assumption that innovation activity develops according to the principle of agglomeration of innovation activity, relying on global tools of innovation activity and on the development of an ecosystem that allows generating meaningful results [18]. Innovation clusters definitely have a significant impact on the development of industries in the economies of all countries and regions. Creating clusters of innovative enterprises has a positive impact on various spheres of life.

Clusters are a vivid example of the fact that with the high significance of global processes, it is the internal conditions of business vision at the micro level that are of greater importance for business development: the presence of permanent and proven contractors, streamlined infrastructure, qualified personnel, production processes improvement.

When building up and developing innovative territorial production clusters it is necessary to rely on the existing base of science cities, special economic zones, closed administrative and territorial entities. Further research will focus on the study of factors and the identification of its capabilities in the framework of cluster development, taking into account the development of digitalization of the economy.

Russia's first national cluster support program was launched by the Ministry of Economic Development in 2012. Its goal is to improve the interaction of enterprises with research and educational organizations of pilot innovation clusters, as well as to stimulate the development of territories with the greatest scientific, technical and production potential.

The program was developed taking into account the experience of previous support of cluster development centers, as well as the best international practices.

In particular, the following programs were used as benchmarks:

- The program focuses on bridging the gap between science and industry by supporting the strategic development of advanced clusters in knowledge-intensive sectors and the environments in which they are located;
- The program aims to ensure synergy and promote the best national joint public-private R & D projects, as well as commercialization and marketing of their results.

A competitive selection of cluster projects that will receive subsidies is carried out within the framework of both programs. Clusters receive government support over a five-year period.

Pilot innovation clusters were selected on a competitive basis: out of 94 applicants, 25 clusters became the winners. Subsequently, their number increased to 27. All of them were located in regions where, among other factors, science cities, special economic zones (SEZ) and off-limits areas (OLA) are located. From 2013 to 2015, subsidies from the federal budget in the amount of 5.05 billion rubles (113.64 million US dollars) were provided to the regions of the Federation hosting pilot innovation clusters. Funding was provided for the following events:

- developing innovative and educational infrastructure;
- strengthening cooperation, promoting a products cluster, including in foreign markets (business missions, fairs, exhibitions, promotional events);
- staff training, advanced training and retraining, as well as the provision of methodological, organizational, expert and information services;
- development of engineering and social infrastructure.

According to the Ministry of Economic Development, from 2013 to 2015, the total volume of cluster production increased by 429 billion rubles and amounted to almost 2 trillion rubles (32.26 billion US dollars).

Against the background of negative economic trends, these companies showed an increase in various performance indicators: the number of new high-performance jobs grew by more than a third (from 27.2 thousand in 2013 to 36.1 thousand in 2015); 40 thousand workers were trained or improved their professional qualifications. The

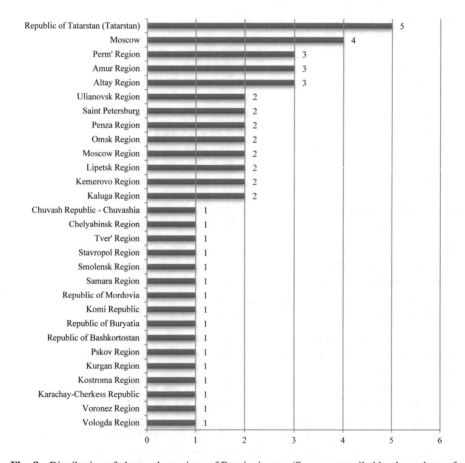

**Fig. 8.** Distribution of clusters by regions of Russia, in pcs. (Source: compiled by the authors of article based on the data of "Geoinformation system of industrial parks, technology parks and clusters of the Russian Federation" https://www.gisip.ru/)

development of pilot innovation clusters gave a significant impetus to investment activity: in just three years, public and private investments exceeded 98 and 360 billion rubles (2.2 and 8.1 billion US dollars), respectively.

The territorial and sectoral profile of the currently operating clusters in Russia are presented in Fig. 8.

At present, we have 50 clusters in the Russian Federation, including 38 at the initial stage, 10 at the medium stage, and only 2 at a high degree of readiness. Table 2

**Table 2.** Number of created regional clusters in the Russian Federation

| Direction | Stage of the organizational development | | | Participants in total |
|---|---|---|---|---|
| | Initial | Medium | High | |
| Aircraft industry | 1 | | 1 | 2 |
| Automotive and automotive parts manufacturing | 2 | 1 | | 3 |
| Environmental Protection and Recycling | 1 | | | 1 |
| Information and communication technology | 1 | | | 1 |
| Space industry | | 1 | | 1 |
| Forestry and processing, pulp and paper production | 3 | | | 3 |
| Medical industry | 8 | | | 8 |
| Metallurgy, metalworking and production of finished metal products | 1 | | | 1 |
| Microelectronics and instrument engineering | 1 | 3 | | 4 |
| New materials | 4 | 1 | | 5 |
| Optics and photonics | 1 | 1 | | 2 |
| Manufacture of railway locomotives and rolling stock | 1 | | | 1 |
| Manufacture of machinery and equipment (including machine tools and special equipment, lifting and hydropneumatic equipment, robots) | 4 | 1 | | 5 |
| Production of food, beverages and tobacco | 1 | | | 1 |
| Production of electricity and electrical equipment | 1 | | | 1 |
| Jewelry production | 1 | | | 1 |
| Industrial biotechnologies (production of products based on enzymes and microorganisms for subsequent use in the chemical industry, health care, food and feed production, detergents, paper and cellulose, textiles, and also in bioenergy) | 1 | | | 1 |
| Shipbuilding | 1 | | | 1 |
| Pharmaceuticals | 2 | 1 | | 3 |
| Chemical production | 2 | | 1 | 3 |
| Nuclear and Radiation Technologies | | 1 | | 1 |
| **Total** | **37** | **10** | **2** | **49** |

Source: compiled by the authors of article based on the data of "Geoinformation system of industrial parks, technology parks and clusters of the Russian Federation" presents the distribution of established regional clusters in the Russian Federation according to their areas of activity. Table 2 shows that currently there are 49 regional clusters in the Russian Federation at various stages of implementation (37 at the initial level, 10 at the middle level and 2 with a high degree of development).

According to the "Geoinformation System of Industrial Parks, Technoparks and Clusters of the Russian Federation", the Ministry of Industry and Trade of Russia, in 2017, there was information about 290 structures, of which 149 are industrial parks, 46 are technology parks and 50 are regional clusters, which include more than 1,765 private owners. Most industrial parks, technology parks and clusters are located in the Moscow Region - 36, Republic of Tatarstan - 27, Kaluga Region - 14, Leningrad Region, Republic of Bashkortostan, Sverdlovsk Region and Ulyanovsk Region - 10 each, Jewish Autonomous Region - 8, in Vladimir Region, Voronezh Region and St. Petersburg - 7, Moscow, Penza Region, Samara Region - 6, Lipetsk Region and Stavropol Region - 5, Amur Region - 4, Kaliningrad Region - 4. Other regions of the Russian Federation created less than 3 organizational structures from the number of industrial parks, technology parks and clusters.

As the cluster policy of Russia improves, the focus shifts from testing formats (pilot innovation clusters) and implementing (cluster development centers) to innovation cluster projects where comprehensive measures are developed, the goal of which is focused on attracting investments to the cluster at a level not lower than world experience. The application of the project approach to the formation and implementation of cluster policy is expected to achieve the following goals:

- reduce the time required to produce results;
- use resources more efficiently;
- increase transparency, soundness and timeliness of decision making;
- increase the effectiveness of cooperation between government agencies and administrative levels.

The best results from the cluster support mechanisms should be achieved through the implementation of intercluster projects involving the sharing of equipment and infrastructure, joint procurement and product promotion to foreign markets, and the upgrading of the skills of team leaders in clusters. An important factor in accelerating the development of clusters will be the constant exchange of advanced experience in cluster cooperation, including in the areas of attracting investment, developing innovation infrastructure and commercialization mechanisms, promoting exports, and developing promising R & D projects.

One of the conditions for the successful implementation of the entire complex of planned activities is that the regions, in which the leading clusters are located, carry out project management synchronously with the roadmap of the priority projects of the Ministry of Economic Development, in particular, through operational monitoring of the leading clusters.

## 5  Conclusions

Thus, the analysis showed the prospects for the development of innovation clusters in Russia in connection with the stimulating function of state bodies and the development of the clustering policy specified in strategic documents. The number of identified problems that should be overcome in the course of the development of cluster formations can be attributed to the shortage of qualified specialists in a whole range of industries, which will be resolved in the future as part of the state program on the modernization of education.

In addition, the problem remains low susceptibility to innovations, depreciation of fixed assets, and, as a result, inadequate quality and cost of products, which requires investments in the renewal of fixed assets and the development of technology, in particular through government subsidies. Low organizational level of development of innovation clusters, problems of cooperation and communication between participants, lack of domestic demand and difficulties in accessing international markets can be compensated by involving regional authorities in coordinating the cluster, introducing incentives to stimulate innovation development, updating the legislative base and correct product positioning. Problems of infrastructure development for innovation clusters can be solved through subsidies under the federal target program "Digital Economy of the Russian Federation", as well as regional development programs, the entire system of grants, concessional loans and the introduction of public-private partnership mechanisms. In general, the state policy and strategic development programs of the Russian Federation contribute to enhancing the clustering processes in the innovation sphere in the digital economy era.

## References

1. Yigitcanlar T (2010) Making space and place for the knowledge economy: knowledge-based development of Australian cities. Eur Plann Stud. https://doi.org/10.1080/09654313.2010.512163
2. Yigitcanlar T, Velibeyoglu K, Baum S (2008) Knowledge-based urban development: planning and applications in the information era. Knowledge-based urban development: planning and applications in the information Era. https://doi.org/10.4018/978-1-59904-720-1
3. Yigitcanlar T, O'Connor K, Westerman C (2008) The making of knowledge cities: Melbourne's knowledge-based urban development experience. Cities 25(2):63–72. https://doi.org/10.1016/j.cities.2008.01.001
4. Bitokova ZH (2010) Cluster education as an aspect of sustainable development of the region. Terra economicus 2–3(8):162–167
5. Afonichkin AI, Mikhalenko DG (2016) Strategies for coordinated development of cluster economic systems Bulletin of the Volga University. V.N. Tatischeva. № 24, pp 244–252
6. Vertakova YV, Polozhentseva YS, Klevtsova MG (2017) Vector analysis of regional cluster initiatives. Scientific and technical statements of the St. Petersburg State Polytechnic University. Economics. № 1 (211), pp 43–50

7. Batov GH, Kandrokova MM, Kumysheva ZH (2016) Organization of sustainable development of the regional economy based on cluster formations. Reg Econ Theor Pract 12:8–14
8. McKinsey Global Institute (MGI) (2014) Global flows in a digital age: how trade, finance, people, and data connect the world economy. MGI, San Francisco
9. Chakravorti B, Chaturvedi RS (2017) Digital planet 2017. How competitiveness and trust in digital economies vary across the world. The Fletcher School, Tufts University
10. Smartphone Connections Forecast (2016) Ovum. https://www.ovum.com/research/smartphone-connections-forecast-2016-21. Accessed 12 Oct 2018
11. More than half of Russians choose e-government services (2017) The Ministry of Communications and Mass Communications of the Russian Federation. http://minsvyaz.ru/ru/events/36563/. Accessed 16 Oct 2018
12. Internet World Stats (2018) Miniwatts Marketing Group. https://www.internetworldstats.com/stats.htm. Accessed 16 Oct 2018
13. Federal State Statistics Service (2017). http://www.gks.ru/. Accessed 14 Oct 2018
14. Collection of Legislation of the Russian (2017) Federation Presidential Decree of 09.05.2017 N 203 "On the Strategy for the Development of the Information Society in the Russian Federation for 2017–2030", 15.05.2017, N 20, Art. 2901
15. Legislation of the Russian Federation (2017) Order of the Government of the Russian Federation of 28.07.2017 N 1632-p "On approval of the program" Digital economy of the Russian Federation", 07.08.2017, N 32, Article 5138
16. Legislation of the Russian Federation (2018) Order of the Government of the Russian Federation of 29.03.2018 N 528-p "On the allocation in 2018 from the reserve fund of the Government of the Russian Federation budget allocations for the implementation of activities under the state programs" Economic Development and Innovative Economy", "Information Society (2011 – 2020)", "Development of the transport system "and "Development of the electronic and radio-electronic industry for 2013–2025", 09.04.2018, N 15 (Part V), Art. 2183
17. The Global Competitiveness Index Historical Dataset (2007–2017) World Economic Forum. Geneva. Version 20180712. http://www3.weforum.org/docs/GCR2017-2018/GCI_Dataset_2007-2017.xlsx. Accessed 14 Oct 2018
18. The Global Innovation Index 2018: Energizing the World with Innovation (2018) Suggested citation: Cornell University, INSEAD, and WIPO. Ithaca, Fontainebleau, and Geneva
19. The Global Innovation Index 2017: Innovation Feeding the World (2017) Suggested citation: Cornell University, INSEAD, and WIPO. Ithaca, Fontainebleau, and Geneva
20. The Global Innovation Index 2016: Winning with Global Innovation (2016) Suggested citation: Cornell University, INSEAD, and WIPO. Ithaca, Fontainebleau, and Geneva
21. The Global Innovation Index 2015: Effective Innovation Policies for Development (2015) Suggested citation: Cornell University, INSEAD, and WIPO. Fontainebleau, Ithaca, and Geneva
22. The Global Innovation Index 2014: The Human Factor In innovation, second printing (2014) Suggested citation: Cornell University, INSEAD, and WIPO. Fontainebleau, Ithaca, and Geneva
23. The Global Innovation Index 2013: The Local Dynamics of Innovation (2013) Suggested citation: Cornell University, INSEAD, and WIPO. Geneva, Ithaca, and Fontainebleau
24. The Global Innovation Index 2012: The Local Dynamics of Innovation (2012) Suggested citation: Cornell University, INSEAD, and WIPO. Geneva, Ithaca, and Fontainebleau
25. The Global Innovation Index 2011: The Local Dynamics of Innovation (2011) Suggested citation: Cornell University, INSEAD, and WIPO. Geneva, Ithaca, and Fontainebleau

26. Doing Business (2007–2018) The World Bank Group. http://russian.doingbusiness.org/ru/data. Accessed 10 Oct 2018
27. Legislation of the Russian Federation (2011) Resolution of the Government of the Russian Federation of 07.02.2011 N 61 (as amended on December 25, 2015) "On the Federal Target Program for the Development of Education for 2011 – 2015". 07.03.2011, N 10, Art. 1377

# The Competitiveness of Single-Industry Cities in the Digital Transformation of the Economy

M. S. Guseva[✉] and E. O. Dmitrieva

Samara State University of Economics, Samara, Russia
{n_econ, n_econ}@sseu.ru

**Abstract.** At the present stage, successful cities are experiencing a digital transformation. Information and communication technologies (ICT) contribute to the enhanced competitiveness, improving the quality of life, establishing the interaction between citizens, business and government. These issues are considered through the concept of "Smart City". The demand for intelligent technologies "Smart City" is gradually taking shape in Russia. In this part, Russia is significantly inferior to leading economies, and the projects being implemented are mainly related to point digitalization. Single-industry cities are in great demand for "smart" solutions. Problems of single-industry cities are considered as one of the main priorities of national policy. For decades, single-industry cities have been the basis of Russian economy and, at the same time, the most vulnerable elements of the socio-economic space. In economy digitalization, single-industry cities do not need measures to overcome the temporary decline in production. They strongly need a new development model. The main goal of this model should be to enhance the efficiency of the economic mechanism based on the introduction of digital technologies and platform solutions into the economy and management of the city. Taking into account the features of socio-economic development and the existing needs, the company has the opportunity to choose the appropriate model. We believe that the model of a single-industry city should be based on a comprehensive assessment of its competitiveness. The purpose of the contribution is to determine the specific nature of assessing the competitiveness of single-industry cities under digital transformation of the economy and to develop guidelines based on the results of theoretical and empirical data. The authors of the contribution attempted to select the optimal scenario for digital transformation of the single-industry city depending on the model of its development.

**Keywords:** The competitiveness · Digital transformation · Digitalization ·
Evaluation · Single-industry city

## 1  Introduction

For decades, single-industry cities have been the basis of Russian economy. They account for about 40% of the country's total GRP. 319 municipalities with a total population of about 13 million people (about 9% of the population) have the status of "single-industry municipality (single-industry city)" in Russia. About 1/3 of single-industry cities are in a difficult socio-economic situation. The main difference of single-industry cities is the strong dependence of the socio-economic situation of the city on activities of one or several

S. Ashmarina et al. (Eds.): *Digital Transformation of the Economy: Challenges,*
*Trends and New Opportunities*, pp. 216–226, 2019.
https://doi.org/10.1007/978-3-030-11367-4_20

enterprises producing homogeneous products. The production base of these enterprises is obsolete today. The products have a low competitive advantage, which causes a reduction in production. Financial problems of a city-forming enterprise gradually develop into socio-economic problems of the whole city. The continued dependence of single-industry cities on city-forming enterprises and the poor quality of the urban environment necessitate the search for effective tools to support and develop single-industry territories.

Solving the problems of single-industry cities is considered as one of the main priorities of state policy in the Russian Federation. The goals of state support for single-industry cities are shifted from "withdrawing a single-industry city from an uncontrolled risk zone" to "ensuring stable and, in some cases, priority development", as evidenced by the continuous improvement of measures to manage the development of single-industry municipalities of Russia.

In economy digitalization, single-industry cities do not need measures to overcome the temporary decline in production. They need a new development model. The main goal of this development model should be to enhance the efficiency of the economic mechanism based on the introduction of digital technologies and platform solutions into the economy and management of single-industry cities. The Center for Strategic Research [7] of priority directions for introducing "smart city" technologies in Russian cities considers three possible scenarios for digital development: centralized, decentralized, and a model of local measures for digital transition. Earlier, the Center for Strategic Research [6] conducted a study in 18 single-industry cities of Russia and proposed three models of single-industry cities and a set of measures and mechanisms for their implementation: controlled compression, stable single-industry city, industrial diversification of a single-industry city.

Each company in Russia, taking into account the socio-economic situation and existing needs has the opportunity to select the appropriate development model and determine the scenario for digital transformation. We believe that the selection of a development model for a single-industry city should be based on a comprehensive assessment of the competitiveness.

The competitiveness refers to a measurable category, which implies the presence of a subject (who assesses), an object (what is assessed), and objectives (criteria) of assessment. Of particular scientific interest is a comparative analysis of single-industry cities of the Samara region upon the integrated index that takes into account various spheres of life. The single-industry city as an object of the competitiveness is poorly studied, which requires a deep theoretical and methodological study.

The purpose of the contribution is to determine the specific nature of assessing the competitiveness of single-industry cities in economy digitalization and to suggest guidelines based on the results of theoretical and empirical data.

## 2   Materials and Methods

The authors used the methods of system analysis, the method of comparisons and analogies, the method of generalizations, the methods of dialectical and statistical analysis, the method of expert estimates, the method of retrospective estimates and structural and dynamic analysis, the use of which allowed ensuring the reliability of the study and the validity of theoretical conclusions and suggestions.

From a methodological point of view, this contribution suggests the following approach to assessing the competitiveness of a single-industry city:

Stage 1. Standardization of index values upon the variation range to one numerical meter (from 0 to 1). To do this, we use the following formula:

$$X = (Xf - Xmin)/(Xmax - Xmin) \qquad (1)$$

If we consider the index, the growth of which is a negative trend (unemployment, the share of the subsistence minimum in the average monthly wage, the mortality rate, the share of the city-forming enterprise in the city-wide production, the share of employed in the city-forming enterprise in the total working-age population), then (1) is the following:

$$X = (Xmax - Xf)/(Xmax - Xmin) \qquad (2)$$

where X is the standardized index value of the company;

Xf - the actual index value;

Xmin, Xmax - the minimum and maximum index value from a set of single-industry cities, respectively.

Stage 2. Formation of six sub-indices of the competitiveness by spheres (Table 1). It is necessary to calculate the arithmetic average of standardized indices included in their composition.

**Table 1.** Indices of the competitiveness of single-industry cities

| Spheres of activity | Index |
| --- | --- |
| Economy (E) | Volume of production of goods and services per capita |
| | Retail trade per capita |
| | Share of profitable enterprises |
| | Balanced financial result |
| Finance (F) | Local budget revenues per capita |
| | Volume of investment in fixed capital per capita |
| | Budget deficit (surplus) |
| Labor resources (L) | Number of working-age population |
| | Unemployment |
| Demographics (D) | Fertility rate |
| | Mortality rate |
| | Migration growth rate |
| | Natural growth rate |
| Single-industry profile (S) | Share of city-forming enterprise in citywide production |
| | Share of employed in the city-forming enterprise in the total working-age population |

Source: compiled by the authors.

Stage 3. Calculation of the integrated index of the competitiveness. It is calculated as the arithmetic mean of sub-indices of the selected spheres.

Step 4. Decoding index values of the competitiveness [16].

Stage 5. Selection of the digital transformation scenario depending on the level of the competitiveness of the single-industry city.

## 3    Results

At the legislative level, Togliatti and Chapayevsk are included in the list of single-industry cities of the Samara region. In accordance with the right of subjects of the Russian Federation to independently determine the criteria for assigning cities to single-industry ones, Chapaevsk, Oktyabrsk and Pokhvistnevo are assigned to single-industry cities (Fig. 1).

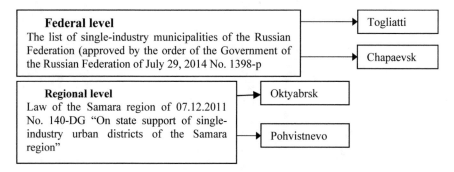

**Fig. 1.** State support for single-industry cities of the Samara region (Source: compiled by the authors)

In the territory of each single-industry city of the Samara region there is a city-forming enterprise. In Oktyabrsk, this is Autocomponent Engineering-2 LLC, specializing in the production of technical and specific products from thermoplastics by injection molding. Its share, according to data for 2017, accounts for 64.6% of the total city production. In Pokhvistnevo - Samaraneftegaz PJSC (94.1%), in Chapayevsk - Promsintez PJSC, the production of explosives (21.9% of the industrial production), in Togliatti - AvtoVAZ PJSC, the production of vehicles and equipment (51%).

Based on systematized data and the adapted index to the specific nature of single-industry territories, taking into account the availability of statistical data, the authors calculated the index of the competitiveness of the Samara region for the period 2013–2017. The results of calculations are presented in the table (Table 2).

The calculations showed a different level of the competitiveness of the studied single-industry cities. A relatively favorable situation in the economic sphere is typical

**Table 2.** Dynamics of indices of the competitiveness of single-industry cities of the Samara region

| Single-industry city | Year | Spheres of activity | | | | | | Index |
|---|---|---|---|---|---|---|---|---|
| | | Economy | Finance | Labor resources | Social sphere | Demographics | Single-industry profile | |
| Oktyabrsk | 2013 | 0,0 | 0,3 | 0,0 | 0,2 | 0,1 | 0,9 | 0,2 |
| | 2014 | 0,2 | 0,5 | 0,0 | 0,2 | 0,0 | 0,9 | 0,3 |
| | 2015 | 0,2 | 0,7 | 0,1 | 0,2 | 0,2 | 0,8 | 0,4 |
| | 2016 | 0,2 | 0,6 | 0,0 | 0,2 | 0,3 | 0,8 | 0,3 |
| | 2017 | 0,2 | 0,4 | 0,0 | 0,2 | 0,3 | 0,7 | 0,3 |
| Pohvistnevo | 2013 | 0,4 | 0,7 | 0,2 | 0,2 | 0,6 | 0,5 | 0,4 |
| | 2014 | 0,7 | 0,7 | 0,5 | 0,2 | 0,6 | 0,5 | 0,5 |
| | 2015 | 0,6 | 0,7 | 0,4 | 0,2 | 0,6 | 0,5 | 0,5 |
| | 2016 | 0,7 | 0,7 | 0,5 | 0,3 | 0,7 | 0,5 | 0,6 |
| | 2017 | 0,7 | 1,0 | 0,4 | 0,3 | 0,6 | 0,5 | 0,6 |
| Togliatti | 2013 | 0,9 | 0,4 | 1,0 | 0,7 | 0,8 | 0,3 | 0,7 |
| | 2014 | 0,6 | 0,4 | 0,8 | 0,7 | 0,8 | 0,3 | 0,6 |
| | 2015 | 0,6 | 0,5 | 0,5 | 0,6 | 0,8 | 0,3 | 0,6 |
| | 2016 | 0,6 | 0,5 | 0,7 | 0,7 | 0,8 | 0,3 | 0,6 |
| | 2017 | 0,6 | 0,2 | 0,9 | 0,6 | 0,7 | 0,3 | 0,5 |
| Chapaevsk | 2013 | 0,3 | 0,5 | 0,3 | 0,2 | 0,4 | 0,9 | 0,4 |
| | 2014 | 0,6 | 0,4 | 0,5 | 0,3 | 0,5 | 0,9 | 0,5 |
| | 2015 | 0,6 | 0,3 | 0,5 | 0,3 | 0,5 | 0,9 | 0,5 |
| | 2016 | 0,6 | 0,4 | 0,5 | 0,3 | 0,5 | 0,9 | 0,5 |
| | 2017 | 0,7 | 0,4 | 0,5 | 0,3 | 0,3 | 0,9 | 0,5 |

Source: compiled by the authors.

for Pohvistnevo, Chapaevsk and Togliatti. Oktyabrsk is characterized by relatively low values of such indices as "Volume of production of goods and services per capita", "Retail trade turnover per capita" and "Share of profitable enterprises".

The absolute leader in the financial sphere is Pokhvistnevo, which demonstrated the maximum index values under consideration. This trend is characteristic only for 2017, when significant volumes of investments of oil producing enterprises (PAO Transneft Volga Region, Samaraneftegaz PJSC) were attracted to the economy. Togliatti became an outsider in this direction. Since 2013, Togliatti's fiscal capacity has been steadily declining. About 51% of revenues in the budget structure of Togliatti accounted for non-repayable income from other levels of the budget system.

The assessment of the component "Labor resources" showed the inconsistency of Oktyabrsk: the city is characterized by a reduction in the number of the permanent population, a negative natural increase. Pokhvistnevo and Chapaevsk also received low scores in terms of the number of working-age population. Stable migration growth of the population covers the natural decline in the population of these urban districts, but does not ensure the growth of the working-age population.

On the contrary, Togliatti is characterized by a high value of labor potential assessment, where the lowest unemployment rate is 0.7%.

The indices of "Social Sphere" did not bring high scores to single-industry cities. The index value for this component was higher than the average for Togliatti. The situation is similar in the demographic sphere of single-industry cities, in Togliatti and Pokhvistnevo.

We note a wide differentiation in terms of mono-profile indices: Chapaevsk is among the leaders, negative index values considered in this group were observed in Tolyatti and Pokhvistnevo, which testifies to the results of economic diversification.

For clarity, Fig. 2 presents indices of the competitiveness of single-industry cities of the Samara region by spheres of activity in 2017.

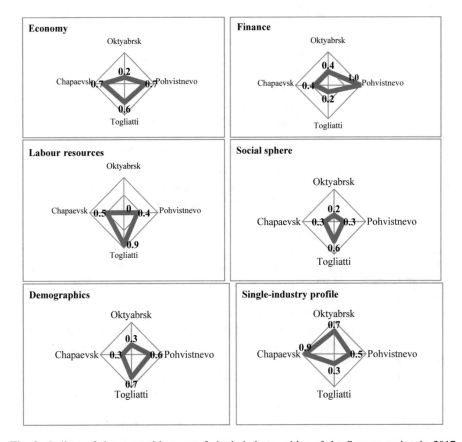

**Fig. 2.** Indices of the competitiveness of single-industry cities of the Samara region in 2017 (Source: compiled by the authors)

According to values of the relevant indices obtained in 2017, Oktyabrsk is among the outsiders in four of the six surveyed spheres (economy, labor, social sphere, demography), which determined the lowest index values of the competitiveness of the single-industry city (from 0.26 to 0.5). The resulting trend is observed throughout the study period (Fig. 3).

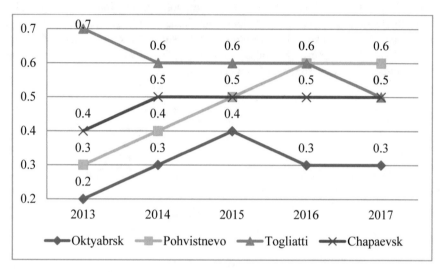

**Fig. 3.** Dynamics of the final index value of the competitiveness of single-industry cities (Samara region in the period 2013–2017) (Source: compiled by the authors)

The cities of Pokhvistnevo, Togliatti and Chapaevsk are in relatively equal conditions, their competitiveness is "above average" (from 0.5 to 0.75). However, for Pokhvistnevo and Chapaevsk, there is a positive trend: the final index value of the competitiveness for the period under review is gradually increasing. The position of Togliatti deteriorates over the years, which is largely due to the decline in indices for such components of the competitiveness as "Economy" and "Finance". On the contrary, there is the improvement in index values for these positions and the growth of labor potential in Pohvistnevo (it was the leader among the surveyed cities upon the index of the competitiveness in 2017).

Thus, on the basis of calculated indices of the competitiveness, it seems possible to determine the most effective model for the development of single-industry cities of the Samara region in economy digitalization (Table 3).

To enhance the competitiveness of single-industry territories, it is necessary to implement a digital transformation scenario that corresponds to its development model, which presupposes a concentration of high-tech production, knowledge, competencies and technologies at this stage.

**Table 3.** Models for the development of single-industry cities of the Samara region

| Single-industry cities | Conditions | Model | Scenario for digital transformation |
|---|---|---|---|
| Oktyabrsk Pohvistnevo | is not included in federal support programs for single-industry cities; does not have enough internal potential to diversify the economy: city-forming enterprise in the medium term maintains stable production volumes; all aspects of vital activity depend on the state of production of a city-forming enterprise | Stable single-industry city | Local measures |
| Togliatti | is included in federal support programs; the beginning of the strategy implementation on economy diversification; city-forming enterprise of the city is stable and is characterized by positive dynamics of industrial production rates | Sustainable development of a diversified city with a city-forming enterprise | Decentralized model |
| Chapaevsk | | | Centralized model |

Source: compiled by the authors

# 4   Discussion

A number of regional studies present various methodologies for assessing the competitiveness of a territory. It has been proven that the same tools and methods for analyzing national and regional competitiveness can be adapted to cities. The combination of econometric and qualitative assessment methods gives a comprehensive picture of the competitiveness of cities and their future prospects [5]. Some authors assess the competitiveness of cities upon one or several indices. Others have developed theoretical models of urban competitiveness, combining quantitative and qualitative indices, aimed at forming an integrated assessment of the economic and non-economic aspects of the competitiveness of a city [10, 11]. In modern works, the competitiveness is assessed by a multi-level system of index values of city development [21]. Studies of the competitiveness of Chinese cities [8, 14, 20] are also based on a multi-level hierarchical model built on the index method and spatial analysis. Major cities of Korea are assessed upon innovative indices proposed in the empirical study. It is proposed to assess the competitiveness of cities from three positions: formation of clusters, human capital and creative economy [12]. Comprehensive studies of cities were carried out in the European Union [2, 3, 15, 18]. In assessing the competitiveness, its sources are identified, which makes it possible to choose the ways of the most efficient allocation of resources in the development and implementation of government programs [1].

The authors of [13] study the conditions under which the concept of a "smart city" will ensure the competitiveness, which seems relevant in the framework of this contribution. A global study of 500 cities from 5 continents [17] analyzes the existing scientific and practical groundwork for assessing the competitiveness in order to develop a conceptual framework and indicator system for sustainable urban competitiveness. In the future, it is planned to build a global sustainability index for urban competitiveness using this assessment system.

In recent decades, the competitiveness of spatial socio-economic systems has been developing very actively at different levels (countries, regions, cities). The contribution [19] showed that the most important aspects of a scientific problem such as the consistent conceptualization of the phenomenon of regional competitiveness, the identification of objects which this concept is applied to, as well as the selection of assessment methods, are not sufficiently developed. In earlier domestic works, it is proposed to measure the competitiveness of the territory [9] by the rank method and the method of potential assessment: resource potential, potential of the quality of life, financial potential, ecological potential and organizational potential. To assess the competitiveness of a city, researchers [4] distinguish the following categories: economy, social sphere, culture, economic activity, production, administrative resources. The study of the competitiveness of single-industry cities, conducted and tested on the example of single-industry areas of the Central Federal District [16], was of particular interest. This methodological approach is taken as a basis for assessing the competitiveness of single-industry cities of the Samara region to select the model and development scenario in economy digitalization.

## 5 Conclusions

The theoretical significance of the contribution is to develop the methodology for assessing the competitiveness of single-industry cities in economy digitalization. The proposed method allows identifying promising areas and tools to improve the efficiency of their development in economy digitalization. The findings and conclusions of the contribution develop the theory and methodology for management of single-industry territories. The obtained scientific results and practical recommendations to enhance state support tools for single-industry cities of the Samara region can be used by federal and regional authorities.

The most important factor in the competitive strength of single-industry cities in current conditions is the introduction of digital technologies and platform solutions in all spheres of life. Digitization of business entities is becoming an indispensable tool that can stimulate the susceptibility of state and business structures to innovations, as well as increase their innovative activity. In this regard, it is necessary to improve the existing methodologies for assessing the competitiveness of cities, taking into account the impact of digital transformations, and determines directions for further research.

# References

1. Akpinar M, Can Ö, Mermercioglu M (2017) Assessing the sources of the competitiveness of the US states. Compet Rev 27(2):161–178. https://doi.org/10.1108/CR-02-2016-0014
2. Annoni P, Dijkstra L (2013) EU regional the competitiveness index: RCI 2013. Publications Office of the EU, Luxembourg
3. Annoni P, Dijkstra L (2017) Measuring and monitoring regional the competitiveness in the European union. In: Huggins Robert (ed) Handbook of regions and the competitiveness: contemporary theories and perspectives on economic development. Edward Elgar Publishing, Massachusetts
4. Bogomolova IV, Mashentsova LS, Sazonov SP (2014) Sustainable development of large cities from the standpoint of assessing the competitiveness of the territory [Ustoychivoye razvitiye krupnykh gorodov s pozitsii otsenki konkurentosposobnosti territorii]. Basic Res 9 (11) 6:2506–2510
5. Bruneckiene J, Cincikaite R, Kilijoniene A (2012) The specifics of measurement the urban the competitiveness at the national and international level. [Miestų konkurencingumo vertinimo nacionaliniu ir tarptautiniu mastu ypatumai]. Eng Econ 23(3):256–270. https://doi.org/10.5755/j01.ee.23.3.1272
6. Center for Strategic Research (2013) Single-industrial cities: reboot. Search for new models of the functioning of single-industry cities of Russia in the changed economic conditions. http://www.ladoga-park.ru/content/2014/04/140426152728/140426152728140426152938.pdf/ Accessed 25 Mai 2018
7. Center for Strategic Research (2018) Smart city in Russia. How to make the cities to mind? https://www.csr.ru/wp-content/uploads/2018/06/Report-Smart-Cities-WEB.pdf/. Accessed 21 Mai 2018
8. Du Q, Wang Y, Ren F, Zhao Z, Liu H, Wu C, Shen Y (2014) Measuring and analysis of urban the competitiveness of Chinese provincial capitals in 2010 under the constraints of major function-oriented zoning utilizing spatial analysis. Sustainability (Switzerland) 6 (6):3374–3399. https://doi.org/10.3390/su6063374
9. Grinchil BM, Korosteleva NE (2003) The most important factors for increasing the the competitiveness of regions [Vazhneyshiye faktory povysheniya konkurentosposobnosti regionov]. Eurograd, St. Petersburg
10. Hu R, Blakely EJ, Zhou Y (2013) Benchmarking the competitiveness of Australian global cities: Sydney and Melbourne in the global context. Urban Policy Res 31(4):435–452 https://doi.org/10.1080/08111146.2013.832667
11. Hu R (2015) Sustainability and the competitiveness in Australian cities. Sustainability (Switzerland) 7(2):1840–1860. https://doi.org/10.3390/su7021840
12. Kwon S, Kim J, Oh D (2012) Measurement of Urban The competitiveness based on innovation indicators in six metropolitan cities in Korea. World Technopolis Rev 1(3):177–185. https://doi.org/10.7165/wtr2012.1.3.177
13. Monfaredzadeh T, Berardi U (2014) How can cities lead the way towards a sustainable, competitive and smart future? WIT Trans Ecol Environ 191:1063–1074. https://doi.org/10.2495/SC140902
14. Ni P, Wang Y (2017) Urban sustainable the competitiveness: a comparative analysis of 500 cities around the world. In: Huggins Robert (ed) Handbook of regions and the competitiveness: contemporary theories and perspectives on economic development. Edward Elgar Publishing, Massachusetts

15. Prokop V, Stejskal J (2015) Impacts of local planning to the competitiveness index change – using approximate initial analysis of the Czech regions. WSEAS Trans Bus Econ 12:279–288
16. Rastvortseva SN, Manaeva EV (2016) The development of methodological support for the assessment and forecasting of the socio-economic state of a monocity [Razvitiye metodicheskogo obespecheniye otsenki i prognozirovaniya sotsial'no-ekonomicheskogo sostoyaniya monogoroda]. Econ-Inform, Moscow
17. Sáez L, Periáñez I (2017) Measuring urban the competitiveness in Europe. Handbook of regions and the competitiveness: contemporary theories and perspectives on economic development. Edward Elgar Publishing, Massachusetts, pp 463–491. https://doi.org/10.4337/9781783475018
18. Sáez L, Periáñez I, Heras-Saizarbitoria I (2017) Measuring urban the competitiveness: Ranking european large urban zones. J Place Manag Dev 10(5):479–496. https://doi.org/10.1108/JPMD-07-2017-0066
19. Ukrainskiy VN (2018) Regional the competitiveness: methodological reflections [Regional'naya Ukrainskiy konkurentosposobnost': metodologicheskiye refleksii]. Quest Econ 6:117–132
20. Wang L, Shen J (2017) Comparative analysis of urban the competitiveness in the yangtze river delta and pearl river delta regions of China, 2000–2010. Appl Spat Anal Policy 10 (3):401–419. https://doi.org/10.4337/9781783475018
21. Zhang W, Deng F, Liang X (2015) Comprehensive evaluation of urban the competitiveness in Chengdu based on factor analysis. In: Ninth international conference on management science and engineering management, pp 1433–1440. https://doi.org/10.1007/978-3-662-47241-5_119

# Problems of Digital Technologies Using in Employment and Employment Relations

M. K. Kot[(⊠)], F. F. Spanagel, and O. A. Belozerova

Samara State University of Economics, Samara, Russia
{mkroz,Belozerovaoa}@mail.ru, shpanur@yandex.ru

**Abstract.** The paper deals with the use of digital technologies in the relations of labour organization and labour management in the Russian Federation. On the basis of the current legislation and judicial practice conclusions about the possibility of introducing information technology in the personnel document, in the process of registration of the employment contract, as well as in the sphere of labour organization for the development of remote employment. The authors come to the conclusion that there are legislative prerequisites for the refusal of the use of material carriers in employment and employment relations, provided a positive approach of the judicial system and the exclusion of a number of legal and technical risks. At the same time, it is noted that some forms of information technology use contradict the requirements of the Russian labour legislation and therefore cannot be implemented in practice.

**Keywords:** Digital technologies · Electronic document · Labour law of russia remote work

## 1 Introduction

1.1. The modern paradigm of development of society as an open information space involves the introduction of digital technologies in all spheres of life, including employment and employment. The computerization of public administration systems, the introduction of information technologies in human resources management processes are real challenges for the restructuring of the current model of legal regulation of labour relations in Russia. The complexity of the restructuring of this legislative body lies in the features of Russian labour law, the concept of which was formed under the influence of bureaucratic practices of the Soviet period, which gave rise to numerous requirements for the design of the processes of internal organization of labour in the enterprise, as well as individual labour relations with employees. The formalization of literally every stage of interaction between an employee and an employer is connected, in addition, with the imperativeness of most regulations in the field of labour law, a high level of violations of workers' rights, low legal culture of the population, methods of state control and supervision in this area.

The interest in the most complete documentation of labour relations is due to the fiscal interests of the state, the need to control the occupancy of the budgets of social security and pension funds. These circumstances demonstrate the need to move to a new technological level of processing and storage of information in the field of employment and employment, which would reduce the time and material costs of the

S. Ashmarina et al. (Eds.): *Digital Transformation of the Economy: Challenges, Trends and New Opportunities*, pp. 227–234, 2019.
https://doi.org/10.1007/978-3-030-11367-4_21

employer, as well as public authorities for the preparation, manufacture, transfer, storage, provision of such documents.

1.2. In addition to the above-mentioned technological problems, the labour market in Russia is also undergoing structural changes: previously uncharacteristic informal, one-time employment, remote or remote employment, and home-based work are on the rise. The usual forms of labour organization for the production economy are lost through direct management of the employee, subordination of his power to the employer. The use of information and communication networks can reduce the managerial impact, the level of control of employees, which generally has a positive impact on the development of human resources, makes it possible to combine work with other forms of social activity (family, education, sports).

1.3. In support of the idea of forming an information society in the Russian Federation, the state is taking certain actions. Thus, the President of the Russian Federation issued a decree "On the Strategy of development of the information society in the Russian Federation for 2017–2030" [1], which, in particular, involves:

> creation of various technological platforms for distance learning in order to increase the availability of quality educational services;
> improvement of mechanisms for providing financial services in electronic form and ensuring their information security;
> stimulation of Russian organizations in order to provide employees with conditions for remote employment;
> development of technologies of electronic interaction of citizens, organizations, state bodies, local governments along with preservation of possibility of interaction of citizens with the specified organizations and bodies without application of information technologies;
> application in public authorities of the Russian Federation of the new technologies providing improvement of quality of public administration.

Federal law No. 60-FZ of 05.04.2013 "On amendments to certain legislative acts of the Russian Federation" [2] added a Chapter regulating the work of remote workers to the Labour code of the Russian Federation. The Chapter provides for the registration of labour relations with such employees through the exchange of electronic documents with the use of electronic digital signature, the use of postal services to send the necessary documents.

The state authorities (the Ministry of labour and social protection of the Russian Federation) are implementing the Digital economy Programme (order of the Government of the Russian Federation of 28 July 2017 No. 1632-R), which envisages the development of a draft law on the accounting of data on the employment of workers in electronic form (e - work book) (December 2018, the introduction of a draft law to the state Duma-February 2019), the project on the introduction of a personnel document management system in electronic form (December 2018) [3]. Since 2020. The Ministry of labour plans to switch to the electronic form of entering and storing information about the labour activity of workers for all first-time entrants to work. At the same time, the paper form of the work book for already working citizens is preserved.

The bodies of state control and supervision in the field of labour (Rostrud) carried out a pilot project to check the state of labour relations in PJSC "Sberbank" and NAO "Yulmart" using a system of protected legally significant exchange of electronic documents through telecommunication channels. The Agency plans to completely abandon the traditional methods of documentary and on-site inspections in favor of the "Automated control and Supervisory activity management system". In this case, the approach to the implementation of state supervision is changing: planned inspections give way to system monitoring with an emphasis on the prevention of offenses in the field of labour [3].

Another area of use of digital technologies in Russia is the replacement of certain labour functions traditionally performed by people with higher professional education, robots, whose activity is to process and analyze information: legal databases, customer information. The innovator in this area was also PJSC "Sberbank", in which the work on drafting contracts with customers was entrusted to robots [4].

The current need for the introduction of digital technologies in the field of labour regulation, as well as the willingness of public authorities to support this process are only a General prerequisite for a qualitatively new format of interaction between employees, employers and the state. Systemic problems and constraints remain that keep the country within the traditional approach to understanding employment and maintaining personnel records. These problems and limitations include:

- technological problems: underdevelopment of information communications in General, uneven development of digital technologies in the country (city-village, "Central" Russia and geographically remote settlements), the use of outdated technology, incomplete and uneven coverage of the "Internet" in the country, the need for continuous monitoring of information threats;
- "human" problems: lack of skills in working with information systems in the middle and older generation, their unwillingness and inability to learn new methods of work; accumulation of fatigue from working exclusively with information resources, blurring the lines between working time and rest time (especially for remote employment), reducing the possibility of live communication, weakening of social ties, sociopathy;
- economic problems: the need to increase the cost of updating information systems, staff training, the creation and updating of information security systems;
- legal problems: inflexibility of the current labour legislation, the judicial system (requirements for the provision of paper media), restrictions on rights or denial of protection of rights in the absence of technological possibility to use electronic documents, technical failure or loss of data;
- organizational problems: time costs for the creation and implementation of information systems, recovery of lost data due to technical failure, the need for duplication of information on paper and electronic media;
- system costs: as a result of the introduction of remote employment and information technologies, the structure of employment will be restructured, associated with the death of a number of working positions, the development of temporary employment options, the growth of informal employment, which can lead to a reduction in the budgets of social funds and social insecurity of such citizens.

This article provides an analysis of solutions to these problems on the basis of an assessment of the existing provisions of labour legislation and judicial practice. To do this, it is necessary to identify specific areas of legal impact that require the use of digital technologies, as well as to address the following issues.

- Does the current legislation allow the introduction of information technologies in the process of labour organization without making significant changes?
- Is it possible on the basis of the existing approaches of judicial practice effective protection of the rights of the parties to the employment relationship: employees and employers?

The purpose of this work is to assess the possibilities of the current legal system for the perception of the requirements of the digital economy, which consists in the rejection of the use of paper media, the introduction of information methods of labour organization and supervision of the state, remote employment.

## 2   Materials and Methods

The authors use the methods of system analysis, generalization, formal legal method and different ways of interpretation of normative acts. These methods made it possible to comprehensively consider the labour and civil legislation norms, judicial practice and they make conclusions about the possibility of widespread use of digital technologies in the labour organization sphere in the existing social and economic conditions in Russia.

## 3   Results

Since the current labour legislation already provides for the use of digital technologies in the design of labour relations with a remote worker (Chapter 49.1 of the Labour code of the Russian Federation), the authors propose to use the rules on remote labour as norms of General impact, extending them to the following areas.

1. Sphere of labour organization: familiarization with local regulations of the organization, orders, orders of the employer can be carried out by exchanging electronic documents using electronic digital signature (part 5 of article 312.1 of the Labour code). Similarly, other mandatory documents may be issued: notifications of change and termination of the employment contract (Art. 74, 79, 81 of the Labour code), agreement on termination of the employment contract (article 78 of the Labour code), liability (article 244 of the Labour code), explanatory notes (article 193 of the Labour code), pay sheet (art. 136 of the Labour Code RF), offers a different existing works, while reducing the number or staff of employees, the change of certain parties to the terms of the contract, etc. (articles 180, 74, 83 of the Labour Code RF).

The ability to use enhanced electronic signature in various respects, including labour, is allowed by law, subject to the obligations under article 10 of the Federal law of 06.04.2011 № 63-FZ "On electronic signature". However, in the opinion of law enforcement officers and scientists, these obligations are excessive, in fact, not allowing ordinary users of electronic means to use the opportunities established by law, forcing them to produce most of the documents on paper.

The court practice related to contesting the legality of actions of employers in terms of familiarizing the employee with the internal documents through electronic media as well as ambiguous and contradictory [5]. It is necessary to form a common position on this issue at the level of the Supreme court of the Russian Federation. Currently, the Appeal ruling of the Moscow city court of 02.12.2015 in case No. 33-45452/2015 is used as a General approach.

Due to the unavailability of services for obtaining a qualified electronic signature for the majority of the population of the Russian Federation, we propose to amend part 4 of article 312.1 of the Labour Code RF in terms of application instead of enhanced simple electronic signature (article 9 of the Federal law "On electronic signature"). At the same time allowed the application by analogy of paragraph 2 of article 434 of the Civil code of the Russian Federation, according to which "the contract in writing can be concluded by drawing up one document signed by the parties, as well as by exchanging letters, telegrams, telex, Telefax and other documents, including electronic documents transmitted through communication channels, allowing to establish reliably that the document comes from the party to the contract". It is proposed to extend this provision both to the conclusion of the contract and to situations where it is necessary to confirm the facts of familiarization with the documents, their receipt, transfer, etc. To confirm these facts, the system can be used, or similar information systems.

The sufficiency of a simple electronic signature is based, inter alia, on the presumption of the good faith of the parties and the reliability of the information provided by the parties. In case of ambiguity in the interpretation of the will of the party, the second party has the right to demand or confirm or duplicate the document on paper.

In any case, the organization of document circulation by means of electronic communication should be based on a special local normative act of the organization and a special instruction on the possibility of applying an electronic signature in the employment contract. The latter condition, in our opinion, will reduce the risk of challenging employers' actions taken on the basis of electronic documents.

2. Scope of the employment contract and additional agreements to it: currently, the Russian labour legislation is quite tough on the form of the employment contract and additional annexes to it: in accordance with the labour code of the Russian Federation, the employment contract must be concluded in writing. In addition, the employer must prove the issuance of a copy of the contract to the employee's hands. To do this, it is necessary to make a note on the copy of the employer on the issuance of the contract, which is confirmed by the signature of the employee (article 67 of the Labour code).

This rigidity of the law is due to the prevalence of informal labour relations in Russia, the evasion of employers from the performance of their duties on social insurance, the lack of knowledge of workers in the field of protection of their rights.

We propose to extend as a General rule the currently existing as a special rule on the possibility of remote conclusion of an employment contract "through the exchange of electronic documents" (article 312.2 of the Labour Code). The legal basis for this is the Federal law of 27.07.2006 № 149-FZ "On information, information technology and information protection". The law provides that "for the purpose of concluding civil contracts or registration of other legal relations (highlighted by us), which involve persons exchanging electronic messages, exchange of electronic messages, each of which is signed by an electronic signature or other analogue of the sender's hand-written signature of such a message, in the manner prescribed by Federal laws, other regulatory legal acts or agreement of the parties, is considered as an exchange of documents" (part 4 of article 11 of the Federal law). At the same time, in order to legalize this method of registration of labour relations, it is necessary to adopt an appropriate local normative act in which the employer-organization to approve the procedures of electronic document flow.

At the same time, based on the current position of the legislator and judicial practice, the introduction of digital technologies in the recruitment process should not worsen the legal status of workers, create obstacles to employment. Therefore, apparently, even with the introduction at the Federal level of the possibility of using electronic documents as a General rule, the employer will retain the obligation to duplicate electronic documents on paper or at the request of the employee to make out the relationship only in the form of ordinary paper documents.

As for the application of the so-called smart contracts or blockchain technology in the sphere of labour legislation, in our opinion, this area of professional activity does not need such changes. As correctly noted in the literature, "the prospects for the development of blockchain technologies and smart contracts in a single state, including in Russia, directly depend on the legal status of the cryptocurrency and the degree of freedom of its use allowed by the rule of law" [6]. Since the calculation of wages according to art. 131 of the Labour code in the Russian Federation should be made in "monetary form in the currency of the Russian Federation (rubles)", the use of cryp-tocurrency is a direct violation of Federal law.

3. The sphere of remote employment: involves remote employment of citizens through the Internet using personal computers. This form of employment is an exception under the current legislation, and is currently not widespread in comparison with traditional methods of employment of citizens. New forms of remote employment based on the use of Internet platforms, such as "crowdwork" and "work-on-demand via apps", which are gaining popularity in Europe, by their legal nature, are more in line with civil law contracts for the provision of services [7–9]. In the conditions of presentation of increased requirements to the employer by the Russian legislator, including the obligation to carry out social insurance, customers of such temporary services are likely to prefer to conclude civil contracts instead of labour contracts with the executors. In the context of the rigidity of the Russian labour legislation, the market of such temporary services will be formed either as an informal, "black", or as a market of civil services.

In addition, remote employment provides for a reduction or complete lack of control of the employee by the employer. This provision is also not peculiar to the labour legislation of the Russian Federation.

Judicial practice shows that the traditional understanding of labour relations is based on the implementation of management and control of the labour process by the employer. Therefore, in the absence of conditions in the employment contract on the performance of work remotely, through the Internet, the employer has the right to require the actual presence of the employee in the workplace in accordance with the rules of the labour regulations (see, For example, the decree of the Presidium of the St. Petersburg city court of 16.11.2016 № 44 g-132/2016).

## 4 Discussion

In the European literature, the issues of the digital economy in labour relations are disclosed in detail [10–12]. The use of information systems in legal practice is the subject of discussion of experts in various fields of law, especially in the areas of civil law (the use of electronic signatures, e-Commerce), legal proceedings. At the mono-graphic level, these problems are reflected in the works of Zharova (for example, Law and information conflicts in the field of information and telecommunications [13]), Savelyev (for example, E-Commerce in Russia and abroad: legal regulation) [14]).

As for labour law, it should be said that the Russian legal discourse on the use of digital technologies in labour law has not yet emerged. It can be noted the lack of a common approach to this issue, the fragmentary nature of research. Thus, more attention is paid to specific aspects of the use of information systems:

- local legal regulation of labour relations (see: Rogaleva IY, Rogaleva GA [15]);
- protection of personal data and other secrets protected by law (see: Petrykina [16]);
- features of the application of remote work (for example, Vasilyeva, Shuraleva, Brown [17]).

The General approach of the participants of the scientific debate on the issues is based on the recognition of the possibility of wide dissemination of information technologies in the processes of labour organization under the condition of a com-prehensive interpretation of labour legislation and the accumulation of positive judicial practice (see: Kourennoj, Kostyan, Khnikin [18]).

## 5 Conclusions

On the basis of the analysis it can be concluded that there are formal legal grounds for distribution as a General rule of the possibility to use electronic documents for regis-tration and maintenance of labour relations under the following conditions:

- ensuring the possibility of using a simple digital signature to confirm the legal facts provided for by the labour legislation of the Russian Federation;
- providing opportunities for employees who do not own information technology to obtain information and documents on tangible media;
- establishment of procedures for the use of digital technologies by employers at the local level;
- inclusion in employment contracts of conditions on the use of electronic documents;
- loyalty of the judiciary to the information ways of organizing labour relations.

At the same time, in Russia, due to the inflexibility of labour legislation, some remote forms of employment can not yet be widespread. Technological problems also persist in the widespread development of information technology.

# References

1. The official Internet portal of legal information. http://www.pravo.gov.ru, 10 May 2017, meeting of the legislation of the Russian Federation, 15 May 2017, No. 20, article 2901
2. The official Internet portal of legal information. http://www.pravo.gov.ru, 08 Apr 2013, meeting of the legislation of the Russian Federation, 08 Apr 2013, №. 14, p 1668
3. Materials of Parliamentary hearings "Features of registration of labor relations in the digital economy" (2018). http://komitet2-7.km.duma.gov.ru/Kruglye-stoly-seminary-soveshhaniya-i-dr/item/15496727/. Accessed 03 Oct 2018
4. "Sberbank" has hired robots (2017). https://www.iguides.ru/main/other/sberbank_prinyal_na_rabotu_robotov/. Accessed 03 Oct 2018
5. Kossov IA (2015) Employment contract with a remote worker: features of drawing up and conclusion. http://www.top-personal.ru/officeworkissue.html?356. Accessed 03 Oct 2018
6. Buterin V (2017) Smart contracts based on blockchain technology. https://vk.com/video-77014399_456239078. Accessed 07 Oct 2018
7. Arbeitsrecht im Betrieb (2016). https://www.bmas.de/SharedDocs/Downloads/DE/PDF-PublikationenDinA4/gruenbuch-arbeiten-vier-null.pdf?__blob = publicationFile. Accessed 01 Oct 2018
8. Crowdwork - die neue Form der „Arbeit"? (2017). https://www.iurratio.de/crowdwork-die-neue-form-der-arbeit. Accessed 01 Oct 2018
9. De Stefano V (2016) The rise of the "just-in-time workforce": on-demand work, crowdwork and labour protection in the "gig-economy". http://www.ilo.org/wcmsp5/groups/public/—ed_protect/—protrav/—travail/documents/publication/wcms_443267.pdf. Accessed 01 Oct 2018
10. Dagnino E (2017) People analytics: work and labour protection in the era of HRM through big data. Labour & Law Issues. https://doi.org/10.6092/issn.2421-2695/6860
11. Jašarević Senad R (2016) Influence of digitalization on labour relations. Zbornik Radova: Pravni Fakultet u Novom Sadu. https://doi.org/10.5937/zrpfns50-12356
12. Morel L (2017) The right to disconnect in French law. Effectiveness of the right to rest in the digital age. Labour & Law Issues. https://doi.org/10.6092/issn.2421-2695/7570
13. Zharova AK (2016) Law and information conflicts in the field of information and telecommunications. Yanus-K, Moscow
14. Savelyev AI (2016) E-Commerce in Russia and abroad: legal regulation. Statute, Moscow
15. Rogaleva IY, Rogaleva GA (2018) Features of registration of labor relations in the digital economy. https://doi.org/10.21686/2413-2829-2018-4-184-189. Accessed 01 Oct 2018
16. Petrykina NI (2011) Legal regulation of personal data turnover. Theory and practice. Statute, Statut, Moscow
17. Vasilyeva Yu V, Shuraleva SV, Brown EA (2016) The Legal regulation of remote work: theory and practice: monograph. State University of Perm, Perm
18. Kourennoj A, Kostyan I, Khnikin G (2017) Digital economy of Russia. Electronic records management of labor relations. https://www.eg-online.ru/article/355018/. Accessed 02 Oct 2018

# Peculiarities of Unstable Employment in the Era of a Digital Economy from Data of Social Media of Russia

V. N. Bobkov[1], M. V. Simonova[2(✉)], N. V. Loktyuhina[3], and I. A. Shichkin[1]

[1] Plekhanov Russian University of Economics, Moscow, Russian Federation
bobkovvn@mail.ru
[2] Samara State University of Economics, Samara, Russian Federation
best-samara@mail.ru
[3] Academy of Labor and Social Relations, Moscow, Russian Federation
cloktn@mail.ru

**Abstract.** The article presents the results of references analysis related to precarious employment in Russia found in social networks, blogs and public sources of information. Using thesaurus compiled on scientific articles and public discussions, automatic monitoring of news, topics, reviews and also debates which dedicated to precarious employment issue has been carried out. The search for subjects of precarious employment was conducted by using Big Data technology. There was revealed the connection of individuals affected by precarious employment to a specific age, gender, occupation, as well as their location in Russian Federation. An estimates of precarious employment scale in Federal districts of Russia has given. In result of the study, a socio-demographic profile of an individual affected by precarious employment in Russia was defined. There were factors and components of precarious employment related to social media users are determined, whereby expanded knowledge concerning manifestations of their precarious employment. A list of key problems faced by precarious workers is established.

**Keywords:** Digitalization · Precarious work · Social guarantees ·
Social networks · Social protection · Unemployment

## 1 Introduction

Global changes in national and international political, economic systems, as well as transformations in world of work highlights a discussions connected with "precarity" and "precarious employment" issues in format of public and academic debates within social media [5].

Nowadays, rapid scientific and technological progress, which led to fast development of media platforms and social networks, such as Facebook and Vkontakte, as well as search engines such as Google and Yandex, enables the opportunity to process data related to labour activity. Media platforms users often upload information or enter queries related with current work or looking for a new job.

© Springer Nature Switzerland AG 2020
S. Ashmarina et al. (Eds.): *Digital Transformation of the Economy: Challenges, Trends and New Opportunities*, pp. 235–243, 2019.
https://doi.org/10.1007/978-3-030-11367-4_22

Internet users search for information about vacancies which implies permanent or temporary employment in certain localities, or they can "tweet" during job search, as well as in cases of starting work [7]. Analysis of social network profiles facilitate to determine the socio-demographic, professional portrait of job applicants and also their location. Web users enter textual reviews and leave comments containing information that can be used in order to obtain specific explanations on various social and labour issues.

Currently precarious employment has become widespread phenomenon in international and national contexts. It represents the forced loss by workers of labour and social rights caused by narrowing of standard employment based on permanent labour contracts with a full working week and high protection of labour and social rights guaranteed to workers [14]. In transition to flexible forms of employment is accompanied by individualization of work, the proliferation of short-term contracts, non-standard working conditions, manifested in part-time employment, volatile wages, expansion of informal employment and shadow economy [12]. This problem is widely explored in scientific literature [11, 15]. Along with mentioned above sources, the importance of such global phenomenon highlighted by its widespread discussion in social networks [10].

The authors identified the problems faced by social media users in the world of labour and manifestations of precarious employment among them. Besides it was revealed the socio-demographic, professional features and also residence of individuals affected by precarious employment.

Using the results of references analysis to precarious employment in Russia found in social networks, blogs and public sources of information it is necessary to draw up an individual's profile affected by precarious employment comprising residence, socio-demographic and professional attributes. It also requires to figure out a key problems faced by precarious workers.

## 2    Materials and Methods

This research was conducted in November 2016. Its empirical base was a sample of users from social networks identified through an automatic search using a thesaurus, compiled on the basis of scientific articles, discussions, debates and public sources of information dealing with precarious employment, informal economy and hidden incomes. During the research there was performed an automatic monitoring of references related to precarious employment in Russia found in social networks, blogs and public sources of information. The search for subjects of precarious employment was conducted by using Big Data technology.

In total, over 2 million web pages were processed during the research. Purposive sample covered more than 4,200 unique users. The research methodology is based on an algorithm consisting of the following steps:

1. Drawing up an initial thesaurus including a list of keywords to further search for discussions in automatic and semi-automatic modes. This thesaurus contains more than 190 original queries.

2. Conducting an automated search based on initial thesaurus, as well as enriching it with phrases (terms) similar in semantics and word forms using machine learning technologies. As a result, the list of keywords was increased by 23.6%, and the total number of identified mentions to precarious employment in 2016 exceeded 43 thousand.
3. Semi-automatic segmentation of found references to precarious employment enable to identify association with a specific age and gender group, professional area, as well as to region and federal district. Spatial boundaries to explore precarious employment in Russia is based on quota sampling (Figs. 1, 2).

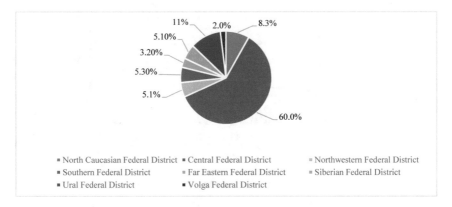

**Fig. 1.** Distribution of social media users in the sample by federal districts of Russia (Source: compiled by the authors)

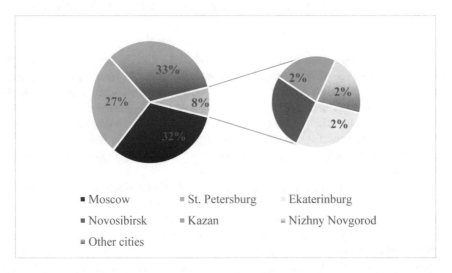

**Fig. 2.** The share of social media users in Russian cities covered by research of precarious employment (Source: compiled by the authors)

## 3  Results

As a result of semi-automatic segmentation of references to precarious employment found in social networks, blogs and public sources of information, it was identified relation of precarious workers to a specific age and gender group, professional area, as well as to region and federal district of the Russian Federation. The chart shown in Fig. 3 illustrates the frequency of age-related cohorts in the sample (bar graph), and curved line indicates the proportion of precarious workers.

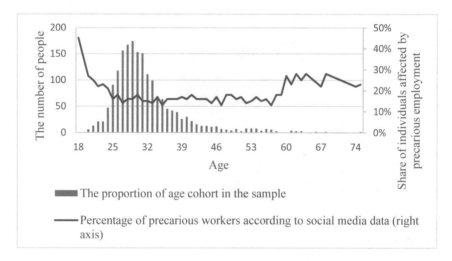

**Fig. 3.** Age structure of individuals affected by precarious employment (Source: compiled by the authors)

According to the research results, youth under 23 years, as well as elderly people refers to the most vulnerable groups influenced with risks of precarious employment. It's determined by low demand on the labour market and insufficient competitiveness relatively to other age cohorts drives youth and people of pre-retirement age (employed pensioners) to occupy labour market segments with high risk of precarious employment [4]. Thus, for these socially vulnerable groups of population, the fact to have payable employment is more important than have decent work including high remuneration and social guarantees [9].

Inner circle of the diagram shown in Fig. 4 implies a distribution of the sample by gender. The external circle depicts the proportion of men and women at risk of precarious employment. In gender aspect of Internet users who have touched on the topic of precarious employment, there is a male's predominance (Fig. 4). This circumstance suggests that advantages and benefits provided by formal employment are more important for women [3]. Salary amount is a most considerable factor encourages men in labour relations [8].

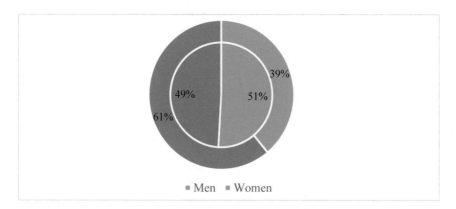

**Fig. 4.** Gender structure of individuals affected by precarious employment (Source: compiled by the authors)

In result of study the educational level of individuals affected by precarious employment, it was revealed a quite high proportion of precarious workers with higher and secondary vocational education (Fig. 5).

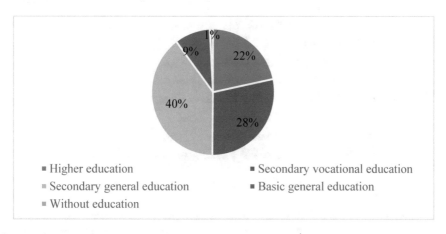

**Fig. 5.** Educational level of individuals affected by precarious employment (Source: compiled by the authors)

Despite the fact that higher education increases worker competitiveness and possibilities to formal employment with a high level to secure of labour and social rights, in Russian reality higher education is not a guarantee of protection against precarious employment [16].

The research found, in economic sectors the majority of precarious employees belong to trade, agriculture and construction, that is explained mass character of occupations they represent, the features of production and labour in these areas, as well low employer requirements to workers education (Fig. 6) [1].

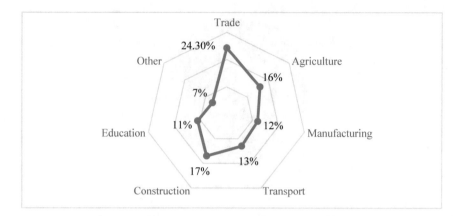

**Fig. 6.** Sectoral distribution of precarious workers (Source: compiled by the authors)

Social media analysis showed that largest scale and spread of precarious employment are observed in North Caucasus, Far Eastern and Volga federal districts (Fig. 7).

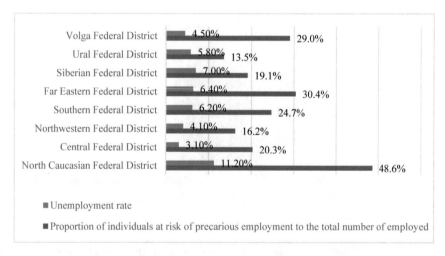

**Fig. 7.** Precarious employment to unemployment ratios in federal districts of Russia according to Rosstat [13] data and social media analysis (Source: compiled by the authors)

Modern information and communication technologies enables to monitor references to precarious employment in social networks, blogs and public sources of information. To search for individuals affected by precarious employment is possible due to usage of Big Data technology. Thus, analyzing the profiles and messages of social media users, it is possible to get data describes the scale, forms and manifestations of precarious employment. This information is a valuable resource to fight such considerable issue in the social and labour sphere as precarious employment.

# 4  Discussion

It occurs because in regions with high unemployment rates and growing labour market tension, people who do not intend to migrate have to accept informal labour relations, lack of social and labour guarantees [17], overtime work or temporary employment and other forms of precarious employment [6].

In addition, the research found that distance between Central federal district and other federal districts of Russia contributes to increase relevance of precarious employment issue to the local population. This is mainly due to the inter-regional disparities in socio-economic development of Russian Federation [2].

# 5  Conclusions

1. According to results of the conducted research based on social media data, a socio-demographic, professional and regional profile of the individual affected by risks of precarious employment in Russia was defined (Table 1).

**Table 1.** Profile of Individual Affected by Precarious Employment in Russia

| Gender | Age | Education | Economic sector | Location |
|--------|-----|-----------|-----------------|----------|
| Male | Up to 23 and over 58 years old | General secondary | Trade, agriculture and construction | North Caucasian, Far Eastern and Volga Federal Districts |

Source: compiled by the authors

2. As a result of social media data analysis it has been concluded that the main problems faced by individuals influenced by risks of precarious employment in Russia are:

   - Lack of social security and confidence in work and life;
   - Harmful or insecure working conditions;
   - Variable and non-competitive wages;
   - Unreasonably low wages;
   - Delays or non-payment of wages, social benefits;
   - Unpaid overtime work;
   - Failure to pay taxes or insurance payments by the employer;
   - The employer does not pay for vacation or sick leave;
   - Difficulties to finding a new job.

**Acknowledgments.** This article has been written under support of the Russian Science Foundation (RNF) "Precarious employment in Russian Federation: the state and ways to reduce № 16-18-10140".

# References

1. Benavides FG, Benach J, Diez-Roux A, Roman C (2000) How do types of employment relate to health indicators? Findings from the second European Survey on working conditions. J Epidemiol Community Health 54(7):494–501. https://www.ncbi.nlm.nih.gov/pubmed/10846191. Accessed 8 May 2018
2. Bobkov VN, Kvachev VG, Novikova IV (2018) Precarious employment in the regions of Russian Federation: sociological survey results. Ekonomika regiona [Econ Reg] 14(2):366–379. https://doi.org/10.17059/2018-2-3
3. Bobkov VN, Kvachev VG, Shichkin IA (2017) Determinants of precarious employment of labour migrants in Russia estimated by respondents of sociological survey. What human resources are needed for the Russian economy?: research papers of the VII International scientific and practical conference "Abalkin readings" held on May 23–25, 2017/edited by Valentei SD. Moscow: FSBEI of HE "REU named after G.V. Plekhanov 131–142
4. Bobkov VN, Novikova IV, Odintsova EV (2017) Profiles of precarious employment in Russia. The standard of living of population in Russian regions, vol 4, pp 26–35. https://www.hse.ru/data/2011/04/06/1211703897/Mode_of_life-ENGL.pdf. Accessed 8 Mai 2018
5. Bobkov VN, Shichkin IA (2017) Priority directions of reducing the unstable employment of labor and forced migrants in Russia. The standard of living of the population of the Russian regions, vol 4(206), pp 45–51. CyberLeninka: https://cyberleninka.ru/article/n/prioritetnye-napravleniya-snizheniya-neustoychivoy-zanyatosti-trudovyh-i-vynuzhdennyh-migrantov-v-rossii. Accessed 8 Mai 2018
6. Puig-Barrachina V, Vanroelen C, Vives A, Martinez JM, Muntaner C et al (2014) Measuring employment precariousness in the European working conditions survey: the social distribution in Europe. Work 49(1):143–161. https://doi.org/10.3233/wor-131645
7. Castells M (2010) The rise of the network society. Blackwell Publishing Ltd., Oxford
8. Cranford C, Vosko LF, Zukewich N (2003) Precarious employment in the Canadian labour market: a statistical portrait. Just Labour 3:6–22. http://www.yorku.ca/julabour/volume3/cranfordetal_justlabour.PDF. Accessed 8 Mai 2018
9. Druzhilov SA (2015) Precariat and informal employment in Russia: social and psychological aspects. HumItarian Sci Res 1(2) [Electronic resource]. http://human.snauka.ru/2015/01/9491. Accessed 8 Mai 2018
10. Golenkova ZT, Goliusova YV (2013) New social groups in the modern stratification systems of global society. Sociological science and social practice, vol 3. http://teoria-practica.ru/rus/files/arhiv_zhurnala/2015/24/sociology/orekhova.pdf. Accessed 8 Mai 2018
11. Kalleberg AL (2009) Precarious work, insecure workers: employment relations in transition. Am Sociol Rev 74:6–8. (Feb 1–22). https://doi.org/10.1177%2F000312240907400101
12. Murgia A (2014) Representations of precarity in Italy. J Cult Econ 7(1):48–63. https://doi.org/10.1080/17530350.2013.856336
13. Number of unemployed according to ILO methodology (2018) Federal State Statistics Service (Rosstat). http://www.gks.ru/wps/wcm/connect/rosstat_main/rosstat/ru/statistics/wages/labour_force/#. Accessed 8 Mai 2018
14. Bobkov VN et al (2017) Precarious employment: Russian and international contexts of the future world of work. Publishing House "RealPrint", Moscow
15. Gorshkov MK et al (2015) Russian society and the challenges of time. Publishing House "Ves Mir", Moscow

16. Benach J, Vives A, Amable M, Vanroelen C, Tarafa G, Muntaner C (2014) Precarious employment: understanding an emerging social determinant of health. Annu Rev Public Health 35:229–253. https://doi.org/10.1146/annurev-publhealth-032013-182500. Accessed 8 Mai 2018

17. Simonova MV, Zhukova AA (2018) Informal employment in the labor market: trend analysis. Bulletin of VSU. Series: Economics and Management, vol 2, pp 82–89. http://www.vestnik.vsu.ru/pdf/econ/2018/02/2018-02-11.pdf Accessed 8 Mai 2018

# The Use of Identical IP Address by Tender Participants as an Indication of Cartel Scheme

Ju. A. Dorofeeva and M. A. Tokmakov[(✉)] [ID]

Samara State University of Economics, Samara, Russia
logl612@yandex.ru, maxim.tokmakov@gmail.com

**Abstract.** The relevance of the study is explained by the increase in the number of contracts entered into by tender, which has caused an increase in the number of cases of illegal (cartel) schemes disclosed by the state body with the use of digital technologies. In this regard, the article is aimed at identifying methods for establishing the presence of a cartel at tenders, including by means of disclosing the use of identical IP addresses by cartel members. The leading approach to the study of the problem of identifying cartel scheme is that its participants do not show the connection between them and the intent of the collusion. Digital technologies make it possible to identify such connection and make it possible, in context with other evidence of scheme presence between parties to identify a cartel, to stop its illegal activities and to not allow an agreement to be entered into at tenders abstracting the law. The article presents an analysis of the norms of legislation, departmental acts of the antimonopoly body and judicial practice in order to identify signs of cartel scheme, including with the use of digital technologies, accounts the grounds for objections of cartel participants to the use of methods to establish illegal collusion by disclosing the fact of identical IP use by tender participants. The materials of the article are of practical value for specialists in the field of antimonopoly law, legal practitioners, government officials specializing in the organization of tenders and the protection of competition.

**Keywords:** Bid rigging · Bidding · Bidding conspiracy · Cartel ·
Competition protection · Digitization · Digital technologies · IP address

## 1 Introduction

Certain peculiarities of organizations' participating in tenders behavior (such as pairwise alternate participation, identical price offers with a minimum decrease from the initial maximum contract price when other bidders allegedly enter into competition, failure of one of the subjects to continue the fight, alternate victory in the absence of other price offers participants, entering into a contract as close as possible to the initial price) can be regarded by the antimonopoly authority as agreements leading to reduction or maintenance of prices at the auctions, and become a ground for initiating a violation of clause 2 of part 1 of article 11 of the Federal Law of 26.07.2006 No. 135-FZ "On Protection of Competition" (hereinafter - the Law on Protection of Competition).

© Springer Nature Switzerland AG 2020
S. Ashmarina et al. (Eds.): *Digital Transformation of the Economy: Challenges,
Trends and New Opportunities*, pp. 244–252, 2019.
https://doi.org/10.1007/978-3-030-11367-4_23

Despite the fact that in the literature about the benefits or harm of cartel different positions are expressed [1–4], it should be assumed that Russian law protects competition as a rivalry between economic entities, where separate actions of each exclude or restrict the capability to affect the general conditions of circulation of goods in the relevant product market by independent actions of each of them.

Considering the negative consequences of recognizing participants in cartels (Article 14.32 of the Code on Administrative Violations of the Russian Federation) and their officials (Article 178 of the Criminal Code of the Russian Federation), defendants in a case of violation of antimonopoly laws stretch to different ways and tricks to avoid liability for violation of anti-cartel law, get anti-cartel amnesty, or at least reduce the amount of liability.

Analysis of the arguments used by the cartel participants in their defense showed that they all come down to one of the main ones, or they are combined with others that are not general but relative and peculiar to a particular situation. Among the general arguments about the absence of a cartel collusion at tender, for example, the argument about the failure to prove the fact of the conclusion of a cartel agreement can be attributed: it is based on the absence of the fact of proper will of the parties to enter into an agreement.

The content of this argument consists in the absence of facts confirming the entry of bidders into the AGREEMENT. Of course, the starting point is the thesis about the need to comply with the written form of the contract (Art. 160 of the Civil Code of the Russian Federation).

Meanwhile, paragraph 18 of Article 4 of the Law on Protection of Competition cartel agreement is defined as an agreement that can be made both in writing, including made in one or several documents, and orally. Thus, p.18 Art. 4 of the Law on Protection of Competition levels the requirement on the will of the parties to the agreement, established by Art. 160 of the Civil Code of the Russian Federation (written form of the transaction), since the rules of civil and antitrust law governing the question of the form of a transaction are in this case correlated as general and particular.

In the absence of a written contract, the cartel participants, their actions are evaluated on the basis of all the circumstances relevant to the determination of the existence of an agreement to commit the actions specified in paragraph 2 of Part 1 of Art. 11 of the Law on Protection of Competition. As explained by the Supreme Court of the Russian Federation in paragraph 9 of the Review on judicial practice issues arising during the consideration of cases on the protection of competition and cases of administrative offence in this area (approved by the Presidium of the Supreme Court of the Russian Federation on March 16, 2016), the fact of an anti-competitive agreement does not depend on its conclusion in the form of a contract according to the rules established by civil law, which does not determine and cannot determine which evidence exactly confirms such an agreement, as well as requirements for the form of supporting documents are not established.

The head of the anti-cartel division of Anti-monopoly service of the Russian Federation, A.P. Tenishev notes that "law enforcement agencies and courts were able to find a common language in the question of how to prove cartels and other anti-competitive agreements": the courts, realizing that antimonopoly authorities cannot find direct evidence of the cartel, often conclude that "an anticompetitive agreement can be proved by a sufficient set of circumstantial evidence" [5]. The conclusion about the sufficiency of the totality of circumstantial evidence to establish the fact of the cartel agreement A.P. Tenishev confirms by reference to the Review of judicial practice and to a number of examples from judicial practice, in particular, to judicial acts in cases No. A40-14219/13-94-135, A40-125017/14, A42-2564/2014, A19-14411/2015, A58-4634/2015. Despite the fact that there exists controversial judicial practice, which refutes the fact that the courts fully accepted the position of the antimonopoly authority on the sufficiency of indirect evidence to qualify the presence of a cartel and the condition of bidders in it (for example you may consider judicial acts on cases No. A47-12654/2016, A36-9469/2016, A54-3726/2015), yet most judicial acts issued in this category of cases confirm the conclusion of A.P. Tenishev, that the courts support the antimonopoly authorities that determine the cartel by qualifying the actions of bidders in the absence of direct evidence [5].

Such indirect evidence is the establishment of such facts as, for example, specifying one contact person, one phone number, using one computer, issuing certificates of electronic digital signatures for one person, the same properties of an electronic file (time and place of creation, author), unique electronic fonts, the number of electronic signs, etc., such establishment is possible through the use of various digital technologies [6]. In this article, we focus on issues of using a single IP address by participants as evidence of a cartel agreement between them.

## 2    Materials and Methods

The methodological basis of the study is set on general scientific methods of cognition, including the principles of objectivity, classification, induction, deduction, etc. Along with general scientific methods of cognition, particular scientific methods were used: descriptive, linguistic and comparative legal methods.

In order to obtain the results of the following study, comparison and contraposition methods were used to compare and contrast the positions of government bodies and cartel participants to substantiate the argument about the absence or presence of cartel collusion in tenders.

In order to achieve the best results of research into the problems of detecting cartel collusion at tender, it would be most reasonable to apply the methods of legal and case comparison to identify ways to solve problems in different legal systems and the possibility of applying those to the Russian legal system, yet the application of these methods and the study of this aspect of the problem is of further research.

## 3    Results

IP address (short for Internet Protocol Address) means the unique identifier (address) of a device connected to a local network or the Internet. An IP address is a 32-bit (IPv4 version) or 128-bit (IPv6 version) binary number. A convenient form of recording an IP address (IPv4) is writing it in the form of four decimal numbers (from 0 to 255), separated by dots, for example, 192.168.0.1. (or 128.10.2.30 that is the traditional decimal form of address representation, and 10,000,000 00001010 00000010,00011110 that is the binary form representing the same address). IP addresses are the primary type of address, based on which the IP network layer transfers packets between networks. The IP address is assigned by the administrator during the configuration of computers and routers. The use of a common infrastructure by tender participants to submit applications for participation in tenders (the use of an identical IP address, one account) confirms the existence of a cartel agreement aimed at price manipulation in tenders. This conclusion of the antimonopoly authority, made on the basis of the analysis of documents of participants in trading using digital technologies, is also reflected in numerous judicial acts issued on the basis of the consideration of applications for the recognition of acts of the antimonopoly authority as illegal. In particular, in case No. A36-2298/2012, the court indicated that the submission of price proposals from one computer and one IP address testifies to the coherence of actions of economic entities, since all actions of the defendants in the case of violation of antimonopoly laws were known to everyone of them in advance, the actions of each of these subjects are caused by the actions of other economic entities and are not the result of circumstances that equally affect all economic entities in the relevant product market. A similar approach was applied by the courts when considering a number of arbitration cases, including in cases No. A42-2564/2014, A40-323/2015, A18-541/2014, A40-170978/2015, A79-3330/2015, A57-2980/2014, A52-706/2015, A58-4634/2015, A03-19328/2016, A41-29471/2017, A40-37124/2017, which indicate the uniformity of the judicial practice on this issue.

The use of information that the violators sent information related to participation in tenders (applications for participation in tenders, documents confirming information about the applicant, as well as the bid itself) using digital technologies by the public authority as evidence of the cartel is based on a particular IP-addresses belonging to a particular person (citizen, organization), according to the contract for the provision of communication services. At the same time, there is no legal definition of the concept "IP address". At present, the Ministry of Communications and Mass Communications of the Russian Federation has developed amendments to the Federal Law of 07.07.2003 No. 126-FZ "On Communications" (hereinafter - the Law on Communications). It has several regulatory objectives, and they are amplified by one more - ensuring the integrity, continuity, stability, sustainability and security of the functioning of the Russian national segment of the Internet; in the aforementioned draft law it is proposed to amplify the conceptual apparatus of the Law on Communication with the definition of IP address. It will be considered as a unique identifier of an intercommunicator or other technical means in the Internet. The application of that law shall serve to overcome the uncertainty in the definition of the concept of IP-address.

But even today, in the absence of a legal definition of an IP address, its significance for establishing communication between various individuals and the existence of certain relations between them, including agreements, is established by the courts not only when a cartel is detected (in tenders), but also in other cases.

Thus, in tax legislation regarding the use of an unjustified tax benefit both tax authority, and as a result of consideration of a complaint against a tax authority's decision, the court often take the conclude that the coincidence of the taxpayer's IP addresses and other persons involved in the transactions indicates actions aimed at reducing the tax burden. Thus, in cases No. A51-5012/2016, No. A40-183993/2014, A56-13035/2010, A56-2141/2010, No. A56-4647/2010, A56-33391/2010, No. A56-13035/2010, the court like the tax authority, using the fact that the counterparties use the same IP address, concluded that these organizations use the same computer when performing transactions in the Client-Bank system, since the presence of a single IP address indicates the management of settlement accounts of organizations from one computer and the formality of funds between the taxpayer and its counterparties, as well as the consistency of actions of such unscrupulous taxpayers aimed at obtaining unjustified tax benefits.

Given the prevalence of the practice of requesting information about the ownership of an IP address to a computer, a natural question arises about the legitimacy of such requests from people who have this information - from the tender participants themselves, providers, Internet providers, other citizens and organizations, as well as digital evidence obtained can be used to establish the fact of an offense. Note that the price of the issue (income from contracts entered into at the auction) is quite high. Thus, in case no. A23-2306/2017, the court found that BuildTechMainanance LLC was reasonably prosecuted under Part 1 of Article 14.32 of the Administrative Code of the Russian Federation in the form of a fine of 1,355,201.04 rubles for cartel collusion at an electronic tender for the maintenance and operation of the engineering infrastructure of the military camp and the shooting range of the military unit with an initial (maximum) contract price (IMCP) in the amount of 54,200,867.00 rubles. Interestingly, in this case, the submission of proposals from the participants of the cartel agreement was carried out not only from the same IP address, but also from the same MAC address (equipment identifier).

The answer to this question should be sought in the substantive rules establishing the rights and obligations of both organizations providing communication services and their subscribers.

Part 2 of Article 23 of the Constitution of the Russian Federation guarantees confidentiality of correspondence, telephone conversations, postal items, telegraph and other messages transmitted via telecommunication networks and postal networks on the territory of the Russian Federation. A similar guarantee is ensured in clause 1 of Article 63 of the Law on Communications, which establishes that the restriction of the right to confidentiality of correspondence, telephone conversations, mail, telegraph and other messages transmitted over telecommunications networks and postal networks is permitted only in cases provided for by federal laws and only on the basis of a court decision. Article 63 of the Law on Communications defines: communication secrets are information about messages, information and documentary correspondence transmitted through telecommunication networks and postal networks, postal items and postal orders, as well as the messages themselves, postal items; the list of information relating to the secret communication is not exhaustive. The duty to ensure confidentiality of communications is stipulated by clause 2 of Article 63 of the Law on Communications to communication operators. In accordance with paragraph 1 of Article 53 of the Law on Communication, information about subscribers includes the last name, first name, patronymic or nickname of the subscriber-citizen, the name (company name) of the subscriber being a legal entity, the last name, first name and patronymic of the head and employees of this legal entity, as well as subscriber's address or address of installation of terminal equipment, subscriber numbers and other data allowing identification of the subscriber or its terminal equipment, information on the databases of billing systems for communication services rendered, including the connection Barrier-, traffic and subscriber payments. According to this standard, information about subscribers and communication services rendered to them, which have become known to communication operators due to the execution of a communication services agreement, are information of limited access and are subject to protection in accordance with the legislation of the Russian Federation.

As a general rule, the provision of information on citizens-subscribers to third parties can be carried out only with their consent, with the exception of cases provided for by this Federal Law and other federal laws. But there is an exception to this rule: federal law provides for cases of mandatory submission of personal data in order to protect the foundations of the constitutional order, morality, health, rights and legitimate interests of others, ensure the defense of the country and the security of the state. The specified is also regulated in the legislation on relations on the provision of communication services, in part of the obligation to provide telecom operators to authorized state bodies carrying out operational investigative activities or ensuring the security of the Russian Federation, information about users of communication services and communication services provided to them, as well as other information necessary to carry out the tasks assigned to these bodies, in particular, in cases of carrying out operational search activities (Article 64 of the Law on Communications). The contents of articles 53, 63 of the Law on Communications and the provisions of the Federal Law of July 27, 2006 No. 152 "On Personal Data" do not allow us to conclude directly that information on the IP address belongs to personal data of users of communication services, but even if the law protects such information about its disclosure to third parties, at the request of the authorized bodies, organizations providing communication services are obliged to disclose it.

The refusal to provide such information is illegal and entails for violators the responsibility established by article 19.7 of the Code of Administrative Offenses of the Russian Federation. Nonetheless, the exemption of liability reason of telecommunications operator who evaded the obligation to provide the authorized state body with information concerning the owner of the IP address cannot be based on the fact that the IP address on which the request was received is dynamic. Thus means that the IP address is assigned to the end user equipment automatically while the device is connected to the Internet. As indicated by the Supreme Court of the Russian Federation in the resolution on 30.03.2016 No. 82-AD16-1, the subscriber's detection at a dynamic address on the basis of a detailed file, which is information about subscribers' connections (connection protocols), and therefore the provision of data about a user of communication services established both by the statistical IP address and by the dynamic IP address to the authorized state body is the responsibility of the telecommunications operator; by providing such data, the organization providing communication services does not intrude into the secret of communication protected by law.

This position is also consistent with the opinion of the Constitutional Court of the Russian Federation set forth in the definition of the Constitutional Court of the Russian

Federation No. 345-O of October 2, 2002, according to which information constituting the secrecy of telephone negotiations protected by the Constitution of the Russian Federation and laws operating in the Russian Federation is information transmitted, stored and installed using telephone equipment, including data on incoming and outgoing signals connecting telephone sets of specific connection. To access such information it is necessary to obtain a court order. Otherwise it would cause non-compliance with the requirements of Article 23 (Part 2) of the Constitution of the Russian Federation on the possibility of limiting the right to privacy of telephone conversations only on the basis of a court order. Therefore, to provide information to the state authorities exercising control in the field of competition protection about the user of communication services set as a statistical IP address (assigned by the communication operator as a permanent address assigned to the user's end user equipment when entering into an agreement for the provision of Internet access services) and by dynamic IP address, no court order is required; Such information must be provided at the request of the antimonopoly authority, and the courts must accept the information as competent evidence.

## 4  Discussion

In both Russian and foreign modern legal literature there is a lot of research devoted to the problems of cartel collusion at tender. Most of them concern the general issues of the notion of collusion as a type of monopolistic activity [7–9], in which it is noted that collusion in bidding is one of the most common forms of monopolistic activity. Many works are devoted to the situational analysis of schemes of cartel collusion and its features [10–12]. Also proposed methods for identifying collusion in the auction [13, 14].

A number of works is devoted to studying the sanctions for violation of the rules on the inadmissibility of collusion and possible ways to prevent such violations [15, 16]. There are also studies of the negative economic effect of cartel collusion [3].

All these numerous publications, of course, indicate the existing interest in the literature to the problems of preventing collusion during tenders. At the same time, in the conditions of the digitalization of the economy, antimonopoly authorities face new challenges (for example, Petrov points out the use of robot-bots for cartel collusion by tender participants [17]), on the other hand they provide new opportunities of detecting collusion using digital technologies. A deep review of the challenges to antimonopoly regulation in the digital age is presented in the monograph with the same name [1].

With all this, it should be noted that insufficient attention has been paid to the issues of studying the evidence of cartel collusion revealed using digital technologies. In most cases, we are talking only about their enumeration [6]. A detailed study of the discovery of the fact of the use of the same IP address by tender participants, as evidence of a cartel agreement, is completely absent. In this regard, the presented study clarifies and develops existing developments in this area.

# 5   Conclusions

Today, largely due to the consistent and reasonable policy of the anti-monopoly authority on the application of digital technologies in the work of territorial bodies, the practice of requesting information about the IP address of a particular computer has become widespread. The provision of data about the user of communication services does not require a court order and is carried out at the request of the state body exercising control in the field of competition protection.

The share of singular infrastructure by tender participants to submit applications for participation in tenders (using identical IP address, one account) is reasonably recognized by the courts as reliable (admissible, relevant) evidence of illegal collusion of tender participants, and therefore, violators of competition law suffer reasonable negative consequences of their illegal behavior at tenders.

# References

1. Tsarikovskiy A, Galimkhanova N, Tenishev A, Khamukov M, Ivanov A, Voynikanis E, Semenova E (2018) Antimonopoly regulation in the digital age, 311 s. Higher School of Economics, Moscow
2. Kloosterhuis E, Mulder M (2013) Competition law on coal-fired power plants. In: 9th ACLE Seminar, Amsterdam, December 12. http://acle.uva.nl/events/competition–regulation–meetings/conference-papers-9th-cr-meeting-2013.html. Accessed 12 Mai 2018
3. Connor J (2014) Cartel overcharges. In: Research in law and economics, vol 29, pp 249–386. https://doi.org/10.1108/S0193-589520140000026008 (March)
4. Shastistko AE (2018) Allow cartels? http://institutiones.com/general/2597-razreshit-karteli.html. Accessed 12 Mai 2018
5. Tenishev A (2016) Practice of the FAS of Russia on cartel cases and other anti-competitive agreements: events of 2016 and plans for 2017. http://www.consultant.ru/law/interview/tenishev. Accessed 12 Mai 2018
6. Khamukov MA (2017) Electronic evidence of cartels at auction. Bulletin of the University O.E. Kutafina, № 9(37), pp 57–61. https://doi.org/10.17803/2311-5998.2017.37.9.057-061
7. Andrews Ph, Gorecki P, McFadden D (2015) Modern Irish competition law: a practical guide. Kluwer Law International. https://lrus.wolterskluwer.com/store/product/modern-irish-competition-law/
8. Franskevich OP (2017) Collusion in the auction as a form of monopolistic activity. J Bus Corp Law 1(5):16–19
9. Khabarov SA (2014) Collusion as a form of coordination at the auction. Lawyer 2014 (15):33–37
10. Morselli C, Ouellet M (2018) Network similarity and collusion. Soc Netw 55:21–30. https://doi.org/10.1016/j.socnet.2018.04.002
11. Posthuma H (2004) Eyes wide open… To catch bid riggers. SUMMIT 3(7):26
12. Spagnolo G, Albano G, Buccirossi P, Zanza M (2006) Preventing collusion in procurement: a primer. Handbook of Procurement, Cambridge University Press. https://ssrn.com/abstract=896723. Accessed 12 Mai 2018
13. Ingraham AT (2005) A test for collusion between a bidder and an auctioneer in sealed-bid auctions. http://dx.doi.org/10.2139/ssrn.712881. Accessed 12 Mai 2018

14. Morozov I, Podkolzina E (2013) Collusion detection in procurement auctions. Higher School of Economics Research Paper No. WP BPR 25/ EC/ 2013. http://dx.doi.org/10.2139/ssrn.2221809. Accessed 12 Mai 2018
15. Mosunova N. (2015) anti-bid rigging policy. Russ Law J, №4. T.3, pp 32–74. https://doi.org/10.17589/2309-8678-2015-3-4-32-74
16. Heimler A (2012) Cartels in public procurement. J Compet Law Econ, 8(4):849–862. https://doi.org/10.1093/joclec/nhs028. Accessed 12 Mai 2018
17. Petrov DA (2018) "Robotization" at the auction in the digital economy era: a business process or a way to circumvent the law? Civil Law 2018(5):12–15. https://doi.org/10.18572/2070-2140-2018-5-12-15

# Structural and Functional Analysis of Requirements to Managers of Innovative Companies in the Conditions of the Digital Economy

V. V. Mantulenko[1]([✉]), A. V. Mantulenko[2], E. P. Troshina[1], and M. V. Vorotnikova[1]

[1] Samara State University of Economics, Samara, Russia
mantoulenko@mail.ru, troshina@yandex.ru,
Maria.Vorotnikova@inbox.ru
[2] Samara University, Samara, Russia
mantulenko83@mail.ru

**Abstract.** In the conditions of the digital transformation of the economy, the demand for highly qualified intellectual potential of employees increases, which objectively requires the integration of educational systems of various countries, the creation of strategic alliances in the field of higher and additional education, educational and business organizations. The rapid development of information technology and technological breakthroughs contribute to the development of innovation in every way. The purpose of this contribution is a structural and functional analysis of requirements to managers of innovative companies in the conditions of the digital economy.

**Keywords:** Innovation · Innovation manager · Digitalization · Professional education · Value added chains

## 1 Introduction

The economy adapts to changes associated with the main trends of the modern world development. Digital technologies are being increasingly introduced into the functioning of organizations, and in some cases determine the success of business projects. For business, modern digital technologies open new opportunities for its development, changes of the market position, and profitability increase. For the state as a whole, digitalization can be considered as one of the tools for innovative development too and, as a result, for increasing the effectiveness of its participation in global value-added chains. The participation in ascending relationships within the global value-added chains prevails over the participation in descending relationships, that means that Russia exports primary products and services, which it acquires later in the form of finished products, so we can see loss of profits associated with the possibility of increasing the level of involvement in downward links within global value chains.

In this regard, there are new requirements for managers of innovative enterprises. The main goal of this contribution is to describe one of the tools for creating an

© Springer Nature Switzerland AG 2020
S. Ashmarina et al. (Eds.): *Digital Transformation of the Economy: Challenges,*
*Trends and New Opportunities*, pp. 253–259, 2019.
https://doi.org/10.1007/978-3-030-11367-4_24

effective system for implementing innovations using advantages of the digitalization. To achieve this goal, first of all, it is necessary that innovative education becomes one of the most important components of this system, and a difficult and not yet formed profession of "innovation manager" becomes attractive and prestigious.

Digitalization is not limited to a simple automation of business processes in the organization or the use of online channels to attract customers, but it also reduces the number of actions that are necessary to complete various business tasks, paper work, enables to avoid mistakes and increase the employees' work productivity. In Russian business practice, such electronic products as 1C (availability and inventory management), OroCommerce (contract management), CRM and others are already widely used. A special issue, that has to be analyzed, is the degree of technical readiness of enterprises to work with such programs. The ability to conduct distance learning (video tutorials, online simulators) is also one of the products of digitalization, allowing to distribute the load in the learning process more rationally.

## 2 Materials and Methods

The theoretical and methodological basis of this research was the works of domestic and foreign scientists who consider the effect of digitization of the economy on the requirements to innovative managers. The study was carried out according to international methodologies based on the analysis of the actual professional activity of small innovative firms, infrastructure of business support in the scientific and technical sphere and universities. The authors also used general logical methods (analysis, systems approach), empirical methods (interviewing, comparison).

## 3 Results

The innovative activity plays the most important role in the implementation of the production of an industrial enterprise. A lot scientists associate innovations with the use of new (high) technologies. Enterprises are trying to carry out innovation activity in one or another form independently or through subsidiaries or innovation centers created specifically for this purpose, designed to promote and use of innovative products [4]. Here, an important component of the survival and development of enterprises and educational institutions in the digital economy is the creation of qualification requirements (professional standards) for the profession of the manager of innovative companies.

As the experience of various countries shows, in the conditions of the modern competition, only professionally trained managers are able to provide the transformation of scientific ideas into a successful innovative project, which is attractive both for investors, producers, and end consumers. The innovative way of the economic development requires, first of all, the susceptibility of the whole society to innovations, as well as a sufficient number of personnel that can manage the innovation process and implement innovations. This development is impossible without such a category of professionals as innovative managers. These specialists should be proficient in the technology of searching and implementing innovations, as well as the principles of their commercialization, the theory and practice of protection of the intellectual property, be able to manage business projects and high-tech enterprises. This aspect is particularly relevant from the standpoint of the latest approaches to the training of highly qualified scientific personnel in the direction of ensuring the innovative development of the economy.

At the same time, in the professional training of innovative managers, we can face a number of the following problems (difficulties) that need to be overcome:

- the absence of the traditions "to commercialize knowledge", "to earn money based on knowledge and skills" in our society;
- the sustainable perception of innovations as real threats to well-being and status positions;
- the undeveloped motivation systems of scientific and educational communities, representatives of government and business sphere in the process of creating a "new economy" and changes in educational systems;
- the absence of the necessary organizational and legal conditions for the development of innovation activity and, above all, its financial support in the real business practice.

All these difficulties are surmountable. However, a favorable environment for innovations in the economy and education can arise only as a result of successful activities of all the participants interested in the innovative transformations.

A new generation of leaders should have modern knowledge in the field of marketing, system analysis, technological knowledge which is necessary for the relevant industry. Obviously, scientists and developers themselves should also have the necessary skills in the field of innovations. They should possess, in addition to the above, competences in the field of forecasting and evaluating the commercial significance of new products and technologies at an early stage of their promotion to the market. In addition, in the current period of the economic development, that is characterized by the activation of the human factor, the growing economic role of all types of knowledge and the development of new forms of management oriented towards this, specialists should have the appropriate personal qualities that enable them to become leaders of changes who can guide the company to the direction of the increased competitiveness of their organization. In this regard, one of the main state ideological tasks should be the formation of an innovative focus of the economic thinking.

The authors have a reason to assert that many traditional methods of providing and evaluating knowledge, mastering special professional skills do not correspond to the principles of the formation of the professional consciousness and do not reflect the modern needs of the quality of education. It requires a lot of work in the development of innovative teaching methods. In universities, the serious revision of the heuristic component is necessary, as well as, if not practical (because of known difficulties in the organization), then as close as possible to the practical activities, of innovative training of future specialists. At the state level, serious work is needed in the direction of developing training programs for managers of the innovation sphere. Practice also needs an answer to the question about the infrastructure of personnel management in an innovative organization, about the content, methods and tools of increasing the personal competence of managerial personnel of various qualification levels that provides certain stages of innovation processes in scientific, technical and industrial organizations. In this regard, it is necessary, first of all, to organize high-quality training and carry out organizational work on the preparation of innovative managers with compulsory practical training. In addition to conducting scientific and practical seminars and modifying training programs, it is necessary to develop projects for the preparation of innovative managers, first at the regional and then at the state level. Preparatory work and the development of special training programs should be based on the following principles:

- a well-developed concept of training an innovative manager is necessary (the main stages should be clearly identified, ranging from the determination of the need to the employment of a specialist with the subsequent monitoring of his work);
- new training program (well-designed and agreed with specific specialists);
- strict professional selection of candidates for the position of innovative managers;
- compulsory internship at enterprises that are interested in such specialists.

At the present stage, in our opinion, the strategy of training innovative managers should be based on the retraining of specialists in the existing training centers and the existing infrastructure of innovation activities. The training system should involve, first of all, highly qualified scientists, practical researchers, entrepreneurs working in the field of innovations. The main "starting" task is the elaboration at the state level and the earliest realization of the mechanism "concept - training - effective practical work".

An innovation manager is a person who is able to realize a new idea, initiate its practical implementation and, in the end, turn it into a viable cost-effective product. The success of most innovative projects largely depends on the professionalism and skills of such people.

Depending on the activity type of a company, the tasks of such a specialist can be very different: from the development of high-tech products to the creation of consumer goods with unique properties. An innovative manager is more a vocation, calling, mission, than a profession. First of all, an appropriate type of character is necessary, as well as the ability to start working on something that seems to be impracticable, unrealizable and then turn the "disembodied" idea into a famous brand.

The innovation manager should be a professional in the technological field in which he implements projects. In addition, it is desirable to have additional education in the field of management (but only in this order).

There are several ways to "enter" this profession. Conventionally, the existing specialists are divided into two groups: "professionals" and "amateurs". "Professionals" are managers who receive the specialized education (managerial, financial, economic) and, not really understanding the subject area, try to manage innovative projects.

"Amateurs" are specialists who come from the scientific sphere and try to "push" their development to the market. These can be scientists involved in research and creation of new products (technologies, drugs, software systems, etc.). Those graduates of technical universities, who are interested not in doing research and development work, but promoting what has already been created by their colleagues, master this profession.

They come from the science, as a rule, with the working experience in fundamental foreign research centers. They can be specialists engaged in business development (business development manager) with the background of the professional training and relevant qualifications. Especially often, this profession involves specialists from the technological or engineering departments of Western companies who were involved in investment projects and the development of new products. However, in its pure form, both ways are flawed. It is necessary either close cooperation of "professionals" and "amateurs" in the framework of one project, or the acquisition by specialists of additional knowledge and missing skills.

To get into the world of innovation management, it is better for potential candidates to maintain constant contact with recruitment agencies that search for professionals for the innovation structures of big international companies.

Functional responsibilities of the innovation manager are:

(1) market research, identification of needs for new materials, products, services, technologies, etc.;
(2) calculation of potential market capacity;
(3) generation, search and preliminary assessment of ideas;
(4) monitoring and timely improvement of technological processes;
(5) legal protection of innovation projects;
(6) development of the project concept, marketing strategy and business analysis;
(7) search for equipment for the production of materials, negotiating with suppliers, conclusion of supply contracts;
(8) work with the manufacturer to develop a technology for the production of new goods.

Innovation management is an art that combines knowledge of the subject area in which projects are implemented, management skills from the branch of high-risk enterprises, as well as the ability to assemble a team and rally it around a common idea.

The main qualities of an innovation manager are an extensive outlook and receptivity. He should be competent in a fairly wide range of issues, as well as work skills in various positions, and be able to learn quickly to convert new experience into new knowledge and competences. Since innovation involves work in the field of new technologies and products, it is necessary to understand the issues of managing intellectual property rights, as well as protecting these rights, to know the specifics of domestic and foreign legislation in this area. A big plus for applicants is the experience of organizing and conducting scientific research, the ability to model various processes and manage them, foreign language proficiency, the basic skills of financial planning and accounting.

Positive and negative aspects of the profession are determined by the specific features associated with the implementation of a new project. On the one hand, there is a chance to grow together with the project and the organization, on the other - there is always a risk that the innovation will fail. Therefore, it is necessary to formulate in advance a description of the attributes of events, conditions, risks that signal the necessity to close the project.

The undoubted advantage of the profession is that, in the process of work, you become an expert in one or several areas and begin to distinguish between projects that have good prospects for implementation and unpromising ones.

The advantages of the profession also include the fact that large and successful companies are primarily engaged in innovations (at least in Russia), these companies can offer professionals not only worthy salaries but also unlimited opportunities for their creativity development. In addition, the need for such specialists is increasing constantly. This means a very interesting and varied work, challenges and the constant search for chances and opportunities, successful career development.

The disadvantages of the profession include unsecured employment. In Russia, there are still a small number of companies with R&D centers (mainly large Western corporations, focused on the creation and production of local products and trademarks), so the vacancy rate for specialists is very limited. And in situations of cost reduction, people who deal with innovation programs are the first candidates for firing. In addition, because of the constant pressure - the need for creating something new every time - there is a high risk of professional burnout.

A manager who has one or several successful innovation projects behind him/her is extremely in demand in the labor market. Especially if we consider that in our country they have just now begun to pay attention to the training of specialists in this field, and it is not known yet what quality of education of the first professional innovation managers we will have. In the future, an innovation project specialist can apply for the position of a quality director, an executive director, a director of a particular direction, or a development director.

# 4   Discussion

The studied problem was considered in the works of Shirokova et al. [5] who examined the strategic need for introducing intrapreneurship into an organization. The functions of innovation managers were studied by Maier and Brem [3]. The entrepreneurial orientation of managers in the context of innovative economy is analyzed in the study of De Souza et al. [1]. The issue of the development of the innovation component in Russia was studied by A. Ivanov, director of the Institute of Law and Development of the High School of Economics (HSE) - Ivanov [2]. In his opinion the "development" of this aspect should be positioned in the state system itself. It is innovative technologies that determine the place within the value chains. Ivanov also presented statistics according to which solutions containing an innovative component in Russian exports are only 0.04%. The Russian economy needs to develop a new category of specialists - innovative managers. The vision of specialists adapted to modern realities will be able to further ensure the economic development of the country.

# 5 Conclusions

In the future, digitization will only facilitate work, including the work of an innovative manager, will open new opportunities for developing organizations, changing market positions, and increasing profits by managing the process of creating value added. Products of digital transformation will allow not only to facilitate and improve the quality of managers' activities, but also to change the quality of training of specialists who are necessary for the modern economy.

# References

1. De Souza LA, Ramos I, Esteves J (2016) The influence of the entrepreneurial orientation of project manager's intention to adopt platforms of crowdsourcing innovation [A influкncia da orientaзro empreendedora dos gestores de projetos na intenзro de adotar plataformas de crowdsourcing innovation]. Atas da Conferencia da Associacao Portuguesa de Sistemas de Informacao 16:237–244
2. Ivanov A (2018) Who in the value chain is higher, and he is an innovator, September 2018. http://pltf.ru/2018/09/10/aleksej-ivanov-br-kto-v-cepochke-sozdanija-stoimosti-vyshe-tot-i-innovator/. Accessed: 12 Sept 2018
3. Maier MA, Brem A (2018) What innovation managers really do: a multiple-case investigation into the informal role profiles of innovation managers. RMS 12(4):1055–1080. https://doi.org/10.1007/s11846-017-0238-z
4. Troshina EP, Mantulenko VV, Shaposhnikov VA, Anopchenko TY (2016) Specific features of entrepreneurial departments in Russian companies. Int J Environ Sci Educ 11(14):6837–6852
5. Shirokova GV, Sarycheva VA, Blagov EYu, Kulikov AV (2009) Intrafirm entrepreneurship: approaches to the study of the phenomenon. Bulletin of St. Petersburg University, Series 8, Issue 1, pp. 3–32

# Problems of the Development of the Digital Economy at the Regional Level

V. A. Savinova[1], E. V. Zhegalova[1]($\boxtimes$), J. V. Semernina[2], and A. S. Kozlova[2]

[1] Samara State University of Economics, Samara, Russia
savinovava@yandex.ru, zhegalova@rambler.ru
[2] Saratov Socio-Economic Institute (Branch) of Plekhanov Russian University of Economics, Saratov, Russia
{ysemernina, alina.brosalova}@yandex.ru

**Abstract.** In order to identify and study existing problems objectively hindering the development of digitalization of regional economies, in the article the analysis of the intensity of the development of the digital economy in Russia for the period 2010–2017 years.

On the basis of the expert survey and subsequent procedure results in ranking as the most important problems for the development of the digital economy at the regional level were identified: (1) the outflow of skilled personnel in the field information and communication technologies from the regions, characterized by relatively low level of digitalization, occurs at both the national and the global level; (2) limited the amount of demand at the regional level, to a greater or lesser extent manifested in all market segments (corporate, Government, households).

**Keywords:** Digitalization of regional economy · Effective demand · Efficiency of economic processes · Financial technologies · Information and communications technologies · Labor resources

## 1 Introduction

Globally, the formation and further development of the digital economy due to the developing over the last several decades, trends in the global economy, in particular, the development of globalization processes, facilitating the integration of national economy in the world, and the rapid increase of the competition's level, which is manifested, including at the level of national economies. In general case the development of the digital economy can and should be considered as one of the key conditions for enhancing the competitiveness of the Russian economy, because the optimum usage of the opportunities are provided by the digital economy, allows to achieve a significant increase in the level of efficiency of various economic processes (both at national level and at the level of individual economic agents).

In turn, the intensity of development of digital economy on the national level is determined primarily by the speed and quality of the digitization's processes occurring at the regional level, which is crucial in terms of significant differentiation of subjects

© Springer Nature Switzerland AG 2020
S. Ashmarina et al. (Eds.): *Digital Transformation of the Economy: Challenges, Trends and New Opportunities*, pp. 260–268, 2019.
https://doi.org/10.1007/978-3-030-11367-4_25

of the Russian Federation directly by the level of digitalization (as its quantitative indicators, it is advisable to use a fraction of the cost of information and communication technologies in total spending on research and development; the share of the sector of information and communication technologies in the gross domestic product; the proportion of households with Internet access, etc.) and other socio-economic characteristics (population of the region; level of urbanization; level of income; industrial production indices; investment structure in fixed capital, etc.).

The overall dynamics of the development of digital economies in Russia, expressed through the index of information's value and communication technologies, is presented in Fig. 1.

The aim of the study is the identification and to describe existing problems objectively, impeding the development of the digital economy at the regional level, due to the influence of economic factors and a general nature for the subjects of the Russian Federation.

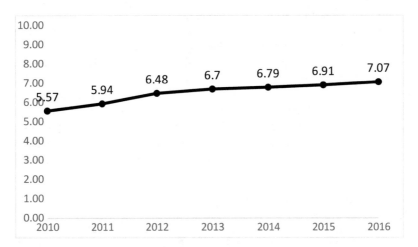

**Fig. 1.** Index of development of information and communication technologies in Russia in 2010–2016. Source: compiled by the authors.

## 2  Materials and Methods

In the making of the study, in September 2018, 30 experts who are working in the area of information and communication technology were interviewed. The method of expert survey was used, (boundaries of analyzed areas was determined in accordance with the methodology developed by the staff of the Higher school of Economics) [12] only one expert from each company was interviewed and selected at random.

Expert survey was conducted by the standard algorithm: each expert was asked to allocate 10 main problems hindering the development of the digital economy in the Russian Federation at the regional level. The formulation of the question presupposed open answers to questions, i.e. some of the problems were formed directly in the survey.

In the next stage of the expert survey was conducted a ranking of all the identified issues, the most important issue was assigned a rating equal to 10 points, and the least significant – rating equal to 1 point. If the equivalence of several selected problems (from the point of view of an individual) of experts used a standard procedure for determining the associated ratings, assuming their average. Opinions of interviewed experts recognized equivalent.

Then using the methods of summary and grouping the results obtained were systematized, and each highlighted the problem of the development of the digital economy at the regional level was rated overall rating is equal to the sum of individual ratings assigned by the experts, i.e. the formula was used:

$$R_{S_i} = \sum_{i=1}^{n} r_i, \tag{1}$$

where $R_{S_i}$ – total rating of the $i$-$^{\text{th}}$ selected experts of the problems impeding the development of the digital economy at the regional level;

$r_i$ – individual rating of the $i$-selected experts of the problems impeding the development of the digital economy at the regional level

$n$ – is the number of experts contributing the $i$-$^{\text{th}}$ problem hindering the development of the digital economy at the regional level.

As the most important development challenges of the digital economy in Russia was highlighted the problems that the value of the total rating was included in the top 10% interval of the possible range of values (from 30 points (minimum) to 300 points (the maximum value)), i.e. the top 10% interval includes values from 273 points to 300 points, inclusive (compliance ratings by $10^{\text{th}}$ %- intervals are presented in Table 1).

**Table 1.** Distribution of rating scores under $10^{\text{th}}$ % of intervals

| Used relative intervals, % | Intervals of ratings, points |
|---|---|
| 90–100 (upper interval) | **273–300** |
| 80–90 | 246–273 |
| 70–80 | 219–246 |
| 60–70 | 192–219 |
| 50–60 | 165–192 |
| 40–50 | 138–165 |
| 30–40 | 111–138 |
| 20–30 | 84–111 |
| 10–20 | 57–84 |
| 0–10 (lower interval) | 30–57 |

Source: compiled by the authors.

The usage of traditional economic-statistical methods in the study of problems of development of the digital economy in Russia at current time, in our opinion, it is not possible due to a very low quality of the existing information base generated at the regional level. The main difficulty encountered during its use, is the lack of necessary information: in the vast majority of cases it is presented in aggregate form (without separate allocation of the sector of information and communication technologies at the regional level) that does not allow for specialized study.

## 3 Results

### 3.1 The Outflow of Qualified Personnel from the Regions

The first problem of regional development of the digital economy is the outflow of qualified personnel in the sphere of information and communication technologies.

The problem of outflow of qualified personnel (total ranking of this problem is the highest according to the results of the expert survey and is 281 points) in the area of the digital economy can be considered at two levels:

(a) national level where the movement of qualified professionals within the national market, and the most common is the outflow of specialists in the leading regions, in terms of the development of the digital economy. In this case, the regions characterized by a relatively low level of development of the digital economy act as a kind of "personnel donors" for the leading regions in this respect (among the latter we should particularly note such regions as Moscow, Moscow region and Saint-Petersburg, and as one of the main indicators of development of the digital economy in these regions is advisable to use level of Internet penetration (Table 2)).

(b) the global level at which the movement of specialists is carried out at the level of national economies. Note that in recent years, recorded an increase in the outflow of specialists in the field of the digital economy from developing countries to developed countries.

**Table 2.** Subjects of the Russian Federation, characterized by a maximum specific weight of population using Internet in total population from age 15 to 74 years in 2017, %.

| № | Subjects of the Russian federation | Specific weight of population using internet in total population from age 15 to 74 years % |
|---|---|---|
| 1 | Moscow region | 92,8 |
| 2 | The Republic of Tatarstan | 92,8 |
| 3 | Saint-Petersburg | 92,6 |
| 4 | Chukotka Autonomous region | 91,7 |
| 5 | Moscow | 91,3 |
| 6 | Tyumen region | 89,7 |
| 7 | Murmansk region | 88,9 |
| 8 | Republic of North Ossetia—Alania | 88,8 |
| 9 | Samara region | 87,1 |
| 10 | Sevastopol | 86,5 |

Source: compiled by the authors.

The fundamental difference between these levels from each other is that in case of "overflow" of specialists in the field of information and communication technologies at the national level, we are talking about raising the level of unevenness in the development of the digital economy. A natural result of skilled emigration is a "gain" of the regions are initially characterized by a higher level of digitization development, and "weakening" of the regions that initially had a low level of digitalization, and the development of the analyzed process at the global level, there is a decrease in the level of digitalization of the Russian economy. In other words, in the first case we are talking about the territorial redistribution of qualified staff, accompanied by increased levels of concentration and in the second case – the "care" of the national economy, which at sufficient scale, this process will lead to an increase in the digitization rate of those economies that were able to attract these professionals and, consequently, to improve their competitiveness on a global scale. It should be noted that the movement of skilled labour in the area of information and communication technologies can have quite complicated "trajectory", i.e. the model described above "movement" of labor are not only possible, in particular, professionals with a high level of professional competence, can either directly move to the global level, and indirectly (first national, then global).

In addition, under certain conditions it is possible and "backward movement" in which experts in the area of digital back in the region, characterized by a lower level of digitalization, however, in soviet practice, such "movements" almost not spread. This is largely due to the fact that in regions with a higher level of development of the digital economy, professionals have the opportunity not only to obtain a higher income when performing similar duties, but also to systematically develop professional competence.

The satisfaction of such material and professional needs of specialists in the area of digital economy in regions with a relatively low level of digitalization is significantly hampered by: working in such regions, the core companies are not able to offer qualified professionals a competitive level of income, and the ability to solve non-standard professional tasks on a permanent basis (generally, specific activities of regional companies working in the area of information and communication technologies, is defined and limited to the simultaneous needs of their customers).

It must be emphasized that this problem threatens the realization of the task of building a territorial development strategy based on the principles of new industrialization, especially in small towns [14].

### 3.2    Limited Effective Demand at the Regional Level

As a second problem, a regional digital economy development in Russia it is possible to provide a limited amount of effective demand. Most clearly, this problem (its overall rating is equal to 274 points) manifests itself in the corporate sector, in which there is an understanding of the importance and necessity of digitalization, but the practical use of the possibilities of the digital economy limited financial capabilities of companies. It seems to us that such financial limitations can be quite clearly divided into several groups:

(1) financial constraints due to the scale of businesses (for a relatively smaller business companies, especially characterized by simplicity of technological operations and a small amount of revenue, large-scale introduction of digital technology (at least at their current levels) does not cause substantial economic effect);

(2) financial constraints caused by the level of profitability of companies (a number of industries and sectors of the Russian economy for various reasons is characterized by a rather low level of profitability, in particular, one of the main reasons is quite low profitability may be inappropriate price situation of the relevant market, which does not allow companies, including major to invest in the digitalization of business);

(3) financial constraints due to the presence of companies have alternative priorities (one of the most common financial priorities are the implementation of investment programs or high level of debt, identifying the main directions of use of financial resources).

The fact should be noted that at the regional level is certainly the predominant relatively small scale of the company's business, demand for information and communication technologies are defined to solve their business challenges, and objectively these needs below the needs of large companies. For example, for large major companies in the last few years is extremely relevant to the problems of forming a unified system of normative-reference information to ensure the comparability of data from different information systems used by them, while for medium-sized companies at this point in time, this issue is not business critical, and in small companies it is non-existent.

In the public segment, the level of effective demand is defined as the General condition of the state budget and placed in the framework of priorities. Given the fact that over the past few years, the state budget in Russia was scarce, the development of the digital economy at the state level largely also been limited by the financial possibilities of the budget. In the segment of households, the problem of insufficient effective demand expressed significantly weaker, however, in this segment, consumers are willing to spend money only on solutions that allow them to solve specific application tasks and at the same time to save certain types of resources (for example, the vast majority of a variety of mobile applications focused on saving time for users). In the Russian market of individual "digital solutions" for households does not actually exist: at the current income level of the vast majority of them just can't afford it (the commercial value of developing such solutions is too large compared to the average budget of households).

Need to focus on the fact that in the regional markets there is a demand for scalable application solutions (products) to digital technology (in this respect, it is the measure of a particular industry startups in the area of automation, financed by big companies), but the vast majority of regional companies working in the area of information and communication technologies, has its own resources for the systematic development of new products demanded by the market, thus attracting loan financing in the field of information and communication technologies is very problematic because of the low level of development of venture investment and the unwillingness of traditional financial institutions, primarily commercial banks increase the volume of unsecured lending.

# 4   Discussion

The development of the digital economy raised in the modern economic literature quite often, foreign researchers usually prefer to analyze either the best practices of those countries which are the indisputable leaders in the field of digital economy at a global scale (thus, among the major economies leaders in the field of digitization are the US and Japan, and among the relatively small economies of the impressive results achieved, in particular, Norway, Switzerland, Singapore, South Korea), or to conduct industry-specific research (for example, recently a lot of the popularity of research in the field of creation of ground and air unmanned vehicles).

Quick development of digitalization, IT- of technologies is the necessary condition of perfection of financial technologies, such as mobile payments and translations, management by the personal finances, crediting. Financial technologies are determined crowd funding and other as the "innovative and front-rank financial services given by companies, a key factor in development of that is information technologies" [7].

In the context of regional aspect of development of digital economy it is necessary to use of digital technologies is important for a management by the financial streams of the state or budgetary management [2].

A level of interference with economic processes from the side of the government of territories bodies can be different in different periods of economic development, but the quality of this interference and its formal results must be based on advantages that are provided by the development of digital economy at the level of territories [16].

The consideration of this subject is taking into account the specifics of the Russian economy is carried out exclusively by local experts. Thus, despite a significant increase in scientific interest during the last few years, the most common remain common theoretical approaches, among which the most popular was the study of the digital economy under the state control.

For example, it is proposed to consider the digital economy as a new direction of economic theory, interpreting it as a complex system of interrelated institutional categories, which, in particular, are information and communication technologies [6].

Also, the digital economy is interpreted in the context of public administration, considering it as the necessary conditions for improving its quality [3].

In a similar vein, are exploring the digital economy and the other scientists. However this group of researchers to a greater extent focuses on the issues of information security [10].

Miroshnichenko [11] mentions the necessity of application of multiregionalist to improve the efficiency of public administration.

Garas [5] already understands the digital economy: in the framework of the public administration system, the author focuses on the study of e-government, noting that one of the factors preventing its development is an information inequality. The study of the digital economy in a similar vein, is typical of the publications of a number of other scientists [1, 4, 8, 13, 15].

As a result we can say that in the specialized scientific literature virtually ignores the regional aspect of the formation of the digital economy, including to identify and systematize the problems hindering its development. Only in some scientific work generally refers to "regional inequality" is holding back the development of the digital economy in Russia [9].

# 5  Conclusions

On the basis of the expert survey and subsequent ranking procedure of the results obtained as the major development challenges of the digital economy at the regional level was highlighted:

(1) outflow of qualified personnel in the sphere of information and communication technologies from regions with relatively low levels of digitization that takes place both at the national and global level;
(2) a limited amount of effective demand at the regional level, in varying degrees, note in all market segments (corporate, government, households).

One of the solutions to the problems is the balanced state policy aimed, in particular, at increasing the level of digitalization of regions of the Russian Federation through the use of distributed production methods of goods and services in the field of information and communication technologies, practically does not impose restrictions on the geographical location of the companies operating in this field.

# References

1. Amanzholova TA (2016) Public administration. Hum Inf 10:47–53
2. Bernatska N (2018) Formation of modern model of budget management based on methods of public administration. Baltic J Econ Stud 4(1). https://doi.org/10.30525/2256-0742/2018-4-1-39-48
3. Bolshakov SN, Leskov IV, Bolshakova YuM (2017) Digital economy as part of the technological platform of public policy and management. Management Issues, No 44, pp 64–70
4. Cheremnykh VYu, Yakovlev LS (2017) E-government: models and perspectives. Bulletin of the Volga Institute of Management, No 1, pp 68–74
5. Garas LN (2017) E-government: a regional perspective. Society: politics, Economics, law, No. 7, pp 9–13
6. Gasanov, GA, Gasanov TA (2017) The Digital economy as a new direction of economic theory. Regional problems of transformation of the economy, No 6, pp 4–10
7. Ryu HS (2018) Understanding benefit and risk framework of fintech adoption: comparison of early adopters and late adopters. In: Proceedings of the 51 Hawaii international conference on system sciences, vol 3864–3873
8. Khvatov, AE, Vatoropin AS (2017) Barriers to obtaining public services in electronic form. Management Issues, No 46, pp 53–61
9. Kravchenko NA, Kuznetsova SA, Ivanova AI (2017) Factors, results and prospects for the development of the digital economy at the regional level. World of Economics and Management, No 4, pp 168–178

10. Merkulov AV, Avdeeva, IL, Golovina TA (2018) Information support of system of public management taking into account modern challenges and threats. Central Russian messenger of social sciences, No 1, pp 153–165
11. Miroshnichenko MA (2017) Development of a system for providing electronic state and municipal services with the use of multiregionalist. Sci J Kuban State Agrar Univ 7(131), 1665–1675
12. National research university higher school of economics [Electronic resource]: official site. Access mode. https://www.hse.ru/primarydata/iio. Accessed 18 Oct 2018
13. Pavlyuchenkova MY (2016) E-government as a vector of innovative state reforms. Bulletin of the Russian university of friendship of peoples. Series: Political Science, No 4, 121–129
14. Pipan T (2018) Neo-industrialization models and industrial culture of small towns. GeoScape 12 (1): 10–16. https://doi.org/10.2478/geosc-2018-0002
15. Stepanova AM (2018) Information policy of the state for the promotion of e-government. Bull Sib Inst Bus Inf Technol 2(26):65–73
16. Walczak D, Rutkowska A (2016) Project ranking for participatory budget based on the fuzzy TOPSIS method. Eur J Oper Res 260(2017): 706–714. https://doi.org/10.1016/j.ejor.2016.12.044

# Digital Farming Development in Russia: Regional Aspect

A. V. Shchutskaya$^{(\boxtimes)}$ ⓘ, E. P. Afanaseva, and L. V. Kapustina ⓘ

Samara State University of Economics, Samara, Russia
avs2020@yandex.ru, parus82@mail.ru, lkap@inbox.ru

**Abstract.** An active use of digital technologies is a characteristic of the current stage of world farming development. Today, the share of Russia's digital economy in GDP is not large. Russia ranks 15th in the digital farming world. Having a huge resource potential, Russia tends to increase its competitive position in the agricultural market. The authors have developed a hypothesis that Samara region, which is one of three Russian leaders in the implementation of digital farming innovations, will be able to make a great contribution into Russian agribusiness that will allow the Russian Federation become a worthy competitor in the digital farming market. The purpose of the study is to identify the promising areas of Russian agribusiness in the context of the world digital farming development. The authors have used general scientific and special methods and techniques of an economic research. The study has showed that, on the one hand, Russia lags far behind advanced economies, but on the other hand, the experience of Samara region has proved that the greatest effect in the implementation of digital innovations is provided by combining the efforts of farmers, scientific institutions and the state. The result of the research is that due to the digitalization, Russia will be able to take advantage of the enormous resource potential, and increase and strengthen its competitive advantages in the world agricultural market.

**Keywords:** Digitalization · Digital economy · Digital farming ·
Information and communication technologies · Innovations · Internet of things ·
Precision farming

## 1 Introduction

The modern world economy is characterized by high innovation activity. The most important area of innovations is the use of information and communication technologies, that is to say, the digitalization that is increasing in all industries and activities. Digital solutions are becoming an indispensable element of the work of enterprises and organizations in the world.

In Russia, the official policy to the digitalization started on 1 December 2016, when the President of the Russian Federation, in his message to the Federal Assembly, proposed to launch a large-scale system program to develop the economy of a new technological generation, so called the digital economy. In 2017, the Digital Economy in the Russian Federation Program was adopted to improve competitiveness of the

© Springer Nature Switzerland AG 2020
S. Ashmarina et al. (Eds.): *Digital Transformation of the Economy: Challenges,*
*Trends and New Opportunities*, pp. 269–279, 2019.
https://doi.org/10.1007/978-3-030-11367-4_26

Russian economy, increase the population's quality of life, provide economic growth, and maintain the national sovereignty [4]. In July 2017, at the meeting of the Council for strategic development and priority projects, Vladimir Putin stressed the importance of digital economy development in Russia. He noticed that the digital economy is not a separate industry, but it is a way of life, a new basis for the development of public administration, economy, business, social sphere, and the whole society [17]. The President compared this project with the electrification of the country in the first half of the XX century.

The development of the digital economy causes the transformation in all aspects of social and economic activities, and farming is not an exception. By the aid of the upcoming reforms, Russia's farming should become a high-tech industry that surely holds a competitive position in the global market. In this regard, it is necessary to realize the current level of the digitalization in Russian farming and the level to achieve in this process.

According to the German experts' analysis of the world farming development, since the 2010s, agriculture has entered the development phase 4.0, that is to say, the digital farming (by analogy with Industry 4.0). The precision farming technologies (agriculture 3.0), supplemented by intelligent networks and data management tools (the Internet of Things), are the basis of agriculture 4.0. The next stage of development will be agriculture 5.0 based on robotics and artificial intelligence, which is able to make decisions and perform operations without human intervention [5].

The USA, Australia, Canada and the EU hold the leading positions in the application of farming digital technologies [15]. Scientific researchers have shown that about 90% of farms in the EU and about 70% of farms in the US use precision farming technologies [31]. In recent years, these countries have been characterized by an accelerated pace in introduction of farming information technologies. Thanks to the use of these technologies, it has become possible to control the full cycle of crop farming and farm animal production through smart devices, that transmit and process the current parameters of each object and its environment (equipment and sensors that measure the parameters of the soil, plants, microclimate, animal characteristics, etc.), as well as seamless communication channels between them and external partners. Farming is gradually turning from a traditional industry with conservative technologies and a business organization into one of the sources of the modern technological revolution [11]. In the EU countries, they believe that digital technologies can make farming smarter, more efficient and sustainable. Thus, farmers who have adopted innovative technologies can count on government support [19, 25]. The efficiency of the applied technologies is expressed in a significant increase of labor productivity, reducing the cost of production and sales, improving the quality and the potential of consumer characteristics of products and services.

In Russia, interest in digital farming technologies is growing exponentially. The results of the research on digital farming techniques, precision farming and precision livestock farming have been reflected in the papers of such authors as Balabanov [2], Fedorenko [10], Korotchenya [15], Kozubenko [16], Oborin [22], Truflyak [31], Zavodchikov [34], and others.

Having a huge resource potential, Russia tends to increase its competitive position in the agricultural production. Only an active engaging in the digitalization of the agricultural economy can make a breakthrough in this area. In this regard, the assessment of the current level of Russian agricultural digitalization, both in general and in some Russian regions, to identify the main problems and promising areas of the development is of a particular interest, and it has become the purpose of this study.

To achieve this goal, the authors have set the following objectives:

- to study the development level of Russian and international digital farming,
- to highlight the territorial differentiation of the digitalization in Russia, and distinguish advanced regions, and
- to summarize the experience of digital technology implementation in farming, and develop recommendations for its use.

## 2  Materials and Methods

The authors have used general scientific and special methods and techniques of an economic research. The methods of theoretical analysis have allowed identifying the special aspects of the digital farming development in Russia and in advanced economies. By means of economic and statistical analysis methods, the grouping of Russian sub-federal units, according to their digital technologies level in farming, is carried out. There is a lack of official statistical data on the application of information and communication technologies in Russian farming. Due to this, the information of the Ministry of Agriculture of the Russian Federation, the Ministry of Agriculture and Food of Samara region, scientists and practitioners' published papers, as well as information from the Internet portals devoted to agribusiness have served as the basis for the analysis.

## 3  Results

Currently, the share of Russia's digital economy in GDP is not large and is only 2.8% that is 4.4 times less than in the UK, 2.5 times less than in China and 2 times less than in the US [31]. Russia ranks 15th in the digital farming world. According to the statistics of the Ministry of Agriculture of the Russian Federation, announced at the Precision Farming Conference in Skolkovo (February, 2018) by Igor Kozubenko, the Head of State and Information Development Department in the Ministry of Agriculture, only 10% of tillable land is processed with the use of digital technologies. The information and computer technologies market in farming is estimated at 360 billion rubles, and is one of the most promising [27].

To intensify the introduction of farming digital technologies, Russian Ministry of Agriculture has begun to develop Digitalization in Farming Program. The regional ministries have created informational support departments in farming, developed the Digital Economy Research Center. Moreover, digital farming departments have been established in the largest agricultural universities since 2017, and training of new employees to work in farming has started.

For a long time, Russian farming has not been an attractive business for investors due to the long production cycle, natural risk exposure, large crop losses during cultivation, harvesting and storage, the inability to automate biological processes, and the lack of progress in productivity and innovations. With the use of information and communication technologies, the agricultural sector will become more attractive for investors, and in the near future, perhaps, one of the most popular.

The largest companies from different regions in Russia are already taking interest in innovative farming technologies. According to the scientists of Kuban State Agrarian University (Monitoring and Forecasting Centre of scientific and technical development in farming: precision farming including automation and robotics), 28 regions out of 40 surveyed, use the precision farming elements in Russia. The precision farming leaders are Lipetsk (812 farms), Orel (108 farms), and Samara (75 farms) regions [31]. These regions have a significant advantage in tillable lands surface, processed with digital technologies. The precision farming is the most widespread in the Southern and Central Black Earth regions, as well as in the Volga region (see Table 1).

**Table 1.** Ranking of Russian regions that used precision farming in 2017

| Russian regions | Precision farming | | Precision livestock farming (number of farms in units) | Precision farming development, support and implementation programs (in units) |
|---|---|---|---|---|
| | Number of farms (in units) | Tillable land surface (thousands of hectares) | | |
| Lipetsk region | 812 | 2352 | 51 | – |
| Orel region | 108 | 684 | – | – |
| Samara region | 75 | 704 | 1 | 1 |
| Kurgan region | 55 | 387 | 2 | – |
| Voronezh region | 54 | 336 | – | – |
| Tyumen region | 54 | 241 | 4 | – |
| Nizhny Novgorod region | 50 | 158 | 4 | – |
| Krasnoyarsk region | 44 | 300 | – | – |
| Tambov region | 41 | 315 | 1 | – |
| Krasnodar region | 32 | 609 | – | 1 |
| Orenburg region | 31 | 218 | – | – |
| Tomsk region | 31 | 177 | 13 | – |
| Republic of Crimea | 30 | 98 | – | – |
| Leningrad region | 24 | 32 | 46 | – |
| Tula region | 23 | 143 | 1 | – |
| Republic of Bashkortostan | 21 | 192 | – | 3 |
| Kaliningrad region | 17 | 94 | 8 | 6 |
| Perm region | 15 | 72 | – | – |

(*continued*)

**Table 1.** (*continued*)

| Russian regions | Precision farming | | Precision livestock farming (number of farms in units) | Precision farming development, support and implementation programs (in units) |
|---|---|---|---|---|
| | Number of farms (in units) | Tillable land surface (thousands of hectares) | | |
| Ryazan region | 15 | 33 | 4 | – |
| Kursk region | 13 | 71 | – | – |
| Amur region | 12 | 110 | 9 | – |
| Adygeya Republic | 7 | 13 | 1 | – |
| Volgograd region | 6 | 101 | – | – |
| Ivanovo region | 6 | 15 | 16 | 1 |
| Kostroma region | 5 | 6 | 24 | – |
| Smolensk region | 4 | 10 | 2 | 1 |
| Republic of Komi | 3 | 35 | 7 | 1 |
| Karachay-Cherkess Republic | 3 | 15 | – | – |

Source: compiled by the authors.

The development of precision livestock farming is more concentrated in the North-Western and Central part of Russia. The largest number of enterprises using these innovations are located in Lipetsk, Leningrad and Kostroma regions.

It should be noted that not all Russian regions have precision farming development, support and implementation programs. The leader is Kaliningrad region that has been using Agrar-Office, AlPro, DelPro and other software for more than 7 years. Samara region has implemented InGEO (geo information system) software since 2009. Smolensk region have introduced Exact Farming since 2016.

The elements of the precision farming that are currently in practical use in Russia include: definition of the field boundaries, remote sensing, parallel driving systems, local soil sampling in the coordinate system, mapping of soil electrical conductivity, mapping yields, differentiated technologies (fertilizer, lime, plant protection products, growth regulators, soil treatment, sowing), crop state monitoring, and crop quality monitoring. In addition, field maps are being digitized, and information databases are being developed in farming.

Identification and monitoring of an individual animal, satisfaction of its needs, automatic regulation of the microclimate and deleterious gas control, monitoring of the herd's health, monitoring of the livestock products quality, electronic database of the production process, and robotization in the milking process are the most widely applicable among the precision livestock farming components.

Samara region is one of three leaders in the digital technologies development in farming. The precision farming is being used on 704,000 hectares of tillable lands (26.7%) from total 2636, 8 thousand hectares. There is a robotic dairy complex LLC Radna in the region.

Describing Samara agribusiness, it should be noted that it is a diversified production and economic system with 559 farming organizations, 1951 farms, more than 290.3 thousand private subsidiary farms, and about 780 food processing organizations and enterprises that provide services to supply and maintain agricultural machinery work [1]. There are five agricultural machinery organizations, such as CJSC Eurotechnika, LLC Eurotechnika MPS, LLC PEGASUS-AGRO producing machinery and equipment for the precision farming in the region. Samara region also has a strong scientific potential. There are 27 research institutes, universities, experimental stations, and research and experimental enterprises: Research Institute of Agriculture in Samara named after N. M. Tulaikov, The Volga Region Research Institute of Selection and Seed Farming named after P. N. Konstantinov, Samara Research Veterinary Station of the Russian Academy of Agricultural Sciences, The Volga Region State Zone Machine-Testing Station, and others.

The activity of Samara region agribusiness focus on an increase of the agricultural production to meet the needs of the population in food, and develop the export potential. In 2017, Samara farmers produced agricultural products for 96.4 billion rubles [8]. Samara region proportion was 1.7% in Russian agricultural production, and 7.2% in the Volga Federal District. The structure of gross agricultural output of Samara region is dominated by crop production (63% of the total in 2017). The livestock production proportion was 37%.

Due to the specialization of Samara region, digital technologies are more used in farming agriculture. Large agricultural holdings and advanced organizations, such as LLC BIO-TON Company with 300 thousand hectares, the agricultural holding Grain of Life with 89 thousand hectares, LLC Steppe Expanses with 6.3 thousand hectares, the Orlovka Agro - Innovation Center with 3.5 thousand hectares, the family farm Tsirulev E.P. with 6.0 thousand hectares, and others are the most active. These enterprises use the precision farming methods and some of its elements, cultivate land using satellite communication, put to use drone technologies and other modern systems in soil monitoring. All heavy equipment has been equipped with GLONASS in LLC Steppe Expanses that allows working with an accuracy not exceeding 5 cm. Heavy machinery has been equipped with a board computers and thruster that are operated by the satellite [7]. Full development of precision farming technologies requires significant investments that is why many agricultural organizations in Samara region have begun introducing individual elements. For example, in some organizations, parallel driving devices have been installed on self – propelled sprayers, and in other companies, navigation aids have been fixed on sowing machines, combine harvesters, cars, and other equipment.

The largest holdings can no longer do without digital technologies in their daily work. The authors have analyzed the use of the precision farming in two Samara agricultural holdings LLC BIO-TON Company and Grain of Life. We have revealed that they account for 55.3% of the total tillable lands cultivated with precision farming technologies in Samara region. According to the dimensional analysis, the land areas of the analyzed farms in municipal districts of Samara region are located unevenly. Therefore, the proportion of the tillable lands cultivated with the use of navigation equipment and other elements of the precision farming, also varies from 1.5% in Stavropolskiy District to 58% in Koshkinskiy and Chelno-Vershinskiy Districts (see Fig. 1).

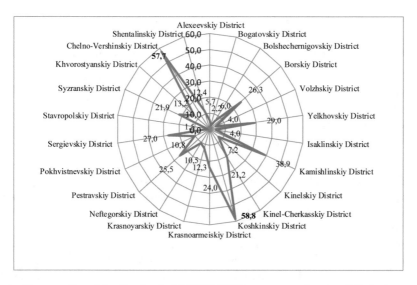

**Fig. 1.** The use of precision farming in LLC BIO-TON company and grain of life in the context of Samara region municipalities in 2018 (in percentage correlation with the tillable land volume) (Source: compiled by the authors)

The effect from the use of innovative technologies has been confirmed by the facts. The average cereal productivity was 26 centners per hectare in Samara region in 2017, but in was 50–74 centners per hectare in LLC BIO-TON Company, and 38–44 centners per hectare in the Agricultural Holding Grain of Life. Previously, there was not such crops productivity.

The interest of Samara farmers in precision farming is growing every year, and there is a need for widespread introduction of these technologies. Market conditions make every farmer think about the efficiency of the technologies used. Practice shows that using the precision farming, farmers have the opportunity to save on fertilizers, plant protection chemicals and irrigation (through their use in accordance with the needs of plants), seeds (through more accurate seeding), oil products (by optimizing machinery and vehicles movement, as well as improving production discipline). The crop productivity is increasing, due to the development of the optimal regime of plant nutrition and timely agricultural efforts. As a result, the cost of goods manufactured has been reduced by a mean of 14–30% [13, 24]. According to V. Orlov, the Director of LLC Eurotechnika MPS, the equipment used in the precision farming is compensated within one year, and more expensive and complex equipment is compensated within two years [29]. Such a short payback period has been confirmed by both Russian and foreign researchers [20, 26].

Digital technologies bring significant benefits in the ecological farming and product quality control.

One of the digital technologies used in farming in Samara region is the GIS system (geo information system in agribusiness). This system has been created in the region since 2008. Currently, the GIS database of agribusiness in the region includes information about 23.977 fields with a total area of 2.433.963 hectares. The total number of agricultural producers registered in GIS, is 3.954 [28].

The main functions of GIS in agribusiness in Samara region are:

– maintenance in the data registry on crops, location of crop acres, and crop condition,
– implementation of space monitoring of acreage,
– monitoring of unused land,
– creation of statistical and customer reports, and thematic maps of agricultural land.

The use of GIS technologies gives the opportunity to provide the necessary cartographic materials with various temporal and spatial generalization to the authorities and agricultural producers. This fact makes it possible to use GIS technologies as a modern system of agricultural management, contributing to the adoption of effective management decisions.

The development of the digital economy in agribusiness was intensified in Samara region in 2018. To develop and implement the measures for the digitalization in agribusiness, Ministry of Agriculture and Food in Samara region has established Information in Agribusiness and Agricultural Land Monitoring Department. It is planned to found an Expert Council on digital farming transformation. All these steps will undoubtedly give positive results in the near future.

## 4 Discussion

Today, interest in information and communication farming technologies is growing both at the level of the Federal and regional authorities, and among farmers. For example, I. Fazrakhmanov, the Minister of Agriculture of the Republic of Bashkortostan, has proclaimed 2018 as the year of the precision farming and satellite monitoring for automated accounting of performed mechanical work and energy consumption [9].

To stimulate the use of digital farming technologies, some Russian regional authorities have granted a subsidy to farmers to reimburse the costs on the purchase of precision farming equipment (satellite navigation stations, on-board computers, monitors, displays, elements of the parallel driving and autopilot system, on-board sensors for accurate fertilizer application, crop monitoring, soil properties measurement, soil moisture monitoring stations), and control and accounting systems (monitoring terminal, fuel level sensors, bulk products sensors, readers). In Samara region, the amount of subsidies is 40%, in the Republic of Tatarstan 50%, in the Republic of Bashkortostan 60%. Subsidy assistance will make the equipment purchasing more affordable not only for large organizations but also for small farms. The staffing problem will be solved by the professional development and training personnel.

Digital technologies in Russian farming are based on the State strategic approach, and the systematic development of innovative programs aimed at the development of agricultural sectors in sub-federal units of the Russian Federation. The effectiveness of innovations is proved by economic, production and environmental indicators in our country and in farming practice in the US, EU, Australia, China, Brazil, Japan, and others.

The role of digital technologies in the farming development has been analyzed in the foreign scientific papers. Eastwood, Klerkx, Nettle [6] and Finch, Samuel, Lane [12] have studied the use of new farming technologies (precision farming technologies: implementation, adaptation and efficiency). Myers, Ross, Liu [21] and Van Hertem, Rooijakkers, Berckmans, Peña Fernández, Norton, Berckmans and Vranken [32] have examined the informatization of agricultural production. Souza, Bazzi, Khosla, Uribe-Opazo, Reich [30], Abd El-kader and El-Basioni [26] have considered job automation and increase of the crops yield. Bansod, Singh, Thakur, Singhal [3], Oliver, Robertson and Wong [23] have done research on production efficiency and optimization of the resource potential.

Only with a close cooperation of science and practice, it is possible to achieve good results in the innovative industrial development. We can give some examples. Agro-terra Company, the resident of the SKOLKOVO Foundation, in cooperation with the Israeli SMART Fertilizer, has successfully conducted an industrial experiment to install sensors on fields with a total area of one thousand hectares in Tula and Kursk regions. This technology has increased the yield of soybeans by 11.5%, wheat by 6.5% for one year [18]. In Menkovskiy branch of the Agrophysical Research Institute in Leningrad region, farmers have obtained yields of wheat of about 60–70 kg/ha, potatoes about 600–650 kg/ha using the precision farming technology. At the same time, there has been a saving of fertilizers and plant protection means of about 35% for 5 years [33].

Today, in Russia, regions leading in the crop output (Krasnodar region, Lipetsk region, Rostov region, the Republic of Tatarstan, Samara region) have been achieved the greatest efficiency in the use of digital technologies, and their experience should be used in other regions. The most popular and promising areas of farming digitalization are navigation equipment, automated control systems, robotic systems, and unmanned vehicles. According to the experts' forecasts, the Internet of Things solutions and farming digitalization will bring a total economic effect of 4.8 trillion rubles per year or 5.6% of Russia's GDP growth [14].

## 5 Conclusions

In Russia, there is a significant potential for growth in production and competitiveness of agricultural products with the introduction of digital technologies. National and international experience in the application of these technologies in farming proves their production, economic and environmental efficiency.

The study has showed that according to the scale of the introduction of digital farming technologies, Russia lags far behind advanced economies. At the same time, there is a number of positive examples in the use of digital technologies in various Russian regions. Large agricultural enterprises and agricultural holdings are the most active in the digitalization process. The experience of Samara and other regions has shown that the greatest effect in the implementation of these innovations is provided by combining the efforts of farmers, scientific institutions and the state.

# References

1. Agribusiness in Samara region (2018). http://mcx.samregion.ru/apk/. Accessed 28 Apr 2018
2. Balabanov VI (2018) Overview of innovative developments in precision farming. Agric Mach Technol 2:24–27
3. Bansod B, Singh R, Thakur R, Singhal G (2017) A comparison between satellite based and drone based remote sensing technology to achieve sustainable development: a review. J Agric Environ Int Dev 111(2):383–407. https://doi.org/10.12895/jaeid.20172.690
4. Digital economy in the Russian federation program (2017) 28 July 2017, 1632 p
5. Digital farming: what does it really mean? European agricultural machinery. CEMA (2017). http://cema-agri.org/sites/default/files/CEMA_Digital%20Farming%20%20Agriculture%204.0_%2013%2002%202017.pdf, Accessed 28 April 2018
6. Eastwood C, Klerkx L, Nettle R (2017) Dynamics and distribution of public and private research and extension roles for technological innovation and diffusion: case studies of the implementation and adaptation of precision farming technologies. J Rural Stud 49:1–12 https://doi.org/10.1016/j.jrurstud.2016.11.008. Accessed 28 Apr 2018
7. Farmers in Bolsheglushitskiy District harvest via satellite. http://volga.news/article/481722.html. Accessed 18 Aug 2018
8. Farming. https://www.samregion.ru/economy/farming/. Accessed 28 Apr 2018
9. Fazrakhmanov I (2018) This year should be the year of mass introduction of precision farming system https://agriculture.bashkortostan.ru/presscenter/news/32615/. Accessed 28 Apr 2018
10. Fedorenko VF (2018) Farming digitalization. Machinery and equipment for the village, 6:2–9
11. Filina FV (2018) Socio-economic conditions of economy digitalization in farming. Management of socio-economic development in the regions: problems and solutions. In: Collection of scientific articles of the 8th International scientific conference. Kursk, pp 364–367
12. Finch HJS, Samuel AM, Lane GPF (2014) Precision farming. Lockhart & Wiseman's Crop Husbandry Including Grassland (Ninth Edition), pp 235–244 https://doi.org/10.1016/B978-1-78242-371-3.50028-X. Accessed 28 Apr 2018
13. Golokhvastova SA (2017) Agribusiness becomes digital. Agric News 4(111):3
14. IT in Russian agribusiness. http://www.tadviser.ru/index.php/Статья:ИТ_в_агропромышле нном_комплексе_России#.D0.98.D0.BD.D1.82.D0.B5.D1.80.D0.BD.D0.B5.D1.82_.D0. B2.D0.B5.D1.89.D0.B5.D0.B9_.D0.B2_.D1.81.D0.B5.D0.BB.D1.8C.D1.81.D0.BA.D0. BE.D0.BC_.D1.85.D0.BE.D0.B7.D1.8F.D0.B9.D1.81.D1.82.D0.B2.D0.B5_.28IoTAg.29. Accessed 18 Mai 2018
15. Korotchenya VM (2018) Russia and agriculture 4.0. Econ Agric Russia 6:98–103
16. Kozubenko IS (2017) Precision farming and Internet of things. Machinery and equipment for the village 11:46–48
17. Latukhina K (2017) Arithmetic of the future. Russian Newsp 7312(146). https://rg.ru/gazeta/rg/2017/07/06.html. Accessed 28 Apr 2018
18. Medvedeva A (2017) What kinds of digital technologies are coming to farming. https://www.agroxxi.ru/stati/kakie-cifrovye-tehnologii-prihodjat-v-selskoe-hozjaistvo.html. Accessed 28 Apr 2018
19. Medvedeva A (2017) Digital technologies are set to revolutionize farming. AGROXXI. Agro-industrial portal. https://www.agroxxi.ru/stati/cifrovye-tehnologii-nastroeny-na-revolyuciyu-v-selskom-hozjaistve.html. Accessed 28 Apr 2018

20. Mungalov D (2016) Precision farming. http://sk.ru/news/b/articles/archive/2016/03/31/tochechnoe-tochnoe-zemledelie.aspx. Accessed 28 Apr 2018
21. Myers D, Ross C, Liu B (2015) A review of unmanned aircraft system (UAS) applications for agriculture. In: Paper presented at the American society of agricultural and biological engineers annual international meeting, vol 5, pp 3598–3612. https://doi.org/10.13031/aim. 20152189593, www.scopus.com. Accessed 28 Apr 2018
22. Oborin MS (2018) The farming development based on digital technologies. Bull Samara State Univ Econ 5(163):38–48
23. Oliver Y, Robertson M, Wong M (2010) Integrating farmer knowledge, precision agriculture tools, and crop simulation modelling to evaluate management options for poor-performing patches in cropping fields. Eur J Agron 32(1):40–50. https://doi.org/10.1016/j.eja.2009.05. 002. Accessed 28 Apr 2018
24. Petrov K, Grigoriev N (2016) Organizational and economic mechanism to stimulate the introduction of precision farming technologies (on the example of Saratov region). Agric Sci J 10:96–100
25. Precision Agriculture: an opportunity for EU farmers. Potential support with the CAP 2014–2020: Study (2014). http://www.europarl.europa.eu/RegData/etudes/note/join/2014/529049/IPOL-AGRI_NT%282014%29529049_EN.pdf. Accessed 28 Apr 2018
26. El-kader SM, El-Basioni MM (2013) Precision farming solution in Egypt using the wireless sensor network technology. Egyptian Inf J 14(3):221–233 https://doi.org/10.1016/j.eij.2013. 06.004. Accessed 28 Apr 2018
27. Shustikov V (2018) Digital technologies are coming into agriculture. http://sk.ru/news/b/pressreleases/archive/2018/02/21/cifrovye-tehnologii-prihodyat-v-selskoe-hozyaystvo.aspx. Accessed 28 Apr 2018
28. Shutko SK (2018) The use of GIS technologies in agribusiness in Samara region. Youth's knowledge: science, practice and innovation. In: Collection of scientific papers of the XVII International scientific-practical conference of students and young scientists, pp 310–314
29. Skachkova A (2018) Online fields. Agro-Inf 4(234):2–3
30. Souza EG, Bazzi CL, Khosla R, Uribe-Opazo MA, Reich RM (2016) Interpolation type and data computation of crop yield maps is important for precision crop production. J Plant Nutrition 39(4):531–538 (2016). https://doi.org/10.1080/01904167.2015.1124893
31. Truflyak EV, Kurchenko NY, Kramer AS (2018) Precision farming: state and prospects. Krasnodar, Kubgau, p 27
32. Van Hertem T, Rooijakkers L, Berckmans D, Peña Fernández A, Norton T, Berckmans D, Vranken E (2017) Appropriate data visualization is key to precision livestock farming acceptance. Comput Electron Agric 138:1–10. https://doi.org/10.1016/j.compag.2017.04. 003. Accessed 28 Apr 2018
33. Yakushev V (2018) The introduction of precision farming technologies in Russia will raise the average yield of grain crops twice. http://xn–80abjdoczp.xn–p1ai/novosti/v-rossii/3215-vyacheslav-yakushevvnedrenie-tehnologiy-tochnogo-zemledeliya-v-rossii-podnimet-srednyuyu-urozhaynost-po-zernovym-kulturam-v-dva-raza.html. Accessed 28 Apr 2018
34. Zavodchikov ND, Larina TN, Shakhov VA (2018) The digital economy of Russian agriculture: regional aspect. Drukerovskiy Vestnik 2(22):216–226

# Analysis of the General Development Trends and the Level of Digitization of the Pharmaceutical Market in the Russian Federation

A. L. Beloborodova[1]([⊠]), N. G. Antonchenko[1,2], A. V. Pavlova[1]([⊠]), A. D. Hajrullina[1], and A. A. Soldatov[3]

[1] Institute of Management, Economics and Finance of Kazan Federal University, Kazan, Russian Federation
{a-beloborodova,halbi}@mail.ru,
anton4enkonataly@gmail.com, 930895@list.ru
[2] Volga Region State Academy of Physical Culture, Sport and Tourism, Kazan, Russia
[3] Russian State Social University, Moscow, Russia
sold123@yandex.ru

**Abstract.** The paper presents a brief overview of companies that form analytical databases on the Russian pharmaceutical market. On the basis of information from four such databases and independently conducted field research, which involved Russian and foreign manufacturers of originator drugs and generics in Russia and abroad, the main trends in the development of the Russian pharmaceutical market are presented. Particular attention is paid to the analysis of trends implementing innovative technologies and digital tools in the activities of companies targeting the pharmaceutical market.

**Keywords:** Innovative technologies · Pharmaceutical market ·
Russian pharmaceutical market · Pharmacy · Digital technologies

## 1 Introduction

The pharmaceutical market is one of the most profitable and rapidly developing sectors of both the world economy and individual countries. Today, Russia is among the ten largest pharmaceutical markets in the world. The demand for drugs in the Russian market continues to grow. This is confirmed by the results of statistical observations which are analyzed in detail in the works of Sokolov et al., Denisova, Bondarev et al. and Sukhorukova [3, 5, 13, 14].

Fedorova and Fedotova [6] in their article "Import Substitution as a Factor in the Development of the Pharmaceutical Market in Russia" describe and analyze in detail the reasons for the observed growth and distinguish the main ones.

- The majority of the country's population is regular buyers of various drug dosage forms;

S. Ashmarina et al. (Eds.): *Digital Transformation of the Economy: Challenges, Trends and New Opportunities*, pp. 280–290, 2019.
https://doi.org/10.1007/978-3-030-11367-4_27

– A high proportion of aging population;
– The growing desire of the population to take medicines and treat themselves;
– Increased attention of Russians to their health.

In many studies, there are also negative trends in the development of the Russian pharmaceutical market, so the main dependence is import dependence. The issue has been studied in detail in the articles of Lin et al. [8] and Fedorova and Fedotova [6], so the authors state this fact and argue that the replacement of imported goods in the pharmaceutical industry is significant not only in terms of its financial well-being, but also from the point of view of strengthening national security.

Much attention in the research of the pharmaceutical market in Russia is paid to the issue of pricing in general and the possibility of transferring world pricing models to Russian reality in particular. The work of Narkevich [9] considers this issue.

The work of Olkhovskaya [10] "Consequences of WTO Accession for the Pharmaceutical Industry in Russia" is very interesting, because the author came to the conclusion that there are no growth prospects for the Russian pharmaceutical market due to WTO accession, since the competitive advantage of Russian drugs is rather low. Today, the fact of the predominance of imported drugs is confirmed by the findings of Olkhovskaya [10].

The Russian pharmaceutical market consists of two main segments: commercial and state. The commercial segment of the pharmaceutical market includes sales of drugs and parapharmaceuticals excluding sales under the pharmaceutical benefits scheme (PBS). The state segment of the pharmaceutical market includes sales of drugs under PBS, as well as sales through health facilities. The study analyzes the dynamic sales of drugs in the commercial segment of the pharmaceutical market for 2012–2017. The purpose of the study is to study the general trends of development and the level of digitalization of the pharmaceutical market in the Russian Federation.

## 2 Materials and Methods

The pharmaceutical market is a dynamic, rapidly changing and strategically significant industry for the economy of any country, in particular for Russia. Therefore, in the pharmaceutical industry, important attention is paid to the analysis of the pharmaceutical market. The study presents a brief overview of the companies that provide consulting services in the field of research of the pharmaceutical market, and on the basis of these databases the authors have analyzed the Russian pharmaceutical market and made the main conclusions.

Today, there are four major players in the market of analytical services in the pharmaceutical sector: DSM Group, RNC Pharma Analytical Company, Deloitte Company and ALFA RESEARCH AND MARKETING Company.

The result of the work of research teams in these companies are analytical databases on the pharmaceutical market. These companies own their own sources of primary data, including exclusive and unique methods of their analysis and interpretation. DSM Group, RNC Pharma Analytical Company and ALPA RESEARCH AND MARKETING Company specialize exclusively in collecting pharmaceutical market analytics. Deloitte also provides services and forms databases for enterprises operating in various sectors of the economy. In particular, for the pharmaceutical market, it conducts annual research devoted to the analysis of trends in its development.

In order to identify the features and patterns of the pharmaceutical market development of the Russian Federation in 2012–2017, marketing research methods were used: a desk research method for secondary data, which is to collect and analyze the main key indicators for the Russian pharmaceutical market.

More than 30 respondents from almost 10 Russian manufacturers of originator drugs and generics in Russia and abroad took part in the study.

## 3 Results

In the course of the study, the following results were obtained. Figure 1 shows the dynamics in the volume of the Russian pharmaceutical market from 2012 to 2017. The volume of the pharmaceutical market grew during all five years; as a percentage, the market grew by 85% and amounted to about 63 million rubles. The annual increase was observed at around 12% and only in 2017 it was 22% compared to 2016.

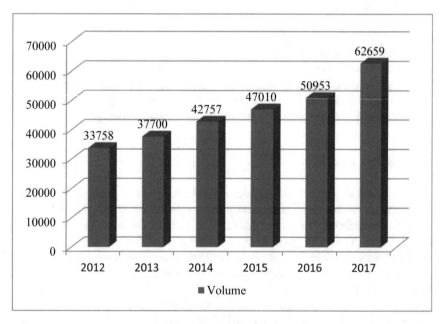

**Fig. 1.** Volume of the pharmaceutical market, 2012–2017, million rubles * (Source: compiled by the authors)

The authors also analyzed the dynamics of the total sales of drugs depending on dispensing requirements and belonging to the list of vital and essential drugs (VED). In 2017, sales continued to grow in value terms among both prescription (Rx-drugs) and non-prescription (OTC-drugs) drugs. The main results are presented in Fig. 2.

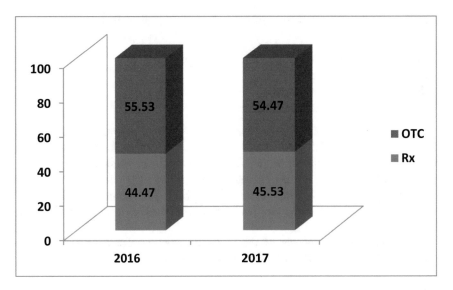

**Fig. 2.** Rx/OTC sales ratio in the retail sector of the pharmaceutical market, 2016–2017 * (Source: compiled by the authors)

As can be seen from the diagram presented in Fig. 2 significant changes in the market in the ratio of prescription and non-prescription drugs are not observed. As for specification of VED, the largest share of drug consumption is non-VEDs: their share in rubles was 67.6%, in volume - 55.34%, which is clearly shown in Fig. 3.

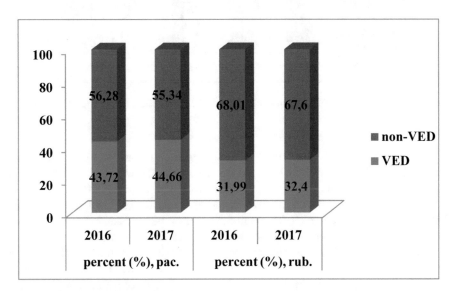

**Fig. 3.** Ratio of sales of VED/non-VED in the retail sector of the pharmaceutical market, 2016– 2017 * (Source: compiled by the authors)

As for the distribution of sales of drugs by ATC - groups, then for 2017 there are three leaders:

- 18.91% in the total volume of commercial drugs belongs to A - gastrointestinal tract and metabolism drugs;
- 16.08% - R- respiratory drugs;
- 14.12% - C- cardiovascular drugs.

The increase in sales in monetary terms in the group of gastrointestinal tract drugs was 7% compared to 2016, in terms of the number of packages, the increase was about 5%. Based on the dynamics of these indicators, it can be concluded that the observed increase in sales in rubles is due to the increase in the average prices of drugs. Indeed, the average price of this group increased by about 3% compared with 2016 and amounted to almost 205 rubles per package. A similar situation can be observed in other groups.

Let us turn to the analysis of the weighted average cost of packaging in the dynamics from 2012 to 2017, which is presented in Fig. 4.

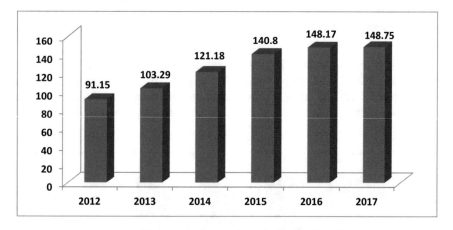

**Fig. 4.** The weighted average cost of packaging from 2012 to 2017, rubles * (Source: compiled by the authors)

In 2016–2017 the weighted average cost of packaging was 148 rubles, this figure increased by 63% compared to 2012. At the same time, the weighted average cost of packaging practically did not change in comparison with 2016 and 2017.

Next, the authors analyzed the relationship of domestic and imported drugs in the Russian pharmaceutical market. The ratio of pharmacy sales of domestic and imported drugs in Russia in December 2012 and 2017 is reflected in Figs. 5 and 6, respectively.

The results of 2012 showed that the share of domestic drugs in the total volume of the pharmaceutical market was 25% in terms of total value, and 56% in volume. Accordingly, foreign-made drugs prevailed in the total value of the market - 75%, and

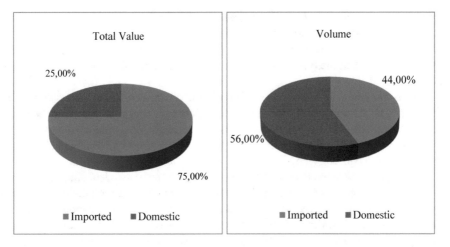

**Fig. 5.** Ratio of sales volumes of imported and domestic drugs in the Russian pharmaceutical market in December 2012 * (compiled by the authors)

accounted for 44% in volume. Despite the fact that Russian-made drugs are sold more in volume, the market share in terms of the total value is significantly higher for imported drugs.

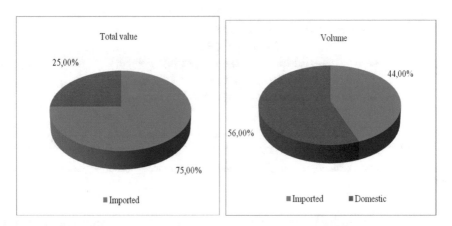

**Fig. 6.** Ratio of sales of imported and domestic drugs in the Russian pharmaceutical market in 2017 * (Source: compiled by the authors)

In 2015, the leaders among importers of Russian pharmaceutical products are the countries of the former Soviet Union: Uzbekistan (19%), Ukraine (15%), Kyrgyzstan (6%), Azerbaijan (5%) (according to RNC Pharma 2015).

The program "Development of the Pharmaceutical and Medical Industry for 2013–2020" sets out goals for active import substitution. So, analyzing the statistics of 2017, one can already draw the first conclusions about the effective implementation of the above program (Beloborodova et al. 2016).

If you look at the ratio of imported and domestic drugs in the pharmaceutical market in 2017, then there is an increase in the market share of domestic drugs both in total value and in volume.

This is due to one of the observed trends in the pharmaceutical market in the 2016–2017 - this is the switching of consumers to domestic drugs. In the pharmaceutical market, the increase in the popularity of Russian drugs is due, among other things, to the vigorous activity of domestic companies using innovative digital technologies (Antonchenko et al. 2014).

In 2018, the research center of Deloitte conducted an independent study, one of the directions of which is to identify the level of implementation of digital technologies in production and sale processes of pharmaceutical companies. The respondents of the survey were the following: 50% - representatives of foreign companies without localization of production in Russia, 25% - Russian companies, and 25% - foreign companies with localization of production in Russia.

In the course of the study, the following results were obtained (Fig. 7).

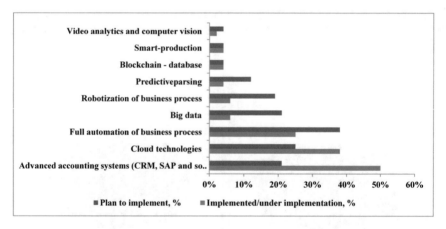

**Fig. 7.** Innovative technologies in pharmaceutical companies * (Source: compiled by the authors)

The figure shows that the most popular technological solution in the pharmaceutical industry is the implementation of advanced accounting systems - 50%, cloud technologies - 38% and automation of individual business processes - 25%. Only 4% of organizations implemented innovative technologies into production processes. After examining the full report presented in the figure, we can conclude that innovative technologies are more embedded in analytical processes than in production processes.

Also in the report you can find materials relating to the analysis of the level of implementation of digital tools in activities of pharmaceutical companies while interacting with end users. The results obtained are analyzed and presented in Table 1.

**Table 1.** Digital tools for the interaction of the pharmaceutical business with end users, 2018

| Indicator | Survey results, % |
|---|---|
| Informing consumers through the site, official pages on social networks | 78% |
| Informing consumers through doctors' blogs and forums | 73% |
| Medical representatives are using modern methods of data visualization (digital dashboard) | 62% |
| Using SEO-complex measures | 59% |
| Online sales of non-prescription drugs | 57% |
| Informing consumers through applications | 56% |
| Using big - data analysis | 56% |
| Developing and using a registry of patients with certain diseases | 47% |
| Online sales of prescription drugs | 43% |
| Using digital devices/applications to collect patient data | 42% |
| Electronic prescriptions | 29% |
| Telemedicine | 25% |

\* (Source: compiled by the authors)

The results of the analysis showed that the most popular digital tool for interacting with end users for pharmaceutical companies is informing about the company's products and activities through official websites (78%), as well as medical forums and blogs (73%). Analyzing the data presented in the table, we can conclude that the development of all digital tools for interaction with consumers listed in it is at an acceptable level.

Summarizing, we can say that the development of innovations and digitalization of business processes of pharmaceutical companies affect, first of all, analytical processes and communication processes of these companies with the end user, while production processes are still poorly affected in both Russian and foreign companies.

We also conducted a small study aimed at asking the representatives of the pharmaceutical industry which government measures, in their opinion, to a greater degree influence the decrease in the share of imported drugs in the Russian market.

The study involved more than 30 respondents from 10 Russian manufacturers of originator drugs and generics in Russia and abroad, and was in a form of interview.

The following data was obtained:

1. Restricting or banning the participation of foreign manufacturers in the field of public procurement - 57%;
2. Encouraging foreign producers to switch to full-cycle production by providing various benefits and preferences - 56%;
3. State support for Russian producers through the provision of state subsidies - 49%;
4. Indexation of registered prices for VED, produced in Russia and the absence of such indexation on VED, produced abroad - 29%;
5. Simplified procedure for registration of drugs manufactured in Russia - 24%;
6. Measures to combat corruption and increase competition - 15%.

For 2017, the main players in the Russian pharmaceutical market can be combined into the TOP -50, they account for more than 76% of the total value of sales of drugs. The first 10 players are presented in Table 2.

**Table 2.** Top 10 corporations in the retail segment of the Russian pharmaceutical market in 2017

| Rating | Rating change 2016/2017 | Corporation | Total value, mln. rubles | Share | Growth 2017/2016 |
|---|---|---|---|---|---|
| 1 | 1 | Bayer Group | 36 948 | 4,48% | 12% |
| 2 | −1 | Sanofi | 36 786 | 4,46% | −1% |
| 3 | – | Novartis | 32 823 | 3,98% | 5% |
| 4 | – | OTCPharm | 28 811 | 3,50% | 5% |
| 5 | – | Servier | 27 251 | 3,31% | 5% |
| 6 | – | GSK | 25 337 | 3,07% | 1% |
| 7 | – | Gideon Richter | 24 188 | 2,93% | 2% |
| 8 | – | Stada | 23 809 | 2,89% | 9% |
| 9 | – | Berlin-Chemie | 22 921 | 2,78% | 6% |
| 10 | 2 | Takeda | 21 985 | 2,67% | 7% |

\* (Source: compiled by the authors)

Top 50 leading brands account for 22.18% of the total value of the pharmaceutical market. In 2017, the top ten was as follows (Table 3).

**Table 3.** Top 10 brands by sales in the retail sector of the Russian pharmaceutical market in 2017

| Rating | Rating change 2016/2017 | Brand | Total value, bln. rubles | Share | Growth 2017/2016 |
|---|---|---|---|---|---|
| 1 | – | Cardiomagnil | 6,63 | 0,80% | 6% |
| 2 | 4 | Nurofen | 6,12 | 0,74% | 12% |
| 3 | 8 | Detralex | 5,71 | 0,69% | 17% |
| 4 | 8 | Miramistin | 5,68 | 0,69% | 19% |
| 5 | −2 | Theraflu | 5,59 | 0,68% | −2% |
| 6 | 3 | Pentalgin | 5,28 | 0,64% | 4% |
| 7 | 1 | Concor | 5,27 | 0,64% | 1% |
| 8 | −4 | Actovegin | 5,23 | 0,63% | −8% |
| 9 | −7 | Kagocel | 5,21 | 0,63% | −13% |
| 10 | −5 | Ingaverin | 5,18 | 0,63% | −8% |

\* (Source: compiled by the authors)

As can be seen from the data of the rating change 2016/2017 that a dozen leaders among corporations are the most stable compared to a dozen leaders among brands.

# 4   Discussion

Theoretical and methodological basis of the study was the work of Hotelling [7] and Chamberlin [4] 1920–1950, devoted to the theory of sectorial markets. From 1950 to the present, the formation of the economy of sectorial markets has taken place as an independent division of economic theory. In the mid-1950s the first textbook on the economy of sectorial markets was published. Based on the formed theoretical base, the rapid development of empirical research took place. Research affects all sectors of the economy. Today, individual authors and entire research organizations devote considerable attention to the study of industries that are strategically important for the economy. Thus, in Russia one of such industries is the pharmaceutical industry and the study is based on the basic principles of the economy of sectorial markets.

# 5   Conclusions

In the course of the study the following conclusions are made:

1. Over the past 5 years, the volume of the commercial market of drugs has grown by 85%, 22% of which - over the period from 2016 to 2017.
2. Market analysis, depending on the conditions of supply from pharmacies and belonging to the VED list, showed that in the total sales volume of the commercial market of drugs in 2017, VED - 54.47% and non-VED - 67.6% prevail. This suggests that the pharmaceutical market is quite independent of the health sector as a whole.
3. After analyzing the ratio of domestic and imported drugs in the total volume of the pharmaceutical market, we noted signs of the implementation of measures prescribed in the program "Development of the Pharmaceutical and Medical Industry for 2013–2020". So, for 5 years, the market of domestic drugs in total value grew by 4.8%. In our opinion, it is worth developing the chosen support measures, achieving a growth of at least 20% within the next five-year plan.
4. Analyzing the level of digitalization of the pharmaceutical industry, it was concluded that the development of innovations and digitalization of business processes of pharmaceutical companies affected, first of all, analytical processes and communication processes of companies with the end user, while production processes are affected slightly in Russian and in foreign companies. The most popular digital tool for interacting with end users for pharmaceutical companies is informing about the company's products and activities through official websites (78%), as well as medical forums and blogs (73%).
5. In order to identify support measures that give the maximum effect, we conducted a survey of representatives of the pharmaceutical industry. Thus, according to respondents, the most effective measures are: limiting or banning the participation of foreign producers in public procurement (57% of respondents), encouraging foreign producers to switch to full-cycle production by providing various benefits and preferences (56% of respondents) and government support of Russian producers by providing state subsidies (49% of respondents).

# References

1. Antonchenko N, Kalenskaya N (2014) Developing a methodology for assessing the efficacy of managerial decisions in entrepreneurial establishments. Life Sci J 11(7):365–369
2. Beloborodova AL, Martynova OV (2016) The development of the trade industry in Russia: trends, problems and prospects. Bull Volga State Technol Univ 1(29):53–60
3. Bondarev AV, Linnikova OV, Novikova NB (2016) Overview of the Russian pharmaceutical market. New Sci Curr State Ways Dev 8:117–120
4. Chamberlin EH (1951) Monopolistic competition revisinted. Economica 18:343–362
5. Denisova OV (2016) Key trends in the development of the pharmaceutical market in Russia. Econ Entrepreneurship 4–1(69–1):1163–1168
6. Fedorova YuV, Fedotova LS (2015) Import substitution as a factor in the development of the Russian pharmaceutical market. Bulletin of Udmurt University. Ser Econ Law 6:55–58
7. Hotelling H (1921) Stability in competition. Econ J 39:41–57
8. Lin AA, Sokolov BI, Slepnev DM (2013) Pharmaceutical market: the production of drugs in Russia. Prob Mod Econ 1(45):191–195
9. Narkevich IA, Lin AA, Orlov AS (2015) The level and price dynamics in the commercial retail sector of the pharmaceutical market in Russia and St. Petersburg. Pharmacy 4:28
10. Olkhovskaya MO (2013) Consequences of WTO accession for the pharmaceutical industry in Russia. Space Time 1(11):125–131
11. Orlov AS (2014) Features of pricing on the Russian pharmaceutical market. New pharmacy. Effective Manag 10:18–23
12. Orlov AS (2014) Improving state regulation of prices on the Russian pharmaceutical market: using foreign experience in reference pricing. Econ Manag Sci Pract J 2(118):14–19
13. Sokolov BI, Lin AA, Orlov AS (2012) Pharmaceutical market: structural features in Russia. Prob Mod Econ 4:336–341
14. Sukhorukova MG (2016) Analysis of the pharmaceutical industry: problems and development prospects. Guide Businessm 29:254–262

# Clusters, Digital Economy and Smart City

L. V. Ivanenko[1]([✉]), E. A. Karaseva[2], and E. P. Solodova[1]

[1] Samara University, Samara, Russia
{ivanenko_lv, se-11.83}@mail.ru
[2] State University of Economics, Samara, Russia
elenakaraseva063@gmail.com

**Abstract.** Currently industry modernization is carried out in cities, the digital economy is implemented resulting in smart plants, companies and organizations which represent almost completely computer-aided manufacturing. Then these companies are integrated into innovative cluster structures where required conditions for introducing the digital economy elements in urban space were created. The paper covers the concept content of a "smart" city, the digital economy, the industrial cluster. The examples of operating clusters are given. The problems and ways of forming a smart city are demonstrated.

**Keywords:** Cluster development · Cluster interaction · Cluster structures · Digital economy · Digital economy implementation · Industry modernization · Smart city · Smart management · Smart people

## 1  Introduction

Nowadays we live in the conditions of the fourth industrial revolution which is characterized by certain directions – activation of clustering and cluster functioning, development of the digital economy. As a resultant of these two processes we have the possibility of organizing smart cities in the country. However, it should be noted that one of the vital tasks under current conditions is training staff that will be ready to create smart cities and manage them effectively as well. Besides, not less complex problem is forming smart population owning skills and abilities to use all innovations and digital economy elements in a smart city.

## 2  Materials and Methods

The paper considers dynamic development of the digital economy in smart cities by means of the set of methods and techniques. First of all, when writing the paper theoretical development of this problem was applied. For this purpose, the authors used the scientific research of theorists and economists, referred to the analysis of the particular academic literature and the findings of the empirical research using the method of the system approach as well.

S. Ashmarina et al. (Eds.): *Digital Transformation of the Economy: Challenges, Trends and New Opportunities*, pp. 291–295, 2019.
https://doi.org/10.1007/978-3-030-11367-4_28

# 3   Results

The statistics demonstrates that nearly 75.0% of all the population of the country lives in the cities. Therefore, improvement of the existing cities, organization of modern urbanized computerized arrangement and their transformation into smart cities with the digital economy, serve as the relevant purpose of urban area development.

We will consider the content of the "digital economy" definition. It is activity with the use of the digital technologies in real life. It is supposed that in the long term the digital economy and its technologies will completely change the existing business models and correct conventional economic communications [7].

The "smart" city term has been used for a long time, but the single, agreed-upon definition has not been provided so far.

A smart city is the city with smart management and smart people. A smart city is characterized by the following parameters: high quality of life; availability of innovative technologies; city infrastructure with the elements of the digital economy; environmentally- friendly living environment [1–3].

Recently the industry has been intensively modernized in cities, the digital economy is developed resulting in smart plants, enterprises and companies which represent almost fully automated production. Then these businesses are integrated into innovative cluster structures where there are all prerequisites and necessary conditions for implementing the digital economy elements in urban areas. It should be noted that the industrial or economic cluster can be such a structure in the city. We should refer to the content of authors' definition of this concept: the industrial or economic cluster is a specific form of territorial self-organization integrated by means of the control action into strategic alliances of businesses, organizations, institutions, plants creating unified chains of value added accumulation in the conditions of the particular innovative environment [2, 3, 6].

The cluster integration of smart businesses leads to significant expansion of introducing innovations, implementing intellectual modernization, upgrading production automation, cutting energy and other resources costs, but all these processes can be carried out only by using the digital economy elements.

It seems appropriate that there can be the organization of common space, unified atmosphere for cooperation and interaction of smart businesses –cluster participants in a cluster. Thus, in the city the share of the digital economy elements introduced in the industry increases. In addition, the smart businesses will be served mainly by robots and engineers who will monitor manufacturing equipment performance, operation of installation and machines, computers and hardware equipment.

The management of the industrial or economic cluster demands the use of innovative opportunities of the digital economy elements: firstly, the use of information communication system; secondly, building up data-processing center and information management.

Taking into account all the above, we came to the conclusion that a cluster can form a qualitative basis for introducing the digital economy and forming a smart city which by definition has to possess all the components of the innovative system.

The potential opportunity of creating the digital economy and smart cities in the region is confirmed by the fact that the Samara region is included in the structure of Association of Innovative Regions in Russia. It is developed further in the paper that the examples of some industrial clusters functioning in Samara city and the Samara region are suggested considering. The following industrial economic clusters concern them: automobile, aerospace, petrochemical, oil-producing, IT as well as, agro-industrial, innovative and some other clusters. These clusters are operating actively [6, 9].

The Samara region clustering began with the organization of an automobile cluster. The agreement on the Volga region automobile clustering in 2010 was signed by 15 companies. It was important not only for suppliers of auto parts and car manufacturers, but for the educational institutions training specialists for this branch as well. It should be taken into consideration the participation of the local authorities representatives in the cluster that promotes special state support [10].

The Volga region automobile cluster contributes to the regional economy significantly: it provides 14% of GRP, it employs 50000 people, the members of the cluster are 52 participants and more than 100 businesses operate in its structure as well. They include automobile assembly plants, major integrating units of components and assemblies: OOO VALEO Service, OOO TPV RUS, OOO Rulevye sistemy, OOO Broze Tolyatti Avtomotiv, OOO DSK, AO AD PLASTIK, AO AKOM, OAO TZTO, OOO Nobel Avtomotiv Rusia, OOO Metalloproduktia and other component suppliers.

The cluster, being an innovative structure, constantly recognizes the importance of transformation, optimization, innovation, changes which demand challenges of digital economy today. In this regard the opportunity of production digitalization is realized actively in the automobile cluster, i.e. new products in a digital form, for example, auto parts are developed. It allows production making a certain economic breakthrough and get new competitive advantages [5, 6].

The special place in the innovative development of the region, takes the hi-tech territorial aerospace cluster which has marked impact on introducing the digital economy elements. The aerospace cluster represents the set of the main industrial productions: rocket-and-space, engine-building and aircraft-building. Moreover, the services in development, test and operation of aircraft are provided in the aerospace cluster. The participants of the territorial innovative aerospace cluster include the following enterprises and companies: OAO Kuznetsov, FGUP GNPRKTS TsSKB - Progress, OAO Aviaagregat, OAO Aviakor-Aircraft Factory, OAO Agregat, FGUP NII Ekran, OAO Metallist-Samara, OAO Salyut. Also two leading educational institutions: Samara National Research University named after academician S.P. Korolev and Samara State Technical University are members of the aerospace cluster. These higher educational institutions have enormous scientific research potential. The obligatory members of the territorial aerospace cluster are the local authorities: The Ministry for Economic Development, Investments and Trade of the Samara region, and the Ministry for the Industry and Technologies of the Samara region [5].

Considering the importance of the aerospace cluster for the region, its structure is supplemented by such organizations as: the Public Autonomous Institution of the Samara region Center for Innovative Development and Cluster Initiatives, Chamber of Commerce and Industry of the Samara region, Regional Center of Innovations and Transfer of Technologies non-profit partnership, OOO Eko Energy, OOO Research and

Production Company Razumnye resheniya, OOO Akvil, Regional Employers' Association Union of Employers of the Samara Region.

The coordinator of aerospace cluster activities is the public autonomous institution of the Samara Region the Center for Innovative Development and Cluster Initiatives. Analyzing the list of the members of the aerospace cluster, their high scientific and innovative level, it is advisable to evaluate the digital transformation which takes place constantly in the cluster that, first of all, assumes the use of new innovative technologies for improving enterprise competitiveness and significant rise in labor productivity in the cluster.

Unambiguously, a lot of new technologies are actively used in the aircraft industry and astronautics. This exactly confirms that the innovative aerospace cluster can also form a qualitative basis for introducing the digital economy elements and forming Samara smart city [5, 7].

The complicated problem is training staff obtaining knowledge of clustering, digital economy, smart cities and managing them effectively [5, 8].

In addition, it is necessary to prepare smart population that will use all the innovations and digital economy elements in a smart city as well. Thus, the personnel problem is of high priority: who has to form a smart city, the digital economy, a cluster and how? What managers, engineers, economists will be in demand? Where will they be found? Who can and has to train them? These issues have to be dealt with expeditiously and require special, careful and well-thought-out decision.

## 4   Discussion

It should be noticed that in the Samara region the opportunities are considered and discussed, the measures for introducing digital economy in activity of state authority are undertaken and work on implementing the projects on digital transformation of municipal government is intensified as well. Besides, the pilot projects on digitalization are provided to introduce in educational institutions of the higher education, the need for development and planning implementation of digitization projects and digital economy elements is proved for industrial production and enterprises as well as digitalization of agricultural industry.

Except for several effectively working clusters (the innovative IT - cluster, the agro-industrial one, the territorial cluster of medical and pharmaceutical technologies, the recreational and tourist one, etc.), there are the forming cluster structures being almost ready to work in the Samara region, for example transport and logistics, construction, housing and utilities clusters, etc. In addition, it is possible to identify potential clusters which can be transformed in forming ones, and then – in actively operating clusters in the Samara region. It is possible to refer to such structures the following clusters: the cluster of municipal solid waste management, the cluster of physical education and sport, the education cluster, the cluster of culture and arts, etc.). Various changes and economic digital transformations are taking place in the city so that Samara will become a really smart city with smart people living in it [4, 6].

# 5   Conclusions

The paper considered the industrial or economic clusters as the specific form of the territorial organization of integrated enterprises, organizations, institutions, plants creating unified chains of value added. The opportunities for introducing the digital economy elements in cluster structures are presented. The parameters characterizing a smart city are stated: high quality of life; availability of innovative technologies; city infrastructure with the digital economy elements; environmentally-friendly living environment. It is proved the interrelation between clustering and the digital economy and formation of a smart city on their basis.

# References

1. Antonov I (2012) Innovative development of the Russian economy: the current state. Integration, Moscow, p 251
2. Bochkarev A (2018) To the question of the connection of wave and digital economy. Bulletin of the Volga Region state University of service, series "economy". Issue № 2 (52) Togliatti: publishing and printing center PVGUS Togliatti, pp 26–30
3. Chelyshkov P, Sedov V (2015) The concept of "smart city": monograph. Ministry of Education and Science of the Russian Federation, National Research University, Moscow "MGSU"
4. Ivanenko LV (2018) Implementation of the project to create a "smart city" as a result of effective innovative technological development of the territory. Trends and prospects, Russia. Yearbook, vol 13, INION.Otd, pp 955–960
5. Ivanenko LV, Ivanenko AA (2015) Aerospace cluster. Innovation sphere. Bulletin of Samara State University. Series Economics and Management. Publishing House Samara University, vol 9/2 (131), pp 165–172
6. Ivanenko LV (2016) Clusters and innovation. Standard Cluster. Hugo's Messenger, Science, Education. Economy Series: Economy. Ufa State Academy of Economics and Service 1 (15):7–22
7. Kanabeeva RA (2018) Development of social innovations in digital creative economy. J Creative Econ Soc Innovations Int Inf Anal J 8(22):67–86
8. Timoschuk NA (2013) Personnel problems of social cluster formation in the region materials of the 4th all-Russian personnel forum "Innovative personnel management", Samara State Technical University, Samara, pp 17–21
9. Strategy of innovative development of the Russian Federation for the period up to 2020 [Electronic resource]. Innovative Russia-2020. Access mode: http://innovation.gov.ru/taxonomy/temp/586. Accessed 11 March 2018
10. Churkina I (2011) Influence of clusters on the development of the region. Bulletin of the Volga State University of Service. Econ Ser 4(18):26–30

# Human Capital Accounting Issues in the Digital Economy

O. Yu. Kogut[1]([✉]), R. Es. Janshanlo[1], and K. Czerewacz-Filipowicz[2]

[1] Al-Farabi Kazakh National University, Almaty, Kazakhstan
kogut.1108@gmail.com, ramazan1951@mail.ru
[2] Bialystok University of Technology, Bialystok, Poland
czerewacz.k@gmail.com

**Abstract.** The article discusses the issues of accounting for human capital in the digital economy. In the accounting community, company staff is recognized as an object of personnel records, but is not recognized as an object of accounting, cost accounting. The structure of accounting objects is proposed to include not only the property of the enterprise, but all the resources over which the company exercises control. The accounting methodology must respond appropriately to the development of accounting in the digital economy. At the same time, IT technologies cause significant modifications both in the methodology and in the applied direction of the science of accounting.

Many researchers believe that the main directions of transformation, in terms of improving the theory of accounting and reporting in the digital economy, are the areas of transformation of the digital economy: manufacturing, financial services, education, health care, and the media. As the analysis of theoretical sources showed, human capital management is not limited solely to economic studies of its value and value, but also includes the study of external social and mental effects. It became possible to estimate human capital by calculating the possibility of replacing an employee, including the cost of training or hiring a new employee. The general accounting model of intellectual capital considers the objects of accounting (structural, consumer, human capital, goodwill) from the point of view of the overall construction of the research process of a complex problem. A human asset can be identified, has evidence of its existence, can arise or cease to exist in an organization at an identifiable time point. In order to create effective accounting and management reporting in an organization, elements of human capital should be classified and accounted for as separate and independent human capital assets. In the context of focusing on the information needs of the stakeholders in the digital economy, the requirements for disclosing information, reducing the time needed to prepare and submit accounting (financial) statements are increasing.

**Keywords:** Accounting · Assets · Auditing · Digital economy ·
Human capital · Intangible assets · Intellectual capital · IT technologies

© Springer Nature Switzerland AG 2020
S. Ashmarina et al. (Eds.): *Digital Transformation of the Economy: Challenges, Trends and New Opportunities*, pp. 296–305, 2019.
https://doi.org/10.1007/978-3-030-11367-4_29

# 1   Introduction

Accounting issues of human capital are not theoretically developed, there is no clear definition (definition) of human capital in the scientific literature, researchers have not come to a common opinion regarding the methods of its financial evaluation. In the accounting community, company staff is recognized as an object of personnel records, but is not recognized as an object of accounting, cost accounting.

The development trend of accounting is the accounting of assets and the provision of information about the resources of the company, which cannot be fully controlled, which cannot be reliably estimated financially, but the use of which will lead to economic benefits. The structure of accounting objects is proposed to include not only the property of the enterprise, but all the resources over which the company exercises control.

Property is represented mainly in the form of intangible assets (methods of identification, valuation, accounting and reporting), while intellectual property (human, structural and consumer capital, goodwill) has not been comprehensively considered comprehensively.

Among the areas of development of the world economy stands out the transition to digital technology, where information is the main resource. This resource is of great value and acts within organizations as an intangible asset.

The accounting methodology must respond appropriately to the development of accounting in the digital economy. Particularly relevant is the problem of rethinking and developing a methodology for cost accounting and calculating processes and products. Solving the problem is caused by the need to analyze past activities and plan future ones.

One of the most important parts of the accounting system is the information component of the two subsystems - financial and management accounting. The reorientation from the control function to the informative one, based on the organization of points of digital transformation of the enterprise, is noted. It requires the development of new indicators, methods of collecting and processing not only financial information, but also the sufficiency of its integration with information about other aspects of business and the external environment. Analyzing the content of literary and scientific sources devoted to this problem, it can be argued that the development of the theory and improvement of accounting practices is metaphysically connected with the expansion of the information potential of the existing economic space. At the same time, IT technologies cause significant modifications both in the methodology and in the applied direction of the science of accounting.

Many researchers believe that the main directions of transformation, in terms of improving the theory of accounting and reporting in the digital economy, are the areas of transformation of the digital economy: manufacturing, financial services, education, health care, and the media. Particularly, it is possible to single out a study of the possibilities for evaluating new accounting objects, which are intellectual human capital, customer base, innovative products, R&D results, etc. [8].

Managing human capital in modern conditions is the management of humanity in the direction of enhancing the humanization of human relations, which makes additional political relevance of the topic of human capital internationally.

In theoretical terms, the clarification of features, principles of human capital management in modern highly competitive conditions can contribute to the development of forecast scenarios for the development of both a single enterprise and society as a whole in its macroeconomic scale.

An assessment of the degree of elaboration of the topic of human capital management in modern highly competitive conditions made it possible to reveal the theoretical lacunae available in studies of human capital.

As the analysis of theoretical sources showed, human capital management is not limited solely to economic studies of its value and value (future value), but also includes the study of external social and mental (social and psychological) effects that are important both for social relations and for internal political relations, society, as well as for international politics, if we talk about international relations in the field of production, distribution and use of knowledge and skills that make up nova human capital.

## 2  Materials and Methods

The growing importance of human capital has always been associated with the emergence of a new type of workers - the creative class in the conditions of a corporation of knowledge, characteristic of a new type of economy.

According to the definition of human capital [2], the following components can be distinguished:

- opportunities of workers (education, professional skills, experience, network of contacts, values and ideas);
- the growth potential of employees (the potential of the employee to achieve the goals of the organization outside of its current role);
- employee motivation (firm values as a basis for employee motivation);
- Innovation of employees' activities (contribution to the creation of new products and services of the company, loyalty to the changes of the company, the desire to learn).

Of course, it is almost impossible to evaluate these components. However, in order to measure work with human resources, a large number of various performance indicators have been created, including in terms of costing and pricing. Thus, it became possible to estimate human capital by calculating the possibility of replacing an employee, including the cost of training or hiring a new employee. As an alternative way to assess human capital, researchers B. Lev and A. Schwartz in 1971 proposed [10] to estimate the current value of all the income that an employee can bring to the company in the future.

However, these methods when trying to meet the requirements of financial statements face the problem in the form of the ambiguous nature of such a component of intellectual capital as human capital, since its essence is dynamic and unstable for being able to be accurately assessed.

The terms "intellectual capital" and "intangible assets" are often used interchangeably. However, this approach is wrong, since there is a huge difference between these concepts. While intellectual capital implies almost all intangible sources of value creation, intangible assets are an identifiable non-monetary asset that does not have a tangible form. An intangible asset is a resource created as a result of an organization's past actions (including purchasing or direct creation) and managed by an organization, with which the organization plans to gain economic benefits in the future.

According to this approach, intangible assets have the following characteristics:

- identifiability;
- the possibility of legal protection;
- the established ownership right;
- material evidence confirming the existence of the asset;
- the creation as a result of a specific event or an identifiable time interval;
- the possibility of destruction or termination of existence as a result of a specific event or the expiration of time.

Obviously, only a small part of the intellectual capital can meet the above requirements. Patents and trademarks in most cases satisfy these conditions. However, many other objects of intellectual capital are not suitable for these requirements. It is almost impossible to determine the moment when customer loyalty to an organization arises, the innovative spirit of a company arises or when an employee's computer literacy ends. This is the main reason why, when working with IC assets, it is not the usual formulas for calculating cost that are used, but rather integrated circuits for determining approximate values.

The advantages of modeling include:

(1) repeated use for the analysis of various situations;
(2) modeling systems with a complex solution that cannot be expressed by one or several mathematical relationships;
(3) requires mathematical skills much lower level than optimization models (optimization models) [13];
(4) accounting is fractal, that is, there is an infinite variety of models that never repeat exactly as a result succeeds the scientific or practical school that concentrates the model that most closely matches the stochastic conditions of a market economy, which led to the use of several hundred models in accounting.

All these positions have led to the need to develop and test a general accounting model for intellectual capital.

Accounting objects in the general model with respect to recommended accounting systems are considered as follows:

- structural capital (reflected in financial and transaction accounting: managerial and strategic);
- consumer capital (not reflected in financial accounting and reflected in transaction accounting: managerial and strategic);

– human capital (not reflected in financial accounting and reflected in transaction accounting: management and strategic);
– goodwill (business reputation) is reflected in financial accounting and is not reflected in transaction accounting, that is, in order to avoid double counting, since transaction accounting regulates the value of business reputation from the moment of acquisition of goodwill and the result is related to the predicted goodwill. Goodwill is subject to regular revaluation.

The general accounting model of intellectual capital considers accounting objects (structural, consumer, human capital, goodwill) from the point of view of the overall construction of the research process of a complex problem, reflected in the form of blocks: accounting and analytical support; intellectual capital model; intellectual capital pattern; complex of defining problems of financial accounting (structured chart of accounts; traditional financial accounting; international model of financial accounting); transaction accounting; change accounting; expression of accounting results in the form of net assets and net liabilities; audit; control.

The general accounting model of intellectual capital is built in terms of efficient use of structural, consumer, human capital and goodwill from the standpoint of identifying growth assets and treating liabilities as an expected value to be created by subsequent investments, that is, based on the requirements of fair valuation and determination of net liabilities in market and fair valuations [7].

## 3  Results

International financial reporting standards typically include intellectual and human capital as internally created goodwill—expenses incurred by an enterprise to generate future economic benefits that do not meet the conditions for recognizing intangible assets such as identifiability (possibility of separation from other assets) and control (the opportunity to secure and restrict to others the right to receive economic benefits). Expenses arising from the creation of internal goodwill are not subject to capitalization and are recognized in the period they are committed according to IAS 38 Intangible Assets.

A human asset can be identified, has evidence of its existence, can arise or cease to exist in an organization at an identifiable time point.

Signs of an asset:

(a)  An asset brings economic benefits. Labor is a factor of production, allowing the employer to produce and sell products, works, services produced by workers. The company is able to receive future economic benefits from the use of a human asset throughout the term of the employment contract.
(b)  The presence of control over the asset. The acquisition of labor by the company is in the process of hiring, executed by the employment contract and the order to work. An enterprise has authority over an asset, it has the ability to influence income by exercising its authority. The employer has the right to impose requirements on the level of education, work experience, experience in the specialty and other requirements. Set the date of commencement of work, place of

work, working hours and rest time, the size and order of payment of wages. The employer assigns the employee specific work, monitors its implementation, in the event of non-compliance, the employee is disciplined. Thus, the employer controls the asset object.

(c)  Asset valuation. At the stage when labor activity has not yet begun, the person directs his efforts towards receiving upbringing and education. In order to adapt to the constantly changing conditions of the labor process, the employee repeatedly improves skills, improves skills and abilities. The financial valuation of a human asset is the monetary expression of the accumulated knowledge of the skills and abilities of an individual to obtain aggregate data on the organization when preparing financial statements.

The object is able to bring economic benefits in the future, the company has the right to receive economic benefits, the initial cost is determined and a human asset can be allocated as an object of accounting.

Accounting for a human asset is an orderly system of collecting, recording and summarizing information in monetary terms by means of complete, continuous and documentary accounting.

The organization discloses in accounting policy the ways of keeping records of human assets, namely: primary observation, cost measurement, current grouping, final summary.

In order to create effective accounting and management reporting in an organization, elements of human capital should be classified and accounted for as separate and independent human capital assets. Moreover, the classification of human capital is to some extent feasible: the number of new products, the assessment of customer loyalty, the percentage of employees with a degree - these quantitative indicators are almost identical to the assets of human capital (human, structural and relational capital). However, human capital assets are often interdependent, which excludes the possibility of separation from each other and subsequent independent classification. It is because of the complex nature of human capital that the approach to the inclusion of all assets of human capital in one asset - goodwill [14] has spread. If the contract price exceeds the market value of all assets, positive goodwill arises. Conversely, if the contract price is lower than the market value of all assets, negative goodwill arises [10].

However, these quantitative indicators are still difficult to translate into monetary value in order to include them in reporting or use them as a management mechanism. Moreover, even if monetary valuation becomes possible, management is much more difficult to convince management of its necessity. So, for example, if you try to evaluate the company's unique business methods, their patenting, valuation and valuation can lead to the loss of their uniqueness, for example, as a result of the publication of patent documents, which will lower their "cost" and destroy the competitive advantage of the company.

At the same time, the initiative to create a system of strict reporting, valuation and asset management of human capital should come primarily from investors and top management of the company, who are willing to invest their resources in the creation and development of human capital, and are interested in making informed decisions based on human capital asset reporting system. The development of such a reporting system requires certain investments from investors, which, however, will pay off many times as a result of the implementation of this reporting system.

Considering human capital as a working example, let us try to imagine that this asset is human capital and its characteristics will be reflected in the company's reporting system. Suppose that employees of the company in question have higher computer literacy than competitors, and that this feature is a competitive advantage of the organization and, therefore, the cause of value creation. Of course, in this case, the investors of the company will be ready to invest in measures to improve computer literacy of employees, which means that developing an accounting system for the organization is necessary, because if computer literacy of employees is a competitive asset of the company, then it must be reflected in the financial statements.

However, it is difficult to assess the growth of computer literacy of staff, which has arisen due to investments of investors, and to separate it from the previous level of computer literacy, as well as from the entire set of IT skills of employees (such as programming, system administration, etc.) and the overall organizational culture of the company. This raises the problem of the monetary valuation of such a single asset, as computer literacy of staff, and therefore will not allow to determine the necessary investments of investors for growth and management of this asset.

At the same time, the company cannot pay off the debt to creditors using IT skills of employees or part of the organizational culture that encourages computer literacy, the founders cannot eliminate the IT skills of employees as a result of the company's bankruptcy, and sales management is unable to sell this human capital is an asset to persons wishing to purchase it. Thus, given that IT literacy of employees is an asset that does not have liquidity in the market, investors and founders of a company do not always support the development of a special reporting system for accounting for human capital assets.

In this regard, many investors are not ready to invest in the creation of such a system of reporting human capital assets, and many financial analysts cannot benefit from the existing reporting system, due to the problematic monetary valuation of human capital assets.

# 4   Discussion

The company's intellectual resources are the main value generators in the knowledge economy [15]. In this regard, such a knowledge management initiative, such as the creation of IC reporting for management purposes, can help expand the list of opportunities for assessing human capital at the organizational level. This initiative, linking the analysis and evaluation of human capital, will structure the priorities and goals of the organization, which in the conditions of the knowledge economy becomes a top priority.

J. Fitts-ents complains: "As far as accounting ignored human capital, you can see in almost any book on business ratios …… indicators related to employees appear only once, and then as an expense, and not profit" [6].

E. Flamholtz, identified three main criteria for recognizing human resources as an asset: potential gain, existence of rights of ownership or control by an economic entity, measurability in monetary terms [5].

Opponents of the concept of HRA believe: human resources cannot be reflected in the account as an asset due to the fact that the company does not have ownership rights to the person.

Researchers strive to take into account the fullness of the factors affecting the value of human capital, health, culture, the intellectual qualities of the individual, etc., propose calculation formulas for which the value of human capital is "not calculated".

For the purposes of investment projects, business plans, it is important to have a database for computing, so that all corporations calculate the cost uniformly (using a uniform methodology), in this case the indicator will be informative for investors.

G. Becker determined the value of human capital on investments for special training. This is the time spent and the efforts of the student himself, the teaching activities carried out by others, and the equipment and materials used. G. Becker also referred to investment in human capital as payment for the services of employment agencies, the cost of finding a new job, the time spent on interviewing, testing, inquiring and clerical work, investments in health that increase productivity [1].

B. Newman argues that to a certain extent the capitalization of human capital is reflected in a reporting item such as "goodwill" and believes that the value of human intellectual capital is estimated by the buyers of the acquired company [11].

J. Fitts-ents refers to human capital expenditures (HCCF) on employee records and benefits, costs for non-permanent workforce, losses due to the lack of employees in the workplace, and employee turnover losses [6].

Competencies determine the skills needed to do the job effectively, while human capital addresses a global problem. That is, competencies relate to the decomposition, disintegration and sale of human capital. Thus, human capital is considered as a whole, or rational, with the value and uniqueness of knowledge. Taking into account the relationship between competencies and human capital, the next question arises: what do we consider, what competencies determine human capital in a company? Is it enough to say that it is valuable or unique? Or: is it possible to determine what conditions of competence and determine the value and uniqueness of human capital? The value and uniqueness of human capital are two concepts that differ in content, and in this regard, they are related to the company's strategy, which clearly indicates the definition of these two values [3].

Human capital in auditing is fundamental to the audit industry. Does human capital in audit activity (on education, on the level of customers of auditors) affect the likelihood of financial distortion? Is there a link between professional experience, level of education and financial distortion? However, the question remains whether human capital in auditing affects the quality of auditing. Human capital can be formed through education and professional experience gained in the workplace, training, consumption and in-depth knowledge of the economic system. Among these factors, education and professional experience are considered as the two most important aspects of human capital. Education, an important component of human capital, can help people improve their cognitive abilities and solve ethical problems of human capital in auditing [16].

Organizational learning is becoming increasingly important for the strategic renewal of organizations in the digital and high-tech economy. Organizations with a huge variety, especially successful in the current environment, when firms must be effective and quickly adapt to change. Research results show differences in training between marketing and production units, as well as various methods of personnel management and types of human capital. Human capital mediates between the practice of HRM and learning [4].

In the context of focusing on the information needs of the stakeholders in the digital economy, the requirements for disclosing information, reducing the time needed to prepare and submit accounting (financial) statements are increasing. "Regulators in many countries of the world and far-sighted leaders of modern business are already actively thinking through the changes necessary to succeed in high-risk conditions when more demands are made on their business" [9].

Human capital is a collection of relatively stable qualities of individuals that determine the ability of people to generate effective solutions and allow them to create and disseminate innovations in the outside world. The most important functions of this component of intellectual capital as a single economic value of any business are the creation and dissemination of innovations, the creation and use of unique solutions to the problems encountered. The functioning of human capital serves the renewal, development and progression of economic entities [12].

# 5   Conclusion

1. Currently, the need for an effective system of reporting intangible assets remains high, and the concept of human capital, together with ordinary financial reporting, is most suitable for these purposes in the new economy, where knowledge is the source of wealth. We have investigated the reasons for the need to create a system of human capital-reporting, as well as factors that prevented the introduction of human capital-reporting in most economic entities.
2. Human capital assets are difficult to classify, evaluate, and identify and can often be used only in conjunction with other human capital assets, making it difficult to evaluate them separately and ultimately affect the reliability of the proposed human capital reporting systems. At the same time, an inefficient human capital-asset market does not allow an accurate valuation and increases the contract price many times in a transaction involving human capital assets, which, in turn, affects market participants and owners of assets who don't want to invest their resources in setting up a human capital-asset reporting system due to high costs. However, the most important problem in this area is the disagreement between the practice and the theory of human capital, since until the accounting of intangible assets and human capital accounting are separated from each other, the possibility of creating an accurate system of reporting assets of human capital is controversial.
3. It is necessary that the theoretical results be used in the development of methodologies and specific recommendations, the practical implementation of which will contribute to the effective implementation and development of the human capital accounting system in the digital economy. In addition, the situation suggests that

there is a need to accumulate experience in keeping records and disclosing economic information in financial statements based on fundamental modifications in the field of receiving, exchanging and processing economic information in the digital economy.

4. Observation and measurement of human capital today is possible only in the management accounting system of the organization. But even here, a conceptual framework for the financial assessment of human capital has not yet been created, which would have been widely used to manage it: there is no unequivocal judgment about the approaches to the financial assessment of human capital, to the disclosure of financial indicators of intellectual and business qualities, labor results.

# References

1. Bekker G (1993) Human capital. Impact on earnings of investments in human capital. SShA. EPI. № 11, pp 109–119
2. Bykova AA, Molodchik MA (2011) The impact of intellectual capital on the results company activities. Bulletin of St. Petersburg University. Series 8 "Management". No 1, pp 27–55
3. Díaz-Fernández M, López-Cabrales A, Valle-Cabrera RA (2014) Contingent approach to the role of human capital and competencies on firms strategy. Bus Res Q 17:205–222
4. Díaz-Fernández M, Pasamar-Reyes S, Valle-Cabrera R (2017) Human capital and human resource management to achieve ambidextrous learning: a structural perspective. Bus Res Q 20:63–77
5. Flamholtz EG (2012) Human resource accounting: advances in concepts, methods, and applications. The Jossey-Bass management series. Springer Science & Business Media, Heidelberg
6. Fitts-ents J (2006) Return on investment in staff. Measuring the economic value of staff. Vershina, Moscow
7. Grafova TO (2011) Model of financial, transactional management and strategic accounting of intellectual capital. Manag Account J 4:75–89
8. Karpova TP (2004) Accounting management accounting of production: the concept of improvement: thesis abstract for the degree of doctor of economic sciences, Moscow
9. Kaspin L (2013) Possibilities of using XBRL in the formation of integrated reporting. Innovative development of the economy. No 1(13), pp 148–149
10. Lev B, Schwartz A (1971) On the economic concept of human capital in financial statements. Account Rev, 46–52
11. Newman BH (1999) Accounting recognition of human capital assets. Pace University Press, New York
12. Pronina IV (2008) Intellectual capital: essence, structure, functions. VES, Moscow
13. Rzhanitsyna VS (2007) Goodwill: accounting and taxation. Accounting 20:18–29
14. Tseng C, Goo YJ (2005) Intellectual capital and corporate value in an emerging economy: empirical study of Taiwanese manufacturers. R&D Manag 35(2):187–201
15. Wang WY, Chang C (2005) Intellectual capital and performance in causal models: evidence from the information technology industry in Taiwan. J Intellect Cap 6(2):222–236
16. Xingqiang D, Jingwei Y, Fei H (2018) Auditor human capital and financial misstatement: evidence from China. China J Account Res 11(4):279–305. https://doi.org/10.1016/j.cjar.2018.06.001

# Risk Management of Innovation Activities in the Conditions of the Digital Economy

T. N. Syrova[✉]

Samara State University of Economics, Samara, Russia
ts260679@yandex.ru

**Abstract.** In this contribution, the author analyzed the identification of risks of innovation activities and the development of methods for assessing and managing these risks. The main problems of identifying risks and difficulties in developing and implementing management programs are determined when it is impossible to use statistical data processing methods. The author considers the possibility of increasing the efficiency of identification and risk management programs for the innovation activity through the use of digital technology. A model of using digital technologies in the methods of risk assessment of innovative activities is proposed.

**Keywords:** Digital economy · Innovations · Innovation activity ·
Risks of innovation activity · Risk management of innovation activity

## 1 Introduction

Innovation activity is the basic development strategy at any economic level. The competitiveness and well-being of not only companies, but also the state as a whole, depend on innovations: their availability and implementation. The relevance of this contribution is determined by the fact that the requirements of the consumer group in any sphere are constantly increasing on a progressive scale depending on the satisfaction of their needs. And what once seemed a science fiction, now, in this age of technological break through, that seems to be ordinary. Therefore, companies need to constantly modify and transform their services and products in order to retain their customers. This need is one of the foundations for innovations [1, 3, 5].

Despite the importance of the innovation activity, this activity is always associated with a high level of risk. At the same time, the risks of the innovation activity themselves, if we take into account its main property – uniqueness, are difficult to identify and analyze. As a result, there is no clear formulated classification of risks yet, as well as any programs to manage them. When assessing risk, we use qualitative and quantitative analysis methods. Heuristic methods are characterized by their inaccessibility and certain subjectivity. Quantitative assessment methods that are based on the analysis of information, are also difficult to implement when assessing risks because of the lack of information in companies. Thus, the process of innovation activity becomes costly and difficult to implement already at the stage of analysis, which undoubtedly affects the final cost of the innovation product, and, most importantly, its consumers [2, 3, 11].

S. Ashmarina et al. (Eds.): *Digital Transformation of the Economy: Challenges,*
*Trends and New Opportunities*, pp. 306–311, 2019.
https://doi.org/10.1007/978-3-030-11367-4_30

In this work, it is proposed that possibilities of the digital economy can help to optimize this stage of development, which will positively affects the expansion of the innovation activity and the increase in the number of innovative companies [4, 7, 13].

## 2 Materials and Methods

The used research methods include comparative and system analysis, expert assessment methods, modeling. At the initial stage, we analyzed the existing approaches to risk assessment and management in the sphere of innovations, as well as the main directions and opportunities of the digital economy. Next, we carried out a study of the main problems that impede the reliable and full consideration of risks, as well as the creation of programs for their management and overcoming the consequences of the risks' occurrence. At that stage, we also modeled a unified risk management concept using the capabilities of the digital economy. At the final stage, the obtained results were processed and analyzed and tasks for further research were formulated.

## 3 Results

Any innovative project is planned with the aim of increasing profitability, competitiveness, improving the performance of the company, that is, welfare in general. The innovation activity leads to a planned positive result if a clear analysis of the risks of this innovation type is carried out, as well as a program to manage these risks is developed [17].

Under the risk in the innovation activity the following is commonly understood: the potential possibility (danger) of the occurrence of an event (events) causing a certain material damage or a shortfall in the project's profit (income) [12, 19–21].

In the modern science there are many different research methods. In their work, Rustemov and Dosymbekov [16] divide them into 2 main groups: heuristic methods and formal research methods. Based on the work of these authors and other scientists (Sklyarov [18] and others), who are engaged in the analysis of risk assessment methods, we compiled a summary table of methods taking into account their advantages and disadvantages.

Based on the data presented in the Table 1, it can be noted that in assessing the risks of innovation activities, to obtain representative results, it is equally important to use both qualitative and quantitative methods of risk assessment (since in fact, none of the currently developed methods provides an unambiguous assessment of the level of risks for various innovations). To increase the effectiveness of the innovation activity, it is urgently necessary to develop a program that generalizes all the studied parameters. In this context, we see a way out only in using the opportunities of the digital economy [9, 12, 19–21]. At this stage, the digital economy itself is still in its formation and early development stage everywhere. But already now its possibilities are endless.

**Table 1.** Methods for assessing the risks of the innovation activity

| Heuristic methods - qualitative risk assessment | Formal methods - quantitative risk assessment |
|---|---|
| Logical conclusions, deduction, induction, analysis, synthesis, comparison, expert evaluation method, Delphi method, pattern, etc.<br>Using these methods, experts, either on the basis of personal knowledge and experience or on the basis of available information, make their assumptions about the probability of risks occurring and can give a characteristic of the level of uncertainty, by which an innovative project is further characterized and possible risks are assumed | Analytical methods, probabilistic-statistical methods, methods of operations research, methods of the theory of choice and decision-making, methods of mathematical logic, modeling<br>Using this methods, it is possible to determine the correlations between factors, evaluate and analyze random variables, identify the best way to implement an innovative project in conditions of limited resources, and make a choice of alternatives to implement an innovative project |
| The main advantages of these methods:<br>– no need for accurate data and software development;<br>– simplicity of evaluation and the possibility of its implementation to calculate the effectiveness of an innovative project | The main advantages of these methods:<br>– the ability to assess and analyze the nonlinearity of uncertainty and risks; the dynamics of their behavior; the probability of the nature of certain processes and external factors, their study;<br>– assessment of a larger volume of information;<br>– digitized evaluation of information |
| The main disadvantages of this analysis are:<br>– the difficulty of attracting independent experts;<br>– a high degree of subjectivity in the assessment;<br>– the lack of qualitative assessments of the risk level when making decisions on innovation activities | The main disadvantages of this analysis are:<br>– the need for deterministic information about the internal and external environment of the company;<br>– the lack of reliable data in the company for analysis;<br>– the complexity and labor intensity of the process |

Source: compiled by the author.

There are many approaches to the definition of "digital economy":

– economic activity based on digital technologies;
– a system of economic, social and cultural relations based on the use of digital information and communication technologies;
– communication environment of the economic activity on the Internet, as well as forms, methods, tools and results of its implementation.

Based on the definition that the basis of the digital economy is communication and information exchange, we believe that it is important to create a model for assessing, identifying and managing risks using the digital economy's opportunities. This model can be presented in the form of a single database formed on the results of evaluating innovations, with risks combined according to certain criteria and the possibilities of their identification in the project, as well as experts connected to this system who can evaluate on-line discussions or other projects. For clearer and more individual calculations, it will be difficult to develop such a program, but for the initial assessment of an innovative project and the definition of its guidelines, this will be a very popular and useful project. In addition, this system will contain operational information on directions of the innovation activity, which will help the state to develop programs to support innovation activities based on the interests of companies, actual business needs [4, 8–10, 14, 15].

## 4   Discussion

Despite the fact that the issues of innovation, risk management and digital economy are currently being actively studied, a clear structure of the relations between innovations and risk management in the innovation activity has not defined in the science in the conditions of the digital economy yet.

Contributions to the analysis and development of innovations, innovation activities, classification of risks and risk management of innovation activities have been made by Bobrova [2], Brockman, Khurana, Zhong [3], Suraeva [1, 17], Du-binyak, Olekhnovich [5], Edwards-Schachter [6], Kahn [8], Kaluzhskiy [9], Liu et al. [10], Mottayeva [12], Rustemov, Dosymbekov [16], Sklyarova [18], Sergeeva, Nekrasov [19], Yusupova [21] and others. As for the digital economy, it is important to note the following scientists: Gasanov GA, Gasanov TA [4], Carlsson [7].

Studying their works, it can be emphasized that almost all the authors agree on one thing: that both the digital economy and innovations are the basis for the development of the state. The functioning of these systems is aimed at increasing the efficiency of social production, maintaining sustainable economic growth rates in order to improve the welfare and quality of life of citizens of the country.

According to the Order of the Government of the Russian Federation, the program "Digital Economy of the Russian Federation" [14] was adopted, according to which the development and use of the digital economy is assumed in all spheres of social and economic activity.

In their work, Gasanov G. and Gasanov T. [4] give the following definition of the digital economy: "it is a system of institutional categories (concepts) in the economy, based on advanced scientific achievements and progressive technologies, primarily in digital information and communication technologies ... The result of the digital economy is a specific product (or service), getting this service by citizens in the socio-economic activities: scientific and educational, in the field of health care and the organization of medical care, the provision of effective business management and its control, legal services, in the field of advertising ...".

According to Sklyarov [18], innovation is the process of creating a new product, service, or improvement of already existing goods and services. An innovative project is an investment in the creation, implementation, and production of the results of fundamental and applied research in order to extract profit and social benefits.

The interrelation of these two directions, points of view is obvious. In part, it means that the digital economy itself (its development and implementation) is an innovative activity.

## 5 Conclusions

In the world of information technology, scientific and technological progress, the basis of well-being is the latest developments in all the spheres. In this context, the relevance of the innovation activity increases daily. Moreover, it concerns not only state projects, but also individual enterprises, including small and medium-sized businesses. But, unfortunately, any innovation is directly related to the risks that, in opinion of all researchers, are inevitable. In this connection, in order to create a favorable environment for the development of the innovation activity, programs for identification, assessment and risk management are necessary. The existing programs are very laborious and costly, which hinders the development of innovations. And here, innovative enterprises can be helped by the "digital economy" that is actively developing in our country, using its opportunities it is possible to optimize, standardize and summarize all the information necessary for the risk assessment. Moreover, this program will allow the state to create targeted programs to support innovation activities based on the needs of citizens, thereby increasing their well-being.

## References

1. Abuzyarova MI, Suraeva MO, Zhabin AP (2013) Principles of implementation of the economic policy of the state based on the management of innovations in modern Russian conditions. Bull Samara State Univ Econ 11(109):54–58
2. Bobrova NM (2012) Risks of innovation activity: types and classifications. Russian Entrepreneurship 8(206):44–48
3. Brockman P, Khurana IK, Zhong R (2018) Societal trust and open innovation. Res Policy 47 (10):2048–2065. https://doi.org/10.1016/j.respol.2018.07.010
4. Gasanov GA, Gasanov TA (2017) The digital economy as a new direction in economic theory. Reg Probl Econ Transform 6:4–9
5. Dubinyak TS, Olekhnovich SA (2016) Risks of the innovation project. "Naukovedenie" Online J 8(5). http://naukovedenie.ru/PDF/22EVN516.pdf. Accessed 11 Mai 2018
6. Edwards-Schachter M (2018) The nature and variety of innovation. Int J Innov Stud 2(2):65–79. https://doi.org/10.1016/j.ijis.2018.08.004
7. Carlsson B (2004) The digital economy: What is new and what is not? Struct Chang Econ Dyn 15(3):245–264. https://doi.org/10.1016/j.strueco.2004.02.001
8. Kahn KB (2018) Understanding innovation. Bus Horiz 61(3):453–460. https://doi.org/10.1016/j.bushor.2018.01.011
9. Kaluzhskiy ML (2014) Marketing network in e-Commerce: an institutional approach. Direct-Media, Moscow, Berlin
10. Liu Y, Lv D, Ying Y, Arndt F, Wei J (2018) Improvisation for innovation and innovation factors. Technovation 74(75):32–40. https://doi.org/10.1016/j.technovation.2018.02.010
11. Vasiliev P (ed) (2018) Management of innovations: a manual. Business and Service, Moscow

12. Mottayeva AB (2017) Principles of management of innovative risks of entrepreneurial structures. Internet magazine "Naukovedenie" 9(5). https://naukovedenie.ru/PDF/93EVN517.pdf. Accessed 11 Mai 2018
13. Prange Ch, Schlegelmilch BB (2018) Managing innovation dilemmas: the cube solution. Bus Horiz 61(2):309–322. https://doi.org/10.1016/j.bushor.2017.11.014
14. Program the digital economy of the Russian Federation (2017) No 1632-r, Moscow
15. Rauter R, Globocnik D, Perl-Vorbach E, Baumgartnerd RJ (2018) Open innovation and its effects on economic and sustainability innovation performance. J Innov Knowl 1–8. https://www.sciencedirect.com/science/article/pii/S2444569X18300325, https://doi.org/10.1016/j.jik.2018.03.004. Accessed 11 Mai 2018
16. Rustemov AA, Dosymbekova DM (2016) Methods for assessing and reducing the risks of innovation activity. Stat Account Audit 4(63):134–139
17. Suraeva MO (2016) Features of the process of strategic innovative development of organizations. Trends in the development of modern society: managerial, legal, economic and social aspects. Collection of scientific articles of the 6th international scientific and practical conference. In: Gorokhov AA (ed) JSC "University book". The South-West State University, Kursk, pp 140–144
18. Sklyarova VV (2011) Features of assessment and management of innovative risks. Finance and credit. Manag Risks 13(445):72–79
19. Sergeeva IG, Nekrasova OA (2016) Classification of the risks of innovative activity. Adv Mod Sci Educ 3(11):191–193
20. Frishammar J, Richtner A, Brattström A, Magnusson M, Björk J (2018) Opportunities and challenges in the new innovation landscape: implications for innovation auditing and innovation management. Eur Manag J 1–14. https://www.sciencedirect.com/science/article/pii/S0263237318300653, https://doi.org/10.1016/j.emj.2018.05.002. Accessed 13 Mai 2018
21. Yusupova ER (2014) Risk assessment in the development of innovative development strategies. Economics and management. Economics 4(113):60–63

# Market Paradigm of the Digital Economics

# Transformation of the Institution of Money in the Digital Epoch

M. E. Konovalova[✉], O. Y. Kuzmina, and S. Y. Salomatina

Samara State University of Economics, Samara, Russia
mkonoval@mail.ru, pisakina83@yandex.ru,
salom771@rambler.ru

**Abstract.** The issues of raising the efficiency of functioning of the institution of money deserve close attention in the current conditions of business activity plummeting, dragging recession, and increasing turbulence. This article presents an exploratory research aimed at solving the issue of improvement of some elements of the institutional environment of the financial sector. Additional relevancy and novelty of the research are stemming from the growing processes of digitization, virtualization and technocratization of the monetary economy resulting in the transformation not only of the objective economic reality but also of its subjective perception by businesses. The originating new space-time paradigm of the monetary reality characterized with absence of any territorial or state borders of the financial flow movement, substantial escape of money from the material world to the virtual one, requires a fundamentally new approach to the understanding of the establishment and functioning of the institution of money. It is of great importance to research the mechanisms and means of coordination of economic agents in the business environment stipulating a transfer from a vertical management type to a horizontal one primarily based on the fiduciary character of cooperation. The results received within the framework of the article will enable raising the efficiency of functioning of the monetary institutions of the Russian economy to entail growth of their resilience to the global economic challenges of the modern times and facilitate an increase in their competitiveness in the global world community.

**Keywords:** Digital money · Digital paradigm · Electronic money ·
Institution of money · Money

## 1 Introduction

A peculiarity of the development of the institutional system of the modern Russian monetary economy is its extreme dependence on a number of exogenous factors making it highly vulnerable to global financial, economic and geopolitical challenges. Instability and low efficiency of the Russian institution of money are also determined by internal (endogenous) factors lending themselves to control and adjustment unlike the external ones. Imperfection of the institutional structure of the money and credit sector leads to mass appearance of institutional traps leading to inefficient functioning of monetary institutions [2]. The issues of institutionalization of the money sphere are profoundly studied in the economic theory; however, the modern conditions of the

© Springer Nature Switzerland AG 2020
S. Ashmarina et al. (Eds.): *Digital Transformation of the Economy: Challenges,
Trends and New Opportunities*, pp. 315–328, 2019.
https://doi.org/10.1007/978-3-030-11367-4_31

genesis of a new digital paradigm of global development trigger the need for deeper fundamental and exploratory research of these issues. This is especially relevant as one can observe a change not only in the objective reality but also in the subjective perception of the monetary reality by businesses [3, 5]. The increasingly monetarizing and virtualizing economy starts functioning in the 4D space-time format generating new monetary and meta-monetary effects bringing about the need for study of in-depth mechanisms of the transformation of the institution of money. There are arising conceptually new technologies such as customization, commoditization, gamification covering inter alia the money market, where crowdfunding, crowd lending, scoring and digital banking are now actively applied [4, 6]. The growing asymmetry of information determines instability of the monetary system raising its entropy, which displays itself in the permanently arising global world crises [8, 10, 25]. In these conditions, adaptive expectations of subjects produce their own economic reality, which in its turn affects objective economic processes including monetary institutions transforming their content and the mechanism of functioning. Solution of the issue of improvement of the institutional environment of the monetary sector will ensure an increase in the resilience of the Russian monetary institutions to global economic and geopolitical challenges.

## 2    Materials and Methods

A typology of basic approaches to the study of the economic essence of monetary institutions is proposed on the ground of application of the institutional and evolutionary methodology of research based on the genetic method. The meaning of fiduciary principles in the process of the establishment of the institutional environment of the monetary economy is shown. The available methodology of a qualitative assessment of the level of trust towards the institution of money as the key monetary economy transformation factor is adjusted.

## 3    Results

Money is one of the oldest and still most highly demanded inventions of the humanity. It plays the key role in the development of a civilization broadening communications in the course of social and economic interactions. Forms of money are changed upon the amendment of socioeconomic conditions, and transformation and expansion of its forms alter the social reality. The present-day society cannot discard money as it forms the basis for the social construction of the environment.

The character of the environment is subject-object as on the one part a man is an object of the environment, a part of the objective reality, and on the other part, an individual perceiving the surrounding world acts as a subject creating its own worldview based on his experience, mentality, trust, socioeconomic conditions money is the main attribute of. Money has the same basis within the framework of the space-time continuum, i.e. is a product of the real economy as well as a result of the human perception of reality.

But does it mean that money appeared at the genesis of a human society?

There is no categorical answer to this question as each sphere of scientific knowledge is based on its own methodological provisions, has its own space-time dimensions, i.e. the system of coordinates forming various perceptions of the essence of money, its forms, types and content.

From the standpoint of the sociogenesis theory, money appeared only when human existence acquired a conscious character. Primitive people developed inter-personal connections following the principles of loyalty stipulating performance of specific rituals (gift, present, sacrifice, etc.) not connected to equivalent exchange, paternalism, nominal impersonal transactions or agreements. For example, money was given as a gift to the deceased, which speaks to the presence of pralogical, primitive thinking defining monetary relationships.

This type of thinking does not exclude the fact that a primitive communal man had a specific conceptual framework, could define the quantity of the received benefits and the number of Kula rings passed by money [19]. However, all these presentations were based on the idea of magical performatism, i.e. on the belief of ritual influence on functioning of a separate man and a tribe in general. Mystical attributes were granted to the objects acting in the form of money, they were animated. Moreover, things were given not only mystical, but social force absolutely strange to common goods. Possession of money was a sort of a social elevator for persons on the lower step of the social ladder, who acquired respect and friendship of highly ranked individuals.

Thus, from the sociological point of view, primitive money forms originated within non-commercial exchange types, where the financial side was not of primary importance. The key role of primitive money lied in the creation of a social environment, securing its integrity and dynamics.

The economic science assumes a completely different position in respect to the money origination time and the evolution of money forms. Economists are sure that money is a historical category, which originated due to the appearance of exchange of commodities preceded by the emergence of private property, deepening of the social division of labour, separation of product manufacturers. Money should obviously have only commercial nature if such approach is applied.

K. Marx paid special attention to the commercial nature of money in his evolution theory. In his opinion, money expresses specific industrial relationships acting as a commodity. The theory of money is a reflection of the theory of value having strict space-time frames of a market mechanism. There is no value and thus no money beyond the boundaries of a market. The evolution of money is a change of the forms of value. Initially, there existed a simple or accidental form of value arising out of spontaneous, non-regular exchange. A self-subsistent form of economy dominated in conditions of the primitive communal system where every community used its own forces to produce everything needed for its own consumption. Products were exchanged in exceptional cases, most often these were the gift exchanges described above. Within that period, the connection between various owners of benefits occurred only in the course of the exchange process, and the things subject to exchange performed social functions rather than economic ones. These benefits cannot be viewed as money as they are neither an instrument of circulation nor a measure of value.

Exchange acquires a regular character upon the development of the commodity production, more and more products get drawn into it, the accidental form of value is replaced with an expanded one, where the presence of one equivalent commodity is unlikely. As a result of the first large social division of labour (separation of animal breeding from crop farming), livestock started to be exchanged for other goods systematically rather than accidentally, while other goods (most often grain) became an equivalent to cattle. The importance of accuracy of quantitative correlations between exchanged goods increased manifold in conditions of the expansion and strengthening of production ties leading to the change in the form of value.

One equivalent commodity appears within the framework of the universal form of value when one product performing the function of an equivalent is singled out from the overall mass. In various countries small cowry shells, kuna (marten skins), salt, tulip bulbs, etc. could be such equivalent commodity. Origination of a universal equivalent is explained by the fact that exchange became vitally important for the existence of the economic system bringing about the need for decreasing the transaction costs of such exchange.

The transfer to the monetary form of value became possible as a result of an increase in the extraction of gold and silver and expansion of the commodity exchange beyond the boundaries of the local market. This is when modern commodity-money relationships started forming. Precious metals were given the equivalent commodity function due to their natural properties. They are easy to transport, store, and exchange since they are homogeneous, durable, divisible, do not spoil, are portable and rather rare. The use of precious metals as a monetary asset underwent long-term evolution, so it is impossible to state an exact period when precious metals acquired the function of money. It is traditionally considered that money in form of precious metals originated 2.5–2 thousand years before Christ. The monetary circulation acquired a finished form after the Age of Discovery and opening of banks in Western Europe.

Money was cast primarily of gold and silver, i.e. there existed a bimetallic monetary system stipulating simultaneous circulation of two metals. This system was unstable; presence of two measures of value hindered the development of exchange operations. The need for maintenance of fixed gold and silver exchange rates raised transactions costs, led to outflow of metal from the country and failed to facilitate the economy development in general.

An increase in the extraction of silver in the late XIX century resulted in the drop of its purchasing power, which contributed toward forcing silver out of the monetary circulation. A system of monometallism established; along with it, there was paper money in circulation, which appeared in the VI–VII century A.D. The gold standard era began.

The evolution of the gold standard system passed three stages: gold coin, gold bullion and gold exchange. The gold coin standard originated in Great Britain in 1816 (was established in 1837 in the USA, in 1895 in Russia). Within the framework of this standard, each gold holder had a right to strike gold coins out of metal at the state mint; such money was in free circulation and considered as units of account for any goods. Gold acted as a measure of prices as banknotes could be freely exchanged for gold at face value. Hoarding suppressed any inflation processes; excess banknotes were exchanged for gold and withdrawn from the monetary circulation system. From the standpoint of Marxism, in the gold coin standard period, national currencies "dropped their national garments" and easily moved to the international payment cycle; and then could serve internal monetary circulation again after being recoined.

After the First World War there established a distinct state monopoly on the emission of money in the form of gold, there appeared a gold bullion standard (banknotes were exchanged for bullions of 12–14 kg). Bullions were used only in international settlements since they could not be applied in the internal circulation in view of their significant weight and high cost. However, this standard did not last as most countries did not have a substantial gold reserve after the First World War and could not afford carrying out of international transactions with gold bullions.

Gold left not only the internal circulation, but was almost withdrawn from the world monetary circulation system in 1922. The appeared gold exchange standard stipulated presence of reserve currencies (foreign exchange bills) designed for international settlements and serving as bank reserves as they were still secured with gold.

The Bretton Woods Conference of 1944 established that gold maintained the world money function; however, US dollar was acknowledged as a reserve currency for the international circulation along with gold; US dollar was set equivalent to gold at the ratio of 35 US dollars/troy ounce; as a result, the gold exchange standard is often called a gold dollar standard.

The gold exchange standard system was rather stable, lasted for about 50 years, but in the early 1970s there appeared a misbalance between the US gold reserve volume and the number of issued dollars. Gold got demonetized also because dollar holders demanded exchange for the precious metal. As a result, gold was withdrawn from the international monetary circulation in 1971 and transformed from money to common goods. In addition to the above listed reasons, there is a number of objective factors responsible for the transfer to the paper monetary circulation and abandoning gold as a payment means. Among such factors are the ones determined by the natural evolution of the monetary circulation: the growth of the role of the state in the gold reserve regulation and control; increase in gold reserve storage costs; inflexibility of gold-related exchange rates; limited opportunities of emission of gold-secured monetary units; obvious advantages of paper money use (acceleration of settlements, decrease in the cost of emission, difficulty to counterfeit, etc.). Apart from the economic factors, there are some political ones facilitating transfer to the Jamaica currency system: absence of a coordinated state economic policy of countries supporting the gold standard system and establishment of hegemony of the USA in the world system through the leadership of dollar as a reserve currency with no relation to gold. The Jamaica currency system states that IMF stops transactions with gold and should not have it in its reserves. A new international currency, SDR (special drawing rights) is introduced in the world monetary circulation acting as a basis for international settlements. The Jamaica currency reform completed the demonetization process only from the position of law; in fact, there was imposed no direct prohibition against the use of gold as an international reserve asset. In practice, this means that monetary gold is still in the gold and currency reserves of many states. After the cancellation of the gold standard, the precious metal has not fully lost all money functions, e.g., as a means of accumulation and maintenance of wealth. However, gold as an intermediary at goods exchange has been replaced with a paper carrier.

The economic publications carrying out a comparative analysis of *real* (gold) money and paper money view the latter only as a symbol of value *as it represents the known quantity of gold, while the quantity of gold is at the same time the quantity of value similarly to any other quantities of goods.* In the opinion of many authors, even the cancellation of the gold security of paper money does not mean the loss of its commercial nature. The modern paper money has both the use value (utility) and the exchange value. The utility of money is the quantity of benefits, which can be acquired for one monetary unit, i.e. represents its purchasing power. The exchange value is a result of the market mechanism, the process of interaction between the demand for money and the offer. The issue of the "internal" value in its traditional interpretation by the economic theory remains open for discussions. From the standpoint of the labour theory of value, some benefits have price but no value, money refers exactly to such benefits. Mismatch between the value and the price may contain a qualitative contradiction, as a result, price stops expressing the value of objects, which cannot be goods by themselves, e.g., conscience, honour, etc. may become a subject of sale for their owners and thus acquire a commodity form thanks to their price. The institutional theory views the internal value of money differently. Using the tools of the contractual approach, institutional scientists suggest that functioning of the monetary system entails high transaction costs; the amount of such costs determines the value of money [9]. Paper money reflects the institutional nature to the highest extent regulating the means of organization and the rules for cooperation between the market participants. The modern money reflecting contractual relationships is based on the high level of trust between individuals in the society, which emphasizes their fiduciary character [16, 17].

Scientific publications often call paper money fiduciary, fiat, credit money often using these concepts interchangeably. In fact, these characteristics reflect only separate aspects of functioning of the paper form of money. Thus, money should be called fiduciary assets if one believes trust to be the key factor of functioning of the monetary circulation system, and fiat money with legislatively consolidated circulation guaranteed by the state if one is talking about institutionalization of the monetary circulation. Being fiat, i.e. representing obligations of the state, money functions as a means of payment, which emphasizes its credit nature, so no wonder it is called credit money in economic publications.

A planned transfer from real to paper money takes place parallel to the transformation of the space-time continuum causing substantial escape of money from the material world to the virtual one. The paper form of money is of dual nature; on the one hand, money acts as a title serving the intangible sphere of reality, on the other hand, presence of a paper carrier makes it a part of the tangible world. A revolution in the information technology sector facilitated the process of transformation of the space-time paradigm of the socioeconomic reality. The increasingly monetarizing and virtualizing economy starts functioning in the new 4D space-time format generating new monetary and meta-monetary effects, in the new paradigm of the monetary reality of the society in the conceptual framework of which the dichotomy of *active-passive* in relation to the *space* and *time* categories becomes dialectically inevitable reality similar to the dichotomy within the wave-corpuscle light theory.

One may single out the following characteristic traits of the appearing new space-time paradigm. The speed of information exchange increases manifold leading to the growth of the speed of circulation of money, corrupting the dependence between the amount of money in circulation and the price level. The growing asymmetry of information determines instability of the monetary system raising its entropy, which displays itself in the permanently arising global world crises. In these conditions, adaptive expectations of subjects produce their own *economic reality*, which in its turn affects objective economic processes transforming their content. Inhomogeneity of the monetary environment is reflected in the presence of various monetary forms (monetary polyformism), many of which perform traditional functions of money in the classical interpretation (Table 1).

The modern economic environment is more transparent, permeable from the position of information and almost eliminates the factors of physical dimension. Movement of money as an information quantum is nothing more than the movement of substance having neither territorial nor state borders.

**Table 1.** Monetary polyformism

| Forms | Secured (good) money | Non-secured (bad) money | Non-secured quasi-money |
|---|---|---|---|
| Characteristic traits | Commodity money | Fiat, credit, fiduciary money | Fiduciary money |
| Varieties | Primitive quasi-money Animalistic money Vegetabilistic money Giloistic money | Paper money and fractional coin Non-cash money Electronic money | Digital money |

Source: compiled by the authors.

This space-time paradigm of the monetary reality possesses high phenomenological capacity reflected in the distribution of new money forms. Development of the information technology of information storage and transfer has led to the transformation of the institution of money; the process of replacement of paper banknotes with records on electronic carriers has started. The electronic money concept is gaining popularity in research works; scientists interpret it in different ways. Pursuant to Directive 2009/110/EC of the European Parliament and of the Council of 16 September 2009 On the Taking up, Pursuit and Prudential Supervision of the Business of Electronic Money Institutions amending Directives 2005/60/EC and 2006/48/EC and repealing Directive 2000/46/EC, electronic money is understood as value in the monetary form stored electronically including on a magnetic carrier, represented as receivables from the issuer, imitated at receipt of money for carrying out of payment transactions… and applied by individuals and legal entities not being the electronic money issuer. Money simultaneously transforms into value stored on an electronic carrier at purchase of a pre-paid instrument. International standards regulating monetary circulation do not view electronic money as deposit money as the latter involves attraction of deposits and is secured with state guarantees. The key characteristics of electronic money can be singled out following the review of the legal grounds for the electronic money circulation. Electronic money is credit money in its nature as it represents the issuer's obligations before money holders. Electronic money is circulating with no use of bank accounts. Being pre-paid, electronic money can be used to make payments in favour of third persons, not the issuer. Judging from these attributes, electronic money is not non-cash money means (deposit money), although the latter also represents an electronic recording format. It is worth noting that from the legal standpoint only compliance with all above named characteristics gives money the status of electronic money; otherwise, not only digital, but also non-cash money can be referred to electronic (Table 2).

**Table 2.** Classification of money from the standpoint of the legal approach

| Criterion | Varieties |
|---|---|
| Pre-payment | – pre-paid money (electronic)<br>– non-pre-paid money (cash, non-cash, digital) |
| Use of a bank account | – money within the banking system (cash, non-cash)<br>– money outside the banking system (electronic, digital) |
| Issuer | – "private" money (partially non-cash, electronic, digital)<br>– "state" money (cash, partially non-cash) |

Source: compiled by the authors.

The *electronic money* term raises considerable disputes among the economic community representatives as the concept of *electronic* is borrowed from natural sciences and an attempt at its incorporation in the institutional field does not facilitate unambiguous interpretation of this definition. The following contours of the modern monetary reality paradigm can be segregated from the standpoint of the institutional and legal approach (Fig. 1).

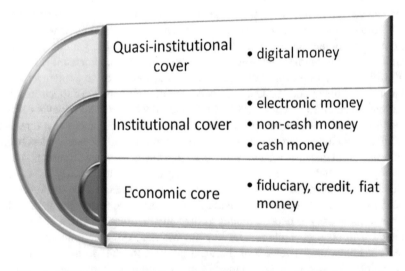

**Fig. 1.** Modern monetary reality paradigm (Source: compiled by the authors)

The attributes enabling interpretation of money as electronic, declared by legal provisions, may suffer considerable changes as opposed to the technological approach establishing more strict frames for reference of money to electronic money. Cash as well as non-cash money without reference to a bank account opening can on the modern stage be represented in form of a digital record on an electronic carrier (Table 3).

**Table 3.** Structure of money from the standpoint of a technological approach

| Electronic money | Non-electronic money |
| --- | --- |
| – Cash (electronic cash)<br>– Non-cash (deposit) money<br>– Non-cash money issued with no use of a bank account<br>– Digital money | – Cash and fractional coin |

Source: compiled by the authors.

The differentiating attribute of cash and non-cash money is money being "on hand" at the money holder, which is equivalent to a tangible information carrier being in the actual possession of the holder; if third persons are in such possession, this is non-cash electronic money. Electronic cash (tokens issued by system administrators) acts as payment means in a system of cash electronic money (WebMoney, e-port, CiberPlat, etc.), their circulation in the network stipulates no connection to the central server; while systems of non-cash electronic money (PayCash, PayPal, CiberCash developed by VeriSign, DirektPay developed by Yahoo, online software of commercial banks, etc.) use payment means in form of electronic receipts or payment orders transferred to the electronic money issuer, who is subsequently making a payment. Issue of cash and non-cash electronic money appears as issue of an electronic analogue of traditional currency.

The only doubtful factor from the position of the technological approach is reference of cryptocurrencies to electronic money as the *technological aspect* itself can be interpreted differently. We believe it to be not only a mechanism of information recording on an electronic carrier, but the electronic money creation technology itself, which is absolutely different for cryptocurrencies. Digital money (cryptocurrencies) do not transpose to national or other currency, i.e. an independent quasi-money form appears. The absence of security, specific creation mechanism (cryptography), limited sphere of application and use, potentially predetermined emission scope do not allow identification of cryptocurrencies as electronic money. Moreover, digital currencies are now performing the functions of money only partially, being more of a speculative asset than a monetary one.

The future of quasi-money in form of cryptocurrencies is uncertain; being an unstable monetary form it can be withdrawn from the monetary circulation sphere or become its full-fledged element [14, 21, 24]. Digital money becomes the most relevant form of carrying out of monetary transactions in conditions of the establishment of the post-industrial production mode characterized by changes in the business entity coordination mechanisms, systems of search for information and decision-making, space-time continuum.

Broadening of the monetary form profile gives an opportunity to mark the establishment of the new stage of the development of the world currency system (see the annex). This is especially relevant in the times of the crisis of the Jamaica currency system, reformation of which is proposed by many economic research works. Among such proposals are creation of a dollar-euro bi-currency system, a multi-currency system using regional currencies most important on a worldwide scale, increase in the degree of SDR involvement in international settlements, return to the gold standards or transfer to a multi-commodity standard. The largest part of the outlined proposals is not in line with the post-industrialization principles and the principles of establishment of a polycentric economic system.

Certain institutional consolidation of the originating world monetary and financial structure is required on the current stage. It is important to take into account peculiarities of the modern process of social reproduction, growing financial globalization of the economy, increasing misbalance between the real and the monetary sphere. The Jamaica currency system has to suffer radical changes by replacement of the outdated principles with the new ones able to appear only in case of historical continuity, high level of adaptation to modern economic conditions and correspondence to the interests of the majority of countries.

The modern currency system as distinct from its preceding stages is more flexible, agile and dynamics. Thus, e.g., the currency liberalization process has covered almost the whole world consolidating the transfer from fixed to floating currency exchange rates. One of the tasks of the IMF reform started in 2008–2009 is strengthening of functions of the monetary regulation of debtor countries.

Transformation of the space-time paradigm of the monetary reality concerns not only the formally consolidated institutional contour but also informal quasi-institutional elements making the system more complicated and raising the degree of its entropy. The modern stage observes strengthening of monetary poliformism facilitating diversification of risks of business entities, reduction of transaction costs, acceleration of exchange transactions possible only in conditions of the growth of actors' trust towards financial institutions.

# 4   Discussion

The issue of research of the institution of money has always been the centre of attention of the economic science. Evolution of scientific representations on the essence of money is observed in the works by Hodgson [9], Keynes [11], Mitchell [17], von Mises [18], Moss [19], Miller [16] and others. These research works review money as a systemic macroeconomic phenomenon making a considerable influence on the proportions of social reproduction. Many Russian authors also partially cover this issue and review the problems of the content of money discussing their institutional nature as well as the essential issues of functioning of the money and credit sphere.

Although the economic science pays attention to the study of the monetary circulation category, in some cases the issue at hand refers to the institutional nature defining the development of money as a systemic basic economic institution. Fiduciary aspects of functioning are almost not reflected, which is an obvious gap in the scientific cognition of this category.

Traditionally, trust as a socioeconomic phenomenon is reviewed in the works of sociologists, philosophers and psychologists, such as Aglietta [1], Simmel [23], Fukuyama [7], Polanyi [22], Keynes [11], Mises [18]. Paying considerable attention to the psychological and social nature of trust, the authors leave out the issues of the influence of trust on the adoption of economic decisions by individuals. This gap has been successfully filled in by representatives of the economic science such as Hendrickson [8], Mario [15], Lawlor-Forsyth [13], Fontana [6], Kudrryashova [12], Vymyatnina [25], Paech [20], Holt [10]. However, these researchers analyze fiduciary relationships in general, with no consideration of the institutional specifics and their influence on the money and credit sphere. The issues of qualitative definition of institutional trust, inter-dependence of the level of institutional trust and the key macroeconomic parameters, establishment of the fiduciary rating of business entities for increase of their wealth, remain underdeveloped.

## 5   Conclusions

Money within the framework of the space-time continuum is a product of the real economy and a result of the human perception of reality, which explains why social reality alters upon change of the socioeconomic conditions, transformation and expansion of monetary forms. It is worth noting that the principle of monism cannot be applied to the review of genesis of monetary forms as each area of scientific knowledge has its own space-time dimensions (coordinate system) where there forms a different presentation of the essence, forms, types and content of money. The evolution of the monetary circulation system observes an increase in the number of money rather than modification of the money forms and replacement of one for the other. Monetary poliformism becomes a distinctive trait of the modern world currency system. A planned transfer from real to paper money takes place parallel to the transformation of the space-time continuum causing substantial escape of money from the material world to the virtual one. Movement of money as an information quantum is nothing more than the movement of substance in the modern economic environment, which is transparent, permeable, has neither physical dimension nor territorial or state borders. Scientific publications do not currently have any exact criteria of referring of money to a particular form. Contours of the modern paradigm of the monetary reality will differ significantly depending on the standpoint (institutional and legal or technological approach) used to research the issue. In our opinion, digital money (cryptocurrencies) is a quasi-monetary form as it performs the functions of money only in part and the technology of its creation is quite specific. However, it does not mean that cryptocurrencies will not become an institutionalized, more relevant form of carrying out of monetary transactions in the future society in conditions of the establishment of the post-industrial production mode. A new world monetary and financial structure is arising requiring institutional consolidation. High flexibility, agility and dynamics of the modern currency system is largely achieved by the expansion of monetary forms, which is able to raise prosperity of business entities in conditions of the growth of personified and de-personified trust.

# Annexure A

See Table 4.

**Table 4.** Transformation of money forms in the course of the evolution of the world currency system

| Period | Currency system | Standard type | | Distinctive traits |
|---|---|---|---|---|
| Since 1867 | Paris currency system | Gold coin standard | Gold standard | – gold is used as world money<br>– national banknotes are secured with gold<br>– currency exchange rates are freely fluctuating within gold points<br>– money is represented in form of coins, banknotes secured with a gold reserve |
| Since 1922 | Genoa currency system | Gold bullion standard | | – gold acts as a reserve and a payment means<br>– USD and GBP are secured with gold and act as gold equivalents on the international arena<br>– currency exchange rates are freely fluctuating with no reference to gold points<br>– money is represented in form of coins, banknotes secured with a gold reserve |
| Since 1944 | Bretton Woods currency system | Gold exchange standard | | – gold acts as a reserve and a payment means<br>– USD becomes the key reserve currency, the only one secured with gold<br>– currency exchange rates are fixed as well as parities<br>– money is represented in form of coins, banknotes; only some of them are secured with a gold reserve |
| Since 1976 | Jamaica currency system | Special drawing right (SDR) standard | | – gold almost loses its monetary attributes and is viewed only as goods (gold is officially demonetized)<br>– SDR become a new international currency serving as a basis for international settlements<br>– free selection of the currency exchange rate regime<br>– money is represented in form of coins, banknotes not secured with gold, electronic money appears |
| Since 2000 | Multi-currency system[a] | Multi-currency standard | | – not only reserve currencies but also currencies of the most developed countries are used in international settlements<br>– transfer from the monopolar world monetary system to the multipolar system<br>– originating tendency of decentralization of the world monetary system<br>– money is represented in form of coins, banknotes, electronic money |

Note: [a]Not officially recorded on the level of the world community
Source: compiled by the authors.

# References

1. Aglietta M (2006) Currencies: between violence and trust. State University Higher School of Economics, Moscow
2. Cattelino JR (2018) From locke to slots: money and the politics of indigeneity. Comp Stud Soc Hist 60(2):274–307. https://doi.org/10.1017/S0010417518000051
3. Chadha JS (2018) Of gold and paper money. Manch Sch 86:1–20. https://doi.org/10.1111/manc.12242
4. Coelho F, Pereira MC (2018) Mindfulness, money attitudes, and credit. J Consum Aff, 34–45. https://doi.org/10.1111/joca.12197
5. Cutts TRS (2018) Modern money had and received. Oxf J Leg Stud 38(1):1–25. https://doi.org/10.1093/ojls/gqy002
6. Fontana G, Passarella MV (2018) The role of commercial banks and financial intermediaries in the new consensus macroeconomics (NCM): a preliminary and critical appraisal of old and new models. Alternative approaches in macroeconomics: essays in honour of John McCombie, pp 77–103. https://doi.org/10.1007/978-3-319-69676-8_4
7. Fukuyama F (2004) Trust: the social virtues and the creation of prosperity. AST, Ermak, Moscow
8. Hendrickson JR, Salte AW (2018) Going beyond monetary constitutions: the congruence of money and finance. Q Rev Econ Financ 69:22–28. https://doi.org/10.1016/j.qref.2017.11.013
9. Hodgson J (2003) Economics and institutions: a manifesto for a modern institutional economics. Business, Moscow
10. Holt JP (2017) Modern money theory and distributive justice. J Econ Issues 51(4):1001–1018. https://doi.org/10.1080/00213624.2017.1391584
11. Keynes JM (1930) A treatise on money. The pure theory of money. MacMillan, London
12. Kudryashova (2018) IV World currencies: comparative analysis in context of monetary functions. World Econ Int Relat 62(8):26–34. https://doi.org/10.20542/0131-2227-2018-62-8-26-34
13. Lawlor-Forsyth E, Gallant MM (2018) Financial institutions and money laundering: a threatening relationship? J Bank Regul 19(2):131–148. https://doi.org/10.1057/s41261-017-0041-4
14. Liu Q, Li K (2018) Decentration transaction method based on blockchain technolog. In: Proceedings of 3rd international conference on intelligent transportation, big data and smart city, ICITBS 2018, pp 416–419. https://doi.org/10.1109/icitbs.2018.00111
15. Mario S, Eugenia C (2018) Rethinking money as an institution of capitalism and the theory of monetary circulation: what can modern heterodox economists/institutionalists learn from Karl Polanyi? J Econ 52(2):422–429. https://doi.org/10.1080/00213624.2018.1469889
16. Miller RR (2000) Modern money and banking. Infra-M, Moscow
17. Mitchell WC (1896) The quantity theory of the value of money. J Polit Econ 4(2):139–153
18. Mises L (2005) Human action: a treatise on economics. Society, Chelyabinsk
19. Moss M (1996) Ocherk o dare. Vostochnaya literature Publisher, Moscow
20. Paech P (2017) The governance of blockchain financial networks. Mod Law Rev 80(6):1073–1110. https://doi.org/10.1111/1468-2230.12303
21. Penkova IV, Korolev VA, Butenko ED, Glazkova IY, Eldarov SK (2018) Crypto currencies as a modern financial tool of digital economy: global experience of state regulation. Adv Intell Syst Comput 726:326–334. https://doi.org/10.1007/978-3-319-90835-9_38
22. Polanyi K (2006) The great transformation. State University Higher School of Economics, Moscow

23. Simmel G (1978) Philosophy of money, Boston
24. Vujičić D, Jagodić D, Randić S (2018) Blockchain technology, bitcoin, and ethereum: a brief overview. In: 17th International symposium on INFOTEH-JAHORINA, Infoteh. Proceedings 2018, pp 1–6. https://doi.org/10.1109/infoteh.2018.8345547
25. Vymyatnina YV, Grishchenko VO, Ostapenko VM, Ryazanov VT (2018) Financial instability and economic crises: lessons from Minsky. Ekonomicheskaya Politika 13(4):20–41. https://doi.org/10.18288/1994-5124-2018-4-02

# Business Analytics of Supply Chains in the Digital Economy

T. E. Evtodieva[1]([⊠]), D. V. Chernova[1], N. V. Ivanova[1],
and O. D. Protsenko[2]

[1] Samara State University of Economics, Samara, Russia
{evtodieva.t,nataliaivanova86}@yandex.ru,
danacher@rambler.ru
[2] Russian Presidential Academy of National Economy and Public
Administration, Moscow, Russia
procenko@ranepa.ru

**Abstract.** Within the conditions of economy digitalization, the intellectual and information potential of society as a main renewable resource ensuring sustainable progressive development is being used intensively, which can significantly increase the efficiency compared to the material production of industrial society and, as a result, has a significant impact on the dynamic development of economic relations between subjects. The need for high-qualitative and structured information is increasing if market participants are in the integrated interaction, in particular, are integrated into the supply chain, allowing for the planned movement of material flows from the place of production to places of consumption. These are required amounts of timely relevant information that ensure adequate management decisions. Information forms the basis that provides all management decisions in the field of supply chain logistics, contributes to its sustainable development and achieving goals. In this regard, it is necessary to consider the impact of change on the possibilities of analytics in supply chains in the digital economy. The purpose of the contribution is to determine the impact of digitalization on new and modified analytical capabilities in logistics supply chains and to assess the market evolution for business analytics in Russia. The main results of the contribution are identified directions for business analytics in logistics supply chains and determined growth prospects for business analytics in Russia.

**Keywords:** Business analytics · Digital economy · Digitalization ·
Logistic integration · Logistics · Predictive analytics · Supply chain

## 1 Introduction

The urgency of the problem under study is due to the fact that a dynamically changing world makes the participants of supply chains take operational decisions with minimal time and high productivity. Studies of evolutionary changes in supply chains [5, 11, 13, 27, 28] indicate that the analytical complexity of logistics processes and supply chain management is due to a number of features inherent in these entities, in particular: plurality of legally independent participants; complexity of the organizational

© Springer Nature Switzerland AG 2020
S. Ashmarina et al. (Eds.): *Digital Transformation of the Economy: Challenges,*
*Trends and New Opportunities*, pp. 329–336, 2019.
https://doi.org/10.1007/978-3-030-11367-4_32

interaction; contradictory economic interests, goal-setting and tasks of organizations; instability of relations between enterprises; multiplicity and complexity of logistics operations and functions performed by participants in the supply chain; the qualitative nature of relationships and criteria for the functioning of enterprises that are difficult to formalize; the stochastic nature of most factors affecting logistics processes. In addition, the increasing complexity of market and competitive relations leads to the need to control the time and speed of logistic flows through the supply chain and to complicate products and expand their assortment list. It was determined that the consumer is willing to pay 30% more for the delivery of products during the day than for a longer time frame. In this regard, there is an urgent need for active use of logistic tools that allow obtaining objective information for a short time lag, based on past or future trends of the phenomenon under consideration. As a result, it is necessary to determine the awareness of intensifying analytical tools to handle big data by Russian business structures in order to make management decisions adequate to the market environment in logistics supply chains. In this regard, the importance of the contribution is to identify the main opportunities provided by business analytics in supply chain management and the elimination of theoretical gaps in the theory of logistics in current conditions.

## 2   Materials and Methods

As the main methods for studying the problem of business analytics of supply chains in the digital economy the authors used general scientific methods of knowledge, allowing objectively and comprehensively studying the conceptual framework of business analytics in the digital economy, and private techniques, including: systematic and integrated approaches to analytical capabilities in supply chains within the conditions of economy digitalization, comparative analytical and situational analysis.

## 3   Results

The source of long-term economic growth in current conditions is economy digitalization, which has innovation background and is based on active automation and robotization, information exchange in real time, additive production and technologies that provide cyber security and augmented reality. Attempts to systematize and unify the categories and concepts of the digital economy have been made by many authors, who have revealed its essence to some extent. In the most general form, the digital economy is defined as an economy based on new methods of generating, processing, storing, transmitting data, as well as digital computer technologies.

The leading analytical agencies of the world show the prospects of the new economy. According to McKinsey Global Institute estimates, by 2025, due to digitalization GDP growth will increase dramatically in a number of major economies of the world. For example, in China such an increase can be up to 22%, in the USA, about 1.6–2.2 trillion US dollars, in Russia, from 19 to 34% of the total expected growth of GDP [4].

What does new digitalization introduce in the economic structure? In the digital economy, new forms and properties of post-industrial policy are built on the basis of extensive informatization of society and practical implementation of information and telecommunication mechanisms of self-organization and harmonization of the economic system [9, 33]. Speed is becoming a key factor in the company's development, including the rate of change in the business, the speed of managing business processes, changes in the lifestyle of consumers and their requests, influenced by the increasing availability of information. Information becomes an objective competitive advantage. In this regard, the digital economy, which is a synthesis of the real and the virtual [15], intensively uses intellectual potential and the accumulated data set that ensures sustainable progressive development, and as a result, influences the system of economic relations between market subjects. The processed, structured and systematized information forms the basis for the development and evaluation of managerial influences that can be aimed at identifying possible negative trends and threats and way to prevent their consequences.

Within digitalization, business analytics in any field of activity, including logistics, contains three aspects presented in Fig. 1.

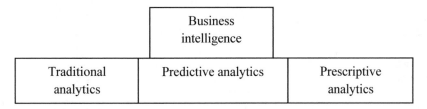

**Fig. 1.** Components of business analytics in the digital economy (Source: compiled by the authors)

Traditional analytics is based on historical data, allows for an analytical assessment of existing phenomena and processes, which can later be read depending on the objectives of business analysis. The set of statistical analysis methods used to obtain new information by processing previously accumulated (historical) and current information and identifying explicit and implicit factors, dependencies, anomalies, is usually called predictive, "smart", "advanced" analytics. Prescriptive analytics is responsible for choosing the best solution to the problem and making management decisions to improve their effectiveness.

At the present stage of economy digitalization, the greatest interest is shown in predictive analytics. Predictive analytics is not a new phenomenon in business. Back in 1960, many foreign corporations tried using non-linear programming and solving heuristic problems in order to develop the shortest itinerary of cargo movements in logistics deliveries and making decisions about credit risk. Already in the 70s–90s of the last century, statistical reports were drawn using predictive analytics and allowing evaluating the company's activities over the past period of time and forming a strategy of behavior in the market in the future. Currently, the capabilities of predictive analytics are expanded, due to data mining (DataMining), machine learning (MachineLearning), neural networks and other mechanisms. Today, smart analytics allows the company both receiving information on the current state of affairs in a systematic way and forming development forecasts. These forecasts can be made by the objectively existing trends and influence factors, and by solving problems using a non-predetermined algorithm, and by solving many other problems, identifying implicit links between events or phenomena in the business sphere, and forming analytical innovations and successfully competing with them.

With regard to supply chain management, we can define the following uses of predictive analytics (Table 1).

**Table 1.** The main business problems solved by predictive analytics in supply chains

| Business challenge | Opportunities |
|---|---|
| Sales forecast | With regard to historical conversion ratio and external factors, determine expected sales |
| Demand forecast | Mathematical modeling with regard to external factors and historical data on the volume of sales provides short-term, medium-term and long-term sales forecasts with regard to seasonality and consumer behavior |
| Analysis of the client base and supplier base | Possibility to systematize, cluster and rank the database, revealing the behavior of each client (customer group) or supplier (supplier group) |
| Optimization of material flows in space | Solve the "shortest path task" with regard to economic, temporal and other factors, as well as a given level of logistics service |
| Determination of the optimal level of logistics service | Deep analysis of consumer requirements and preferences and determination of the optimal level of service from the client's point of view with regard to costs of this service |
| Optimization of operational processes in production | Analyze the process in real time, forecast the further optimality range and the possibility of changing control parameters and give recommendations to optimize them |
| Technological maintenance and repair of equipment | Visualize the data on the current state of equipment and forecast equipment failure scenarios |
| Optimization of warehouse stocks | Solve a multi-criteria task on structuring reserves, determine the time and size of their replenishment, options for optimal placement in the warehouse |
| Effective relationships determination | Determine cost parameters for interaction, determine return on investment in maintaining and managing the relationships and give recommendations to optimize them |
| Behavior strategy determination | Develop supply chain scenarios and determine pragmatic behavior strategies |
| Logistics risk management | Identify, analyze and prevent risks by determining proactive effects and assessing negative consequences |
| Reduction of returnable logistic flows | Develop a scenario model optimizing returnable logistics with regard to an array of data on real and potential demand |

Source: compiled by the authors

Obviously, the use of predictive analytics provides the following opportunities for a company participating in the supply chain:

(1) Improve product quality and increase the productivity of the enterprise logistics system;
(2) Identify weak areas of the enterprise and other actors in the supply chain;
(3) Receive time for the prevention of negative consequences from various logistic efforts;
(4) Form high loyalty on the part of the subjects of interaction;
(5) Develop an individual customer service policy;
(6) Develop a strategic initiative of behavior and provide competitive advantages.

In general, supply chain management in logistics can be divided into three main areas for application of analytics [12]:

– Improving operational efficiency by making critically weighted and informed decisions, optimizing resource consumption, improving the quality of logistics processes and their productivity;
– Increasing customer satisfaction by identifying customer needs in each consumer segment and building an individual logistics service system;
– Development of a new business model for logistics operations, which leads to additional profits, either by reducing costs or by increasing sales.

The range of tasks and opportunities explains why 57% of respondents participating in research conducted abroad by MHI Annual Reporting Repetition Key Survey Findings in 2017 noted a high potential of predictive analytics, giving way to robotization and automation (61%) [21]. Studies by Transparency Market Research show the rapid growth of the market for forecast analytics. If in 2012 the volume of this market reached $2.8 billion, by 2019 it is expected to increase to $6.5 billion. Thus, the average annual growth of the international market for analytics systems will be 17.8% [24]. In general, the market for business analytics, according to IDC research, is also characterized by an upward trend. If in 2015 it amounted to 122 billion dollars, in 2016 it was already 130 billion dollars, and by 2020 the market volume is expected to grow to 203 billion dollars [8]. The domestic market for business analytics is characterized as highly concentrated, in which both manufacturers (vendors) of analytical systems and system integrators are represented. There is a transition from multi-vendor to highly specialized manufacturers. Due to the substitution policy, conditions have been created for the active development of the market for business analytics by Russian enterprises. However, Russian solutions have not reached the level of complexity and opportunities offered by foreign systems yet. However, the market leaders are precisely domestic companies (Softline, Glow Byte Consulting, AT Consulting, Krok, Sapran, Bars Group and others), geographically concentrated in Moscow and St. Petersburg.

Systems are gaining increasing popularity, allowing for the daily addition of new dimensions, indicators, changes in the report. Relying on demand behavior, demand for reporting analytics is declining and experts note an increase in demand for predictive analytics in the future. Therefore, it can be stated that Russian market for predictive analytics is in the formative stage. Business entities strongly need it. Today, the network of retail and process industries with well-established supply chains are more interested in it. The estimated size of the market for predictive analytics in Russia is up to $100 million with a growth prospect of up to 100% [23]. A sufficient number of predictive analytics solutions are emerging on the market, which are undergoing pilot projects. For the most part, they are aimed at systematizing data. However, in the nearest future, there is a shift in emphasis towards generating action able insights and automating logistical decisions (digital advisors).

## 4   Discussion

The conceptual foundations for new conditions of economic activity were considered in the works of such foreign scientists as Anderson, Druker, Castells, Poratm & Rudinm, as well as domestic scientific representatives Avdushin et al., Lazarev, Melyukhin, Popkova EG (ed), Swan [1, 3, 10, 14, 16, 17, 20, 25, 26, 29, 32]. The main emphasis in the presented works was made on the definition of driving forces and pathways for the development of the post-industrial society and the description of technologies and methods of their application in the economy of a new type. The issues of digital society and economy digitalization were reflected in the works of Andiyev and Filchakova, Bakin, Babkin and Chistyakov, Laudon, Tim, Tapscott, Webster [2, 6, 7, 18, 22, 31, 33], who identified the tools that form the basis of digital society, and present the possibilities of its use, as well as the problems of the digital economy. However, despite the systematic vision of basic technologies of the digital economy, their application to logistics is not considered enough. In this connection, the scientific works of Afanasenko and Borisova [5], Evtodiev et al. [13], Lynch [19], Stock [30] deserve attention, in which attempts are made to describe the possibilities of digital technologies for logistics operations. In order to theoretical understand and determine the possibilities of using such a tool of the digital economy as a business analyst, we consider it relevant and timely to study the problem of changing the existing analytical capabilities in supply chains and assess the market evolution for business analytics in Russia.

## 5   Conclusions

The volume of data received and moved in logistics supply chains is constantly increasing, which makes participants in the market space focus on the possibility of obtaining detailed qualitative information from them to make management decisions. Information forms strategic assets that must be integrated into logistics activities and provide for the analytical needs of supply chain participants.

Thus, business analytics is a new trend of the modern economy and an effective tool to ensure the ongoing sustainable development of the supply chain, optimize costs and minimize losses of both material and temporary nature.

# References

1. Avdokushin EF, Sukhova VS (2009) New economy. Master, Moscow, p 584
2. Andiyeva EY, Filchakova VD (2016) Digital economy of the future, industry 4.0. Appl Math Fundam Inform 3:214–218
3. Anderson C (2011) Makers: new industrial revolution. Gardners Books, p 272
4. Aptekman A, Kalabin V, Klintsov V (2017) Digital Russia: a new reality. https://www.mckinsey.com/. Accessed 23 Sept 2018
5. Afanasenko ID, Borisova VV (2018) Digital logistics. Peter, St. Petersburg, p 351
6. Bakin A (2017) Trends in the development of the economy and industry in the digital economy. Polytechnic Publishing House University, SPb, p 658
7. Babkin AV, Chistyakova OV (2017) Digital economy and its impact on the competitiveness of business structures. Russ Bus 18(24):4087–4102. https://doi.org/10.18334/rp.18.24.38670
8. Business analytics and big data in Russia in 2017 (2018) http://www.cnews.ru/reviews/bi_bigdata_2017. Accessed 10 Sept 2018
9. Brown LR (1081) Building a sustainable society, New York
10. Druker PF (1993) Post-capitalist society. HarperCollins, New York
11. Dybskaya VV, Sergeev VI (2018) Global trends in the development of supply chain management. Logist Supply Chain Manag 2(85):3–15
12. DHL (2013) Big data in logistics. A DHL beyond the hype
13. Evtodieva TE, Chernova DV, Voitkevich NI, Khramtsova ER, Gorgodze TE (2017) Transformation of logistics organizational forms under the conditions of modern economy. Contributions to economics, pp 177–182
14. Castells M (2010) Informational epoch. Economy, society and cultures. HSE, Moscow, p 318
15. Keshelava AV, Budanov VG, Rumyantsev VU et al (2017) Introduction to the "Digital" economy. VNII geo-system, p 28
16. Lazarev IA (2005) Information economy. MEPI, Moscow, p 542
17. Lazarev IA (2015) New information economy and network development mechanism. DashkoviK, Moscow, p 248
18. Laudon K, Laudon J (2012) Management information systems: managing the digital firm, 12th edn. Pearson, p 630
19. Lynch K (2005) Logistics network of the next generation. Ascet 7:176–179
20. Melyukhin IP (1997) Problems of the foundations of information society in Russia. Informatics Problems. Moscow, p 120
21. Optimization modeling of the logistics network (2018) http://www2.deloitte.com. Accessed 12 Aug 2018
22. Tim O (1998) Information society "vulnerable to criminals". Inf World Rev 13:103
23. Predictive analytics for Russian business (2018) https://iot.ru/promyshlennost/prediktivnaya-analitika-dlya-rossiyskogo-biznesa. Accessed 15 Sept 2018
24. Predictive analytics in production (2018) https://iot.ru/promyshlennost/vozmozhnosti-prognoznoy-analitiki-keys-ot-beltel-datanomics. Accessed 1 Oct 2018
25. Porat M, Rubin M (1997) The information economy: development and measurement. Coordin Printing Office, Washington, DC
26. Popkova EG (ed) International crisis management, Contributions to economics. https://doi.org/10.1007/978-3-319-60696-5_20
27. Sergeev VI, Dutikov IM (2017) Digital supply chain management: a look into the future. Logist Supply Chain Manag 2(85):87–97

28. Sergeev VI, Kokurin DI (2018) The use of innovative technology "Blockchain" in logistics and supply chain management. Creat Econ 12(2):125–140. https://doi.org/10.18334/ce.12.2.38833
29. Svon M (2017) Scheme of the new economy. M. Olimp-Business, p 348
30. Stock R, Lambert M (2011) Douglas strategic logistics management. Irwin, McGraw-Hill, p 438
31. Tapskott D (2003) Electronic-digital society. Refl-beech, Moscow, p 432
32. The new digital economy. How it will transform business (2016) A research paper produced in collaboration with AT&T, Cisco, Citi, PwC & SAP. Oxford Economics, p 34
33. Webster F (2004) Theory of the information society. The Aspect of Press, Moscow, p 258

# Monitoring as a Tool to Ensure the Quality of Services Provided in the Interaction of Service Organizations and Municipal Authorities in Economy Digitalization

M. V. Vesloguzova[1]([⊠]), L. S. Petrik[1], K. M. Salikhov[2],
and O. A. Bunakov[2]

[1] Volga State Academy of Physical Culture, Sports and Tourism, Kazan, Russia
{mariaves,petrikls}@mail.ru
[2] Kazan Federal University, Kazan, Russia
{hafizms,oleg-bunakov}@mail.ru

**Abstract.** Monitoring is a tool that ensures the quality of services provided in the interaction of service organizations and municipal authorities in economy digitalization. In this regard, this contribution is aimed at identifying monitoring tools as an integral part of the management cycle. These tools make it possible to adjust the management of an object or process through the use of digital services. The results of this process determine the functional importance in providing "feedback": the possibility of identifying the needs of the controlled object, monitoring the efficiency of the chosen methods and instruments of influence on it by the subject. To achieve the goal, the authors relied on theoretical and empirical methods of scientific research. At the same time, objective monitoring results characterizing the quality of services provided in the interaction of service organizations and municipal authorities should be determined by applying an interaction model of all participants in this process based on the method of ensuring the quality of services provided. This methodology is a leading approach to this problem, which allows a comprehensive review when monitoring the quality of services provided upon five general criteria, as well as additional indicators formed by the results of this monitoring. The contribution revealed that these monitoring criteria allow determining the efficiency of subordinate organizations, which is justified by conducting independent service quality monitoring provided in the region by the municipal committee and organizations subordinate to it according to common criteria such as: openness and availability of information about the organization, comfortable conditions for services, customer waiting time, friendliness, politeness, care when providing services, satisfaction with the quality of services. The materials of the contribution are of practical value for service organizations when interacting with municipal authorities, since they determine the results of independent monitoring conducted through digital services. Thus, the advantages and disadvantages of monitoring within the framework of this interaction model were identified. On the one hand, the interaction between them is in opposite directions to each other, which illustrates a well-structured and well-coordinated process of information transfer, where everyone understands his role and knows his functions. On the other hand, it is necessary to note, as a model deficiency, that information acquired when conducting independent monitoring, is not brought to the notice of subordinate organizations, the population, for whom this independent service quality monitoring was conducted.

**Keywords:** Digital services · Economy digitalization · Monitoring · Monitoring criteria of the quality of services provided · Municipal authorities · Quality of services · Quality monitoring of services · Service organizations

## 1 Introduction

Monitoring is a tool that ensures the quality of services provided in the interaction of service organizations and municipal authorities in economy digitalization. In view of increasing control over the provision of services, it is necessary to form and use tools that ensure its efficiency, where monitoring has a special place. Monitoring is understood as a process of systematically monitoring an object, monitoring, analyzing, evaluating and predicting its state or method and system of monitoring the state of a particular object or process, making it possible to monitor them in development, evaluate, quickly identify the effects of various external factors. Monitoring results provide an opportunity to adjust the management of an object or process [8]. In this connection, the purpose of the contribution is to identify monitoring tools as an integral part of the management cycle. These tools make it possible to adjust the management of an object or process through the use of digital services, and these issues were considered by Candiello, Albarelli, Cortesi [4], Wulan [13]. The information platform of this monitoring method is digital services of official committees and departments located on the Internet telecommunications network with the transfer of documents and communications to digital media, due to the dynamic development of the digital economy. For any organization, regardless of its size, type of product or service, an integrated approach to ensure the quality and safety of information becomes a strict, logical and useful necessity [9, 12].

## 2 Materials and Methods

To achieve the purpose, the authors relied on theoretical and empirical methods of scientific research. It is about continuous monitoring of any process in order to identify its compliance with the desired result. In other words, if the diagnostics of a situation is carried out systematically with certain predetermined frequency and using the same (in any case, basic) indicator system, we are dealing with monitoring [10].

Monitoring as an instrument of control is an integral part of the management cycle through which municipal authorities and service organizations interact on the basis of digital services. The functional importance of this interaction is to provide "feedback": the ability to identify the needs of the controlled object, monitor the efficiency of the selected methods and tools to influence it by the subject [10]. In this regard, the authors attempted to determine the impact of these aspects on service visitors, satisfaction and their behavioral intentions [4, 7].

## 3 Results

Municipal authorities in the face of relevant municipal committees use in their practice quality monitoring of services provided based on the criteria for monitoring the performance of subordinate organizations (Federal law of Russia 2014). So, on the territory of the Republic of Tatarstan, the municipal committee conducted independent service quality monitoring provided according to common criteria by organizations under its jurisdiction (the Order of the Government of Russia 2013). Independent service quality monitoring was conducted in socially-oriented organizations:

(1) Teenage clubs;
(2) Social Service Center;
(3) Rehabilitation, Recreation, Employment Center for Children and Young People;
(4) Social and Rehabilitation Center for Children with Deviant Behavior.

The results of monitoring upon the criterion - openness and availability of information about the organization are presented in the diagram (Fig. 1).

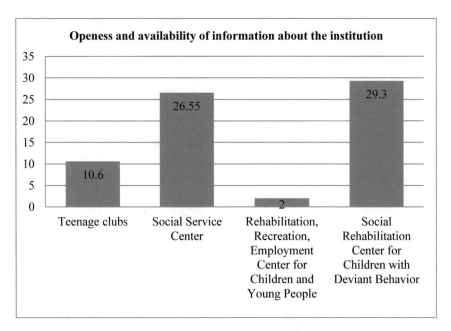

**Fig. 1.** The result of independent monitoring of organizations upon openness and availability of information about the organization (Source: compiled by the authors)

Based on the chart, you can draw several conclusions. The highest score for this criterion was the "Social Rehabilitation Center for Children with Deviant Behavior", while the "Rehabilitation, Recreation, Employment Center for Children and Young People" received the lowest score, although it deals with package tours for children and young people in summer health camps. Openness and availability of information about the organization is necessary for the center to attract more attention to the children and their parents, which will increase the selling of package tours, and, consequently, increase the amount of profit received for package tours in certain sessions, which, in turn, will increase more qualitative process of organizing the session. Adherence to these principles is an important factor in improving the research conducted, obtaining objective and reliable information [3, 6].

The next criterion is comfortable conditions for services (Fig. 2).

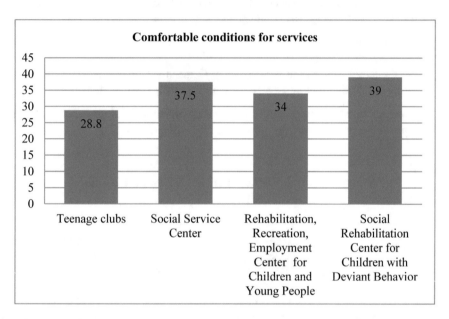

**Fig. 2.** The result of independent monitoring of organizations upon comfortable conditions for services (Source: compiled by the authors)

According to this criterion, the organizations received fairly close ratings to each other. However, the Social Rehabilitation Center for Children with Deviant Behavior again received the highest rating. It is necessary to note that the Social Center works with children with deviant behavior, children and adolescents who are in a difficult life situation and socially dangerous situation, as well as with children who are without parental care and awaiting transfer to orphanages. For this category of persons, a high degree of comfort and availability of the service is important, which is taken into account in the work of the organization, judging by a high result for this criterion.

The next criterion is the customer waiting time (Fig. 3).

According to this criterion, the organization "Teen clubs" received the lowest rating. This suggests that the management of the organization should pay close attention to the availability of a preliminary appointment to specialists of the organization, as well as to the possibility of providing the service on the same day.

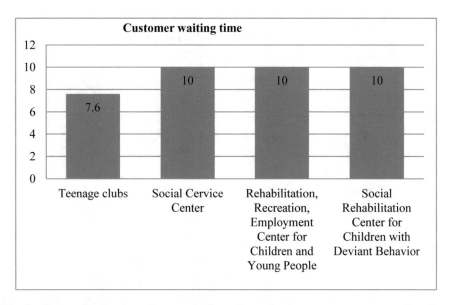

**Fig. 3.** The result of independent monitoring of organizations upon customer waiting time (Source: compiled by the authors)

The next criterion for independent monitoring is friendliness, politeness, care when providing services (Fig. 4).

When working with children and young people, the results of independent monitoring upon this criterion should be with the highest rating. However, the "Teenage clubs" gets a low rating again compared to other organizations. Friendliness, politeness and care are the main components of the general opinion about the degree of satisfaction with the service provided by the organization, which directly affects the number of consumers of this service.

The last criterion is satisfaction with the quality of the service provided. Satisfaction with the quality of service is one of the most important indicators of the organization. As can be seen from the diagram, the consumers of services of these organizations are not fully satisfied with services. The management of organizations needs to analyze in more detail the work of the organizations' specialists for further changes that will lead to an increase in the level of customer satisfaction with services provided (Fig. 5), which are necessary for making changes in the organization [8].

After monitoring the organizations subordinated to the municipal committee, we have to note that according to the general criteria of independent quality service monitoring, namely, openness and availability of information about the organization, comfortable conditions for services, customer waiting time, friendliness, politeness, care when providing services, satisfaction with the quality of services, "Teenage clubs" ranks the lowest in the rating of subordinate organizations according to these criteria,

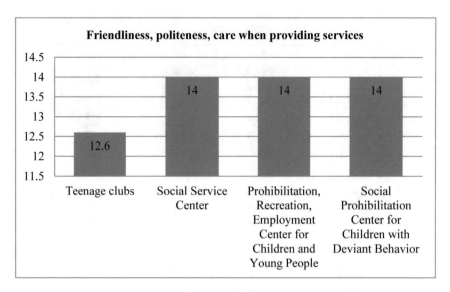

**Fig. 4.** The result of independent monitoring of organizations upon friendliness, politeness, care when providing services (Source: compiled by the authors)

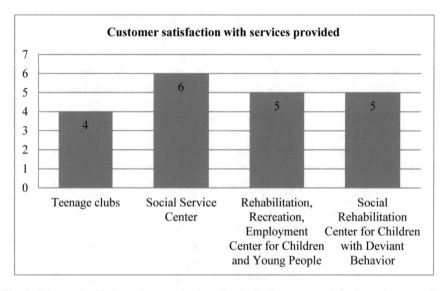

**Fig. 5.** The result of independent monitoring of organizations upon satisfaction with the quality of services provided (Source: compiled by the authors)

while "Social Service Center" and "Social and Rehabilitation Center for Children with Deviant Behavior" take the first lines. "Rehabilitation, Recreation, Employment Center for Children and Young People" takes the middle line among subordinate organizations.

# 4   Discussion

Although, the problem was studied in the works of Simakhin [10], Pavlova [8], Candiello et al. [4], Mel et al. [7], the authors consider the interaction model between service organizations and municipal authorities as an integrated approach that allows monitoring its efficiency (in this case, they are subordinate organizations that provide social services to the population, and the quality of these services are independently monitored by municipal authorities). Municipal relevant committees create an organizational framework for conducting independent service quality monitoring. In relation to a public committee, a municipal committee forms its composition, approves independent monitoring work plan, establishes indicators characterizing general criteria for monitoring the quality of services, examines information on the results on the quality of services provided, considers recommendations on the development of measures to improve the activities of service organizations based on the results of independent monitoring, approves the terms of reference for independent monitoring.

In relation to the Operator Organization, the municipal committee approves organizations, allocates a budget for the provision of service for collecting and processing the results of independent service quality monitoring. The Community Committee, in turn, in relation to the Committee directly organizes independent service quality monitoring, and provides results of service quality monitoring of organizations, as well as proposals for improving the quality of their activities when interacting with municipal authorities and service organizations [11].

The Public Committee determines the list of organizations in relation to which independent monitoring was conducted, draws up proposals on the development of the terms of reference, which spells out the methodology for conducting independent monitoring. The Operator collects, compiles and analyzes information on the quality of services provided, provides information on the results of independent monitoring to the Public Committee in accordance with the terms of the municipal contract concluded with the authorized body [5].

# 5   Conclusions

Thus, as a result of interactions between the participants of this model, when conducting service quality monitoring upon five general indicators and additional indicators, monitoring results are generated through digital services. At the moment, the activities of participants of this process look this way. We seen the interaction between them is in opposite directions to each other, and it means a clearly structured and well-coordinated process of information transfer, where everyone understands his role and knows his duties. The implementation of this interaction is enhanced by digital resources, allowing up-to-date information processing and presenting results achieved.

However, it should be noted that the model is incomplete, and information acquired when conducting independent monitoring is not brought to notice of subordinate organizations, the population, for whom this independent service quality monitoring was conducted.

# References

1. About modification of separate legal acts of the Russian Federation concerning conducting independent service quality monitoring provided by the organizations in the field of culture, social service, health protection and education: the Federal law of July 21, 2014 No. 256-FZ. http://www.consultant.ru/document/cons_doc_LAW_165899. Accessed 11 Mai 2018
2. About the approval of the plan of actions for forming of independent monitoring system of the quality of social services provided by organizations for 2013–2015: the order of the Government of the Russian Federation of 30.03.2013 N 487-p, item 9. http://www.consultant.ru/document/cons_doc_LAW_144318. Accessed 11 Mai 2018
3. Basarangil I (2018) The relationships between the factors affecting perceived service quality, satisfaction and behavioral intentions among theme park visitors. Tour Hosp Res 18(4):415–428
4. Candiello A, Albarelli A, Cortesi A (2012) Quality and impact monitoring for local e-government services. Transform Gov People Process Policy 6(1):112–125. https://doi.org/10.1108/17506161211214859
5. Colakovic A, Bajric H (2016) Assessing customer satisfaction based on QoS parameters. Int J Qual Res 11(1):221–240. http://www.ijqr.net/journal/v11-n1/14.pdf. Accessed 11 Mai 2018
6. Goryaeva SE (2012) Methodological and methodical bases of the study of public opinion on the quality of public services. Bull Inst Complex Stud Arid Territ 1(24):50–57
7. Mele C, Colurcio M, Russo-Spena T (2014) Research traditions of innovation: goods-dominant logic, the resource-based approach, and service-dominant logic. Manag Serv Qual 24(6):612–642. https://doi.org/10.1108/MSQ-10-2013-0223
8. Pavlova VV (2011) Typology of changes in the organization and their characteristics. Actual Probl Econ Law 2:102–106
9. Pop AB, Țîțu AM (2018) Implementation of an integrated management system: quality-information security in an industrial knowledge-based organization. Qual Access Success 19(166):87–93
10. Simakhin OG (2011) Development of the system of management of state customs services on the basis of customs monitoring: thesis of candidate of economic sciences, Moscow, p 153
11. Vesloguzova MV, Zemskova OA (2017) Methods of assessing customer satisfaction with the quality and comfort provided by the municipal services. In: Problems and prospects of development of social and economic potential of the Russian regions materials VI all-Russian electronic scientific-practical conference, pp 195–200
12. Wickramasinghe V (2015) Effects of human resource development practices on service quality of services offshore outsourcing firms. Int J Qual Reliab Manag 32(7):703–717. https://doi.org/10.1108/IJQRM-03-2013-0047
13. Wulan IR (2017) The role of regulations on administrative and practices in improving quality of services in public organizations. Cogent Bus Manag 4:1–16. https://www.tandfonline.com/doi/pdf/10.1080/23311975.2017.1396952?needAccess=true. Accessed 11 Mai 2018

# Prospects for the Integration of Environmental Innovation Management on the Platform of Information and Communication Technologies

I. V. Kosyakova[✉], N. Yu. Zhilyunov, and Yu. V. Astashev

Samara State Technical University, Samara, Russia
iv-kos@mail.ru

**Abstract.** The management structure of a large commodity company is very complex, which causes problems in communications, in coordinating management actions, and there is a danger of distortion and loss of information. Particular difficulties begin when combining environmental and innovation management. Solving the problems of environmental innovation management is much easier on the information and communication technology platform. The goal of the contribution is to develop an information flow management algorithm when introducing environmental innovation management. The leading method for the study of this problem is the method of modeling, which allows considering this problem as a focused and organized process of integrating environmental innovation management, subject to the application of information technology. As a result of the study, it is possible to single out the conclusions obtained in the course of analyzing the state and trends in the development of the digital economy in Russia and abroad, as well as its role in the implementation of environmental innovation management. An algorithm is proposed for the use of information flows in the implementation of environmental management at large oil companies, highlighting four levels of negative environmental impact. The practical significance of the contribution is the possibility of using the proposed approach in practical activities of large oil companies in implementing innovation integrated management practices at all levels, including top management.

**Keywords:** Digital economy · Digitalization · Ecology · Information and communication technologies · Innovations · Vertically integrated companies

## 1 Introduction

World economic growth today is associated with the development of the digital economy. So by 2025, due to Internet technologies, in China the gross domestic product (GDP) is expected to increase to 22%, in the US it can reach $ 1.6–2.2 trillion. According to McKinsey Global Institute [2], by 2025 the digital economy will increase Russia's GDP from 19% to 34% of the total expected GDP growth. The basis of these forecasts is the positive effect from the introduction of breakthrough technologies and business models, the development of digital platforms and ecosystems, the new level of

© Springer Nature Switzerland AG 2020
S. Ashmarina et al. (Eds.): *Digital Transformation of the Economy: Challenges,*
*Trends and New Opportunities*, pp. 345–355, 2019.
https://doi.org/10.1007/978-3-030-11367-4_34

information processing, the introduction of Industry 4.0 [14], the spread of the Internet of Things, the widespread introduction of robotics, etc. The basis of the new digital economy is intellectual activity. Information technologies penetrate deeply into all spheres of human activity, the face of society is changing, and new approaches to managing the economy both in Russia and abroad are forming [1].

The digital economy is linked with innovation management, whose task is to implement innovations, since the success of any business is impossible without them. Innovation management is designed to spread new thinking at all levels of business management, to maintain high intellectual potential and creativity of employees, to accelerate the growth of the company, to increase profits from innovation [3].

In the context of sustainable development, it is impossible to bypass the aspect of environmental innovation [13]. Ecological innovations always take place if a decrease in the negative impact on the environment is recorded as a result of the introduction of innovation in any sphere - a product, a technological process, management, social sphere, etc. The environmental aspect of the company is controlled by environmental management, which, like the innovation one, is part of the overall business management. The integration of environmental innovation management is a natural process that, however, should be managed. First of all, the mutual diffusion of these two types of management is necessary for companies that are significant environmental polluters [9], which require targeted, sound investment policy in the field of environmental innovation.

## 2 Materials and Methods

The authors used the following methods: theoretical (analysis; synthesis; specification; generalization; method of analogies; modeling); empirical (the study of foreign and Russian experience in the application of digital technologies, the study of the possibility of introducing environmental innovation technologies into the work of industrial companies, the study of Russian legislation in the field of digital technologies); experimental (stating, forming, control experiments); methods of mathematical statistics and graphic images of the results. The experimental base of the research was the information of large oil companies of Russia, presented in the open press.

The methods used made it possible to study the indicated problems in three stages:

- At the first stage, the authors analyzed the level of development of the digital economy in Russia and abroad, as well as general approaches and methods for studying economic problems, and drew a study plan;
- At the second stage, the authors developed a methodology to enhance the effectiveness of introducing environmental innovation management using digital technologies;
- At the third stage, the authors refined the theoretical and practical conclusions, and summarized and systemized the results.

# 3 Results

## 3.1 Digital Economy and Environmental Sustainability

Russia is not a leader in promoting the digital economy yet.

Since 2014, the European Commission has been publishing the Digital Economy and Society Index (DESI), a composite index that measures the progress of EU countries in the direction of the digital economy and society. It combines a set of relevant indicators of current European digital policy. DESI consists of five main policy areas that are regrouped by 34 indicators [18]:

1. Digital technologies: fixed broadband access, mobile broadband access, fast and ultra fast broadband.
2. Human capital: basic skills and use of the Internet, advanced skills and development.
3. Internet service: the use of content by citizens, communications and online transactions.
4. Integration of digital technologies: business digitalization and e-commerce.
5. Digital government services: e-government and healthcare.

According to DESI [18], four leading EU countries (Denmark, Finland, Sweden and the Netherlands) are among the world leaders. Figure 1 shows the international indices of the digital economy.

The figure shows that Norway and New Zealand are lagging behind Korea and have higher scores than the United States and Japan. Russia has a lower rating and is lagging behind in the electronics segment, in terms of digitalization, mastering digital technologies and computer-aided design systems, as a whole in the share of the digital economy in GDP. However, as noted by The Moscow Times, Russia is starting to look like a normal country, and 2018 will take a few more steps to keep its status as a developing market [18]. It is worth noting that the volume of the digital economy increased by 59% from 2011 to 2015. The infrastructure of information and communication technologies (ICT) is developing rapidly in Russia. It is a profitable business.

As positive features of ICT in Russia, it is possible to highlight the active development of mobile communications, a significant increase in the role of ICT in the educational process and scientific research, the almost universal introduction of software into the practice of both large organizations and small firms.

In 2010, V.V. Putin signed the Order of the Government of the Russian Federation "On the Plan for the Transition of Federal Executive Bodies and Federal Budget Institutions to the Use of Free Software (2011–2015)". This made it possible to create a national software platform, which included such software as 1C, Crypto-pro, Alt Linux and others [11].

The program "Digital Economy of the Russian Federation" was approved in 2012. By Order of March 29, 2018 No. 528-p, the Government of the Russian Federation planned to allocate up to 3040836.86 thousand rubles from the government reserve fund in 2018, and this funding is specific, targeted for activities within the framework of "Information Infrastructure", "Formation of Research Competences and Technological Groundwork" and "Information Security" [12]. Due to private investment these

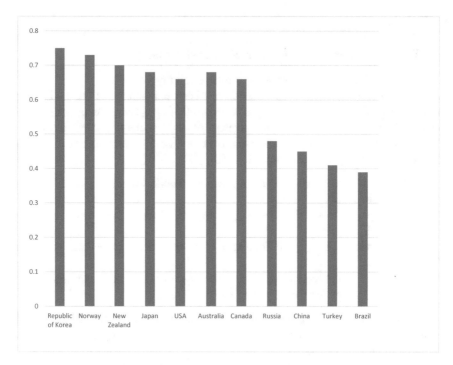

**Fig. 1.** Countries' rating according to DESI (Source: compiled by the authors)

volumes may increase. Thus, the financial base for the development of the digital economy is sound. But this is only one of the most important aspects for modernization of Russian society.

Russia has long lagged behind North America and Europe in the field of techno-logical innovation. As a result, a significant proportion of industrial equipment requires modernization, especially in primary industries. On the other hand, Russia has one of the most developed telecommunication infrastructures. And this is an excellent base for the advancement of innovative technologies. It is known that primary industries in the Russian economy play a significant role. For example, the share of oil and gas revenues in the federal budget reached 40% by the end of 2017. The mining and oil refining sector accounts for over 60% of all export earnings [7]. But the raw materials industry cannot be considered innovatively capacious. So the Nelson index for the Russian oil refining and petrochemical complexes is only 4.3, while in the US it is 9.5. As a result, primary industries are also lagging behind in the development of the green economy.

### 3.2    Basics of Combining Environmental and Innovation Management

According to Ciocoiu [3], the idea of the development of the global economy and society is based on two issues: the potential of information and communication tech-nologies (ICT) and the problem of environmental sustainability. The green economy model is a global trend of sustainable development, which involves the preservation of natural capital with simultaneous economic growth. The green economy has achieved significant success in most developed countries.

In Russia, the transition to the green economy is slow, usually dotted. It relies on technological developments of past decades mainly in the field of energy consumption, production of greener products, in the field of waste management. Until now, the driving force here has been the initiative of firms, public organizations, and informal associations engaged in innovation activities. Obviously, the free market does not have the properties of environmentally sustainable self-regulation. According to the rating of countries in The Environmental Performance Index, determined by the Center for Environmental Law and Policy at Yale University, Russia occupies only 52 place in the rating [17]. The transition to the green economy requires government measures. And above all, Russia's lag in the green economy is associated with the limited development of environmental innovations in both technology and management.

Russia has a resource-oriented economic structure. It is dominated by the fuel and energy complex. Due to the fuel and energy complex Russia receives tangible benefits from carbon regulation. Other sectors of the economy have rather low specific indicators relative to international competitors. So metallurgy, mining, and others require long-term investments in order to meet international ratings. According to the Ministry of Economic Development, the annual volume of investments should exceed € 6 billion (0.18% of GDP). There is no doubt that the existing problems cannot be solved without the intensive development of management, both environmental and innovation, on a common information and communication technology platform. Both environmental and innovation management are part of the overall management system, which is directly related to the production and management structure.

In large commodity companies, for example, the fuel and energy complex, vertical integration with a large number of management links is mainly used. The management structure of the company becomes complex, it has not only horizontal and vertical, but

**Table 1.** Objects of the negative impact of a vertically integrated oil company as a structural unit of environmental management

| Name | Definition |
| --- | --- |
| Local object (LO) | Production facility as part of one installation (or one production process - stationary or mobile), located within the same land plot and having a direct negative impact on the environment |
| Industrial object (IO) | The aggregate of local objects that can be identified within a single geographic boundary and for which a single environmental aspect can be identified |
| Segmental object (SO) | The aggregate of industrial objects of one functional activity of the company, comparable in significant environmental aspects and types of the negative impact on the environment |
| Corporate object (CO) | The aggregate of the company's corporate objects, which determines the environmental component of the company's sustainable development |

Source: compiled by the author

also spatial differentiation. There are problems in communications, in the coordination of managerial actions, and a danger of distortion and loss of information. Solving these problems is easier if ICT is available as a platform for management, both environmental and innovation. But an even greater need for ICT arises when these two components are integrated into environmental innovation management (EIM).

As for environmental management, in a number of works [9] it is proposed to highlight the objects of environmental management of vertically integrated oil companies in accordance with the dominant criterion, which today in the raw materials industry is understood as the negative impact on the environment. In accordance with this, for example, for vertically integrated oil companies (VIOC) it is proposed to identify 4 levels of objects of the negative impact (NI) on the environment (E), which are presented in Table 1 [9].

The need to differentiate the objects of the negative impact of a vertically integrated oil company is due to the fact that it:

- Allows identifying the structural units of environmental management at different levels of the hierarchy of vertically integrated oil companies;
- Allows considering, comparing and developing management actions for objects of the negative impact of the same scale of this impact, but having different economic, social aspects, technological base, etc.
- Can serve as the basis for the environmental management system (EM) of vertically integrated oil companies;
- Manages information within the ICT Integrated Management Platform.

Using the approach suggested above, it is possible to use the algorithm for managing the information of a vertically integrated oil company. In the proposed algorithm, the ordering of information accompanying control actions of environmental management is carried out "from top to bottom" (Fig. 2).

Information in the form of a set of indicators {IS1} is formed at the level of the top management of the VIOC. These indicators are policy for the next level of environmental management. At the level of the segment with the object of the negative impact of SO, a set of indicators {IS2} is formed, which is formed by top managers of main activities of the vertically integrated oil company: oil production, oil refining, etc. This information is used to formulate targeted environmental programs with respect to SO. In turn, {IS2} are taken into account when developing strategies for environmental management at the level of PO, for example, at the level of oil refinery. The objects of the negative impact of PO are local structural units - installations, technological processes. They are the direct sources of emissions. The level of emissions is determined through the monitoring of the environment, and the requirements limiting these emissions are included in the set of parameters {IS4}. The exchange of information is carried out through feedback, which allows adjusting management actions almost in real time. It is possible to implement such an algorithm only if there is an ICT platform, i.e. practically when digitizing the management actions of the VIOC.

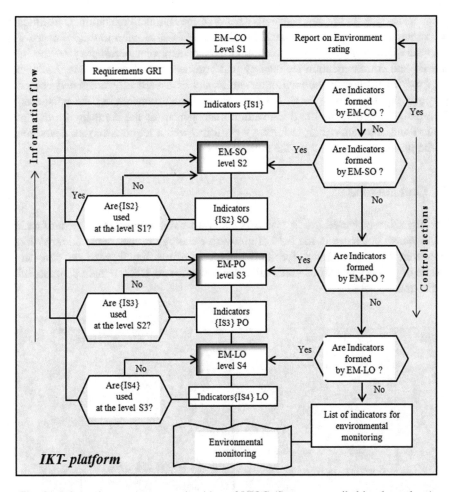

**Fig. 2.** Information management algorithm of VIOC (Source: compiled by the authors)

## 4   Discussion

Studies of the impact of the digital economy on the development of modern society are conducted by a number of organizations, for example, the European Commission on Communication Networks, contentand technology - the European Commission on Communication Networks, Content and Technology [15], the Federal State Statistics Service of the Russian Federation [5] and others. Some problems are being investigated in the area of the impact of the digital economy on the environment and the development of the green economy. The works of such authors as V. N. Lopatin [11], C. N. Ciocoiu [3], Gunkov et al. [7], Pakhomov et al. [13], Earnhart et al. [4]. A review of the current scientific literature on the problem under study shows that the coverage of the problems of integrating environmental and innovation management of the company with a complex hierarchical structure, which are also significant environmental polluters and are part of raw materials sectors of the Russian Federation, is insufficient. In particular, the aspect considered in this contribution concerning the prospects for integrating environmental and innovation management of the company on the information and communication technology platform deserves close attention. Methodical approach to managing the environmental aspect of a vertically integrated company, proposed by Kosyakova et al. [9] can be used when considering the prospects for integrating environmental and innovation management of the company on the information and communication technology platform, which requires further development of the problem considered by the authors.

## 5   Conclusions

The algorithm proposed in Fig. 2 allows solving a rather important problem of making management decisions in the field of innovative ecology in the "here and now" mode. However, this is not enough for the development of the digital green economy in the sphere of primary industries - after all, these industries are still the main polluters of the environment.

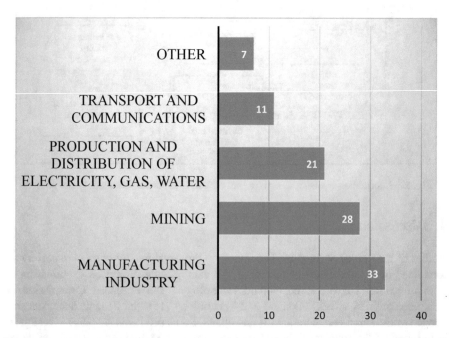

**Fig. 3.** Contribution of industries to environmental pollution, % (Source: compiled by the authors based on the data of the Analytical Center under the Government of the Russian Federation at the end of 2016)

Figure 3 shows information on the level of environment pollution by various sectors of the national economy, provided by the Analytical Center under the Government of the Russian Federation in 2016 [5]. The Fig. 3 shows that the greatest environment pollution is caused by manufacturing and mining companies. At the same time, indicators of eco-innovation for polluting industries are low [4, 5, 16]. This can be seen in Table 2.

One of the ways to improve the situation is related to the integration of environmental and innovation management. First of all, an integrated EIM should ensure the identification of "bottlenecks" with respect to the negative impact on the environment in order to optimize the investment policy for all levels of the hierarchical structure of the VIOC. This will allow a systematic approach to the management of innovation and investment activity, ensuring the implementation of the company's goals in the field of environmental policy, from limiting emissions of LO to the introduction of innovative practices of integrated management at all levels, including top management.

**Table 2.** The share of organizations that carried out innovations that ensure the improvement of environmental safety using innovative products, works, services by the consumer, in the total number of organizations that carried out environmental innovations: 2015,%.

| Industry, which organizations belong to | Reduction of energy consumption or loss of energy resources | Reduction of air pollution, land, water resources, reduction of noise level | Improvement of opportunities for recycling of goods after use |
|---|---|---|---|
| Production and distribution of electricity, gas and water | 50,8 | 55,6 | 6,3 |
| Extraction of fuel and energy minerals | 45,0 | 65,0 | 40,0 |
| Production of coke and oil products | 29,4 | 70,6 | 17,6 |

Source: compiled by the authors based on the data of the Federal State Statistics Service 2018

Thus it is proved that environmental innovation has its own distinctive features. Eco-innovation in technology is usually secondary, supplementary, such as in the "clean production" model. Such innovations are gradual, increasing the efficiency of existing technologies without a qualitative change in the technological process [4, 6, 13]. Uncertainty arises in estimating the payback period of the environmental component of innovation. Breakthrough environmental innovations are used much less frequently, their payback period may be more than 10 years, and sometimes it is just difficult to define. The long-term payback of environmental investments is one of the reasons why companies prefer to use borrowed environmental innovations. These facts, as well as the increasing complexity of technological processes and the limited resources of the company, bring open innovation to the forefront, attracting third-party organizations, both domestic and foreign. That is why the integrated EIM needs a unifying ICT platform [1].

Objective monitoring of the negative impact at the level of LO, taking into account the characteristics of the territory of functioning of LO (including the level of environmental degradation), can be the source material for the formation of basic models for identifying "bottlenecks" relative to the negative influence on environment.

# References

1. Berisha-Shaqiri A (2015) Information technology and the digital economy. Mediterr J Soc Sci 78–83. https://doi.org/10.5901/mjss.2015.v6n6p78
2. Aptekman A, Kalabin V (2017) Digital Russia: a new reality. McKinsey Global Institute, p 133. https://www.mckinsey.com/media/McKinsey/Locations/Europe%20and%20Middle%20East/Russia/Our%20Insights/Digital%20Russia/Digital-Russia-report.ashx. Accessed 30 Mar 2018
3. Ciocoiu Carmen Nadia (2011) Integrating digital economy and green economy: opportunities for sustainable development. https://ideas.repec.org/a/rom/terumm/v6y2011i1p33-43.html. Accessed 30 Mar 2018
4. Dietrich EH, Khanna M, Thomas LP (2017) Corporate environmental strategies in emerging economies. Rev Environ Econ Policy 8:164–185. https://doi.org/10.1093/reep/reu001. Accessed 30 Mar 2018
5. Federal State Statistics Service. Indicators of innovation activity: 2018: statistical compilation (2018) Gorodnikova NV, Gokhberg LM, Ditkovsky KA and others; Higher School of Economics. - Moscow: HSE, ISBN 978-5-7598-1742-0
6. Forrester SV, Ustinova GH, Kosyakova IV, Ronzhina NV, Suraeva MO (2016) Human capital in innovative conditions. Int Electron J Math Educ 11(8):3048–3065
7. Gunkov AG, Kholopov YA (2017) Environmental management of the enterprise. Bull SamGUPS 1(35):80–83
8. Korolev I (2018) The state will spend 436 billion dollars on the ICT infrastructure. http://www.cnews.ru/news/top/2018-07-05_na_iktinfrastrukturu_gosudarstvo_potratit_436. Accessed 30 Mar 2018
9. Kosyakova IV, Kudryashov AV (2016) Methodical approach to managing the environmental aspect of a vertically integrated company. Prof Sci 5:14–24
10. Kudryashov AV (2016) Managerial approach to assessing the negative environmental impact of a vertically integrated company. UEKS 12:94. https://cyberleninka.ru/article/n/upravlencheskiy-podhod-k-otsenke-negativnogo-vozdeystviya-na-okruzhayuschuyu-sredu-vertikalno-integrirovannoy-kompanii. Accessed 30 Mar 2018
11. Lopatin VN (2018) Information security risks in the transition to a digital economy. State Law 3:77–88
12. Medvedev D (2018) Order of March 29, 2018 No. 528-p. http://static.government.ru/. Accessed 30 Mar 2018
13. Pakhomova NV, Sergienko OI (2016) Innovations of environmentally sustainable development: the situation in Russia in the context of international experience. In: Problems of the modern economy, vol 1–2, pp 247–254
14. Pukha Y (2018) Industry 4.0: global digital operations Study 2018. https://www.pwc.com/gx/en/industries/industry-4.0.html. Accessed 30 Mar 2018

15. Review of the digital agenda in the world (2018) Website of the Eurasian Economic Union. http://www.eurasiancommission.org. Accessed 30 Mar 2018
16. Sahin F, Koksal O, Garbus S (2014) Revisiting of theory X and Y: a multilevel analysis of the effects of leaders' managerial assumptions on followers' attitudes. Manag Decis 10:1888–1906. https://doi.org/10.1108/md-06-2013-0357
17. The Environmental Performance Index (2018) http://epi.envirocenter.yale.edu/. Accessed 30 Mar 2018
18. Van der Linden N, Hols M, Berends J, Bos B (2018) Digital economy and society index (DESI) 2018. http://europa.eu/rapid/press-release_MEMO-18-3737_bg.html. Accessed 30 Mar 2018

# Social Media Marketing as a Digital Economy Tool of the Services Market for the Population of the Republic of Tatarstan

S. Shabalina[1]([✉]), E. Shabalin[1], A. Kurbanova[1], A. Shigapova[1], and R. Vanickova[2]

[1] Kazan Federal University Kazan, Kazan, Russia
sshabalina74@gmail.com, kurbanovaas84@gmail.com,
metallicana.metin@gmail.ru
[2] Institute of Technology and Business, České Budějovice, Czech Republic
vanickovaradka@gmail.com

**Abstract.** The authors analyze the popularity of social media as one of digital economy tools among users of the Republic of Tatarstan. It contains the comparative analysis on age categories of users and activity of respondents. The research purpose is analysis of social media and behavioral characteristics of their users for determination of weak spots of use of social networking sites for studying of the services market as a channel for obtaining tourist information and their elimination. The most popular social networking services are identified according to the number of hits. There are also defined key positions, on which potential clients plan their vacation and which have the greatest credibility. The conclusions on the increasing role of SMM in the service sector were drawn on the basis of respondent responses. The age categories of users and the most reliable information channels are determined. They are regarded by marketing specialists as effective distribution channels of products and, in particular, of any services, including tourist ones. Rapid information dissemination and its real tracking requires from suppliers more accelerated strategy in provision of information for clients and decrease in time for order processing, when there is shift of accent from advertizing and direct marketing to marketing of directions and sales in the Internet.

**Keywords:** Behavioral model · Digitalization · Marketing · Services market · Social media · Targeting

## 1 Introduction

Currently social media occupy a significant place in life of the majority of people. According to the statistics, 93% of Russians, who use the Internet, look through social media about three hours a day. It has been a long time since social media are not just a platform for communication. For users it is an additional source of information, a place of exchange of views, entertainment, work and inspiration. For some people it is their image.

© Springer Nature Switzerland AG 2020
S. Ashmarina et al. (Eds.): *Digital Transformation of the Economy: Challenges, Trends and New Opportunities*, pp. 356–367, 2019.
https://doi.org/10.1007/978-3-030-11367-4_35

Social media in Tatarstan have high integration in life of inhabitants of the Republic. The most popular social networking sites in Tatarstan are: VKontakte, Facebook, Instagram, Odnoklassniki. According to the tag "Tatarstan", there are about 2 million posts in Instagram, and 4 million with the tag Kazan. By the request "Kazan tours", there are about 338 communities in the social networking service VKontakte, and in the largest communities the number of subscribers reaches 40 thousand.

The contribution of Tim O'Reilly "Web 2.0" [4], describing the existing information space, can be considered as the starting point of the present development stage of the Internet where social networking sites occupy an extremely important niche. According to the Forbes magazine, equal distribution of social media with widespread replacement of television by the Internet became one of twenty two main trends of the first decade of the twenty first century [2].

## 2  Materials and Methods

The analysis of social networking services provides more accurate and wide information on the target audience. 80% of the Russian Internet audience daily use social media. Social media generate targeted traffic under satisfaction of information needs of users and formation of credibility to a brand [13]. Formation of the portrait of the brand audience, reviews and estimates of users of SMM help to evaluate strengths and weaknesses of work for its further improvement.

## 3  Results

One of the most popular social media site in Russia is VKontakte (Fig. 1). It was launched on October 10, 2006. This website is an analogy with Facebook, a social networking service for students and graduates of the Russian higher education institutions. According to the data as of August 2017, the average daily audience of the portal makes more than 82.5 million visitors. The total amount of users, registered in the site, is more than 460 million. According to the SimilarWeb service, VKontakte ranked 7th in popularity in the world in September 2017 [7].

The second popular social networking website in Russia is Odnoklassniki. At the same time, it is the first social media site in the popularity rating of Armenia, the fourth – in Kazakhstan and Azerbaijan and the twenty fourth – in the universal rating. As well as VKontakte, Odnoklassniki has a prototype. It is an American social networking service "Classmates", which mission is to unite schoolmates from the entire planet. This service is more "domestic" and many companies avoid it. Because of superficial opportunities, settings of advertizing campaigns and acquisition of statistical data are not in great demand for product promotion [11].

The leader among the world social networking services is Facebook. It ranks third in Russia: its audience is twice smaller than VKontakte. Since 2006 the service is available for all Internet users of 16 years worldwide.

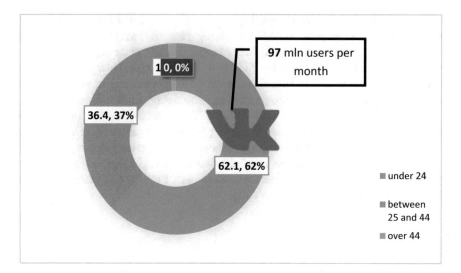

**Fig. 1.** The age of users of the social media service Vkontakte (Source: compiled by the authors)

Facebook ranks among top - 5 the most visited web sites of the world. As of April, 2017, the monthly audience of the social media service makes nearly two billion people. As of June 2017, there were more than 2 billion users, who visited the website at least once per month, and who were traced by the system with the help of the Like button and the cookie technology. In March 2017, the daily active audience made 720 million people. The mobile application Facebook for smartphones is monthly used by 1,03 billion people. Every day users leave 6 billion "likes" and comments in the site and publish about 300 million photos (Fig. 2).

Currently, it is one of the key tools of a SMM specialist [1]. It should be noted that the Russian-speaking segment of Facebook strongly differs from the universal one. It is caused by the development mechanics, behavioral features of the audience, and nuances of social and demographic characteristics of users. The focus of interest of Facebook users also strongly differs from the interests of users of VKontakte and Odnoklassniki, therefore it is more difficult to hold their attention. It resulted in more expensive, but well-designed advertising campaigns [8].

In recent years, the visualization trend is the leading one in SMM. It is considered that photo and video content enjoys bigger popularity than text, and is more effective [5]. The best platform for implementation of such campaigns is Instagram, a social media service, which is based on exchange of photos and videos, which ranks fourth in Russia (Fig. 3). By December 2010, one million users were registered in the social networking service. By June 2011, this figure increased up to five million. In two months the number of users was 10 million, and by March 2012, the amount of users reached nearly 30 million. In April 2014, more than 200 million people were registered in Instagram.

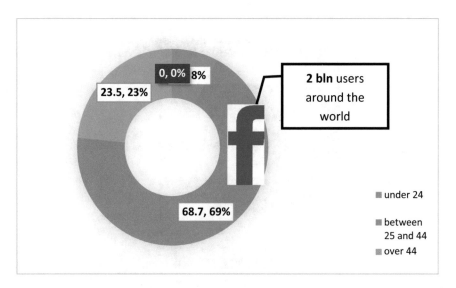

**Fig. 2.** The age of Facebook users (Source: compiled by the authors)

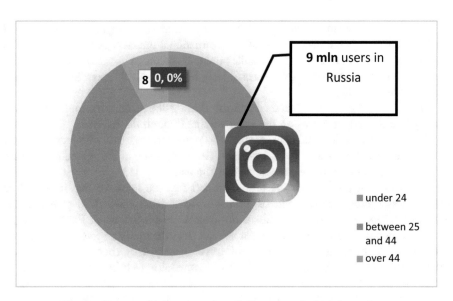

**Fig. 3.** The age of Instagram users (Source: compiled by the authors)

In July 2011, the Instagram team announced 100 million loaded photos. By November 2013, users loaded 16 billion photos. Currently, Instagram belongs to the Facebook corporation.

SMM-experts conducted the research, in which they allocated the gender ratio of social media users in Russia. In Fig. 4, it is shown that the biggest share of users of social networking services is women.

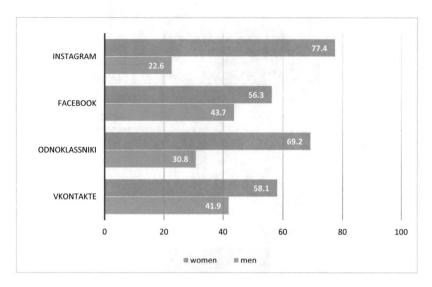

**Fig. 4.** The gender ratio of social media users in Russia (Source: compiled by the authors)

The World Tourism Organization UNWTO recognizes an important role of social media in tourism. More than a third of the international travelers gain access to social media sites by means of smartphones – they publish content in blogs, on pages in social media, recommend these or those places to visit, leave comments about locations while they are travelling [9].

Since 1995, the Internet market is on the rise. At the same time, tourist products occupy one of leading places on the sales volume through the Internet.

This tendency is sustainable in Europe. The number of tourists, who prefer to book in advance only flight and a hotel, increases. Other elements of the program are made on the spot, according to their tastes, a schedule and plans [13].

Consumers of products of the tourist market feel the need for live information: detailed descriptions, photos, videos, reviews. Advertising catalogues, core journals and newspapers, television and radio is not capable to fill this need. For this reason, social media are important for the tourist sphere. Besides, tourists appreciate their time and instead of spending several hours in a travel agency, they choose tours by themselves according to their own parameters on aggregator websites, tourist portals or social media.

Currently, there are two main ways of use of the Internet by tourist firms:

– Advertisement of tourist services and the firm image on its own or someone else's webpage;
– Sale of tourist services through web pages (Internet-store) [14].

There are specialized social media for the tourism sphere. In the Russian market they are:

– Foursquare - a social networking site with geolocation function, which is mostly designed for work with mobile devices. On the basis of his geolocation, an author checks-in` and submits a review about this place. For each check-in, the user gets points and achieves ranks.

- Tourister - a community of experienced travelers and guides in a format of diaries. Here it is possible to fill in a travel map, to keep your own blog, to get advice from tourism experts. Besides, the service users participate in cashback-programs. The monthly audience is 220 000 people.
- Enjourney combines all functions of a social networking service and a platform for blogging. There is no limitation on loading photos and videos. Here is a good atmosphere for communication, ratings of community members, and an opportunity to create your own route map.
- TourOut is one the first social media services in tourism. It enables to communicate, to upload photo reports and travel maps, to conduct thematic forums. Its feature is that except social loading it bears possibilities of aggregator websites. The audience is more than 500 000 people per month.
- Turbina.ru is one the largest Russian-language tourist community, which feature is a big amount of professional photographers and travelers among its members. The audience is more than 400 000 visitors per month [11].

## 4  Discussion

The international metasearch service momondo.ru conducted the research and found out that for the last five years online services for tourists and social networking sites became one of the most popular sources of information for travelers from Russia. Judging by the research results, for Russians publications in the content feed of social networking services are more important than advice from staff of travel agencies. The news feed, which is formed by social networking services by certain algorithms, became the most effective motivator and inspirer for 51% of young tourists between 18 and 22 years old. The need for social media also increases among people of the senior age group. This year the number of tourists of this age category, which plan their trips according to publications in social media, has doubled in comparison with 2014. It makes 34%. At the same time, interest in tourist information in social media grows more slowly among respondents of the middle age (36–55 years). In comparison with 2015 the difference made 7%, while only of 33% of the interviewed tourists of this age category choose social media as the source of information [10].

## 5  Conclusions

The top list of the most popular sources of information for travelers was made in the course of the research. Consequently, stories of friends (48%) are in the first place, the second place is shared by tourist websites (41%) and social media (40%), TV and radio (24%) and travel agencies (22%) – the most unpopular channels for obtaining information on travelling (Fig. 5).

In comparison with 2016, a bigger amount of users began to rely on blogs and travel applications while planning their vacation. At the same time, the number of fans of applications has grown from 8% to 16% within five years. Following the results of the poll, the amount of fans of travel blogs makes 16% (a 3 per cent increase). Young people between 18 and 22 years old (30%) prefer blogs, respondents of 23–35 years old and 36–55 years old look for information in mobile applications.

However, a universal tendency is that social media – an important part of any trip. "If you did not post anything – you were not there", - a popular slogan in the Internet. According to Mark Sharron, the President of TripAdvisor for Business, such involvement of social media into the travelling processes is called «a phenomenon of an independent traveler, who is always online» [3]. In his opinion, it is a chance for all representatives of the tourism and hospitality sphere for sharp rise, as it became much simpler to influence on potential clients.

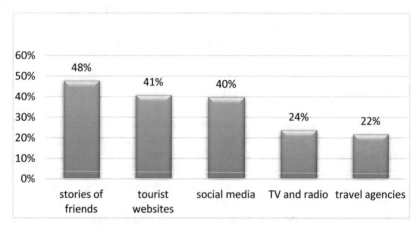

**Fig. 5.** The most popular sources of information for travelers (Source: compiled by the authors)

The portal TripBarometer1 conducted the research, devoted to the phenomenon of an independent traveler, who is always online. These studies showed how strongly the content, which was created by other users, influences the processes of planning a trip and booking. Comments on the website are of great importance for 65% of Russian tourists, who find inspiration while planning a trip in the Internet. Then follow ratings, which influence on the choice of 64% of compatriots [10].

The author conducted the survey research among potential clients of the travel agencies, who live in the territory of the Republic of Tatarstan. The age of the main part of respondents is 20–25 years old (78,6%). At the same time all the respondents in equal shares travel less than once a year, once a year and every six months (Fig. 6).

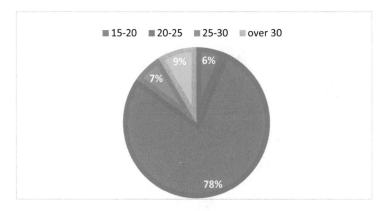

**Fig. 6.** The age of the respondents (Source: compiled by the authors)

According to the poll, the travel geography of the respondents extends to the territory of the Republic of Tatarstan and Russian destinations (47,8% and 63% respectively). Asia is visited by 28% of the respondents, Europe – 21%, Turkey – 17,4% (Fig. 7).

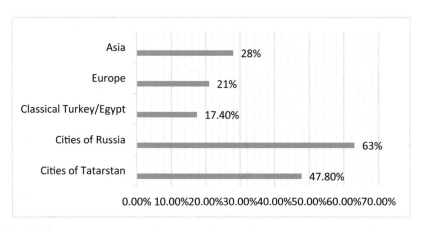

**Fig. 7.** The most popular regions according to the survey results (Source: compiled by the authors)

However, only 37% of the respondents use package tours. Others plan their vacation individually, using several options of obtaining information.

Figure 8 shows that in spite of the fact that the majority of the respondents book hotels, tickets, plan time and entertainments by themselves, the package last minute tours still arouse their interest.

In the course of the poll, the respondents distinguished several priority channels for

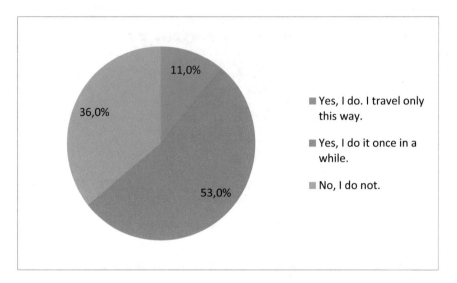

**Fig. 8.** Do you monitor last minute tours? (Source: compiled by the authors)

**Table 1.** The rating of the priority communication channels

| Communication channel | The amount of respondents, who are satisfied with communication channels, people (%) |
|---|---|
| Website of a travel agency | 11 (23,9) |
| VKontakte communitites | 29 (63) |
| Instagram accounts | 16 (34,8) |
| Telegram channels | 6 (13) |
| e-mailing | 2 (4,3) |
| Actually visit travel agencies | 14 (30,4) |

Source: compiled by the authors

obtaining tourist offers (Table 1). The table shows that visiting a travel agency does not hold the leading position, and it is not attractive to 70% of the respondents. It allows drawing a conclusion that integration of the Internet and social media into work of travel agencies in particular and into the tourism industry in general influences the overall involvement of the population into investment of the capital into the industry [6].

According to the statistical service "Wordstat. Yandex", about 35 273 people per month search for "tours from Kazan" [11]. It confirms the fact that the tourism industry gains popularity in the Internet. In the course of the poll, it was found out that the most preferable way of obtaining information on tours, on opportunities for trips, promotional offers is social media: the first one in the rating is VKontakte, then follows Instagram, the second place is taken by telegram bots, the third one is e-mailing.

In the course of the research, the author analyzed travel agencies, which are the most popular in social media according to the number of subscribers (Fig. 9). By

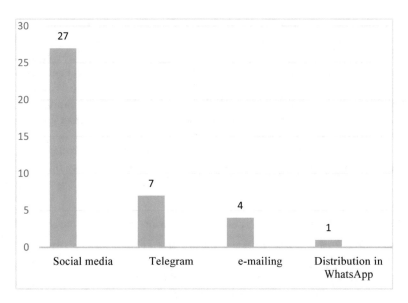

**Fig. 9.** The priority communication channels for obtaining information (Source: compiled by the authors)

request "tours Kazan" [11], the social networking service VKontakte gives top 3 communities:

- The travel agency «Four Season» – amount of subscribers: 37 900 people. There are 400 new subscribers each day. On average, 5 posts with offers are published in the community per day. The group has hashtag navigation for simplified use. It also has bright graphic content, and you have a possibility to leave request for an individual tour. The column "Contacts" contains valid numbers, and administrators of the community answer within 4 h. There is also distribution of offers in WhatsApp and Telegram.
- The travel agency «Tuzemets» – amount of subscribers: 34 104 people. The daily audience increase – 100 people. On average, there are published 11 posts with various offers each day. There is also good interaction with the audience and beautiful graphic content. The average response time – 40 min. The column "Contacts" contains valid numbers. Navigation is in the form of the clickable menu.
- The travel agency «Bounty Tours» – amount of subscribers: 27 106 people. The daily audience increase – 6 people. On average, there are published 20 posts with various offers each day. At the time of the research, the administrators of the group held a contest for audience capture. It means that they are familiar with SMM tools and actively implement them in work. There is no community navigation. The average response time – 15 min.

The situation with Instagram is similar: by the same request, the account TA "Bounty tours" has the biggest number of subscribers—26,2 thousand people, the account Pegas tours – 24,9 thousand subscribers. On average, three offers are published in the account "Bounty tours" per day. There is a reference to the chat in WhatsApp and the phone number of the agency. The account Pegas tours is absolutely similar.

These figures show that interactive planning of vacation is relevant both for audience of VKontakte, and for users of Instagram. However, travel agencies put a lot of effort into interaction with clients in the social networking service VKontakte.

Nevertheless, if we turn to the conducted research, 68% of the respondents do not consider social media as a convenient way to search for favorable offers for vacation for several reasons:

Social media become one of the most popular channels for promotion of services, including the travel ones (Table 2). The share of social media among interaction tools of representatives of the tourism industry with potential clients rapidly grows from year

**Table 2.** Refusal of social media as a communication channel with travel agencies

| The reason for refusal of social media as a channel of acquisition of travel offers | The amount of respondents, who gave the reason, % |
|---|---|
| The graphic content of communities distracts from offers | 15 |
| It is impossible to follow new offers without independent monitoring of numerous groups | 45 |
| Glut of the news feed resulting in loss of needed publications | 12 |
| Rare updating of the content | 28 |

Source: compiled by the authors

to year [2]. Currently, 4 social networking services take the leading positions in Tatarstan: VKontakte, Facebook, Instagram and Odnoklassniki. They are united by similar forming algorithms of the news feed and by the diversity of the generated content. The feature of Instagram is its graphic content, which strongly influences on motivation of tourists. Except for classical social media, there are specialized tourist social networking services such as Tourister, Forsquare, TourOut, Enjourney, Turbina. ru, which contribute to development of tourism in the Internet. In the course of the research it was found out that the most popular sources for obtaining information from travel agencies are communities in VKontakte and accounts in Instagram.

# References

1. Bernd W Wirtz et al (2010) Strategic development of business models. Long Range Plan 43 (2–3):272–290
2. Fakhrutdinova LR, Syradoev DV, Terehova TA, Antonova NV (2015) Organizational forms and methods of production management control. Mediterr J Soc Sci 6(3):727–773
3. Forbes magazine (2017) http://www.forbes.ru/mneniya/. Accessed 25 Mai 2017
4. Gitomer J (2012) Business in social media. How to sell, lead and win. Saint-Petersburg, Piter
5. Interview of Mark Sharron President of TripAdvisor for Business (2016) http://www. hotelstop.ru/. Accessed 31 Mar 2017

6. Khalilov D (2016) Marketing in social media. Mann Ivanov and Ferber, Moscow
7. Krainova OS (2014) Innovation marketing tools in promotion of tourism enterprises: spare capacity of social media. In: Materials of the conference "Tourism and service industry: condition, problems, efficiency, innovations", pp 22–35
8. Kuzmin O (2014) Social media «Vkontakte». RBC References. https://ria.ru/spravka/20140916/1024355414.html/. Accessed 27 Mai 2017
9. Novruzova E (2017) Social media in Russia, summer 2017: figures and trends. Brand Analytics. http://blog.br-analytics.ru/sotsialnye-seti-v-rossii-leto-2017-tsifry-i-trendy/. Accessed 30 Mai 2017
10. Popova AI (2014) Role of social media marketing in the tourism industry. In: Modern directions of theoretical and applied researches, pp 21–23
11. Portal TripBarometer1 https://www.tripadvisor.com/TripAdvisorInsights/w783/. Accessed 1 Aug 2017
12. Shabalina SA, Shabalin ES (2017) Innovations in the tourism sphere: neural networks in development of the regional tourism sphere. In:Tourism as a factor of modernization of economy and development of Russian regions. Yoshkar Ola, pp 67–69
13. Smirnova I (2017) Social media for tourists and travelers. https://rina-rina.ru/articles/socialnyx-setej-dlya-turistov-bolshoj-obzor/. Accessed 10 Mar 2017
14. Statistical service «Wordstat. Yandex» (2016) https://wordstat.yandex.ru/. Accessed 25 Mar 2017

# Model of the Internet Escrow's Legal Regulation as a Factor of Efficiency of Its Use in E-Commerce

M. A. Tokmakov[(✉)] [iD]

Samara State University of Economics, Samara, Russia
maxim.tokmakov@gmail.com

**Abstract.** The global e-commerce market continues growing year by year. Alongside, low confidence of consumers in sellers on the Internet and electronic platforms remains one of the key issues predetermining relevance of the researches focused on looking for the tools to solve the issue of confidence between the e-commerce participants. Internet escrow is among those tools. In contemporary literature, the focus is mainly on the economic assessment of its use, while the efficiency of using the Internet escrow depends largely on the legal regulation in the relevant legal system. The research deals with the models of the Internet escrow's legal regulation in the USA, where it has been settled long enough, and in the Russian Federation, where the legal and regulatory base in the relevant area is only at the initial stage of development. The aforesaid analysis renders it possible for us to identify the major "strong" properties of escrow, providing for efficiency of its use in e-commerce. The characteristics of various legal escrow models as presented in the research can be used as a criterion for choosing the escrow agent from among both the Russian and international companies offering the relevant services.

**Keywords:** Beneficiary · Confidence · Depositor · E-commerce
Escrow account · Escrow agent · Internet escrow · Legal model
Nominal account

## 1 Introduction

The size of the e-commerce market goes on growing worldwide steadily over the past few years. According to the analytical agency We Are Social, contained in 2018 Digital Yearbook [1], being the report on the global digital market, the total amount of the global e-commerce market for consumer goods in 2017 increased by 16% and made almost USD 1.5 trillion. At the same time, the number of people around the world using the e-commerce platforms grew by 8%, and made almost 1.8 bln persons.

The Russian e-commerce market is even slightly ahead of the world average rates. Thus, according to the forecasts of the research agency Data Insight [16], the growth of the domestic Russian online sales of tangible goods in 2018 will be 18%, and the market size will go up to RUB 1.115 billion (the market growth in 2017 was 18%, in 2016: 23%).

S. Ashmarina et al. (Eds.): *Digital Transformation of the Economy: Challenges, Trends and New Opportunities*, pp. 368–375, 2019.
https://doi.org/10.1007/978-3-030-11367-4_36

Alongside, the consumer confidence in the Internet sellers and electronic platforms remains to be one of the key factors influencing development of e-commerce. The public opinion poll concerning confidence in e-commerce services as carried out by the Regional Public Organization "Internet Technologies Centre" (ROCIT) in early 2018 showed that 63% of respondents chose that they had fear for their security among the reasons for refusing to make purchases on the Internet, as well the major fears were associated with the possibility of the bank card or e-wallet data theft (80%) or with the shop's denial to deliver the already pre-paid goods (80%) [12]. In this light, examination of the tools solving the issue of confidence in e-commerce is rather relevant.

One of the above tools is the Internet escrow, enabling to secure transactions between the network users. The funds of the parties to the obligation are placed under control of an independent third party (escrow agent), which holds and releases them only after the obligation has been completed and both the parties are satisfied.

The escrow agent checks the funds transferred by the buyer and keeps them on the special escrow account. The seller sends the goods only after the buyer has made the full payment to the escrow account. The escrow agent tracks the goods to make sure they are delivered on time. The seller does not receive the payment until the buyer accepts the goods and is able to check them comprehensively. As soon as the buyer has checked and accepted the goods, the funds from the escrow account will be transferred to the seller. If the goods fail to arrive to the buyer within the time specified or they have got defects, the funds will be returned to the buyer after sending the goods back to the seller. During the above period, none of the parties is entitled to dispose of the funds on the escrow account.

The aforesaid services become more and more popular all around the world in recent times. Among the largest global players, one can distinguish Escrow.com, a company which as of 2017 reached the amount of USD 2 billion in secured transactions [11]. Successful use of the Internet escrow in Russia would contribute to further development of e-commerce through reducing the risks on the part of its participants, as well as through the overall enhancement of confidence in Internet transactions. However, it requires, above all, confidence particularly in the applicable model of escrow, which might be ensured by the transparent legislative regulation and high level of professionalism of the escrow agents providing the relevant services. Otherwise, the Internet escrow's use can achieve an adverse effect, instead of the positive one, by becoming another fraudulent tool.

There has been almost no escrow legal regulation in Russia until recent times. Some regulations concerning escrow accounts were fragmentary introduced into the Russian law as late as in 2014; and the regulations in the Civil Code of the Russian Federation governing the escrow agreement took the effect only from 01.06.2018. However, the features of rendering escrow services on the Internet have not been separately recorded in the legislative regulation of the Russian Federation yet. With that, the above situation was not a hindrance for domestic companies to offer sundry services of the secured or protected transactions on the Internet actively: SafeCrow, Escrow.WebMoney, "Safe Deal" from Yandex, "Guaranteed Settlement Service" from Sberbank and some others.

Within the framework of globalization of the world, especially through the Internet, competitiveness of the legal system may become crucial for choosing the most reliable escrow agent, and, as a consequence, become a factor in formation of the unique "safe havens" for global players of the Internet escrow services. It is not too much of a leap to imagine that the international e-commerce participants will opt for, when choosing a mediator in their transactions, the escrow agent, the activity of which is settled as transparently as possible by the state, where such agent operates, and the legislative model of which allows the escrow agent's clients to be protected, to the greatest degree, from potential abuses of that agent.

It appears that assessment of such models is vital for building an efficient and competitive escrow structure in Russia. In this regard, it may be of relevance to assess the Russian escrow services market's conditions as for its compliance with the Russian law in this area, which law has emerged almost from scratch, and also to compare it with the escrow models on the Internet in other legal systems, which models are being successfully implemented internationally.

## 2   Materials and Methods

The major research method is the comparative-legal analysis of the US law, where the escrow, including on the Internet, has been settled long enough and with most details, thereby making it possible to successfully use escrow in e-commerce, and the law of the Russian Federation regarding escrow, the formation of which has just begun. The aforesaid analysis enables identification of the key "strong" properties/features of escrow, determining its active use in the USA.

Through the prism of the US Internet escrow legal regulation model, the research deals with the Russian law in this area, identifies similar and missing elements. Further, the existing Internet escrow services in Russia are assessed in the context of the identified major structural characteristics both in the Russian and US legal systems.

From a practical point of view, the characteristics of various escrow legal models presented in the research can be used as a criterion for choosing an escrow agent from among both the Russian and international companies offering the relevant services.

## 3   Results

### 3.1   Internet Escrow in the USA

One of the most successful escrow models is the model developed in the United States or more precisely in the State of California. Other countries frequently take the rules proposed by the Californian lawmakers as a basis for their own legislation. The Californian escrow law (the so-called Escrow Law) means the regulations contained in the Division 6 (commencing with Section 17000) of the California Financial Code and Subchapter 9, Title 10, California Code of Regulations commencing with Section 1700 (10 C.C.R. § 1700, et seq.).

The active regulation of Internet escrow in the State of California began as early as in 2000, when the number of complaints in the area of e-commerce increased significantly and the regulators realized that Internet escrow companies could reduce the risks of consumers ordering goods and services via the Internet.

The logical conclusion was that all the Internet escrow companies had to follow the rules prescribed for ordinary escrow agents [5]. However, the Californian law still contains some clarifications with regard to the Internet escrow. Thus, in accordance with Section 17004.5 of the California Financial Code, the Internet escrow agent means an entity carrying out escrow transactions via the Internet.

Paragraph 1737.1 of the California State Code (§ 1737.1. Internet Escrow Agent Special Accounts) prescribes the basic rules for regulating the special accounts as used by the Internet escrow agents in their activity.

1. The deposited funds shall be placed, within the following working day, by the escrow agent to the special bank account (trust account or escrow account). Only the deposited funds can be available on that account.
2. The Internet escrow agent may not dispose of the deposited funds before placing them to the bank trust or escrow account.
3. The Internet escrow agent shall create the reserve fund in the minimal amount of 10% from the average monthly receipts to cover the losses that may be caused by use of credit cards, including the chargebacks, fraudulent practices, etc.

Besides, it should be noted that escrow services in California are subject to mandatory licensing by the Department of Business Oversight. Section 17210 of the California Financial Code sets the minimum capital of an escrow agent to be no less than USD 50 000, including USD 25 000 in liquid assets. Moreover, in order to obtain the license, an escrow agent shall provide the pledge or guarantee amounting from USD 25 000 to 50 000 in favour of the Department of Business Oversight and also insure liability for at least USD 125 000 against damages by fraudulent or wrongful actions of any official or "trustee" of the escrow agent.

Among other things, the requirements have also been made for the escrow agent's employees: they all have to pass through tests as for previous involvement in fraud, embezzlement or misappropriation of property. Further, the escrow-agent's manager should have minimum 5 years of experience in the escrow area.

The escrow agent's activity is still supervised by the Department of Business Oversight following issuance of the license: the annual reports of the escrow agent, as well as the regulatory licensing examination once per 2–4 years are required.

Naturally, the above strict licensing requirements to the escrow agent contribute to increase of confidence both in it and in the escrow system as a whole.

The escrow agent is a professional participant of the escrow services market and is subject to compulsory licensing and license control, therefore it has the incremented responsibilities in relation to its customers corresponding to this level (more detailed examination was made by Tokmakov, 2017 [14]). It appears that particularly the aforesaid approach of the American law to the escrow agent has determined, to a large extent, efficiency and relevance of escrow both in e-commerce and in other areas of the economy. Incidentally, the official website of the Californian Department of Business Oversight informs, in free access, about all the Internet escrow companies licensed to provide online escrow services in the State of California: there are only three of them (including the above-mentioned Escrow.com)! And this point emphasizes not monopolization of the Internet escrow market, but the regulator's thorough and careful selection of the entities rendering such services in order to prevent the possible negative consequences for consumers.

Thus, the US Internet escrow model's efficiency is achieved through the following key components:

1. The escrow agent's activity is subject to licensing. The licensing requirements contain the incremented financial and professional criteria.
2. Escrowing of funds is achieved through entrance into a trilateral agreement between all the parties to the transaction.
3. The funds are deposited on a special account in the escrow agent's bank.
4. The deposited funds have a separate regime: none of the parties (including the escrow agent) is entitled to dispose of the funds before the conditions stipulated in the contract are met and the funds cannot be foreclosed under the claims of creditors of any party.

## 3.2   Internet Escrow in Russia

Analysis of the Russian law renders it possible to conclude that at present there is no special legal and regulatory base for Internet escrow yet. There are not any requirements for the escrow agent among the general regulations concerning escrow either, which fact is, to a large extent, based on absence of the exclusive list of the entities authorized to provide the escrow agent services.

However, with regard to the Internet escrow, there are still some provisions that can be mentioned. The Civil Code of the Russian Federation acknowledges the escrow agreement as an agreement, whereunder one party (the depositor) undertakes to deposit the property to the escrow agent, and the escrow agent is obliged to keep it and transfer it to the other party (the beneficiary) provide that certain conditions are met (clause 1 of the Article 926.1 of CC of RF). Alongside, the escrow agreement is the trilateral one and it shall indicate, as an essential condition, the depositing term, which may not exceed five years.

One can deposit movables, non-cash funds and non-documentary securities (clause 3 of the Article 926.1 of CC of RF). With that, the deposited property shall be separated from the property of the escrow agent (clause 1 of the Article 926.4 of CC of RF). None of the parties is entitled to use and dispose of the deposited property.

The deposited non-cash funds are separated through opening the special bank accounts: escrow account, if the escrow agent is a bank, or nominal account, if the escrow agent is not a bank (clause 3 of the Article 926.7 of CC of RF). Such use of the special accounts ensures safety of funds against foreclosure, arrest or interim measures in respect of the escrow agent's or depositor's debts. Thus, it can be stated that the Russian escrow model has generally perceived the positive global experience. Most of the above escrow efficiency criteria (three of four) are also peculiar for the Russian law. However, it is obvious that probably the most principal and strategic criterion, licensing of escrow agents, is not available.

Under the Russian law, currently the escrow services can be provided by any entity without any additional requirements. This situation has emerged particularly at the same time with commencement of the new regulations of the Civil Code of the Russian Federation concerning the escrow activity from 01.06.2018. Before that time, only the banks within the bounds of the escrow account's agreements could act as an escrow agent. Naturally, such closeness in the circle of escrow agents was out of line with the global experience or with the essence of the escrow's structure, which fact was repeatedly criticized [10]. However, the way this problem is solved today, as it appears, raises even more questions. For successful development of the escrow services market in Russia, this obvious gap in regulation of the escrow agents' activity shall be eliminated as soon as possible. Escrowing as designed to be a reliable tool on reduction of risks and intensification of confidence can on the contrary become, if the subjects acting as escrow agents are chosen wrongfully, a tool replicating the risks and undermining confidence in the entire escrow institution.

Nevertheless, even without taking into account the incremented requirements to the escrow agent on the part of the Russian legislator, it is possible to form the criteria showing the Russian model of the Internet escrow:

1. Escrowing of funds is achieved through the trilateral escrow agreement between all the parties to the transaction (escrow agent, depositor and beneficiary).
2. The escrow agent (if it is not a bank) shall open a nominal account in the bank in its name and indicate the depositor and the beneficiary under the escrow agreement as beneficiaries to the nominal account.
3. Such placement of funds on the nominal account forms a separate regime for them: none of the parties (including the escrow agent) is entitled to dispose of them before the conditions stipulated in the contract are met and the funds cannot be foreclosed under the claims of creditors of any party either.

Only the above structure, combining all the indicated features, can give according to Russian law the legal effect, which is peculiar for escrowing.

Assessment of the escrow services offered by the Russian Internet services at the e-commerce market leads to the conclusion that none of them (of which we know, at least) is in line with the developed escrow model. In most cases, the deposited funds are placed not on the nominal accounts, but on the ordinary settlement accounts of the companies; the escrow agreement is substituted for the service agreements and contracts of agency, etc. With this approach, formally, references to provision of escrow services on the web-sites of such Internet services and as a whole usage of the term "escrow" can mislead consumers, which hardly matches the good faith. Such a situation may undermine confidence not only in those who offer the said services, but, even worse, in the area of escrow services, which is just emerging in Russia.

# 4   Discussion

One should mention a lack of attention to the problems associated with the Internet escrow, not only from the legislator, but also from the science and professional community. A lot of works in the Russian legal literature are devoted to studying escrow, however they, as the institution is young, raise the escrow issues in general and are not focused on a separate examination of the Internet escrow [3, 15]. The American legal literature has the similar situation, oddly enough [4, 8].

At the same time, an active discussion of the Internet escrow's use in e-commerce is held among the economists. Models of the escrow efficient use at online auctions [7, 9] and strategies to form confidence in online platforms [6] are being developed; the impact of using escrow on intensification of confidence and reduction of risks of the e-commerce participants is estimated [2, 13].

However, it seems that efficiency of using the Internet escrow largely depends not only on economic factors, but also on the escrow's legal regulation in the relevant legal system. The properly selected legal escrow model may meet the needs of the e-commerce participants, take into account the balance of their interests and solve the issue of confidence, thereby encouraging development of e-commerce and acquiring a specific economic effect.

## 5   Conclusions

The tools, creating the safe conditions of property exchange for participants, solving the issues of fear of the first performance and of various abuses subject to practical impersonality of the counterparty on the Internet, are among the key factors for development of e-commerce. Internet escrow is one of those tools.

However, the Internet escrow can cope with the problems in a successful and effective manner, provided that only there is a properly developed model of legal regulation in the relevant legal system. The legislator's gaps may invoke a lack of interest on the part of e-commerce participants in the proposed escrow services, or choice preference in favour of the escrow services regulated by another legal system, being a certain economic threat to the relevant state. It is completely inacceptable for Russia under the conditions of the global competition and sanctions policy.

## References

1. 2018 Digital Yearbook. https://wearesocial.com/blog/2018/01/global-digital-report-2018
2. Antony S, Lin Z, Xu B (2006) Determinants of escrow service adoption in consumer-to-consumer online auction market: an experimental study. Decis Support Syst 42(3):1889–1900. https://doi.org/10.1016/j.dss.2006.04.012
3. Bashkatov ML (2013) Legal structure of escrow transactions: problem statement. M-Logos. https://goo.gl/4gJBDU
4. Bhagat S, Klasa S, Litov LP (2016) The use of escrow contracts in acquisition agreements. http://dx.doi.org/10.2139/ssrn.2271394
5. California Escrow Law Protects Online Consumers (2000) News Release 09-13-2000. http://www.dbo.ca.gov/Licensees/Escrow_Law/nr0013.asp
6. Delina R, Drab R (2009) Trust building on electronic marketplaces in the field of escrow services and online dispute resolution. In: Idimt-2009: system and humans, a complex relationship, vol 29, pp 149–156

7. Hu X, Lin Z, Whinston AB, Zhang H (2004) Hope or hype: on the viability of escrow services as trusted third parties in online auction environments. Inf Syst Res INFORMS. https://doi.org/10.1287/isre.1040.0027 (Inst Oper Res Manag Sci)
8. Huber W (2005) Escrow 1: an introduction. Educational Textbook Company, 530 p
9. Nian LC (2006) The viability of Online Escrow Services (OES) in online exchanges. In: Duserick FG (ed) Fifth wuhan international conference on e-business. Integration and innovation through measurement and management, vol 1–3. ALFRED UNIV, pp 439–445
10. Medentseva EV, Tokmakov MA (2017) Escrow: international experience and perspectives of application in Russia. J Adv Res Law Econ 3(25):906–909. https://doi.org/10.14505/jarle.v8.3(25).26 (vol VIII, Summer)
11. PayPal v. Escrow Services: Efficiency and Safety by the Numbers (2018) Byte technology. https://byte-technology.com/blog/paypal-v-escrow-services-efficiency-and-safety-by-thenumbers/. Accessed 1 Apr 2018
12. Rybakova O (2018) Methodological materials for online stores on interaction with citizens. http://files.runet-id.com/2018/rif/presentations/19apr.rif18-zal-2.15-30–ribakova.pdf
13. Shim S, Lee B (2010) An economic model of optimal fraud control and the aftermarket for security services in online marketplaces. Electron Commer Res Appl 9(5):435–445. https://doi.org/10.1016/j.elerap.2009.08.003
14. Tokmakov MA (2017) Peculiarities of the escrow agent's legal status under common law of the USA. Issues of economic and law, № 3, pp 19–22
15. Vasilevskaya LY (2016) Escrow account agreement: legal qualification issues. Electronic Supplement to the "Russian juridical journal". https://goo.gl/RxAbKQ. Accessed 1 Apr 2018
16. Virin F (2018) Online trading in Russia 2018. http://datainsight.ru/sites/default/files/DI-RIF2018.pdf

# Development of the Practice of Sharing Economy in the Communicative Information Environment of Modern Urban Communities

B. Nikitina[1([⊠])], M. Korsun[2], I. Sarbaeva[1], and V. Zvonovsky[1]

[1] Samara State University of Economics, Samara, Russia
belanik@yandex.ru, roxanna@mail.ru,
zvonovsky@gmail.com
[2] Samara National Research University, Samara, Russia
forpost@gmail.com

**Abstract.** In the last decade, the idea of sharing economy has spread widely in the world, inspiring people with the prospects of rapid and widespread practical realization of the values of the future: social solidarity, effective use of industrial and human resources, overcoming the phenomenon of capitalist alienation in the process of social and industrial production. At the same time, not the most progressive types of practices of the share economy get success in real life. A real alternative to the implementation of the ideas of the sharing economy became so-called "Uberization" as increase in the exploitation of workers instead of selfless participation of citizens in joint activities. At the same time, the development of the information and communication environment and the multiplication of the number of Internet communities naturally entail the growth of social solidarity and the strengthening of "weak" links in a wide variety of communities. The article investigates spontaneously emerging practices of donations and exchanges, which are carried out by users of a wide variety of virtual communities, identifies specific types of sharing economy on different web platforms, and examines the differences in the communication of modern citizens involved in the sharing economy, depending on the cultural software that is installed in a particular virtual community.

**Keywords:** Communicative information environment ·
Digital platform for sharing practices · Online community ·
Resource efficiency · Sharing economy · Social solidarity · Urban community

## 1 Introduction

The concept of "sharing economy" is a good theoretical construct that not only described innovative practices in modern society, but also stimulated their implementation. Such umbrella terms make a great contribution to public discussion, stimulates the imagination and encourages skepticism, promoting both deeper theoretical understanding of social changes and it's practical applications.

On the other hand, we agree with the criticism of this term by Wittel [22], who argues that it is impossible to study a term meaning too many different things. In this

S. Ashmarina et al. (Eds.): *Digital Transformation of the Economy: Challenges, Trends and New Opportunities*, pp. 376–394, 2019.
https://doi.org/10.1007/978-3-030-11367-4_37

regard, a recent study "Sharing what? "Sharing economy" in sociological discussions" [1] analyzing a variety of interpretations of "sharing economy" is of interest. Using a multidimensional scaling method, the authors presented the academic publications of the last decade devoted to this topic in the form of a thematic map. The map shows a great variety of context and interpretation of "sharing economy". The structured space of interpretations was divided into 4 parts, noticeably differing in the meanings invested in that concept. If one of the quadrants focuses on the regulatory side of the issue using terms such as "power", "regulator", "city", as well as the terms "sustainable" and "development", the other quarter of the field focuses on the transformation of the labor market where we can find words such as "work", "digital markets", "P2P", "trust" and, by the way, "Uber" and "Airbnb". The third quadrant analyzes the "sharing economy" as a new production/distribution/consumption model (expressed by words "consumers", "suppliers", "design", "access", "use"), and the fourth quadrant deals with the exchange as a form of social innovation using terms such as "social", "local", "community", "participation", "including", "sustainability" and "we".

Some authors focus attention on the essential differences in understanding of "sharing economy", as for example, R. Belk who identifies in his works two totally different types of "sharing economy" [2, 3]. (1) "True participation and sharing", which characterizes the majority of non-market exchange sites, within which the product is reinterpreted as a common resource. Based on this critical discourse, «sharing economy», or "participative economy", is a form of social exchange that takes place between people known to each other, without any profit. But to be inside that system people must establish their affiliation with the some community. Such a "sharing economy" should be called a community-oriented economy or a participatory economy. (2) "Sharing" can be understood as a process that expands the number of people who can use or benefit from the product; in this case, R. Belk describes "sharing economy" as a misleading term that actually identifies the market coordination of users/consumers and their activities in order to access the resource in exchange for a fee. In this interpretation, he was joined by Bardhi and Eckhardt, who spread this criticism, defining sharing practices, based on new information technologies as "access-based consumption" [4]. They regard such relationships as a form of a market-mediated transaction in which no transfer of ownership takes place. Rather, consumers pay for access to other people's products or services for a certain period of time. In an influential article in the Harvard Business Review [5], the authors emphasize that in this form the "collaborative economy" is not actually associated with a change in the type of social relationships, it's a business-oriented economy, and a transition process to this type we will recognize as "UBERization". In fact, such a model assumes the presence of a company that is an intermediary between consumers who do not know each other. This is commercial exchange, and consumers see it as utilitarian, rather than social values, which are paid for in the traditional capitalist regime [5].

J. Kennedy also points to different interpretations of the "sharing economy" when he says that the sharing economy can be viewed as (1) as a type of household, (2) a way of redistribution, and (3) a tool for creating communities [13]. And although different variations in the "sharing economy" can be distinguished into independent terms, she agrees with N. John that the boundary between social phenomena behind the term "sharing economy" is blurred [11, 12].

It would be better to accompany all debates on the issue the specification of the types of social interaction carried out and the type of actors (commercial, non-commercial, individual or institutional) in order to avoid misunderstanding of the reasoning about the sharing economy as a social phenomenon. However, we should accept that the complexity of the problem will not be overcome even if we focus on the actors involved in what we see as the practice of the sharing economy. The fact is that a set of exchange practices in a person is not homogeneous, but heterogeneous, and they can include both business-oriented and community-oriented actions as its shown in some recent articles [21].

Considering the habitus of those involved in the implementation of the sharing economy practice authors argue that visually similar sharing practices, which are episodic cases, may have different motives and triggers, and are not evidence that a person's lifestyle and habitus are changed towards a society with a new type of sociality. We also add that even actively using the sharing economy practices, the respondents are nevertheless included at the basic level in the economic structure of the existing post-industrial society, which is a society of overconsumption. This is due to the fact that the habitus of a person who is fully represent the "sharing economy" society has not yet been described, because it is not yet clear for now how this new sociality society, based on participatory practices, could look as a type of a socio-economic system. Apparently this is the peculiarity of modern practical social constructionism - defining certain social practices as revolutionary and breakthrough, the authors only outline possible options for their impact on the existing social reality, without predicting and not describing the total change in the system.

Some scientists point to the fact that every "sharing economy" practice are useful [21], doesn't matter if that social interactions flow from online space to offline or in the opposite way. But very often such communications are not aimed at either easing the boundaries between unequal social strata in community or reducing inequalities in society in general. On the other hand, in the process of such interactions there is an equalization of the level of material well-being in society, which provides opportunities for the gradual transition of the population to the value system of a post-industrial society oriented towards self-realization, more active social interaction and universalist values.

We agree with opinion that "these activities are having certain success mostly in Western countries" [7]. The hegemony of the idealized vision of "sharing economy" appears to have been established in more mature consumer societies. In general, the habit of a person of a more developed consumer society naturally implies a willingness to share the goods that are available in abundance. From this point of view, next we consider the "sharing economy" via objects, which is most popular to share. Some authors suggest to divide sharable resources into three groups: goods (food, material objects); assets (amenities, facilities, spaces) and services (knowledge, labor) [14]. This means that the listed types of resources have ceased to be scarce to such an extent that sharing them has become quite normal. As for food and material goods, people are ready to give these resources for free, but not for social utility, but of individual benefit associated with getting rid of un-necessary and even hindering further consumption of objects. However, from the point of view of the transfer of property rights, such practices should be characterized as true sharing methods, community oriented. Being a charity visually, in fact, such behavior hides consumer selfishness.

As for the second type of resources, which are more expensive and scarce - amenities, facilities, spaces, we should say that resources joint usage does not imply the transfer of property rights, but only gives temporary access rights—often with partial compensation. Sometimes such a short-term or inexpensive rental or free usage have an economical motives of care about some extra property. Most likely, the sharing of resources of the second type is less often completely community-oriented and more often involves the achievement of commercial benefits.

Situation with the exchange of intangible resources even more complex. Information flows, appearing for the use of people who are interested, can be useful both for the author of the initial information and for the local community. Useful information can be a source of increase in both individual and social efficiency, reducing the costs and costs of the individual and society, for example, information about traffic jams, weather surprises, utility problems, upcoming cultural events, etc. In this case, the information contributes to improving the efficiency of life activity in a particular place or region. Millions of videos on YouTube, the development of wiki-resources - all of these are examples of free information services, the motivation for which can be both commercial and non-commercial, or be mixed. Solving complex tasks on crowd-sourcing platforms also brings more moral than commercial satisfaction to participants, but definitely develops their human capital, i.e. brings them undoubted individual benefits.

General growth of public good as a target and side effect of people's life which is difficult to define as a job, household behavior or hobby, turns into a trend that appeared due to the wide spread of a variety of sharing practices. That looks very close to Marx's idea of the collective use of resources, if interpret it correctly, without usual stereotypical mistakes [18]. Marx interpreted the socialization of production as the use by many people of the same common resource, which was not exhausted, but increased in the process of use. This kind of resource can only be knowledge, understanding, as well as creativity, which allows achieving socially significant results without overexploitation of material and human resources.

Despite the great importance of increasing creativity in the process of consumption of things and their production, more pronounced today is the trend of more complete and effective use of existing consumer value typical not only for socially-oriented initiatives, but also for business, where it succeeded. The development of various management systems such as «lean production», «just in time» etc. began to reduce production costs, and the development of the Internet and artificial intelligence became another powerful lever for intensification of industry. Emergence of various types of sharing (sharing of equipment, space, energy, and even labor) actually became only a new stage in the development of capitalist production, the purpose of which is to maximize profits.

Simultaneously consumers also began to respond on economic crises by rationalization of consumption and cost reduction, became much more selective, economical and reasonable. That could be viewed as a signal of a decrease in their level of well-being, but can be viewed as a change in the way of thinking associated with the phase transition of industrial society to a different stage, which is described in R. Inglehart/

K. Welzel concept of value modernization [10]. Within the framework of this concept, the system of values in society gradually changes depending on the level of satisfaction of basic needs and sustainability of well-being. Survival values recede into the background, while universalist values come to the fore: existential values, values of self-expression and belonging to a group.

As for Russia, it is a society with a predominance of materialistic values, whose population judging by the results of Ingleheart's research still feels itself in economic and social insecurity, the "sharing economy" is unlikely will develop rapidly and in the same forms as in the countries with a developed consumer society.

Realizing that the Russian Federation does not belong to the countries with the most mature capitalist economy, but rather the opposite - it has many features that cast it to the previous stage of initial capitalism, however, we would like to consider the nature of the penetration and dissemination of the idea and practice of "sharing economy" in Russia, taking into account the limitations of our country

In this paper, we proceed from the following features: (a) Modern Russian society is limited in its political self-expression, as well as in public activity, expressed publicly. The existing attempt to activate grassroots social activity often creates fake activities. Any public activity is immediately assigned to its results by the state. (b) Despite the fact that modernization is tight at the level of infrastructure [roads, sewage, electricity] at the level of the individual, it is fast, as consumption has led to the introduction of innovative technologies in life, which has sneaked in with a development of capitalism. This was reflected in both overconsumption and the penetration of digital technology. (c) Generations of the end of the 20th - beginning of the 21st century did not know hunger and an extreme degree of need. According to Inglehart's theory, they have a more postmaterialistic value system. In addition the emergence and wide spread of the Internet has changed not only the amount of available information, but also influenced the deeper foundations of society, including the system of values. (d) Here we should recall the theory of sociologist M. McLuhan that a change in the prevailing social media and communication entails a change in the type of society [17]. (e) Aggressive propaganda plays a "repulsive" role for consumers of TV content, who, in a situation of expanding possibilities of accessing the Internet, successfully use it. A special role was played by a breakthrough in the transition from computers to smartphones, which democratized access to various Internet resources. It is obvious that such a simple way of accessing the Internet plays a dual role in the development of society. Layers appeared that cannot work on a computer - i.e. work with large texts, meaning both writing and reading, as well as editing and compilation, i.e. in some way, they lose the horizon of thinking, make it more immediate and narrow. On the other hand, these same groups are distracted from propaganda TV media, are much more involved in the selection of consumed information, as well as in network interaction. Thus, for them, earlier there comes a stage of greater individual choice, more active social communication. (f) No wonder that one of the first books of R. Inglehart was called "Silent Revolution" [9]. In Russia, judging by how much the regime has succeeded in suppressing opposition activity, a revolution is more possible at the grassroots level and not at all in the form in which it is expected by the authorities, arming special units for physical counteraction. (g) The revolution is likely to sneak through a consumer society, the existence of which is welcomed by the state, as it allows distracting the population from political activity. The consumer society in Russia has certainly developed. But in the near future serious social changes will happen associated with both economic constraints caused by the specific political conditions of Russia's existence, and with the change of generations and increase of the digital population share in society.

While scientists from developed countries are talking about how "sharing economy" as a whole bunch of innovative practices can make a difference at the political level, challenging capitalism [8, 15, 19], in the case of Russia, we would like to examine the situation in the opposite direction - can we see in the sharing economy practices the potential for grass-roots changes imperceptible to the authorities, which would help move the countries towards increasing solidarity of the population, quality his life without participation in this process of state institutions.

Going back to the core of "sharing economy" we must emphasize one more time that this is a system of social interaction aimed consciously on more efficient use of resources belonging to individuals, a group of individuals or a community, which is carried out to improve the quality of life of the population and its environment. Thus, the "sharing economy" is viewed by us mainly as a social phenomenon closely related to the development of local communities in modern industrial/ post-industrial society, which is especially important for Russia, a country that represents a specific arena for the development of innovative technologies in a desert landscape of social disunity and mistrust [23]. Development of grass-root practices that reinforce sociality and sustainability regardless of the political and economic situation is especially important for Russia in a situation where social initiatives visible to the state become objects of authoritarian control that inhibits their development.

In connection with a fairly basic level of Russian sociality, we will consider mainly simple sharing practices, which are accessible for implementation, but go beyond the economy of collective consumption of the past because of Internet mediation. We are supporting idea, that Internet activity is not just a means of communication at a distance, but also contributes to the development of interpersonal communication and civic activity [6]. We suggest that the very structure of Internet communities – so called "cultural software" [16] – actually provokes truthful interaction between users, when they got an inspiration with a new ideas of sharing practices.

In that discourse we want to keep our frame – to determine how the need for physical contact and the transfer of material objects from hand to hand, stimulates the unification of real and virtual communities in a single offline entity. How the idea of uniting people into virtual communities can contribute to the emergence of real off-line communities, make them qualitatively more perfect, strengthen social ties between people not only in the virtual space, but also in real life. On the other hand, we want to trace how this same idea, without being axial for the formation of territorial communities, penetrates into the interaction of their members in a natural way. We are interested in the following question: people who meet physically for exchanges and gifts, as a result of the existence of virtual communities that clearly mark their territoriality - are they a "local community".

In general, our task is to describe and study the spectrum of social practices of a sharing economy implemented by members of virtual and/ or real communities, which are aimed at (1) increasing the level of social integration and solidarity; (2) the intensification and optimization of the use of resources in the social, economic and environmental aspects, as well as the non-description of the sharing communities themselves. At the same time, we will try to identify the main types of social communities that exist today in the virtual or real social space of a modern Russian city.

## 2  Materials and Methods

The purpose of this study is to analyze the existing information and communication environment used by modern Russian citizens as a potential tool for the development of sharing economy within local communities.

To achieve the goal, the following tasks were set: (1) to identify and describe the most popular Internet platforms used to implement sharing practices in the modern Russian-speaking Internet sector, to conduct their analysis and typology; (2) to describe the practices of developing social solidarity and increasing the efficiency of using material and human resources associated with the development of online platforms for sharing economy in the modern Russian-language Internet space; (3) to determine the specifics of the information and communication environment of communities tied to the localities of the city of Samara and their difference from citywide online communities.

As mentioned above in the theoretical part of this article, communities that implement sharing practices have not just a network structure, but have the characteristics of a rhizome. Since the analysis of a rhizome can be started from any of its points, we began the search for practical examples describing the phenomena of a sharing economy in the virtual space of local communities from the most convenient entry point for researchers: a semantic series related to the practice of donation, exchange, joint activity. This strategy was implemented on the basis of the search engine Google, as one of the most popular in the Russian-speaking sector of the Internet.

The study has a 3-step character. In the framework of the first stage, the practice of sharing economy was considered as a donation of resources from person to person in order to use them most effectively. To search for thematic online communities focused on donations and the redistribution of things, we used the keywords "give it away", "give it away for free", "accept it as a gift", "give it away for free".

At the second stage of the study, Internet resources focused on more complex types of interaction, such as the exchange of resources, the offer of services in exchange for services, services in exchange for resources, or vice versa were considered. Accordingly, the second wave of the search was based on the phrase "free exchange", "exchange for free" "free exchange of services". As a result, we received a large number of websites on which not only the practices of the exchange of services were presented, but also donations of things - repeatedly. Since the results of the second query partially overlapped with the results of the first query, we took for analysis only non-repeating sites that appeared in the top list of queries for the above-mentioned key phrases. As expected, a significant proportion of selected Internet platforms for the implementation of sharing practices were based on social networks, where we reviewed not only the main pages, but also the most interesting and relevant threads of the discussion, as well as partner sites that were advertised by the administrators themselves.

Another large group of sites was represented by independent platforms created directly for the organization of exchange processes (https://avito.ru, https://vse-zadarma.ru, https://yaotdam.ru, https://barahla.net, https://vamotdam.ru/ https://www.bashnabash.org, https://otdamtak.ru, https://beru-davai.org, etc.). In the part of the list that the urban Internet forums were composed with thematic sections on the proposal of various gifts and exchanges, the most active and relevant groups were also selected by us for consideration.

Separate pages of major Internet forums related to the topic of motherhood and childhood (https://deti.mail.ru, https://babyblog.ru, https://baby.ru) appeared on our list quite often, which forced us to refer to their main pages to find out the general rules of interaction. And finally, let us point out that the Internet platform https://DaruDar.org, which was one of the first sharing communities in RU-Net, was studied by us separately, because it has a unique, very progressive and promising system of rules for user interaction.

The most recent at the time of viewing the ad on these Internet platforms were taken by us for description and analysis. In general, search queries found a relatively stable set of sharing sites with sufficiently high users activity. As for social network resources, it is necessary to note the colossal advantage of groups hosted on the "VKontakte" social network, unlike such social networks as "Odnoklassniki" and "Facebook", which have some similar groups but were far from our top-list. Here, of course, one should take into account that any pages and public places hosted on the "VKontakte" social network today have high priority in search queries due to the fact that the network is almost a universal information and communication space for a huge number of users of the Russian Internet.

Based on the fact that the virtual space of social networks, as well as the Internet space as a whole, has a rhizome structure, we followed the channels of the "VKontakte" network that looked thematically relevant and promising for our topic. Thus, it turned out that an additional search with the help of internal navigation through the most developed resources within the "VKontakte" network makes it possible to more clearly see both the subject area of sharing and the localization of sharing practices.

It was revealed that inside these pages there are links to friendly sites on the same social network that deal with similar sharing practices only in another geographical region or another thematic area.

Thematic Internet communities taken as basic for description and analysis were considered in various aspects. We described and analyzed (a) the rules proposed by the administration of Internet platforms (b) the degree of their compliance by users (c) the objects generating the non-commercial communication that we are studying, as well as (d) the nature of the interaction of the actors and (e) the "habitus" of the users, which predetermine their motivation to participate in sharing practices.

As for the third task of the study, it was focused on the search for sharing practices within stable virtual local communities in Samara. In the search for this kind of interaction, we looked at the largest and most active websites found at the requests of "Samara community", "Koshelev community" and "Uznyi gorod community". These communities, as it turned out, were again based primarily on the social network

"VKontakte". Another variant of Internet groups that duplicate real offline communities are the online groups called very strange - "Overheard". Such groups began to appear in the social network "VKontakte" about 5 years ago and today in almost any territorial or professional community such online mirrors exist.

Online groups that duplicate offline communities have more subscribers than online sharing communities, because their content is more diverse: here community members discuss local problems, post information about finds/ losses, ask for tips on how to solve a particular issue, relying on nearby resources. In such communities, the discussion takes place in the form of comments under the published post. Due to the fact that these communities have several tens of thousands of users and very high activity, we took only 4 groups, but in each of them about 500 messages were scanned, which turned out to be enough so that the subsequent message would not give an increase in the original information.

In the framework of the third stage of our study, not only the facts of donation and exchange, but also higher-level sharing practices were analyzed. These include "search and request for services" on demand "at the place of residence", "tool rental" and "rent of technical premises for sharing", practice of organizing joint activities to solve local problems, as well as information exchange between community members matters important for everyday life. We did not begin to search for sites that develop such social interaction, since such practices are rarely found in modern Russian society. However, we hoped that such examples could be found in virtual online communities in isolated cases.

In any case, the activities carried out according to the principle of "the sharing economy" are designed to increase social solidarity and the efficiency of resource use, which should benefit both the local residents and the environment. Therefore, it is important to seek and manifest such examples.

The research strategy described above allowed us to collect a vast array of data that can be considered as sufficient to solve the set tasks.

## 3   Results

Before presenting the results of the study, we recall that the search and selection of target Internet platforms took place for 1, 5 months [08–09. 2018], in the process of which it was noted a constant change in the list of top sites and their places in the ranking. Such dynamics is explained not only by the number of visits to these sites, but also by the policy of the software developer or Internet provider. Nevertheless, during the first stage of the study we were able to identify several main types of Internet resources, on the basis of which the sharing practices are implemented.

As the first type of Internet platforms, we consider commercial sites developed for the sale of second-hand items. These are such services as "Avito", "From hands to hands", "1000 boards" and similar virtual bulletin boards. In addition to the main array of announcements of sales, a section "Free market" was found on all the platforms listed, in which the subject subcategories "Animals", "Personal things", "For home and cottage", etc. are distinguished.

There are no separate or special rules for submitting ads to the free transactions section on these sites. Any visitor who wants to register must agree to follow the rules adopted on the site. These rules relate to compliance with state laws when placing ads [the absence of prohibited or inappropriate content in ads, particular privacy policies, etc.] and limit the responsibility of resource owners for conducting a transaction using their service.

The structure of these platforms is focused on the most effective interaction between the participants in the exchange: the rules for submitting ads are clearly described, and specially designed forms that are required to fill in; communication between participants of the exchange is possible only at a private level, it is impossible to comment on the announcements or ask clarifying questions. After transactions ads are marked by the author as closed and is stored in the history of the activity of the personal profile, but the identity of the recipient remains undisclosed.

Despite the presence of a sufficiently large segment of donated items on these sites, in essence, transactions here are not a form of charity, but are dictated by the desire to get rid of unclaimed things. And if in some cases people want to save themselves from the remorse of conscience associated with placing objects in good condition in the garbage, many just want to save themselves from transporting heavy, bulky and unnecessary items.

Web sites of the second type are represented by online platforms focused on non-commercial transactions. In fact, these resources are similar in structure and mechanics to virtual bulletin boards, but suggest the absence of a monetary equivalent for the goods being placed. In addition, the developers of these resources often post information about the purpose or "mission" of their creation to potential users. For example, the resource "All for nothing" and positions itself as "a site where a freebie is often and honestly distributed". To fill the site with content, its employees search for all possible options for free receipt of goods or services as part of ongoing advertising campaigns or product promotion. However, on the same site there is an opportunity to get something for free from people who want to give their things as a gift. The site also provides for very simple registration, but there are no warnings about personal data or references to the need to comply with laws and regulations. But there is an owner's attempt to describe changes in the world that allow modern people to meet their needs for free. One feels that the author is a bit naive and inexperienced in writing texts of this kind. However, the free gifts page works quite actively, includes several tens of thousands of ads, which are divided into more than 40 headings. As for most often free gifts here - there are a lot of animals, especially cubs, but various household items are also given away: home furniture, appliances, etc.

There are also resources, on the main page of which there is a categorical statement about the non-commercial nature of the interaction on the platform. These sites contain descriptions of the environmental and social benefits of a peer-to-peer economy. Some of them focus attention on saving nature from pollution, some on helping people, some calling for self-improvement and implementing morally positive actions. Of interest is the attempt to completely suppress any market relations, which is implemented on the website BeriDavai.ru. In particular, the site is completely forbidden to offer something in return or for money, it is forbidden to even indicate the money equivalent of the gift. In addition, the administration even points out the fact that they do not accept the economy of merit, stating that website intentionally does not keep records of ratings and statistics among donors. Somewhat more mundane is the motivation of users by otdamvam.com site administrators, who focus on the benefits of acquisitions or donations of unnecessary things. And vice versa - the Internet platform vamotdam.ru differs in that the administration in the rules immediately takes into account the possibility of charitable transfers to users of things both to public organizations and foundations.

However, despite the orderly structure of the posted ads, the ability to search for the right product by category or keyword, as well as the motivational component, was not found to be significant in these resources. Ads are rarely posted and do not find a response for a long time. In addition, the platforms do not allow potential participants in the exchange to discuss the features of the further exploitation of the «gift» or other aspects that are similar thematically. In some cases, commercial ads were found on such - presumably purely non-commercial sites. For example, there were cases when titles contain the phrase "giving away for free", but ad texts themself indicates the value of the goods in rubles. Sometime there were barefaced commercial ads.

Separate and completely exclusive in this series is the Internet platform DaruDar.org, which is also created as a base for posting gifts, but has a uniquely designed user interaction system. It is important that this project aims to actively introduce the practice of donation into everyday life and foster selflessness in the people around them. The ideology of this site is focused on making the world more open and friendly. This is especially relevant in the conditions of a low level of public confidence in various institutional practices, individualization of people's lifestyle, as well as the emergence of an increasing number of things and objects in everyday life. The site explains in detail the difference between a gift and a present. Presents are involuntary, they are foreseen for certain events, their list is limited, they require a reciprocal step. The gifts are selfless, have a free nature, they are not regulated in time, are spontaneous and do not require anything in return. This is an occasional celebration for that person who needs in particular things much more than donors.

The main difference of this service from those considered earlier is that gifts here must be valuable, although they have become useless to donors. In general, the format of interaction on this website is fundamentally different from the rules on other platforms: self-organization prevails here and plays an important role; participants themselves regulate problems through instruments of disapproval and public feedback about so called "accomplices" [special term used on that platform].

The service has its own traditions and code of conduct. Exchange in this project under strict prohibition, as it violates the very promise to disinterested communication. DaruDar is also different in that the donor can choose the donated person to whom his gift is most needed. To do this, you need to read the reasons in the comments and look at the profiles of those who wanted a gift. In other previously reviewed services, the process of selecting the recipient of a thing is more primitive: usually it is received by the one who first marked the item as liked, booked it in comments or personal messages. In this case, the thing can be given even to a user with deliberately false information about himself in the profile.

This project calls to express gratitude and to continue donating the received item so that a greater number of individuals can take advantage of it. A similar idea like no other expresses the principles of a circular economy and rational consumption, a reduction in the number of wastes and the development of a network of partnerships.

The next type of resources on which sharing practices are implemented is represented by city and parent forums. Practices of donating things or searching for the necessary arise here spontaneously, as a result of the concentration of people having the same interests. Children's things have always traditionally been transmitted within the family and between families. But in the face of declining marital and neighborhood communication, for some women, the emergence of Internet resources, where you can share not only information, but also things, has become very relevant. Things in such forums are often given to each other for free, but we have noted cases when users express their wishes not only to donate, but also they want to change things to something specific ("sweets", specific products that are desirable buy). Often, the proposed objects or requests are discussed, and also accompanied by a description of life situations, describes the features of how it is necessary and not necessary to use objects, their service life and other details. It is on forums of this kind that discussions are most often found with the aim of moral support, advice, and opinions. It expresses to a greater degree the principles of the «Sharing Economy», which consist in the fact that goods cease to be alienated from the seller, and the act of transfer becomes personalized. The reason for this is the fact that such sites grow in principle on information resources aimed at discussing issues that concern parents in connection with the process of raising children.

Note that the rules governing activities at such sites usually do not directly relate to issues related to the transfer of gifts or exchange. The ways of interaction are generated by the users themselves, who send their contact information to each other in different forms - in the form of email addresses, phone numbers and accounts in social networks. At the same time, the presence on these sites this implies acceptance of the rules of the site as a whole, which usually require complying with laws and ethical norms. There are some problems with that type of sites: specific forum-like format complicated a search of the desired product by keywords and practically excludes the possibility of categorizing the goods placed for exchange.

As a separate type of resources, we have identified thematic exchange groups in social networks (basically, the communities of the social network "VKontakte" were analyzed, the most relevant to the search query).

Each community organized on this social network has its own rules for posting, which include the requirements for the ad text and its design, as well as restrictions (for example, to place ads only from your own page on the social network, do not put links to the ad text and etc.). Each group describes the mechanism for selecting the recipient of the item: the advertiser may choose the person to whom he will pass the item, from those who have marked the advertisement as liked or made a repost on his page. The analysis showed that the larger the community, the more strictly regulated and formalized the exchange procedure. In the largest community there is no platform for communication between the donor and the recipient. The donor sends the advertisement to the administrator for consideration, after checking for compliance it is placed on the "wall" of the group on behalf of the group with the donor's signature, which cannot be written about their desire to receive this gift. Comments to the post are disabled. Prospective recipients may mark this record as liked ("like"), thereby expressing interest in acquiring this thing and deanonymizing to the donor, because "Likes" are personalized. After a certain period of time [most often, a day], the donor looks at the list of "likes" from the recipients and decides to whom to transfer the thing through personal communication with the recipient in personal messages. In smaller communities, it is possible to comment on the posting with the ad. This mechanism is used by recipients to clarify the presence of a thing, requests to "book" a thing up to a certain time, or to clarify the characteristics of a thing. However, donors often transfer communication into a private message space, directly communicating with the recipient who left a comment ("answered in a personal"). Also, this mechanism is used by the donors themselves, fixing the transfer of things in the comments ("taken away").

It should be noted that all the studied communities are much more focused on informing the donation of things than on their request. Only in local communities, we recorded announcements like "I will accept as a gift" or direct requests. Often, the very presentation of the site on the main page orients users to get rid of unnecessary junk, but not to redistribute resources as a manifestation of social solidarity ("After all, each of us has an unnecessary thing? Get rid of them!"). The main categories of products offered as "gifts" on these resources are children's and women's clothing, accessories, furniture and household items. Often you can see ads for animals donated, which are placed by activists, volunteers shelters for stray animals.

One of the important observations is the discovery of an obvious division of roles among the participants of the studied communities: cases of combining the roles of donor and recipient by one member of the community are extremely rare [the exception here is the platform DaruDar.org]. It is noticeable that donors are more affluent and high-status participants, the socio-economic situation of recipients is much worse. In fact, sharing practices are replaced by charity practices, when donors do not see themselves as potential recipients within a given community, and the role of the recipient is, by default, perceived as an object that benefits. Practitioners of asking for help with things or food can be called reactive, since the need for exchange arises when there is a sharp deterioration in financial status due to job loss, divorce, increased treatment costs or diseases, which he often briefly informs about in self-description. That way recipients trying to stimulates donations by articulating his request. Donors can also act in accordance with the reactive strategy; posting information about the items given away with the wording "will give to the needy".

Proactive exchange strategies are observed in communities where the socioeconomic status of participants and their values are mostly equal, or in those communities where exchange items do not belong to the category of goods that a person needs for survival (first of all, items related to hobbies).

Note that the situation is quite typical when one local Internet community engaged in the exchange and redistribution of things is divided into several more highly specialized ones. So, for example, the community "Give free/ Samara for free" has a whole pleiud of subcommunities "Samara free technique", "Samara free furniture", "Samara free Building Materials", "Samara free Tableware", "Samara's free Medicines" etc. Interaction in such groups is not very intense [about 5 ads per day], but they have a deeper communication of an equal nature, where the donor and the recipient feel mutually beneficial. Thus, people understand each other's motivation when they give away an unused portion of an expensive medicine or diapers, they often explain the reasons why the medicine was not used in full, discuss the conditions of storage and administration of drugs, etc. Passing each other building materials, people are less sociable, but pay more attention to practical issues of physical transportation of objects. In the community "Samara Free Tableware" recipients place requests for a gift of kitchen utensils, explaining their request not with a difficult financial situation and emotionally colored appeal to a difficult life situation, but with a calm explanation of the situation ("they gave us an apartment from the state, help with some dishes will accept unnecessary porcelain, earthenware, ceramic dishes for crafts in school"). Thus, it can be said that highly specialized communities for exchange demonstrate an atmosphere not of charity, but of social solidarity.

The second stage of the research involved the search for higher level sharing practices. As already mentioned, the search for communities was carried out using the keywords "free exchange", "exchange for free", "free exchange of services". The most relevant search results for these keywords largely coincided with the results of the first stage of the research search (virtual bulletin boards, VamOtdam.org platform). References to resources offering car exchanges or fake users accounts and "likes", which, in essence, are commercial marketplaces, were also excluded.

Among the remaining sites, books [fiction and educational literature] and exchange of services became the predominant content for exchange. The practice of exchanging books discovered as a result of this research search can hardly be called really sharing, since the books are given in one direction, their further fate is not of interest to those who give them, communication is practically absent. In fact, the main motivation for the group members is to get rid of unnecessary items, because supply in these segments significantly exceeds demand.

There are some specific with Internet platforms promoting the exchange of services. Here, groups organized on the basis of territorial communities have been especially active, as the exchange of services requires the physical presence of participants. As it turned out, most of the proposals in the groups for the exchange of services come from novice masters of the beauty industry. Offering free services in return for receiving something similar, beginner masters gain experience and reputation, and also develop professional connections. For example, a manicure master begins a collaboration with a makeup artist. Subsequently, both are likely to rely on joint projects to prepare clients for significant events, advertise each other to their clients and other bonuses from the interaction. Such requests can be designed for long-term inter-action, and not focus on a one-time benefit.

Offers from nannies, photographers, tutors and pastry chefs are also found in service exchange groups, but not so often. Sometimes participants are offered repairs of various kinds, tailoring, walking pets. Such services are already close to the category "work on demand", but being free, they look like more socially oriented and humane. In contrast to the simple exchange of things, which is carried out as a brief non-binding episode, the service implies a longer interaction time between actors, which include, in addition to the time for rendering the service itself, the stage of preparation and discussion of the results. In the process of this, it is precisely communication that takes shape and the first steps for further interaction, the formation of "weak" ties.

This group is interesting for us because the participants are trying to define their geographical community in order to make the exchange of plants and interaction easier and more enjoyable. Although these attempts do not look successful yet, we can record the fact that users of these groups are in demand for such a tool that would combine thematic and geographical proximity of users as a basis for the formation of the community

The third stage of the study was to discover the practice of a sharing economy in the activities of local communities that have mirror sites on the Internet. During the previous stages of the study, it became clear that the most active exchange processes happen where communication itself is most intense - on forums, in groups of local urban communities. In our study we investigated communities created for residents of remote areas of the city of Samara: "Koshelev Club", "Koshelev Project "Krutye Kluchi - Samara" and "Uzhnyi Gorod". Samara - the official group" as well as the online community "Overheard. Samara". The last group in the social net "VKontakte" does not have a clear link to any neighborhood, but has a large audience of residents of the city of Samara as a whole.

These communities are very active, dozens of ads from residents are published on the day, which only rarely remain without comment. Posts on the "wall" are very diverse. The most frequent are requests for information. An hour after a request is made about finding a master for the repair of residential premises 3–4 comments usually appear, where could be both: information about the organizations involved in the repair or direct answer from the masters who are ready to take on the job for a moderate payment. Often you can find questions about purchase of a high quality products: directly from local residents or in "reliable" stores. In the comments, residents often share their experience in solving different kind of problems or offer to unite to solve if problems are not individual (for example, to place barrier blocks on the sidewalk in order to prevent car parking in an inappropriate place).

Also we found exchange/donation practices that turned out to things which are unnecessary (building materials, children's products) in these communities. Once we met with post suggesting sharing a rental garage for storing sports and automotive equipment. Sometimes announcements about the loss/ finding of things or pets were posted in these communities which made a help. Considering the territorial remoteness of the districts, practices of searching for travel companions for work trips are being implemented in these communities, and the ads are posted by both resident drivers and resident fellow travelers. The described interactions can be both situational in nature (joint trip in the near future), and suggest the possibility of long-term interaction (joint trips on an ongoing basis).

Thus, it can be said that on the basis of such communities (originally not oriented to the practices of the sharing economy) processes spontaneously arise that have a direct relationship to it. At the same time, there is no separation between donors and recipients among residents, communication is carried out at the same status level. In addition, unlike specialized exchange groups, in these communities we can observe the practices of such information exchange, which enrich not only the immediate recipient, but also potential recipients, who also take advantage of network.

# 4   Discussion

Comparing the activity of users on Internet platforms created specifically for the exchange, and exchange practices that developed spontaneously on existing non-thematized Internet platforms [mainly in social networks], we came to the conclusion that the more structured the exchange process is, the less intensive the associated communication between the exchange participants. Thus, on platforms like vse-zadarma.ru, where the infrastructure is optimized for exchanges (there is a classifier of goods, a catalog search with the ability to set custom criteria, strict requirements for the form of an ad, etc.) there was no communication between the participants in the exchange (including the inability to ask clarifying questions, give advice, etc.). The situation with exchange practices arising spontaneously in local Internet communities is completely different: a posted announcement about the possibility of exchange or a request for exchange is often accompanied by comments containing tips, personal experience in similar situations, jokes or reprimand. At the same time, the communication carried out in the comments may contain links to other announcements about the exchange or even go beyond the originally designated topic, enriching (at least in terms of information) all communication participants, including outside observers who have not directly entered into the discussion. At the same time, it is practically impossible to effectively search for the necessary thing or a potential exchange partner, because the information is not structured, dynamic and variable. However, the search for a necessary thing or an exchange partner in this information field can bring pleasure by itself (akin to shopping without purchases) or encourage the implementation of sharing practices for those goods that were not originally considered by a person as suitable for exchange.

Thus, the analysis of exchange practices in the information and communication environment partly confirms the statements of the theory of McDonaldization expressed by G. Ritzer: an increase in formalized, rational practices in a given system reduces its level of humanism and orientation to the development of the person himself [20]. Although there are exceptions: among the practices we found, there were also those who, subject to the formal requirements for the absence of a search for commercial gain, were in essence commercial. So, on the described by us Internet platform "DaruDar" you can see users who received a gift much more than they gave themselves, for such users on the slang of this platform even the special name "vacuum cleaners" appeared.

It should also be noted that the effectiveness of sharing practices does not increase with an increase in the number of members of the themed Internet community or users of the Internet platform. During the analysis of the exchange communities in the social network "VKontakte", we found that the degree of formalization of communication and the rigidity of the rules are the higher, the more participants in the community. The most numerous communities are organized in the form of so-called "public", where any information that a community user would like to place is tested by a moderator who makes the final decision on publication. If a community is less numerous, it is most often organized in the form of a "group", where all participants have equal status and can post information on their own, moderation can be carried out immediately after publication. On the one hand, large communities use a form of "public" to ensure information security, since it is problematic to track the quality of content produced by a community of several tens of thousands of users. Often, the ability to place information on a resource that has a significant audience reach provokes some users to use these platforms for advertising or drawing attention to a particular problem, completely ignoring the original motivation. On the other hand, such a community organization demonstrates a lower level of trust of the community administration to the participants. More "chamber" thematic communities regulate to a lesser extent the content posted and the activities of the participants, which allow them to engage in dialogues and discussions, resulting in live communication between them. At the same time, the presence of common interests or values shared by the participants is the key to the adequacy of the information posted.

In general, it can be stated that "sharing" practices arise spontaneously and reinforce an atmosphere of social solidarity in a communicative-information environment, naturally formed by local communities. Examples include forums or communities in social networks created by people living in close territorial locations, sharing similar interests or engaging in similar practices. On the other hand, the actual environmental effect of reducing the amount of waste due to the optimal use of resources is made precisely by the narrow thematic communities that branched off from large, but less productive communities that emerged to implement the idea of a sharing economy.

It is also worth mentioning the serious differences in the habitual characteristics of the participants in groups, which manifest themselves both in different motivations for joining the sharing community, and in the differences in their role repertoire and intensity of participation. In the thematic exchange communities in social networks there are practically no permanent activists, "regulars", and the participants are active situational, until the issue of interest is resolved (give something up or find something necessary). The situation is different in the Internet mirrors of local urban communities (or communities of interest) - there are always the most interested members of the group who are actively involved in all sales, one might say, they "hold the box". Such participants often communicate with others, give advice or links to necessary information, or simply comment on posted ads, which can stimulate further discussion. Formally useless, such a position makes the community more "alive" and interesting for users. The weakest attempts to implement the idea of a peer economy can be attributed to specially created Internet platforms that are devoid of lively communication and close ties with their members, so they are poorly integrated into the daily practices of using the Internet environment. The activity of their users is as situational as that of members of the exchange communities in social networks, and the forms for implementing long-term communication practices by developers are not provided.

## 5  Conclusions

The results of the study showed that sharing practices have a wide variety of manifestations that are recorded in a communicative information environment. These practices were discovered by us on virtual bulletin boards, second-hand items platforms, subject and local forums, social networks. Sharing practices on the above-described Internet platforms are polarized: either the resource is created in accordance with the values of the sharing economy for the redistribution of goods and services, or the resource is created for other purposes, and sharing practices arise there independently during the communication of users.

The effectiveness of sharing practices also has a dual nature, since they combine both material exchange and emotionally colored social interaction. The study showed that the sharing practices recorded on the Internet can be represented in the form of a continuum, where practitioners are oriented at one end, oriented towards efficient material exchange and not involving close communication between participants in the exchange. On the other hand, there are practices focused both on communication and on information sharing, but thereby creating information noise, biding effective material exchange. However, such practices are more correlated with the values of a true community-oriented sharing economy.

Note that sharing practices, accompanied by emotionally colored social interaction, can contribute to increasing solidarity among community members, even those not included in the implementation of the principles of sharing. The intensification of the use of resources is realized by the participants of the exchange communities often regardless of the fact of the presence of personal communication between the participants.

To implement the most effective sharing practices, it is necessary to form convenient and easy-to-use platforms that allow the user to easily and timely satisfy their need for both the products exchange and the appropriate communicative support, as well as to satisfy the internal request for the implementation of universalist values. In the course of the study, we found that the quality of the communicative-information environment actively used by local communities in Samara has certain disadvantages. At the same time, its undoubted advantage is rather ample opportunities for the development of local communities independent of the state and business. These advantages include strengthening universalist values oriented towards a socially and environmentally friendly lifestyle, developing humanistic communication and strengthening weak ties, which, of course, will never become strong ties of the traditional and family type, but can support the population in the transition to a new, uncertain economy of the future, the outlines of which today excite the imagination, causing both fears and hope for the best.

**Acknowledgements.** The research conducted for this article is part of the project "Readiness of local communities to develop joint consumption and management of household waste through the development of information technologies as a strategic factor influencing the socio-economic development of Samara", funded by Competition of fundamental research conducted in 2018 by RFBR together with the Subjects of the Russian Federation, grant number 18-411-630003.

# References

1. Arcidiacono D, Gandini A, Pais I (2018) Sharing what? The 'sharing economy' in the sociological debate. Sociol Rev Monogr 66(2):275–288. https://doi.org/10.1177/0038026118758529

2. Belk R (2010) Sharing. J Consum Res 36(5):715–734
3. Belk R (2013) You are what you can access: sharing and collaborative consumption online. J Bus Res 67(8):1595–1600
4. Bardhi F, Eckhardt GM (2012) Access-based consumption: the case of car sharing. J Consum Res 39(4):881–898. https://doi.org/10.1086/666376
5. Eckhardt GM, Bardhi F (2015) The sharing economy isn't about sharing at all. https://hbr.org/2015/01/the-sharing-economy-isnt-about-sharing-at-all. Accessed 2 Apr 2018
6. Hampton K, Wellman B (2003) Neighboring in Netville. City Community 2(4):277–311
7. Herbert M, Collin-Lachaud I (2017) Collaborative practices and consumerist habitus: an analysis of the transformative mechanisms of collaborative consumption. Recherche et Applications en Marketing (English Edition) 32(1):40–60
8. Hult A, Bradley K (2017) Planning for sharing–Providing infrastructure for citizens to be makers and sharers. Plan Theory Pract 18(4):597–615
9. Inglehart R (1977) The silent revolution. Princeton University Press, Princeton
10. Inglehart R, Welzel C (2005) Modernization, cultural change, and democracy: the human development sequence. Cambridge University Press, Cambridge
11. John N (2012) Sharing and web 2.0: the emergence of a keyword. New Media Soc 15 (2):167–182
12. John N (2013) The social logics of sharing. Commun Rev 16:113–131
13. Kennedy J (2018) Theorising sharing practice in relation to digital culture. Media Int Aust 168(1):108–121
14. Kennedy J, Nansen B, Meese J, Wilken R, Kohn T, Arnold M (2017) Mapping the melbourne sharing economy. Networked Society Institute Research Paper, Melb, p 5
15. Loh P, Shear B (2015) Solidarity economy and community development: emerging cases in three Massachusetts cities. Community Dev 46(3):244–260
16. Manovich L (2013) Software takes command. Bloomsbury Academic, New York
17. McLuhan M (2011) Understanding media. Understanding Media (trans. from English 3rd ed. Kuchkovo Pole). Moscow
18. Mezhuev VM (2007) Marks against Marxism. Cultural Revolution. Moscow
19. Moon MJ (2017) Government-driven "sharing economy": lessons from the seoul metropolitan government. J Dev Soc 33(2):223–243. https://doi.org/10.1177/0169796X17710076
20. Ritzer G (1983) The "McDonaldization" of society. J Am Cult 6(1):100–107
21. Rufas A, Hine C (2018) Everyday connections between online and offline: imagining others and constructing community through local online initiatives. New Media Soc 20(10):3879–3897
22. Wittel A (2011) Qualities of sharing and their transformations in the digital age. Int Rev Inf Ethics 15:93–98
23. Zvonovskiy VB (2007) Daily interindividual impersonal trust as a factor in economic activity. World Russia 16(2):133–151

# The Internet of Things: Possibilities of Application in Intelligent Supply Chain Management

T. E. Evtodieva[1]([✉]) [ID], D. V. Chernova[1] [ID], N. V. Ivanova[1] [ID],
and J. Wirth[2]

[1] Samara State University of Economics, Samara, Russia
{evtodieva.t,nataliaivanova86}@yandex.ru,
danacher@rambler.ru
[2] HEG - Haute école de gestion Arc, Neuchâtel, Switzerland

**Abstract.** Topicality of the research based on the statement that modern Economy changes rapidly according to consumer's requirements and technology progress: informational technology, robotics, Internet technologies, business automatization, AR and VR technologies and others. These conditions called Digital Economy or Industry 4.0. It provides a huge amount of requirements to Supply Chain Management (SCM) by rising expectation of consumers to service level and delivery time. In this regard, this article aims to disclose an importance of the informational technologies in the sphere of logistics. That was noticed in the 1970's while an information and data rate become a competitive advantage. Since that time technology's progressed deeply and nowadays Intelligent systems transform a paradigm of business and Supply Chain. The leading approach to the study of this problem is synthesis of the different researches views and practice analysis allowing a comprehensive review of the actual informational technologies in SCM. The article presents challenges facing the SCM Industry, disclosed the Internet of Things in SCM definition, discovered that IoT-based SCM take a middle part between customer IoT and Industrial IoT. The materials of the article are of practical value for the Internet of Things applications in SCM.

**Keywords:** Digital economy · Industry 4.0 ·
Intelligent supply chain management · Intelligent systems · Internet of Things ·
Supply chain management

## 1 Introduction

Topicality of the theme approved by the modern SCM applications such as in the late 1980's it was a new paradigm for business, and nowadays intelligent supply chain management is able to reach the same goal such as reduce costs, improve profitability, and enable competitive advantage for organizations [18, 24, 25]. The problem is that increasing variety of goods and services and multi-channel and omni-channel sales has made it needed to forecast precisely to predict demand in supply chain. All aspects of SCM became more complicated and we shouldn't rely only on past experience. As

© Springer Nature Switzerland AG 2020
S. Ashmarina et al. (Eds.): *Digital Transformation of the Economy: Challenges,*
*Trends and New Opportunities*, pp. 395–403, 2019.
https://doi.org/10.1007/978-3-030-11367-4_38

modern researchers state, intelligent supply chain management is changing the game's rules for traditional warehouses, retailers, consumers, and employees alike by using Blockchain in Trucking Alliance (BiTA) to the use of the IoT [17, 18].

Thus the research goal is to research of the Internet of Things applications in SCM and the research problems are:

1. To discover the Iot-based SCM definition and complexion;
2. To declaim conceptual model of IoT in SCM;
3. To define an application spheres of IoT in SCM.

First steps for IoT are made by analysis of the customer behavior. It became possible to track customers purchasing and on-line activities. Benefits of these activities for SCM regarding products to be tracked more correctly, delivered more promptly and stored more efficiently.

According to Business Insider's premium research it's estimated that there will be more than 24 billion IoT devices on Earth by 2020 [14]. That's four times more than modern people population on the Earth. Therefor about $6 billion will flow into IoT solutions and these investments will generate $13 trillion by 2025 [14].

The IoT opportunities and competition advantages it gives to business require adapting current business models or find new ones to succeed. Today is a new era for using intelligent technologies in SCM and it's time for invest in this technologies. But at first we should understand the key points of its application and implementation challenges we will face to.

## 2    Materials and Methods

Methodological base of this paper consist of General scientific methods and particular scientific methods such as analysis of existing theoretical approaches, synthesis of the different researches views. A systematic approach that considers the objects of study from the standpoint of systems theory and integrated approaches according to IoT application in the SCM sphere let us made conclusion about IoT implementation and future trends in business. The set of applied methods confirms the objectivity and validity of the research.

## 3    Results

Modern consumer's requirements to SCM rising rapidly. Customers use different sales channels and expect to get the same service level by using all of them. Rising material flows (including returns flows) and informational flows volumes make it complicated and much more expensive to satisfy the demand. In this conditions suppliers, retailers, dealers and other supply chain links faces new challenges (Fig. 1) that they have to overcome. These challenges relate to different aspects of SCM: future sales volume, products price, promotion offers, inventory volume, safety stock volume etc. Modern technologies are able to be a tool of optimal decisions in supply chains management by using big data analytics, deep learning algorithms, machine learning, IoT. As Jason

Rosing states Intelligent supply chain management means merely using data, lots of it, to make better decisions, use more advanced technology, and gain actionable insights into operations, says Intelligent supply chain management requires a connection between anything and everything in the warehouse, distribution center, storefront, and e-commerce portal [18].

As it was mentioned below "Internet of Things" is the most perspective base for Intelligent SCM. The definition "Internet of Things" started life as the title of a presentation K. Ashton made at Procter & Gamble (P&G) in 1999 linking the new idea of RFID in P&G's supply chain to the then-red-hot topic of the Internet [2]. The maim idea of the IoT meaning by K. Ashton is that we need to empower computers with their own means of gathering information, so they can "see", "hear" and "smell" the world for themselves. RFID and sensor technology enable computers to observe, identify and understand the world—without the limitations of human-entered data [2].

IoT based on the machine to machine (M2M) technology that means that devices "communicate" with each other without human. IoT is a next step of automatization using TCP/IP protocols to data exchange via Internet. IoT brings an advantage by gathering nets with each other and constructing the "network of networks", moreover it helps to change business models of different economy spheres. IoT forming new economy rules that called "shared economy", excluding dealers out of the business model.

Researches define two ways of IoT application – "customer IoT" and "Industrial IoT" (IIoT or Industry 4.0). We can state that difference between the IoT variants concern the main goal and final user of the IoT. Customer IoT incudes' smartphones and tablets, wearable fitness and outpatient medicine sensors, smart house technology etc. Industrial IoT is used widely then customer IoT. IIoT combines' M2M technology, BigData analysis and production automatization technology. IIoT is much more

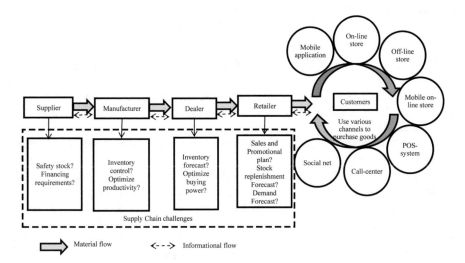

**Fig. 1.** Challenges facing the Supply Chain Management Industry (Source: compiled by the authors according to Senergies Intelligent Systems. http://synergies.ai/scm.html)

advanced then the human being because of it continuous, accurate and error-free data collection and rating. IIoT raises' product quality control, arranges' lean production process, provides' high reliability of supply chain and optimizes' manufacturing business processes. The main goal of IIoT is business processes automatization and optimization, reducing material costs and time. Both of IoT applications have the same technical basis and ecosystem.

Defining "customer IoT" and "Industrial IoT" is not really the true way of IoT development. Traditionally SCM linking manufacturing and sphere of circulation. According to mentioned below we can state that IoT-based SCM take a middle part between customer IoT and Industrial IoT like a chain it links two spheres of IoT

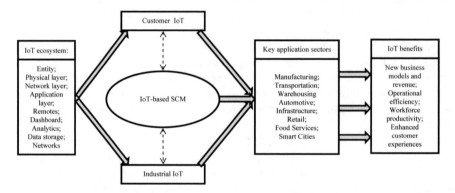

**Fig. 2.** The conceptual model of IoT-based Supply Chain Management application (Source: compiled by the authors)

together. Conceptual model of IoT in SCM shown at the Fig. 2.

IoT joins' together a number of components forming IoT ecosystem. A. Meola describe all them as follows:

- Entity: different groups of IoT users and providers such as businesses, governments, and consumers.
- Physical layer: material part of IoT - devices, sensors and other hardware.
- Network layer: needed for transfer data between devices that was collected by physical layer.
- Application layer: consist of protocols and interfaces used by devices communication.
- Remotes: various of remotes using mobile application to connect and control IoT devices. That could be smartphones, smartwatches, tablets, smart TVs etc.
- Dashboard: part of control system displaying information about IoT ecosystem. Usually it's a part of remote.
- Analytics: different software programs needed to analyze data collected by IoT devices. This data analysis could be used for predictive analytics;
- Data storage: venue (server) for IoT data collected by IoT devices to be stored;

- Networks: communication layer realized via Internet IoT devices to communicate with entity and each other [14].

One of the most important part of the IoT is utilized technology. All of IoT technologies could be used in SCM such as application programming interfaces (APIs), Big Data management tools, predictive analytics, AI and machine learning, the cloud, and radio-frequency identification (RFID) (The Internet of Things definition 2018). The main thing that slow down the IoT expanding is absence of the unit protocol and platform which will organize IoT utilization.

A. Meola mentioned the spheres that will mostly benefit from the IoT (Fig. 3). IoT

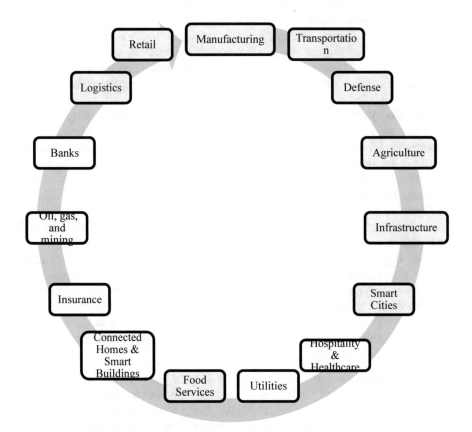

**Fig. 3.** The beneficiaries spheres of IoT applications (Source: compiled by the authors)

systems will gain efficiency for all application spheres and all of them depend on SCM. We marked the part of them which closely tied to SCM and must be researched first for IoT implementation.

IoT-based SCM in the nearest future will realized by following AI technologies:

- Predictive analytics for Demand Forecasting. Predictive analytics combine numerous of advanced analytics tools: ad-hoc statistical analysis, predictive modeling, data mining, text analytics, optimization, real-time scoring and machine learning [16]. These tools assist companies to predict different future events precisely;
- AI for Warehouse Management. Warehouse automatization has being increasing fast because this sphere of SCM is the most costly one and could be automated easily than the others. Swisslog company created a concept of "learning warehouse". It appears to use intralogistics systems which can help improve efficiencies autonomously by learning the most-optimized workflows from data such as customer ordering behavior, a company's machine use and its use of resources [4];
- Chatbots in Procurement. Chatbots are conversational interfaces that potentially could bring several benefits to businesses, including reduced cost of transactions and sales cycle time by replacing humans in the procurement processes [4];
- Transportation. Intelligent transportation systems are moving people and cargo all over the world. Thousands of IoT sensors are used on planes, trains, ships, and vehicles to optimize all transport processes and costs: from engine performance and safety to logistics and supply chain management [23].
- Navigation systems. The finest example that already exists' is Yandex.Navigator. Yandex service collect information received from users and analyzed on the company's server. An application automatically offers the driver detour options and displays the route on the screen of the phone or tablet. Artificial intelligence Yandex will begin to redistribute the load on the roads of cities soon. Taking into account the accumulated statistics, it will offer such routes that optimally load the highways and minimize traffic jams [26].
- Location Tracking. Revealing a real-time information about cargo location provide possibility to monitor and manage incoming and inside facility flows [17];
- Sensitive goods monitoring. It's important to make perishable or sensitive goods stay fresh. This goods transportations and storage require specific environmental conditions to maintain quality. IoT application allow to monitor the temperature of the goods and temperature and humidity in the storage area. Also the shock and vibration levels could be measured during shipment. This data could be used to inform supplier or driver about damaged shipments and make decisions according the damaged goods [17].

Thus we mentioned most likely spheres of SCM to applicate IoT-based technologies. And by IoT devices expanding one of the important themes is information security and its ethical use. There is a real risk of cyber attacks to get information of connected cars, critical infrastructure, and even people's homes. Because of this growing threats a number of companies are focusing on cyber security. Security technologies are needed in IoT-based SCM as it concentrate a huge information flows of consumers private data and companies commercial secret. Data also shouldn't be used to make harm to humanity and environment and that is a question of the ethics' field. All users must be informed of their data use and these statement mustn't be violated.

# 4 Discussion

SCM paradigm was studied and discussed by a lot's of scientists. Among them could be mentioned Bowersox [5], Closs [5], Stock [21], Lambert [21], Martin [12], Shapiro [20], Christopher [7], Harrison [8], Van Hoek [8], Collins [6] and others. The main idea of these author's papers is integration of supply chain links and gaining economic effect through all the supply chain.

The aspects of the digital economy and digital society were stated by Bell [3] in the 1970's. The ideas of informational technology application in logistics and SCM was outlined by Arntzen [1], Brown [1], Harrison [1], Trafton [1], Kumar [11], Minis [15], V. I. Sergeev [19] and nowadays ideas of intelligence technologies in business and supply chains are expanded. There are variety of intelligence technologies and its applications in supply chain which now called Intelligent Supply Chain Management. Intelligent SCM ideas were launched in the researches of following authors: Ho [9], Lau [9], Yücesan [28], Van Wassanhove [28], Zubair Khan [29], Al-Mushayt [29], Alam [29], Ahmad [29]. According IoT in SCM we can state, that the definition "Internet of Things" was first used by Kevin Ashton in 1999 during a presentation he made at P&G [13]. Almost 20 years passed but there is no common opinion of this definition. Researchers from all over the world discuss this theme: Ashton [2], Meola [14], McFarlane [13], van Kranenburg [25]. Corporates and other companies also pay their attention to the IoT definition, for example Microsoft, Google, IBM, SAP, Amazon etc. There were also established public organization issuing IoT application in different spheres such as The Internet of Things Council in Europe and Internet of Things Russian research center.

All these institutions and scientists are looking for the best way to applicate IoT and other intelligence technologies in economy and SCM but there is still a strong requirement for further researches.

# 5 Conclusions

The future has come already. There are millions devices becoming "smarter" day to day. Successful use of smart devices and technologies bring new possibilities to business to get competition advantages and reduce cost, provide high service level and rise accuracy in customers responds by letting businesses manage devices, analyze data, and automate the workflow. SCM links consumers and manufacturers and dealers through material flow. Planning and managing the flow in any way is a challenge. Today SCM is a critical aspect of almost every industry and unfortunately hasn't received as much focus from AI startups and vendor companies compared to healthcare, finance, and retail [4]. The supply chain links collect a vast amount of data which provide benefits by using it correctly. By this reasons SCM must be the pilot sphere of IoT implementation to rise supply chain efficiency and reducing costs.

# References

1. Arntzen BC, Brown GG, Harrison TP, Trafton LL (1995) Global supply chain management at digital equipment corporation. Interfaces 25:69–93
2. Ashton K (2009) That "Internet of Things" Thing. https://www.rfidjournal.com/articles/view?4986. Accessed 4 Oct 2018
3. Bell D (1999) The coming of post-industrial society. A venture in social forecasting. Basic Books, New York
4. Bharadwaj R (2018) Artificial intelligence in supply chain management – current possibilities and applications. Retrieved from https://www.techemergence.com/artificial-intelligence-in-supply-chain-management-current-possibilities-and-applications/. Accessed 5 Oct 2018
5. Bowersox DJ, Closs DJ (1996) Logistical management: the integrated supply chain process. McGraw-Hill Companies Inc., New York
6. Collins J, Arunachalam R et al (2004) The supply chain management game for the 2005 trading agent competition. School of computer science, Carnegie Mellon University, Pittsburgh
7. Christopher M (2011) Logistics and supply chain management: creating value-adding networks, 4th edn. Financial Times Prentice Hall, New York
8. Harrison A, Van Hoek R (2011) Logistics management and strategy: competing through the supply chain, 4th edn. Pearson Education Prentice Hall, New York
9. Ho GTS, Lau HCW (2004) An intelligent information infrastructure to support the streamlining of integrated logistics workflow. Expert Syst 21(3):123–137
10. Internet of Things Russian research center (2018). http://internetofthings.ru. Accessed 4 Oct 2018
11. Kumar K (2001) Technology for supporting supply-chain management. Commun ACM 44 (6):58–61
12. Martin Ch (2005) Logistics and supply chain management: creating value-adding networks. Prenties Hall, New York
13. McFarlane D (2015) The origin of the internet of things. https://www.redbite.com/the-origin-of-the-internet-of-things/ Accessed 6 Oct 2018
14. Meola A (2018) What is the internet of things (IoT)? Meaning & Definition. https://www.businessinsider.com/internet-of-things-definition. Accessed 7 Oct 2018
15. Minis I (2011) Supply chain optimization, design, and management: advances and intelligent methods, 1st edn. IGI Global, Hershey
16. Predictive analytics (2018). https://www.ibm.com/analytics/predictive-analytics. Accessed 7 Oct 2018
17. Ray B (2018) 5 IoT applications in logistics & supply chain management. https://www.airfinder.com/blog/iot-applications-in-logistics. Accessed 8 Oct 2018
18. Rosing J (2018) What is intelligent supply chain management? https://cerasis.com/2018/03/15/intelligent-supply-chain-management/. Accessed 7 Oct 2018
19. Sergeev VI, Grigoriev MN, Uvarov SA (2008) Logistics. Information systems and technology. Alfa-Press, Moscow
20. Shapiro JE, Singhal VM, Wagner SN (1993) Optimizing the value chain. Interfaces 23 (2):102–117
21. Stock R, Lambert MD (2011) Strategic logistics management. McGraw-Hill, New York
22. The Internet of Things Council (2018). https://www.theinternetofthings.eu/what-is-the-internet-of-things. Accessed 8 Oct 2018

23. The Internet of Things definition (2018). https://www.sap.com/uk/trends/internet-of-things.html. Accessed 8 Oct 2018
24. Turban E et al (2018) Electronic commerce 2018. Springer Texts in Business and Economics, vol 636, pp 270–273. https://doi.org/10.1007/978-3-319-58715-8_7
25. Van Kranenburg R (2014) Internet of Things (IoT); a new ontology and an engineering challenge. http://internetofthings.ru/internet-of-things-and-russia/92-internet-of-things-iot-a-new-ontology-and-an-engineering-challenge. Accessed 8 Oct 2018
26. What is the Internet of things: existing technologies (2018). https://strij.tech/publications/tehnologiya/chto-takoe-internet-veschey.html. Accessed 5 Oct 2018
27. Your IoT business needs the right business model (2018). https://content.microsoft.com/iot/business-models. Accessed 7 Oct 2018
28. Yücesan E, Van Wassanhove LN (2004) Supply-Chain. Net: the impact of web-based technologies on supply chain management. Springer US, New York
29. Zubair Khan M, Al-Mushayt O, Alam J, Ahmad J (2010) Intelligent supply chain management. J Softw Eng Appl 3(4):404–408. https://doi.org/10.4236/jsea.2010.34045

# The Use of Technology of Digital Economy to Create and Promote Innovative Excursion Products

G. V. Aleksushin[(⊠)], N. V. Ivanova, and I. J. Solomina

Samara State Economic University, Samara, Russia
{gva3, ivanovanvl971}@yandex. ru, sui@bk. ru

**Abstract.** The relevance of the topic is caused by the growing trends in the Internet environment and computer innovations aimed at the modernization of the excursion business. The aim of the study is to analyze the possible directions of the commercial component in the innovative technologies of modern excursion activity as a tourist product and the most important component of the economy of the tourist and recreational cluster and the electronic economy. The object of research is innovative excursions – audio and virtual excursions. The subject of the study was the technology of organization of modern excursions. The materials of the article are of practical value for the staff of the tourism and recreation cluster in the era of digitalization of the economy. The article based on the results of research work №47-18 characterized audio tour as a product of the electronic economy, investigated through a comparative analysis of the development of this type of tourist product in Russia and Samara.

**Keywords:** Audio excursion · Virtual excursion · Guide · Innovation · Internet · Customer · Promotion · Market · Website · Social network · Tourism · Tourism product · Tourism and recreation cluster · Service · Economy · Excursion · Excursionist · Tour guide · Electronic economy

## 1 Introduction

Currently, the market of tourist services is becoming a space for the active introduction of digital economy technologies. The rapidly growing demand for tourism products and the orientation of tourism firms or the individual needs of the client led to the transition to digital technologies in the creation and promotion of tourism products. The most promising in this direction is the use of digital economy technologies. It is important not only to create innovative content of excursions, but also to bring its offer to the consumer.

Before modern designers of excursion services the e-economy challenges:

(1) optimization of automatically arranged tours and excursions routes based on logistics;
(2) the ability of designers to choose a excursion objects to electronic databases with regard to their value to the tourists from different countries, of different ages and interests;

(3) payment for services provided to customers will be made through the Internet, and prices will be lower, as they exclude part of the actions usually carried out by people and attracted resources.

Currently, the form of organization of the Audio guide is actively developing all over the world. Audio guide is the soundtrack used to explore the exposition of the museum, exhibitions, countryside, device to play it. By type of technological basis Audio guides can be divided into 2 main groups: on the basis of special devices ("iron", or hardware, Audio guides) and on the basis of standard mobile devices – Smartphone, tablets, PDAs and other (mobile Audio guides).

Typically, an Audio guide recording consists of a series of audio clips. Each fragment is numbered and tied to the scheme (map) of the inspected area, museum, or to the numbers of exhibits. If an Audio guide involves a coherent and complete story from fragments, it is called an Audio excursion. Audio excursions are widespread abroad, more and more they appear in Russia. They are created by museums and independent developers. Audio guide services are provided by some mobile operators.

Audio guide is a direct and convenient way to make your own tour of a new city or place, to discover them, that is, you can travel without visiting a specific place. Just download the material ordered from the website (for example, Your AulioGid). You can identify the following reasons for using the audio guide:

- no need to spend time with strangers, the consumer decides where and when to listen to the excursion;
- it is impossible to get lost, because the application has a map;
- you can start and finish listening to the information on any object.

Of course, an Audio excursion will not replace a real excursion in a certain place with a real tour guide, but can provide information that will determine the choice of the consumer. Audio tour, on the one hand, has the properties of self-education, as the user receives the necessary information, and on the other hand – prepares the tourist for a real journey, when the perception of the material will fall into consciousness on the prepared ground. In addition, having a sound application with excursion material, the tourist can independently lay the routes of his journey.

From the point of view of the economy, the Audio tour is not free, the payment of downloaded excursion applications will contribute to the replenishment of the budgets of the developer and the region.

## 2 Materials and Methods

As the sources of the study were used:

(1) materials of economic statistics;
(2) materials of websites and social networks of tourist firms offering tourist product in the form of Audio excursions;
(3) results of the survey of employees and customers of travel companies.

The survey was conducted to identify the role of e_communications in the creation and promotion of excursion services. 25 Russian tourist companies from Samara region took part in the survey. The study of the results of the survey revealed the ratio of different forms of communication with customers in the creation and promotion of tourist products (Fig. 1):

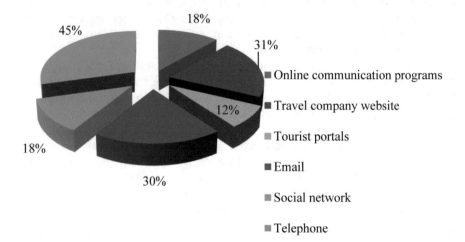

**Fig. 1.** The ratio of different forms of communication with customers in tourist firms (Source: compiled by the authors)

The analysis of the role of different forms of communication with clients of tourist companies has led to the conclusion that currently more than 63% of all interactions with customers account for social networks, websites and tourist portals. Presentation of excursion services to customers through the electronic environment is possible on several levels:

(1) simple (low-tech) – creation, support, development and modernization of con- solidated information bases of tour guides, their products (excursions, quests, quizzes, etc.), materials about excursion sites;
(2) complicated (medium-tech) – formation based on the previous-level information databases commercial sites which retain the physicality of tour guides and guides who conduct the real excursion due to electronic their sale – something like booking.com in selling excursions;
(3) complex (high-tech) – the formation of new electronic formations, which will not only contain a ready-made information base and specific excursion objects, but also contribute to the creativity of consumers who will be able to design their own electronic excursions and tours of the available elements, using the capabilities of modern digital technologies – in the case of pre-recorded audio or video support without the participation of a real tour guide or guide;
(4) promising (innovate-tech) – the use of the latest achievements in these proposals – AR, VR, gadgets, geolocation

The analysis of offers from different websites allows to estimate commercial potential of sales of an Audio tours through the Internet – see Table 1.

**Table 1.** Offers Audio tours from the site to different places in the world

| № | The name of the tour and direction | Characteristic | Time | Language | Price |
|---|---|---|---|---|---|
| 1 | Package tours: sightseeing Tours around Prague | Contains 7 audio tours. Contains text descriptions; photos. Support for Apple; Android devices | 07:33:59 | Russian | $9.99 (499 rub.) |
| 2 | Package tours: Five tours of Vienna | Contains 5 audio tours. Contains text descriptions; photos. Support for Apple; Android devices | 05:08:58 | Russian | $10.99 (549 rub.) |
| 3 | Package tours: Five tours around of St. Petersburg | Contains 5 audio tours. Contains text descriptions; photos. Support for Apple; Android devices | 10:52:41 | Russian | $8.99 (449,5 rub.) |
| 4 | Package tours: Around of St. Petersburg | Contains 7 audio tours. Contains text descriptions; photos. Support for Apple; Android devices | 14:24:02 | Russian | $13.99 (699, rub.) |
| 5 | Package tours: Roman holiday | Contains 4 audio tours. Contains text descriptions; photos. Support for Apple; Android devices | 05:00:49 | Russian | $8.99 (449,5 rub.) |
| 6 | Package tours: Moscow tours | Contains 8 audio tours. Contains text descriptions; photos. Support for Apple; Android devices | 15:00:29 | Russian | $15.99 (799,5 rub.) |
| 7 | Package tours: All tours in Russia | Contains 32 audio tours. Contains text descriptions; photos. Support for Apple; Android devices | 66:22:26 | Russian, english | $19.99 (999 rub.) |
| 8 | Package tours: All tours in Europe | Contains 42 audio tours. Contains text descriptions; photos. Support for Apple; Android devices | 68:32:25 | Russian | $29.99 (1499 rub.) |

Source: compiled by the authors.

Also on this website you have the materials for the cities of Alupka (1 guide), Anapa (1 guide), Volgograd (1 guide), Dmitrov (1 guide), Evpatoria (1 guide), Kazan (1 free Audio tour, 1 guide), Kerch (1 guide), Kolomna (1 guide), Moscow (25 Audio tours, 4 free, 1 guide), a neighborhood of Saint Petersburg (8 Audio tours, 1 free, 1 guide), Omsk (1 free Audio tour), Saint Petersburg (11 audio tours, 1 tour guide), Sevastopol (1 Audio tour), Sochi (1 Audio tour), Suzdal (1 guide), Yalta (1 guide). The analysis of this website shows that Audio tours in the information market are mainly represented by the cities of Moscow and St. Petersburg, other cities are poorly represented, most cities in Russia are not displayed at all. This suggests that the Audio tours are poorly represented in the market of tourist services in Russia, although there is a demand for this service.

The Internet space is a favorable environment for the development and promotion of virtual tour content through groups in social networks, configured to actively include subscribers in joint activities to create an excursion product. Speaking about the promotion of finished products, among the many ways to demonstrate the tourist product (colorful texts, impressive photos, bright logos, etc.), video content can attract more attention of potential consumers. It is necessary to develop and substantiate the principles and technologies of interaction in social networks of all participants of social groups to create and promote virtual excursion products.

## 3   Results

In 2018, Samara, along with other Russian cities such as Moscow, St. Petersburg, Sochi, Kazan, Yekaterinburg, Kaliningrad, Nizhny Novgorod, Volgograd and Saransk, became a co-organizer of the world Cup. As part of the research work Samara State Economics University signed a contract №47-18 dated 4 June 2018 with ANO KDR "U-Ra" for the performance of research work "Development of methodological support in the organization of excursion service for participants and guests of the world Championship on football of 2018", the execution of which the Professor of Department of service and tourism Commerce G.V. Aleksushin, associate professors N.V. Ivanova and I.J. Solomina has developed informational materials on 60 cultural and historical objects, located in the historic part of the city, for inclusion in the Audio excursion. The results of the study were the following aspects of formation and promotion:

(1)   trends in the development of e-tour and e- excursion tourist product;
(2)   identification and characteristics of 4 levels of development of the principles of organization of electronic excursion and tourist products;
(3)   study of the role of social networks in the formation of virtual excursion content;
(4)   evaluation of the commercial potential of Audio excursions sales via the Internet;
(5)   conclusions from the research work №47-18, in which the authors of this article were developed and provided for the audio recording of 4 excursion routes: 2 guest, "river excursion route", excursion route "Stalin's Bunker".

The excursion guest route was designed for 2 routes of the world Championship guests: from the airport to the "Samara-Arena" stadium, and from the railway station to the "Samara-Arena" stadium. The route lasts 40 min, contains historical information about Samara and Samara region, represents the main brands of the city and region. The features of the guest excursion route are the lack of attachment to specific excursion objects and informational nature.

The river route is designed for offline use on river boat tours. At the time this route takes about 1 h. The route contains historical information about Samara and Samara region, and is minimally tied to the excursion objects located on the banks of the Volga river, along which this route passes.

Excursion route "Stalin's Bunker" is intended for autonomous use through audio recording or in the work of guides inside the bunker. By time, this route lasts 1 h, contains historical information about the bunker and Kuibyshev as the capital of the USSR in 1941–1943, and is tied to the excursion objects located inside the bunker along the route through it.

Of course, the format of Audio excursions in Samara in the initial state, but the work done in the framework of research work №47-18 "Development of methodological support for the organization of excursion services for participants and guests of the 2018 world Cup" allows you to continue to work in this direction and use the capabilities of the audio guide more widely, not only locally, as it was during the 2018 FIFA world Cup, where guest routes were broadcast in buses that follow from/ to the airport and train station, on ships during the river excursion, during the excursion to the museum of Stalin's Bunker.

## 4 Discussion

Exploring the information age, M. Castells writes that in the information economy, the productivity and competitiveness of factors or agents depend on their ability to generate, process and effectively use knowledge-based information. The main driving force of the information economy is the production and consumption of information, both in material form (e.g. high-tech products) and in non-material form.

The concept of electronic economy was first formulated by the American scientist Nicholas Negroponte in 1995. He presented it in the form of transition from the movement of atoms to the movement of bits.

In the "Concept of formation and development of the unified information space of Russia" the information space is defined as a set of banks and databases, technologies of their maintenance and use, information telecommunication systems, operating on the basis of general principles and providing information interaction between the organizations and citizens, as well as meeting their information needs.

The analysis of the interests of guests from different countries on the formation of a valuable complex of excursion objects is taken from the study of different countries [1, 3, 4, 13]. Theoretical analysis of the audio tours [2, 6] and the problems of their information saturation during the organization and holding of the 2018 world Cup are displayed in the article [15] and presented in the materials of the research work 47-18 "Development of methodological support for the organization of excursion services for the participants and guests of the 2018 world Cup".

In fact, innovative technologies in the excursion business have been actively discussed in the last few years [9, 10], but not in the framework of e-commerce. Many researchers have noticed that every year the Internet environment is becoming more popular, the placement and promotion of tourism products in social networks and other Internet resources is becoming more popular [5, 7, 8, 11, 12, 14]. Some tours and excursions in the form of virtual tours are presented on the Internet as separate types of tourist products, among which Internet users are most interested in interactive tours [16]. The review of modern scientific literature on the problem under study showed insufficient elaboration on the development of electronic tourism technologies, and there are no studies of practices at all.

# 5 Conclusions

Summarizing the materials presented in this article, it is necessary to develop excursion and tourist activities in the format of electronic economy, which is an important element of the information society. To do this, there are many developments analyzed in the article.

# References

1. Aleksushin GV (2015) Perspective of Samara-Canadian relations in the tourism. Mediterr J Soc Sci 6:323–327, №6S3
2. Barichev SG (2007) Audio tours as a means of self-education. Open Educ, 28–30 №3
3. Hudson L, Hudson S (2017) Marketing for tourism, hospitality & events: a global & digital approach, Sage Publications, London, Gurhan Aktas. J Hosp Tour Manag 36:127–128
4. Hur K, Kim T, Karatepe O, Lee G (2017) An exploration of the factors influencing social media continuance usage and information sharing intentions among Korean travelers. Tour Manag 63:170–178
5. Jansson A (2018) Rethinking post-tourism in the age of social media. Ann Tour Res 69:101–110
6. Jing L, Pearce P, Low D (2018) Media representation of digital-free tourism: a critical discourse analysis. Tour Manag 69:317–329
7. Karpova DN (2013) Internet communication: new challenges for young people. Bull MGIMO Univ №5(32):208–212
8. Kofler I, Marcher A, Volgger M, Pechlaner H (2018) The special characteristics of tourism innovation networks: the case of the regional innovation system in South Tyrol. J Hosp Tour Manag 37:68–75
9. Kyrgina SO, Koptseva MG, Surzhikov VI (2017) Quest-excursion as an innovative form of excursion product. Azimut Scientif Res Econ Manag
10. Liah AA, Likhanova VV (2017) Innovation in excursion activities. Scientif Notes Zabaikalje State University. Series: Social Sciences
11. Liu Z, Park S (2015) What makes a useful online review? Implication for travel product websites. Tour Manag 47:140–151
12. Narangajavana Y, Fiol L, Tena M, Artola R, García J (2017) The influence of social media in creating expectations. An empirical study for a tourist destination. Ann Tour 65:60–70
13. Paniagua FJ, Huertas A (2018) The content of tourist destinations in social media and the information search of users [El contenido en los medios sociales de los destinos turísticos y la búsqueda de información de los usuarios]. Cuadernos de Turismo, (41):513–534 +723–725
14. Sariev SS, Chernova DV (2014) Social networks as a tool of marketing communications in commercial activity. Bull Samara State University of Economics. №11(121):99–103
15. Solomina IY (2013) The influence of the social memory of the city on the formation of tourist space of Samara. Postgrad Bull Volga Reg №3–4:61–67
16. Volobueva MV (2017) Virtual tour as an effective marketing tool. Education and science without borders: social sciences and humanities: Orel state university of economics and trade. № 8:180–183

# Intellectual Algorithms for the Digital Platform of "Smart" Transport

T. B. Efimova[1(✉)], V. A. Haitbaev[2], and E. V. Pogorelova[1]

[1] Samara State Economic University, Samara, Russia
TB_Efimova@mail.ru, jour.ru@gmail.com
[2] Samara State Transport University, Samara, Russia
vhaitbaev21@mail.ru

**Abstract.** In conformity with the concept of a "smart" city, resources of all municipal services are to be used in an optimum manner thereby ensuring maximum comfort for all city dwellers. It particularly concerns city passenger transport, the basic part of a "smart" city, for which development of an intellectual digital platform allowing operational management of transport processes as well as reaction to events in real time is essential at present.

Development of proposals and recommendations in improving and developing the existing intellectual digital solutions is essential for the municipality of Samara, as available software products use simplified optimization models and do not take into account the existing restrictions, therefore leading to results which do not meet the demands of city dwellers in the full extent.

This research is devoted to the development of the digital transport platform allowing city dwellers to obtain up-to-date information about city transport and about the possibility of optimizing their routes. In this research analysis of various models suitable for the solution of the mentioned aim is conducted, models suitable for the formulated tasks under the conditions of the most significant restrictions are chosen for the purpose of optimization of passenger flows and creating an intellectual system.

Passenger flow processes based on dynamic modeling which are analyzed in this research envisage using models based on the concept of metaheurustic, the latter being the particle swarm method, the ant colony algorithm, iteration technique, the combinatorial method, etc. An algorithm of using these methods with a variant of modified transport infrastructure worked out with due regard to the changing requirements to passenger traffic is also proposed. The intellectual system created on the basis of the chosen models and algorithms will allow obtaining the necessary predictive information for city dwellers using public transport.

**Keywords:** Digitization · Dynamic programming · Intellectual system ·
Municipality · Optimization of passenger flows · Passenger flow processes ·
Passenger traffic · "Smart" city

© Springer Nature Switzerland AG 2020
S. Ashmarina et al. (Eds.): *Digital Transformation of the Economy: Challenges,
Trends and New Opportunities*, pp. 411–418, 2019.
https://doi.org/10.1007/978-3-030-11367-4_40

# 1 Introduction

The concept of a "smart" city contains various aspects of life activity; acceleration of the processes of development and enlargement of conurbations, the growth of population in cities and in the rural area require solution of the herewith arising problems. Up-to-date conurbations are an intellectual product of engineers, architects and designers in which almost all social and cultural, industrial, technological and infrastructure facilities are related to the implementation of the plans of city development [12].

Efficient and reliable operation of urban and suburban passenger transport is an essential factor of realization of "smart" city solutions. Urban passenger transport provides the basic part of workmen's rides of the population, which directly influences the efficiency of functioning of municipal services, enterprises, organizations, institutions and all industries of regions and the country as a whole. In order to optimize operation of city passenger transport, transport transfer junctions in particular, it is necessary to develop a unified digital platform of "smart" transport possessing a friendly and clear user interface oriented towards the needs of city dwellers [13].

Passenger transport is one of the major factors of municipal maintenance of more than 1300 municipal settlements in Russia. It carries more than 120 million passengers daily. Transport mobility of a city dweller in Russia is averaged to about 450 journeys a year and it is still increasing. City passenger transport amounts to 85–90% of all city journeys. The state of the transport sector of most cities, especially density of road traffic, is being influenced by the process of active increasing of vehicle-to-population ratio [4].

Development of a transport system is a sophisticated complex of interrelated solutions concerning economic activity in the municipality of Samara and directly influencing the quality of life of all city dwellers and guests, as well as operation of institutions and enterprises. Prescript of administration of the municipality of Samara, 30, October, 2015 On approval of the municipal programme "Development of urban passenger transport in the municipality of Samara" for the period of 2016–2020 states that the measures taken in regard to the development of urban passenger transport do not allow providing transport services corresponding to the up-to-date level of quality to the population in full extent, because the average length of time consumption for travel from residential districts to job placements in the municipality of Samara exceeds one hour (the norm is 40 min). Moreover, conditions of passenger transportation during peak hours on separate routes are unsatisfactory; there is no sufficient number of transport infrastructure facilities designed for people with limited mobility. Available means of control of movement of vehicles on the line do not duly allow ensuring control of fulfilling agreements in carrying out transportation by municipal routes in the municipality of Samara.

Thus, the aim of this research is analysis and assessment of the instrumentarium of economic and mathematical methods of research of the urban passenger transport system, as well as working out proposals in regard to probability of their use on the city level of Samara forming the basis of the corresponding intellectual system.

## 2   Materials and Methods

In developing a new digital transport platform of Samara peculiarities of city facilities are taken into account, e.g. length and carrying capacity of roads, use of various modes of transport, transport boarding junctions, stops of public transport, etc. The necessity of transition to "smart" city solutions, involvement of city dwellers into them determines new approaches, methods, algorithms of analysis and planning of passenger flows [16].

It is obvious that the task of development of the city transport platform of the municipality of Samara is essential, however it is rather complex due to a great number of factors applied; among them one may single out the structure of routes, location of stops (proximity to areas of active transport flow, transport junctions), diversity of modes of city passenger transport, amount of rolling stock, its service and technical characteristics, types of passengers (students, pensioners, working population, etc.) and their distribution along the transport network, fare, headways (depending on modes), time of the day, a weekday, season, etc. [1]. These issues are now being actively studied within the framework of creation of "smart" cities as a part of the world community [2, 10, 14, 15].

When developing the digital platform let us represent as initial data the transport network of the municipality of Samara in the form of a graph $\Gamma(V,E)$, where $V$ denotes transport junctions (locations significantly changing characteristics of passenger flow), and $E$ denotes arcs and edges (modelling ways), each is assigned time of travel, length, carrying capacity for each mode of urban passenger transport and its types. Thus, it is preferred to assign its own graph to each mode of transport, i.e. $\Gamma_n$ is a graph for the mode of transport $n$. Transport junctions $S \subseteq V$ are generating flows (these are residential areas, as a rule), $D \subseteq V$ are absorbing flows (e.g., city downtown). The route for $\Gamma$ will be a succession of edges:

$$e_1 = (i \rightarrow k_1),\ e_2 = (k_1 \rightarrow k_2),\ \dots\ e_m = (k_{m-1} \rightarrow k_m),\ e_{m+1} = (k_m \rightarrow j)$$

The worked–out digital platform contains a group of optimum routes for delivery of passengers to the key points of the municipality. When analyzing the existing methods, the first one to be considered was the method of branches and edges, however it demands significant time consumption when dimension of the problem is large-scale. Metaheuristic algorithms are considered to be the most suitable ones for solution of complex problems of combinatorial optimization [2, 3, 5, 6].

"Ant algorithms" relating to tasks of optimization are widely used at present; the first algorithms were successfully applied for solving the travelling salesman problem [9].

AS-algorithms use artificial agents acting on the basis of constantly reviewed information [3]. Algorithms in which the classical form is improved in a certain way, e.g. by means of using elite strategy [11] or application of the idea of establishing upper and lower edges of parameters of renewing the pheromone, are well-known.

## 3   Results

When developing the digital transport platform at the first stage a map of transport junctions of the municipality was drawn up in the form of an undirected graph. Then the system of agents which configure the main routes of a graph (i.e. optimize the key part of city directions, i.e. terminals, central main lines, large trade centers and enterprises) is activated. Every next junction is chosen at random; apices which the agent has already passed are of special importance. At every step adjustment of the amount of "pheromon" on each passed arc of the initial graph takes place.

In this way a map of key optimum routes is drawn up.

Further a matrix of correspondences $\|a_{xy}\|$ is formed in the following way. The map of the municipality is divided into squares of small size (1000 by 1000 m); in each square the number of dwelling houses and the number of floors are evaluated; according to special tables the number of dwellers in each house is determined. Further the figures obtained are summed up and written into the corresponding cell of the matrix of correspondences. Let us consider each square to be the one both generating and absorbing flows. In squares corresponding to a park zone, for example, "0" is written instead of the corresponding element of the matrix.

The following services may be used as sources of information for the purpose of practical realization of the digital platform: maps Google, Yandex. Maps, 2GIS, OpenStreetMap, BingMaps, HERE, CityMapper, the service of the Transport operator of Samara. Notwithstanding the convenient user interface, they have a number of drawbacks rendering their application in this research difficult: Google maps and 2GIS use information about transport movement in the form of the time-table provided by transport operators, which lowers precision of information; application of Yandex is encumbered as there is no access to API in applications of outside developers; OpenStreetMap does not allow configuring a route upon request.

Transport operator of Samara is developed by the OJSC "Samara-Informsputnik" within the framework of realization of the program of the Government of Samara region "Electronic Government" under the order of the Management of information resources and technologies of the Administration of the municipality of Samara, under control of the Department of transport of Administration of municipality of Samara and the Municipal Enterprise "Passenger transservice" within the framework of "The new approach to passenger information" included into the Strategy of development of the transport complex of the municipality of Samara for the period of 2011–2020. Useful services are provided by the Internet portal, such as: prediction of arrival of transport at a stop, a guide of routes of all modes of transport with live maps of movement, search of the most suitable route of travel from A to B and a lot of others [7].

The use of more precise information about the supposed time of movement along the configured optimum route as initial data will allow improving realization of the developed digital transport platform.

Apart from the above mentioned data the developed software product takes into account carrying capacity of rolling stock, number of enlarged types of urban passenger transport in the municipality of Samara, volumes of transportation (the mean value for the last 3 years), the average service speed, the average headway, the total length of routes of each mode of transport.

When drawing up a model let us consider the routes including transfers, as well as the shortest ones (a criterion to be considered is a temporal one which takes into account the time of approaching the junction, the time of awaiting a means of transport, as well as fare).

Let us single out an enlarged transport section with indicated points of transfer and the respective transport routes (Fig. 1).

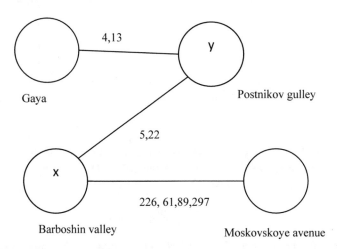

**Fig. 1.** Enlarged transport section (source: compiled by the authors)

Let us make the points "Barboshin valley" (x) and "Postnikov gulley" (y) providing an opportunity to board trams 4, 13 ("Gaya street") and 5, 22 ("Barboshin valley") places of transfer (Fig. 1).

In order to apply the real situation (matrix of correspondences $\|axy\|$) to the optimal (the shortest) route obtained at the previous stage using AS-algorithms (based on heuristics and optimization applied for solving many combinatorial problems) let us consider some section of the route not presupposing transfer (xy). For the junction x a number of passengers axm having arrived at it to board another tram (1) going to the junction y (m – mode of transport by which a passenger has arrived) is known:

$$\sum_{m} a_{xm} = a_{xy} \tag{1}$$

In the model of area (xy) all routes crossing this area, their length (initial data are provided by the official map of the municipality, as well as by the resource http://www. samaratrans.info, where headways, duration, length, points passed-by are given for each route), time spent by a passenger to cover this distance on foot are taken into account. The speed of all means of transport and the average headway are taken from Table 1.

**Table 1.** Characteristics of city passenger transport in the municipality of Samara

| Name of a transport means | Capacity according to the registration certificate, people | Amount in the municipality of Samara, units | Volumes of traffic, million passengers a year (the mean value for the last 3 years) | Average service speed, km/h | Average headway, minutes | Total length, km |
|---|---|---|---|---|---|---|
| Tram | 224 | 847 | 115,1 | 18 | 17,3 | 166 |
| Trolleybus | 108 | | | 19 | 19,2 | 208,4 |
| Metro car | 250 | | | 45 | 8 | 12,7 |
| City bus | 120 | 32893 | 169,7 | 21 | 39,8 | 41900 |
| Articulated bus | 180 | | | 20 | | |
| Bus of Pavlovsk Bus Plant | 43 | | | 21 | | |
| Minibus | 13 | | | 100 | 6,4 | |

This table is compiled by the authors on the basis of Samara statistical yearly periodical 2017 and data from http://samarastat.gks.ru [8]

It is reasonable to use probability-based choice (2): a passenger chooses a mode of transport and the corresponding route (l - a type of route a passenger chooses, n – a mode of transport going by this route):

$$P_{lmn}^{xy} = P_{ml}^{xy} P_{nl}^{xy} \tag{2}$$

Consequently the number of passengers of this route may be calculated according to the formula (3):

$$a_l^{xy} = \sum_m a_m^x p_{ml}^{xy} p_{nl}^{xy} \tag{3}$$

To take into account the criterion of time let us divide "peak hours" (4) into intervals (let us subtract the initial time of the peak period from the final one):

$$t = t_{кон} - t_{нач} \tag{4}$$

The starting point of the "peak hour" is taken earlier than a working day begins (the average value is 7.00); besides reserve time is taken into consideration for the passenger not to be late (when calculating this parameter it is necessary to take into account a number of random factors; in the case under consideration let us take an averaged value equal to 15 min). In order to load the network during each time interval it is necessary to take into account movements into the points of absorbing passenger flow.

For this purpose it is necessary to distribute correspondences when introducing a new parameter, i.e. the number of passengers passing the section (xy) towards the junction with the number j (beginning of work - tj) - . The time of movement between areas is taken from official sources on the basis of the optimum route configured earlier.

Let us consider practical realization of the formulated task – the result of introduction of the created digital transport platform:

1. A variety of stops and routes is formed automatically.
2. A user chooses a number of key directions (from the microdistrict Krutie Klyuchi along the Moscow highway, from the microdistrict Yuzhnyi Gorod along Avrora street, from Tashkentskaya street to the center of the city, to key industrial enterprises on Zavodskoye highway, hospitals, universities).
3. With the help of AS-algorithm the digital platform configures optimal key routes.
4. Matrices of correspondences, departure and arrival are configured.
5. A point of starting or termination of movement is chosen (coordinates are stored; it is possible to specify them by using a geocoder).
5. An optimum route is suggested.
6. Receiving recommendations for improving the choice of a route on account of parameters of a model (the developed intellectual system gives a user the corresponding recommendations).

## 4 Discussion

The most significant investigations in the field of organization of city passenger traffic are presented by proceedings of the following scientists: Madar [2], Turpisheva et al. [4] and others. In these works the most general principles of planning and organization of efficient operation of transport enterprises relating to passenger traffic are substantiated. The problems of economic and mathematical modeling of city passenger transport are investigated in the works of Krushel [1], Parluk [3], Tutigin and Popov [5], Fomicheva and Barzikov [6] and others. These problems are touched upon in the proceedings of conferences connected with the theme of "smart" cities.

## 5 Conclusions

The worked out digital platform based on the analyzed algorithms and models is designed to enable provision of comfortable conditions for life activity of the population of the municipality of Samara by means of development a consistently functioning system of city passenger transport which is economically effective, attractive and accessible for all social groups. The proposed model will allow optimizing the existing transport system by reducing the time of awaiting city public transport and the level of overcrowding passenger compartments during "peak hours"; it will also allow improving ecological parameters of the transport system of the municipality of Samara.

**Acknowledgments.** To Kopeikin Sergei Vladimirovich – Doctor of Engineering Science, Professor of Samara State University of Transport for valuable advice in modeling.

# References

1. Krushel EG et al (2015) Management of passenger transport in a small town. Models and algorithms. Volgograd State Technical University, Volgograd
2. Madar ON (2017) Simulation of demand for transport services in the suburban gravity zone of passenger flows for long-distance passenger trains. Scientific trends: issues of exact and technical Sciences. https://doi.org/10.18411/spc-12-10-2017-13
3. Parluk EG (2014) Imitation modeling of passenger flows in the system "Railway-city public transport". Bulletin of Rostov State University of Transport. Rostov State Transport University (Rostov-on-Don). ISSN: 0201-727X №3(55):78–82
4. Turpisheva MS, Nurgaliev ER, Jahyaeva SB (2017) Research of the processes of passenger transportation by automobile transport. Bulletin of Astrakhan State Technical University. Astrakhan State Technical University (Astrakhan). ISSN: 1812–9498. №1(63):56–61
5. Tutigin RA, Popov VN (2016) Measurement of parameters of passenger flows for the purposes of modeling. Advanced topics of metrological assurance of scientific and practical activity. Northern (Arctic) Federal University named after M.V. Lomonosov (Arkhangelsk). ISBN: 978-5-261-01201-6. №2.:155–160
6. Fomicheva OE, Barzikov KV (2013) Agent and the model of its behavior in the multi-agent environment of modeling passenger flows. Mining information-analytical bulletin (scientific and technical journal). Ltd «Mountain book»(Moscow). ISSN: 0236–1493. №S5:193–195
7. Samara transport operator. http://tosamara.ru (date of reference: 03.05.2018)
8. Samarastat. http://samarastat.gks.ru (date of reference: 03.05.2018)
9. Anagnostopoulou E et al (2018) From mobility patterns to behavioural change: leveraging travel behaviour and personality profiles to nudge for sustainable transportation. J Intell Inform Syst. https://doi.org/10.1007/s10844-018-0528-1
10. Byun J (2016) Smart city implementation models based on IoT technology. Adv Sci Technol Lett. https://doi.org/10.14257/astl.2016.129.41
11. Huang L et al (2018) A multi-objective optimization model for determining the optimal standard feasible neighborhood of intelligent vehicles. Trends Artif Intell. https://www.springer.com/gp/book/9783319973036. August 28–31 2018
12. Massobrio R et al (2016) Towards a cloud computing paradigm for big data analysis in smart cities. Trudy ISP RAN/Proc. ISP RAS. https://doi.org/10.15514/ispras-2016-28(6)-9
13. Parr G (2015) Smart city architecture and its applications based on IoT. Proced Comput Sci. https://doi.org/10.1016/j.procs.2015.05.122
14. Pasquale GD (2016) Innovative Public Transport in Europe, Asia and Latin America. A Survey of Recent Implementations. Transportation Research Procedia. https://doi.org/10.1016/j.trpro.2016.05.276
15. Tiwari et al (2018) Scalable prediction by partial match (PPM) and its application to route prediction. Appl Inform 20185:4. https://doi.org/10.1186/s40535-018-0051-z
16. Zheng X et al (2016) Big data for social transportation. IEEE Trans Intell Transp Syst. https://doi.org/10.1109/tits.2015.2480157

# Networks of Competences of Subjects of the Local Food Market in the Conditions of Formation of Digital Economy

N. V. Sirotkina[1], O. G. Stukalo[1]([✉]), N. V. Nikitina[2], and A. A. Chudaeva[2]

[1] Voronezh State University, Voronezh, Russia
docsnat@yandex.ru, stukalo_oksana@mail.ru.ru
[2] Samara State University of Economics, Samara, Russia
nikitina_nv@mail.ru, chudaeva@inbox.ru

**Abstract.** The network research of the local food markets participants is based on author's methodology. At the same time, the methodology is considered generally as the form of knowledge and self-knowledge provided by three expressed aspects: world outlook, cognitive, technological. In general the methodology represents the multi-level system of receipt of converting and statement of knowledge about any object. Notice that it is about methodology of a specific research, but not about methodology as the area of knowledge which study means, prerequisites and the principles of the organization of cognitive and practical and reformative activity. In article, allowing equivalence of methodology and a method and considering that the methodology represents a peculiar method which establishes the relation between the theory and practice, we will consider author's understanding of management of network development of participants of the local food markets, believing that the world outlook part of methodology (which is in focus of our attention) isn't subject to the strict proof, only argumentations of an author's line item. The world outlook aspect influences statement of a research problem and interpretation of results. World outlook aspects of methodology are dualistic as they represent unity objective and subjective. We suggest to understand set of the methods allowing to synthesize theoretical views on a problem of management of network development as a method of regulation and activization of opportunities, resources and knowledge of the subjects of the local food market showing interest in development of regional economy and showing responsible behavior taking into account the situation determined from a line item "role status" as world outlook aspect of methodology of the considered area of a research. The approaches created during formation and development of the classical economic theory, the neoinstitutional economic theory, network theories and digital economy became theoretical basis of a research namely: system, resource, institutional, network, evolutionary, behavioural approaches. Research methods: cognitive modeling, method of analogies, abstract and logical method, system approach.

**Keywords:** Digital economy · Food market · Network of competences · Regional economy subject

S. Ashmarina et al. (Eds.): *Digital Transformation of the Economy: Challenges, Trends and New Opportunities*, pp. 419–425, 2019.
https://doi.org/10.1007/978-3-030-11367-4_41

# 1 Introduction

Spatial-economic transformation of subjects of the local food market is shown in forming of the structures uniting the milti-branch participants of economy of the region having convergent lines. At the same time the processes answering to objective laws, leading to consecutive change of formations are observed: from chaos to hierarchy and to more perfect frictionless structures. The research of such processes is in focus of attention of the scientists (Stepnov et al. [7], Tallman [12], Wernerfelt [14]) showing reasonable scientific interest in network form designing of development of the food sector of the region economy considering nature of placement and interaction of participants, allowing to use effectively their resource opportunities, experience and knowledge. The research purpose consists in development of ideas of a network interaction as which authors consider network of competences. According to the purpose, in this work the following tasks were set and solved: to determine network of competences, to provide its structure, to allocate imperatives of forming and development of network of competences.

# 2 Materials and Methods

The original approach consisting in a combination of a morphological method and a method of analogies was applied to obtaining and converting of ideas of network of competences. "Having spread out" the word network on letters, the terms determining nature of behavior of various groups of regional economy subjects were found. At the same time an analogy to users of social networks was drawn, and the group behavior was determined according to type of behavior social networks users. 7 types of groups (behavior) were as a result received that became the basis for search of the directions of ensuring effective interaction of the subjects of regional economy entering this or that group. Except the specified methods, authors applied resource, institutional and network approaches and developments of the classical economic theory, neo-institutional economic theory, theory of network economy, the theory of digital economy and the management concept development of network of competences.

# 3 Results

Being guided by the principle of ontologic engineering, we will determine new economic category: the network of competences of participants of the local food market is a dynamic great number of the interconnected economic agents (Table 1) having the resources, resource potential, experience and knowledge necessary for implementation of production of quality food products in the amount and an assortment satisfying to requirements of the population of the region.

**Table 1.** The subjects creating network of competences of the food sector of region economy

| Group of subjects | Structure of group |
|---|---|
| **N** (*newcomer*) | "BEGINNERS": the organizations which are engaged in conversion of waste, rendering transport, logistics, information and telecommunication services, the organizations of the sphere of public catering |
| **E** (*enthusiast*) | "APOLOGISTS": the overworking and agricultural enterprises; organizations of wholesale and retail trade |
| **T** (*troublemaker*) | "TROUBLEMAKERS": mass media; trade professional publications; the organizations exercising control and supervision |
| **W** (*watcher*) | "OBSERVERS": the organizations performing environmental and social and economic monitoring; population; the organizations which are engaged in public examination |
| **O** (*originator*) | O (originator) "CREATORS": organizations of education, research and development; organizations of the sphere of physical culture and sport |
| **R** (*redactor*) | "MODERATORS": executive bodies of the government; population; institutes of civil society |
| **K** (*keen*) | "CONSUMERS": population; the entities and the organizations showing interest in development of the food sector and with enthusiasm helping it |

Source: compiled by the authors.

In other words, the network of competences is a concentration of subjects (participants) having special knowledge, experience, and, above all resources. Resource providing plays a crucial role as, quite often, acts as the factor limiting development. In the conditions of digital economy to resources as special requirements are imposed on the basic element determining competences. The most relevant is security innovative and information [10].

## 4   Discussion

The network of competences is a set of experience, knowledge and resource opportunities of participants. Among all variety of necessary resources, in the conditions of digital economy the special importance is purchased by innovative and information resources. Innovative resources in works of domestic economists are considered as "the phenomenon capable to remove society on a new stage of economic development without the mass social conflicts". A. Shishkin with coauthors consider that activization of innovative resources happens in the conditions of openness to the new ideas and vision of prospects, elimination of institutional traps, overcoming the traditions and prejudices complicating innovative development, destructions of the barriers established by the previous level of knowledge and information, creations of new examples of culture of society [4]. As a rule, innovative resources are understood as set of the social and economic relations developing between split level economic subjects (households, firms, the state) concerning forming and implementation intellectual productive and powers of consumption and qualities of the person providing the continuity of creation process, use and preserving innovations in modern knowledge economy [8].

The special role for forming and development of network of competences is played by information resources whose importance is extremely highly appreciated by modern economists from among those who consider the third industrial revolution the "digital revolution" promoting forming of post-industrial society. In modern researches the concept of the third industrial revolution was transformed to the Industrie 4.0 program developed and realized in Germany as the public and private project of creation of completely automated production assuming the interaction happening within the concept of the Internet of things allowing to adapt flexibly production processes under new requests of consumers [11]. The main benefit of the concept "the Industry 4.0" is that core competencies of subjects of network are possibilities of handling of big data arrays. Paying attention to this circumstance, Stepnov et al. [7] specify that "the Industry 4.0" "provides end-to-end digitalization of all physical assets and their integration into a digital ecosystem together with the partners participating in a value creation chain". The quoted authors consider the main differences of this concept the shift of emphasis on investment ensuring engineering procedures, change of competences and qualifications of workers, institutional transformation of approaches to the organization of interaction of accounting entities, including the population which on projections will shortly prefer creation of workplaces of "house" that quite corresponds to "spirit" of digital revolution especially as the generation which learned to maneuver in flows of the redundant information [2] already grew.

The prospects of use of information resources are offered with transition to the new (fifth) industrial revolution allocated with P. Marsh, afraid of loss of the USA of the line items as leader in world industrial production. The recommendations offered by P. Marsh [3] are universal for economy of any level which implementation will allow to keep its industrial potential. The food sector of region economy on scales of production and specific weight of innovative products can be considered the industrial sector that actualize in relation to it the following components of the fifth industrial revolution: active development new mainly complementary technologies; use of new opportunities for automation and robotization of production; development of capabilities of social and economic systems to adaptation and individualization of production; placement of production in the territory close to the place of sale (on condition of technological feasibility); the expansion of global networks use which is expressed in consolidation of supply chains of products and information flows; clustering of economic space of the territory; growth of influence of ecological factors; expansion of a range of services, the accounting entities accompanying implementation of a production profile of interaction [1].

## 5   Conclusions

Ideas of forming of competences network supplement the following imperatives:

1. Forming of competences network is always connected with accommodation of subjects of regional economy. Really, the problem of a setization is a problem of placement. We will draw an analogy to expansion of the network companies which

as regions, priority for the placement, consider the territories with availability of the centers of high concentration of the consumer demand [14]. Forming of network of competences is the same way observed in the regions differing in high concentration of participants, co-creating value, but at the same time positively competing as the lack of the competition deprives of subjects of regional economy of an opportunity and aspiration to development. It is obvious that it is totally pointless to compete with a mirror. That is why subjects of regional economy need availability of the organizations initiating their interest in promotion and enhancement [8].

2. The network of competences in fact is a cluster with the institutional agent as whom the state acts. Taking into account new endogenous and exogenous challenges the task of development and implementation of state economic policy consists in regulation of interaction of regional economy subjects in the directions limiting regional growth. The solution of the specified task is possible because of fixing to executive authorities in competences network of function of the institutional agent performing the initiative responsible creative activities in the system of global risks and uncertainty aimed at forming of the conditions favoring to the fair competition [5].

3. Competences networks have regional specialization. The necessity of isolation of regional competences networks is caused by presence of federal level of economic management which functionality shouldn't be duplicated at the level of the region.

4. Executive bodies of the government perform function of provider of the fair competition in the regional market as regulate resource allocation between networks of competences and federal networks in the region. The tools allowing executive bodies of the government to support balance of competences, to promote completion of missing competences, to redistribute resources and to regulate implementation of resource opportunities are project management and the system of state programs.

5. Criterion of forming and development of network is justice because all that limits opportunities isn't fair. At the level of regional economy, the problem of justice is directly connected with distribution of means of the Government budget, the interbudget relations, implementation of state programs and use of other instruments of state regulation. The economic space of our country is characterized by essential differentiation of regions on the level of economic development that is extremely negative fact [9]. However, the tendencies which are qualitatively worsening the current situation, being manifestation of "injustice" are noted in recent years. So, the regions which were traditionally acting as donors of the government budget (The Kaluga, Lipetsk, Belgorod regions) appear in group of subsidized regions that is a consequence of inefficient policy of the state. According to us, the regional policy shall be directed to support of strong regions to create conditions for implementation of their potential. Unfair redistribution of budgetary funds leads to decrease in national level of economic development and doesn't promote forming of the competitive environment which is a basis of forming of network interaction.

6. Coordination of competences network is provided due to forming of the internal information space organizing flows of innovative knowledge. The building-up of

amounts of the circulating information promoting diffusion of innovations leads to creation of collective intellectual property items and increase in external effect of activities of network.

7. The optimality of network structure is determined: cost efficiency of its functioning; speed of knowledge generation; speed of accumulating of a good practice; speed of detection of missing resources.

8. Stability of network is provided due to preserving independence and isolation of partners in implementation process of the competences by them as joint activities are resulted by convergence and the subsequent unification of competences. The lack of original competences deprives of participants of network of appeal, doesn't allow to be created to a competitive environment and inevitably leads to disintegration of network, creating a vacuum on economic space of the region that is not admissible in a case with the food sector of regional economy which violations in functioning cause loss of a food and economic safety of the country.

9. The effective motivational mechanism for participants of network is forming of need for a cross-functional cooperation on the terms of parity and equality because "neutralization of a priority of one participant of network before the others guarantees protection against manifestation of opportunistic behavior".

10. Prepotent signs of competences network are: each participant of network has a certain functional competence (a set of competences) which are considerably localized; the number of participants of network is rather high, at the same time it is comparable to quantity of technological stages of a cycle co-creation realized by participants of network, determined by its subject content and less number of the projects realized by participants of network is incomparable; participants of network are autonomous functioning organizations involved in interaction by the aspiration to satisfy the resource problems of each other and interest in implementation of resource opportunities.

Thus, we selected the provisions disclosing the world outlook aspects of methodology of management of development of competences network of consisting in the following. First, evolution of a methodological basis of the management concept development of competences network of the food sector of region economy became possible thanks to development of system approach as universal methodology of a research. Secondly, networks of competences should be considered from a line item of the subject, a subject and an object what allows them deeply and to research comprehensively to offer the efficient acceptances of managerial interaction considering features of their forming and development.

# References

1. Doroshenko SV, Shelomentsev AG, Sirotkina NV, Khusainov BD (2014) Paradoxes of the "natural resource curse" regional development in the post-soviet space. Econ Reg №4:81–93. https://doi.org/10.17059/2014-4-6
2. Goncharov AY, Sirotkina NV (2015) The mechanism for managing a balanced development of regions with dominant economic activities. News of higher educational institutions. Technol Text Indus № 4(358):35–43

3. March P (2015) New industrial revolution. Consumers, globalization and the end of mass production, Moscow, Publishing house of the Gaidar Institute, 420 p
4. Shishkin AF, Shishkina NV (2013) Economic theory, Tom 1. LLC KDU Publishing House, Moscow, p 1160
5. Sirotkina NV, Golikova GV, Romashchenko TD (2018) Policy technologies, and approaches to management of organizational changes. Stud Syst Decis Control, №135:31–38. https://doi.org/10.1007/978-3-319-72613-7_4
6. Sirotkina NV, Stukalo OG (2015) Clustering of the economic space of a region in the context of the formation of the food industry. Terra Economicus, T 13(3):99–109. https://doi.org/10.18522/2073-6606-2015-3-99-109
7. Stepnov IM, Kovalchuk YA, Demochkin SV (2016) Structural and functional analysis of theories of economic and industrial development. Part 1. Sci Time № 9(33):232–244
8. Stigler GJ (1961) The economics of information. J Political Econ 69(3):213–225
9. Stiglitz J, Akerlof G, Spence M (2001) L'asymetrie au coeur de la nouvellemicroeconomie. Problemes Econ, N 2734:19–24
10. Stukalo OG (2015) Conditions and prerequisites for the formation of the region's food industry. Bull Kursk State Agric Acad T 135:31–38
11. Stukalo OG (2016) Methodical approach to the formation and development of the food industry in the region. Mod Trends Dev Sci Technol № 11–9:111–113
12. Tallman SB (2000) Forming and managing shared organizational ventures: resources and transaction costs. In Faulkner D, DeRond M (ed) Cooperaive strategy. Oxford University Press, Oxford, pp 96–118
13. Wernerfelt B (1984) A resource-based view of the firm strategic management. Journal 5:171–180
14. Wiliamson O (1996) Comparative economic organization: the analysis of discrete structural alternatives. Adm Sci Quartely № 36(2):269–296

# Education of the Future: New Professional Frames and Jobs in Conditions of Digital Economy

# Readiness to Changes as One of Educational Values of Innovation-Oriented Procurement

L. V. Averina🅳, E. P. Pecherskaya$^{(\boxtimes)}$🅳, and A. R. Rakhmatullina

Samara State University of Economics, Samara, Russia
Alv94@ya.ru, pecherskaya@sseu.ru, sseu_ar@mail.ru

**Abstract.** The article deals with the essential characteristics of the willingness of contract managers to professional activities in the procurement system, the possibility of their change in the learning process and some challenges connected from educational values' influence point of view. The purpose of this article is consideration of the existing approaches to understanding of the phenomenon of contract managers' training from the point of view of value orientations, which are the basis for business activity in conditions of innovative development of the global economy. The results of the conducted research on the readiness to changes in procurement are presented in the study. They characterize the economic and social environment in procurement in Russia at present and in Samara region of the Russian Federation, exactly. Readiness to changes is crucial for creation of a new mode of thinking of the modern business community. By results of the conducted research, the authors offered a model of this integrative characteristic, which can be the basis for purposeful process of training and retraining of contract managers for development of the procurement and, furthermore, economics of the region and country in the context of requirements of the innovative economy.

**Keywords:** Contract manager · Innovative economy · Procurement
Training · Values in education

## 1 Introduction

Contract managers' training, acting as one of the leading factors in the successful functioning of the contract system in the procurement sphere for state and municipal needs, is one of the most important tasks facing educational organizations [1–3]. Despite the urgency and relevance of professional training in the field of procurement, the scientific thoroughness of this topic remains, as before, at low level.

At the same time, the existing methodological and didactic material allows to formulate the main approaches to the training process for contract managers.

Thus, when designing and organizing the educational process, it seems expedient to be guided by the principles of andragogy, since the educational process and its results must correspond both to the structure of the learner's experience and to develop (stimulate its development) in the desired direction [4, 17, 18, 21].

According to the principles of andragogy formulated by Hadjar and Samuel [11], an adult learning person has a leading role in the learning process.

© Springer Nature Switzerland AG 2020
S. Ashmarina et al. (Eds.): *Digital Transformation of the Economy: Challenges,*
*Trends and New Opportunities*, pp. 429–436, 2019.
https://doi.org/10.1007/978-3-030-11367-4_42

The learning process should be organized in the form of a joint activity of the learner and the trainer at all its stages, be, first of all, practical-oriented.

In the choice of methods it is necessary to give preference to active ones, including project training, dual training, case study, interactive technologies [20].

In theoretical training, disciplines containing integrated material on several related fields of knowledge come first [12–16].

Particular attention should be paid to the principle of electivity (variability, flexibility), as one of the principles of androgogy. The implementation of the principle should provide a real opportunity to optimally take into account the contingent of trainees, their profile, the level of readiness for mastering the content of education, create the possibility of correcting the content of professional education for contract managers, depending on changing conditions in the labor market and in society. The reform of the contract system is far from completing [3], which requires the constant updating of both the content content of educational programs, and approaches, teaching methods [6, 19]. That is why authors in the process of professional training and retraining of the contract manager, pay special attention to the availability of experience in this area (the initial level of knowledge), which allows to correct the educational process in a timely manner taking into account this factor.

## 2  Materials and Methods

The research methods are the system analysis, expert evaluation methods, polling and interviewing, modeling.

The main research methods

- the first stage was implemented by means of analysis of the existing approaches to educational values and competency of contract managers based on self-reflection method and progress monitoring;
- the second stage included monitoring and diagnostic work, development of the model of readiness of contract managers to changes, and the complex of the most significant values in the procurement environment is investigated;
- at the third stage the results of diagnostics are processed and analyzed, objectives for a further investigation phase are made.

## 3  Results

Considering the essence of values, it should be noted that understanding of both key purposes and ways of their achievement is the basis of this category. These are the fundamental standards, principles, guiding ideas, which provide development of any society, direct certain individuals, help us to choose this or that behavior in vital situations [9].

In conditions of the innovative oriented economy, readiness to changes should be an essentially important value orientation of the business community [5]. Exactly this quality can become fundamental in the new values system of the business world, considering the turbulent nature of the external environment both at the individual level and at the level of organization in general.

Readiness consists of knowledge, skills, practical actions, formed motives, cognitive activity, ability to self-esteem and reflection as showed on the Fig. 1 below.

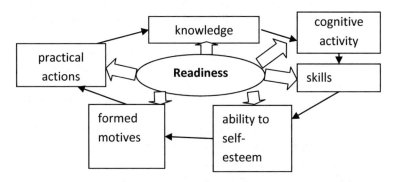

**Fig. 1.** Structure of contract managers' readiness (Source: Authors, 2018)

When modeling the structure of readiness for professional activity, we took into account the specificity of the activity in the context of the labor function (the processing of large amounts of information in the context of the EIS (information procurement system), the specificity of activities in terms of competencies (knowledge in the field of economics and law (synthesis of knowledge, interdisciplinarity) analytical skills, planning and forecasting skills).

The structure of readiness for professional activities of contract management activities includes such interrelated components presented in Table 1, which are mutually determined with functions (planning, rationing of procurement, search, information processing, analysis, procurement system management).

**Table 1.** Components of contract managers' professional activity

| Component name | Component content |
| --- | --- |
| The technological component | implies the use of analytical operations in professional activities: the ability to analyze, classify and summarize information, group facts, draw conclusions based on available data, predict the situation, and make judgments about the findings |
| | This also includes the ability to use various information sources |
| The cognitive component | Presupposes knowledge of the theoretical foundations of mental operations |
| The information component | Includes the ability to work with information, including, to receive, process and interpret large amounts of data |
| The reflexive-evaluation component | Assumes that the contract manager should be able to analyze his actions, evaluate the results of his activity, and gain experience. So, the reflexive component contributes to an objective assessment of the results of their actions and self-improvement through the analysis of their professional activities |

Source: compiled by the authors

Indicator level of readiness of the learner for activities is the formation and severity of these components.

At the same time, despite the monitoring of the pace of mastering the material and the results of the training conducted throughout the entire educational process (input, intermediate, final testing, preparation of design works), there are goal-setting risks and intellectual risks that affect the willingness of contract managers to professional activities.

Under the risks of goal-setting within the framework of this study we mean purposefulness and level of satisfaction with the result achieved, the level of reflection. Under intellectual risks within the framework of this study we mean congruence, readiness to perceive new information, mastering new skills, the influence of negative patterns.

As the study showed in a representative sample of trainees - contract managers (150 respondents), the purposefulness and level of satisfaction with the result achieved are in inverse correlation from age and professional experience (work experience in procurement) and the lower the above mentioned indicators:

46% out of 100% are satisfied with the achieved result among students with three years of experience in the field and above, 34% of 100% are satisfied with the result in the age category from 45 and above. Among the students with an experience of 0 to 3 years, these indicators were 89% of 100%.

At the second stage we studied the development of readiness of contract managers to changes, and the complex of the most significant values in the procurement environment for trainees. The results are given below in Chart 1.

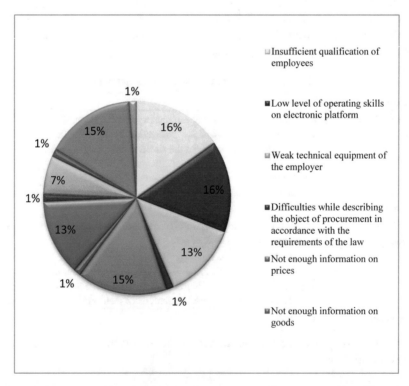

**Chart 1.** The values and attitude to challenges of contract managers while studying procurement (Source: Authors, 2018)

At the initial stage of training, only 32% of contract managers had an average level of formation of project competency, that is more significant level of professional qualities, while 16% have a higher level of formation of project competency. At the final stage of training 60% of trainees have a higher level of formation of design competence, whereas 11% of contract managers have the base level of formation of such competency. Analysis of the dynamics of formation of professional competence in the experimental group gives reason to believe that the implementation of the developed model of formation of project competency makes it possible to significantly improve the efficiency of the process of formation of professional competence of contract managers.

Experimental accuracy had been ensured by the fact that the initial characteristics of the groups under study were similar. This is proved by the data ascertaining experiment, during which it was found out that contract managers had virtually the same starting performance by the level of knowledge and development of monitoring actions. In addition, the results of our research on a representative sample – contract managers as trainees on refresher courses of "Contract system in the procurement of goods, works and services for state and municipal needs" on the basis of federal innovation platform FGBOU VO "Samara State Economic University" in 2014–2015 showed, that the new employees of companies who have been studied the above-mentioned course, at rates faster got accustomed to their professional duties and more effectively took on responsibilities, since such contract managers had mastered their knowledge in the field of operation of the contract system, they also possess practical skills in the field of procurement, they know the actual models of procurement management and ready to solve specific practical problems in their field of activity.

## 4   Discussion

Modern realities in the education system requires theoretical understanding and identifying pedagogical conditions that would ensure the formation of practice-oriented competencies of contract managers in the process of professional education and practice. In addition, due to the recent introduction of Russian training programs for contract managers as CPE (from 2013 only), the lack of systematic analysis of professional competencies and, in particular, project competency formation of contract managers, in domestic and foreign academic literature should be mentioned. Meanwhile, analysis of pedagogical activity of tutors shows that only a few of them can properly control the results of the educational process, to determine the factors and means of its successful implementation, assess the extent and effectiveness of its implementation, to identify problems and difficulties to make the appropriate adjustments.

Due to the study there were eliminated some challenges that reduce the effectiveness of additional vocational education or training in procurement. First of all, it is inadequate preparation for pedagogical activity within the framework of a practice-oriented approach and the design of such activities, to which all new requirements are made. For trainer who trains contract managers, it is necessary to own and use in his activity the ability to design it on the basis of an integrative (hlistic) approach that takes into account differences in the types of knowledge; technology of problematic, inter-active learning, learning through cooperation [7]. Secondly, development of an inno-vative component of the economy enhances intensive interaction of universities and business communities, that is why universities should take into account the develop-ment of those professional competencies that are most in demand for employers. To solve the above-mentioned problems the authors conducted a longitudinal study on a representative sample of contract managers as trainees of training courses in the field of procurement.

Innovative technologies proved to be effective in approaching the methodological problems in the course of authors' work as lecturers at the university and as tutors on business training programs of contract managers.

## 5   Conclusions

As the result of mastering new forms and methods of professional activity in the field of procurement will be formed critical thinking and a system of values that promote active participation in the sustainable development of society and business [8, 11]. However, one should take into account the impact of the risks in the study on the successful formation of the readiness of contract managers to professional activities.

Monitoring of educational phases on each module included diagnosis, analysis of the success of the implementation of training stages (modules) and timely correction of the structural components of the modules.

Educational activity in the monitoring mode and the training session using the project-learning technique allow to implement a quality approach to the educational process. Obtaining timely information on the occurrence of each stage of the educa-tional process allows tutors to analyze how tasks developed at different stages of the educational process, consistent with the objectives, and how they are focused on the verification of knowledge and level of skills in accordance with their castings on reproductive and productive levels. In the event of deviations it allows tutors make adjustments quickly and to improve the quality of educational process in a short period of time. The ability to design training sessions on the subject of a monitoring system to determine the location of each session in order to address common goals and objectives of training, organize the effective interaction with the trainees provided the achieve-ment of high quality education.

Monitoring activities had been formed successfully by all subjects of the educa-tional process in contract managers' training model implementation based on their readiness to changes. During the joint analysis and evaluation of the work performed by both the tutor and by the trainee, both conducted a discussion of difficulties, gave substantial and detailed description of learning outcomes, analyzed strengths and weaknesses of the work performed. It allows tutors to relate his assessment of the achievements of trainees with their self-esteem and, therefore, make adjustments to it in time. All of the above-mentioned technologies and methods helped to successfully implement in the structure of educational process the "feed-back" system and quality management of educational process.

Blended learning has become increasingly popular, when interactive training is combined with personal communication with teachers [10, 17]. In this regard, when developing the training process of contract managers, it is considered to be useful taking into account the requirements for vocational education by the social and economic environment.

# References

1. Abakumova OA (2014) Legal status of the contract manager. Acad Herald. N2 (28):269–276
2. Averina LV, Pecherskaya EP (2015) Actual problems of preparation of a specialist in the field of procurement. Materials of the international scientific-practical conference of students, graduate students and young scientists "Social behavior of youth on the Internet: new trends in the era of globalization." SSEU, pp 39–41
3. Averina LV (2017) Models of organization of dialogical interaction in the implementation of DPO programs in the field of procurement. Sb. articles of the International scientific-practical conference "European research" pp 266–268
4. Axelrad H, Luski I et al (2016) Behavioral biases in the labor market, differences between older and younger individuals. J Behav Exp Econ 60:23–28
5. Berry A (2008) Tensions in teaching about teaching: understanding practice as a teacher educator. Springer, Dordrecht
6. Clandinin DJ, Huber J (2010) Narrative inquiry. In: McGaw B, Baker E, Peterson PP (eds) International encyclopedia of education, 3rd edn. Elsevier, New York, NY, pp 436–441
7. de Vries Marc J (2018) Handbook of technology education. Springer International Publishing, DOI. https://doi.org/10.1007/978-3-319-44687-5
8. Dufek L (2015) Public procurement: a panel data approach. Proced Econ Financ 25:535–542
9. Enke J, Kraft K, Metternich J (2015) Competency-oriented design of learning modules. Proced CIRP 32:7–12
10. Garbett D, Ovens A (eds) (2017) Becoming self-study researchers in a digital world. Springer, Dordrecht
11. Hadjar A, Samuel R (2015) Does upward socila mobility increase life satisfaction? A longitudinal analysis using British and Swiss panel data. Res Soc Stratification Mobil 39:48–58
12. Karanatova LG, Kulev AYu (2015) Organization of university innovative platforms as factor of development of competences of innovative business [Organizatsiya universitetskikh innovatsionnykh ploshchadok kak faktor razvitiya kompetentsii innovatsionnogo predprinimatel'stva]. Administrative consulting [Upravlencheskoe konsul'tirovanie]. N 12:15–23
13. Kirichenko YuA (2012) Modern information technologies in additional professional education. Vestnik TISBI. № 2(50):18
14. Larchenko Yu.G (2015) Game technologies in additional professional education. Additional vocational education. Modern Stud Soc Probl 1(21):195–200
15. Menshenina SG (2018) The structure of knowledge to the professional activity of information security specialists.Bull Samara State Technical University. Series Psychol Pedagogical Sci 2(18):100–107
16. Pecherskay EP, Averina LV, Kochetckova NV, Chupina VA, Akimova OB (2016) Methodology of project managers' competency formation in CPE. IJME-Math Educ 11(8): 3066–3075

17. Pellegrino JW, Hilton ML (ed) (2012) Education for life and work: developing transferable knowledge and skills in the 21st century NAS Press
18. Shamakhov VA, Karanatova LG, Kuzmina AM (2017) The research directed to results of activity of the federal innovative platforms forming professional competences of the sphere of the state and municipal procurement assessment. Administrative consulting [Upravlench-eskoe konsul'tirovanie]. N 12, pp 8–21
19. Sklyarova IM (2014) Functionality of contract managers in the field of public procurements. Labor and social relations. №8:26–32
20. Sullivan KPH, Czigler PE, Hellgren JMS (2013) Cases on professional distance education degree programs and practices: successes, challenges, and issues. IGI Global
21. Zhang Yu (Aimee) (2015) handbook of mobile teaching and learning. Springer. https://doi.org/10.1007/978-3-642-54146-9

# Gaps in the System of Higher Education in Russia in Terms of Digitalization

S. I. Ashmarina[1]($\boxtimes$), E. A. Kandrashina[1], A. M. Izmailov[1],
and N. S. Mirzayev[2]

[1] Samara State University of Economics, Samara, Russia
asisamara@mail.ru, kandrashina@sseu.ru, airick73@bk.ru
[2] Lankaran State University, Lankaran, Azerbaijan
mirzoev.n@mail.ru

**Abstract.** The authors study the main aspects of economy digitalization as a global megatrend, its trends and directions, as well as the impact of economy digitalization on gaps in the system of higher education in Russia. The digital economy affected processes in the main social and socio-economic systems (SES). The purpose of this contribution is to study the impact of the digital economy on the processes in the system of higher education in Russia, namely the impact on gaps. The main range of research tasks can be outlined by analyzing such a phenomenon as gaps in socio-economic systems, the nature of their occurrence, as well as the role and influence of digitalization on them.

**Keywords:** Digitalization · Gaps · Gaps model · Higher education
Socio-economic system · System of higher education system

## 1 Introduction

Digitalization is a global trend spawned by the widespread dissemination of information technology. The depth of penetration and the relevance of information and communication technologies today form a new environmental reality, transforming the classic social and socio-economic systems to requirements of the time [10].

Considering the fact that the main part of the business sector has shifted to a "number", today we have an economy that is significantly different from the one that was only a quarter of a century ago. Today we live in the era of the digital economy, which forms a new reality.

The term digital economy appeared relatively recently, in 1995, and was first introduced by American scientists from the University of Massachusetts [5]. The digital economy is a new stage in the development of the economy as a whole, and its main driver is the rapid development of information and communication technologies.

Considering the significance and level of penetration of digitalization into the life of modern society, specialized state and regulatory acts and regulating aspects of the

---

The original version of this chapter was revised: Last author information have been corrected. The correction to this chapter can be found at https://doi.org/10.1007/978-3-030-11367-4_69

© Springer Nature Switzerland AG 2020
S. Ashmarina et al. (Eds.): *Digital Transformation of the Economy: Challenges, Trends and New Opportunities*, pp. 437–443, 2019.
https://doi.org/10.1007/978-3-030-11367-4_43

digital economy began to be adopted not so long ago. The Strategy Development for Information Society for 2017–2024, approved in Russia, defines the digital economy as follows: "The digital economy is an economic activity in which digital data, processing large volumes of information are the key factor in production. Compared to traditional forms of management, new forms can significantly improve the efficiency of various types of production, technologies, equipment, storage, sales, and delivery of goods and services" [19].

The digital economy has had a significant impact on many social and economic processes, the results of which were expressed in the positive dynamics of the main macroeconomic indicators. If we talk about Russia, the country's GDP from 2011 to 2015 increased by 7%, and the volume of the digital economy over the same period increased by 59%, to 1.2 trillion rubles (upon prices in 2015). Thus, the digital economy accounted for 24% of the total GDP growth. From the point of view of analysts at McKinsey, by 2025 economy digitalization in Russia will make it possible to increase the country's GDP by 4.1–8.9 trillion rubles [17].

## 2   Materials and Methods

The main methods used in the contribution should include analysis and modeling. The analysis is a tool for studying the system of higher education in Russia, which is a complex socio-economic system. In terms of the impact of digitalization on this system, it is an analytical approach that allows assessing the influence of the external environment on the structural elements of the education system and studying gaps. After studying the impact of digitalization on the socio-economic system, the most successful method is modeling, which allows you to visually see the formation of gaps.

## 3   Results

The system of higher education is closely interconnected with the digital economy. In accordance with the Program for the Development of the Digital Economy in the Russian Federation for a period until 2024, one of the priority areas is "Personnel and Education". In the framework of this priority direction, it is intended to solve such tasks as:

– Develop key conditions for training in the digital economy;
– Improve the education system, which should provide the digital economy with competent personnel;
– The labor market should be based on the requirements of the digital economy;
– Develop a system of motivation for the development of necessary competencies for personnel in economy digitalization in Russia.

However, the interaction of the education system with the digital economy is focused not only on the aspect of adaptation to changing conditions through the development of new educational programs and the training of qualified personnel with a set of key competencies. The HE system as a large and significant SES requires the training of qualified personnel that is able to meet the requirements of a dynamically changing environment. At the same time, it must be said that the digital economy is able to influence the processes taking place within the HE system. For example, gaps in the education system are deeply affected by the digital economy.

To understand the subject matter of the question under consideration, we first consider the concept of discontinuities, their nature and specific features within the framework of the HE system.

The scientist who first applied the concept of strategic gaps is considered to be Ansoff [1], who at one time identified two types: the gap between the environment and strategy, and the gap between strategy and capabilities. Another Western scholar, Post et al. [3] interprets a strategic gap as the difference between a strategic plan and an operational execution. These gaps, according to Coveny, can be caused by such reasons as: the strategy isolation from life, the lack of strategic focus, the lack of realistic planning and the lack of control over implementation.

The issues of gaps are partially covered, including in the scientific works of domestic scientists. For example, in the work of Kramin and Kramin [19] the gap is represented as a deviation of the desired state, determined by strategic plans, from the current state. The main parameters that can characterize the gap phenomenon are such as: declarations (high expectations), expert assessments (maximum achievable indicators), forecasts (realistic indicators), and results.

The declaration is official statements within the framework of which the basic principles of the state's external or internal policy and its program provisions, and the main forms for the system development are proclaimed for the near and long term. The formation of official documents (declaration) is the responsibility of the relevant ministry and the Government of Russia.

An expert assessment is the conclusion of one person or a group of persons who are competent authorities in a particular environment. In relation to the HE system, expert assessments are formed in relation to the possibility and degree of achievement of the planned actual development results in a given period.

The forecast, in the context of the problem under consideration, is a probabilistic judgment about the state of any phenomenon in the future, based on a special scientific research (forecasting).

Declarations, expert assessments and forecasts can be grouped into the "expected results" (desired) group, within which the main development directions are formed, parameters and vectors are set, development forecasts are developed, and opinions of authorities are taken into account. However, the expected results may not coincide with the actual results (achieved) and differ from them a lot.

The strategic gap can be called the difference between the desired and the achieved, the elimination of which will be associated with large time and resource costs. The causes of strategic gaps can be the leadership, planning processes, and the information technology used.

Understanding the causes of gaps is extremely important in the analysis process. From our point of view, when analyzing strategic gaps, it is necessary to identify discrepancies: firstly, between desires, secondly, between those indicators that can be achieved with the realization of different development pathways.

However, despite the fact that there are declarations, expert assessments and forecasts, they may differ from each other to a greater or lesser degree. Goals and objectives with their inherent indicators, officially declared in program documents may be different from those presented in the materials compiled by experts and voiced in forecasts from various research institutes and scientific schools. Projections for the development of different indicators of the HE system may differ greatly from those presented in Development Programs. Experts can predict that it is impossible to achieve the planned "heights" while scientific institutions predict the opposite.

So, based on all of the above, we can say that there are two types of discontinuities. The first type is the gap between declarations, forecasts and expert assessments. The second type is between expected and actual results.

It is reasonable to assume that the external environment is actively influencing the occurrence of one or another type of gaps. If we present the described situation in the form of a model, we get the following picture (see Fig. 1).

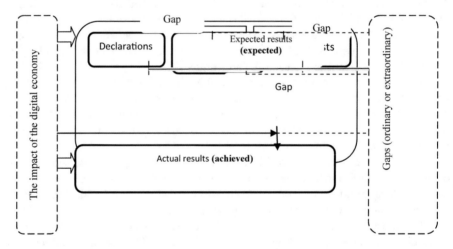

**Fig. 1.** The impact of the digital economy on gaps between parameters (Source: compiled by the authors)

As can be seen from the figure, there are gaps both between the elements of the expected results and actual results. Often, the results of the development of the HE system at different stages differ from those planned in declarations noted in expert assessments and formed in official development forecasts. In both cases, the influence of digital factors can be noted.

We present the key digital factors that influence the elements of the group of expected results according to the figure below.

Consider the possible reasons for gaps between the expected results and actual results in economy digitalization in the context of three elements: declarations, expert assessments and forecasts.

If we talk about declarations, then we can say that the goals in declarations are formed, firstly, on the level of ICT development for the period of producing official documents, secondly, based on forecasts for the development of economy digitalization for a given period. However, it is difficult to predict the dynamics of digitalization, so the accuracy in this aspect is very conditional (Table 1).

**Table 1.** The impact of digital factors on the group of expected results for the development of the HE system in Russia

| Impact of digital factors | | |
|---|---|---|
| Declarations | Expert assessments | Forecasts |
| – The dynamics of digital development, the depth of its penetration and the response of SES may differ significantly from those declared in official documents. In this regard, the expected results for the development of the HE system indicated in declarations in any case are conditional | – High specificity and dynamics of digital development, which requires an expert to have a massive amount of accurate information, which is often impossible due to a high level of secrecy (including trade secrets) on information technologies that are being prepared for implementation;<br>– It is difficult to predict the impact of digitalization on individual elements of the HE system | – High dynamics of digital development and it is very difficult to assess the level of development of SES affected by the technology in a given period of time;<br>– Deep penetration of digitalization both in the HE system and in adjacent systems, thereby forming a picture of the world based on the response of SES to the penetration of new technologies into them |

Source: compiled by the authors

On the one hand, it is difficult to have deep knowledge in the field of development and the impact of ICT on the emerging picture of the future when producing official documents. On the other hand, the dynamics of technologies and their penetration into the life of society does not always have the same speed. According to Moore's law [9], the dynamics of technologies should change their speed downward in the foreseeable future. However, nobody knows when this moment comes. It is also necessary to say that digitalization (digital economy), influencing the HE system, causes its reaction expressed in adaptation to the requirements of the time. Adaptation is expressed in the penetration and adoption of new technologies by the education system. However, even if there are new technologies in the digital economy that affect the HE system, then the rate of change in HE also depends on many factors and it is difficult to predict them.

The gap between expert assessments and actual results can be justified, first of all, by the level of competence and expert knowledge of the impact of digitalization on HE. If knowledge does not allow assessing the influence of digitalization on the HE system, then the gap between the expected and actual results will be higher.

Forecasts for the development of HE are taking into account the development of technologies both worldwide and in Russia. However, due to the fact that the level of development of technologies affecting HE in Russia and in the world is different, then it is difficult to forecast this impact. First of all, it is difficult to assess the development within the framework of the HE development. There are no education systems based on digital technologies in Russia. This is accompanied by both the lag in the development of the national HE in general, and difficulties in the development of the digital economy.

Speaking about the HE system in economy digitalization in Russia, one can say that the key result of the gap between expected results and actual ones is the discrepancy between the demands of the labor market formed under the influence of modern requirements and actual training results. The main reason for the discrepancy is the lag in the level and quality of training from the real needs of the labor market.

## 4   Discussion

Certain aspects of gaps in socio-economic systems are studied and presented in the works of Western scientists [1, 14, 18] as well as domestic researchers [6, 7, 11, 13, 14, 16, 18, 19]. Key elements of the influence of contact audiences on the development of socio-economic systems are presented in many scientific works [2, 4, 12, 15, 13]. The degree of environmental influence on the system is also very important. In the light of the active diffusion of digitalization into all existing spheres of life, the studies of theoretical aspects of the influence of digitalization on socio-economic systems are relevant [8, 9]. However, taking into account the diversity and varying degrees of complexity of the existing socio-economic systems, the issue of gaps in the HE system, which has specific features and a complex structure, are still little studied.

## 5   Conclusions

The HE system in Russia is closely interconnected with the digital economy. Key aspects of the relationship are reflected in regulatory acts at the state level. The HE system as one of the forms of SES is subject to internal gaps expressed in the discrepancy between the expected results of development and actual results. The influence of the digital economy also plays an important role in this aspect. The impact of the digital economy as a HE environment on certain elements of the expected results is expressed primarily in the discrepancy between the level of trained personnel and emands of the labor market.

**Acknowledgments.** The study was carried out within the framework of the state assignment of the Ministry of Education and Science of the Russian Federation, project No. 26.940.2017/4.6 "Managing Changes in the System of Higher Education Based on the Concept of Sustainable Development and Coordination of Interests".

# References

1. Ansoff I (1989) Strategic management. 3.2.6. Four stages of evolution. M.: Economics, pp 258–259
2. Coveney M, King D, Hartlen B, Ganster DG (2003) The strategy gap: leveraging technology to execute winning strategies. John Wiley & Sons, Inc., New York, NY, USA
3. Coveny M (2010) Strategic gap. Technology implementation of corporate strategy in life, etc. Alpina Business Books, Moscow, pp 69–73
4. Izmailov AM, Ashmarina SI, Kandrashina EA (2016) Features of change management in socio-economic systems, taking into account structural changes in the development of entrepreneurship. Financ Manag 3:46–52
5. Izmailov AM, Bazhutkina LP, Yakhneeva IV (2016) Features of change management in pharmaceutical business structures. Financ Manag 4:112–123
6. Keshelava AV, Budanov VG, Rumyantsev VU (2017) Introduction to the "digital" economy. In: Keshelava AV, Budanov VG, Rumyantsev VYu, others (eds) under total ed. A.V. Keselava; Ch. "Numbers". - VNIIGeosystem, p 28
7. Kramin MB, Kramin TV (2011) Meeting stakeholder interests as a strategic aspect of corporate governance. Actual Probl Econ Law 4:164–171
8. Mendelow A (1991) Environmental scanning: the impact of the stakeholder concept. MA, Cambridge, pp 407–418
9. Moore E (2003) No exponential is forever: But "forever" can be delayed! [semiconductor industry]. In: 2003 IEEE international solid-state circuits conference, 2003. Digest of Technical Papers. ISSCC., San Francisco, CA, USA, 2003, pp. 20–23, vol 1. https://doi.org/10.1109/issee.2003.1234194
10. Negroponte N (1995) Being digital. Knopf, N. Negroponte. – NY, p 256
11. Post JE, Preston LE, Sachs S (2002) Redefining the corporation: stakeholder management and organizational wealth. Stanford University Press, Stanford, CA
12. Presidential Decree of May 9, 2017 No. 203 "On the Strategy Development of the Information Society in the Russian Federation for 2017–2030"
13. Samuelson PE (2015) Economy. In: Samuelson PE, Williams M, p 1358
14. Sanin VV (2009) Balance and conflict of interests of stakeholders in the strategic and business plans of the company [Electronic resource). Electron J Corp Financ 2(10):112–132. https://cyberleninka.ru/article/n/balans-i-konflikt-interesov-steykholderov-v-strategicheskih-i-biznes-planahkompanii. Accessed 21 Apr 2018
15. Smilek NI (2017) The gap as a form of quantitative and qualitative change in the economic system: the agricultural sector. Bull Chuvash GSHA 2:94–97
16. Tazhitdinov IA (2013) Application of the stakeholder approach in strategic development of the territory. Econ Reg 2:17–27
17. The official portal of RBC. The digital economy will increase Russia's GDP by 8.9 trillion by 2025. https://www.rbc.ru/technology_and_media/05/07/2017/595cbefa9a7947374ff375d4. Accessed 21 Apr 2018
18. Toynbee AD (1991) Comprehension of history/ AD. Toynbee. - M.: Progress, p 731
19. Zenkina IV (2012) Analysis of strategic gaps as a tool for strategic analysis and the potential for its use in the strategic management of an organization. Audit Financ Anal 4:107–112

# Training Employees in the Digital Economy with the Use of Video Games

L. V. Kapustina$^{(\boxtimes)}$ (iD) and I. A. Martynova (iD)

Samara State University of Economics, Samara, Russia
lkap@inbox.ru, martynov1998@rambler.ru

**Abstract.** The growth in competition across the digital world means that today's employees need to be better trained than ever. The skills and knowledge provided by traditional training are already insufficient in the digital environment. The flexibility and interactivity of new digital technologies assist people in meeting the needs of the growing digital economy. Using digital technologies in training promote creativity and high order thinking that are highly valued by employers. The purpose of this paper is to show the correlation between the application of video games in training workforce in the digital economy and the increase of their motivational efficiency and competitiveness in the digital business environment. The methodological approach taken in this study is a mixed methodology based on the experiment, mathematical statistics and survey instruments. The experiment was held on E-Commerce refresher courses for employees in Samara State University of Economics. Two groups of staff from Samara Region companies were taught by two different methods: one group was taught by a trainer at a traditional training, while training a second group included the use of video games. The results confirm the benefits of this approach from a motivational perspective and competitive growth of employees in the digital market. This suggests that the application of video games in the process of computer assisted training for at least some of the time available can be recommended to train staff in the digital economy.

**Keywords:** Competitiveness · Digitalization · Digital economy ·
Training employees · Video games

## 1 Introduction

Digitalization is the processes associated with the introduction of information and communication technologies in production, service and management. Digitalization is global and covers all levels of the economy, from large industrial enterprises to small firms and individual, including virtual organizations.

The digital economy is an innovative part of information technology that allows industries, infrastructure, government, social institutions and individuals to have integrated functioning and development. In this economy, the Internet and other networks, including computer social networks, contribute to the business globalization.

© Springer Nature Switzerland AG 2020
S. Ashmarina et al. (Eds.): *Digital Transformation of the Economy: Challenges,
Trends and New Opportunities*, pp. 444–454, 2019.
https://doi.org/10.1007/978-3-030-11367-4_44

McKinsey estimates that three quarters of the value produced by the Internet is attributed to manufacturing, financial services and other industrial organizations. The report, presented by McKinsey Institute for globalism, shows how trade, finance, people and global economic data form a network of interaction, based on information and communication technologies and, as a result, economic growth [24].

In December 2016, OECD countries (Organization for Economic Cooperation and Development) adopted the document "Digital Economy Coverage of Developed Countries", which notes the following effects of information and communication technologies:

(1) cross-border data flows increased the world GDP by 3% annually from 2003 to 2014, and gave an increase equivalent to $2.2 trillion,
(2) the Internet economy in the G20 countries grew by more than $4 trillion, showing growth from $2.5 trillion in 2010,
(3) the Internet covered 1 billion users in 2005, and 3 billion at the end of 2016. Internet coverage is 40% of households in the medium-developed countries, 80% in the developed countries. Less than 10% of enterprises have access to the Internet in the underdeveloped countries,
(4) developed countries now have more than 5.5 billion mobile phones, developing countries - 1.5 billion,
(5) Internet and mobile phones contribute to the new technologies implementation. Scientific and technological progress is accelerating. The time gap between the emergence of global innovations (phone, personal computer, mobile communication) is reduced,
(6) the emergence of Facebook and WhatsApp illustrates new innovative ideas [7].

The growth of the digital economy in the GDP of the G20 countries from 2010 to 2016 is shown in Eurasian Economic Commission [11].

According to the analytical company Boston Consulting Group, the volume of the digital economy in the world will have reached $16 trillion by 2035 [14]. The Digital Economy of the Russian Federation Program was adopted by the Russian Government in June 2017 [9]. The ways for the development of Russian economy up to the level of developed countries with the help of digital platforms have been presented in this program. In the context of the transition to the digital economy, the problem of imbalance in the labor market is becoming more sensitive because of the need to train personnel to work in the digital economy from a new angle.

The current state of the labor market is focused on increasing the intensity of information processes of the productive forces of society, the active use of information resources and the need for their application in the modern digital economy. The discrepancy between the ways of staff training and the actual and future needs of the labor market, in terms of qualification level and professional structure, has led to a shortage of qualified personnel to work in the digital economy. On the one hand, the demand for specialists directly related to the elements and labor functions of the digital economy is increasing in the labor market but on the other hand, there are not enough ways to train such employees.

It should be taken into account that in the digital economy training should focus on the modernization of production, changing the application speed of new technologies in production, the expansion of the use of high technologies in the production process and development. Therefore, training is considered as one of the main values, without which it is impossible to carry out the formation of the labor force in the conditions of economy digitalization.

In order to eliminate the dysfunctions between the needs of employers and the existing labor market, a system of retraining is used. Investment management in human resources significantly neutralizes the loss of professional and qualification capital. Therefore, training staff should not only follow the demand of the current labor market, it should be developed according to the advanced needs.

GALLUP has conducted the research aimed at analyzing the work behavior of employees in the United States based on their level of interest and involvement. In the period from 2010 to 2012, the results are presented in the research of Sorenson and Garman [33].

The authors of this paper have put a question if video games can be a solution to the problem of labor apathy and lack of interest in the professional development. If it is possible to turn video games to the benefit and to include them in the process of employee's evolution?

The authors have noticed that video games are widely used in the HR field, where they help in the preselection, adaptation and training of employees, as well as generally able to help in the organizational culture.

As an example, we can mention the corporate gaming solution Business Start in Sberbank, developed by Insight One (at that time—Social Insight) by McCann Moscow Agency [28]. The gaming solution has been developed to familiarize employees of the Bank with new banking products (including the purchase of the franchise). A feature of the game is that the employee was offered to visit the role of the person taking a loan. The game was built in such a way that the employee, firstly, got acquainted with the way of thinking of the borrower and secondly, interacted with the ideal (virtual) employee and, building a dialogue with him, got acquainted with the preferred model of communication.

While considering the foreign projects, the authors have distinguished startup Knack [21]. They produce two versions of video game applications for HR Directors: one, mobile and desktop for business representatives, the second, strictly mobile for potential applicants. Knack has pioneered the use of video games to identify a candidate's strengths, abilities and inclinations. In the game (one of the many), Wasabi Waiter players are invited to try on the role of a sushi restaurant waiter, that is, take orders, keep orders and sculpt rolls.

Another video game, Balloon Brigade, offers strongly reflect the attacks of a legion of fast-moving burning enemies with the water balls, which also needed to be filled. In this case, during the game Knack keeps record of the player's actions, including non-obvious, and translate them into a special system of assessments and characteristics that determine the intellectual development of the player, emotional involvement and stability, and the ability to take risk and recognize images. It helps applicants to identify their strengths, career paths and if desired, the opportunities not only to share the results on social networks, but also to contact employers who are interested in such a combination of characteristics [17].

The authors have studied the game Human Resource Machine [18]. This is a logic puzzle game, the gameplay in which is built on creating a series of teams. At each level, a new command is introduced. This approach allows players to better understand the purpose of each of the tools, operate and explore new combinations without unnecessary confusion. It is important that the game teaches abstract thinking, problem analysis, data processing and transformation.

To develop social, psychological and physical potential, video games have been being used for many decades. Nevertheless, the authors of this study have noticed that there is a certain research gap. There are a few scientific researches on the use of video games for training employees in the digital economy. That is why, this paper seeks to explain if video games are able to motivate staff to get new knowledge and work more efficiently in the digital economy. It traces the 'immersion' potential of computer games and their potential to motivate employees for searching right decisions in a diverse digital business. It has been hypothesized in the study that the application of video games in training workforce in the digital economy to achieve the higher motivation and knowledge, and to boost staff's competitiveness represents an approach with excellent prospects.

There are two primary aims of this study:

(1) to conduct the experiment,
(2) to ascertain whether the application of video games is an effective computer assisted training strategy from a motivational perspective and competitive growth of employees in the digital economy.

## 2   Materials and Methods

The methodological approach taken in this study is a mixed methodology based on the experiment, mathematical statistics and survey instruments. Participants, who formed the cohort, belonged to employees in the digital economy from Samara Region companies such as Sberbank, Alfa Bank, Eldorado, M Video, Rostelekom. The cohort consisted of two groups of around 25 people each (N = 50). From the total number of employees, 25 people were randomly assigned to the experimental group and 25 people to the control group. Therefore, we had one experimental and one control group in each training context (traditional training and blended learning). The experiment was held on E-Commerce refresher courses in Samara State University of Economics. Table 1 summarizes information about participants, both of the experimental and control groups.

**Table 1.** Descriptive statistics of the sample

| Educational Setting | Group | N | % |
|---|---|---|---|
| Traditional training | Control | 25 | 50 |
| Blended learning | Experimental | 25 | 50 |
| Total | | 50 | 100 |

Source: compiled by the authors.

The study used a pre-test-post-test survey design. Based on the methodological framework proposed by Il'ina [19] a diagnostic questionnaire was developed. All the items diagnosed are detailed in the Results section. Data was collected during the period from September 2017 until January 2018. The pre-test questionnaire took place at the beginning of September and post-test questionnaire was presented to the employees on the last week of the period. The training program for the experimental group occurred between those dates. Figure 1 illustrates the design and procedure of our study.

**Fig. 1.** Experimental design. (Source: compiled by the authors)

The E-Commerce training program with the application of video games for the experimental group was developed based on the model proposed by the authors of the paper. Table 2 demonstrates the skills that were trained through the video games application and the types of games employed.

**Table 2.** Experimental material

| Game type | Skills trained |
|---|---|
| Graphical games | Comprising digital documents, graphs, diagrams and tables |
| Personalized games | Describing business inside the digital business environment |
| Cultural games | Getting the knowledge on country-specific digital realities |
| Historical games | Getting the knowledge on country's economic history |
| Creative games | Developing metacognitive skill and creativity |
| Informational games | Reading skills in digital business environment |

Source: compiled by the authors.

## 3    Results

At the first stage of the experiment, 50 employees in the digital economy from Samara Region companies such as Sberbank, Alfa Bank, Eldorado, M Video, Rostelekom were chosen and then were randomly assigned to one of the groups either control or experimental on E-Commerce refresher courses in Samara State University of Economics. At the second stage, the staff of both groups were assessed to determine the level of their motivation and professional skills. The employees in both groups demonstrated surprisingly low level of motivation and skills in the digital business environment. The results are shown in Table 3.

**Table 3.** Indicators of motivation and professional level before the use of video games in training

| Criteria | Number of employees that meet the criteria in the control group | Number of employees that meet the criteria in the experimental group | Total (N = 50) |
|---|---|---|---|
| The desire to acquire new knowledge in digital economy with minimal cognitive efforts | 19 | 21 | 40 |
| Aspiration to master professional skills and knowledge in the digital economy | 9 | 8 | 17 |
| A firm grasp of e-market conditions | 7 | 8 | 15 |
| The ability to make independent decisions in a risky and uncertain digital business environment | 10 | 10 | 20 |
| The ability to find original digital solutions | 9 | 10 | 19 |
| Communicative and social-adaptive skills | 13 | 12 | 25 |

Source: compiled by the authors.

As Table 3 shows, there is no significant difference between the two groups. It is apparent from this Table that at the beginning of the experiment, the employees of both groups demonstrated low level of motivation. It is clear that the desire to get new knowledge in the digital economy without much cognitive efforts outweighs the aspiration towards mastering professional skills and knowledge. Surprisingly low level of adaptation to changing e-market conditions is revealed. Moreover, the staff in both groups can be characterized by inability to find original digital decisions and lack of communicative and social-adaptive skills.

At the final stage of the experiment the groups were tested again and this time the results were quite different (see Table 4).

**Table 4.** Indicators of motivation and professional level after the use of video games in training

| Criteria | Number of employees that meet the criteria in the control group | Number of employees that meet the criteria in the experimental group |
|---|---|---|
| The desire to acquire new knowledge in digital economy with minimal cognitive efforts | 15(−4) | 11(−10) |
| Aspiration to master professional skills and knowledge in the digital economy | 11(+3) | 16(+8) |
| A firm grasp of e-market conditions | 10(+3) | 17(+9) |
| The ability to make independent decisions in a risky and uncertain digital business environment | 14(+4) | 17(+7) |
| The ability to find original digital solutions | 11(+3) | 19(+9) |
| Communicative and social-adaptive skills | 17(+5) | 20(+8) |

Source: compiled by the authors.

The results have indicated the correlation between the application of video games in training employees and motivation variables growth. As hypothesized, the use of video games to achieve the higher motivation and knowledge and to boost staff's competitiveness in the digital economy represents an e-approach with excellent prospects. Furthermore, the data of the experiment may serve as additional evidence in favor of substantial efficacy of this approach for computer assisted training for staff to work more efficiently in the digital economy.

# 4   Discussion

The modern training system increasingly obliges its participants to interact with the virtual space and tools. At the same time, video games remain one of the key elements of the training space. According to the sociological survey conducted by TNS, 82 million people visited the Internet [13].

Since 2003, the number of scientific articles on video games has grown rapidly. Around the same time, a scientific debate began in this area. The question of this discussion is how to understand computer video games. Some researchers interpret video games as types of narratives or texts of transmedia stories. They conclude that games should be studied using narrative theories [20, 26]. Their opponents argue that video games must be seen as the games themselves-in-itself [2].

After this debate, the study of the material aspect of video games began. Apperley and Jayemane [5] have identified three important trends in the research of current and future games: ethnographic studies of players and their communities, the study of game artifacts and codes, and the study of the relationship of work motivation and computer games. Productive activities in this context mean activities that not only create economic, social or spiritual value, but also develop a person's personality.

Shmeleva, Pavenkov [31] consider the multi-semiotic analysis as the perspective analysis of computer games from the point of view of patterns formation for productive activity.

Nicola Whitton has proposed the typology that differentiates educational games not only by the internal structure of the activity from the game to the player, but by the specifics of the formation of the educational space in the virtual gaming space:

(1)   use of commercial entertainment games,
(2)   modification of existing games,
(3)   use of commercial educational games,
(4)   use of virtual worlds,
(5)   creating custom games,
(6)   creating games by teachers and students [32].

According to McGonigal [23], the average age of the players is 35 years, and 69% of the heads of families and 97% of young people devote time to games. She also has mentioned the data concerning the quality and qualification characteristics: 61% of top managers allow themselves to make small "game breaks" during the working day.

Recent examples of researches into computer assisted training also include Deubel [8], Frydenberg [12], Gtinther and Mentzel [15], and Holm et al. [16]. Aarseth [1],

Eskelinen [10] and Wolf Mark [36] have studied the difference between video games and other types of media. Wolf [36] have mentioned that watching films involves people's imagination to read movies, while playing video games needs practical actions limited in time, such as entering some data. Eskelinen [10] has noticed that the player always understands the difference between the narration on the screen and the action in the video game. Aarseth [1] believes that other kinds of media would like to colonize video games.

The majority of modern researchers have mentioned four traditional genres of computer games:

1. simulations: America's army [4], The Sims [25],
2. multi-users on-line role games: Ever Quest 2 [29], Ragnarok Online [30], Nori School [34], World of Warcraft [35],
3. multi-user virtual environment: Second Life [22, 27],
4. 3D quests: Bone [6].

According to Aldrich [3], virtual worlds, simulations and serious games are the properties of the same training domain. Training simulations represent structured scenarios with thoroughly developed rules, strategies and tasks. They are created with the definite goal of training a skill that could be applied then in a real life context. All in all, games are seen as a valuable tool for the development of human potential.

The authors have analyzed the received results and obtained one evidence that generally video games, applied both in formal and informal training settings for developing skills, improve human potentiality for learning and working.

However, up to now, far too little attention has been paid by researchers to the role of video games in training employees in the digital economy, and data about the efficacy and motivational potential of video games are limited.

We consider that further studying of the use of video games in training staff in digital economy is necessary as well as providing the today's digital business environment with well-trained employees.

## 5 Conclusions

Life in the era of the digital economy is cutting not only the need in some professions but also is changing the labor market. The new digital era should be provided with new professionals. Human Resources Directors used to find staff with part-skills, in other words, professional skills. Now employers want to hire employees with soft skills (ability for studying, working in teams, increasing motivation to reach companies' goals). In near future there will be a trend for changing jobs regularly, though keeping the same job position, employees will have to develop their skills during all their lives.

Thus, staff and the training are becoming the key factors to develop and improve the digital business environment.

The digital economy has influenced an introduction of computer assisted training. Trainers and instructors used to teach with paper medium representation. To date, they have become guides to the digital world using all kinds of Internet technologies to help workers be competitive in the e-market.

The authors of this study believe that the modern process of training staff in the digital economy could not be imagined without the application of video games.

The results of our experiment provide important insights into the application of video games into the process of computer assisted training.

To summarize, we must say that the use of video games in training employees can:

1. boost employees' motivation for getting new knowledge in the digital economy,
2. strengthen the desire to acquire new knowledge in the digital business environment,
3. eliminate the formal approach towards knowledge acquisition,
4. assist employees in their adaptation towards constantly changing e-market environment,
5. develop employees' autonomous decision- making process,
6. stimulate employees' creativity,
7. develop staff's communicative and social-adaptive skills.

This subject would be a fruitful area for the further research. The potential of using video games in training staff in the digital economy is huge, and abandoning it would be illogical.

The findings of this study will be of interest to academics, training instructors, managers, employers and employees. By using video games in training staff, instructors will give them a competitive advantage in this ever-evolving digital economic environment.

# References

1. Aarseth E (2001) Computer game studies. Game Stud 1(1). http://www.gamestudies.org/0101/editorial.html. Accessed 14 Apr 2018
2. Aarseth E (2001) Videogames of the oppressed: videogames as a means for critical thinking and debate: a thesis presented to the academic faculty. Atlanta. http://www.ludology.org/articles/thesis/FrascaThesisVideogames.pdf. Accessed 30 Apr 2018
3. Aldrich C (2009) Virtual worlds, simulations, and games for education: a unifying view. Innovate 5(5). http://www.innovateonline.info/index.php?view=article&id=727. Accessed 14 Apr 2018
4. Anderson et al (2008) Video games in the english as a foreign language classroom. In the second IEEE international conference on digital game and intelligent toy enhanced learning, pp 188–192
5. Apperley TH, Jayemane D (2012) Game studies' material turn. Westminst Pap Commun Cult 9(1):5–25. http://doi.org/10.16997/wpcc.145
6. Chen HHJ, Yang C (2011) Investigating the effects of video game on foreign language learning. In: Proceedings of the 6th international conference on E-learning and games, Edutainment, pp 168–175
7. Dahlman C, Mealy S, Wermelinger M (2016) Harnessing the digital economy for developing countries. OECD Developing Centre. Working paper 33 No. 334. December 2016. http://www.oecdilibrary.org/development/oecd-development-centre-workingpapers_18151949. Accessed 14 Apr 2018

8. Deubel P (2003) Learning from reflections issues in building quality online courses. Dist Learn Administr 7(3). https://pdfs.semanticscholar.org/b6b0/21e4311e91beb8f8cf1de8576a 8521309575.pdf. Accessed 12 Mar 2018
9. Digital economy of the Russian Federation Program. http://d-russia.ru/wp-content/uploads/ 2017/05/programmaCE.pdf. Accessed 13 Mar 2018
10. Eskelinen M (2001) The gaming situation. Game Stud 1(1). http://www.gamestudies.org/ 0101/eskelinen/. Accessed 2 Apr 2018
11. Eurasian Economic Commission. http://www.eurasiancommission.org/ru/act/dmi/ workgroup/materials/. Accessed 27 Apr 2018
12. Frydenberg J (2002) Quality standards in e-Learning: a matrix of analysis. Int Rev Res Open Dist Learn 3(2). http://www.irrodl.org/index.php/irrodl/article/view/109/189. Accessed 12 Feb 2018
13. Granic I, Lobel A, Engels RCME (2013) The benefits of playing video games. Am Psychol Assoc 69(1):66–78. https://doi.org/10.1037/a0034857
14. Gref GO (2018) None of us will be able to sit in ambush, and we need to be ready for this. About the trends of the new digital era. https://www.sign.com/2017-06-29/glava_sberbanka_ german_gref_o_trendah_novoy_cifrovoy_epohi. Accessed 29 Apr 2018
15. Gtinther A, Mentzel B (2003) E-Learning bei der Volkswagen Coaching GmbH / A.Gtinther & B.Mentzel. E-Learning: Einsatzkonzepte und Erfolgsfaktoren des Lernens mit interaktiven Medien/U. Dittler (Hrsg.) Miinchen: Oldenbourg: 291–300
16. Holm C, Franzen M, Grohbiel U (2002) Grundlagen fur nachhaltiges E-Learning: Erhebung von Einstellungen und Erwartungen bei Dozierenden und Studierenden der FH Nord-westschweiz. C.Holm, M.Franzen & U.Grohbiel. Vortrag Web-Based Training, Olten
17. How It Works. Knack. https://www.knack.it/how-to/index.html. Accessed 24 Apr 2018
18. Human Resource Machine. Tomorrow Corporation. http://tomorrowcorporation.com/ humanresourcemachine. Accessed 25 Feb 2018
19. Il'ina TI (2017) Methodology for Study Motivation of University Students. http://testoteka. narod.ru/ms/1/05.html Accessed 12 Jan 2018
20. Jenkins H (2004) Game Design as Narrative Architecture. In: Wardrip-Fruin N, Harrigan P (eds) First person: new media as story, performance, and game. Cambridge (Mass.) pp 118–130
21. Knack in the news. Knack. https://www.knack.it/media/index.html. Accessed 19 Mar 2018
22. Liang MY (2011) Foreign lucidity in online role-playing games. Comput Assist Lang Learn 25(5):455–469
23. McGonigal J (2011) Reality Is Broken: Why Games Make Us Better and How They Can Change the World. Penguin Press, 400 p
24. McKinsey Global Institute. Digital globalization: the new era of global flows. March, 2016. www.mckinsey.com/mgi. Accessed 15 Apr 2018
25. Miller M, Hegelheimer V (2006) The SIMs meet ESL: incorporating authentic computer simulation games into the language classroom. Interact Technol Smart Educ 3(4):311–328
26. Murray JH (2005) The Last Word on Ludology v Narratology in Game Studies. Vancouver, Canada, June 17, 2005. https://inventingthemedium.com/2013/06/28/the-last-word-on-ludology-v-narratology-2005/. Accessed 16 Apr 2018
27. O'Connor E (2004) Using Second Life (a Virtual Reality) in Language Instruction: Practical Advice on Getting Started. Informational Technologies in Linguistics, lingodidactics and intercultural communication: 52–63
28. Press release. Insight One. http://www.insight-one.ru/pr1_05_08_14.html. Accessed 18 Apr 2018
29. Rankin YA, Gold R, Gooch B (2006) 3D role-playing games as language learning tools. In: Proceedings of Eurographics, pp 33–38

30. Reinders H, Wattana S (2011) Learn English or die: the effects of digital games on interaction and willingness to communicate in a foreign language. Digital Educ Cult. http://www.digitalcultureandeducation.com/cms/wp-content/uploads/2011/04/dce1049_reinders_2011.pdf Accessed 15 Feb 2018

31. Shmelev I, Pavenkov O (2006) Coping behaviour and difficult life situations of university students in Russia. Soc Stud 8(1):63–77. https://doi.org/10.13165/SMS-16-8-1-4 Accessed 14 Apr 2018

32. Slavin RE et al (2009) What works for struggling readers? The University of York. Institute for Effective Education, 12 p

33. Sorenson S, Garman K (2018) How to tackle US Employees' stagnating engagement. GALLUP. http://www.gallup.com/businessjournal/162953/tackle-employees-stagnating-engagement.aspx. Accessed 28 Feb 2018

34. Suh S, Kim SW, Kim NJ (2010) Effectiveness of MMORPG-based instruction in elementary english education in Korea. J Comput Assist Learn 26(5):370–378

35. Thorne SL (2008) Transcultural communication in open internet environments and massively multiplayer online games. In: Mediating Discourse Online pp 305–327

36. Wolf Mark JP (2008) The video game explosion: a history from pong to playstation and beyond. In: Mark JP Wolf (ed) Westport, Connecticut, London: Greenwood Press, p 23

# Legal Education in Conditions of Digital Economy Development: Modern Challenges

M. A. Yavorskiy, I. E. Milova, and V. V. Bolgova(⊠)

Samara State University of Economics, Samara, Russia
{yavorm, vvl976}@mail.ru, irina.milova@ro.ru

**Abstract.** Digital economy has become today's reality. Its implementation in our country is the realization of the state programme approved by Vladimir Putin, the President of the Russian Federation. The complete transition to digital economy is going to be finished by 2030. The creation of the information environment is systemic; it encompasses various segments of public life. In particular, electronic technologies are being actively introduced into education in Russia, significantly increasing its quality and efficiency. Similar innovations appeared in the legal cluster of education. Of course, they can have a significant impact on the change of approaches to the formation of professional competences of future lawyers. The analysis of the practice shows that at present social order is articulated specifically for specialists who are fluent in electronic technologies and are able to use them to solve complicated legal problems. We found it interesting to study the approaches to this problem in foreign literature. The positive experience accumulated abroad could be useful for optimizing the domestic educational environment and would make it possible to move much faster in this direction. The goal is to analyze the opportunities digital economy can provide for the training of lawyers specializing in criminal law and criminal process; to understand to what extent such innovations help students to prepare for independent work, allowing to reduce the period of adaptation entry into investigative and judicial activities. The leading approach to the research of this problem is the understanding that the transition to digital economy should not be formal. The implementation of the program should give us a lawyer of a new type who is going to be broad-minded, ready to solve complicated practical problems and take non-standard decisions. Also, taking into account digital technologies, these processes should be much faster, without intermediate links, with greater accuracy of the results obtained. Results: analyzing achievements of a number of foreign countries, the authors have shown the prospects for introducing digital economy into legal education in Russia. At the same time, they have identified a number of peculiarities that require increased attention when introducing this kind of technologies into Russian education. On the whole the implementation of digital technologies in education in general and in legal education in particular is positively evaluated by the authors. They have no doubts that such innovations are inevitable and will have a practical effect in the near future when students who received the appropriate skills will be able to use them in their professional activities.

The analysis of the possibilities of digital economy for the training of lawyers was conducted for the first time in the article on the basis of the use of a complex of scientific methods. Practical significance: the main provisions and conclusions of the article can be used in scientific and scientific-educational activities

S. Ashmarina et al. (Eds.): *Digital Transformation of the Economy: Challenges,*
*Trends and New Opportunities*, pp. 455–462, 2019.
https://doi.org/10.1007/978-3-030-11367-4_45

when considering conceptual issues related to the creation of a digital educational environment, including the field of jurisprudence. The materials of the article may be of practical importance for students of law universities, as well as for teaching and professorial staff, whose interests include issues of criminal law, criminal procedure and forensic science.

**Keywords:** Digital economy · Information safety · Cyber-crimes · Internet services · Electronic evidence · Digital education · Online courses · Digital jurisprudence

# 1  Introduction

Singapore, Israel, Japan and the United Kingdom are considered leaders in transition to digital technologies. The United States, Scandinavian countries and South Korea slowed down the pace of informatization, although they had previously shown a tendency of faster growth of such processes.

Egypt, Greece, Pakistan are lagging behind in digitalization, demonstrating that they are not ready to face modern challenges in this segment [13].

Russia along with China belongs to the group of countries that are considered promising in terms of the development of digital environment, where corresponding achievements are used in educational environment as well. Statistics show that the vast majority of students in these countries use the Internet resources. At the same time, young people show articulated interest in digital technologies. In many universities distance education and online courses are being introduced. In this regard, we can believe that our country will be able to take a respectable place among the global digital elite [16].

So, the development and implementation of measures aimed at distributing digital mechanisms in educational programs, as well as the development of students' practical skills of using electronic technologies, are becoming very acute. Despite the fact that these issues are widely spread in scientific community today, there are practically no research works devoted to studying the processes of introducing elements of digital economy into legal education.

Consequently, it turns out that national education lags behind the real social demand. It is still oriented towards analog education, emphasizing the development of standards and the preparation of bachelor and magistracy educational programs. Modern realities urgently demand specialists ready to work with digital material, including investigation of crimes and trial of criminal cases [3]. That is why it is necessary to change the approach to training radically, and lawyers are no exception to the general rule.

The attempts to find fundamental research in this segment have been unsuccessful. With a sufficiently wide range of textbooks on information technologies in various fields of activity, there is practically no monographic base for the digitalization of education and there are practically no serious scientific research works. The study of the relevant issues is conducted at the level of individual publications in the format of articles, does not have a systemic nature and gives only general recommendations.

The issues of digital economy importance for the educational process were raised to a certain extent in the works of Averyanov [1], Aniskina [2], Bulgakova [3], Dzhatdoev [4], Elistratova [5], Voroshilova [6], Evtushenko [7], Kupriyanovskiy [8], Latyshev [9], etc.

A number of problems concerning information technologies and their importance for training specialists including lawyers were raised in the works of foreign authors.

Most scientists highlighted the tendency of the educational environment digitalization growth and its importance for training of lawyers of a new type. On the plus side there is indicated the possibility of fast obtaining of a large amount of reliable information and timely reaction to global changes through monitoring mechanisms with the aim of adjusting the educational process and optimizing it. The scientific community is convinced that the time has come for the formation of a special sphere - information culture [5]. The authors studying the issue believe that electronic technologies can help the participants of the educational process to develop creative thinking; to give teachers an additional creative impulse and to increase the level of students' motivation to gain knowledge in their professional activities.

All the researchers agree that in modern conditions informatization of education will make it more individualized and differentiated, and this, in turn, will make it possible both to develop self-education skills and to work directly with each student [4].

At the same time, there is a certain concern that the national education system is not ready for a radical restructuring. The logic of acquiring knowledge is now built on the receipt of exclusively textual material that is proposed for memorization. In conditions of digitalization it is complemented by a video sequence containing drawings, graphics, animation elements; this is accompanied by sound, light effects. It cannot be excluded that the above-mentioned things will make it difficult for future lawyers to understand the essence of the issue, especially taking into consideration the fact that people do not have the same perception of information. In addition, a large degree of freedom leads to the problem of its use [21]. Not all students are able to concentrate on getting specific information. When working independently, many of them will follow web links to other segments and it can distract their attention from the main issue.

The scientific community has no doubts in the objectivity of the process of digitization of the educational environment, stating that it is time to move from the stage of understanding of its significance to the stage of specific actions [24]. The declaration of its usefulness should be a practical reality giving a clear and predictable result.

Goal: to analyze the potential of digital economy for the education of lawyers with the aim of giving them practical skills that will enable them to get engaged in relevant activities as quickly as possible as an independently functioning members.

## 2 Materials and Methods

In the process of research, the following methods were used: analysis and comparison as the main methods, system-structured and specific-sociological methods in the context of studying the state of social relations in the modern educational environment when training future lawyers.

The research was based on scientific works, publications of Russian and foreign scientists studying various aspects of the informatization of the educational process.

The study of the problem was carried out in two stages:

The first stage: the analysis of the existing scientific literature on the subject of the study, as well as the normative base in the field of legal regulation of the digitalization of the educational cluster, were carried out.

The second stage: conclusions obtained from the analysis of scientific literature and legislation were formulated, the publication of the article was prepared.

# 3   Results

Distant learning is very up-to-date. Such a mechanism makes education more affordable as it makes constant physical presence of students at the university unnecessary and at the same time preserves their contact with teachers. Russia is a country that has a vast territory, so distant learning can solve a great number of problems. A lot of universities in Russia, including Samara State University of Economics, actively use this educational format and obtain rather good results.

Legal education effectively uses information educational resources that cover educational literature, teaching materials, visual teaching aids and technical equipment. Thus, it is difficult to imagine the training of a future lawyer without business games (for example, educational process for studying criminal cases) [8]. During such a game students have an opportunity to play various procedural roles: a judge, a prosecutor or a court counsel. According to results they draw up legal documents [10]. The information environment allows you to translate this format from real to virtual, to make it more operational and interesting and to involve a large number of students in it [22].

The portfolio method is widely used in the educational cluster. It allows estimating the achievements of students in education and science quickly. It is also a kind of electronic storage of information, giving the idea of what practical skills they get during the training. In the portfolio all written works of the student are saved - course projects, essays, final qualification (bachelor) works, master theses. It contains reports on all student practices [25]. The student saves scientific publications in it. The teacher, using his account, is able to be in constant dialogue with students, encouraging them for self-education, assessing their progress in time and pointing out mistakes.

Electronic technologies give the teacher the possibility to check and review any written work quickly, to make specific recommendations and comments and orient the student towards an independent searching of additional information.

Using the results of this continuous work, we get positive results of pedagogical monitoring that allow us to improve the quality of the educational process and to determine the future directions of its progressive development. Information technologies greatly facilitate the procedure of intermediate and final control over the knowledge gained. Various tests are very effective [12]. In Samara State University of Economics they are used in all training courses for bachelors, masters and specialists as part of training of future lawyers.

A very promising way of learning is communication of students and teachers in the format of online conferences and round tables, including those with foreign partners. This direction makes it possible to reveal the range of opinions on the most controversial issues, articulates their informal discussion and gives students a vector for self-studying of them.

Digital videos that can be used in the study of criminal process and forensic science can be considered an interesting format for obtaining knowledge [7]. In addition, students can make digital educational films themselves. This practice is used in Samara State University of Economics, where several digital videos were made showing a series of investigative activities. For example, there is a film on the inspection of the scene, which clearly demonstrates its consistency and cognitive abilities.

Nowadays the world community is facing the problems of cyber-crime, as well as the fact that the possibilities of modern engineering are quite actively used by criminals [1]. Unfortunately not enough attention is paid to this direction of training future lawyers in Russia; theoretical knowledge has rather general character and practical skills of detecting signs of similar crimes, peculiarities of their investigation and proceeding are not taught.

In addition, the national system of evidence does not include electronic evidence as a separate type of evidence and does not give its legal definition [15]. Meanwhile, in a number of foreign countries positive experience has been gained in using it in criminal proceedings.

It should be noticed that video recording is used quite effectively in our country during investigation of criminal cases [6]. However, its resolution quality is not always high as recording is made in the process of carrying out operational activities (for example, when fixing the transferring of a bribe to an official or the moment of selling drugs) [19]. In some US states in such cases it is allowed to improve the quality of video recording by digitizing the image. This evidence is assessed by the courts as valid, provided that digitizers confirm its authenticity and the correctness of the procedure [18]. This kind of precedent took place in the Court of Appeal of Georgia.

There is no doubt that it would be very useful to use digital records in the national criminal proceedings as an official source of evidence. It is obvious that corresponding changes should be introduced in the criminal procedure legislation. At the same time, universities should teach appropriate practical skills to law students at the classes of criminal procedure and forensic science right now [2].

For example, Samara State University of Economics has a forensic laboratory, the technical capabilities and equipment of which allow working with low-quality video recordings. Through digitization it would be possible to improve an image, which can be followed by giving it evidence value by procedural forms. Along with such practical activities it would be very useful to teach students to draw up expert opinions on the course of this process and on preserving authenticity of an image [9].

Digital expertise, as an independent type of research, could become another interesting innovation. Now it has a fragmented character, as part of technical expertise. In addition, digital issues would form a layer of research when conducting various complex forensic expertises [11]. For example, such innovations, of course, are in demand with ballistics, traceological analysis and fingerprint analysis. A significant amount of evidence information would be given by ballistic-digital, trace-digital, chemical-digital, biological-digital, genetic-digital, commodity research-digital and construction-digital expertises [20].

During the process of future lawyers training it is necessary to teach skills of conducting the above-mentioned types of expertise. Bachelor and master students of Samara State University of Economics have the opportunity to work in this direction when studying the special course "Expertise and Expert Activities" and using the technical capabilities of the forensic laboratory.

The majority of expertises are carried out with the use of information technologies [14]. They are used when making a research itself and in drawing up conclusions on its results. The students should get practical skills during the educational process.

In addition, it would be advisable to create digital banks to accumulate information on the most interesting expert research including typical and non-standard methods and on the analysis of investigative and judicial practice, both of federal and regional levels [26].

Digital technologies would be very useful in creating a criminal case model, where the student could gain skills of independent participation in legal proceedings as well as practical experience in preparing various kinds of procedural documents such as resolutions, protocols and conclusions in online mode.

One more aspect is the use of information opportunities for educational purposes, for example, in counteractions to extremism and terrorism. It seems that the further development of the field of informational education in general and media education in particular will contribute to counteractions to extremism among young people. The main task in this direction is to prepare young people for life in modern information conditions, for correct perception and evaluation of various kinds of information [23].

## 4  Discussion

The number of scientific studies and scientific publications based on them, methodological and practical recommendations does not reduce the scientific novelty and, on the contrary, indicates the relevance and controversial nature of the problems of transition to digital technologies in domestic education in general, and in training of future lawyers in particular [17].

Despite the fact that a significant number of publications have appeared in this segment recently, the content analysis of scientific works allows us to conclude that many aspects of the topic stipulated by current circumstances are still insufficiently studied and need further development.

At the present stage, the priority tasks are the increase of the digital literacy of students and teachers, informatization of the training itself and the accompanying processes, creation of information culture of education, rational use of electronic resources to form professional competencies, and, as a result, getting a lawyer fully prepared for independent activities in various fields of public life, able to find information quickly, analyze it and take a legitimate and reasonable decision based on the analysis of this information.

The use of the obtained results has a big practical importance. It can contribute to solving of the scientific problem in the national educational environment.

## 5  Conclusions

Russian education is in the process of transition from traditional acquisition of knowledge to a systemically different one, built on information technologies. Measures taken for the informatization of education pursue the goal of training personnel capable of working effectively in the new economic conditions. For a future lawyer, it is important to obtain not only a certain amount of theoretical knowledge, but also practical skills. There is no doubt that digital processes contribute to this. However, specific electronic formats require development. Of course, the well-proven portfolio method, distance learning, intermediate and final tests of various types are to be preserved but they need a number of innovations: digital videos; digital models of various forensic situations and digital descriptions of legal cases; digital data banks (for example, on fingerprint analyses materials; on judicial acts and procedural documents); digital video archives of various kinds of legally relevant information (including that of regulatory and statistical nature), digital expertise and digital opinions of experts on various legal issues.

## References

1. Averyanov MA, Evtushenko SN, Kochetkova EY (2016) Digital society: new challenges. Econ Strat 141:90–91
2. Aniskina EV (2017) Informatization of education in the Russian Federation at the present stage. Mod Sci Res Innov, no.12 [Electronic resource]. http://web.snauka.ru/issues/2017/12/85225
3. Bulgakova EV, Bulgakov VG (2013) Repository of video archives of data on dynamic characteristics of a person for solving forensic tasks. Leg Comput Sci 4:28–31
4. Dzhatdoev AH (2018) Information technologies in jurisprudence. Young Sci 6:20–24
5. Elistratova NN (2012) Information culture as a criterion of informatization of higher education in modern reforming conditions. Mod Sci Res Innov, no. 7 [Electronic resource]. http://web.snauka.ru/issues/2012/07/15770. Application date: 29 Oct 2018
6. Voroshilova OS, Mironova MN (2016) Problems of the use of innovations in professional training and methods of solving them. Mod Sci Res Innov, no. 10 [Electronic resource]. http://web.snauka.ru/issues/2016/10/72934. Application date: 26 Oct 2018
7. Evtushenko KN, Rannih VN (2013) Didactic audit of E-education in high school: monograph. Tula State University Publishers, Tula
8. Kupriyanovskiy VP, Sinyagov SA, Lipatov SI (2016) Digital economy – a clever method of working. Int J Open Inf Technologies 4:26–32
9. Latyshev DS (2017) A brief overview of information technologies used in legal activities. Innov Sci 10:14–17
10. Masyuk MA (2011) Analysis and visualization of the relationship of normative-legal documents in reference and legal systems. Sib J Sci Technol 35:40–45
11. Rak IP (2016) Information technologies in the activities of law enforcement agencies. Innov Sci 14:132–135

12. Robert IV (2004) Interpretation of words and phrases of the conceptual apparatus of informatization of education. Comput Sci Educ 6:63–70
13. Sakovich SI, Pavlova YV (2015) Informatization of education. Mod Sci Res Innov, no. 11 [Electronic resource]. http://web.snauka.ru/issues/2015/11/59010. Application date: 28 Sept 2018
14. Habliyev SR (2015) Information and educational environment in various educational systems. Mod Sci Res Innov, no.12 [Electronic resource]. http://web.snauka.ru/issues/2015/12/60716. Application date: 25 Sept 2018
15. Chernenko VV Piskorskaya SY (2012) Expert systems. Actual Probl Aviat Cosmonaut 8:322–323
16. Chernyshev PM (2016) Information technologies in legal activities. Mari Judic Bull 18:34–36
17. Yudina TN (2016) Understanding digital economy. Theor Econ 3:12–16
18. Yakusheva NM (2012) Didactic principles of creating E-learning tools and problems of their implementation. Bulletin of Moscow State Humanitarian University named after M.A. Sholohov, Pedagogy and Psychology Series, vol 1, pp 87–96
19. Brown Dina, Warschauer Mark (2006) From the university to the elementary classroom: students'experiences in learning to integrate technology in instruction. J Technol Teacher Educ 14:599–621
20. Judson Eugene (2006) How teachers integrate technology and their beliefs about learning: is there a connection? J Technol Teach Educ 14:581–597
21. Pereira CMG, de Carvalho e Silva Afonso M, da Cunha Santos DF (2015) Education for Entrepreneurship – an experience report in a higher education institution. JETT 6(1). ISSN-e 1989-9572
22. Boulet Maude, Boudarbat Brahim (2015) The economic performance of immigrants with Canadian education. Estudios Economicos Regionales y Sectoriales: EERS: Regional Sect Econ Stud 15:23–38
23. Bouhajeb M, Mefteh H, Ben Ammar R (2018) Higher education and economic growth: the importance of innovation. Atlantic Review of Economics: Revista Atlántica de Economía 1 (2):2174–3835
24. Uetela P (2015) Higher education and the challenges for economic growth in Mozambique: some evidence 4:276–294
25. Rice Kerry Lynn (2006) A comprehensive look at distance education in the K-12 context. J Res Technol Educ 38:425–448
26. Rasinen Aki (2003) An analysis of the technology education curriculum of six countries. J Technol Educ 15:31–47

# Potential of the Education System in Russia in Training Staff for the Digital Economy

E. A. Mitrofanova[1], M. V. Simonova[2], and V. V. Tarasenko[3(✉)]

[1] State University of Management, Moscow, Russia
elmitr@mail.ru
[2] Samara State University of Economics, Samara, Russia
m.simonova@mail.ru
[3] Orenburg State Pedagogical University, Orenburg, Russia
tarasenko56@mail.ru

**Abstract.** The contribution presents the research results of the potential of the education system in Russia and ways of their updating in order to improve the quality of training for the digital economy. The authors of the contribution disclose answers to such questions as: 1) what qualification requirements are applicable to labor resources; 2) what potential does the education system in Russia have in training the workforce for the digital economy; 3) what are the risks in the education system in Russia in training for the digital economy. The conclusions substantiate the potential of the education system Russia in training for the digital economy and determine the leading role of managerial staff of the educational organization in its updating.

The authors used general scientific research methods: systemic and functional approaches; special research methods: sociological (expert assessment method, observation and analysis of educational practices, questioning and interviewing participants in educational relations, interviewing employers) and statistical methods. The methods used allowed scientifically characterizing the potential of the education system in Russia in training for the digital economy. The source base for the contribution was materials from the Organization for Economic Cooperation and Development, the International Telecommunication Union, the World Economic Forum, the Boston Consulting Group, Russian legislative and executive bodies, the Institute for Statistical Studies and Higher School of Economics and etc.

**Keywords:** Education · Educational organization · Education system ·
Digital economy · Digitalization · Labor resources · Management ·
Managerial staff

## 1 Introduction

In the modern information society, according to a number of scientists [1, 6, 9, 13], knowledge and information become the most important resource of socio-economic development. At the same time, according to the research of the International Telecommunication Union (ITU), Russia occupies only 45th place in the ranking of countries in terms of information and communication technology development, which

© Springer Nature Switzerland AG 2020
S. Ashmarina et al. (Eds.): *Digital Transformation of the Economy: Challenges, Trends and New Opportunities*, pp. 463–472, 2019.
https://doi.org/10.1007/978-3-030-11367-4_46

is significantly lower than the position of countries with developed and developing digital economies. The main reason for a low value of the indicator is the problem of digital literacy and digital competencies, when citizens cannot fulfill their needs and interests in the network due to the lack of basic digital skills. The research results of the consulting company Boston Consulting Group (BCG), Gartner, show that more than 80% of Russian employees do not have skills necessary for the digital economy and are not ready for the role they will have to play in the future. Only 17% of Russian specialists are engaged in highly qualified labor in the category of "knowledge". This is 1.5 times less than Japan and the United States, Germany - 1.7 times, Singapore - 2 times, Great Britain - 2.6 times. And, according to the report "Russia 2025: from Human Resources to Talents" (prepared by the order of the consulting company BCG), Russia can catch up with the leading countries if the system of education is not retargeted to train human resources in accordance with the requirements of the digital economy.

Taking into account the trends in the digitalization of society and economy, the state program Digital Economy of the Russian Federation has been developed and is being implemented in Russia. The main goal of the program is to create the necessary conditions for the development of the digital economy and the preservation of Russia's competitive advantage in world markets. The education system has a key role in training for the digital economy. At the same time, the analysis of theoretical and empirical data [2, 5] made it possible to identify the contradiction between the need of Russia in the development of the digital economy and the lack of the efficient education system in training staff for the digital economy. The controversy necessitated a study of the potential of the education system in Russia, and system upgrading will improve the quality of training for the digital economy. In accordance with this goal, the tasks of the contribution were as follows:

1. Clarify the qualification requirements of the digital economy to labor resources.
2. Assess the state and potential of the education system in Russia in training staff for the digital economy.
3. Determine the existing risks in the education system in Russia in training staff for the digital economy.

The solution of the tasks assumed the search for answers to the following questions: 1) what qualification requirements are applicable to labor resources; 2) what potential does the education system in Russia have in training the workforce for the digital economy; 3) what are the risks in the education system in Russia in training staff for the digital economy.

## 2  Materials and Methods

The following scientific research methods were applied in the contribution: systemic and functional approaches; special research methods: sociological (expert assessment method, observation and analysis of educational practices, questioning and interviewing participants in educational relations, interviewing employers) and statistical methods.

# 3   Results

With the development of the digital economy, labor resources with the current level of digital competence/literacy are in demand [9, 14]. Digital competence refers to the ability and willingness of an individual to confidently, effectively, critically and safely select and apply info-communication technologies in various areas of life: working with content (information and media competence), communication (communicative competence), techno-sphere (technical competence) and consumption (consumer competence) [17]. The modern requirements of Russian economy, applicable to the content of the digital competence of labor resources, have been clarified in professional standards. The Labor Code of the Russian Federation defines a professional standard as a characteristic of qualifications required for an employee to perform a certain type of professional activity. In turn, the labor code describes qualifications of an employee as the level of knowledge, professional skills and work experience of an employee.

At present, more than 1,100 professional standards have been approved, which employers use when developing staff policy and staff management, organizing training and certification of employees, concluding labor contracts, forming job descriptions and establishing wage systems. It is worth noting a high efficiency of professional standards in the field of professional activity "Communication, Information and Communication Technologies", which corresponds to the current trends in the digitization of the economy and society as a whole. Along with IT specialists, the key role in the development of the digital economy is played by the subjects of professional activity in the following areas: education and science, finance and economics, industry, service and the provision of services to the population, administrative, management and office activities, transport, etc. Analysis of professional standards for relevant areas of professional activity allowed forming a generalized list of knowledge and skills necessary for the digital economy (Table 1).

The authors assessed the state and potential of the education system in Russia in training staff for the digital economy. The contribution made it possible to establish that, in accordance with state requirements, the principal educational programs contain compulsory academic subjects such as "Economics" and "IT", which are aimed at building the knowledge and skills demanded by the digital economy. In accordance with the requirements of the federal state educational standards, the planned results of mastering "Economics" are:

- Formed economic thinking; the ability to navigate current economic events in Russia and in the world;
- Skills to search for relevant economic information in various sources, including the Internet; the ability to independently analyze and interpret data to solve theoretical and applied problems;
- The ability to develop and implement projects of economic and interdisciplinary orientation based on basic economic knowledge and value orientations;
- The ability to personal self-determination and self-realization in economic activities, including in the field of entrepreneurship; knowledge of the modern labor market, knowledge of the ethics of labor relations, etc.
- The planned results of mastering of "IT" are:

**Table 1.** Generalized list of knowledge and skills required for the digital economy

| Generalized professional requirements | |
| --- | --- |
| You need to know: | You must be able to: |
| • international and national legislation, standards, ethical norms and principles governing info-communication activities;<br>• fundamentals of computer science and computing, structural construction of information systems and programming;<br>• requirements for the technical equipment of the workplace, rules for working with equipment and its software;<br>• criteria for assessing the quality of information;<br>• technologies for systematization of large amounts of information, development and maintenance of electronic databases;<br>• functional features of electronic documents and information protection, etc. | • develop information requirements in accordance with professional tasks;<br>• use organizational and technical means in working with large amounts of information and when digitizing it;<br>• work with information systems and electronic databases, including local networks and information and communication network "Internet";<br>• carry out the structural construction of information systems and programming;<br>• apply electronic document management and information protection technologies;<br>• carry out activities to upgrade the hardware and software of the workplace;<br>• continuously improve digital literacy, etc. |

Source: compiled by the authors

- Formed ideas about the role of information and information systems in modern society; databases, their structure, means of creating and working with them;
- Computer tools for data presentation and analysis; methods of processing numeric and textual information; programming skills, including testing and debugging programs;
- Formed basic skills and abilities to comply with the requirements of safety, hygiene and resource saving when working with information systems;
- Adoption of norms of information ethics and law, principles of information security, etc.

A theoretical analysis of the state requirements for basic educational programs led to the conclusion that the planned results of graduates of the education system in Russia meet the qualification requirements of the digital economy presented in Table 1. Educational and professional programs implemented with the development of educational programs are also aimed at developing skills of the modern economy. With the digitalization of society and economy, the most demanded vocational training program is "Operator of computing machine" (Fig. 1).

In accordance with the Russian legislation, the implementation of educational programs is possible only if there is a license confirming that staff, material, technical, financial, psychological, pedagogical, and a number of other conditions comply with state requirements. In Russia, more than 46 thousand educational organizations have a license, confirming that the conditions for implementing general educational programs comply with state requirements. Along with this conclusion, the authors assessed the compliance of the staff and material and technical support of the education system in Russia with the digitization of society and economy.

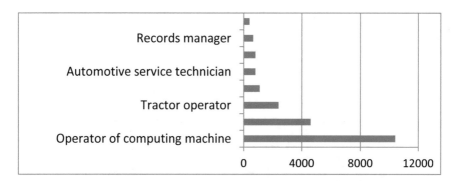

**Fig. 1.** The number of people trained in vocational training programs with the development of educational programs, people (Source: compiled by the authors based on open data from the Ministry of Education of the Russian Federation)

In the 2017–2018 academic year, 66.6 thousand teachers of Economics/Social studies and 29.1 thousand teachers of Information Technology (IT) participated in general educational programs. Most of them have higher pedagogical education, the first or highest qualification category, additional professional education in the field of pedagogical activity (updated at least once every three years) and work experience of over 20 years (Fig. 1). The presented characteristic of the professional portrait of teachers positively characterizes the staff conditions of the education system in Russia.

When studying the education system in Russia, staff risks were also identified in preparing graduates for the digital economy. Staff risk is a situation that reflects the undesirable developments for the organization, staff and society as a whole, the causes of which may be: the inefficient staff management system; behavior, action (inaction) of staff; external environment of the organization [11]. An analysis of educational programs made it possible to establish that an insignificant amount of time is allotted for mastering "Economics" (as a rule, not more than 140 h—2 h per week in grades X and XI). This amount of time is not enough to attract and provide a full-fledged academic load of "narrow" specialists in teaching Economics. Therefore, the discipline "Economics" is often assigned to "conditional specialists" - teachers of History, Social studies, Geography or Information Technology. This can adversely affect the quality of training for the digital economy. The following staff risk is due to a high proportion of young teachers of IT under the age of 30 years (Fig. 2), whose professional competence requires updating through the acquisition of methodological experience in organizing and conducting training sessions.

The staff risk of reducing the quality of training for the digital economy is also a high turnover of teachers of Economics/Social studies - 10.6% and Information Technology (IT) - 11.9%. For comparison, the turnover of teachers of Mathematics, Geography and Biology is 7%, Russian language and Literature - 9%. The staff risks include the classical problem of the education system in Russia - the presence of vacant positions of teachers of Economics/Social studies (377 units) and Information Technology (IT) (394 units). This problem is especially urgent for small rural education organizations (Fig. 3).

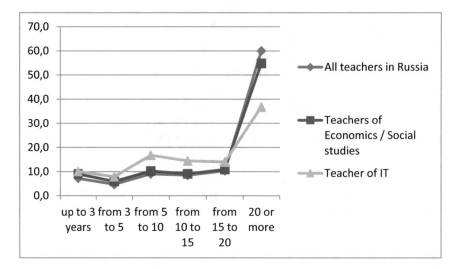

**Fig. 2.** Distribution of teachers by work experience, in % (Source: compiled by the authors based on open data from the Ministry of Education of the Russian Federation)

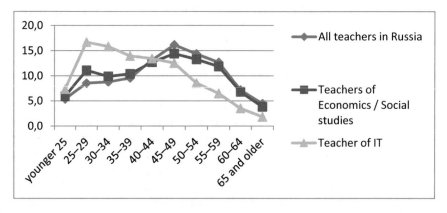

**Fig. 3.** Distribution of teachers by age, in % (Source: compiled by the authors based on open data from the Ministry of Education of the Russian Federation)

The education system in Russia takes into account the trends of the digitalization of the economy, developing the material and technical support of educational programs. The data presented in Fig. 4 indicate an annual increase in the number of personal computers used for educational purposes per 1000 students.

More than 1,465 thousand computers out of 2050 thousand used for educational purposes, have access to the Internet. Active work is underway to increase the maximum speed of access to the Internet to 100 Mbit/s and above. The computers used for educational purposes are equipped with various educational programs (more than 26 thousand), computer testing programs (more than 20 thousand), electronic versions of

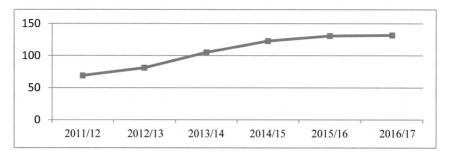

**Fig. 4.** The number of personal computers used for educational purposes, per 1000 students, units (Source: compiled by the authors based on open data from the Ministry of Education of the Russian Federation)

textbooks and manuals (more than 45 thousand), electronic journals and electronic diaries (more than 34 thousand) and other applications. In addition to computers, classrooms are equipped with multimedia projectors (more than 550 thousand), inter-active whiteboards (more than 260 thousand), multifunctional devices (more than 300 thousand) and other office equipment necessary to ensure a high-quality educational process and its result.

At the same time, a significant risk for the education system in Russia is a low applicability of material and technical support in educational activities. In particular, the proportion of students enrolled in programs implemented using distance learning technologies is only 2.5%, e-learning - 13.5%, and network - 1.8% (Fig. 5).

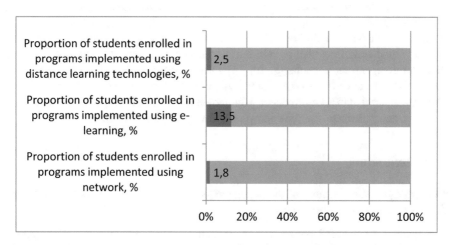

**Fig. 5.** The share of students in educational programs implemented using modern information technologies (Source: compiled by the authors based on open data from the Ministry of Education of the Russian Federation)

## 4 Discussion

The digital economy requires the education system to train the relevant staff. At the same time, to achieve significant results in the education system, as noted by Professor Michael Barber, professionals with the necessary qualities and skills are also required [4]. Such professionals, according to a number of scientists [3, 7, 12, 15], are managerial staff with an actual level of professional development.

In accordance with the roadmap of the "Digital Economy of the Russian Federation", it is managerial staff of educational organizations that are given priority in ensuring:

- Updating the legal framework of the educational organization and its educational programs to meet the requirements of the digital economy;
- Creating a profile of competencies for each student in accordance with the content of the basic competencies for the digital economy;
- Implementing an information system to support the individual profile of students' competencies with the possibility of planning his personal development path;
- Introducing and using distance learning technologies and e-learning in all types and forms of students' activities, including in networking with other educational organizations, including additional education, in accordance with the requirements of the digital economy;
- Implementing vocational orientation of students on the basis of the atlas of new professions, developing and implementing a system of early identification, support of highly motivated and talented students on the basis of competencies and personal development pathways for the digital economy, etc.

The presented list of tasks of managerial staff of the educational organization for the digitalization of society is not final. According to the authors, it can be expanded with the tasks of introducing digital technologies into the management system of an educational organization, as well as into the development system of managerial staff. At the same time, research results [8, 10, 16] show that the digital competence of managerial staff of educational organizations requires continuous development. This will allow managerial staff to more effectively update the potential of the education system in Russia and minimize the risks in training for the digital economy.

## 5 Conclusions

When studying the potential of the education system in Russia in training staff for the digital economy, it was found out that:

1) Principal educational programs contain compulsory study subjects "Economics" and "Information Technology", which are to develop the knowledge and skills demanded by the digital economy;
2) Conditions for educational programs correspond to the state educational requirements established in Russia;

3) Pedagogical staff of the education system in Russia has a high level of qualification, systematically improved in the profile of pedagogical activity;
4) The education system in Russia takes into account the trends of digitalization of the economy in the process of developing the material and technical conditions for the implementation of educational programs;

The study also revealed staff risks and risks of low applicability of material and technical support in educational activities, which significantly limit the potential of the education system in Russia in training staff for the digital economy. According to the authors, the leading role in minimizing the identified risks and updating the potential of the education system in Russia belongs to managerial staff of educational organizations, which has an actual level of professional development. This conclusion actualizes the further study of managerial staff of the educational organization as a condition for increasing the efficiency of the education system in Russia.

# References

1. Abdrakhmanova GI, Plaksin SM, Kovaleva GG (2017) Internet economics in Russia: approaches to definition and assessment. Forsyth 11(1):55–65
2. Abdrakhmanova G, Vishnevskiy K, Volkova G, Gokhberg L et al (2018) Digital economic indicators in the Russian Federation: 2018. Data Book. National Research University Higher School of Economics, HSE. Moscow
3. Adizes IK (2014) Leading the leaders: How to enrich Your Style of management and handle people whose style is different from yours. Alpina Publisher, p 259
4. Barber M, Donnelly K, Rizvi S (2012) Oceans of innovation. The Atlantic, the Pacific, Global leadership and the future of education. Institute for Public Policy Research, London
5. Bobkov VN, Vakhtina MA, Simonova MV (2016) Economic factors of Russian inequality. Int J Environ Sci Educ 11(16):8900–8910
6. Castells M (2000) Information age: economy, culture, society. M.: HSE, p 608
7. Daft RL, Lane PG (2007) The leadership experience. Eksmo, p 480
8. German MV (2012) Continuing education: the evolution of development, objective reality. Bulletin of Tomsk State University. Economy 2:147–154
9. Inozemtsev VL (2000) Modern post-industrial society: nature, contradictions, perspectives: studies. Manual for University Students. M.: Logos, p 304
10. Lenskaya EA, Brun IV (2016) Are the directors of Russian schools ready to work in a transformational mode. Educ Issues 2:62–99
11. Mitrofanova AE (2013) Socio-economic content and structure of staff risks in the organization. Electron J "Bulletin of Moscow State Regional University" 2:1–7. http://evestnik-mgou.ru/vipuski/2013_2/stati/ekonomika/mitrofanova.html
12. Mitrofanova EA, Tarasenko VV (2017) Complex strategy of differentiated development of the managerial personal of educational organization. In: International conference on research paradigms transformation in social sciences. The European proceedings of social & behavioral sciences: future academy, vol 35, pp 1347–1355
13. Naisbitt J (2003) Megatrends, p 380
14. Neustroev SS, Simonov AV (2015) Innovative directions for the development of e-learning. Man Educ 3(44):9–15

15. Savvinov VM, Strekalovskiy VN (2013) Consideration of the interests of stakeholders in the management of educational development. Bull Int Organ Educ Sci New Econ 8(1):87–99
16. Simonova MV, Ilyukhina LA, Romantsev GM, Zeer EF, Khamatnurov FT (2016) Approaches to monitoring of competences and qualifications. IEJME: Math Educ 11 (7):2745–2760
17. Soldatova GU, Nestik TA, Rasskazova E I, Zotova EYu (2013) Digital competence of adolescents and parents. Results of the all-Russian study. M.: Foundation for Internet Development, p 144

# Open Educational Resources in the Digital Economy: Legal Regulatory Framework for Free Software License

N. V. Deltsova[✉]

Samara State University of Economics, Samara, Russia
natdel@mail.ru

**Abstract.** Relevance of the problem under examination is subjected to the active development of the digital economy and introduction of information technologies into the educational environment, formation of the open educational resources, which can be used only in compliance with the intellectual rights of authors and copyright holders of the works. The purpose of this article is to inquire into the free licenses used in the open educational resources: open licenses and Creative Commons licenses in Russia. The key method of research is a particular-scientific formal legal method, rendering it possible to arrive at the concept of the open educational resources, to determine the major characteristics of the open licenses and Creative Commons licenses. Following the research, it is concluded that the open license mechanism is the most convenient for arranging and distributing content of the open educational resource in the digital economy. The research has shown that the modern legal structure of the open license as recorded in the Civil Code of the Russian Federation makes it possible to integrate the Creative Commons license used worldwide into practical activities to form the open educational resources in the Russian Federation. The Creative Commons licenses can be used in the modern law enforcement activity without significant amendments into the Russian law. The article's materials are of practical value for professionals and scholars concerned with the intellectual property protection, as well as for those implementing and using the open educational resources.

**Keywords:** Digital economy · Creative commons license ·
Intellectual property free software license · Information technology ·
Open educational resources · Open license

## 1 Introduction

Technological process in the digital economy is a driving force for the socio-economic development, creation of new jobs and formation of the highly skilled personnel. Contemporary conditions of the social development, involving active introduction of the information technologies, require training of the professionals, being able to meet the acute needs of the society. Development of the information technologies exerts a significant impact on improvement of the global educational system, which is based on the communicative capabilities of the information environment, openness and use of

© Springer Nature Switzerland AG 2020
S. Ashmarina et al. (Eds.): *Digital Transformation of the Economy: Challenges, Trends and New Opportunities*, pp. 473–480, 2019.
https://doi.org/10.1007/978-3-030-11367-4_47

the state-of-the-art Internet-based training technologies. Global informatization has significantly changed the approaches to the content and methods of training and teaching. Currently, the educational standards and training programs in Russia and abroad are focused on use of the e-learning tools available through the specialized educational resources, called the Open Educational Resources [1]. Dissemination of the educational information through the open educational resources is associated with the necessity to respect the intellectual rights for the intellectual products comprising them. Majority of the information as included into the content of the relevant resources is intellectual property: its distribution and use must be subject to the law governing the creation and use of the intellectual products [4]. In this regard, the issue of using the free licenses while creating and implementing the open information resources is relevant. This situation is especially true for the Russian Federation, where the legal institution of open licenses was legally formalized only in 2014 [8]. It appears that the current trends affecting development of the organizational and legal mechanisms for using information technologies based on the open licenses in the educational area require an adequate legal regulation thereof, which also determines relevance of this research.

## 2 Methodological Framework

### 2.1 Research Methods

During the research, the following methods have been applied: general scientific, i.e. the methods of analysis, synthesis, comparison, generalization; particular-scientific formal-legal method, rendering it possible to define the open educational resources, to specify the major features of the open licenses and Creative Commons licenses.

### 2.2 Research Base

The research base includes the scientific surveying, publications of the Russian and foreign scholars and lawyers examining the legal aspects of use of the open educational resources and free licenses. Analytical materials on the issues related to the open educational resources and technologies, open licenses and Creative Commons licenses are also used in the research.

### 2.3 Research Stages

The issue has been examined in two stages:

The first stage: the existing scientific and analytical literature within the research's topic, as well as the law concerning legal regulation of the free licenses for activity is reviewed; the problem, purpose and methods of the research are identified. The second stage: the conclusions obtained following the scientific and analytical literature's and law's analysis have been formulated; the publication has been prepared.

# 3  Results

The concept of the "open educational resources" was first heard in 2002 at the Conference "UNESCO Forum on the Impact of Open Courseware for Higher Education in Developing Countries". It was suggested, at that conference, to understand the open educational resources as the materials placed in public domain, intended for use in the academic process, the authors of which materials had approved the free use and processing thereof.

In the Russian scientific literature, the open educational resources mean the training, academic or scientific resources that are freely available or released with a license authorizing their free use or processing. The open educational resources include complete courses, educational materials, modules, textbooks, videos, tests, software, as well as any other tools, materials or technologies being utilized to provide access to knowledge [13].

The open educational resources offer an innovative approach to development, dissemination and use of knowledge, contribute to implementation of the education-for-all approach and the lifelong learning and encourage a spread of educational programs [6].

The first steps in formation of the open information resources were made by the Massachusetts Institute of Technology in 2001 as a part of the MIT Open Course Ware (OCW) project: more than 9000 academic and training materials were published on a wide range of subjects taught at the university [1].

The non-profit public organization "Creative Commons" played a vital role in promoting the open educational resources. This organization was established in the USA in 2001 with the support of the Centre for the Public Domain. Its purpose was to enhance the number of the works available for legal free distribution and use through the ready-made Creative Commons licenses, having the legal force and being adapted to the laws of many countries of the world.

The Creative Commons licenses are treated by the professionals as the free licenses different from the conventional license agreements that are based on the "all rights reserved" concept, particularly owing to their "free" nature as they are based on the "some rights reserved" concept. By virtue of the above concept, the user is granted with the widest range of powers to use the intellectual property items, the so-called "freedoms". The following four basic freedoms are singled out: to use the work for any purposes, to modify the work, to distribute the work and to distribute the modified work [5]. Content of such licenses provokes discussions with the professionals, which fact testifies to their relevance and development [11].

Alongside, creation of such licenses does not contradict the principal legislative instruments in the area of copyright: the Copyright Act of 1976 and the Digital Millennium Copyright Act (DMCA) of 1998. Moreover, the Creative Commons have become acknowledged in the US judicial practice as well [9]. The Creative Commons licenses mean the license agreements containing various alternatives of use of the protected works that allow users determining the scope of the work to be used. They are conditionally grouped into the full and limited licenses depending on the scope of the conferred authorities.

The first group includes: Attribution and Attribution-Share Alike. The Attribution license authorizes distribution, editing, modification and use of the work as a ground for derivative works, even on the commercial basis with the obligatory indication of authorship. The second group of licenses also authorizes users to process the works, including for commercial purposes, but subject to indication of the authorship and licensing of the derivative works under similar conditions.

Besides, one can single out the four types of the limited Creative Commons licenses: Attribution Non-Commercial, Attribution Non-Commercial Share Alike, Attribution No Dervis, Attribution No Derivatives. The first type of the limited licenses is aimed to authorize the users to reprocess the work on a non-commercial basis with obligatory indication of the author and non-commercial nature of use. As part of this license, it should be noted that there is no obligation to license the derivative works under similar conditions. The second type allows processing of the work on a non-commercial basis as long as the author is mentioned, and the derivative works are licensed under similar conditions. The third type of licenses permits distribution, commercial and non-commercial use of the original work, provided that the work is transferred with indication of the authorship, remains unchanged and preserves its integrity. The fourth type of the limited licenses authorizes the users to download the work and share it with others as long as they mention and provide reference to the author. However, it is prohibited to amend the work and use it for commercial purposes [12].

The Creative Commons licenses have been widely used in practices of many educational and academic institutions worldwide. These include the World Intellectual Property Organization, UNESCO, and educational institutions all around the world.

It is worth noting that Russia is also experienced at using the Creative Commons. In particular, since 2005 the Moscow State Institute of International Relations has been distributing materials through posting on its website: www.mgimo.ru, under the license: Creative Commons Attribution-Non Commercial 2.5. The portal of the New Media and Theory of Communication Department of the Faculty of Journalism at the Moscow State University named after M.V. Lomonosov (www.convergencelab.ru) publishes materials under the conditions of the free license: Creative Commons Attribution 3.0 Unported, authorizing to use them through showing the name of and providing the reference to the author. In addition, the materials of the corporate portal of the National Research University "Higher School of Economics" (www.hse.ru) can be reproduced in any media, in the Internet or in any other media pursuant to the license: Creative Commons Attribution – Share Alike 3.0 Unported [11].

No definition of the "free license" is fixed in the civil law of the Russian Federation. This term is used only in scientific researches. The open license is a legal tool enabling to use the content of the open educational resource in the Russian Federation as covered by the intellectual rights.

The Russian legal doctrine is based on the fact that intellectual rights for the protected works belong to the group of exclusive copyrights, where the author is entitled to prohibit all third parties from using the protected intellectual property items. Besides, the researchers assume that the exclusive right, in addition to the right for prohibiting the use of the intellectual activity deliverable (prohibitive function of the exclusive right), is at the same time the right to allow others using the author's work (positive function). Use of intellectual property is possible only with the copyright holder's consent [6].

According to the Article 1286.1 of the Civil Code of the Russian Federation (hereinafter CC of RF), an open license means that the licensing agreement, whereunder the author or any other rightholder (licensor) provides the licensee with the ordinary (non-exclusive) license for using the work of science, literature or art, can be concluded in a simplified manner.

From the point of view of the Russian law, the open license is an accession agreement, therefore its conditions shall be accessible to the public at large and placed in such a way that the user will be able to read them immediately before the use.

The very process of entrance into such a license agreement can be called simplified, whereas the open license may contain an indication of actions that will be considered as acceptance of its terms. In that case, the written form of the agreement will be considered as complied with (paragraph 2 of clause 1 of Article 1286.1 of CC of RF).

The essential condition of any agreement, failing which condition the agreement is considered not to be concluded, is the one of the subject. The open license's subject is the right to use the work of science, literature or art within the limits provided by the agreement. Consequently, the licensor individually determines which rights it is going to grant to the user under the license. The holder of the exclusive right's capacity to provide the licensee with the right to use the work it owns for creation of the new intellectual activity deliverable is an important feature of the open license.

In that case, unless otherwise is provided by the open license, it is considered that the licensor has proposed to enter into an agreement on use of the work owned by it to anyone desiring to use the new intellectual deliverable created by the licensee on the basis of such work, within the limits and under the conditions contemplated by the open license. Acceptance of such proposal is also considered to be an acceptance of the licensor's proposal to enter into the license agreement for the above work (paragraph 2 of clause 2 of Article 1286.1 of CC of RF).

In view of the aforesaid regulation, one should also pay attention to the current edition of the Article 1266 of CC of RF, which deals with the right for the work's integrity. In the context of the legal regulation of open licenses, the regulations of the above-mentioned Article prescribe the rule that the author may agree to make future changes, reductions and additions to his/her work, to supplement his/her work with illustrations and explanations, if necessary (correction of errors, updating or addition of actual information, etc.), provided that the aforesaid action does not distort the author's intention and does not violate integrity of the work's perception. Thus, the provided version of the regulations of the Article 1266 of CC of RF makes it possible for the author of the intellectual deliverable to confer the right for making changes to his/her work to third parties.

It should be emphasized that, as a general rule, an open license is gratuitous, unless otherwise is provided by the license agreement.

In addition, if the open license does not stipulate the territory, where the use of the relevant work is permitted, such use will be allowed worldwide.

Let's also note that the validity term of the open license is determined by the agreement, and if the open license does not specify the term, the agreement is considered to be concluded for five years (paragraph 2 of clause 2 of Article 1286.1 of CC of RF).

The said characteristics of the open license, which are currently recorded in the applicable civil law, allow us concluding that it can be described as free because it provides the user with quite a vast range of options for using the copyrighted works.

Besides, one may state that the open license is an effective tool for formation of the open educational resources:

- The licensor (rightholder) is able to determine, at its discretion, the rights granted under that agreement, using the broadest approach to granting the rights, reserving only some of the powers. It is the principle of "some rights reserved," which, unlike the concept of "all rights reserved," is based on exercise of the copyright's authorization function.
- The licensor is entitled to authorize using the protected work for creation of the new derivative works, which can also be used further under the terms of the open license. This rule is very important for the works used for educational purposes. Amendments can be made both by other professionals and by the students themselves. Alongside, a violation of the right to the work's integrity and the right to have it protected from distortion is precluded.
- The right of use is granted free of charge, which makes it possible for a vast audience to use the educational content. At the same time, it is not prohibited to distribute the intellectual products on a payable basis, which serves as a worthy incentive for the authors and rightholders at creating the open educational resource. It seems practically expedient to provide the intellectual property rights within the bounds of the single educational resource, both on a payable and free basis.
- The global principle of using the protected items on the basis of the open license, which makes it possible to acquire education from anywhere in the world, is a positive feature.

## 4   Discussion

The open educational resources and certain aspects of their use are being actively dealt with in the contemporary science. The relevance of the issues was highlighted in the researches of Santos-Hermosa et al. [8]. The necessity to regulate the relations in respect of the open educational resources was emphasized by Galkina et al. [2]. The issues related to the concept and content of the open educational resource were examined by Ysupov [13]. The issues of content of the Creative Commons licenses and criticism thereof were presented by Stallman [10]. The issues of free licenses in education were paid attention to by Moiseyechev [5], who surveyed the experience of application of free licenses in foreign countries and prospects of such application in Russia. The scientific concepts in the area of copyright, justifying the possibility of using the open licenses, were presented in the research of Labzin [6]. Some scholars in their works analyze the content of the legal institution of the open license in Russia.

The content, legal characteristics of that agreement were disclosed in the works of Grin [3] and Zdanovich [14]. Rusanova [7] wrote about application of the Creative Commons in Russia. The aspect dealt with in this article is not sufficiently investigated at the level of scientific literature. The above publications show relevance of the research in the area of use of the free licenses: open licenses and Creative Commons licenses in the open educational resources.

## 5  Conclusions

The open license mechanism is the most convenient for arranging and distributing content of the open educational resource in the digital economy.

The research has shown that the modern legal structure of the open license as recorded in the Civil Code of the Russian Federation makes it possible to integrate the Creative Commons license used worldwide into practical activities to form the open educational resources in the Russian Federation.

The Creative Commons licenses can be used in the modern law enforcement activity without significant amendments into the Russian law. Unlike the traditional license agreements, which are based on the "all rights reserved" concept, the open licenses and the Creative Commons licenses are based on the "some rights reserved" principle, providing the user with more expansive opportunities to deploy the open educational resource and its substantial content.

Depending on the type of the agreement based on its content, the open licenses and the Creative Commons licenses provide the user with a number of opportunities: to use the protected works for educational purposes, to distribute them, to distribute modifications of the protected works (derivative works) that constitute educational content and, in some cases, to modify the works. This enables to the highest extent to ensure compliance with the student's interests, on the one hand, and to observe the interests of the copyright holder, on the other hand.

## References

1. CIS on its way to open educational resources. Analytical Review. UNESCO Institute for Information Technologies in Education. UNESCO (2011) https://iite.unesco.org/pics/publications/ru/files/3214683.pdf. Accessed 20 Mar 2018
2. Galkina AI, Burnasheva EA, Grishan IA, Komarova MV, Bobkova EY (2015) Statistics of productivity and effectiveness of experimental support of the educational system (For scientists and education experts). Mediter J Soc Sci 6(5) S3:62–70
3. Grin ES (2014) On the issue of legal nature of open licenses. Acute Probl Russian Law 11:2411–2416
4. Intellectual Property Rights: Textbook/Badulina EV, Gavrilov DA, Grin ES, etc.; Under general editorship of L.A. Novoselova. M.: Statut, Vol 1: General Provisions, 512 p (2017)
5. Labzin M (2014) Scientific concepts of understanding the intellectual property law. Patents and Licenses. Intellect Rights 8:57–68

6. Moiseyechev E (2014) International experience and Russian perspectives of open educational resources. Information resources of Russia http://www.aselibrary.ru/press_center/journal/irr/irr5924/irr59245927/irr592459276096/irr5924592760966103/. Accessed 21 Mar 2018

7. Rusanova YV (2018) Creative Commons Licenses. In: Law in the Area of Internet. M.: Statut, pp 415–430

8. Santos-Hermosa G, Ferran-Ferrer N, Abadal E (2017) Repositories of open educational resources: an assessment of reuse and educational aspects. Int Rev Res Open Distribu Learn 18, 5 Aug. http://www.irrodl.org/index.php/irrodl/article/view/3063/4300. Accessed 2 Apr 2018

9. Sobol IA (2014) Free licenses in the Russian Copyright. M., 196 p

10. Stallman RM (2005) Fireworks in Montreal. The official website of the free software foundation. http://www.fsf.org/blogs/rms/entry-20050920.html. Accessed 2 Apr 2018

11. Khokhlov YE (ed) (2011) Use of the creative commons licenses in the Russian Federation: analytical report. M: Information Society Development Institute, 94 p. http://creativecommons.ru/sites/creativecommons.ru/files/docs/ispolzovanie_ss_v_rf.pdf. Accessed 21 Mar 2018

12. Web-site of Creative Commons. http://creativecommons.ru/faq/. Accessed 19 Mar 2018

13. Ysupov IF (2017) The application of open electronic educational resources in the conditions of implementation of information security professional education http://scipress.ru/pedagogy/articles/primenenie-otkrytykh-elektronnykh-obrazovatelnykh-resursov-v-usloviyakh-realizatsii-informatsionnoj-bezopasnosti-organizatsii-professionalnogo-obrazovaniya.html. Accessed 1 Mar 2018

14. Zdanovich GV (2014) On the issue of content of the free license. Business Education Law. Bull Volgogr Inst Bus 2(27):280–286

# Matrix Model of Cognitive Activity as One of the Meta Basis of Digital Education

N. A. Tymoschuk[1(✉)], E. N. Ryabinova[1], O. A. Sapova[2], and V. Oddo[3]

[1] Samara State Technical University, Samara, Russia
{7.60n, eryabinova}@mail.ru
[2] Samara State University of Economics, Samara, Russia
loli_air@mail.ru
[3] University Nice Sophia-Antipolis, Nice, France
virginie.oddo@unice.fr

**Abstract.** Due to the transition to informatization and digital education society focuses on self-education. Digitalization development will demand the search for meta basis in strategic and tactical approaches to education: there will be formalization of educational information adoption by step-by-step implementation of logical schemes of an approximate basis of actions. It is shown that the educational environment has a considerable impact on the preparation of competitive engineers. The education guidance, a workbook and the reference media on the studied subject are modern educational methodological support for creation of necessary conditions to provide sufficient achievement of theoretical material and practical skills to organize effective students' self-work. The technology of the organization of self-educational students' activity on the basis of matrix model of cognitive activity is rather simple and clear both for teachers and students, and can be applied to studying of any subject. By means of cognitive and activity matrix all volume of preparation material breaks into four modules, each of which has various level of complexity. The first module contains the simplest tasks of the first level of complexity, the second one includes the tasks of the second level of complexity, etc. Each module contains the optimum number of theoretical data and tasks, both with detailed stage-by-stage analysis and an explanation and for self-work. The presented discretization and formalization of development of the subject by means of matrix model of cognitive activity opens only one of possible ways to digital education.

**Keywords:** Cognitive and activity matrix · Competitive engineers ·
Digital education · Metaeducation · Metabases · Self-educational activity

## 1 Introduction

Meta-education is a new strategic goal of modern educational policy based on mental-action integration of studying material and the principle of reflexive thinking. New standards of basic education are formed according to the metasubject approach, the requirements to metasubject results of studying are given there. Such concepts as "metaobjects", "metaproject training", "metasubject" (sub-subject, polysubject)

© Springer Nature Switzerland AG 2020
S. Ashmarina et al. (Eds.): *Digital Transformation of the Economy: Challenges,*
*Trends and New Opportunities*, pp. 481–493, 2019.
https://doi.org/10.1007/978-3-030-11367-4_48

abilities, metasubject competences (universal educational actions – UUD), metasubject communications, the metabases, etc. are included in the concept of formation of modern cultural and educational space of teaching and educational activity of the technical university [3]. These terms have deep historical roots and have been known since the time of Aristotle. However in the Russian pedagogics meta-education started to develop at the end of the 20th century in Y.V. Gromyko's works, A.V. Hutorsky, etc. [4, 5, 8, 9, 12, 13].

Due to the transition to informatization and digital education society focuses on self-training. The main task of educational institutions is to create the most favorable conditions for self-education and self-development of the personality, mutual education, requirements to continuous education as to the means for strengthening of adaptation opportunities to sharply accelerated pace of life. The process of these tasks' solution is multivariable that corresponds to the main tendencies of studying which M. V. Clarin called "metatendencies" [2]: mass nature of education and its continuity as a new quality; the importance of education both for the individual, and for public expectations and norms; orientation to active person's development ways of cognitive activity; adaptation of educational process to inquiries and needs of the personality; studying orientation to student's identity, providing opportunities for its self-disclosure.

## 2   Materials and Methods

Modern education requires the increase of computer participation in educational process. Only the computer that contains special psychological and pedagogical software can carry out the individual studying process in mass education. In this case it can be considered, along with the teacher, as an equal participant of educational process. It is obvious that the elaboration of psychology and pedagogical computer software will be the content of modern pedagogical science. Digitalization development will demand the search for metabases in strategic and tactical approaches to education: there will be discretization of educational information and formalization of its adoption by step-by-step implementation of logical schemes of an approximate mechanism of actions [6, 10].

N.V. Gromyko connects the sense and purpose of metasubject approach with the formation of mental-action mode in the education [4]. From the Soviet school in modern didactics there is still the mode (knowledge, abilities, skills); the introduction of the Unified State Examinations (USE) brought information-test mode; mental-action mode appeared with the emergence of mental-action pedagogics (since the beginning of 90s). In 2008 metasubject approach was declared as one of the reference points of new educational standards of basic education.

Meta (over, between, for, through, after means something general, integrating) is the system which investigates and describes other system by means and terms which are out of these systems: universal knowledge and methods– mental-action (metaactivity). Metaactivity uses the general methods of knowledge and metaabilities based on flexibility of mind, its dialectics, capacity to transformation and anti-conformism. It includes the theoretical thinking that allows giving the definitions to various concepts, giving necessary evidential base, being able to generalize, systematize, classify, etc.; creative thinking that allows structuring objects, finding solutions in unusual and new situations; critical thinking that allows distinguishing facts from opinions, seeing ambiguity of the statement, etc.; skills of information processing (analysis, synthesis, assessment, interpolation, extrapolation, etc.) and regulatory abilities (right definition of criterion function, rational planning, optimum choice of strategy and tactics, etc.).

Metasubject approach to the organization of self-educational students' activity allowed authors to develop cognitive and activity matrix providing the process of step-by-step performance of algorythmic intellectual actions and measurability of mastered studying material [10].

# 3 Results

## 3.1 Theoretical Basis of the Considered Issue – Theoretical Bases of Cognitive Activity

By modern psychology it is established that the person in his perception progresses on the levels and forms of activity [1, 14]. The levels of activity depend on the way of expression of the information that is mastered in the course of studying. The person moves ahead from learning to imitation (copying) and further to heuristic and creative actions. Forms of activity are significantly personified and are connected with knowledge processes, such as reflection, perception, attention, memory, thinking, consciousness, speech, imagination, abilities, intelligence, etc.

The carried-out analysis of psychological processes allows allocating the following defining informative levels from a set of knowledge processes: reflection, judgment, an algorithmics and monitoring which can represent one of possible structurization of knowledge process from the psychological point of view.

We define the listed above informative levels through $\psi_i$, $i = \overline{1,4}$. Level $\psi_1$ is the level of reflection (approximate level) which characterizes the perception of studying material by the student and includes such psychological processes as feeling, perception, attention, imagination, memory (as reminiscence), evident and figurative thinking, motive.

Level $\psi_2$ is comprehension. Cogitative function includes processing of accepted educational information, finding ways of objective solution. It is characterized by such psychological processes as memory, consciousness, is evident – effective or conceptual thinking, motive.

Level $\psi_3$ is algorithmic. Algorithm formation of the task solution is an executive function which includes the analysis of the ways applied to realize algorithm and is characterized by such psychological processes as memory, consciousness, attention, imagination, speech thinking, emotions, motive.

Cognitive level $\psi_4$ is monitoring. Control-correcting function is responsible for the correct registration of the result and is characterized by such psychological processes as memory, attention, thinking, speech, motive. The controlling procedure is the means to elaborate students' technique and ability to analyze and correct their own activity (self-examination).

The allocated cognitive levels allow building the structure of cognitive process (Fig. 1) from which follows that different levels of cognitive activity are characterized by identical psychological components. However each of the considered components changes in the course of passing through cognitive levels from $\psi_1$ to $\psi_4$.

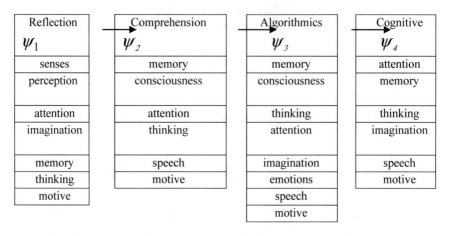

| Reflection $\psi_1$ | Comprehension $\psi_2$ | Algorithmics $\psi_3$ | Cognitive $\psi_4$ |
|---|---|---|---|
| senses | memory | memory | attention |
| perception | consciousness | consciousness | memory |
| attention | attention | thinking | thinking |
| imagination | thinking | attention | imagination |
| memory | speech | imagination | speech |
| thinking | motive | emotions | motive |
| motive | | speech | |
| | | motive | |

**Fig. 1.** Structure of cognitive process (Source: compiled by the authors)

There are two levels of activity depending on the way of expression of information acquired in the course of studying – reproductive and productive. At the reproductive level of activity the acquired information is only reproduced in various combinations – from direct copying before reconstructed reproduction and application in standard situations. Reproductive level of student's activity is copying of teacher's activity, direct reproduction of the acquired action algorithm. There are two types of reproductive activity. The first type corresponds to the reproduction on familiar pattern, and at the same time logical communications between concepts aren't required (activity at the level of learning). The second type of reproductive activity provides logical communications and elementary analogies at the reproduction of studying material (activity at the level of reproduction).

Productive levels of activity are implemented with the use of mastered methods. In the course of these levels of activity the mastered algorithm adapts to a new situation, or it is created from the parts of several other algorithms again. As a result of productive activity regarding the content of studying new information is always created, and this information will be new, as a rule, not objectively but subjectively. Productive activity can be also considered at two levels: application and creativity. The level of application corresponds to the solution of non-standard tasks by the studied methods when existing solution algorithm is simplified. Level of creativity is reached within educational, scientific, research work of the students. The information that is obtained can be often objectively new, and it is published in the press, reported at conferences, etc. This activity has to include creative action, a research element, knowledge transformation or transfer.

## 3.2 Creation of Cognitive and Activity Matrix

The considered cognitive levels of mastering of studying materials $\psi_i$, $i = \overline{1,4}$, and activity levels $d_j$, $j = \overline{1,4}$, can be formally presented in the form of matrix of the size $4 \times 4$ (Table 1) – matrix model of cognitive activity where each combination of match will correspond to a certain number of the mastered studying material that is a structural unit of an educational task which we will call an educational element of cognitive-activity matrix. From Table 1 we can see that the considered structure of cognitive activity which basics are not only psychological processes, but also kinds of activity, and it allows presenting the mastering of studying material as "movement" on the elements of $\psi d$ matrix which is made of the listed above cognitive and activity levels. At the same time certain quantity of mastered studying material corresponds to each of the elements of this matrix $Y_{ij}$, beginning from the most elementary level $Y_{11}$ (learning at the level of reflection) and finishing with the highest level $Y_{44}$, that is the research with control of own actions. The smallest volume of student's knowledge takes place at the level $\psi_1 d_1$. The further a student moves on the educational elements of $\psi d$ matrixes ($i \rightarrow 4$; $j \rightarrow 4$), the more difficult knowledge is mastered as the complexity of elements increases.

The content of any subject can be presented as the system of educational tasks (tasks), and any educational task can be structured by means of the presented matrix model of cognitive activity of students. At the same time the level of complexity of the task is determined by the activity sign $j = \overline{1,4}$, as solution of task of any complexity demands using all four cognitive levels.

**Table 1.** Matrix model of cognitive activity of students

| Cognitive levels | Activity levels | | | |
|---|---|---|---|---|
| | Reproductive activity | | Productive activity | |
| | Learning (introduction) | Reproduction (copies) | Application (transformation) | Creativity (research) |
| | $d_1$ | $d_2$ | $d_3$ | $d_4$ |
| Reflection $\psi_1$ | $Y_{11}$ | $Y_{12}$ | $Y_{13}$ | $Y_{14}$ |
| Comprehension $\psi_2$ | $Y_{21}$ | $Y_{22}$ | $Y_{23}$ | $Y_{24}$ |
| Algorithmic $\psi_3$ | $Y_{31}$ | $Y_{32}$ | $Y_{33}$ | $Y_{34}$ |
| Monitoring $\psi_4$ | $Y_{41}$ | $Y_{42}$ | $Y_{43}$ | $Y_{44}$ |

Source: compiled by the authors.

The scheme of mastering tasks for each j-level ($j = \overline{1,4}$) will be various. For the tasks of the 1st level the scheme of mastering will be defined by the following formula: $Y_{11} \rightarrow Y_{21} \rightarrow Y_{31} \rightarrow Y_{41}$; for the tasks of the 2nd level: $Y_{11} \rightarrow Y_{12} \rightarrow Y_{21} \rightarrow Y_{22} \rightarrow Y_{31} \rightarrow Y_{32} \rightarrow Y_{41} \rightarrow Y_{42}$; for tasks of the 3rd level: $Y_{11} \rightarrow Y_{12} \rightarrow Y_{13} \rightarrow Y_{21} \rightarrow Y_{22} \rightarrow Y_{23} \rightarrow \ldots$; for the 4th level: $Y_{11} \rightarrow Y_{12} \rightarrow Y_{13} \rightarrow Y_{14} \rightarrow Y_{21} \rightarrow Y_{22} \rightarrow Y_{23} \rightarrow Y_{24} \rightarrow Y_{31} \ldots$ (Fig. 1). Thus, the complexity as the formal task characteristic, is defined in our case by the structure of the search process for the solution. To solve test tasks of the 1st level the student consistently has to pass the following stages through the elements of $\psi d$ matrixes: reflection - comprehension –algorithmic - monitoring. At the same time the $\psi d$ matrix will have the size $4 \times 1$. To solve the tasks of the 2nd level of complexity: reflection at the level of learning - reflection at the level of reproduction -comprehension at the level of learning - comprehension at the level of reproduction - algorithmic at the level of learning –algorithmic at the level of reproduction - monitoring at the level of learning - monitoring at the level of reproduction. In this case the size of $\psi d$ matrixes is $4 \times 2$. The stages of mastering tasks of the 3rd and 4th levels of complexity are completed in the same way. Taking into account metasubject approach each element of cognitive- activity matrix defines universal educational action.

### 3.3   Technology of the Organization of Self-educational Activity of the Students on the Basis of Cognitive-Activity Matrix

Modern reality is that the educational environment has considerable impact on the preparation of competitive engineers. It is connected, first of all, with the introduction of new approaches to the preparation at higher school, with new additional requirements to university graduates, with new competences of future engineers, etc. The generating of the educational environment in higher educational institution is defined by joint communication of teachers and students, their joint situational activity, experience transformation and equality of all participants of educational process. The creative educational environment is not only a condition, but also the means of training and education.

The education guidance, workbook and the reference media on the studied subject are a modern educational methodological support to create necessary conditions of sufficient mastering of theoretical material and practical skills at the organization of effective out-of-class students' self-work, self-education and self-checking. The technology of the organization of self-educational students' activity on the basis of cognitive-activity matrix is rather simple and clear both for teachers and students and can be applied for studying of any subjects. By means of cognitive-activity matrix all volume of studying material is divided into four modules, each of which has various level of complexity. The first module contains the simplest tasks of the first level of complexity; the second includes the tasks of the second level of complexity, etc. Each module contains the optimum number of theoretical data and tasks both with detailed stage-by-stage analysis and explanation, and with independent solution.

To create effective knowledge system, students are offered to start mastering studying material with the solution of the simplest tasks – tasks of the first level. The leading principle in performing educational tasks is consequential movement in the levels of mastering of educational information that reflects the hierarchy of the levels of person's activity. Only in this case the organization of experience of educational activity is mastered gradually according to the mastering of cognitive actions of the lowest level of knowledge complexity. At the same time studying actions are carried out with the understanding of formation mechanism of student's knowledge.

At the end of each module the self-examination set is given that consists of the tests of appropriate complexity level and the technique of independent assessment of available knowledge level. By means of offered educational and methodical guide the student can independently study the material with an individual speed and master types of mental performance to solve the tasks of appropriate complexity, to regulate the quantity of solvable tasks, to do self-performance-assessment and estimate the level of mastered studying material, gaining skills of self-education and self-assessment.

It is shown that such approach to the education allows individualizing the system of vocational students' preparation, creating educational and methodical computer software. The development of the presented studying system that represents the model of modern educational environment opens one of possible ways to digital education. The model of the educational environment, in this case, represents not just some essence, but the mode of actions in joint situational activity of all participants: the synergetic effects arising in the system "environment-subject" are expressed in new adaptive qualities of the students which allow them to feel more familiar when they solve arising problems and tasks [11]. The transition from traditional to innovative educational technologies provides active creative interaction of educational process entities, unity of their cognitive, research and practical activities. We give examples of structuring of educational tasks according to cognitive activity matrix.

### Tasks of the First Complexity Level.
In the conditions of mass globalization there is a problem about social inequality in society that affects the level of Earth population, the index of development of human potential. It is necessary to find out how negative consequences influence the personality in the conditions of social inequality (Table 2).

**Table 2.** Tasks of the first complexity level

| Studying elements | Sequence of actions |
| --- | --- |
| $Y_{11}$ – reflection on the level of learning | Social inequality is a form of differentiation in which certain individual, social groups, strata, classes are at different levels of the vertical social hierarchy and have unequal life chances and opportunities to meet needs |
| $Y_{21}$ – comprehension on the level of learning | Personality tries to overcome social inequality by any means |
| $Y_{31}$ – algorithmic on the level of learning | There are attempts to move from one social stratum to another, which are often illegal |
| $Y_{41}$ – monitoring on the level of learning | This means that the individual has the opportunity to move to another social class, but due to existing negative experience this process can be difficult |

Source: compiled by the authors.

### Tasks of the Second Complexity Level.
The problem of culture preservation is urgent in modern society. The source of culture is the cultural heritage as the general property of all people and all mankind. Cultural heritage are both natural, and cultural objects. To define the opportunity to refer this or that object in cultural heritage of UNESCO (Table 3).

**Table 3.** Tasks of the second complexity level

| Studying elements | Sequence of actions |
| --- | --- |
| $Y_{11}$ – reflection on the level of learning | Defining the concept "cultural heritage" |
| $Y_{12}$ – reflection on the level of reproduction | Identification of object's signs (work by the creative genius of the person, certificate of a civilization, connection with traditions, inclusion of unique territories and so on) |
| $Y_{21}$ – comprehension on the level of learning | To define object type (monument, ensemble, creation of person and nature) |
| $Y_{22}$ – comprehension on the level of reproduction | To identify the role of the monument (history, city planning, archeology) |
| $Y_{31}$ – algorithmic on the level of learning | Definition of the object category and its status (federal, regional, municipal) |
| $Y_{32}$ – algorithmic on the level of reproduction | Make a general characteristic of the object |
| $Y_{41}$ – monitoring on the level of learning | To correlate the characteristic with the laws of the Russian Federation and the provisions on cultural heritage |
| $Y_{42}$ – monitoring on the level of reproduction | The decision to assign / not assign the status to the object |

Source: compiled by the authors.

### Tasks of the Third Complexity Level.

The national park "Samara Luka" is the pride of Samara and the Samara region, but anthropogenic pollution and soil cover destruction can lead to the threat of disappearance of many species of plants and animals therefore the task is program development "On Reduction of Harmful Effects on Soil Pollution of National Park Samara Luka" (Table 4).

**Table 4.** Tasks of the third complexity level

| Studying elements | Sequence of actions |
| --- | --- |
| $Y_{11}$ – reflection on the level of learning | Definition of the concepts "program", "environmental protection" |
| $Y_{12}$ – reflection on the level of reproduction | Definition of specific problems of the Samarskaya Luka national park |
| $Y_{13}$ – reflection on the level of application | Justification of the relevance of the program "On reduction of harmful effects on soil pollution of the Samarskaya Luka National Park" |
| $Y_{21}$ – comprehension on the level of learning | Definition of the concept of "harmful effects" |
| $Y_{22}$ – comprehension on the level of reproduction | Defining the goals and objectives of the program "On reduction of harmful effects on soil pollution of the Samarskaya Luka National Park" |

*(continued)*

**Table 4.** (*continued*)

| Studying elements | Sequence of actions |
|---|---|
| $Y_{23}$ – comprehension on the level of application | Defining the structure of the program " On reduction of harmful effects on soil pollution of the Samarskaya Luka National Park" |
| $Y_{31}$ – algorithmic on the level of learning | Defining management methods of the program "On reduction of harmful effects on soil pollution of the Samarskaya Luka National Park" |
| $Y_{32}$ – algorithmic on the level of reproduction | Organization of mechanisms for program implementation "On reduction of harmful effects on soil pollution of the Samarskaya Luka National Park" |
| $Y_{33}$ – monitoring on the level of application | Defining program funding "On reduction of harmful effects on soil pollution in the Samarskaya Luka National Park" |
| $Y_{41}$ – monitoring on the level of learning | Justification of the social efficiency of the program implementation "On reduction of harmful impact on soil pollution of the Samarskaya Luka National Park" |
| $Y_{42}$ – monitoring on the level of reproduction | The expected results of the program "On reduction of harmful effects on soil pollution of the Samarskaya Luka National Park" are described |
| $Y_{43}$ – monitoring on the level of application | Compilation of applications to the program " On reduction of harmful effects on soil pollution of the Samarskaya Luka National Park" |

Source: compiled by the authors.

## Tasks of the Fourth Complexity Level.

The young family made the decision to buy the apartment in a mortgage. They want to warn themselves against typical mistakes in this situation. It is necessary to make the plan according to which they will be able to take the apartment in a mortgage and will protect themselves from unnecessary risks (Table 5).

**Table 5.** Tasks of the fourth complexity level.

| Studying elements | Sequence of actions |
|---|---|
| $Y_{11}$ – reflection on the level of learning | Contact a mortgage broker who knows everything about a mortgage and will help to resolve many issues |
| $Y_{12}$ – reflection on the level of reproduction | Sign the contract with a competent realtor to ease the procedure of buying an apartment |
| $Y_{13}$ – reflection on the level of application | Collection of employment references, photocopies and original documents for applying to the bank to obtain a loan |
| $Y_{14}$ – reflection on the level of creativity | Choosing a bank and its positive decision on granting a loan |
| $Y_{21}$ – comprehension on the level of learning | Search for a suitable apartment to be approved by the bank |

(*continued*)

**Table 5.** (*continued*)

| Studying elements | Sequence of actions |
|---|---|
| $Y_{22}$ – comprehension on the level of reproduction | The choice of the term of the loan, taking into account income and other opportunities of the family |
| $Y_{23}$ – comprehension on the level of application | Appraisal of the apartment. Transferring evaluation certificate to the bank |
| $Y_{24}$ – comprehension on the level of creativity | Documents are sent to be checked to the bank and the insurance company. The security service of the bank, together with the legal department considers the selected apartment. If they are satisfied with everything, then permission is given to purchase an apartment |
| $Y_{31}$ – algorithms on the level of learning | Signing credit agreement |
| $Y_{32}$ – algorithms on the level of reproduction | Transferring money to the seller |
| $Y_{33}$ – algorithms on the level of comprehension | Signing the contract of sale |
| $Y_{34}$ – algorithms on the level of creativity | Registration of home insurance and the borrower |
| $Y_{41}$ – monitoring on the level of application | Drawing up a full repayment schedule |
| $Y_{42}$ – monitoring on the level of learning | State registration. The transferring of rights is implemented at the time of state registration. Registration of transactions lasts from 3 days to a month |
| $Y_{43}$ – monitoring on the level of reproduction | Insurance, due to which the bank wants to reduce its risks |
| $Y_{44}$ – monitoring on the level of creativity | The last step is the registration of real estate in the registration chamber and receiving the documents confirming the right to housing, property ownership certificate |

Source: authors' development
Source: compiled by the authors.

## 4   Discussion

The presented formalization of mastering the subject by means of cognitive activity matrix by step-by-step implementation of logical schemes of an approximate basis of actions can be applied to assess students' educational activity [15, 16]. Step-by-step performance of the test task is the interconnected sequence of educational elements defined by an algorithm and solution logic of an educational task. Specially developed forms of tests answers which show quality of academic performance demonstrate which quantity of educational elements is not acquired and what quality they are. It allows using tests' results not only for quality and quantitative standard of educational achievements of students, but also to realize their studying potential. The objectivity and measurability of education quality open a great number of opportunities for management of educational process – from the correction of content of educational standards and programs to the improvement of teaching techniques and increase in efficiency of students' self work [7]. According to the research, for the pedagogical theory of performance assessment the concept "pedagogical task" is initial. Multiple choice are widespread in test practice, they are simple, traditional and convenient to check knowledge automatically. In the systems of test tasks the probability of the correct answer to the subsequent task can depend on the probability of the correct answer to the previous tasks. Such tasks are significant for the organization of students' self-work and also for implementing of final graduates' assessment of professional educational institutions.

## 5  Conclusions

The presented adaptive system of professional training of technical universities has the personified character and fully satisfies to all system properties: integrity, divisibility, organization and communications, integrative qualities.

The matrix model of students' cognitive activity in the adaptive system of personified professional training provides the mechanism of systematization of educational tasks; content of the subject can be considered as the system of the educational tasks which have their structure that means studying the essence of the objects; the structural analysis of the systems of educational tasks in different subjects revealed the violation of tasks' percentage ratio of the complexity levels: with the increase of complexity the number of the appropriate tasks sharply decreases; in the systems of tasks their hierarchy on complexity is broken; the specified shortcomings interfere with effective formation of students knowledge system that eventually leads to the complications of the principle of the developing studying;

The basic principles of construction and structure of matrix model allow presenting any block of educational information in the form of final number of elementary information chunks; students master them moving on the elements of cognitive-activity matrix. Similar structuring of studying material makes mastering process measurable, at the same time the concept "measurabilities" becomes solvable from the quantitative point of view (how many structured elements are mastered with specified quality), and in qualitative (which cognitive-activity elements are mastered, which are not).

The developed tests for a periodic qualimetry of educational information mastering are the interconnected sequence of educational elements of cognitive-activity matrix. It is defined not only by an algorithm and logic of the solution of an educational task of exact level, but also by the dependence of receiving the correct answer at mastering subsequent educational element of cognitive-activity matrix. The organization of qualimetry follows the principle of subsequent movement on the levels of complexity of studying material which shows the hierarchy of opportunities for a person. Educational tasks in a test form by means of cognitive-activity matrix can be used as means of monitoring and studying.

The matrix model of cognitive process is invariant concerning the content of the studied subject, and can be equally applied both to basic, mathematical and natural-science subjects, and to all-professional and special subjects and also to humanitarian, social and economic subjects.

The algorithm of realization of the presented learning system in educational process can be used as the metabasis of psychology and pedagogical providing computer learning systems. Concepts of metasubject activity of the technical university are formulated on the basis of the concept of functional educational systems and regularities of development of subject and metasubject knowledge [17]. The satisfaction of personal students' needs in individual educational trajectories, in obtaining certain educational results, in the introduction of internal monitoring system and quality assessment of knowledge and competences demands the development of metasubject students' abilities. For this reason formation of the metasubject competences (MC) become the main task of modern education, they are applied both within the educational process, and in real life situations. Metasubject approach provides complete perception of the world, contributing thereby to the development of the complete personality, transferring from the informed person to a thinking person.

The presented formalization of development of educational subject by means of cognitive -activity matrix by step-by-step implementation of logical schemes opens only one of possible ways to digital education.

# References

1. Bespalko VP (2010) Nature corresponding pedagogics. National Education, Moscow
2. Clarain MV (1997) Innovations in training: metaphors and models. The analysis of foreign experience Nauka Publishing house. Moscow
3. Concept of federal state educational standards of the general education: projects/Russian academy of education; under the editorship of Kondakov AM, Kuznetsov AA (2008) Prosveshchenie, Moscow, 39 p
4. Gromyko YV (2000) Mental pedagogics: theoretical-practical guidance on the development of the highest models of pedagogical art. Minsk, edition "Technoprint", p 376
5. Kolesina KY (2009) Metaproject training: the theory and technologies of realization in educational process. Dissertation. Rostov-on-Don: UFU
6. Lord FM (1980) Application of item response theory to practical testing problems. Lawrence Erlbaum Ass. Publ, Hillsdale, N-J
7. Mayorov AN (2000) The theory and practice of tests' creation for the systems of education (How to choose, create and use tests for education). Moscow, edition SPUTNIK, p 352
8. Ozerkova IA (2010) Metasubject approach: ways of realization. New educational standards. Metasubject approach. Materials of pedagogic conference, December 17, 2010. http://eidos.ru/shop/ebooks/220706/index.htm. Accessed 13 Mar 2018
9. Prokopenko ML (2010) Metasubject content of education in primary school. New educational standards. Metasubject approach. Materials of pedagogical conference, Moscow, December 17, 2010. http://eidos.ru/shop/ebooks/220706/index.htm. Accessed 5 Apr 2018
10. Ryabinova EN (2009)The adaptive system of the personified vocational training of students of technical colleges. Mechanical Engineering. Moscow

11. Ryabinova EN (2007) Synergetic approach to the construction of individual- corrected educational technology. Synergetrics of natural, technical and social and economic systems: collection of articles of the International scientific and technical conference. (8–9 November), Issue 1, chapter 2, Togliatti, edition TGUS, 2007, pp 183–189
12. Safonova OY (2010) Possibilities of realization of metasubject approach at informatics lessons. New educational standards. Metasubject approach. Materials of pedagogical conference, Moscow, December 17, 2010. http://eidos.ru/shop/ebooks/220706/index.htm. Accessed 25 Mar 2018
13. Skripkina, YV (2010) Metasubject approach in new educational standards: issues of realization. New educational standards. Metasubject approach. Materials of pedagogical conference, Moscow, December 17, 2010. http://eidos.ru/shop/ebooks/220706/index.htm. Accessed 2 Mar 2018
14. Talyzina NF (2003) Pedagogical psychology. Academy. Academy, Moscow
15. Tymoschuk NA (2017) Modern strategy of assessment of educational activity K.A. Kirsanov, S.V. Rykov. Scientific reflection. On materials I of the Volga region pedagogical forum "The system of continuous pedagogical education: innovative ideas, models and prospects". Togliatti, No. 5–6 (9–0):142–146
16. Tymoschuk NA (2016) Innovative approach in assessment of educational activity. New strategy of assessment of educational activity: collection of conference articles "The new strategy of assessment of educational activity". Samara, 337–342

# Information and Education Means of Social Educational Networks

A. S. Zotova[1], V. V. Mantulenko[1(✉)], N. A. Timoshchuk[2], and L. Stašová[3]

[1] Samara State University of Economics, Samara, Russia
azotova2012@gmail.com, mantoulenko@mail.ru
[2] Samara State Technical University, Samara, Russia
7.60n@mail.ru
[3] University of Hradec Králové, Hradec Králové, Czech Republic
leona.stasova@uhk.cz

**Abstract.** The publication defines various aspects of the concept of "information and education space", the key features of modern educational process are revealed. Authors represent social networks as one of the means of formation of such space outlining positive and negative aspects. In order to study the effectiveness of using the social nets in education process the survey has been performed in the frame of the research. The survey studied to what extent the students and lecturers are being equipped, the types of information they used, time costs for the information search, their experience of learning and teaching with the help of thematic internet sites and social nets, the wish to use the possibilities of the Internet in their own practice, the ways of accepting and implementation the information online and the attitude of the students and the lecturers to education through social nets.

**Keywords:** Digitalization · Education · Information and education space · Internet recourses · Social networks

## 1 Introduction

Digital technologies are creating new opportunities for higher education. One of these new technologies are social media, which enable the participants of educational process to stay connected regardless of time and space boundaries. Studying perspectives of using social media in the higher education, the researches emphasize some advantages and disadvantages of these resources for the educational process [3, 6, 9]. Among positive aspects are mobility and efficiency, technological adaptability, recognizability, the difference in the methodology of the teachers' work from the traditional, familiar, well-known communication environment for students, accessibility, interaction, etc. [6, 16].

On the other hand, the use of social media in the university educational process has a number of difficulties: among them the lack of a developed concept and scientific and methodological support for the effective application of these technologies in the learning process; lack of access to social networks from the university's classrooms as a result of their blocking or lack of necessary technical support, as well as a large amount of entertainment content, distracting from training activities, psychological barriers, etc.

The use of media resources, on-line technologies in education is becoming more and more spread among Russian universities. The HEIs introduce various on-line platforms in their activities (Blackboard, Edmodo, Joomla, Moodle) to develop educational communications between students and professors. As a rule materials generated at these resources are presented as the educational ones or having the educational potential but very often they lack some essential characteristics. Off-line education concentrates on such processes as knowledge adoption, skills training, experience development, motivation, control and results assessment and such processes are practically not presented in on-line educational environment if we take it in complex [1, 10, 11, 14].

The monitoring of innovative behavior of Russian population performed by the researchers from National research university "Higher School of Economics" (Moscow, Russia) in order to find out the main channels of life-long learning showed that in 12% of cases the interviewed students used at least one option of the Internet to get knowledge. This is 12 times higher than the previous indicator of 2006 [17]. The analysis of the results describes various forms of getting new information: communication at thematic sites; participation in distance learning seminars; webinars; audio and video lectures.

But the research doesn't give any answers to the following questions: whether the information they got can be assessed as knowledge which might be useful in the future? What are special characteristics and features of online-education in comparison to "classical" way of getting knowledge?

The research of Graham C.R., Hilton J., Rich P. & Wiley D which compares the use of online technologies in the classroom and distance learning through online technologies shows that effective change of traditional approach of learning process organization demands extra resources and knowledge such as the communication time evaluation, elaboration of rules for net communication, creating of extra communication between students online and offline, definition of roles and responsibilities between the instructor and the lecturer [5].

## 2   Materials and Methods

The methods of empirical observation were used to obtain the missing data. Using the methods of comparison and chronological analysis, groups of Internet educational resources users are identified and systematized. The synthesis of information was carried out by means of formalization and graphical methods of data presentation. In the course of the study, methods of analysis of literary sources and scientific reasoning, abstraction, and graphic modeling were used.

## 3   Results

Today, according to the last Eurobarometer survey [12], the use of online social networks is rising, reducing the gap between this medium and the written press: 58% of Europeans use them at least once a week. Over four in ten Europeans do so every day or almost every day (42%). Although the use of online social networks is rising, remarkable is, that simultaneously has been also rising Europeans' distrust in the Internet and online social networks.

In Russia, the annual growth of the online education market is 27%, the traditional - 5%. In 2016, more than 50 million people around the world studied using the largest online platforms (EdX, Coursera, Udacity, etc.), 800 thousand of them are Russians. In addition to private projects, a number of states launched their own national platforms - the United Kingdom, Australia, Brazil, and Russia joined to the countries in 2015.

The main innovation of such platforms is the design of the courses: they consist of small video fragments for 5–10 min, due to which the listeners kept concentration. 2015 was a new milestone in the development of digital education. LinkedIn for $1.5 billion bought the project Lynda.com, which revised the training process. The basis of the program was not mini-lectures, but tutorials, short video instructions for representatives of different professions. For several years, the world's leading experts have created more than 25 thousand tutorials, and for access to them it is enough to buy a monthly, semi-annual or annual subscription. There was a new type of educational project - a kind of "intellectual fitness room", where at any time you can buy a ticket.

The development of information technology entailed a boom of educational startups. Since the beginning of the 2000s, a new class of projects has appeared in the field of education. For the first time significant financial resources were attracted not by universities with their centuries of history and brand, but young teams of entrepreneurs. Ten years ago, the global volume of venture investments in educational startups barely reached $100 million, and in 2016 it exceeded $3 billion, which is comparable to the annual costs of education of individual countries.

Profiles of educational startups are diverse. For example, the most successful start-up, attracted more than $300 million, - TutorGroup - teaches English on the Internet. And the project Achieve3000, which collected a quarter of a billion dollars, develops the ability to read and understand the text. One of the most vivid examples of the new era is the Minerva program, which claims to train world leaders and innovators of the future. Its founders managed to attract more than $25 million at the idea stage.

All new educational projects unite one thing - they are built on modern technologies, use a synthesis of advanced developments in the field of computer intelligence, digital technologies and behavioral psychology. The best educational startups already compete with traditional universities for their investments and talents.

One of the tendencies of modern education is the creation of information-educational space which provides interactive interaction between the entities of educational process. The purpose of the suggested research is to define the possibilities for the synthesis of classical classroom forms of teaching and on-line technologies through distinction of target students groups and creation of unique electronic resource - social educational net, the essence of which is defined in the research of Goryachev and Mantulenko [6].

The information and educational space is a set of systemically organized information, technical, educational and methodical provision that is inseparably linked with the person as an education entity and also information sources (electronic and paper ones), distributed databases, program systems, virtual libraries, etc., and the networks that unite them, technologies that support management process of organizational activity of HEI, taking into account social, economic, cultural, psychological and pedagogical conditions of information implementation and the processes connected with it.

Information in the form of information streams (information transfer) moves in space and in time (storage of information), carrying out at the same time information exchange between entities, who make, collect, transform and store it according to the requirements, getting emotional -semantic coloring, occupying at the same time a certain part of cultural and social memory.

The same subjects at the same time create information and education space and consume it for realization of the set goals including educational. Separately taken information which has been analyzed and comprehended by the entity becomes an information product, property, means of production of information and education space in general. Therefore, it (information and education space) periodically becomes the result of human activity [6].

One of the means that is used to create such space is social networks. The most popular according to the students who participated in the poll which aim is to study the effectiveness of social networks in education, are such networks as VKontakte, Youtube, Instagram, "Odnoklassniki", Twitter, Facebook, etc. use. It is connected with the convenience of communication in the interactive mode, taking into account the interests of community participants, with the variety of working forms with the content, giving the chance to approach creatively to putting their ideas and projects into practice and also the opportunity for prompt informing on the changes in personal information space.

Social networks and Internet communities, being habitual for modern students, have high educational potential which gradually starts to offer the services, ways of interaction, new information and communication channels, special communication technologies and activity regardless of nationality and religion [6].

From 289 students of Samara HEI (Samara State University, Samara State Technical University, Samara State University of Economics) participating in our poll, 85% of respondents wanted to study by means of social networks, and 23% already had positive experience of education with the help of Internet – technologies [6].

Practically all the students who have participated in the poll answered that they need a computer to search for a job, operate in computer programs, communicate, watch videos and movies, download and listen to music, study". Only 3% added that they use the computer for games. All the respondents specified search engines which they use to prepare for classes, they are Google and Yandex.

More than 90% of poll participants use Internet resources for educational purposes. The most popular is Wikipedia (37%), then comes Google search engine (17%), the directory Consultant system (8%), iBooks (7%). 27% of poll participants indicated such websites as lingualeo.com, theoryandpractice.ru, allbest.ru, bibliotekar.ru, the official sites of media, universities. 9% of the respondents did notname any educational resource. The popularity of Wikipedia (37%), and Wiki technologies in general, is that students have an opportunity to create network educational content with their teachers or independent cognitive activity, to comment, edit their own and joint written network projects. It gives students an opportunity to work in their own temporary mode and also to build individual educational trajectories, expanding communicative space of educational process, giving the chance to teachers to assess the contribution of each student into collective educational, creative activity. High level of the interaction provides the continuity of the educational process which is beyond the university classes [6].

The introduction of social networks into the educational process allows not only solving specialized pedagogical problems (for example, control of knowledge and provision of flexible interaction between a teacher and a student), but also increases the interest in the process of knowledge obtaining.

21 teachers from Samara, Moscow, St. Petersburg, Nizhny Novgorod which participated in our poll questionnaire marked that social networks VKontakte and Facebook taking into account their functionality, allow creating courses for students, holding videoconferences, inviting teachers of other HEIs to implement it. Besides, social networks can be used to maintain the relations between conference participants' seminars, webinars, summer schools that allows not only improving emotional group climate, but also increasing quality of the held events by exchange of the ideas and remarks. By social networks it is possible to organize club activity, having united students and teachers of various regions of the country.

Teachers pointed the opportunity to place in blogs curricula, tasks to practical and laboratory research and also thematic references, video topics. As the means of the assessment of students' competences, they recommended the creation of electronic portfolios, conducting testing, placing the topics of the presentations for the seminars or for total check work, specifying the criteria of estimation depending on the performed work or task.

In the Czech survey of Poulová [15] more than 85% of public universities have been using one of the e-learning portals and the mostly used one is being called LMS Moodle.

In the survey conducted in 2017 at 5 Czech universities (233 questionnaires distributed among students of environmental courses were evaluated), the mostly used open online source was English and Czech Wikipedia. Otherwise, students declared that they usually use the open online sources only as additional sources for their studies. The English Wikipedia was seen as the valuable and high-quality source. Despite that the Czech Wikipedia had been used more often [13]. To the advantages in the educational process the respondents referred mobility and efficiency, technological effectiveness, recognizability, difference in techniques from traditional ones, the habitual, habitual students' environment, availability, interaction, etc. [7].

Together with indicated positive aspects the respondents marked that the use of social networks in the educational process has a number of difficulties: lack of developed concept and scientific methodological support of these technologies in the educational process; lack of access to social networks from educational classrooms as the result of their blocking or lack of necessary technical provision and also the large volume of entertaining content that distracts students from educational activity, psychological barriers, etc. [18].

Solving of existing problems is possible only by deep studying of the capacity of educational social networks, development and approbation of effective techniques of their application in educational space, elaborations of specialized educational networks. It will allow using them as a unique powerful tool to create information and education space.

## 4  Discussion

Monitoring of innovative behavior of the Russian population, organized by the researchers of the Higher School of Economics [17] with the aim of identifying channels of life-long-education, showed that in 12% of cases receiving new knowledge was realized through using at least one of the Internet services. This figure is twelve times greater than that was registered in 2006. Based on these monitoring data, scientists identify forms of obtaining new information - from communication on thematic sites to participation in distance seminars, webinars, listening to audio lections, watching videos.

However, the above analyst does not provide answers to the following questions. Is the received information a kind of knowledge that can be used in the future? Is online learning really a learning process as an independent form of study? What are its characteristics, properties and differences from the "classical" form of the knowledge acquisition?

The study of the online technologies application in higher education in the classroom and in a distant form [5] shows that for the changing the traditional learning process organization, in addition to development of an activity plan, it is necessary to involve additional resources and knowledge. That means regulation of such aspects as communication time, rules for building communication in the network, formation of additional types of interactions between students, online and offline, defining the roles between instructor and teacher.

## 5  Conclusions

Synthesis of "classical", classroom forms of teaching and modern computer technologies, clear determination of the target students group (for example, correspondence students, students with disabilities) create conditions for using electronic programs and online resources in the professional education.

The use of information and communication technologies as additional means of learning, organizational forms of the education process creates conditions for the modernization of the Russian higher school education, providing its accessibility, democracy, the high quality of the academic preparation of future bachelor students, master students and specialists. The introduction of social networks into the educational process allows not only solving specialized pedagogical problems (for example, control of knowledge and provision of flexible interaction between a teacher and a student), but also increases the interest in the process of knowledge obtaining.

**Acknowledgment.** The research is done in the frame of the state task of the Ministry of Education and Science of the Russian Federation №26.9402017/PC "Change management in high education system on the basis of sustainable development and interest agreement".

# References

1. Altbach PG (2013) Advancing the national and global knowledge economy: the role of research universities in developing countries in Studies in Higher Education 38(3):316–330
2. Baikova AA, Shakleina TA (2014) Megatrends. Main evolution trends of global character in XXI century, Aspekt Press, p 448
3. Bathmaker AM (2016) Higher education in further education: the challenges of providing a distinctive contribution that contributes to widening participation in Research in Postcompulsory Education, 20–30
4. Conner TW, Rabovsky TM (2011) Accountability, affordability, access: a review of the recent trends in higher education policy research. Policy Stud J 39(1):93–112
5. Graham CR, Hilton J, Rich P, Wiley D (2010) Using online technologies to extend a classroom to learners at a distance in distance education 31(1):77–92
6. Goryachev MD, Goryachev MM, Ivanushkina NV, Mantulenko VV.(2014) Application of network resources in modern education. Bull Samara State Univ 5(116):220–227
7. Harrington HJ (2006) Change management excellence: the art of excelling in change management. Paton Press
8. Hiatt JM (2006) ADKAR: A model for change in business, government and our community. Prosci Learning Center Publications, Loveland, Colorado
9. Lemoine PA, Hackett PT, Richardson MD (2017) Global higher education and VUCA-Volatility, uncertainty, complexity, ambiguity in Handbook of Research on Administration, Policy, and Leadership in Higher Education. IGI Global, Hershey, PA, pp 549–568
10. Levy M (2009) WEB 2.0 implications on knowledge management. J Knowl Manag 13 (1):120–134
11. Marginson S (2007) Prospects of higher education: globalization, market competition, public goods and the future of the university. Sense Publishers, Rotterdam, The Netherlands
12. Media use in the European Union. Standard Eurobarometer 88, Autumn 2017. European Union. 2018, pp 80
13. Petiška E (2018) Usage of Open Educational Resources (OER) by Students of Environmental Disciplines in the Czech Republic. https://www.envigogika.cuni.cz/index.php/Envigogika/article/view/548. Accessed 10 Oct 2018. https://doi.org/10.14712/18023061.548
14. Prigozshin AI (2003) Innovations: motives and obstacles. In: Social problems of innovation management. Polit Press, pp 271
15. Poulová P (2010) Uplatnění eLearningu na českých univerzitách - desetiletá historie. Application of eLearning at the Czech universities – ten years of history. Proceedings: 6th international konference on distant learning – DisCo 2010. Plzeň: Západočeská univerzita v Plzni, Ústav celoživotního vzdělávání
16. Sher PJ, Lee VC (2004) Information technology as a facilitator for enhancing dynamic capabilities through knowledge management in Information & Management 41(8):933–945

17. Sobolevskaya OV (2014) The Internet will become the main assistant in the matter of self-education. http://www.opec.ru/1715182.html. Accessed 20 Feb 2018
18. Teichler U (2016) Changing structures of the higher education systems: the increasing complexity of underlying forces. Higher Educ Policy 19(4)

# Information Modeling of the Students' Residual Knowledge Level

S. I. Makarov and S. A. Sevastyanova[✉]

Samara State University of Economics, Samara, Russia
matmaksi@yandex.ru, s_sevastyanova@mail.ru

**Abstract.** The training of specialists for the economy of the new format will be more effective if all components of the education system are brought in line with the new paradigms. Digitization of economy changes the goals, principles, forms, means and methods of the learning process, including its control measures. Fixing the level of residual knowledge is one of the procedures used in the Russian education system to monitor the quality of education at various levels. Typically, this procedure is implemented in the form of computer testing on the previously learned disciplines. The analysis of test results gives grounds for making decisions on the implementation of corrective measures. This determines the importance of the task to get reliable, accessible and informative monitoring results. The article suggests a method of visualization and interpretation of monitoring results of retained knowledge, and substantiates the possibility of its use in order to analyze the problems of training at the individual and group levels. It is offered to use a vector form of representation of the data set to process statistical information. In this case, the visualization of monitoring results has the form of a spatial vector or a cut-out shape of a multidimensional vector onto a plane. The possibilities of the visual assessment to reach threshold values of knowledge are discussed. Examples of using a vector model for monitoring the level of retained knowledge in three or more disciplines are given. An approach to solve the problem of comparability of monitoring results carried out in various assessment systems is proposed. The main conclusions and results can be used directly in the educational process, in the field of education management, in psychological and pedagogical work.

**Keywords:** Transition to digital economy ·
Mathematical methods in pedagogics · Monitoring results ·
Monitoring of retained knowledge · Processing of monitoring data

## 1 Introduction

The active introduction of digital technologies in various spheres of social communications poses new challenges for the system of training specialists. For their implementation a change of paradigms in education is required. The necessary qualities of a specialist of a new formation are: the ability to search and process information, qualified data analysis, structuring and presenting information, the ability to formulate a problem and find non-standard approaches to solving, the ability to use digital technologies of production and communication. The formation of these and other necessary

© Springer Nature Switzerland AG 2020
S. Ashmarina et al. (Eds.): *Digital Transformation of the Economy: Challenges,*
*Trends and New Opportunities*, pp. 502–509, 2019.
https://doi.org/10.1007/978-3-030-11367-4_50

competences should be the goal-determining factor of the whole learning process in any of its forms. Accordingly, all elements of the didactic system must be aligned with the goals set.

One of the mandatory procedures in education is to assess the residual knowledge of students. It is necessary for pedagogical diagnosis, correction of the educational process, the choice of forms and methods of teaching disciplines, assessing the quality of education, effective management of education. To a greater extent, the procedure for monitoring retained knowledge is used in higher education, since this system has more opportunities to make changes in the educational process, both at the individual level and at the level of the education institution and the education system as a whole. The monitoring results of student retained knowledge are one of the factors that are important in the licensing of the education institution and the accreditation of training programs.

Monitoring of retained knowledge should have signs of purposefulness, objectivity, completeness, adequacy, as well as informative and clear presentation of results. The latter is necessary in order to draw correct conclusions and, if necessary, to organize a correct process using monitoring results. The quantitative assessment, expressed graphically, has advantages in perception and analysis before the digital form to present results. The visualization of statistical data in the form of diagrams and graphs is the standard practice of presenting information in business, management, and economics.

In the works of teachers-researchers, the modern experience of conducting, analyzing and interpreting monitoring of student knowledge is reflected. The specificity of diagnostic work in the university is discussed in the scientific works of Artisticheva [2]. The functional purpose and organization of test knowledge monitoring is deeply discussed [3, 7]. Academic writings [1, 4–6, 14] offer various ways of processing qualimetric procedure results for diagnosis and correction of knowledge. A number of studies devoted to the presentation of the data in various forms [8–11, 13]. The works [7, 12, 15] are devoted to the problems of monitoring retained knowledge.

At the same time, having a significant number of publications on this subject, the problems of processing, analysis and interpretation of the received data remain topical. Based on the analysis of the literature on this issue, we can conclude that in most cases, the analysis of results is carried out by common statistical methods: sample characteristic are calculated, histograms are constructed, various hypotheses are tested, models and forecasts are developed, etc. These scientific approaches well perform their qualimetric functions, but have a significant drawback: the complexity of perception for the unprepared user. This paper illustrates a different approach that helps present results.

## 2 Materials and Methods

The methodology of the research includes general scientific and mathematical methods: theoretical analysis of the problem on the basis of the scientific literature analysis, generalization of experience for determining ways of solving the problem, comparison, mathematical modeling, and fundamental foundations of vector algebra for solving applied problems.

The search for new ways of visual representation of monitoring results, their processing and analysis leads to the idea of using mathematical methods in educational activities. In the practice, the procedure for monitoring retained knowledge is part of internal or external monitoring and, as a rule, is based on subject testing with a certain lag in time. The quantitative assessment of test results takes the form of an ordered set of data, and therefore can be presented in a vector form and analyzed on the basis of a vector criterion. The components of the vector can be generated based on the current academic performance of students or results of later testing.

In this case, the problem of monitoring retained knowledge can be reduced to the task of constructing a ratio that converts each test result from one discipline to some well-ordered set [16], such as the set [0,100]. Moreover, the ratio should be constructed in such a way that the qualitative features increase with the increasing value of each criterion. This will allow geometrically correlating the results.

## 3  Results

We will consider that education in the education institution lasts for $k$ years and every year monitoring of retained knowledge is conducted in $n_r$ disciplines, $1 \leq r \leq k$ и $\sum_{r=1}^{k} n_r = n$. We believe that monitoring is carried out in the form of testing on a 100-point scale. The number of points scored by the student when monitoring the current academic performance of the past period and when monitoring retained knowledge of the $i$-th discipline at the $j$-th year of training will be denoted $x_{ij}^0$ and, $x_{ij}$ respectively.

Thus, the entire set of monitoring results of the current academic performance and retained knowledge in all disciplines is assigned ordered $n$-tuples - vectors for monitoring the current academic performance and retained knowledge.

$$\overline{X}^0 (x_{11}^0, x_{12}^0, \ldots, x_{1n_1}^0; \ldots; x_{k1}^0, x_{k2}^0, \ldots, x_{kn_k}^0) \text{ and}$$

$$\overline{X}(x_{11}, x_{12}, \ldots, x_{1n_1}; \ldots; x_{k1}, x_{k2}, \ldots, x_{kn_k})$$

These objects from the mathematical point of view represent vectors in $n$-dimensional space, from the semantic point of view they are monitoring the current academic performance and retained knowledge. These vectors are located in the $n$-dimensional cube

$$C = [0, 100] \times [0, 100] \times \ldots \times [0, 100].$$

For example, a vector $\overline{X}_I$ - $n$- a dimensional vector, all coordinates of which are equal to 100, corresponds to the "ideal student", who scored the maximum number of points in all disciplines.

For the visual and quantitative assessment of the compliance of the knowledge level with the specified minimum requirements, the following approach can be proposed.

In this cube, you can select two subsets - *CT* and *CN*

$$CT \cup CN = C,$$

where $CT = [50,100] \times [50,100] \times \dots \times [50,100]$ - is the $n$-dimensional cube into which the ends of vectors hit, all coordinates of which satisfy the minimum requirements for the current and retained knowledge of the student; $CN = C\backslash CT$ – is the complement of *CT* to the set *C*. The ends of vectors, in which at least one coordinate does not satisfy the minimum requirements for the current and retained knowledge of the student, will hit in the set *CN*.

## 4 Discussion

In the first year of study at the education institution, monitoring of retained knowledge is the assessment of the initial level of student knowledge upon his/her enrollment to the education institution after finishing school. Test assignments should correspond to the school curriculum and the content of the unified state examination, but they may be supplemented by other questions. The initial minimum acceptable level is established on the basis of assumptions about the amount of knowledge required for successful learning. The resulting numerical data is visualized in the vector form.

Figure 1 shows the vector interpretation of monitoring results of the first year students in the selection of three disciplines of testing. Here the parts of *C* and *CT* set of three-dimensional points of possible and admissible monitoring results, respectively. The vector (1) illustrates the level $\overline{X}_I$ with "ideal features", the vector (2) visualizes results not satisfying the minimum requirements.

If the number of disciplines for monitoring knowledge exceeds three ($n > 3$), then it is not possible to build a cube and vectors of monitoring results in $n$-dimensional space. However, there is a technique that allows working with $n$-dimensional vectors graphically, using a cut-out shape of the vector on a plane in the form of a petal diagram. Consider this technique using the example of the comparative analysis vectors of current academic performance and retained knowledge acquired in the first year of training, provided that the number of monitoring disciplines is seven. Monitoring was conducted in the second and third year of training.

In the diagram above, the external broken line represents the "ideal level" of the respondent's knowledge, and corresponds to the maximum possible score for each of monitoring disciplines. The internal line corresponds to the level of the minimum admissible requirements for each of disciplines and is a kind of barrier, the lower threshold value. Other closed lines of the diagram illustrate the results of monitoring the current academic performance (intermediate testing, exam, offset) and retained knowledge one year and two years after studying these disciplines. From the figure above it is clear that the level of retained knowledge decreases with increasing time and in some disciplines can fall below the permissible level, as happened with discipline $X_{17}$ (Figs. 2 and 3).

The described method of visualizing monitoring results of the current academic performance and retained knowledge can be applied not only individually to each

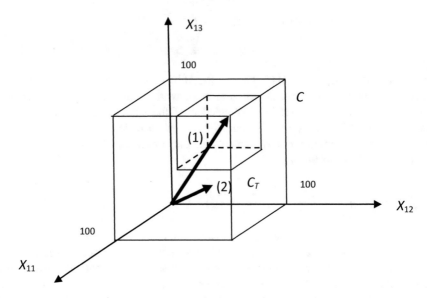

**Fig. 1.** A three-dimensional cube of monitoring results of retained knowledge (Source: compiled by authors)

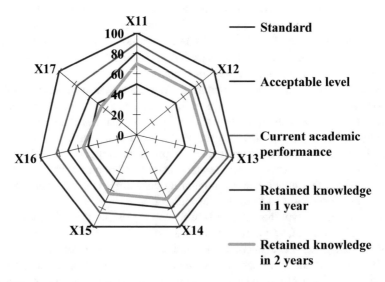

**Fig. 2.** The comparative analysis of vectors of the current academic performance and retained knowledge (Source: compiled by authors)

student, but also to homogeneous groups of students on the same educational programs. For this, a vector $\overline{X}group$ is formed, whose coordinates are equal to the arithmetic mean of coordinates of monitoring vectors of all students.

$$\overline{X}group = \frac{1}{m}\sum_{i=1}^{m}\overline{X}_{i},$$

where $m$ - is the number of students in the group, $\overline{X}_i$ - the monitoring vector of each student.

The presentation of monitoring results in the vector form will allow visually identifying problematic positions, and on the basis of their analysis and more detailed study, organize measures to correct the situation or make changes in the educational process in the future.

The vector interpretation of assessments can also be used, for example, in psychological and pedagogical work on the vocational guidance of students. One aspect of this work is to identify scientific and practical preferences of students in the future profession, the dominant inclination towards humanities or natural sciences, and so on. To implement this task, vector diagrams are constructed based on monitoring results. We will give an example of this approach when the student is choosing the course of study. If the number of disciplines studied in the education institution is $n$, then some of them are social and humanitarian disciplines and $n$-$q$ disciplines are natural-scientific. Consequently, the $n$-dimensional cube

$$C = [0, 100] \times [0, 100] \times \ldots \times [0, 100]$$

can be represented as a product of cubes of dimension $q$ and $n$-$q$:

$$C = Cq \times Cn - q.$$

Using the projections of monitoring vectors of the current academic performance and retained knowledge to the corresponding spaces of dimension $q$ and $n$-$q$, it is possible to obtain monitoring humanitarian and natural-science vectors, based on the results of the study, we can draw conclusions about the preferences of the student, as well as the level of teaching disciplines of the corresponding cycle in this institution.

If monitoring vectors of the current academic performance and retained knowledge are built on different scales or in different scoring systems, then these vectors can be normalized for comparability of results. We will perform a valuation of monitoring vectors $\overline{X}^0$ and $\overline{X}$. To do this, we divide their coordinates by the lengths of vectors. As a result, vectors of the same directions of unit length are obtained, and whose ends are on the unit sphere with center at the origin.

$$\overline{X}_N^0 = \frac{1}{\sqrt{\sum_{i=1}^{k}\sum_{j=1}^{n_k}x_{ij}^2}}\overline{X}^0, \quad \overline{X}_N = \frac{1}{\sqrt{\sum_{i=1}^{k}\sum_{j=1}^{n_k}x_{ij}^2}}\overline{X}$$

The coordinates of $\overline{X}_N^0$ and $\overline{X}_N$ vectors and are equal to direction cosines $\cos\alpha_i$, $i = \overline{1, n}$ of vectors directions - $\overline{X}^0$ and $\overline{X}$. Using the direction cosines, you can compare monitoring results of the current academic performance and retained knowledge of one

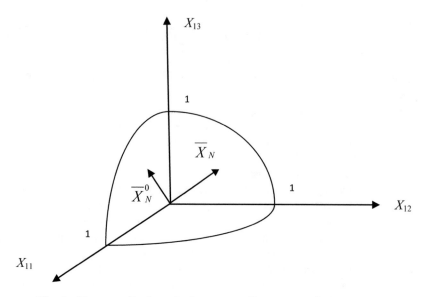

**Fig. 3.** The normalized monitoring vectors (Source: compiled by authors)

student. The more the vector is inclined to one of the axes, the greater the corresponding direction cosine is and the more the student is successful in mastering the corresponding discipline.

## 5    Conclusions

The vector method of interpreting monitoring results of retained knowledge suggested in this paper allows visualizing data, visualizing the situation, and identifying problem areas. The possible areas of application of this method in qualimetric support of education monitoring, vocational guidance work, individual educational activity are described. The vector approach opens new prospects for using mathematical methods in the management of education.

## References

1. Subetto AI, Chernova YuK, Gorshenina MV (2004) Qualimetric support of management processes. St. Petersburg: Publishing House "Asterion", 278
2. Artisheva EK (2015) Pedagogical diagnostics as the basis of the knowledge correction system. Educ Technol 3:85–103
3. Slepukhin AV (1999) Use of new information technologies to monitor and correct students' knowledge in mathematics. Thesis for the degree of Candidate of Pedagogical Sciences (13.00.01). Yekaterinburg, 121
4. Krasheninnikova YuV (2010) Statistical processing of monitoring results for higher mathematics. Bull Pskov State Univ Ser Nat Phys Math Sci (1):116–122

5. Gumennikova YuV, Ryabinova EN, Chernitsyna RN (2015) Statistical processing of students' monitoring results. Bull Samara State Techn Univ Ser Psychol Pedagog Sci 3 (27):78–87

6. Karpinsky VB (2010) Modern mathematical methods of processing pedagogical monitoring results. Alm Mod Sci Educ № 2:82–84

7. Shestova EA (2012) Development of models and methods for analyzing and processing monitoring results of knowledge. News of Southern Federal University. Technical Science. 2 (127):146–152

8. Kim VS (2012) Matrix representation of monitoring results. Bulletin of Moscow State Regional University. Series: Pedagogy (4):114–120

9. Lobova TV, Tkachev AN (2014) An adaptive fuzzy procedure for interpreting results. News of higher educational institutions. North-Caucasian region. Series: Technical Sciences. 5 (180):102–105

10. Yuryev GA (2013) Mathematical model for the interpretation of computer monitoring results using Markov networks: the thesis abstract. Moscow

11. Fedyaev OI (2016) Forecasting retained knowledge of students in separate disciplines with the help of neural networks. News of Southern Federal University. Technical Science 7 (180):122–136

12. Kremer NSh (2016) Diagnostics and forecasting of the level of mathematical preparation of students. Modern Math Concept Innov Math Educ 3(1):263–265

13. Melnikova NN, Shchelokova EG (2012) Career orientation: a vector model of diagnosis and interpretation. Eur Soc Sci J 2(18):270–277

14. Sevastyanova SA (2006) Formation of professional mathematical competencies of students in universities of economics. Samara, Dissertation

15. Artishcheva EK, Bryzgalova SI, Gritsenko VA (2013) Background level of knowledge: essence, analysis, assessment of assimilation: monograph. Publisher: BFU named after I. Kant, Kaliningrad

16. Makarov SI (2003) Methodical foundations for the creation and application of educational electronic publications (using the example of the course of mathematics. Dissertation. Moscow

# Effective Staff Training for the Contract System in the Conditions of the Digital Economy: Opportunities and Limitations

L. V. Averina[1] (ID), E. P. Pecherskaya[1]([✉]) (ID), and O. V. Astafeva[2]

[1] Samara State University of Economics, Samara, Russia
Alv94@ya.ru, pecherskaya@sseu.ru
[2] Financial University under the Government of the Russian Federation,
Moscow, Russia
astafeva86@mail.ru

**Abstract.** Effective staff' training, acting as one of the leading factors in the successful functioning of the contract system, is an urgent need of the society. In this regard, the subject of this study is the formation of integrative model for staff in procurement (contract managers) training that can serve to developing the potential of the Russian economy. The results of this research is the development of a model for training contract managers (syn. procurement specialists), taking into account the aspects of professional experience revealed in the process of approbation of this model in the pedagogical activity of the author on the basis of the Federal Innovation Platform in Procurement, Samara State University of Economics (further referred to as "SSEU") from 2013 till present. Characteristic features of the methodical system for the training of contract managers have been analyzed. The results of the research can be used as a basis for the organization of vocational training and retraining, optimization of existing programs of professional development and implementation of innovative projects in these areas.

**Keywords:** Continuous education · Additional vocational education ·
Contract system · Innovative infrastructure

## 1 Introduction

Despite the urgency and relevance of this area of professional training, there is a low level of scientific elaboration of this topic in the scientific literature. Much attention is paid to Russian legal researchers by regulatory and legislative problems in the field of procurement, while the educational part of the process remains without due attention [7, 16, 18–20]. At present, the main components of the organization of training in general and additional vocational education, in particular, the goals and technology of such training are transformed and acquire a certain specificity, different from traditional didactics and traditional educational strategies.

The purpose of additional vocational training today is considered in the form of a set of universal, general professional and professional competencies formed by students of courses in the programs of additional professional education [8–11, 17, 33].

© Springer Nature Switzerland AG 2020
S. Ashmarina et al. (Eds.): *Digital Transformation of the Economy: Challenges,
Trends and New Opportunities*, pp. 510–517, 2019.
https://doi.org/10.1007/978-3-030-11367-4_51

In the conditions of realization of additional vocational training, the technologies of problem (practice-oriented), interactive training, training through cooperation are gaining popularity. The reform of the contract system is far from over [2, 3], which requires the constant adjustment of both the content of educational programs, and approaches, teaching methods [6].

Despite the urgency and relevance of this area of professional training, there is a low level of scientific elaboration of this topic in the scientific literature. Much attention is paid to Russian legal researchers by regulatory and legislative problems in the field of procurement, while the educational part of the process remains without due attention [4, 14, 15, 22, 23].

Among Russian scientists studying the problem of vocational training in the procurement sphere are Shamakhov et al. [28]; Karanatova [17]. Among foreign scientists we relied on the studies of J. Edler and J. Yeow [11] and L. Dufek [10] in the procurement sphere and on studies on innovative educational approaches of Garbett and Ovens [13]; Axelrad et al. [1]; Sullivan et al. [30]; Pellegrino and Hilton [24].

## 2 Materials and Methods

In the process of research, methods such as retrospective analysis, statistical, dialectical, analytical methods, content analysis of scientific literature were used. The analysis was carried out in two stages using a combined methodological approach. At first, the legislative base of the Russian Federation was considered in relation to professional education and procurement regulations, as well as the materials of scientific publications, official reports of government research institutions, state agencies.

Then, as an example of the practical implementation of staff training in the sphere of procurement, Center for Business Education of SSEU was taken. In addition to the statistical data obtained following the results of the survey of the participants of the refresher courses on the basis of Center for Business Education, the results of other similar studies in the field of applying innovations in education were considered [24, 31].

When creating an innovative training program for staff in procurement, we address systemic, synergistic, personality-activity and axiological approaches and emphasize the prognostic aspect when using the monitoring system for learning outcomes at all its stages (intermediate and final results).

## 3 Results

Training of contract managers as a methodological system for the organization of the educational process should be represented by the unity of the organization of learning and development processes aimed at preparing contract managers for the successful performance of professional activities. The CPE program of contract managers as specialists who should take special training is a specific sphere of business training conducted according to the federal law of the Russian Federation 05.04.2013 № 44-FZ "On contract system in the sphere of procurement of goods and services for federal and municipal needs" and departmental acts and guidelines. This institution has its origin in foreign practice (US and Europe), where it has long existed and is referred to as «contract manager», «contract officer», «contracting officer» [4, 10, 28].

In the process of teaching in advanced training courses in the field of procurement (in the contract system), we have identified the conditionality of the formation of readiness for professional activity in the procurement system by the effectiveness of the system of methodological provision of training, interaction of the internal environment of the individual with the external influence of the methodological system [9, 13].

In addition, due to the recent introduction of Russian training programs for contract managers as CPE (from 2013 only), the lack of systematic analysis of professional competencies and, in particular, project competency formation of contract managers, in domestic and foreign academic literature should be mentioned.

First of all, attention is drawn to a wide range of methodological issues related to the preparation of contract managers. These include the following: not developed or weakly developed system of "feed-back" between the lecturer (tutor) and trainees (contract managers), the inadequacy of presented to contract managers learning material to practical aspects of professional duties and responsibilities (rapid obsolescence of didactic material, the lack of communication theoretical developments with the realities of procurement activities in the modern high-tech conditions), the absence of an explicit model of practice-training due to the novelty and a wide range of competencies of contract manager [3, 26]. In this regard, attention is drawn to the model of project competency of contract managers positively proven in practice of CPE programs implementation and approved in the course of educational activity of the authors.

In order to control the quality of the educational process the authors conducted a longitudinal study on a representative sample of contract managers as trainees of training courses in the field of procurement. there is evidence that the most important skills for the contract managers are considered to be: evaluating procurement bids participants, ability to maintain effective execution of contracts, good skills in IT [3, 19, 32]. Based on the structure, stipulated by the regulations governing the competence of contract managers and the corresponding range of professional duties, the structure of contract managers' competencies could be schematically represented as follows (Fig. 1).

The analysis of the questionnaire data of the participants of the advanced training courses under the program "Contract system in the procurement of goods, works and services for provision of state and municipal needs", carried out on the basis of the Business Education Center of SSEU for 2014–2017 showed, that the most important for the contract manager are the following skills:

- evaluating the applications of the participants in procurement,
- ability to conduct contractual work (on the execution of contracts),
- possibility to operate with information and technology.

Knowledge of information technologies plays an increasingly important role in the competence of the contract manager (contract service worker), since July 1, 2018 all purchases for state and municipal needs have been translated into electronic form. At the same time, strict administrative responsibility is provided for violation of the

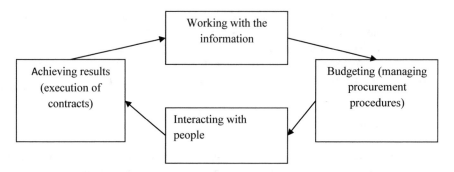

**Fig. 1.** Structure of contract managers' competencies as a cyclic system (Source: Averina & Pecherskaya 2016 [2])

procedure for working with electronic documents in the sphere of procurement and the terms of their placement. That is why the use of information and communication technologies plays a special role in the implementation of educational programs in procurement.

The importance of human resource development and the formation of the information technology industry was emphasized in the Decree of the Government of the Russian Federation of 01.11.2013 No. 2036-r "On the Approval of the Strategy for the Development of the Information Technology Sector in the Russian Federation for 2014–2020 and for the Future to 2025".

For each of the key skills identified by the results of the survey, further monitoring of the pace of formation of competence with the use of IT technologies (in particular, the educational platform Moodle, successfully applied on the basis of SSEU), which allows to manage the learning process using an individually-oriented approach.

We also should mention the meaning of risk-oriented approach that helps to provide more deliberate, well-targeted training [27]. According to our study conducted from 2015 till 2017, the negative expectations regarding the risks of professional activity in the procurement sphere were as follows (Table 1).

**Table 1.** System of risks and negative patterns of contract managers, in %.

| | |
|---|---|
| Risks of violations during planning | 43% |
| Risks of violations in the organization of procurement procedures | 21% |
| Risks of breach of contract | 92% |
| Risks of violations during the evaluation and consideration of applications for participation in procurement procedures | 11% |
| Risks of breach of contract performance | 78% |
| Risks of violations when accepting the results of performance of a contract | 59% |

Source: compiled by the authors.

The obtained data allow to identify the "problem" area of activity, the functions of the contract manager and pay special attention to the development and consolidation of necessary knowledge, skills and skills in the training process in order to reduce such risks and negative expectations of the student. At this stage, we can talk about the introduction of a risk-based approach in training contract managers.

## 4   Discussion

The study of national scientifical literature allows to ascertain the absence of specific studies on the problem of formation of contract managers' project competency, according to the problems of actual practice. However, the general issues of the organization of studying in high school using project-learning techniques are mentioned in the works of Clarin [8], Guryanova et al. [14, 15], Kirichenko [18], Larchenko [19]. To the relation of project activities and competence of the expert, practice-building approach to teaching are devoted research works of Sarkisyan [25], Vvedensky and Donnikova [32].

The foreign research works, for example, Barron [5], Enke et al. [12], Markham [20], Sawyer [26], Tisch et al. [31], and others emphasize that the project-learning method allows to organically integrate knowledge from different areas of learning in the solution of a problem and makes it possible to apply this knowledge in practice. There also observations on a level of positive impact on the motivational sphere and the meaning of it to the development of cognitive abilities and creativity.

In this study, the authors discuss the project methodology in a wider context as a trend in education, and in CPE, in particular, which is based on concepts such as project (material effect), project technology (the process of the training during the project). Thus, the project is presented as a method of purposeful activities within a flexible tutor-led, focused on the solution of the problem and aimed to obtain concrete results in the form of material and / or ideal product. Obtaining timely information on the course of the learning process, made possible in the framework of formation of project competency of the trainee (contract manager), it allows tutor to quickly make adjustments if deviations occur. Thus, the use of such technique of formation of project competency makes it possible to promptly intervene in the mechanisms, patterns of the educational process, and leads to increased efficiency of that. Based on the study of theoretical and methodological, technological and methodological foundations of the implementation of additional professional education (CPE), training model of contract managers is considered by authors through a set of motivational, informative, technological units and, moreover, diagnostic unit, correction and control. Particular attention in the training structure in the system of CPE, according to the authors, should be given to diagnostics, monitoring and correction, based on the determination of the effectiveness of the process of development of professional competencies.

A special role in this process is given to the introduction of innovative technologies in the educational process [28]. Recently, blended learning has become increasingly popular, when interactive training is combined with personal communication with teachers [29]. In this regard, when developing an integrative model for the training of contract managers, we also took into account the requirements for additional vocational education by the social and economic environment. We refer to such requirements:

1. Ability to predict the future needs of business in the staff.
2. Ability to form an alternative personnel pipeline. Training should be closely integrated into the economic environment
3. The ability to form the measured skills. Training should provide a balance of purchased and updated hard and soft skills [2].

## 5 Conclusions

In the process of research, new questions and challenges emerged that need their solution:

1. Among identified limitations in vocational training are the weak motivation of contract managers to improve the existing level of knowledge and skills due to the instability of the legal field in the procurement sphere and a high degree of responsibility in the absence of a quality monitoring system and the effectiveness of contract management procedures. It could be minimized by amends in law on procurement in regard of shaping the limits of responsibility and compensation mechanisms in case of breach more thoughtfully.
2. The problem that reduces the effectiveness of vocational training is inadequate preparation for pedagogical activity within the framework of a practice-oriented approach and the design of such activities which subject to all new requirements (legal, economical and other). A tutor who trains contract managers should use technologies of problematic, interactive training, training through cooperation.

The result of mastering of new forms and methods of professional activity in procurement will be critical thinking and a system of values that promote active participation in the sustainable development of society and business corresponding to the era of the digital economy.

## References

1. Axelrad H, Luski I et al. (2016) Behavioral biases in the labor market, differences between older and younger individuals. J Behav Exp Econ 60:23–28
2. Averina LV, Pecherskaya EP (2015) Actual problems of preparation of a specialist in the field of procurement. Materials of the international scientific-practical conference of students, graduate students and young scientists "Social behavior of youth on the Internet: new trends in the era of globalization." SSEU, pp 39–41
3. Averina LV (2017) Models of organization of dialogical interaction in the implementation of DPO programs in the field of procurement. Sb. articles of the International scientific-practical conference "European research", pp 266–268
4. Abakumova OA (2014) Legal status of the contract manager. Academic Herald. N2 (28):269–276

5. Barron B (1998) Doing with understanding: Lessons from research on problem- and project-based learning. J Learn Sci 7(3&4):271–311
6. Berry A (2008) Tensions in teaching about teaching: understanding practice as a teacher educator. Springer, Dordrecht
7. Clandinin DJ, Huber J (2010) Narrative inquiry. In: McGaw B, Baker E, Peterson PP (eds) International encyclopedia of education, 3rd edn. Elsevier, New York, NY, pp 436–441
8. Clarin MV (1997) Innovations in education: metaphors and models. Analisys of foreign experience. Moscow, Nauka. 263 p
9. de Vries Marc J (2018) Handbook of technology education. Springer International Publishing. https://doi.org/10.1007/978-3-319-44687-5
10. Dufek L (2015) Public procurement: a panel data approach. Proced Econ Financ 25:535–542
11. Edler J, Yeow J (2016) Connecting demand and supply. The role of intermediation in public procurement of innovation. Res Policy 45(2):414–426
12. Enke J, Kraft K, Metternich J (2015) Competency-oriented design of learning modules. Proced CIRP 32:7–12
13. Garbett D, Ovens A (eds) (2017) Becoming self-study researchers in a digital world. Springer, Dordrecht
14. Guryanova A, Khafiyatullina E, Kolibanov A, Makhovikov A, Frolov V (2018) Philosophical view on human existence in the world of technic and information. Adv Intell Syst Comput 622:97–104. https://doi.org/10.1007/978-3-319-75383-6_13
15. Guryanova A, Astafeva N, Filatova N, Khafiyatullina E, Guryanov N (2018) Philosophical problems of information and communication technology in the process of modern socio-economic development. Adv Intell Syst Comput 726:1033–1040. https://doi.org/10.1007/978-3-319-908359_115
16. Hadjar A, Samuel R (2015) Does upward socila mobility increase life satisfaction? A longitudinal analysis using British and Swiss panel data. Res Soc Stratif Mobil 39:48–58
17. Karanatova LG, Kulev AYu (2015) Organization of university innovative platforms as factor of development of competences of innovative business [Organizatsiya universitetskikh innovatsionnykh ploshchadok kak faktor razvitiya kompetentsii innovatsionnogo predprinimatel'stva]. Administrative consulting [Upravlencheskoe konsul'tirovanie]. N 12, pp 15–23
18. Kirichenko YuA (2012) Modern information technologies in additional professional education. Vestnik TISBI. № 2(50):18
19. Larchenko YuG (2015) Game technologies in additional professional education. Additional vocational education. Modern Stud Soc Probl № 1(21):195–200
20. Markham T (2011) Project based learning. Teach Librarian 39(2):38–42
21. Menshenina SG (2018)The structure of knowledge to the professional activity of information security specialists. Bull Samara State Techn Univ Ser Psychol Pedagogical Sci № 2 (18):100–107
22. Naumova TB (2009) Additional professional education in the system of continuing education. Competence № 5:10–16
23. Pecherskaya EP, Averina LV, Kochetckova NV, Chupina VA, Akimova OB (2016) Methodology of project managers' competency formation in CPE. IJME-Math Educ 11 (8):3066–3075
24. Pellegrino JW, Hilton ML (eds) (2012) Education for life and work: developing transferable knowledge and skills in the 21st Century NAS Press, 204 p
25. Sarkisyan UV (2011) Project competency of goverment and municipal officials as an object of sociological study. Authority. from: http://cyberleninka.ru/article/n/proektnaya-kompetentnost-gosudarstvennyh-i-munitsipalnyh-sluzhaschih-kak-obekt-sotsiologicheskogo-analiza

26. Sawyer RK (2006) The cambridge handbook of the learning sciences. Cambridge University Press, New York, p 250
27. Scheglova TM (2013) Proektnaya kompetentnost: suschnost, struktura, soderzhanie. from http://cyberleninka.ru/article/n/proektnaya-kompetentnost-suschnost-struktura-soderzhanie
28. Shamakhov VA, Karanatova LG, Kuzmina AM (2017) The research directed to results of activity of the federal innovative platforms forming professional competences of the sphere of the state and municipal procurement assessment. Administrative consulting [Upravlencheskoe konsul'tirovanie]. N 12, pp 8–21
29. Sklyarova IM (2014) Functionality of contract managers in the field of public procurements. Labor and social relations. № 8:26–32
30. Sullivan KPH, Czigler PE, Hellgren JMS (2013) Cases on Professional Distance Education Degree Programs and Practices: Successes, Challenges, and Issues. IGI Global
31. Tisch M, Hertle C, Cachay J, Abele E, Metternich J, Tenberg RA (2013) Systematic approach on developing action-oriented, competency-based learning factories. Proced CIRP 7:580–585
32. Vvedensky VN, Donnikova TS (2014) Correlation of activity and project competency of specialist. Science. Arts. Culture. from: http://cyberleninka.ru/article/n/sootnoshenie-deyatelnosti-i-proektnoy-kompetentnosti-spetsialista
33. Zhang Yu (Aimee) (2015) Handbook of mobile teaching and learning. Springer. https://doi.org/10.1007/978-3-642-54146-9

# Political Economy Assessment of the Educational System Promotion in Russia in the Conditions of Digital Economy Formation

V. A. Noskov[1]([⊠]) and V. V. Chekmarev[2]

[1] Samara State University of Economics, Samara, Russia
noskov1962@inbox.ru
[2] Kostroma State University, Kostroma, Russia
tcheckmar@ksu.edu.ru

**Abstract.** The article reviews a political-economic analysis of the growing dynamics of the education system in Russia, using the example of higher education in the creative solutions of the digital economy. The author examines the economics of education from the political economy view. The evolution of targets in higher education is presented. When comparing definitions of higher education targets, they provide two mutually related, but competing functions of higher school: the role of higher education in the teaching workforce for all economic sectors and the general intellectual growing of a creative, socialized personality. The nature of economic interests in the higher education system is determined. Market economy and democratic society principles imply models of competitive harmony in economic interests of higher education. The educational product is analyzed from the principle of diglossia as a public and private benefit.

In the near future, a combination of fee and free higher education in Russia is inevitable. However, the share of fee-paying educational services of universities will dramatically increase. The analysis confirms academic steps to create a scientific product in the digital economy. Competition of universities in the market of scientific products for receiving profitable contracts with firms and corporations stimulates universities to increase their rankings. The conclusion shows the inconsistency of the dynamics of the development of the education system of Russia in the digital economy. An important contradiction is in the prevalence of political decisions free from long-term socio-economic consequences.

**Keywords:** Digital economy · Economic interests in higher education system · Education economics · Educational product as a public and private benefit · Higher education targets · Higher school · Political economy approach · Russian education system · Scientific product of universities

S. Ashmarina et al. (Eds.): *Digital Transformation of the Economy: Challenges, Trends and New Opportunities*, pp. 518–528, 2019.
https://doi.org/10.1007/978-3-030-11367-4_52

# 1 Introduction

In economic research of education in general and higher education, in particular, we note [1] several related issues and challenging problems. One of them covers everything that represents the economics of education as one of the economies' sectors. These include issues of financing, taxation, economic costs of the educational process, organization of wages, demand for educational services of different levels and quality, the functioning of the market for implied services, the role of the state in its regulation, schemes for managing educational institutions and appropriate legal support. It is this range of questions that constitutes the content of many monographs and textbooks [2, 3].

And the subject of the economics of education is usually determined in such a way that issues are limited by this approach: "The subject of the economics of education is the driving laws of material, financial, labor resources led to the field of education or used in it to achieve socially and personally determined goals of its functioning and development" [4, p. 382].

This definition of the subject of the economics of education clearly reflects the prevailing view on research concentration around the "driving laws of material, financial, labor resources", and "socially and personally determined goals of functioning and development" is something external, exogenous with respect to the movement of resources in education [5].

Moreover, in the literature there are other definitions of the subject. So A. Maloletko [6, p. 18] notes: "In the economics of the system of higher and postgraduate professional education we see the overall material, technical, financial, labor resources and potential in various organizations".

Despite the evidence in the importance and productivity of studying the mechanisms of economic resources movement in the field of education, it is impossible to ignore other challenges in the economics of education.

# 2 Materials and Methods

The title of this paper associates with the necessity to analyze statistical indicators characterizing changes in the quality of educational organizations for some period, the number of students and teachers, their scholarships and salaries, the ratio of organizational and legal forms, etc. However, statistical analysis is not the subject of this work. The authors intend to focus on political economy analysis of the education system promotion.

We talk about the role and functioning of the education system in general, and higher education in particular, in the socio-economic tenor of the country, the growth of human capital and intellectual potential in the conditions of digital economy. About the special properties of the product of production in the field of education as a public and private good, about a complex system of economic relations in which a higher educational institution operates. We study contradictions, compromises, harmony of economic interests, manifested in higher education, on the ratio of goals in the long- and short-term runs.

The significance of a strategic approach to planning the development of higher education and the management of this elaboration - all that goes beyond the sector economy, and which may be called the political economy of higher education [7]. Within the framework of the study, we use methods of analysis, synthesis, modeling, a method of scientific abstraction, a systematic approach to analysis, analogies, comparisons, and categorizations.

# 3    Results

The framework of the article uncovers any political economic issues of higher education; therefore, we will single out only a few, as we see it, especially relevant and deserving priority issues.

## 3.1    The Objectives of Higher Education in the Formation of the Digital Economy

The definition of targets is a crucial starting point in shaping the concepts of higher education development. Few people, apart from universities, are able to foresee socio-economic changes, and in fact their mission includes both adaptation to these changes (for example, the desire of technical universities to develop humanitarian areas with appropriate connivance of the Ministry, which has administrative rent from this process), and foreseeing the impact on these changes [8, 9].

The political and economic assessment of the changes in the field of higher education in the conditions of the digital economy suggests that the government has formed an opinion about higher education as a subsidiary. Its existence is justified only from the point that it contributes to the functioning of labor markets. Universities must shape the skills required by potential employers. The effectiveness of universities is estimated by how quickly graduates find a job. Hence the conclusion-recommendation to subsidize the employment of qualified holders of diplomas, and not to spend money on training low-quality specialists [10].

National concepts of higher education vary across countries. To a large extent, such differences are associated with different participation of the state budget in financing higher education [11–13]. Different participation is determined by the targets for higher education results.

In our previous published papers [14], we identify four main approaches to the goal setting of higher education.

The first is the orientation of the higher educational institution to the training of highly qualified professional personnel with a clearly marked specialization for further work in battle or another sector of the economy in a particular specialty. With such a definition of the goals of higher education, attention is focused on special disciplines and professional knowledge and experience.

However, as the characteristics of a post-industrial society matured and the digital economy had a go, the lifelong or almost lifelong attachment of a person to a particular profession conflicted with new socio-economic conditions.

First, the increasing variability of the structure of production and economic relations is increasingly raising the role of post-professional education, which allows a person to change the specialization and nature of his activities.

Second, narrow professional specialization is fraught with the danger of limiting the horizons of the socio-economic world outlook. And it is not by chance that in recent years the expansion of humanity education in highly specialized higher educational institutions, which train, for example, doctors, engineers, and geologists, has an urgent demand.

Third, a narrow professional specialization limits freedom of choice, increases a person's dependence on the state of the industry and on the behavior of the employer, that is, limits the freedom of the individual as a whole.

The second approach to defining the goals of higher education and to developing its concept is the opposite of the first. Targets are shifted from specific training to the development of human intellectual, creative potentialities. It is this that makes it relatively easy to move from one type of activity to another, to acquire new information and new special professional knowledge and skills.

This approach has become prevalent in the top-rated universities in the world. With the described approach to defining the goals of higher education and developing its concept, the tasks of intellectualization, enhancement of the creative potential of a significant part of the country's population and the formation of its intellectual elite are much more successful.

The third approach: with a view to higher education, the problems of young people's differentiation in terms of intellectual development coefficient and the ability to assimilate and use increasingly complex information come to the fore. The method involves a series of exams, during which an examination threshold is found for the subject.

The fourth approach is the opposite of the third: no exams, diplomas, or degrees are needed. The student is studying at the university disciplines of his own choice. Whether useful or not for his future career, that is the problem of his personal choice and understanding of his capabilities. With all the seeming exoticism of such target-setting, there are already quite a few universities in the world that build the learning process on such principles.

## 3.2 Economic Interests in Higher Education

The market exists not only for graduates, but also for other products of higher education, including professional opportunities in the universities themselves.

The situation in the education in Russia is characterized by the continuous reorganization of the Ministry of Education and Science. An analysis of this situation using the method known as the Saaty AHP suggests that the government wants to retain the main (if not decisive) role in determining the parameters of functioning of higher education. This method, used in the process of creating an analytical hierarchy and developed by Thomas L. Saaty [15], is a planning and decision-making method in the process of multi-criteria decision. (There is an evidence using this method in the USA in 1985–2000).

The purpose of the method is to set, through a series of steps, a hierarchy of scenarios that should or can lead to the large-scale of a specific goal. In a scenario involving evaluation processes, the influence or significance of individual parties and sub-goals is also measured. The work ends with the construction of a script inscribed in a hierarchical framework with a dimension with its own vector.

The impulse of transformations coming from the government is highly likely to correlate with the influence of market mechanisms (current economic interest) than with concern about the quality of education (long-term economic interest due to the fact that education has such a characteristic as hereditarily).

In a number of works, researchers note that the mechanism for the realization of national economic interests is complex and multi-dimensional. The state calls to advocate their interest in determining long-term and short-term state interests, including concepts and programs for the development of higher education [16, 17].

State interests are realized through power and administrative structures, through the activities of people working in these structures, and civil servants. In moving along these steps, national economic interests are subject to deformations to a greater or lesser extent.

First, state interests are far from matching with national ones. This circumstance is analyzed in detail by the theory of public choice (D. Buchanan) [18], which is famous as the new political economy. In our opinion, a broader and more precise definition of the subject of new political economy is presented in publications in the pages of the journals "Problems of New Political Economy" and "Issues of Political Economy", as well as in [19].

Secondly, departmental interests, competitive opposition of various departments arise within the management structures. Under certain conditions, departmental interests are able to replace and push into the background national interests. Thirdly, in the activities of individuals of public servants, national and state interests are combined with personal economic and personnel interests. And the point here is not only in behavior oriented towards the granting of status rents, but also in the fact that the effect of the realization of long-term economic interests often turns out to be beyond the time limits of the terms of office and responsibility of individuals, civil servants. State interests do not always correspond with the economic interests of universities and with the interests of professors and teachers.

The choice of school and specialty of higher education is poorly associated with the real needs of the economy. Despite, for example, the signals of the market about the overproduction of economists and lawyers, these specializations occupy a priority place in the applicants' choice. And after graduating from university, graduates still find jobs that are acceptable to themselves, although they do not quite coincide with the specialization obtained (another argument in favor of the second approach discussed above to the definition of goals and the concept of higher education) [20].

Another issue is burning: the uneven distribution of the intellectual, educational and informational potential throughout the country. Schools and hospitals, cultural institutions in remote places and especially in rural areas do not receive an influx of young teachers, doctors, cultural workers.

The temptation of a simple solution is great: the activation of an administrative resource. They are implementing the proposal to solve the problem by reviving targeted admission to higher educational institutions and distributing graduates. But it is necessary to take into account social costs. They consist, firstly, in restricting the democratic right of any citizen to receive higher education and to follow up on their own, free choice. Secondly, work in the direction will become a predominantly for young people from poor families, which will reinforce the stratification of the population, socio-economic inequality.

It is necessary to recognize the weakness of institutionalization of competition of universities for obtaining limits on budget financing and for attracting applicants on a budgetary and extra budgetary basis. Determining the number of students admitted to a university with budget funding, without sufficient and publicly known criteria in competition, translates the solution of the problem into the plane of the relationship between negotiating power and personal informal relationships with employees of the Ministry of Education and Science of the Russian Federation.

Institutionalization of inter-university competition and the development of a system of objective criteria in the competition of universities for admission numbers financed from the budget not only makes it possible to limit the space for decisions of an informal and shadow nature, but also contributes to improving the quality of work of universities stimulated by competitive conditions [21].

### 3.3 Educational Product as a Public and Private Benefit

Public goods, as well known, have two characteristic features: they are noncompetitive and non-exclusive. These characteristics apply to university educational products with very significant limitations. No competitiveness means zero marginal cost. If adding the student audience with each additional student did not increase university costs, there would be non-competitiveness. In reality, the increase in the number of students for each additional unit is still associated with an increase in a certain proportion of costs. This is the burden on the teacher in all forms of individual communication with students, the area of audiences, and equipment with computers, other teaching and material means, etc. So the concept of non-competitiveness is only very limited to the educational product of the university [21].

Equally limited is another characteristic of the public good in relation to the educational product of the university - non-exclusivity. A product is non-exclusive, if no one can be excluded from the scope of its consumption. If anyone had free access to a university education, the non-exclusivity of this product would indeed be asserted. In reality, not all young people can become students: admission to budget places is limited by funding, there are applicants' contests, access to higher education for a fee. So the concept of non-exclusivity in relation to the system of higher education remains only within the limits of equality of opportunities and rights in competition for obtaining educational products of universities.

Thus, an educational product, having very limited properties of a public good, is mainly a private good. The social significance of higher education is not determined by the properties of the public good, but by exceptionally large positive externalities. Higher education, firstly, meets the needs of firms, the economy and culture of the country as a whole in highly qualified personnel, ensures the growth of intellectual potential, and secondly, the formation of a socially active, responsible population, high-quality society, and civil society. That is, the most fundamental national economic interests are served and realized. Yes, and within the micro-social neighborhood, communication with highly educated people is comfortable for others.

The prevalence of the properties of private good argues the fee for higher education. In most economically developed countries, it is [22–25].

However, this problem is far from clear, especially in the conditions of modern Russia. First, the payment for higher education is objectively perceived by society as a departure from the usual social achievement, as a deterioration of living conditions. Secondly, the payment of higher education can limit access to it by young people from low-income families, who constitute a very large part of the population so far. Consequently, the conditions for social and economic inequality are reproduced, it even increases. In detail, this thesis is developed in the works of N.G. Yakovleva [26].

Thirdly, as it is quite clearly stated in economic theory, market mechanisms themselves, without additional regulatory influence from the state, can not sufficiently direct resources to the production of goods with large positive externalities, including in the higher education.

Funding higher education will be partly carried out by entrepreneurs and charitable foundations. This perspective is closely linked with the trend of increasing autonomy of universities and the conditions for a significant increase in the salaries of professors and teachers, and the improvement of teaching and material support.

### 3.4    Scientific Product of Universities in the Frame of Digital Economy

Together with the prospects and trends discussed above, new trends are also emerging in Russia. It is about turning universities into scientific and information centers with their research laboratories, not only enriching science with new discoveries, but also providing efficient, new and improved technologies to the economy. This aspect of university activities requires special attention. In Russia, for decades, the division of scientific work has been. Fundamental theoretical research was carried out by academic research institutes.

Applied scientific research was carried out mainly by industry research institutes (research institutes and design bureaus). And universities and colleges were mainly engaged in the training of highly qualified personnel, the translation of scientific knowledge to the student audience. And although in recent years, attention to universities' own research activities has increased, it still remains in the background as compared with the educational process, which is still to a small extent connected with the scientific research of the professors and students themselves.

World practice is moving in a different direction. Universities have become the main medium of scientific and technological progress. Moreover, in the concept of a post-industrial society there are three stages of development. And if in an agrarian society the church and the army were a specific form of social organization, and corporations in the industrial society, in a postindustrial society this role is assigned to universities. Universities are turning into centers of scientific, informational, economic, and cultural life of the country that are system-organizing society.

At the same time, the content of the scientific product of universities has fundamentally changed. The distinction between fundamental and applied research in various institutes remains in the past. In most cases, the university is directly focused on the needs of the economy in the field of technology. Applied science, drawing on ideas from fundamental research, has become the main field for the realization of the research potential of universities. Direct relations with corporations, the sale of the scientific product of the university, the fulfillment of orders for the development of new technologies, the implementation of joint scientific and technical programs and projects with the business - all this causes fundamental changes in the sources of income of universities, in financing their activities.

These processes fundamentally change the methodology of higher education. The participation of students in research and development demanded by economics and business, with their focus on significant applied results, contributes to the formation of highly qualified specialists ready for effective practical work. Often, since the student time, cooperation with certain firms has been established.

# 4   Discussion

A political-economic analysis of the dynamics of the promotion of the education system of Russia on the example of higher education in the conditions of the digital economy gives the interpretation of the content of the economy of higher education from the position of political economy. The tendency of universities to turn into system-organizing centers of economic and social life is very important for determining the place and role of a university in each given region.

This interesting and actual problem is already to a large extent considered in our publications [1].

Universities are socio-economic institutions (in terms of institutionalism) that are not clearly defined. This thesis is unequivocally criticized from the standpoint of the history of the emergence and development of universities. But in Russia in the early twenty-first century, it acquires a special form. The education system in Russia suffers from many shortcomings, the most significant of which is organizational and financial inefficiency. Many universities are called universities, although in fact they do not even have advanced research programs [27, 28].

According to the previously stated position [7], the intellectual potential of the population, its level of professionalism and education, the development of science, technology and culture are an external resource of socio-economic development.

The multiplication and effective use of this resource is the main condition for economic growth, improving the welfare of the people, promoting freedom and democracy, forming a civil society, ensuring social and political stability in the country, occupying a worthy place in the global economy and international relations, that is, in aggregate, and solving the most important problems of modern Russia.

## 5   Conclusions

Summing up our political and economical ideas about the development of the Russian education system on the example of higher education in the conditions of the digital economy, we emphasize that current trends are very contradictory, but the main one, unfortunately, is the prevalence of political decisions without taking into account long-term socio-economic consequences.

When comparing approaches to the definition of higher education goals, essentially two mutually related, but competing functions of higher school are found. This is the role of higher education in preparing qualification personnel for all sectors of the economy, on the one hand, and general intellectual development, the formation of a creative, socialized personality [26]. Finding the optimal combination of these two goals is related to the specifics of the economic relations in which the university operates, with a compromise resolution of the contradictions of economic interests in the higher education system. In other words, we have a problem of a political economy nature.

Corresponding to a market economy and democratic organization of society, are models built on market principles based on these principles, providing for a compromise harmonization of economic interests. The core of such models can be contracts providing for additional, scholarship student support and obligations relating to housing and other living conditions of the young specialist. Another important aspect is related to the economic interests of each individual university, as well as the personal interests of the professors and teachers working in it.

The predominance of the properties of the private good of the educational product of universities argues paid for higher education. For the foreseeable future, a combination of payment and free higher education in Russia is inevitable. However, the prospect looks in such a way that the share of payment for educational services will consistently increase.

Competition of universities in the market of scientific products for the receipt of the most profitable orders and the conclusion of contracts with firms and corporations stimulates universities to increase their ranking. What matters is not whether the capital is a university or a provincial, but the scientific potential, image, brand, obtained patents for discoveries and inventions, scientific publications, the level of informatization of activities, implemented projects and, ultimately, a place in the scientific hierarchy.

## References

1. Institutionalization of Public Goods (2003) Noskov VA, Sviridov NN and etc. under total ed. Deserved Professor of Science of the Russian Federation, Professor VV Chekmarev. Publishing House of SSC RAS, Samara
2. Belyakov SA (2013) Russian higher education: models and development scenarios: monograph SA Belyakov, TL Klyachko. Publishing House "Business" RANEPA, Moscow
3. The future of higher education in Russia: an expert view. Foresight study - 2030: Analytical report (2014) Ed. Efimova VS INFRA M, Moscow; Sib. feder. Univ, Krasnoyarsk

4. Economic theory (political economy) (1997) Under. Total ed. Vidyapina GI Zhuravleva. INFRA, Moscow
5. Noskov VA (2003) Basics of the formation of the market of higher education services. Econ Sci № 5:91–98
6. Maloletko A (2009) The system of higher professional education: management of economic security. Probl Theor Pract Manag № 9:17–22
7. Chekmarev VV (2003) Economic relations in the production of educational services. VV Chekmarev, VA Noskov and etc. Publishing House of SSC RAS, Samara, 168 p
8. Clark BR (2011) Higher education system: academic organization in a cross - national perspective. Trans. from English A Smirnova. Moscow House of the Higher School of Economics
9. Clark BR (2011) Maintaining change in universities. Continuity of case studies and concepts. Trans. from English E Stepkina. Moscow: Izd.dom Higher School of Economics
10. Subetto AI (2015) Scientific and educational society as a basis for the development strategy of Russia in the XXI century. Assertion, St. Petersburg
11. Fifty modern thinkers about education. From Piaget to our days (2012) Trans. From English SI Denikin; under the scientific ed. MS Dobryakova. Moscow: Izd. House of the Higher School of Economics
12. Readings B (2010) University in the ruins. Trans. from English IS Korbut. Izd.dom state, Moscow. University - Higher School of Economics
13. Thomas G (2016) Education. A very brief introduction. Trans. from English A Arkhipova, under scientific. Ed. S Filonovich. Izd.dom HSE, Moscow
14. Noskov VA (2010) The economic role and economic functions of education in the cluster of the regional economy. VA Noskov, EV Bolgova. Econ Sci № 5(66):289–292
15. Saaty TL (2016) Relative Measurement and its generalization in decision making. Why pair comparisons are key in math for measuring intangible factors. Cloud Sci 3(2):171–262. https://cloudofscience.ru/sites/default/files/pdf/CoS_3_171.pdf. Accessed 4 Apr 2018
16. Mass Higher Education. Triumph of the BRIC (2014) Karnoi M, Loyalka P, Dobryakova M, Dossani R, Frumin I, Koons K, Tilak DBG, Wang R. Trans. from English M Dobryakova, L Pirozhkova. HSE, Moscow
17. Ortega y Gasset H (2010) Mission of the University. Trans. with isp. Golubeva MN, Korbut AM. Izd, Moscow. State House University - Higher School of Economics
18. Buchanan James (1999) The logical foundations of constitutional liberty, vol 1. Liberty Fund, Indianapolis
19. Chekmarev VV (2009) New political economy: sources and results (Kostroma initiative). The report at the theoretical seminar "Discussion problems of modern social science and economic thought" (Moscow, October 20, 2009). / VV Chekmarev. Moscow: MSU. M Lomonosov. Center for Social Sciences; Kostroma: KSU them. NA Nekrasov
20. Slobodchikov VI (2017) The systemic crisis of education as a threat to the national security of Russia. Slobodchikov VI, Korolkova IV, Ostapenko A, Zakharchenko MV, Shestun EV, Rybakov SYu, Moiseev DA, Short SN. Free Thought., № 2, № 3
21. Vasilenko NV (2017) Economics of education: textbook. NV Vasilenko, A Ya Linkov. INFRA-M, Moscow
22. Tao BL (2017) The Commercialization of Education/BL Tao, M Berci, W He the New York Times. http://www.nytimes.com/ref/college/coll-china-education-005.html. Accessed 4 Apr 2018
23. Twebaze R M (2016) Commercialization of education in Uganda; causes and consequences. Int J Recent Sci Res 6(7):5107–5112
24. Borgohain S (2016) Critical Analysis. International Research Journal of Interdisciplinary & Multidisciplinary Studies (IRJIMS). –Vol. I, Issue XII. January, pp 71–76

25. Hanushek EA, Machine S, Woessmann L (eds) (2016) Handbook of the economics of education. Elsevier, Boston
26. Yakovleva NG (2017) Commercialization, bureaucratization, education management of post-Soviet Russia: a political economy view. Probl Theor Pract Manag № 3:121–130
27. Chistyakova SN, Gevorkyan EN, Podufalov ND (2016) Actual problems of professional and higher education, Collective monograph. Econom-Inform Publishing House, Moscow
28. Collini Stefan (2016) Why do we need universities? Trans. from English D Kralechkina. Izd. HSE House, Moscow

# Gamification for Handing Educational Innovation Challenges

Z. Yordanova[(⊠)]

University of National and World Economy, Sofia, Bulgaria
zornitsayordanova@unwe.bg

**Abstract.** The papers aims at exploring gamification as a tool for handing some educational innovation challenges. The study summarizes though a literature analysis some educational innovation challenges that have been provoked by the increasing innovations in the area of education and the growth of failures and unsatisfying results from educational innovations implementation and application. On the other hand, gamification proposes a way to handle, with its diverse elements and characteristics, some of these challenges and possibly to impact the effectiveness of implementing and applying educational innovations. The research methodology employs a literature analysis so as to elicit some fundamental educational challenges and then a cluster analysis is done via content research. The result from this step is formulating some common educational innovation-related groups for further analysis. Later in the research, by utilizing the explored and summarized gamification tools and elements, to each educational innovation group, a potential gamification tool is assigned. The results are from interest of both educational and innovation experts and researchers. The outcomes may also be used directly in solving some of the educational issues.

**Keywords:** Education · Innovation management · Educational innovation · Gamification

## 1 Introduction

Innovation in education has been a topic from wide interest during the last decades. Reasons for the extensive research on educational innovations are the growing incapability of humanity to address the future problems and to organize the development direction [32]. On the contrary, the intensity of industries, globalization and innovation in general, make people following and trying to handle the already existing issues rather than to predict them, to steer and to manage the progress. Education per se is the unique concept, which refers to so many aspects of human future, and at the same time it is responsible for it. All these emphasizing the urgency of educational innovation in terms of its management, application and relevance to the future problems in any area.

And while the importance and urgency of educational innovation are clearly recognized by the wide audience, including society, academics, politicians and business, the ways for its application and implementation for delivering the required results are still struggling with inefficiency [41].

© Springer Nature Switzerland AG 2020
S. Ashmarina et al. (Eds.): *Digital Transformation of the Economy: Challenges, Trends and New Opportunities*, pp. 529–541, 2019.
https://doi.org/10.1007/978-3-030-11367-4_53

The current research aims at summarizing some common educational innovation challenges and at exploring the possibility of employing gamification for handling some of them. The research is explorative and provides contribution to the practical management of educational innovations for achieving efficiency.

## 2 Literature Analysis

Intensive processes of blurring boundaries between industries, economic sectors and businesses are being observed recently [18, 43]. A tendency for eliminating borders between countries, overlapping of markets [23, 35], merging of multinational companies [21, 36] has been reported. In this collaborating and globalizing world, knowledge and technology transfer is an essential process that aims at boosting effectiveness, productivity and innovativeness of industries, education, companies and countries [3, 44]. Knowledge transfer between two fields is ordinary and commonplace [7]. It aims at creating synergy and adding value by combining some strengths specific to each of the actors in the interaction or at solving problems in a better way [5]. Pursuing effectiveness and boosting optimization of educational innovations, in the current research, we explore gamification in its role for addressing educational innovation challenges.

### 2.1 Educational Innovation Challenges

Innovations are considered as the main driver for growth and a determinant for organizational and sectoral productivity and competitiveness. They are recognized as the growth engine of modern economy and they ensure growth regardless of the economic environment. Many organizations have declared that improving and increasing innovativeness and the ability to develop innovations are amongst the most substantial factors for growth [4, 12, 29, 31]. Innovations are equally important for the private and governmental sectors, important for the humanity in general. Since it has been clarified that innovations are the most reliable tool for transforming the past and present up to a superior level, the issue how more effectively and successfully innovations should be managed is still valid. Educational innovations are defined by Taylor et al. [39] as any novel teaching technique, strategy, tool, or learning resource that could be used by an instructor to lead to effective (or promising) instructional techniques that benefit student learning and engagement. According to Fullan [17], educational innovation must contain three elements: use of new revised materials (curriculum materials or technologies); use of new teaching approaches (teaching strategies or activities); alteration of beliefs (pedagogical assumptions).

Much research has been done on problems that education is facing. Utilizing the idea of problem driven innovation [10], the current research aims at extracting some commonly identified problems and challenges and their innovative solutions from the literature. These all are called challenges in the research and will be analyzed later from the perspective of solving them with gamification elements. According to OECD (2016) the main issue in education and the starting point for innovation in the sector are productivity and efficiency. Here, efficiency means the balance between resources invested and the outcomes in terms of students' performance and equity.

According to Kozma [25], educational innovation means supporting a shift from traditional paradigms towards emerging pedagogical approaches based on ICT solutions such as fostering learner-centred and constructivist processes, and the acquisition of lifelong learning skills. Hannon [20] refers innovation to a complete shift in the educational paradigm, driven by the four principles of social innovation, i.e. openness, collaboration, freedom, and direct participation of those involved. Innovation has become an essential ingredient in creating and sustaining a culture of performance in higher education and keeps transforming higher education [42].

Cortes-Robles et al. [11] emphasized on the importance of the Information and Communication Technologies (ICT) in education and linking educational innovation and challenges to the integration of ICT into a higher education institution as a tool to innovate teaching practices, as well as providing the possibility of including new didactic strategies that arouse the interest and motivation of students to improve the quality of teaching-learning processes inside and outside of the classroom.

Staley and Trinkle [37] formulated ten trends in managing higher education and respectively referring the educations innovation. These are: Increasing Differentiation of Higher Education; Transformation of the General Education Curriculum; Changing Faces of Faculty; Surge in Global Faculty and Student Mobility; The New "Invisible College"; The Changing "Traditional" Student; The Mounting Pressure to Demonstrate Value; The Revolution of "Middle-Skill" Jobs; College as a Private vs. Public Good; Lifelong Partnerships with Students.

Creativity has been identified as a key ingredient of educational innovation in many researches. But in practice, there is still widespread ignorance of creativity in the formal education's field and a lack of scientific research about creativity and education, particularly in teacher training [19]. According to Nosari [32], exactly creativity would be the criteria of future transformation in education. The aspects of creativity are described by Lee [27] as vividness in physical-physiological sphere, cooperation in social sphere, quest in rational sphere, virtue in moral sphere, beauty in artistic sphere, and belief in religious sphere and are defined as the properties of value ability and presented the educational purpose. Cheung [8] arises the entrepreneurship to the top priority in modern education since entrepreneurs have been so important to our economy, schools should be responsible for cultivating in students a suitable entrepreneurial spirit and skills. Wai [42] analyses a series of facilitators that determine creativity and innovation in teaching learning processes, emphasising the role of assessment, organizational culture and information technologies as relevant and essential elements of the educational processes. Information and communication technology plays a crucial role in the way of learning and allows changes in education for an innovative and creative school environment. These technologies could act as a platform to encourage creative learning and innovative teaching, while providing a variety of opportunities for a constructive change. Clements [9] and Levi [28] have encouraged the trends of educational innovation circles by insisting to apply innovation teaching in the teaching profession. E-learning in university education as a source of innovation is driven by the lack of clear approach for delivering e-learning technologies and is also identified as a challenge for universities. The rapid changes and increased complexity in today's world and dynamics in the education for the future put new challenges and demands on the education system in the perspective of online learning. There has been generally a growing awareness of the necessity to change and improve the existing system towards online learning [2]. Online learning is also examined as a collaborative learning environment and departamental management [40]. Still, the acceptance of e-learning by the employees and students is a challenge in many spheres.

Cortés-Robles, Luis García-Alcaraz and Alor-Hernández [11] pointed out in their research that organizations achieve innovation performance under challenging conditions and these for instance are: the technologies are not always accessible, the workforce does not have the required competences or demand constant training not easily available; the lack of techniques to facilitate the assimilation of new technologies and to learn from others; A limited perception about the technical side of innovation, which produces the impression that this process depends on a random creative effort, and no less important, a low assimilation degree of tools for planning the evolution of a product or an innovation system.

Ebersole [16] has defined the following challenges which higher education leaders face: a trend toward competency based education, tougher accreditation standards, an emphasis on assessment, voids in leadership, and the growing diversity of students as challenges that will plague higher education in the coming years. Wai [42] detected globalization and collaboration as big challenges, which the educational innovation should be, addresses as cross-disciplinary collaboration received increasing attention. Sustainability has also been identified as a crucial factor for as to encompass the different effects of human resources for sustainable development [1].

## 2.2   Gamification

Gamification is often researched for the purposes of adding value to education and training. Despite of the general understanding of gamification to support mainly students' motivation and to stimulate learning processes, gamification may be used in many other issues to cover them partially or fully. In respect to assign gamification characteristics, elements or identified strengths to the formulated educational innovation challenges in the research, in this section of the research gamification is first clarified as a concept and then it is researched for application in different situations for handing diverse problems and challenges.

### 2.2.1   Nature and Purpose of Gamification

Gamification is often confused in the literature with some close concepts as games, serious gaming, game learning, applying game design, utilizing video games for educational purposes. A game is structured play, usually for enjoyment but gamification is a strategy for influencing and motivating the behavior of people [29]. The gamification presents use of gaming-oriented mechanisms to address practical problems or to engagement between specific audiences [30]. According to Zichermann and Cunningham [45], gamification is an application of game dynamism and system to real-life problem-solving. Kapp [22] defines gamification as the usage of the concept of games and game mechanics to engage users and troubleshoot. De-Marcos, Domínguez, Saenz-de-Navarrete, & Pagés [13] formulate gamification as the use of elements of gameplay and game design techniques in non-gaming contexts to engage people and solve problems. Defined by Deterding et al. [14] gamification is the use of game elements in non-game contexts. In addition, it is also defined as a process of using gaming mechanisms and game thinking to solve problems. Lee and Hammer [26] believe that gamification is the use of gaming mechanisms, dynamics, and frameworks to promote desired behavior.

To Koivisto and Hamari [24] the gamification is a technological partner with collector potential phenomenon claimed to provide a multitude of benefits, such as entertainment and education, as well as the social benefits through communities and social interaction. The use of gamification is now considered as an emerging phenomenon, which emanates directly from the popularity of games, and its inherent characteristic to motivate action and its performance, solve problems and enhance learning in the most diverse areas of knowledge and life of individuals [30].

### 2.2.2 Applications of Gamification

Menezes and Bortolli [30] explore how gamification may be applied to the design of innovative tasks related to formative assessment when it comes to evaluation for training purposes. Gamification can be also used in fields such as marketing to encourage customers as well as in employment to motivate employees [15]. The following elements have been identified as part of the application of gamification concept. They are divided into three categories [6]: dynamic, mechanical and component. Dynamic elements consist of: limitations - choices that make them meaningful; emotional reinforcement; storytelling - permanent graphic experiences, creating a sense of flow and use of ideas for history; progressive relationships. Mechanical elements are: challenges - goals to achieve resource acquisition; opportunities for rewarding; cooperation in the field of competition; cooperation; feedback, opportunity for a victory. The components are: achievements; leadership; avatars; levels; badges; points; fights; searches; collections; social graphics; combat the team; unlocking the content; virtual goods, gifts.

### 2.2.3 Gamification in Education

Many researches reveal how exactly gamification benefits education and how it is employed by it. According to Lee and Hammer [26] gamification increases level of engagement and motivation in classrooms which are crucially needed by students. It also provides a great chance for shy students to express themselves. It provokes a number of emotions which are important to the learning and educational process in their wide and common perspective. These are experimenting with powerful emotions, from curiosity to frustration to joy, positive emotional experiences, such as optimism and pride in a controlled environment [38]. Matsumoto [29] discuss the pros and cons of gamification of education and concluded with the importance of well-designed tutorial, task, interface and feedback for the effective game-based e-Learning. As a final result of the study, "a classroom based on gamification" is helpful in improving learners' understanding level and motivation. Some gamification elements are used in a research for educating blind students and the results were promising [33]. Students have freedom to fail and try again without negative impact [34].

## 3  Methodological Approach

The methodology used in the study employs a synthesis analysis and content analysis. The main purpose is first the research to summarize some already raised challenges in education which should be the future drivers of educational innovations. Then the identified challenges are clustered into groups and after it, all the groups are explored for potential gamification elements as a possible tool for handling these challenges. Inductive reasoning is used for developing conclusions from literature review by weaving together new information into theories. The choice of the content analysis is made because of the purpose to organize and elicit meaning from the data collected and draw realistic and practical conclusions. In the case of the study, the collected data are researches on the topic of educational innovations and their implementation. The assignment of a gamification element to each of the educational challenges is made based on the gamification tools described by [6] since they are wide ranged enough for achieving diverse results. These are: limitations; emotional reinforcement; storytelling; progressive relationships; challenges; opportunities for rewarding; cooperation in the field of competition; cooperation; feedback, opportunity for a victory; achievements; leadership; avatars; levels; badges; points; fights; searches; collections; social graphics; combat the team; unlocking the content; virtual goods, gifts. The assignment is done on the bases of relevance and adequacy. The extracted challenges from the performed literature analysis are presented in Table 1.

**Table 1.**  The identified educational challenges as future drivers of educational innovations

| Identified educational challenge for the future | Study |
| --- | --- |
| Productivity and efficiency of education | OECD (2016) |
| Emerging pedagogical approaches based on ICT | Kozma [25] |
| Learner-centred and constructivist processes | Kozma [25] |
| Acquisition of lifelong learning skills | Kozma [25] |
| Openness, collaboration, freedom, and direct participation of all the involved in the educational process | Hannon [20] |
| Creating and sustaining a culture of performance in higher education | Wai [42] |
| Integration of ICT into a higher education institution as a tool to innovate teaching practices | Cortes-Robles et al. [11] |
| Motivation of students | Cortes-Robles et al. [11] |
| Improve the quality of teaching-learning processes inside and outside of the classroom | Cortes-Robles et al. [11] |
| Increasing Differentiation of Higher Education | Staley and Trinkle [37] |
| Transformation of the General Education Curriculum | Staley and Trinkle [37] |
| Changing Faces of Faculty | Staley and Trinkle [37] |

(*continued*)

**Table 1.** (*continued*)

| Identified educational challenge for the future | Study |
|---|---|
| Surge in Global Faculty and Student Mobility | Staley and Trinkle [37] |
| The Changing "Traditional" Student | Staley and Trinkle [37] |
| The Mounting Pressure to Demonstrate Value | Staley and Trinkle [37] |
| The Revolution of "Middle-Skill" Jobs | Staley and Trinkle [37] |
| College as a Private vs. Public Good | Staley and Trinkle [37] |
| Lifelong Partnerships with Students | Staley and Trinkle [37] |
| Creativity | Giménez [19] |
| Creativity for transformation | Nosari [32] |
| Cooperation in social sphere | Lee [27] |
| Entrepreneurship as a top priority in modern education | Cheung [8] |
| Assessment, organizational culture and information technologies as relevant and essential elements of the educational processes | Venera [41] |
| E-learning | Clements [9], Levi [28] |
| Delivering e-learning technologies | Ibezim [31] |
| Rapid changes and increased complexity | Ibezim [31] |
| Dynamics in the education for the future | Ibezim [31] |
| Online learning | Ibezim [31] |
| Online learning | Almarabeh et al. [2] |
| Collaborative learning environment | Tziallas et al. [40] |
| Departamental/faculty management | Tziallas et al. [40] |
| Acceptance of e-learning | Rym et al. [4] |
| Technologies are not always accessible | Cortés-Robles et al. [12] |
| The workforce does not have the required competences or demand constant training not easily available | Cortés-Robles et al. [12] |
| Lack of techniques to facilitate the assimilation of new technologies and to learn from others | Cortés-Robles et al. [12] |
| A limited perception about the technical side of innovation | Cortés-Robles et al. [12] |
| A low assimilation degree of tools for planning the evolution of a product or an innovation system | Cortés-Robles et al. [12] |
| Competency based education | Ebersole [16] |
| Tougher accreditation standards | Ebersole [16] |
| Voids in leadership | Ebersole [16] |
| Growing diversity of students | Ebersole [16] |
| Globalization | Wai [42] |
| Collaboration | Wai [42] |
| Cross-disciplinary collaboration | Wai [42] |
| Sustainability | Al-Khateeb et al. [1] |

Source: compiled by the author

# 4 Results

The first step after identifying the selected educational challenges from the literature analysis is to form some groups for their better presentation and analysis for assigning a gamification element to handle them. The results from the clustering process are presented in Table 2. The clustering process is done according to the respective identified challenges, their content and characteristics.

**Table 2.** Grouping of educational innovation challenges

| Identified educational challenge for the future | Challenge group |
| --- | --- |
| Productivity and efficiency of education | Efficiency |
| Emerging pedagogical approaches based on ICT | ICT |
| Learner-centred and constructivist processes | Process |
| Acquisition of lifelong learning skills | New skills |
| Openness, collaboration, freedom, and direct participation of all the involved in the educational process | Collaboration |
| Creating and sustaining a culture of performance in higher education | Sustainability |
| Integration of ICT into a higher education institution as a tool to innovate teaching practices | ICT |
| Motivation of students | Motivation |
| Improve the quality of teaching-learning processes inside and outside of the classroom | Quality |
| Increasing Differentiation of Higher Education | Educational Model |
| Transformation of the General Education Curriculum | Transformation |
| Changing Faces of Faculty | Faculty management |
| Surge in Global Faculty and Student Mobility | Mobility |
| The Changing "Traditional" Student | Transformation |
| The Mounting Pressure to Demonstrate Value | New skills |
| The Revolution of "Middle-Skill" Jobs | New skills |
| College as a Private vs. Public Good | Educational Model |
| Lifelong Partnerships with Students | Collaboration |
| Creativity | Creativity |
| Creativity for transformation | Creativity |
| Cooperation in social sphere | Collaboration |
| Entrepreneurship as a top priority in modern education | New skills |
| Assessment, organizational culture and information technologies as relevant and essential elements of the educational processes | Culture |

*(continued)*

**Table 2.**  (*continued*)

| Identified educational challenge for the future | Challenge group |
|---|---|
| E-learning | E-learning |
| Delivering e-learning technologies | E-learning |
| Rapid changes and increased complexity | Complexity |
| Dynamics in the education for the future | Educational Model |
| Online learning | E-learning |
| Online learning | E-learning |
| Collaborative learning environment | Collaboration |
| Departamental/faculty management | Faculty management |
| Acceptance of e-learning | E-learning |
| Technologies are not always accessible | ICT |
| The workforce does not have the required competences or demand constant training not easily available | Educational Model |
| Lack of techniques to facilitate the assimilation of new technologies and to learn from others | ICT |
| A limited perception about the technical side of innovation | ICT |
| A low assimilation degree of tools for planning the evolution of a product or an innovation system | Educational Model |
| Competency based education | New skills |
| Tougher accreditation standards | Accreditation |
| Voids in leadership | Leadership |
| Growing diversity of students | Transformation |
| Globalization | Globalization |
| Collaboration | Collaboration |
| Cross-disciplinary collaboration | Collaboration |
| Sustainability | Sustainability |

Source: compiled by the author

From the 45 identified education challenges, 19 groups are formed. These are: Efficiency, ICT, Process, New skills, Collaboration, Sustainability, Motivation, Quality, Educational Model, Transformation, Faculty management, Mobility, Creativity, Culture, E-learning, Complexity, Accreditation, Leadership, Globalization. The groups play the role of a focus for better analysis on educational innovation challenges and gamification knowledge transfer. To each group of educational challenge for further innovation is assigned a potential gamification element from those identified by Boer [6] presented in Table 3.

**Table 3.** Match between education challenge group and possible Gamification elements

| Education challenge group | Possible Gamification elements |
|---|---|
| Efficiency | Virtual goods, emotional reinforcement, opportunities for rewarding, feedback, achievements, gifts, challenges |
| ICT | Virtual goods, opportunities for rewarding, challenges |
| Process | Fights, opportunities for rewarding, feedback, gifts, challenges |
| New skills | Virtual goods, emotional reinforcement, collections, progressive relationships, achievements, gifts, storytelling, challenges |
| Collaboration | Emotional reinforcement, collections, cooperation in the field of competition, progressive relationships, gifts, storytelling, challenges |
| Sustainability | Virtual goods, fights, collections, progressive relationships, challenges |
| Motivation | Emotional reinforcement, opportunities for rewarding, gifts, challenges |
| Quality | Fights, progressive relationships, opportunities for rewarding, achievements, gifts |
| Educational Model | Levels, collections, storytelling, challenges |
| Transformation | Levels, progressive relationships, achievements, gifts, challenges |
| Faculty management | Levels, challenges |
| Mobility | Virtual goods, levels |
| Creativity | Emotional reinforcement, progressive relationships, storytelling |
| Culture | Progressive relationships, challenges |
| E-learning | Virtual goods, cooperation in the field of competition, gifts, storytelling |
| Complexity | Cooperation in the field of competition, progressive relationships, storytelling |
| Accreditation | Levels, cooperation in the field of competition |
| Leadership | Emotional reinforcement, achievements |
| Globalization | Virtual goods, cooperation in the field of competition |

Source: compiled by the author

The presented matching between education challenge groups and possible gamification elements has been done based on observations of the gamification elements in different education-related context and assuming these may be relevant and appropriate for handing the formulated challenges. Further research will employ the results of this study case by case for proving these assumptions via case studies.

## 5   Conclusions

In conclusion, we may summarize that educational challenges could be grouped in 19 groups, namely: Efficiency, ICT, Process, New skills, Collaboration, Sustainability, Motivation, Quality, Educational Model, Transformation, Faculty management, Mobility, Creativity, Culture, E-learning, Complexity, Accreditation, Leadership,

Globalization. The contribution of that grouping is working on the educational innovations by focusing on the different challenges that these diverse groups propose. The second contribution of the study is the examination of a possible knowledge transfer between some separate and self-contained concepts as gamification to educational management and educational innovations in particular. The results shows relevance for searching and trying such a cooperation for handling educational challenges because of the already practiced and successful gamification application in education. The new approach in the research is exploring gamification as a management tool in education not only in the learning process itself bearing in mind that much of the research has proposed only that until the moment.

Future work of the author will be to experience gamification in each of the identified educational management groups and examining via case studies the possible to impact these challenges. The study has an explorative nature and may be a starting point for many new research on the topic.

**Acknowledgements.** This work was supported by the Bulgarian NFS under the grant "Young Scientists – 2017" No DM 15/4 – 2017 and also by University of National and World Economy under the project No NID NI-14/2018.

# References

1. Al-Khateeb M, Al-Ansari N, Knutsson S (2014) Sustainable university model for higher education in Iraq. Creat Educ 5:318–328. https://doi.org/10.4236/ce.2014.55041
2. Almarabeh T, Mohammad H, Yousef R, Majdalawi Y (2014) The University of Jordan E-learning platform: state, students' acceptance and challenges. J Softw Eng Appl 7:999–1007. https://doi.org/10.4236/jsea.2014.712087
3. Argote L, Ingram P (2000) Knowledge transfer: a basis for competitive advantage in firms. Organ Behav Hum Decis Process 82(1):150–169
4. Rym B, Olfa B, Mélika B (2013) Determinants of E-learning acceptance: an empirical study in the tunisian context. Am J Ind Bus Manag 3(3):307–321. https://doi.org/10.4236/ajibm.2013.33036
5. Bagheri S, Kusters R, Trienekens J, van der Zandt H (2016) Classification framework of knowledge transfer issues across value networks. Product-Serv Syst Across Life Cycle Proced CIRP 47:382–387
6. Boer P (2013) Introduction to Gamification. https://cdu.edu.au/olt/ltresources/downloads/whitepaper-introductiontogamification-130726103056-phpapp02.pdf. Accessed 2 Mai 2018
7. Cano-Kollmann M, Cantwell J, Hannigan T et al (2016) J Int Bus Stud 47(3):255–262. https://doi.org/10.1057/jibs.2016.8
8. Cheung C (2012) Entrepreneurship education at the crossroad in Hong Kong. Creat Educ 3:666–670. https://doi.org/10.4236/ce.2012.35098
9. Clements J (2013) British film institute anime: a history. Palgrave Macmillan, London
10. Coccia M (2016) Problem-driven innovation in drug discovery: co-evolution of the patterns of radical innovation with the evolution of problems. Problem-driveninnovationsindrugdiscovery: Co-evolution of the patterns of radical innovation with the evolution of problems, https://doi.org/10.1016/j.hlpt.2016.02.003. Available at SSRN: https://ssrn.com/abstract=2810128

11. Cortes-Robles et al. (2019) Managing Innovation in Highly Restrictive Environments, Management and Industrial Engineering. https://doi.org/10.1007/978-3-319-93716-8_7
12. Cortés-Robles, García-Alcaraz L, Alor-Hernández (2016) Managing innovation in highly restrictive environment: lessons from Latin America and Emerging Markets. Springer
13. De-Marcos L, Domínguez A, Saenz-de-Navarrete J, Pagés C (2014) An empirical study comparing gamification and social networking on e-learning. Comput Educ 75:82–91. https://doi.org/10.1016/j.compedu.2014.01.012
14. Deterding et al (2011) From game design elements to gamefulness: defining "gamification." In: Lugmayr A, Franssila H, Safran C, Hammouda I (eds) MindTrek 2011, pp 9–15. https://doi.org/10.1145/2181037.2181040
15. Domínguez A et al (2013) Gamifying learning experiences: practical implications and outcomes. Comput Educ 63:380–392
16. Ebersole J (2014) Top issues facing higher education in 2014. Forbes
17. Fullan M (2007) The new meaning of educational change, teachers college press, 5th edn
18. Gilsinga V, Bekkersb R, Freitasc I, Steend M (2011) Differences in technology transfer between science-based and development-based industries: Transfer mechanisms and barriers. Technovation 31(12):638–647
19. Giménez NP (2016) Profile of promoters and hindering teachers creativity: own or shared? Creat Educ 7(10)
20. Hannon V (2009) 'Only Connect!': a new paradigm for learning innovation in the 21st Century, Centre for Strategic Innovation
21. Heidenreich M (2012) The social embeddedness of multinational companies: a literature review. Socia-Econ Rev 10(3):549–579. https://doi.org/10.1007/978-1-84457-884-9
22. Kapp KM (2012) The Gamification of learning and instruction: Game-based methods and strategies for training and education. Pfeiffer
23. Kohonen A (2013) On detection of volatility spillovers in simultaneously open stock markets. J Empir Financ 22:140–158
24. Koivisto J, Hamari J (2014) Demographic differences in perceived benefits from Gamification. Comput Hum Behav 35:179–188. https://doi.org/10.1016/j.chb.2014.03.007
25. Kozma RB (2003) Technology, innovation, and educational change. a global perspective: a report of the second information technology in education study module 2. ISTE Publisher
26. Lee J, Hammer J (2011) Gamification in education: what, how, why bother? Acad Exch Q 122:1–5
27. Lee Y (2013) The teaching method of creative education. Creat Educ 4:25–30. https://doi.org/10.4236/ce.2013.48A006
28. Levi A (2013) The sweet smell of Japan: Anime, Manga and Japan in North America. J Asian Pacific Commun 23:3–18
29. Matsumoto T (2016) Motivation strategy using Gamification. Creat Educ 7:1480–1485. https://doi.org/10.4236/ce.2016.710153
30. Menezes C, Bortolli R (2016) Potential of Gamification as assessment tool. Creat Educ 7:561–566. https://doi.org/10.4236/ce.2016.74058
31. Ibezim N (2013) Technologies needed for sustainable E-learning in university education. Mod Econ 4(10):633–638. https://doi.org/10.4236/me.2013.410068
32. Nosari S (2012) Creativity at the crossroad creative education as moral education? Creat Educ 3:63–65. https://doi.org/10.4236/ce.2012.37B015
33. Pinho T, Delou C, Lima N (2016) Origami as a tool to teach geometry for blind students. Creat Educ 7:2652–2665. https://doi.org/10.4236/ce.2016.717249
34. Redfield CL (2013) Gamification and creating game developers. In: 2013 Proceedings of the information systems educators conference, San Antonio, Vol 30
35. Sattinger M (2006) Overlapping labor markets. Labor Econ 13(2):237–257

36. Søderberg AM, Vaara E (2003) Merging across borders: people. CBS Press, Cultures and Politics
37. Staley DJ & Trinkle DA (2011) The changing landscape of higher education. Educ Rev 46:15–31
38. Surendeleg G, Murwa V, Yun H-K and Kim YS (2014) The role of Gamification in education—A literature review. Contemp Eng Sci 7:1609–1616. http://www.m-hikari.com
39. Taylor et al (2018) Propagating the adoption of CS educational innovations, ITiCSE 2018, June 2018, Larnaca, Cyprus
40. Tziallas G, Kontogeorgos A, Papanastasiou C (2016) An E-learning platform for departmental use. Creat Educ 7:1189–1194. https://doi.org/10.4236/ce.2016.79124
41. Venera TA (2016) Aspects regarding the role of facilitators in creative learning and innovative teaching. Annals of the "Constantin Brâncuşi" University of Târgu Jiu, Economy Series, Special Issue, vol II/2016, pp 48–52
42. Wai C (2017) Innovation and social impact in higher education: some lessons from Tohoku university and the open university of Hong Kong. Open J Soc Sci 5:139–153. https://doi.org/10.4236/jss.2017.59011
43. Wuwei L (2002) Industry amalgamation and industry innovation. Shanghai Manag Sci 04
44. Zhelev P (2014) International technology transfer to Bulgaria after its European union accession. Econ Altern, Issue 3, pp 83094
45. Zichermann G, Cunningham C (2011) Gamification by design: implementing game mechanics in web and mobile apps. O'Reilly Media, Sebastopol, CA

# The Model for Meeting Digital Economy Needs for Higher Education Programs

E. V. Bolgova$^{(\boxtimes)}$, G. N. Grodskaya, and M. V. Kurnikova

Samara State University of Economics, Samara, Russia
elena_bolgova@rambler.ru, gngsamara@mail.ru,
mvkurnikova@gmail.com

**Abstract.** The study is carried out in response to the need to provide the economic digitalization of Russian regions with academic programs developed by national universities. In this regard, the paper is focused on researching into the reasons of the digital agenda disregarding higher education programs and their role in training the staff able to create new digital platforms and technologies, progressively introduce existing tools and decisions. The disclosure of the environment enabling to transform digitalization-based academic programs into an impetus of digital economy of a Russian subject, a driver to develop an innovative sector and a regional university network would address the shortage of highly qualified staff and help Russian regions catch up with the leading countries of digital transformation. The authors used a method of modelling to address the issues of providing regional economic digitalization with higher education programs within the framework of the unanimity of Russian and foreign scientific opinions and the world practice of territorial administration and sectoral governance. The paper presents theoretical and methodical basis for providing regional economic digitalization with higher education programs, provides an interpretation of triple-helix model as a method based on an innovative initiative and the equality of subjects in the chain 'universities – businesses – governments', reveals the direction of how to use the model in diversifying academic programs and distributing them spatially. The research has resulted in grounding the theoretical and methodical foundations of modelling as a method to providing regional economic digitalization and regional industries with higher education programs; a model well-balanced with regional and industrial strategies, university research activities; a strategy to implement the model in the economy of Russian regions. The study is scientific and practical value for researching into socio-economic development of Russian subjects and regional university networks, programming educational development.

**Keywords:** Economic digitalization · Higher education · Model · Region · Strategy

## 1 Introduction

The policy in the innovative sphere and the priorities of the scientific and technological development defines strategic directions of Russian development and accounts for a conversion to digital technology and the economy based upon them [1].

© Springer Nature Switzerland AG 2020
S. Ashmarina et al. (Eds.): *Digital Transformation of the Economy: Challenges, Trends and New Opportunities*, pp. 542–556, 2019.
https://doi.org/10.1007/978-3-030-11367-4_54

**Table 1.** The Samara Region's rank in the innovative development of Russian regions according to leading rating agencies

| Indicators | 2014 | 2015 | 2016 | 2017 |
|---|---|---|---|---|
| Rating of Expert RA | 11 | 11 | 11 | 12 |
| Rating of the Agency for Innovative Development of Regions, including the following blocks: | 15 | 14 | 10 | 9 |
| • research and development | 12 | 18 | 15 | 20 |
| • innovative activity | 19 | 17 | 16 | 23 |
| • socio-economic environment for innovative activity | 23 | 12 | 13 | 8 |
| • innovative activity | – | – | 7 | 4 |
| Russian Regional Innovative Index by National Research University Higher School of Economics, including the following blocks: | 25 | 20 | 25 | – |
| • scientific and technological potential of Russian regions | 20 | 21 | 20 | – |
| • innovative activity | 33 | 19 | 33 | – |
| • socio-economic environment for innovative activity in a region | 6 | 6 | 6 | – |
| • quality of innovative policy | 36 | 39 | 36 | – |

*Source:* compiled by the authors based on [2–6]

However, the transition to a digital era in our country is impossible without implementing proper regional scenarios, transforming regional economies based on a developed innovative potential with science and education playing a major role in it. Such regions include the Samarskaya Oblast (the Samara Region) that is a region with ambitious strategic goals and a leading role in national innovative development. However, the current methodology of estimating regional innovative development lacks a unified approach to measurements, which is proved by the ranks of the Samara Region in the innovative development of Russian regions that are highly differentiated depending on various rankings (see Table 1).

The goals of digital transformation require to define fundamentally and quantify the category reflecting the level of the innovative development of regional economy being a necessary condition and a prerequisite for developing digital technologies and digital economy based upon them. We believe that 'regional innovative activity' is the most appropriate concept used nationally and worldwide to estimate the innovative activity of regional systems combining various approaches to quantifying. Thus, the Russian Federal State Statistics Service uses the indicator 'innovative activity f businesses defined as the ratio of businesses implementing technological, organizational and marketing innovations in the reporting year of the total number of businesses across Russian regions [7]. The Expert RA National Rating Agency estimates the level of the innovative activity of Russian subjects within the estimation of regions investment attractiveness measured as the number of patent applications and developed technologies advanced for Russia [8]. The Institute for Statistical Research and Knowledge-based Economy of the National Research University Higher School of Economics makes calculations of the Russian Regional Innovative Index. It uses four groups of subindices with innovative activity of businesses characterizing the one of a Russian subject. This calculation involves the ratio of businesses implementing technological and non-technological innovations, possessing their own internally developed technological innovations, participating in joint projects of doing researches and development. The Center for Strategic Research 'North-West' measures the level of regional innovative activity across various stages of innovative process [9]. The cross-classified statistical data by four groups (human resources; the development of new technologies; knowledge transfer and use; market deployment of innovative production) has inherent difficulties related to limited information bases of current Russian statistics and the principles of classifying indicators by various stages of an innovative process. The Rating of Russian Innovative Regions developed by the Agency for Innovative Development of Regions includes a component 'Innovative Activity of a Region'. The following indicators are measured: the investments from federal budget into the innovative sphere of regional economy in relation to Gross Regional Product (GRP); the financial support of innovative projects by federal development institutes; participation in contests organized by federal executive authorities and federal development institutes; winning contests organized by federal executive authorities and federal development institutes; the presence of innovative infrastructure including the state-founded one; performing public innovative activities.

The system of indicators reflecting the level of the innovative activity of different countries has been developed by a number of international organizations based upon the most widely used approaches.

1. The European Innovation Scoreboard (EIS) has developed the system of indicators estimating the innovative activity of different countries, in 2018 including 27 indicators clustered in 4 classes. Framework conditions, Investments, Innovation activities, Impacts are presented by the indicators of employment in innovative industries and impact of innovative production sale are used in comparative analysis of the EU countries' innovative activities to compare them with US and Japanese similar indicators [10].

2. The OECD Methodology differentiates the proportion of high-tech industries in manufacturing production; the amount of investments into knowledge sector (including spending on higher education, research and development, including software development); the design and production of telecommunications equipment; software and services; science and high-tech employment. The methodology reflects the scope and pace of development of innovative sphere in the development of OECD countries and non-members – Russia, Brazil, India, China, etc. [11].

Ambiguity exists in defining and quantifying the level of innovative development via various factors such as economic digitalization, training the staff for digital economy, the model and strategy of providing digitalization with higher education programs. Digital economy is an economic activity, with digital data being a key productional factor, which processing and use comparing with traditional economies enables to significantly increase the efficiency of various types of production, technologies, equipment, storage, sale, delivery of goods and services [12].

Digital economy in a country and its regions can obviously be created when key digital directions are implemented and staffing needs are provided with graduates of specialized university programs. To carry out this task, Russia developed and tested Digital Economy Country Assessment (DECA) [13] in 2017, which is the World Bank's product elaborated jointly with the Institute for Information Society Development with Russian experts. DECA multivariate model includes the blocks reflecting the readiness, use and impact of digital transformation on national socio-economic processes, which apparent advantage is the possibility to use it on national and regional levels of Russian economy. The methodology was tested in Russian and pilot regions (the Ul'yanovskaya Region and the federal city of Sevastopol'). The level of digital economy development in Russia was estimated 'average +', in the Ul'yanovskaya Region 'average'. An important result of DECA methodology was in defining a crucial problem of Russian economic digitalization – the shortage of highly qualified staff able to meet the challenges of digitalization, develop new digital platforms and technologies, widespread adoption of existing digital tools into productional processes and management.

## 2 Materials and Methods

### 2.1 Theoretical Foundations for Providing Regional Economic Digitalization with Higher Education Programs

The international experience shows that, staffing digital economy implies a straightforward model of providing digitalization with higher education programs and the strategy for its implementing in a specific regional environment, with the 'triple-helix model' being the most wide-spread in theoretical foundation and practical realization. Worked out by Henry Etzkowitz and Loet Leydesdorff, triple helix model (TH-model) as a set of interactions between academia, industry and governments offers opportunities for the administration able to provide the digitalization of regional economy with innovational initiatives [14–16]. In this model, an innovative initiative is the main tool of interaction and equal prerogatives to go with universities, industries located in a region and territorial governments being elements of the helix interacting within a scheme similar to the one of the human DNA.

The basic content of the TH-model may be expressed through the theses: (a) universities, businesses, governments are institutional spheres (or spaces) with their traditional missions playing new role in a knowledge-based economy. In the cource of interaction, each subject functions as another institute and forms a common space previously subjected: the one of knowledge – by universities, the one of agreement – by governments, the one of innovations – by businesses; (b) there are various ways to reach a balanced cooperation of these institutional spheres: administrative command, market, hybrid; (c) innovations result from the interaction (intersection) of these institutional spheres leading to the transformation of universities from educational and scientific and research organization into entrepreneurial ones, hybridization of university missions, the change from traditionally individual education to a group approach.

Separately, we should note the substantive aspects of the TH-model relate to higher education development. The borders of TH-model include intersection of knowledge spaces, agreements, innovations increasing the circulation of human resources – the scholar move from universities to businesses and government authorities – from governments to universities, which creates new ways of academic work and the model of Professor-of-Practice (P-o-P). Universities transform themselves from educational and research organizations into entrepreneurial ones, 'hybridize' their missions in the context of developing a specific resource – regional growth drivers, which is a key foundation in modelling the provision of digitalization with higher education programs.

Managerial capabilities account for both TH-model content and principles: (a) the key role of universities in the collaboration of businesses and governments in an innovative system; (b) the twin objectives of generating a new knowledge in providing innovative development, increasing enterprise competitiveness situated within a territory, regional socio-economic development [17].

The study of TH-model limitations in providing regional digitalization with higher education programs and the strategy of its realization in certain regions is based on scientific insights by Russian and foreign scholars researching into the role of higher education in socio-economic development factors within the theory of knowledge-based economy [18–27], in the light of high-tech innovations development [28].

According to foreign researchers, practical value of TH-model is not well studied yet as it needs further clarifying and development within a certain application – in social and economic development, in the development of higher education and an innovative sector. Thus, Botot and Savinsky [29] emphasize the fact that TH-model is more related to regional economic development rather than national policy of innovative development. Carayannis and Grigoroudis [30] agree with these and expand TH-model with the society as a final consumer of innovations and emphasize territorial management and 'smart specialization' – a national or a regional strategy for innovative activity establishing the priorities in innovative potential development according to business priorities. Runiewicz-Wardyn [31] combines a territorial aspect of TH-model with an innovative one. When analyzing the problems of European regional clusters, the author points out the disadvantages of cooperation within innovative systems, inadequacy of cooperation structure and networks resulting in European clusters backlog from the US ones able to overcome with TH-model. Haines and Olson [32] consider TH-model as an innovative driver. When researching into Silicon Valley structure, the authors develop its ecosystem model of nine TH-model components and emphasize the cooperation between innovations and social legitimacy in creating start-ups, show the driver generating a stream of innovation worldwide extending the model with the forthcoming stages of creating innovations globally and locally. McCray [33] points out the efficiency of TH-model in defining the policy of developing certain sectors of an innovative sphere, particularly nanotechnologies.

A vast group of researchers notes the value of TH-model in improving higher education policy, developing universities as a key source of innovations, creating research units within universities (King and Douglass [34], Schuetzenmeister [35], Maki [36]). Douglass [37] studies knowledge-based industries resulting from knowledge and US and Californian universities as 'flagships' in creating new growth ecosystems. Slota and Bowker [38] research into an effective form of innovative initiative within TH-model – a cyberinfrastructure (CI) as the main change agents in large-scale distributed intensive Big Data processing and joint scientific activity. Russian scholars (Suslova [39], Borisoglebskaya and Mikhailov [40] analyze possible use of TH idea in a wide range of issues related to the development of higher education and knowledge-based, skill and technology intensive industries. Malyshev's paper [41] is focused on a regional aspect of TH-model – the cooperation of government, business and universities in frontier areas. The author emphasizes a growing role of government in the strategy of Russia's economic modernization, programming innovative development within TH-model. Surovitskoy and Gerasimova [42] research into the efficiency of the actors of regional innovative systems based on TH-model. Burdakova et al. [43] study regional technological entrepreneurship.

## 2.2 The Methodology for Modelling and Developing the Strategy

Despite the wide range of research, there is no study reflecting the use of TH-model in managing the development of academic programs of a certain university or a university network.

However, TH-model provides digital economy staffing with broad methodological possibilities. The first among them is the practice of Massachusetts Institute of Technology (MIT) establishing standards for realizing TH-model in developing science as entrepreneurship and demonstrating the trends for academic programs development associated with socio-economic development, urgent economic growth areas including digitalization. MIT has pioneered the development of academic programs shaping the competences in fundamental issues and transforming digital economic ideas into practical products. MIT academic programs for high professionalism of its graduates based on deep theoretical knowledge within the technological chain 'education – research – innovations'. Importantly, higher education programs aimed at digital economy are realized according to the directions of scientific laboratories of Faculty for Electrical Engineering and Computer Science (EECS), Institute for Data, Systems, and Society (IDSS). EECS holds studies within Computer Science and Artificial Intelligence Laboratory (CSAIL), Microsystems Technology Laboratories (MTL) and implements training programs in Information Systems and Schemes, Applied Physics and Instrument Engineering, Biomedicine, Informatics (artificial intellect, information systems, information theory; Big Data, Fundamental Biology and Medicine (bio-EECS), computer networks and computer security, electrical power engineering, multi-core processors and cloud computing, nanotechnologies and quantum information-processing, robotic engineering, wireless networks and mobile computing. The IDSS members are engaged in the study of information and sociotechnic systems, analytical sciences (statistics, information and decision-making theory, social and institutional behavior), general problems (sustainable development and risk, design and architecture, politics, managerial decisions) which enabled to organize academic programs in the scope of energy supply, finances, healthcare, social relations, urbanization [44].

The second example of using TH-model in providing regional economic digitalization with higher education programs is based the practices of European countries, which together with Russia implement a continental model of higher education. In contrast to the Anglo-Saxon one, the European system of higher education measures up multi-layered administrative and territorial structure of EU countries, mixed economy, progressive advance of digitalization. Among all European countries, French higher education programs are constantly attractive for French and foreign students, in high demand of employers for their effective cooperation with local community – businesses and local governments. Their main feature ensuring digitalization of France's economy is equal distribution across the country and relevance to industrial specialization of regions (see Table 2).

**Table 2.** Geographic distribution of Master Degree Programs in France by training programs 'Communication Digital/Information technology and networks' in the national ranking Eduniversal Group in 2017/2018 academic year

| Specialization | Top-3 programs |
|---|---|
| Communications | NEOMA Business School, **Reims**, **Rouen**, **Paris**<br>Université Paris-Dauphine - Paris 9, **Paris**<br>IAE Aix-Marseille Graduate School of Management, **Marseille** |
| Social media, communication management | CELSA Sorbonne-Université, **Paris**<br>IIM - Institut de l'Internet et du Multimédia Léonard de Vinci, **Paris**<br>Ecole des mines d'Alès (en partenariat avec le CELSA-Sorbonne), **Alès Cedex** |
| Big Data | Grenoble Ecole de Management/Grenoble INP – Ensimag, **Grenoble**<br>TELECOM ParisTech, **Paris**<br>Université de Reims Champagne-Ardenne, **Reims** |
| Cyber security; IT-security | CentraleSupélec/IMT Atlantique, **Rennes**<br>INSA Lyon, **Lyon**<br>Groupe ESIEA, **Paris** |
| IT-decisions | EISTI (École Internationale des Sciences du Traitement de l'Information), **Cergy-Pontoise**<br>Université Panthéon-Assas (Paris II), **Paris**<br>Université d'Orléans, **Orléans** |
| IT and systems | CentraleSupélec, **Gif-sur-Yvette**<br>INSA de Lyon, **Lyon**<br>ENSEA et Université de Cergy-Pontoise, **Cergy-Pontoise** |
| 3D games and video | Gobelins - L'Ecole de l'Image, **Paris**<br>IIM - Institut de l'Internet et du Multimédia Léonard de Vinci, **Paris**<br>ESGI - École Supérieure de Genie Informatique, **Paris** |
| Web management, Web-strategies | IIM - Institut de l'Internet et du Multimédia Léonard de Vinci, **Paris**<br>Ionis School of Technology and Management, **Ivry- sur-Seine**<br>IAE Montpellier, **Montpellier** |
| IT systems management | Université Paris-Dauphine, **Paris**<br>ESSEC Business School/TELECOM ParisTech, **Paris**<br>CentraleSupélec, **Gif-sur-Yvette** |
| Telecommunications and networks | TELECOM ParisTech, **Paris**<br>IMT Atlantique, **Cergy-Pontoise** |

*(continued)*

**Table 2.** (*continued*)

| Specialization | Top-3 programs |
|---|---|
| | Aix-Marseille Université, **Marseille** |
| Digital marketing and analysis | Audencia Business School, **Nantes** |
| | NEOMA Business School, **Paris** |
| | EMLV/IIM – Institut de l'Internet et du Multimédia, **Paris** |
| E-commerce | IIM - Institut de l'Internet et du Multimédia Léonard de Vinci, **Paris** |
| | EFAP, Lyon, Bordeaux, **Lille** |
| | SKEMA Business School, **Lille, Paris** |

Source: compiled by the authors based on [45]

The analysis of their distribution across the country implies that France could avoid the concentration of master's degree programs providing staff for digital economy in the metropolitan region. The proportion of programs aimed at staffing various areas of digitalization and realized in Paris accounts for less than 44%. Clearly, the country has developed an equal territorial localization of these academic programs providing the development of both the metropolitan region and the territories beyond Île-de-France [45].

The model for providing digital economy with higher education programs and its worldwide use enables to draw the following conclusions:

(a) successful practices of digitalization in Europe and the US reaffirm the validity of modelling as a method to provide digital development of these countries with higher education programs;

(b) a fundamental requirement for modelling is the use of TH-model as a method based on innovative initiative and equitable cooperation of subjects in the continuum 'universities – businesses – governments';

(c) the strategy for realizing a TH-model should take into consideration both a general trend of innovative development and the environment of developing digital economy in a certain Russian subject.

# 3 Results

## 3.1 The Conditions for Developing a Model: The Directions of Economic Digitalization of a Region and Its Industries

The development of TH-model in providing the digitalization of regional economy with higher education programs is based on an innovative initiative – an equal prerogative of universities, businesses, governments. The form of an innovative initiative of local governments is the Strategy of socio-economic development of a Russian subject – a document of prospective planning defining scenarios, direction, goals of regional digitalization within the system of strategic vision (Table 3).

**Table 3.** Digitalization of economy according to the Strategy of socio-economic development of the Samara region until 2030

| Vision | Strategic priorities |
|---|---|
| Scenarios | **'Transformation'** •the growth of IT sector in regional economy |
| | **'Base scenario'** •outstripping growth of telecommunications industry |
| | **'Target scenario'** •the highest contribution of telecommunications industry into gross regional product |
| Directions | **'The development of info- and telecommunications'** |
| | • virtual reality and business technologies: |
| | • artificial intelligence, robotization; |
| | • Big Data; |
| | • highly-efficient computer systems; |
| | • intelligent interface, visual telecommunication; |
| | • information security; |
| | • information networks |
| Strategic goals | **Digital technologies in administration** |
| | 'digital economy'; 'digital society'; digital megapolis'; 'digital cities'; 'digital government'; 'digital companies'; 'digital production';'digital university' |

Source: compiled by the authors based on [46, 47]

An innovative initiative of businesses is realized through industrial strategies – the schemes and strategies of regional sectors, clusters, economies taking into account the national policy towards digitalization and features of local markets of digital platforms and technologies (see Table 4).

**Table 4.** Digitalization of industrial economic systems of the Samara region (examples of sectoral strategies)

| Sectoral systems | Strategic directions |
|---|---|
| Transport and logistics cluster | • integrated digital space for transport complex: |
| | • intelligent transport systems; |
| | • cross-cutting digital technologies of transport and logistics process; |
| | • integrated Internet-based technologies for traffic systems and flows; |
| | • GLONASS – Global Navigation Satellite System; |
| | • integrated system of digital technological telecommunications 'digital railroad'; |
| | • blockchain technologies in logistics and delivery chains |
| Healthcare | • neuroelectronic interface; |
| | • virtual reality; |
| | • highly-efficient computing and artificial intelligence |

Source: worked out by the authors based on [48–50]

The strategic directions of regional economic and sectoral digitalization specifies parameters for the model of providing digitalization and the strategy of its realization with academic programs. In fact, these directions refer to two key shifts to digital economy: (a) the change in the balance between informatization processes (technologies associated with IT-systems mainly locked onto a corporate level) and digitalization (technologies involving the Internet access) to favour the latter; (b) the selection of the most demanded digital directions within digital platforms and Big Data technologies, neurotechnologies and artificial intelligence, blockchain technologies, quantum technologies and industrial intelligence, robotic science and sensory technologies, DECT, virtual and augmented reality.

### 3.2 The Model and Strategy of Providing Regional Economy Digitalization with Academic Programs

The universities localized within the economy of Russian regions are realizing their innovative initiative in digitalization through academic programs whose graduates develop their professional competencies through researches in university laboratories, engineering and project centers. The efficiency of this approach is proved by a complex experience: MIT practice in the US aimed at organizing academic programs within their research, equal distribution of academic programs of digital communications/information technologies and networks across a country in Europe; the development of a university laboratory base and the content of research in digitalization – in Russia (see Table 5).

**Table 5.** The reflection of the digitalization of economy in the research activity of universities in the Samara region (example)

| University | University research directions in digitalization |
| --- | --- |
| Samara National Research University named after S.P. Korolev | **Research laboratories and engineering centers:**<br>• additive technologies;<br>• energy efficiency technologies;<br>• research automated systems;<br>• geoinformatics and information security;<br>• microelectronics;<br>• breakthrough technologies of remote sensing;<br>• radioelectronic systems and devices;<br>• electronic instrument engineering and automation;<br>• fundamental research and innovative technologies;<br>• intelligent aerospace systems;<br>• biotechnical and radio electronical systems;<br>• strain measurement and telemetering;<br>• optoelectronic measurements;<br>• modelling the processes of data management and computing |
| Samara State University of Economics | **Scientific and research programs:**<br>• improvement of training methods for digital economy;<br>• IT in managing the development of a biosphere reserve, working out spatial strategies, providing executive authorities, measuring market attractiveness of goods and services, traffic flow organization, event tourism (the case of Football World Cup in 2018), household waste recycling, developing communicative strategies;<br>• modelling corporate IT systems, electronic services, intelligent IT systems |

Source: compiled by the authors based on [51]

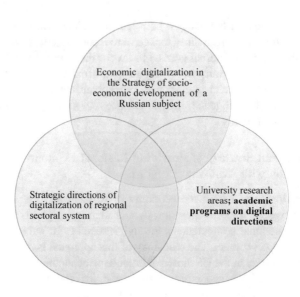

**Fig. 1.** TH-model in providing digitalization of regional economy with academic programs
Source: compiled by the authors

The consensus of the participants in the chain 'universities - businesses - governments ' on strategic objectives in economic digitalization enables to clarify the borders of TH-model and develop on its basis the model of providing digitalization of regional economy with academic programs - the processes of creating and establishing digital platforms and technologies in the economy of a region and its industries (see Fig. 1).

The strategy of realizing the model consists in a clear identification of the needs in higher education programs as a specific resource of economic development created by an innovative initiative of all participants of the chain 'universities - businesses - governments' (see Table 6).

**Table 6.** The strategy of realizing TH-model in providing regional economic digitalization with higher education programs

| Digital directions | The structure of academic programs |
|---|---|
| Computer and IT sciences (block 1) | Computer sciences; fundamental informatics; IT systems and technologies; IT systems administration |
| Computer and Information Sciences, Information Security (block 2) | Program engineering; cybersecurity; applied computing and new industrial technologies; quantum technologies; industrial Internet; robotics and sensorics; neurotechnology and artificial intellect; the technologies of wire-free communication; the technologies of virtual and augmented realities; blockchain systems |
| Economy, management, marketing, law of information systems (block 3) | Digital economy; corporate IT systems; Big Data; Data Analytics; Digital Marketing; Digital Communications; E-Business; IT-decisions; IT of engineering systems; 3D games and video; Web management; Web-strategy; information management systems; cyber security; legal protection of digital technologies |

Source: compiled by the authors

The academic programs staffing digital economy are being realized with a view of digital directions (block 1), the introduction of digital platforms and technologies (block 2), economic, managerial, marketing, legal обеспечение of digital business processes (block 3).

## 4  Discussion

The development of an economy with digital data being a key production factor is a long process with businesses, universities, governments as its subjects. Low efficiency of at least one of them withdraws a necessary support of digitalization and makes impossible the transfer to digital economy. The analysis of global practices in digital development привел к пониманию the fact that the system of higher education may become a weak point in this system with highly qualified staffing of digital economy within academic programs. Russia would not avoid the problem despite the fact that academic courses in informatics and information and communication technologies are being fully and successfully realized, as academic programs in higher education do not fully comply with the needs of digital economy. This risk is emphasized in scientific research, complex decisions within strategies, plans and programs being realized in Russia on a territorial and an industrial level, and the creation of necessary and sufficient conditions for its bridging – in a number of university strategies. The review and systematization of foreign and Russian researches as well as the author's studies of the trends in an innovative development showed that modelling and developing the strategy of realizing a model within specific local environment is a mainstream in solving the problem of providing economic digitalization with academic programs. The analysis of scientific sources enabled to develop theoretical foundations and methodology for working out the model of providing economic digitalization with academic programs based on TH-model and an innovative initiative as a means of установление cooperation within the chain 'universities – businesses – governments'. The aggregation of the strategies of socioeconomic development of the Samara region and its industrial systems allows to state that the model parameters should be defined by digital directions. The analysis of research activities conducted by Samara leading universities enables develop the model and the strategy of its realizing within the blocks of academic programs staffing economic digitalization. As the results obtained in the framework of the research, the authors believe that the methods to solve the problems of staffing digital economy based on modelling and strategizing are quite new. The authors' guidelines to organize higher education programs in view of strategic prospects of digitalization, specific digital directions, economic, managerial, marketing and legal provision of digital business processes may be of practical value.

## 5  Conclusions

In accordance with the urgency of economic digitalization of Russian regions and training within higher education system, the authors have studies theoretical foundations and foreign experience in organizing higher education programs in the context of digitalization. The study resulted in defining the approaches to estimating the level of innovative development and economic digitalization of Russian regions; classifying theoretical and methodological foundations of TH-model and best practices of its application in organizing MIT academic programs; hypothesized high efficiency of modelling the provision of regional economic digitalization with academic programs; while the assessment of France's best practices in spatial organization of academic programs enabled to draw conclusions on the necessity to work out the strategy of realizing TH-model taking into account a specific region. The author's method of modelling resulted in the model of providing digitalization with academic programs and the strategy of its realization within regional economy.

Proved by the authors the method of modelling, the model and the strategy of providing regional economic digitalization with academic programs enable to solve the problem of staffing digital economy.

# References

1. Decree of the President of Russia dated December 1, 2016, No. 642 On the Strategy of scientific and technological development of the Russian Federation. http://www.consultant.ru/. Accessed 12 Apr 2018
2. The Rating of Investment Attractiveness of Russian regions RAEX (Expert RA). https://raexpert.ru/rankingtable/region_climat/2017/tab3/. Accessed 10 Mar 2018
3. The Rating of the Association of Innovative Russian Regions (2017). http://i-regions.org/images/files/airr17.pdf. Accessed 10 Mar 2018
4. The Rating of Innovative Development of Russian Regions (2017) Issue 5, Ed. by Gohberg LM (ed) National Research University Higher School of Economics. 12, 14, 19
5. The Rating by the National Association of Innovations and IT Development, 2015, from http://www.nair-it.ru/pressabout/2015-08-03_kp.php. Accessed 10 Mar 2018
6. The Rating by RANEPA (Russian Presidential Academy of National Economy and Public Administration) 'Innovative Business in Russian regions (2018) http://www.i-regions.org/images/files/presentations/RANEPA_26.12.pdf. Accessed 10 Mar 2018
7. Federal State Statistics Service. http://www.gks.ru. Accessed 10 Mar 2018
8. The Rating of Investment Attractiveness of Russian regions RAEX (Expert RA). https://raexpert.ru/docbank/5e2/a5b/897/dd35c089e004153429d3569.pdf. Accessed 10 Mar 2018
9. Fund 'The Center for Strategic Developments 'North-West". http://csr-nw.ru. Accessed 10 Mar 2018
10. European Innovation Scoreboad (2018). https://www.ewi-vladeren.be/sites/default/files/imce/eu_innovatie_scorebord_2018.pdf. Accessed 02 May 2018
11. OECD. Science, Technology and Industry Scoreboard (2011). http://oecdru.org/zip/9211048e5.pdf. Accessed 02 May 2018
12. Decree of the President of Russia dated May 9, 2017, No. 203 On the Strategy of information society development in the Russian Federation for 2017–2030. http://www.garant.ru/products/ipo/prime/doc/71570570/#ixzz5SsGf9m00. Accessed 10 Mar 2018
13. Yershova TV, Khohlov YuV (2018) Digital Economy: may it be developed in a specific region. http://www.acexpert.ru/events/x-ezhegodniy-kongress-malogo-i-srednego-biznesa-st.html. Accessed 10 Mar 2018

14. Etzkowitz H, Leydesdorff L (1997) Introduction: Universities in the global knowledge economy. In: Etzkowitz H (ed) Universities and the Global Knowledge Economy: a Triple Helix of University-Industry-Government Relations. Pinter, London and Washington
15. Etzkowitz H (2002) Incubation of incubators: innovation as a triple helix of university-industry-government networks. Sci Public Policy 29(2):115–128. https://doi.org/10.3152/147154302781781056
16. Etzkowitz H (2002) MIT and the rise of entrepreneurial science. Routledge, London
17. Etzkowitz H (2011) Tripple-helix model. Innovations 4:5–10
18. Machlup F (1962) The production and distribution of knowledge in the United States. Princeton, New-Jersey
19. Shumpeter J (1982) Theory of Economic Development (Study of Entrepreneurial Profit, Capital, Interest and Conjuncture Cycle), Transl. from Germ. by V. S. Avtonomov et al. Progress, Moscow
20. L'vov DS, Pugachev VF (2001) Mechanism of steady economic growth. Economics of present-day Russia 4:52–58
21. Makarov VL (2003) Knowledge economy: lessons for Russia. Bull Russ Acad Sci 73 (5):450–456
22. Suharev OS (2008) Basic principles of institutional and evolutional economics. Vysshaya shkola, Moscow
23. Gaponenko AL, Orlova TM (2008) Knowledge management: how to convert knowledge into capital. EHksmo, Moscow
24. Innovational Development: Economy, Intellectual Resources, Knowledge Management (2009) ed. by Mil'ner BZ, INFRA-M, Moscow, Russia
25. Popov EV, Vlasov MV (2009) Institutes of knowledge-based minieconomics; ed. by akad. VL Makarov. Academia, Moscow, Russia
26. Nureev RM (2012) On the way to understanding institutional nature of innovations. J Inst Stud 4(2):4–10
27. Nureev RM (2012) Russian economy: problems of innovation mode formation. Hum Cap Vocat Educ 1:18–31
28. Oveshnikova L, Lebedinskaya O, Timofeev A, Mikheykina L, Sibirskaya E, Lula P (2018) Studying the sector of the russian high-tech innovations on the basis of the global innovation index INSEAD. Perspectives on the Use of New Information and Communication Technology (ICT) in the Modern Economy (Editors: Elena G. Popkova, Victoria N. Ostrovskaya). Part of the Advances in Intelligent Systems and Computing book series 726
29. Botot S, Savinsky D (2011) Triple Helix model in Great Britain's, US and Russia's regional development. Innovations 4:43–46
30. Carayannis E, Grigoroudis E (2016) Quadruple innovation helix and smart specialization: knowledge production and national competitiveness. Foresight STI Gov 10(1):31–42. https://doi.org/10.17323/1995-459x.2016.1.31.42
31. Runiewicz-Wardyn M (2009) Innovation Systems and Learning Processes in the EU and US Regions. Economic Research Center Kozminski University in Warsaw. https://escholarship.org/uc/item/6836g9xp. Accessed 02 May 2018
32. Haines JK, Olson JS (2015) Accelerating Innovation in Global Contexts. https://escholarship.org/uc/item/3sv830h5. Accessed 02 Oct 2018
33. McCray WP (2005) Will small be beautiful? Making Policies for our Nanotech Future. https://escholarship.org/uc/item/0vd0v7dx. Accessed 12 May 2018
34. King CJ, Douglass JA (2007) A reflection and prospectus on globalization and higher education: CSHE@50. Results of a Symposium Organized by the Center for Studies in Higher Education University of California. Berkeley. https://escholarship.org/uc/item/6gm168pp. Accessed 02 May 2018

35. Schuetzenmeister F (2010) University Research Management: an Exploratory Literature Review. https://escholarship.org/uc/item/77p3j2hr. Accessed 02 May 2018
36. Maki KM (2015) The Institutional Design for University Knowledge Transfer and Firm Creation. https://escholarship.org/uc/item/1hn0s8pv. Accessed 19 April 2018
37. Douglass JA (2016) Knowledge based economic areas and flagship universities: A Look at the New Growth Ecosystems in the US and California. https://escholarship.org/uc/item/7v10z4gk. Accessed 19 Apr 2018
38. Slota S, Bowker GC (2017) Negotiating science through policy: EarthCube, infrastructure and policy-relevant science. https://escholarship.org/uc/item/7n26p6k1. Accessed 29 Apr 2018
39. Suslova T (2011) The possibilities of realizing the idea of triple helix model. Higher Educ Russ 3:147–149
40. Borisoglebskaya L, Mikhailov V (2015) The study of efficient forms of innovative development of knowledge intensive and high-tech industries based on triple helix model based Latin American countries. Innovations 9(203):78–86
41. Malyshev E (2012) The peculiarities of cooperation of government, business and universities in triple helix model. Bull ZabGU 9(88):103–112
42. Surovitskaya G, Gerasimov S (2014) The efficiency of actors in regional innovative systems. Creat Econ 11(95):74–83
43. Burdakova G, B'yankin A, Vakhrusheva V (2017) The development of technological entrepreneurship in the region on the basis of the triple helix model. St. Petersburg State Polytechnical University Journal. Economics 10 (6):172–181. https://doi.org/10.18721/je.10616
44. Education MIT – Massachusetts Institute of Technology. http://web.mit.edu. Accessed 19 Apr 2018
45. Classement Eduniversal 2018 des Meilleurs Masters, MS&MBA (2018). http://www.meilleurs-masters.com. Accessed 06 Apr 2018
46. The Strategy of socioeconomic development of the Samara region until 2030 (2017). http://economy.samregion.ru/upload/iblock/82a/strategiya-so_2030.pdf. Accessed 10 Mar 2018
47. Working paper of the Skolkovo Centre for digitalization how to understand digitalization (2018). https://cdt.skolkovo.ru/ru/cdt/. Accessed 10 Apr 2018
48. Transport Strategy of the Russian Federation (2018). http://docs.cntd.ru/document/902132678. Accessed 11 May 2018
49. Transport innovations (2018). http://inno-trans.ru/data/documents/IT-31_inet.pdf. Accessed 10 May 2018
50. The paper of the Centre for Breakthrough Researches of the SamSMU "IT in medicine" (2018), Retrieved May, 2018, from http://smuit.ru/about/infrastructure/cpi-it-medicina/. Accessed 02 May 2018
51. Science. Researches. Internet resources by the Samara University (https://ssau.ru/), the Samara State Technical University (https://samgtu.ru/), the Samara State University for Economics (http://www.sseu.ru/). Accessed 03 May 2018

# Digital Economy in the Socio-Economic Development of Enterprises

# Innovation Activities of Enterprises of the Industrial Sector in the Conditions of Economy Digitalization

A. V. Zastupov[(✉)]

Samara State University of Economics, Samara, Russia
oiler79@mail.ru

**Abstract.** The relevance of the contribution is due to the need to study innovative development problems of industrial enterprises in the conditions of economy digitalization. In this regard, the contribution is to identify the development problems of the investment and innovative capacity of industrial enterprises in the region and ways to solve them, taking into account the need to introduce digital forms and approaches. The leading approach to this problem is the development and analysis of ways to improve the efficiency of the innovative business environment and the investment and innovative development of territories in the conditions of economy digitalization. The contribution identifies the factors of innovation activities of enterprises, forms a mechanism for managing the digital innovative development of industrial enterprises. The materials of the contribution are of practical value in the field of improving innovation activities of industrial enterprises and innovation potential of the region as a whole.

**Keywords:** Digital economy · Digital innovative development ·
Digital technologies · Digitalization · Factor · Investment resources ·
Innovations · Industrial policy · Management

## 1 Introduction

In current conditions of development, one of the most important problems is the problem of increasing the investment and innovative development of industrial enterprises in the conditions of economy digitalization in Russia [8], the development of appropriate science-based approaches and mechanisms for improving innovation activities of industrial enterprises [10], which makes the contribution relevant.

Problems related to improving innovation potential of regional enterprises, innovation activities of industrial enterprises in the conditions of economy digitalization were considered in the works of domestic and foreign authors, such as Algina [1], Izmalkova [4], Sannikova [8], Khasaev [6], Duranton [2], Ketels [5], Gawer [3], Sölvell [9], Phillips [7] and others.

According to the analytical review on this issue, the problems of improving innovation activities of enterprises in the conditions of economy digitalization, as well as organizational and economic aspects of the innovative development of industrial enterprises in the regions [11], are not sufficiently studied.

S. Ashmarina et al. (Eds.): *Digital Transformation of the Economy: Challenges, Trends and New Opportunities*, pp. 559–569, 2019.
https://doi.org/10.1007/978-3-030-11367-4_55

In accordance with this, the following goals and objectives are defined:

- Study industrial policy as an element of economic growth of the industrial complex of the country [1];
- Determine approaches to the innovative development of industrial enterprises in current conditions of a "digital" management approach;
- Formulate the main provisions for improving innovation activities of enterprises in current crisis conditions of economy digitalization [4];
- Give proposals for organizational and economic innovative development of industrial enterprises in the regions, taking into account the need to introduce digital forms and approaches.

## 2    Materials and Methods

The author of the contribution used the following methods: theoretical (analysis, classification, specification, synthesis); diagnostic (factor analysis, modeling, forecasting, expert estimates); empirical (observation, comparison, generalization).

The experimental base of the contribution was a complex of enterprises of the automotive industry, oil and gas industry, aircraft industry and the rocket and space industry of the Russian Federation.

The use of these methods contributed to obtaining specific results. From a theoretical point of view, these methods allowed studying industrial policy and from a practical point of view, these methods allowed studying innovation potential of industrial enterprises in the conditions of economy digitalization in Russia.

- Form approaches to the innovative development of industrial enterprises in current conditions of economy digitalization;
- Develop a mechanism for managing the socio-economic development of the enterprise in economy digitalization and organizational and economic measures to improve innovation potential of industrial enterprises in the regions.

## 3    Results

Industrial policy is an essential element for ensuring the economic growth of the industrial complex of the country, its modernization and digitalization in order to ensure compliance with current requirements. The optimal solution of innovation and investment problems of economic growth determine the basis of industrial policy. It should be noted that innovation and investment are in a certain relationship and cannot exist separately. Digitalization of industrial enterprises, digital innovative technical solutions in the manufacturing sector, innovative ideas and developments require considerable investments in production. At the same time, limited investment contributes to the dynamic growth in the volume of innovations required in this case in accordance with the requirements of scientific, technical, industrial, economic, managerial and informational "digital" progress.

An important component of industrial policy is structural policy, which involves certain institutional changes [1]. It is necessary for the restructuring of traditional industries, which form the basis of the industrial complex and for selective support for new promising industries. Industrial policy is implemented through certain mechanisms and tools, including innovative and investment development mechanisms, digital innovations, digitalization, promotion of investments in the form of subsidies, loans, government orders, and tax incentives. The tools of industrial policy include the regulation of foreign investment and foreign trade policy.

Industrial policy acquires great importance in the crisis conditions and in the conditions of economy reforming and digitalization. It allows changing its structure, concentrating resources in terms of their limitations for the implementation of selected target characteristics [2]. The presence of economic cycles implies the need to manage investment and reproduction processes. The implementation of industrial policy will avoid both the "overheating of the economy" and its "excessive fall".

In recent years, there has been a tendency in industrial sectors of Russian economy to reduce research and development costs. It is worth noting that large national companies in a number of industries in the Russian Federation are spending significantly lower amounts of financial resources on research and development than foreign companies of similar industries. According to Industrial Research Institute (IRI), in 2017 research and development costs in the world as a whole amounted to about $2000 billion, while a significant proportion of research and development costs (more than 60%) are accounted for the leading economies of the world - USA, Germany, Japan and China, while the share of Russia was only 2.6%. Experts predict that this trend will continue in the coming years, while in the United States and China, research and development costs will increase, resulting in the US share in global R&D costs at the level of 26–27%, while China's share will grow to 21–22% (Table 1).

**Table 1.** The share of countries in world R&D costs in 2015–2017, %

| Country | The share of countries in world R&D costs (%) | | |
|---------|------|------|------|
|         | 2015 | 2016 | 2017 |
| USA | 26.9 | 26.4 | 26.4 |
| Japan | 9.1 | 8.8 | 8.7 |
| *European union* | | | |
| Germany | 5.7 | 5.6 | 5.5 |
| *BRICS countries* | | | |
| China | 19.1 | 19.8 | 20.2 |
| Russia | 3.0 | 2.7 | 2.6 |
| Brazil | 2.1 | 2.0 | 1.9 |
| India | 3.4 | 3.5 | 3.7 |

Source: compiled by the author.

Innovation industrial policy as part of the socio-economic policy of "digitalization" is able to determine goals, priorities and the mechanism for implementing the innovation strategy in the country. Government regulation of innovation policy is expanding during periods of depression and recovery, as it becomes necessary to form the basis for the transition to the growth stage. In the conditions of market self-regulation, when most innovations of improving character are carried out by entrepreneurs at the expense of their own resources, the state can limit itself to creating a positive innovation climate, maintaining freedom of competition in terms of intellectual property, and protecting it. Basic innovations, especially in periods of depression and recovery, cannot be effectively implemented on a spontaneous market basis. They require government support. The implementation of basic innovations and the formation of new digital technologies radically change the nature of production and its competitive advantage [5]. Therefore, the state has to effectively intervene with its resource support and coordinate organizations that implement innovations.

An important role is played by industrial policy to ensure the success of innovation activities of industrial enterprises, and therefore it seems necessary to determine the factors of innovation activities of enterprises in the conditions of a digital business environment (Fig. 1).

The most important task of industrial policy concerning its coordination with innovation is the formation of demand from industrial enterprises for digital innovations. In the absence of demand for innovation, there is no need to support structures for innovation. Stimulating supply without stimulating it to effective demand is largely becoming unproductive spending. The stretching of a temporary innovation lag is, of course, acceptable, but this can lead to a loss of innovation in development and cash. The results of research and development (R&D) can also be lost to domestic industry and be used by foreign competitors.

In order to accelerate scientific and technological achievements in industrial enterprises and innovation potential in the conditions of economy digitalization, it is necessary to:

1. Improve the scientific and technological investment structure in favor of research and development of technological marketing at the stage of creating scientific and technical reserve [9];
2. Accelerate the pace of investment in R&D, including at the expense of funds from targeted scientific and technical programs of national importance;
3. Implement reserves for the development of dual-use technology in the sub-sectors related to the military-industrial complex;
4. Improve methods for assessing the efficiency of high-tech digital technologies in the leading sub-sectors of the industry [8];
5. Improve pricing methods in digitalization, taking into account the diversity of market factors;
6. Carry out scientific analysis of the production cycle of digital technologies;
7. Develop methods for the integrated assessment of the socio-economic effect of digitalization in production and management of production [7];
8. Introduce digitalization in the nano-industry in order to create new materials and products with predetermined distinctive features for the market.

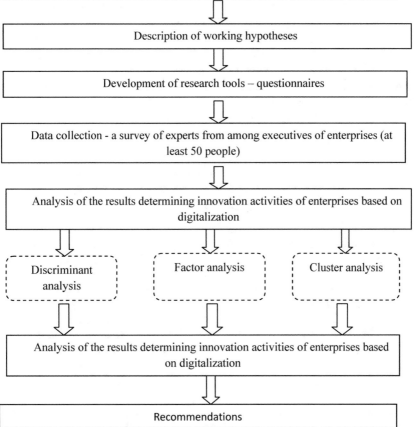

**Fig. 1.** Technology for determining innovation activities of enterprises in the conditions of the digital business environment (Source: compiled by the author)

In current crisis conditions, an important aspect limiting innovation activities of industrial enterprises in Russia is the limitation of financial resources and a high level of risk of innovative projects. The current situation is due to several economic and financial factors, such as instability of demand for industrial products, dynamic legal and tax regulation, currency fluctuations, volatility of the national currency, world energy prices, and others. Trade and economic sanctions on Russia by a number of Western countries, primarily the United States and the European Union, has led to the fact that many niches in the industrial products market have become less competitive and are free that partially eliminates the lack of demand in some sectors.

In this regard, it is important to create a mechanism for managing the innovative development of the enterprise, taking into account economy digitalization and approaches (Fig. 2).

The main production factors in the modern innovative economy include not only labor and capital, but also R&D with the transition from periodic implementation to a continuous process. It should also be noted that innovation activities of economic systems is characterized by their special dynamic relation between demand and supply, in contrast to their dynamic production of standard goods. In the latter case, high demand for goods causes an increase in their production and supply in the market. As innovation demand develops, output also increases, its assortment list expands, but to a certain limit. Any innovation requires long-term R&D, and the time for its development, as a rule, is much longer than for modernization of production facilities. Therefore, demand for innovation, ready for industrial implementation, will always be greater than supply in the sound market economy.

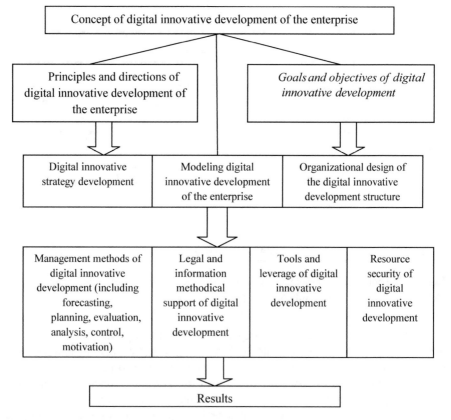

**Fig. 2.** Management of innovative development of the enterprise taking into account economy digitalization and approaches (Source: compiled by the author)

According to experts, the issues of forming innovation technological potential are not being studied sufficiently [6]. The interaction of factors determining dynamic technological resources in the conditions of economy digitalization is poorly studied. Many forms of material incentives for research specialists in engineering departments of enterprises, where applied R&D is performed, do not work. Innovation activities are slowly improving, and this especially applies to the pricing of scientific and technical products in the "developer-producer-consumer" system.

As noted above, the amount of R&D costs by Russian companies is significantly lower than in foreign companies [10]. So, even in the most knowledgeable industries such as the aviation and rocket and space industries, the share of R&D costs from the revenue of the leading Russian companies such as United Aircraft Corporation and Russian Helicopters was only about 1.5% in 2017. This indicator value is seriously lower than R&D costs of Airbus (5.9%), Embraer (5.6%), Boeing (3.5%). This trend is also seen in other industries. In particular, R&D costs of KAMAZ amounted to 2.6% of revenue, and AvtoVAZ - 1.2%, which is several times lower than costs of GM (5.1%), Ford (4.8%) or Renault (4.6%). It should be noted that Russian companies are lagging behind in R&D costs in the oil and gas sector, where, R&D costs of Gazprom amounted to "only" 0.2% of revenues, and Tatneft - 0.3% in 2017, which is also lower than the corresponding figures of leading companies in the oil and gas industry (Table 2).

**Table 2.** Comparative R&D costs of enterprises in selected key industries, 2017

| Foreign industrial enterprises for the following branches | Costs of enterprises (% of revenue) | Russian industrial enterprises for the following branches: | Costs of enterprises (% of revenue) |
|---|---|---|---|
| *Aircraft and rocket and space industry* | | | |
| Airbus | 5.9 | United aircraft corporation | 1.4 |
| Boeing | 3.5 | Russian helicopters | 1.2 |
| Embraer | 5.6 | | |
| *Automotive industry* | | | |
| Ford | 4.8 | AvtoVAZ | 1.2 |
| GM | 5.1 | KAMAZ | 2.6 |
| Renault | 4.6 | | |
| *Oil and gas industry* | | | |
| ExxonMobil | 0.4 | Rosneft | 0.7 |
| Chevron | 0.5 | Gazprom | 0.2 |
| Statoil | 0.6 | Tatneft | 0.3 |

Source: compiled by the author

When studying the development of innovation potential of industrial enterprises, data on the quality of innovations (innovation, "digital" technology) can be used as a tool to improve the incentive mechanism for employees of engineering and technical services, as well as to secure the efficiency of research and development activities related to improving the level of "digital" technological development of enterprises in the industry. The improved quality of basic technologies makes it possible to judge the economic and production potential of enterprises and, to a certain extent, can be a factor in the investment attractiveness of specific enterprises [3]. This improved quality can fairly characterize innovation activity. To study the trends in the quality of basic technologies (depending on the industry specifics), data are needed for a number of years, some of which can be obtained on the basis of expert assessments with the involvement of scientists from research institutes and universities [4]. The creation of a "digital" data bank on the main parameters of technology dynamics is an important condition for managing scientific and technical potential of industrial enterprises in the conditions of economy digitalization.

Currently, the largest amount of funding for economy digitalization is required for the entire cycle of scientific and technical activities, which should be used for important activities in the field of increasing innovation potential of enterprises, which include:

1. Develop information retrieval systems and a data bank on scientific and technological potential, including competitors in the countries of the West, the USA and Japan;
2. Organize targeted scientific missions and training specialists in basic universities of the country and leading foreign universities;
3. Carry out research work in research institutes, universities, industry-specific research organizations and scientific and technical centers of leading enterprises, taking into account the need to introduce digital forms and approaches;
4. Perform scientific and technical expertise of technology for monitoring the rating of critical technologies in the industry.

In the regions of the Russian Federation, which have a high share of high-tech products, it is necessary to carry out the following set of organizational and economic measures that will improve innovation potential of industrial enterprises and the region in current conditions of economy digitalization (Fig. 3).

In the conditions of digital competition in high-tech subsectors, standardization is a significant factor in the output of commodity producers to the external market. Of course, the main role here belongs to the staffing of modern high-tech "digital" technologies, the development of which is impossible without high-class specialists. In current conditions, targeted training of specialists should be funded on a priority basis at the expense of the federal budget for leading state universities. In many ways, this is the most important condition for improving the efficiency of R&D in basic industries that create complex high-tech products to ensure the technological security and defense capability of the country.

R&D takes a long time a large amount of capital investment. Taking into account the modern development features of most domestic industrial enterprises, it can be noted that there is a shortage of resources not only for continuous, but also episodic innovation. Hence the important task of industrial policy is not just to launch a new high-tech product by a certain enterprise, but to recreate the entire R&D chain, ensuring its activities not only with financial resources, but also with staff and experimental base. Therefore, state funds for the development of industry, industrial parks, special investment contracts and other instruments of the Law "On Industrial Policy in the

Russian Federation" are important elements of industrial policy, but they are insufficient. Coordination of organizations is also necessary, often even with different forms of ownership, which are or can be included in a single R&D chain.

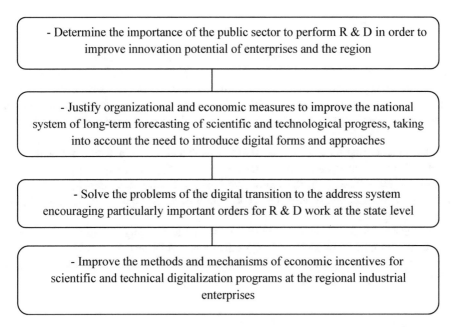

- Determine the importance of the public sector to perform R & D in order to improve innovation potential of enterprises and the region

- Justify organizational and economic measures to improve the national system of long-term forecasting of scientific and technological progress, taking into account the need to introduce digital forms and approaches

- Solve the problems of the digital transition to the address system encouraging particularly important orders for R & D work at the state level

- Improve the methods and mechanisms of economic incentives for scientific and technical digitalization programs at the regional industrial enterprises

**Fig. 3.** Directions for the improvement of innovation potential of industrial enterprises and the region in the conditions of economy digitalization (Source: compiled by the author)

Coordination with the help of state authorities in the framework of "digitization" of industrial policy is necessary not only in R&D, but also in order to recreate and maintain unified production technological processes [11]. The lack of the effective advanced marketing of modern digital technologies, determining the development of the industry, is largely determined by the insufficient level of staff and information support and insufficient use of foreign experience.

Today, at the level of individual regions of the Russian Federation, inter-sectoral scientific and technical centers are being created. By concentrating research and staff potential and investment resources, these centers could carry out inter-sectoral research, based on the needs of leading enterprises, focusing on digital forms and approaches. The purpose of such centers should be the support and promotion of innovative and high-tech projects involving the development and commercialization of unique products and technologies in the conditions of economy digitalization. In particular, in the Samara region, "Zhiguli Valley" High-Tech Technopark has been implemented and operates as a unique digital business environment for the development of high-tech companies and innovative projects in the region, which attract large international leading companies in the field of innovation. This project is aimed at creating a comfortable business environment for innovative digital development and modernization of the economy of the Samara region. To achieve high rates of scientific and technological development of industry in the region, an optimal concentration of public funds in promising areas of high-tech digital technologies is needed.

## 4   Discussion

The contribution studies the theoretical and practical aspects of digital innovative development of enterprises in the industrial complex. The use of theoretical and methodological guidelines and practical recommendations of the contribution will help make informed decisions in the field of digital innovative development of sectoral industrial enterprises, which indicates that the obtained results are consistent with the research hypothesis.

The practical significance of the contribution is that the main results can be used by state and municipal bodies and the business community in developing guidelines for improving the efficiency of the innovative business environment and investment and innovative development of territories, taking into account the need to introduce digital forms and approaches.

Further research in this area will expand the understanding of applying theoretical and methodological provisions and practical recommendations in the field of improving innovation activities of industrial enterprises and innovation potential of the region as a whole, focusing on the results achieved in economy digitalization.

## 5   Conclusions

The contribution presents the main results, conclusions and recommendations on the problems of increasing innovation activities of industrial enterprises and improving innovation potential of the region in the conditions of economy digitalization.

The scientific results obtained fundamentally complement the existing approach to the problem of improving innovation activities of industrial enterprises, assigning the leading role in this process to managing the mechanisms of the digital approach to innovative and investment development of territories.

## References

1. Algina MV, Bodnar VA (2010) Institutional instrument of monitoring and activization of innovative activity of industrial enterprises. Econ Anal Theory Pract 42:18–25
2. Duranton G (2014) California dreaming: the feeble case for cluster policies. Rev Econ Anal 3:3–45
3. Gawer A (2014) Bridging differing perspectives on technological platforms: toward an integrative framework. Res Policy 43(7):1239–1249
4. Izmalkova SA (2013) The choice of the perspective directions of innovative development of regions and their realization by means of formation of technological platforms. Reg Econ Theory Pract Issue 1:29–34

5. Ketels C (2013) Recent research on competitiveness and clusters: what are the implications for regional policy. Camb J Reg Econ Soc 6:269–284
6. Khasaev GR, Mikheev UV (2003) Clusters – modern instruments of increase of competitiveness of the region (through partnership to the future). Part 1. COMPASS of industrial restructuring, pp 5–6. http://compass-r.ru/st-5-03-1.htm
7. Phillips F, Linstone H (2016) Key ideas from a 25-year collaboration at technological forecasting & social change. Technol Forecast Soc Chang 105:158–166
8. Sannikova IN, Tatarnikova EV (2015) Assessment of innovative capacity of the enterprise for management of development. Manag Russ Abroad 3:57–64
9. Sölvell Ö (2012) The multi-home-based corporation: solving an insider-outsider dilemma. In: Heidenreich M (ed) Innovation and institutional embeddedness of multinational companies. Edward Elgar, Cheltenham, pp 59–76
10. Zastupov AV, Streltsov AV, Tatarskikh BY, Poluyanov VB, Murtazova HM (2016) Petrochemical cluster as the engine of the region's economic development. Int J Econ Perspect 10(3):83–93
11. Zastupov AV (2017) Cluster model of development of industrial branches of economy of the region. Bull Samara State Univ Econ 3(149):13–18

# On Collaborations Between Software Producer and Customer: A Kind of Two-Player Strategic Game

I. V. Yakhneeva[1]([⊠]) [iD], A. N. Agafonova[1], R. V. Fedorenko[1],
E. V. Shvetsova[1], and D. V. Filatova[2]

[1] Samara State University of Economics, Samara, Russia
rinadoo@yahoo.com, agafonova.a.n@gmail.com,
fedorenko083@yandex.ru, shvetsova.e@mail.ru
[2] WZiMK Kielce University of Technology, Kielce, Poland
dfilatova@tu.kielce.pl

**Abstract.** The digitalization of economic activity poses new challenges to both enterprises' managers and software developers. Software ceases to be a mass product, acquiring the characteristics required and defined by the consumer. The understanding of customers' needs helps the reducing uncertainty and allows the producer's development strategy selection. In this work, a model of interaction between software developers and its consumers is created. Since the primary goal of this work is an analysis of possible effects of the choice of developer-customer interaction strategies, the model is considered as a specific strategic game, and its interpretation is carried out using the game theory and decision under uncertainty tools.

**Keywords:** Cloud service · Decision making · Game theory · IT solution · Strategic game · Software · Uncertainty

## 1 Introduction

Industrial progress is an integral part of everyday reality. It dictates the necessity to find new, often innovative solutions aimed at improving the efficiency of economic activity. The result of the first industrial revolution, which occurred in the second half of the XVII century and lasted on a global scale during the XIX century, was industrialization changed the socio-economic structure of society, of the second one (from the middle of the XIX century to the middle of the XX century) forever, was electrification, of the third one (from the middle of the XX century to the beginning of the XXI century) was computerization. It should be noted that the observed in our days the fourth industrial revolution is characterized primarily by digitalization, which accelerates transformation processes. The response on the arising question about backgrounds of this phenomenon could clarify the situation and show ways of the adaptation to new and often ambiguous conditions of economic reality.

Any socio-economic transformations require a specific social consciousness background. Thus, starting from the second half of the XXth century, there is an

© Springer Nature Switzerland AG 2020
S. Ashmarina et al. (Eds.): *Digital Transformation of the Economy: Challenges, Trends and New Opportunities*, pp. 570–580, 2019.
https://doi.org/10.1007/978-3-030-11367-4_56

individual perception that business success depends on quick access to information and effective methods of its processing impossible, which are impossible without employment of information technology – IT [1, 9, 16, 24]. The exponential growth of IT, explained by the laws of Guilder, Metcalf, and Moore, makes possible to use significant computer and communication capacities in development of the social and economic interactions infrastructure and particularly in the value chain [5, 11, 17]. The concept of integration of physical and software components arises (despite the heterogeneity of their space-time scales) demonstrating the plurality and the transparency of the behavioral modality in a context dependable environment and often associated with uncertainty conditions.

The context change can be explained by the consumers' behavior, whose opinion becomes decisive in shaping the conjuncture of the market and hence in the production of goods and services. Mass production is personalized, and now it is mass customization production. It is the principal fact that makes producers regularly review not only the methods of management, leading to the passage from "classical" management method to process or project one, but also the choice of an information system [21]. Analyzing the standards and the capabilities of nowadays business-oriented information systems, even such as ERP II (Enterprise Resource Planning), one can conclude that this class of the systems is likely useful in the solution of many problems arising in the context of industrial and socio-economic transformation [19, 20]. Consequently, there is still a particular demand for this kind of software on markets.

Let us pay attention to some more facts. The increasing amount of information is a direct consequence of global digitalization. It is most visible and significant for the unstructured data. For example, productive work with large volumes of information is an essential part of novel risk management strategies [8]. It is clear that under the uncertainty, the ability to manage the generated data, both structured and unstructured, is particularly important. This fact is a keystone to making the proper decisions [26]. The indispensability of new decision-making methods has no doubt. Artificial intelligence (intelligent algorithms, deep learning or Big Data analytics) is beneficial here [2, 12, 27]. In a framework of digitalization era, the cloud services (CaaS, PaaS, IaaS, SaaS, etc.), as well as their practical use for different scale businesses, cannot be neglected [22, 25]. As the demand of the customers, the emergence of new IT solutions promoting digitalization and automatization of business management forms the alternative-proposal market and complicates the customer's decision-making process.

That customer-producer collaborations paradox requires more profound analysis. Therefore, the primary goal of this work is to create a model that describes the mechanism of interaction between the software producer and the potential customer.

## 2  Methodological Framework

### 2.1  Problem Formulation

Let us characterize the software market concerning the capabilities of the software producer and the requirements of the consumer. Standard box pack software is the most popular product sold by software producers in large scale. Small and medium-sized enterprises present the majority of customers; their management processes require the implementation of standard information technologies. The expansion of the enterprise's activities, the information volume increase, and the complication of business processes lead to additional requests which could be hardly satisfied by standard software products. In consequence, the requirement of customized software development appiers. The technology SaaS (Software as a Service) is an alternative approach to customer service. Based on a cloud service it gives the ability to connect any necessary modules to the standard software. The scale of the enterprise is not essential anymore. Moreover, rejecting personalized software development customers have abilities through cloud service to create the information system which fulfills the demand of business processes. The cloud technology market is currently developing at a rapid pace. However, most of the companies prefer to build their own IT capacities.

When distributing the standard box pack software, the producer receives a relatively small income from each customer request, compensated for in the case of large sales volumes and long-term licensed services. As it is possible to see, only high volume sales could guaranty high effectiveness of the producer's activity. The concentration on mass production does not take into account the individual preferences of each consumer. The focus on mass production does not take into account the personal preferences of each consumer, that results in the losses of potential income. Developing software according to customer requirements frees the producer from this deficiency. Besides, the customer-oriented solution is as the basis for the subsequent development of a standard product aimed at a specific segment, as well as for the personalized software development for a new consumer. It is also necessary to take into account the lost profit arising from the producer's unjustified efforts for the unsatisfied customer who prefers to select the other producer. The arising problem corresponds to the following question: how to define the collaboration strategy satisfying antagonistic expectations of decision-makers?

## 2.2   Definition of a Strategic Game

The situation mentioned above is a kind of a two-player strategic game, where "software producer" and "customer" play the roles of the decision-makers who have in the disposition a set of strategies associated with certain payoff functions. Moreover, as it was possible to see, each player's payoff does depend on a list of all other decision-makers' strategies.

In particular, the strategic game consists of

- a set of players - $\{\mathcal{A}, \mathcal{B}\}$, $\mathcal{A}$ stands for the "software producer" called Player $\mathcal{A}$, $\mathcal{B}$ stands for the "customer" called Player $\mathcal{B}$;
- sets $\tilde{A}$ and $\tilde{B}$ describe, respectively, the strategies of Player $\mathcal{A}$ and Player $\mathcal{B}$ ($card\,\tilde{A} = n_A$, $card\,\tilde{B} = n_B$);
- intersections of $\tilde{A}$ or $\tilde{B}$, which are related to the certain payoffs summarized as the matrices $A = (a_{\ell k})$ for Player $\mathcal{A}$ and $B = (b_{\ell k})$ for Player $\mathcal{B}$, $\ell = \{1, \ldots, n_A\}$, $k = \{1, \ldots, n_B\}$.

We assumed that $\mathcal{A}$ or $\mathcal{B}$ chooses the strategy that is best for him or her, given convictions about the other players' strategies. The consideration was limited to the situation such that every player's belief about the other players' strategy was a correct one.

### 2.3  Decision Making Mechanism

In a context of rationality, any decision making is a choice of one strategy other others. While the subjective-objective information is hardly available under uncertainty, each such selection is charged by risk and by the impossibility of outcome's probability distribution assigning. According to decision-making theory, each player can be characterized by own decision-making mechanism. That is to say, taking the strategies of Player $\mathcal{B}$ as potential nature states and those of Player $\mathcal{A}$ as possible strategies, one can deduce the outcomes of payoff function for Player $\mathcal{A}$ (and vice versa for Player $\mathcal{B}$) using the following criteria, namely

- Hurwicz criterion (or optimist–pessimist index)

$$G(\mathcal{A}) = \max_{1 \leq \ell \leq n_A} \left\{ \alpha_A \min_{1 \leq k \leq n_B} \{a_{\ell k}\} + (1 - \alpha_A) \max_{1 \leq k \leq n_B} \{a_{\ell k}\} \right\}, \tag{1}$$

  where $\alpha_A$ is the pessimism rate, $0 \leq \alpha \leq 1$;
- Savage regret criterion with positive flow payoff (MINIMAX)

$$S(\mathcal{A}) = \min_{1 \leq \ell \leq n_A} \left\{ \max_{1 \leq k \leq n_B} \left( \max_{1 \leq \ell \leq n_A} \{a_{\ell k}\} - a_{\ell k} \right) \right\}; \tag{2}$$

- Laplace insufficient reason criterion (MAXIMEAN)

$$L(\mathcal{A}) = \max_{1 \leq \ell \leq n_A} \left\{ \sum_{k=1}^{n_B} p_k a_{\ell k} \right\}, \quad p_k = \frac{1}{n_B} \text{ for any } k \in \{1, \ldots, n_B\}; \tag{3}$$

- Wald's criterion (MAXIMAX)

$$W(\mathcal{A}) = \max_{1 \leq \ell \leq n_A} \left\{ \max_{1 \leq k \leq n_B} \{a_{\ell k}\} \right\}. \tag{4}$$

As it is possible to notice if $\alpha = 0$, then (4) is a particular case of (1). We have to admit that in the static situation (a decision is required only one time) players select only one strategy, in dynamic one (long-time relations) players can infinitely adopt and change the strategy according to the circumstances and beliefs.

## 3 The Collaboration Model of Decision-Makers

### 3.1 Decision-Makers' Strategies

To get the interaction model of decision-makers, first of all, we have to define the sets of the players' strategies.

Player $A$ has the possibility of three actions, namely:

- $A_1$ – to offer standard box pack software;
- $A_2$ – to develop software, focusing on the needs of the particular customer;
- $A_3$ – to provide cloud service to the consumer.

Player $B$ can act as follows:

- $B_1$ – to continue to collaborate with the software producer;
- $B_2$ – to refuse producer's services favoring a competitor or creating his/her own IT service.

As it is possible to notice $\tilde{A} = \{A_1, A_2, A_3\}$, $n_A = 3$, and $\tilde{B} = \{B_1, B_2\}$, $n_B = 2$.

### 3.2 The Action Profile and Payoffs of the Software Producer

The second step in the interaction model definition requires the description of a list of actions of Player $A$ and Player $B$. To get the action profiles the following characteristics of Player are used:

- $i$ is a revenue from the sale of box pack software and its licensed maintenance, $i \in \mathbb{R}_+$;
- $I$ is a revenue from customer-oriented software development, $I \in \mathbb{R}_+$;
- $q$ is a revenue from providing cloud service to customers, $q \in \mathbb{R}_+$, $q < i < I$;
- $n$ is a customer-to-provider call number (it includes the annual renewal of licensed maintenance, personalized software development), $n \in \mathbb{R}_+$, $n > 1$;
- $t$ is a penalty for the lack of new product development and the inability to withstand changing market demand, $t \in \mathbb{R}_+$;
- $s$ is a penalty for the lost client, $s \in \mathbb{R}_+$;
- $T$ is a benefit arising from gained experience in customer-oriented software development under an ability to complete faster further orders of the old or a new one customer, $T \in \mathbb{R}_+$, $t < T$;
- $Y$ corresponds to the costs of maintaining a cloud infrastructure which is sufficient to attract a significant number of customers, $Y \in \mathbb{R}_+$.

We call $\Theta_A = \{i, I, q, n, t, s, T, Y\}$ as the set of the parameters of Player $A$. Let us describe the action profile of Player $A$.

Strategy combination $A_1 B_1$. The producer sells only the standard box pack software. The consumer has been using this services for quite a long time. When using standard products, the producer frees himself from the necessity to solve complex atypical tasks. It does not require additional investments and allows receiving a steady profit quantified by the customers' number. However, issuing the standard software, the producer rejects the specific-needs customers and loses the potential profit. The payoff this strategy combination is $a_{11} = ni - t$.

Strategy combination $A_1B_2$. The producer provides the standard box pack software, and the customer leaves, refusing to renew the licensed service or purchase new versions of the program. Consumer's withdrawal can be caused by the departure to a competitor, by refusal to continue using the software or replacing it with free analogs. Besides, some companies may use unlicensed versions of programs. The producer receives a regular income from the sale of software, but the consumer does not return. The payoff this strategy combination is $a_{12} = i$.

Strategy combination $A_2B_1$. The producer develops customer-oriented software. The customer continues to use the producer's services in the future. The producer receives more significant income from the unique software product development in comparison with that from the implementation of the standard box pack software version. Additional benefits here are the experience of the producer and his improved image. Moreover, these new solution is a platform for the updates of the box pack products attracting new customers. Besides, the successful fulfillment of the order increases the probability of client's re-order. The payoff is $a_{21} = I + T$.

Strategy combination $A_2B_2$. The producer develops the customer-oriented software; however, the customer leaves, refusing to renew the licensed service. The producer increases an income. Nevertheless, these expended efforts are ineffective. The producer failed to develop long-term relationships with the customer, did not receive the additional positive image needed to find new customers. The payoff is $a_{22} = I - T - s$.

Strategy combination $A_3B_1$. The producer provides a cloud service. The consumer refers to it for various tasks, connecting the required modules as needed. The income from a single treatment is lower than that for the sale of the box pack software, but the potential number of such requests is significant. The cost of the cloud service related infrastructure maintenance can be quite large due to the obligation of the assurance of the maximum number of accesses in a real-time. The payoff is $a_{31} = qn - Y$.

Strategy combination $A_3B_2$. The producer provides a cloud service, but the consumer leaves, refusing further cooperation. Investments in the development of the information infrastructure of the cloud service are not justified and do not bring the expected profit. The payoff of is $a_{32} = i - Y$.

The payoffs of Player $\mathcal{A}$ are summarized as the following matrix:

$$A = \begin{pmatrix} ni - t & i \\ I + T & I - T - s \\ qn - Y & i - Y \end{pmatrix}.$$

## 3.3    The Action Profile and Payoffs of the Customer

Let us introduce the characteristics of Player $\mathcal{B}$ in a form of a set $\Theta_B = \{p, P, g, n, W, w, U, u\}$, where

- $p$ corresponds to the payments of the purchase of the box pack software, $p \in \mathbb{R}_+$, $p = i$;

- $P$ corresponds to the payments for the customer-oriented software development, $P \in \mathbb{R}_+$, $P = I$;
- $g$ are payments for cloud service, $r \in \mathbb{R}_+$, $g = q$;
- $n$ is a client-to-producer calls number, $n \in \mathbb{R}_+$;
- $W$ is a benefit from the customer-oriented software adopted to the time-varying external and internal business environments, $W \in \mathbb{R}_+$;
- $w$ is a benefit from the experience of the customer-oriented software ordering used for the own tools development by means of open sources, $w \in \mathbb{R}_+$, $w < W$;
- $U$ is a benefit from the cloud service utilization considered as a possibility to avoid the own IT infrastructure development, $U \in \mathbb{R}_+$;
- $u$ is a benefit from a new producer propositions, $u \in \mathbb{R}_+$, $u < U$.

Since the strategy combinations for Player $\mathcal{B}$ are equivalent to that of Player $\mathcal{A}$, the further description concerns only their interpretations.

Strategy combination $A_1 B_1$. The consumer pays for standard products and their licensed service repeatedly. When using standard box pack software, the customer service level is a standard one. Therefore there are no additional benefits for the customer. Such strategy implementation is the minimum necessary to continue the work. Both the deterioration and improvement of the business situation may induce the client to consider other offers on the software market, which will lead to care from the producer. The payoff is $b_{11} = -pn$.

Strategy combination $A_1 B_2$. The client pays for the producer's services despite the opportunity to benefit from the selection of cheaper software analogs. The gained experience allows the player to create an effective IT service without the obligation to renew licenses annually. The compatibility of particular software tools always allows on the improvement of functionality of the purchased licensed software. The payoff is $b_{12} = u - p$.

Strategy combination $A_2 B_1$. The client pays for the development services and receives the benefits expressed in the maximum adaptation of the software product to the characteristics of the customer's business. At the same time, constant cooperation with the same producer allows the client to receive the most efficient service, including under the time-varying conditions of external and internal environments of the enterprise. The payoff is $b_{21} = W - P$.

Strategy combination $A_2 B_2$. Once paid for the customer-oriented solution the client can interrupt the collaboration due to the unchanged producer's conditions. The client can select free or cheaper software tools such that to keep own satisfaction level. The payoff is $b_{22} = w - P$.

Strategy combination $A_3 B_1$. The client can completely abandon the development of its own IT-infrastructure or significantly reduce it, which can lead to a significant reduction in costs. Also, the client can plug-unplug modules to meet particular requirements optimizing time and costs. The payoff is $b_{31} = U - ng$.

Strategy combination $A_3 B_2$. The client gets the opportunity to test the capabilities of the software in order to understand better the need to use them. As a result, the client can choose free or cheaper software. The payoff is $b_{32} = u - g$.

The payoffs of Player $\mathcal{B}$ are summarized as the following matrix:

$$B = \begin{pmatrix} -pn & u-p \\ W-P & w-P \\ U-ng & u-g \end{pmatrix}.$$

## 3.4   The Payoff Functions as Decision-Making Mechanism

Now we can identify the payoff functions. The optimist-pessimist index (1) is

$$G(\mathcal{A}) = \max \left\{ \begin{array}{l} \alpha_A \min\{ni-t,i\} + (1-\alpha_A)\max\{ni-t,i\} \\ \alpha_A \min\{I+T, I-T-s\} + (1-\alpha_A)\max\{I+T, I-T-s\} \\ \alpha_A \min\{qn-Y, i-Y\} + (1-\alpha_A)\max\{qn-Y, i-Y\} \end{array} \right\}, \quad (5)$$

$$G(\mathcal{B}) = \max\{G_1, G_2\}, \quad (6)$$

where $\alpha_A$ and $\alpha_B$ are the pessimism rate of $\mathcal{A}$ (producer) and $\mathcal{B}$ (client) respectively, $\alpha_A, \alpha_B \in [0,1]$, and

$$G_1 = \alpha_B \min\{-pn, W-p, U-ng\} + (1-\alpha_B)\max\{-pn, W-p, U-ng\},$$
$$G_2 = \alpha_B \min\{u-p, w-P, u-g\} + (1-\alpha_B)\max\{u-p, w-P, u-g\}.$$

Laplace insufficient reason criterion (3) is

$$L(\mathcal{A}) = \max\left\{ \tfrac{1}{2}(in-t+i), I - \tfrac{1}{2}s, \tfrac{1}{2}(qn+i) - Y \right\}, \quad (7)$$

$$L(\mathcal{B}) = \max\left\{ \tfrac{1}{3}(W+U-P-n(p+g)), \tfrac{1}{3}(u+w+u-P-p-g) \right\}. \quad (8)$$

Wald's criterion (4) takes the form

$$\mathcal{W}(\mathcal{A}) = \max\{\max\{ni-t,i\}, \max\{I+T, I-T-s\}, \max\{qn-Y, i-Y\}\}, \quad (9)$$

$$\mathcal{W}(\mathcal{B}) = \max\{\max\{-pn, W-p, U-ng\}, \max\{u-p, w-P, u-g\}\}. \quad (10)$$

We let the reader to deduce the analytical expressions for criterion (2). Here we will give only its abstract numerical illustrations. Let the parameter sets of players characteristics are

$$\Theta_A = \{100, 1000, 10, 5, 10, 50, 20, 100\}$$

and

$$\Theta_B = \{100, 1000, 10, 5, 1100, 25, 50, 25\}$$

The payoff matrix of Player $\mathcal{A}$ is

$$A = \begin{pmatrix} 490 & 100 \\ 1020 & 930 \\ -50 & 0 \end{pmatrix}$$

the payoff matrix of Player $\mathcal{B}$ is

$$B = \begin{pmatrix} -500 & -75 \\ 100 & -975 \\ 0 & 15 \end{pmatrix}.$$

Substituting these data into the expression (2), we get $S(\mathcal{A}) = S(\mathcal{B}) = 0$. That means the compensation of pros and cons bringing by MINIMAX strategies for Player $\mathcal{A}$ and Player $\mathcal{B}$.

### 3.5   Remarks

**Remark 1.** The strategy allocation rules of the players are unknown. Therefore, we can only speculate that the "cautious" strategies (1) or (3) would be the dominative choice of players. The decision-making process of rational players is always goal-oriented. Players perform actions subjected to specific objective functions (e.g., the profit maximization, the leader-image creation).

**Remark 2.** The players' strategies evolve. That is to say since the degree of preference across a set of available alternatives is time-varying, the players do not perform the same behavior. To estimate the degree of preference and its dynamics the empirical study is required.

In our present consideration, we omit these questions leaving a room for further investigations.

## 4   Discussion

The different aspects and barriers of strategic choice of the producer-customer relations were considered in the numerous contributions (see [10, 13, 15, 23]). For example, there are solutions on how to overcome the customers' negative behavior (see [3, 4] or [14]). Pioneered in the works of J. von Neumann and D. Morgenstern [18] and still developed, the game theory is the efficient apparatus for solving complex socio-economic problems [6]. Very often a real-world problem (not only economic one) is a game among players with antagonistic goals; the game theory is capable with a certain probability to ensure the best decision (only concerning players' perception). As the evidence, we can list the decision about the massive international port development supported by Nash equilibrium conditions [7]. However, due to the complexity, the search of the best-way collaboration is still essential and cannot be successful without psychological, economic and mathematical tools.

# 5 Conclusions

How on modern-day IT market to define the best collaboration strategy satisfying antagonistic expectations of decision-makers? – To create and to use the model which is capable of catching the market complexity and uncertainty. In this work, we concentrated only on the aspects of the model selection. For each of the decision-makers, we defined the characteristics and the list of the strategies. That gave the ability to formulate a kind of two-player strategic game, where each player can choose one strategy among the listed. However, the reasons which make players favorite one strategy over others, as well as the mechanism of the preferences aggregation, have not been studied. As it is possible to see, the research poses new questions requiring more detailed theoretical and empirical studies.

# References

1. Bell D (1999) The coming of post-industrial society: a venture in social forecasting. Basic Books, New York
2. Cerchiello P, Giudici P (2016) Big data analysis for financial risk management. J Big Data 3(1)
3. Chaudhry S, Bharatendu NS, Joshi C (2018) Vendor response to customer opportunism in IT service relationships: exploring the moderating effect of customer involvement. Ind Mark Manag. https://doi.org/10.1016/j.indmarman.2018.04.012
4. Cristóbal JRS (2015) The use of game theory to solve conflicts in the project management and construction industry. Int J Inf Syst Proj Manag 3(2):43–58
5. Denning PJ, Lewis TG (2017) Exponential laws of computing growth. Commun ACM 60 (1):54–65
6. Dixit A, Nalebuff B (2010) The art of strategy: a game theorist's guide to success in business and life. W. W. Norton & Company, New York, p 512
7. Do MH, Park G, Choi K, Kang K, Baik O (2015) Application of game theory in port competition between Hong Kong port and Shenzhen port. Int J E-Navig Marit Econ 2:12–23
8. Doszhan R, Mustafina A, Supugalieva G (2015) The empirical analysis of the system of risk management on small and medium-size enterprises. Asian Soc Sci 11(21):240–247
9. Drucker PF (1942) The future of industrial man, a conservative approach. The John Day Company, New York, p 298
10. Fedorenko RV, Zaychikova NA, Abramov DV, Vlasova OI (2016) Nash equilibrium design in the interaction model of entities in the customs service system. Math Educ Res J 11 (7):2732–2744
11. Gilder G (1989) Microcosm: the quantum revolution in economics and technology. Touchstone
12. Jain P, Gyanchandani M, Khare N (2016) Big data privacy: a technological perspective and review. J Big Data 3(1)
13. Khan A, Haasis HD (2016) Producer–buyer interaction under mass customization: analysis through automotive industry. Logist Res 9:17. https://doi.org/10.1007/s12159-016-0144-9
14. Khan A, Keung J, Niazi M, Hussain S, Ahmad A (2017) Systematic literature review and empirical investigation of barriers to process improvement in global software development: customer–vendor perspective. Inf Softw Technol 87:180–205. https://doi.org/10.1016/j. infsof.2017.03.006

15. Li M, Zheng X, Zhuang G (2017) Information technology-enabled interactions, mutual monitoring, and supplier-buyer cooperation: a network perspective. J Bus Res 78:268–276. https://doi.org/10.1016/j.jbusres.2016.12.022
16. Machlup F (1962) The production and distribution of knowledge in the United States. https://press.princeton.edu/titles/1510.html
17. Moore, GE (2003) No exponential is forever: but "forever" can be delayed! International solid-state circuits conference (ISSCC) 2003/Session 1/Plenary/1.1
18. Neuman J, Morgenstern O (2007) Theory of games and economic behavior. Princeton University Press, p 776
19. Nwankpa JK (2015) ERP system usage and benefit: a model of antecedents and outcomes. Comput Hum Behav 45:335–344. https://doi.org/10.1016/j.chb.2014.12.019
20. Olson DL, Johansson B, De Carvalho RA (2018) Open source ERP business model framework. Robot Comput Integr Manuf 50:30–36. https://doi.org/10.1016/j.rcim.2015.09.007
21. Parviainen P, Tihinen M, Kääriäinen J, Teppola S (2017) Tackling the digitalization challenge: how to benefit from digitalization in practice. Int J Inf Syst Proj Manag 5(1):63–77
22. Rahimli A (2013) Factors influencing organization adoption decision on cloud computing. Int J Cloud Comput Serv Sci 2(2):140–146
23. Siddique L, Hussein BA (2016) A qualitative study of success criteria in Norwegian agile software projects from suppliers' perspective. Int J Inf Syst Proj Manag 4(2):63–79
24. Toffler A (1970) Future shock. Random House, New York, p 505
25. Wamuyu PK (2017) Use of cloud computing services in micro and small enterprises: a fit perspective. Int J Inf Syst Proj Manag 5(2):59–81
26. Williams SP, Hausmann V, Schubert P, Hardy CA (2014) Managing enterprise information: meeting performance and conformance objectives in a changing information environment. Int J Inf Syst Proj Manag 2(4):5–36
27. Williams S (2016) Business intelligence strategy and big data analytics. Morgan Kaufmann, p 240

# Digitalization as a Source of Transformation of Value Chains of Telecommunication Companies Using the Example of PAO Megaphone

S. I. Ashmarina[⊠], E. A. Kandrashina, and Ju. A. Dorozhko

Samara State University of Economics, Samara, Russia
asisamara@mail.ru, kandrashina@sseu.ru

**Abstract.** The contribution considers transformation of value chains of telecommunication companies, which is connected with setting out auxiliary business processes into separate business units: shared service centers. The authors considered such a global trend of economic development as digitization of information and business processes, which is a source of these organizational changes. Shared service centers, which can significantly improve the quality of infrastructure support for core business processes, are presented as a variant of internal outsourcing, which has the potential to move to external outsourcing. Based on the concept of the value chain by M. Porter, the authors reviewed the model of the shared service center used by PAO Megafon (a publicly held company under the laws of the Russian Federation), identified sources of added value and identified the development potential of the shared center as an independent business.

**Keywords:** Digitalization · Organizational change · Outsourcing
Shared service center · Telecommunication companies · Value chains

## 1 Introduction

The telecommunications industry is one of key infrastructure sectors of the economy. The development of telecommunications is a necessary condition for the development of modern information and telecommunication technologies. At the same time, traditional telecommunication services are becoming less and less valuable for customers, and the customer value is shifting to the area of additional services. In this regard, telecommunication companies consider the alternative sources of growth that are related to such areas of activity as the development of financial services and electronic document management, commercialization of Big Data and others. There is gradual transformation of traditional mobile operators in companies of integrated digital solutions. Along with this, the matrix solution of tactical and strategic tasks for a large company with a wide network of branches (or for a group of companies) causes to remove supporting business processes and centralize them [9].

© Springer Nature Switzerland AG 2020
S. Ashmarina et al. (Eds.): *Digital Transformation of the Economy: Challenges, Trends and New Opportunities*, pp. 581–589, 2019.
https://doi.org/10.1007/978-3-030-11367-4_57

The object of the research is value chains of telecommunication companies, and the direct subject of the research is processes of their transformation. The purpose of the research is to study the feasibility of setting out auxiliary business processes in separate business units.

## 2  Materials and Methods

The methodological basis of the research is the concept of the consumer value chain, proposed by M. Porter in 1985. Each organization can be represented as a set of various activities aimed at developing value for the consumer, which are combined into a value chain. At the same time, value developing activities are divided into two main types: primary and secondary activities. The main activities are those that are directly related to the physical developing of the product, sales and product movement towards the buyer, as well as to the maintenance and technical support of goods after the purchase. The auxiliary activities are aimed at supporting the main activities. The value chain of the company and the way the company performs certain types of activity is, in the aggregate, a reflection of its history, strategy, its principles towards the implementation of its strategy, as well as economic activities of internal divisions [6].

The materials of the research are data on the activities of PAO Megafon, a federal telecommunication operator that holds leading positions in the telecommunications market in Russia and the world. MegaFon Group of Companies unites all directions of the IT and telecommunications market. In June 2017, a new strategy of MegaFon until 2020 was presented - "Approaching the Digital Future", which is aimed at abandoning price competition and increasing the value of an existing subscriber base through additional services. MegaFon plans to develop as a company of integrated digital communications: the digital client is in the center of the strategy.

## 3  Results

The current trend in the telecommunications market is such that over the past 2 years the mobile market has moved into a state of stagnation. The prerequisites for this were two main reasons:

The first reason is the glut of the market, that is, the lack of potentially new subscribers - mobile users can no longer consume services more than they need;

The second reason is irrational competition, that is, a race among operators for a new share of subscribers by offering more advantageous tariff plans, rather than focusing on the quality of service of existing subscribers and improving the quality of the network.

In this regard, the leadership of telecommunication companies faces the urgent task of making organizational changes in order to extract additional value. Changes in the value chain are for them as one of the main sources for improving the efficiency of companies.

The classic type of value chain implies that all activities, both core and auxiliary, operate within the same territorial unit. The implementation of this approach by telecommunication companies means the distribution of main activities in branches throughout the country, and the coverage of all auxiliary activities directly in them.

The most significant direction to optimize value chains of telecommunication companies is optimization of auxiliary processes, primarily infrastructure. As part of this research, emphasis is on improving the information infrastructure, which is designed to increase the speed and quality of decisions made, improve interaction within the company and improve performance.

Digitalization of information and information exchange provided the development of robotic process automation (RPA) technologies. Office robots can reduce costs by 20–40% in repeatable, algorithmic data processing: HR functions and finance, supply chain management and information technology sales, customer support. It increases the reliability and accuracy of operations, relieves peak loads in processes closing the period, data migration during IT system implementations, improves the quality of the audit and reduces the level of operational risks [2].

These technologies provided an opportunity to develop a SSC. This way of transforming the value chain has become one of the latest trends in organizational changes in telecommunication companies in Russia. As a rule, routine processes for the entire group of companies are highlighted in the SSC. The purpose of such changes is to increase business manageability and productivity, increase the speed of business processes, as well as improve their quality, as the approach to service provision changes, shifting to the interaction between internal suppliers and internal customers. Testing results [8] show that critical success factors for the development of such centers are a clear understanding of strategic goals, the redesign of business processes, the restructuring of the personnel management structure and cost allocation schemes.

Since July 1, 2016 PAO Megafon has allocated a separate division of the SSC located in Samara. When it was developed, the top management of the company set goals:

(1) Improve the quality of internal services;
(2) Optimize and unify the company's business processes;
(3) Improve business manageability;
(4) Increase the potential for business scalability.

The SSC is a multifunctional single service center that provides services to all branches and subsidiaries of PAO Megafon in a number of routine back-office operations. In accordance with them the maximum savings from their automation and scalability should be obtained. That is, in this case, the SSC acts as a functional division, which took over the centralized implementation of auxiliary business processes, the implementation of which is no longer foreseen in branches. The work of the Samara SSC includes such supporting business processes as accounting and reporting, planning and support for orders, treasury and financial control, a number of personnel management functions, administration of receivables and support for corporate business operations.

Due to the fact that auxiliary business processes do not develop added value for consumers, the SSC is a classic cost center in the organizational structure of PAO Megafon. At the same time, due to centralization of these functions, the process of providing internal services becomes more manageable and mature, while its cost is reduced due to cost savings, economies of scale and standardization of processes.

The process of providing services within the framework of the SSC is considered as the interaction of internal suppliers and internal customers. This interaction process is structured in such a way that the result of the work performed is the benefit for the entire organization, that is, the ultimate goals of internal suppliers and internal customers are unified within the framework of the company's implemented strategy (Fig. 1).

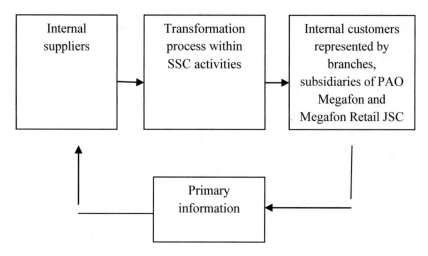

**Fig. 1.** The interaction between internal customers and internal suppliers of PAO Megafon (Source: compiled by the authors)

According to Fig. 1, the SSC is the structure in which transformation of information takes place. The incoming stream will be the information that is to be processed, and the final result has to be sent to consumers. The organizational structure of PAO Megafon has a division - the Federal Documentary Support (FDS), which accumulates the source information and directly sends it to the SSC. Thus, internal suppliers are a division of the FDS, from which the primary information comes. When fulfilling their regular work activities, employees of this company interact with employees of the SSC by placing requests to meet their functional needs.

The final result is necessary for internal customers, which are directly PAO Megafon and Megafon Retail JSC represented by all branches and subsidiaries, for which costs are reflected and profits are generated. These companies perform work that directly develop added value. Accordingly, the branches and subsidiaries of MegaFon are beneficiaries, and they bear the costs of supporting auxiliary processes.

The SSC carries out those operations that provide the process of developing added value. The SSC costs are allocated on the basis of cost drivers (based on the "Cost plus" approach): according to the number of personnel servicing a specific business unit (branch), as well as the number of transactions of this business unit.

Distribution costs are calculated based on the sum of all the costs of operating the SSC and the percentage of marginality for SSC services (management fees) agreed within MegaFon Group of Companies. The cost of operating the SSC includes the direct costs of meeting the needs of the business, as well as indirect costs, such as management and maintenance costs.

It is worth noting that earlier the accounting department or the personnel department were only departments of the company, and now they have become a separate organization and enter into service agreements with customers, implying goals for quality, productivity, and cost. This is a completely different approach to work that forms a different way of thinking. Considering the above, we can say that the SSC, in its essence, is a service organization that provides services for transforming information for the end user, which best affects the quality of service for internal customers and the approach to services provided.

The SSC model provides for a unified IT architecture that allows transition to electronic document management and abandoning the paper format. This means that the interaction with internal suppliers and internal customers is carried out in electronic form (this applies to primary documents and contracts). The work between departments within the company also implies a maximum transition to a paperless format, which significantly increases the speed of processing documents and is a factor in improving the quality of internal services.

Thus, at present, the added value of the SSC for PAO Megafon is in two main aspects:

1. Cost savings due to centralization of support functions and reduction of staff in branches and subsidiaries;
2. Risk reduction through the unification and standardization of business processes, which, in turn, reduces costs in the training of personnel.

The emergence of these effects is due to the fact that the relationship between the SSC, its client departments and headquarters are associated with a constant combination of cooperation and competition, which leads to increased managerial control over labor processes. The competition with clients' departments controlling labor processes leads to the introduction of control mechanisms, norms and standards both in the center and in the departments of clients. These rules, on the one hand, limit uncertainty; on the other hand, they lead to the fragmentation of labor processes, making them less complex [5].

Testing the viability of the SSC as an independent business, bringing additional added value, is carried out by internal outsourcing. And at this stage, due to centralization of auxiliary functions, companies increase the manageability of processes providing internal services, while their cost is reduced due to cost savings, economies of scale and standardization of processes.

Since 2018, the SSC of PAO Megafon has reached its full capacity and will allow saving about 470 million rubles for the year. The contribution of this center to the current capitalization of PAO Megafon is 0.75%. Despite the fact that the ratio in the total value of the company is not great, a positive economic effect is significant for PAO Megafon from the point of view of obtaining additional income from various kinds of auxiliary services.

At this stage of the SSC formation, there are a number of problems that the organization has to solve. One of the weaknesses is that only part of the functionality of the above mentioned directions, which has a transactional flow nature, is transferred to the SSC. From the point of view of entering the external market, this list of services is rather small and, to some extent, is specific only for PAO Megafon and its subsidiaries. The complexity of the functions allocated in the SSC, within the framework of the demand of external customers, is low at this stage.

The philosophy of continuous improvement transforms the SSC model into a model of an integrated service organization, the focus of which is gradually moving from the transaction center to the strategic goals of business development [9]. The setting out of auxiliary activities forms the potential for transforming internal sources of added value into external ones. Accordingly, an important pathway for the strategic development of the SSC as an independent source of added value is commercialization of its services. In this aspect, the process of commercialization can be considered as a way to adapt the services of the SSC in the market environment, aimed at extracting profits as part of transformation of the SSC into an independent source of income. For successful entry into the external market and development of services on it, it is necessary to take into account a number of conditions and factors that are combined in the concept of "strategizing". That is, the SSC will not be able to enter the external market until the company has a clear strategic vision - an understanding of what the company itself should be and what services provided should be.

The further development of the company is considered through the concept of "strategizing". That is, it is necessary to identify the key components of the organizational and economic mechanism and highlight the issues that, within the framework of the development of the SSC, will make decisions about entering the foreign market.

To enter the external market, it is necessary to improve the skills of employees, conduct training both within the company, and cooperate with external organizations, developing additional skills. To make decisions on expanding the functionality of the SSC staff and how to increase the expertise, it is necessary to interact with the Company's Head Office in order to increase the complexity of functions implemented in the SSC.

Taking into account new opportunities for collecting analytics, one can recognize the significant role of the SSC in supporting the development of auxiliary services when entering new markets, as well as the inclusion of valuable analytical information. This suggests that the SSC can act as a business partner for the expertise of owning business processes.

It is also worth transforming the interaction model between domestic suppliers and customers, presented in Fig. 1, and consider it in terms of entering the external market. The transformation takes place when substitute products, competitors and companies that are potentially ready to enter the market are added to the existing model.

Competitors may be both outsourcing companies that provide accounting services, personnel management, etc., and the SSCs of other companies, for example, Sberbank, MTS, VTB, etc. Companies that are potentially ready to enter this market may include various types of businesses that would be interested and profitable to engage in remote service within the framework of identified business processes.

Potential clients may be companies in the corporate sector that are already included in the customer base of PAO Megafon, which will provide an additional way to monetize them. As well as those companies that at their stage of development only think about cost optimization through centralization of auxiliary processes and analyze the options - developing their own SSC or outsourcing the business processes under consideration.

It is worth noting a certain regularity, which implies that the smaller the business that develops the SSC, the longer the payback period of the investment is. This assumption is based on two main factors:

1. A large amount of investments in the infrastructure development on the basis of which the SSC will function: finding a site for location, technical equipment, software, staff selection and training, etc.;
2. A number of operations going through it in the sense that the smaller the company's scale, the fewer transactions occur within its business.

Accordingly, the smaller the company that plans to centralize support functions, the more interested it is to implement this centralization through outsourcing, rather than developing its own data center. If we proceed from the assumption that PAO Megafon is a large company with a minimum payback period, that is, there is reason to assert that the internal price of MegaFon services will be attractive to the market.

And if PAO Megafon attracts additional customers, then the increase in costs will be mainly due to conditionally variables. That is, large-scale additional investments in digital infrastructure will not be necessary taking into account the fact that a sufficient reserve already exists and is functioning. But basically, additional investments will be required related to the increase in the scale of operations, that is, connected with the attraction and training of new employees in order to process these additional transactions.

Accordingly, due to growth in conditionally variable costs, the economies of scale increase, that is, the return on resources used increases, and the growth rate of the number of services provided is higher than the growth rate of costs. This, in turn, affects the reduction in costs per unit of service provided. And, as a result, it will affect the profitability of this type of economic activity for MegaFon. Accordingly, it can be concluded that MegaFon benefits from attracting new customers through additional services and services, which together make a tangible contribution to the company's overall revenue stream.

According to SSON, 79% of owners of SSCs already confirm that SSCs not only reduce costs, but also become a source of additional business benefits (SSON Analytics, 2017). Such a model is more technological and customer-oriented, focuses on knowledge and methodology as the final result, providing a high level of customer satisfaction, and allowing developing new services and bringing added value to the business.

Thus, when business processes transferred to the SSC will be debugged and will be executed as efficiently as possible, errors by employees will be minimized, and there will be an opportunity to expand the functionality, the SSC will be able to enter the open market and function, developing more added values for PAO Megaphone. It will be not just a cost center, but also a profit center.

# 4 Discussion

Although companies in the same industry may have similar value chains, the value chains of competitors are most often different. These differences are sources of competitive advantage.

Currently, the largest corporations are actively engaged in transformation of their business, looking for new and more efficient business management models. This is connected, first of all, with the expansion of the activities of modern corporations, with the increasing complexity of management activities, leading to the need to integrate individual elements of the management system, which allows solving the problems of informational, instrumental, methodological and consulting support of management. In addition, one of the important tasks facing a modern company, especially aggravated in the context of slowing inflation, is to control and reduce costs, primarily for auxiliary activities. Companies can choose different optimization options, one of which is the use of various outsourcing models. G. Mins and D. Schneider in "Metacapitalism and the Revolution in E-business" [4] reveal the essence of the externalization process that is the basis of the outsourcing process by comparing the main assets (factors of activity) in the traditional and a new company. Another trend, caused by the desire of organizations to reduce costs and increase efficiency, is associated with consolidation and optimization of common business functions, and is known as insourcing, in-house services, business services [1].

A specific option for internal outsourcing is the development of shared service centers (SSC), which can significantly improve the quality of service for business units, primarily through the standardization of business processes. According to Deloitte, the financial function, HR, IT, procurement and tax calculation is most often imposed on the SSC.

To date, the choice of options for outsourcing has become a topical discussion in both the academic and practical management environments. However, there is no systematic understanding of management problems in this area: problems that arise and approaches to solve them differ significantly depending on the situation [3].

# 5 Conclusions

So, one of trends in the strategic development of telecommunication companies at this stage is the active development of additional services. The transformation of existing value chains is an important source of such development.

Taking into account the fact that auxiliary activities are supportive of those that create greater added value, the issue of their centralization becomes urgent. In this connection, companies are faced with two main options for optimizing data from business processes - outsourcing or creating their own SSC.

Digitalization as a global trend creates the conditions for centralizing auxiliary activities associated with extensive workflow. The development of shared service centers provides new opportunities for optimizing and transforming processes. This affects the change management system, IT technology, the transfer of new atypical services to the center, the use of robots and artificial intelligence, etc. At the same time, the issues of improving the quality of customer service and continuous improvement of all processes remain unchanged. The combination of financial and IT service centers can be used as a tool for organizing information outside the company, which implies a change in the orientation of such business processes from the domestic market to the external one. Shared service centers are also tools for new business models, especially those based on human capital [7].

The example of PAO Megafon shows that the development of the SSC as a multifunctional single service center that provides services to all branches and subsidiaries of PAO Megafon in a number of routine back-office operations, improves their quality and performance, and also creates savings from their automation and scalability. According to the results of calculations, it was revealed that the potential cost of the SSC as a business line is 0.75% of the current capitalization of PAO Megafon, which is an insignificant share compared to main activities, but reflects the development trend of auxiliary services.

Materials of the contribution can be useful for representatives of IT companies who are solving the problem of optimizing interaction with existing and potential customers.

In the process of research, new questions have arisen that require more detailed study. It is necessary to continue research aimed at optimizing the strategic choice of the customer.

# References

1. Aksin OZ, Masini A (2008) Effective strategies for internal outsourcing and offshoring of business services: an empirical investigation. J Oper Manag 26(2):239–256. https://doi.org/10.1016/j.jom.2007.02.003
2. KPMG. Robotics of business processes. https://assets.kpmg.com/content/dam/kpmg/ru/pdf/2017/11/ru-ru-rpa.pdf
3. Knol A, Janssen M, Sol H (2014) A taxonomy of management challenges for developing shared services arrangements. Eur Manag J 32(1):91–103. https://doi.org/10.1016/j.emj.2013.02.006
4. Means G, Schneider D (2000) Meta capitalism. The e-Business revolution and the design of 21st-century companies and markets. Wiley, New York
5. Mezihorak P (2018) Competition for control over the labour process as a driver of relocation of activities to a shared services centre. Hum Relat 71(6):822–844. https://doi.org/10.1177/0018726717727047
6. Porter ME (1998) Competitive strategy: techniques for analyzing industries and competitors. SSON analytics. Free Press, New York. https://www.sson-analytics.com/. Accessed 10 April 2018
7. Strikwerda J (2014) Shared service centers: from cost savings to new ways of value creation and business administration. https://doi.org/10.1108/s1877-636120140000013000
8. Wang S, Wang H (2015) Shared services management: critical factors. Int J Inf Syst Serv Sect 7(2):37–53. https://doi.org/10.4018/ijisss.2015040103
9. Zaripova DA, Nasyrova VI, Ochaikin KD (2015) The model of a shared service center of the company: selection criteria and directions of transformation. Int Bus Manag 9(7):1714–1717. https://doi.org/10.3923/ibm.2015.1714.1717

# Industrial Enterprises Digital Transformation in the Context of "Industry 4.0" Growth: Integration Features of the Vision Systems for Diagnostics of the Food Packaging Sealing Under the Conditions of a Production Line

R. K. Polyakov$^{(\boxtimes)}$ and E. A. Gordeeva

Kaliningrad State Technical University, Kaliningrad, Russia
polyakov_rk@mail.ru, elena.beznos@klgtu.ru

**Abstract.** The article contains the results of the authors' research the subject of which is the system of technical vision for the diagnostics of the food packaging air-tightness under the conditions of on-line production and their marketing potential.

The article presents the first stage of the research work, which is devoted to the development of the diagnosing method of the air-tightness of the food products package under the conditions of in-line production for a prototype of a self-learning software and hardware vision system that performs the diagnostics of the air-tightness of food packaging in a flow production environment.

Scientific novelty of the solutions proposed in the project is the use of a fundamentally new design of the complex with a self-learning system of technical vision, based on the use of advanced methods in the field of artificial neural networks and machine learning.

The analytical material presented in the article shows the development vectors of modern innovative elaborations, as well as a general trend in the scientific and technical literature.

The authors believe that this research offers a valuable view how innovative systems of technical vision, methods of artificial neural networks and machine learning can influence the digital transformation of industrial enterprises provided "Industry 4.0" growth.

**Keywords:** Artificial neural networks · Digitalization · Innovation · Machine learning · Methods · Quality control · Technical vision

## 1 Introduction

As well as a large-scale digital transformation the "Industry 4.0" growth requires more modern approaches from industrial enterprises which can largely influence the intensification in production. By transiting the basic functions of process control into digital format, enterprise leaders will be able to significantly improve the quality of production, including improvements through modern diagnostic digital methods and advanced

© Springer Nature Switzerland AG 2020
S. Ashmarina et al. (Eds.): *Digital Transformation of the Economy: Challenges, Trends and New Opportunities*, pp. 590–608, 2019.
https://doi.org/10.1007/978-3-030-11367-4_58

management functionality for all production. Resolving such issues, food and processing companies will be able to optimize the production process, reduce costs and improve their efficiency.

Analysis of the food and processing industry showed that the share of domestic food equipment in the domestic market is about 20%, while the market for food and processing industry equipment grew by 2% in 2017 up to 59.7 billion rubles, while import was 48.4 billion rubles.

At the same time, a significant share of demand from food and processing industry enterprises is covered by foreign equipment, which is an order of magnitude higher than that of Russian-made equipment.

At present, technological equipment is modified in all branches of industry, rigged with means of control, diagnostics and automation, which contributes to improving the quality and competitiveness of the products.

The shortcomings of existing technologies include:

1. Presence of manual processes in the process of visual inspection of products and management of technological equipment in the field of canning and tampering.
2. Without attention of manufacturers, there are the remained issues on the existing differences in the weight dimensions of cans, which require automation in the field of self-adjustment and the adaptation of the sealing machine.
3. Until now self-learning software and hardware vision systems are not used to diagnose the air-tightness of food packaging under the conditions of in-line production.

The main criteria of this R & D are as follows: increasing the level of efficiency and accuracy of food packaging air-tightness diagnostics under the conditions of in-line production as compared to the existing one, increasing the accuracy of verification to 99%, resistance to changes, expanding the scope of the self-learning software and hardware vision system at the sites of fish, meat, fruit and vegetable canning industry.

In this connection, the main activity in the field of solving these problems is the improvement of the production process control system, based on the timely introduction of new methods with the help of innovative means for diagnosing the air-tightness of the food products packaging under the conditions of in-line production.

Over the past decades, this problem has been studied by both Russian and foreign scientists. In particular, the following Russian scientists contributed to the development of the theory of vision systems: Potapov et al. [1], Lukyanitsa and Shishkin [2], Veremeenko et al. [3].

There are foreign authors studied the problems and applied modern machine learning algorithms are as follows but not limited to: Redmon et al. [4], Ren et al. [5], Dai et al. [6], Liu et al. [7], Hong et al. [8], Huang et al. [9], Szegedy et al. [10], Wu et al. [11].

The choice of methods and means for diagnosing the air-tightness of food products packaging under the conditions of in-line production is determined by the design features of the flow lines, temperature regimes, operating conditions, and many other factors.

Therefore, the main purpose for which this study is directed is the development of new methods for diagnosing the air-tightness of food packaging under the conditions of in-line production. After all, it is the main self-learning software and hardware system of technical vision.

## 2   Materials and Methods

The foundation of the self-learning software and hardware vision system for diagnosing the air-tightness of food packaging under the conditions of in-line production is the algorithm that will be incorporated into it.

The authors of the article consider that under the control and management algorithm it is necessary to understand the process of converting the initial information received from one or several light-sensitive cameras into a control panel of the technological object and means of displaying information.

Analysis and classification of images is a key process in the task of managing the integrity of canned food.

The general scheme of the algorithm for rejecting substandard packaging is presented in the STAGE 5 article 1.2. The development of the algorithm for identifying damaged objects includes the selection of the main decision algorithm. It is necessary to test and decide on the use of a number of architectural options that may include:

1. Classical machine learning;
2. SSD (Single Shot Multibox Detector);
3. YOLO (you look only once);
4. Faster RCNN (Regional Convolution Neural Network);
5. Mask RCNN (Regional Convolution Neural Network).

With the development of convolutional neural networks and self - training, the scanning window method, the Viola-Jones algorithms and further analysis using SVM are somewhat outdated. There are networks that at their output give not only an object class, but also a bounding rectangle object: x, y, w, h. This, although requires a slightly faster hardware, provides a better result at the output.

Neural networks for detection tasks have developed relatively recently starting from 2016. One of the main trends in 2016 in the field of object detection was the transition to faster and more efficient detection systems. In such approaches as YOLO, SSD and R-FCN, a step is taken to joint computation on the whole image. This is where it differs from the resource-intensive subnets associated with the Fast/Faster R-CNN techniques. This technique is commonly called "end-to-end training/learning".

In fact, the idea is to avoid using separate algorithms for each of the sub-problems isolated since this usually increases learning time and reduces the accuracy of the neural network. Such an adaptation of neural networks for work from beginning to end usually occurs after the operation of the original subnets and, thus, represents a retrospective optimization [12]. However, the Fast/Faster R-CNN techniques remain highly efficient and are still widely used for object detection, in addition to the latest developments in speed, SSD and YOLO are highlighted.

# 3 Results

## 3.1 Market Potential

The food and processing industry of Russia is a system-forming element of the agro-industrial complex. Being the most important branch in the country it ensures food security of the Russian Federation. At present, it unites over 20 thousand enterprises from 30 industries (about 60 sub-sectors). They account for about 16% of the country's total industrial output. According to experts, the share of raw materials in the cost structure of food production exceeds 72%.

The dynamics of the industry development is characterized by the attractiveness of the market, significant capacity, stable demand and extremely high competition in the domestic market of the Russian Federation with foreign partners.

The quality of products, goods and services is controlled by state regulatory documents: over 800 state standards, more than 350 types of industry standards and over 3,000 technical specifications developed. At the same time, thanks to the federal law "On the quality and safety of food products", producers have the right to develop and implement technical conditions for their own products independently [13].

The machine-building market for the food and processing industry in 2015 amounted to 56.1 billion rubles, which is 6.5% more than in the same period in 2014. The volume of manufactured machinery for the food and processing industry in the Russian Federation in 2015 amounted to 10.6 billion rubles (9.7% more than in 2014).

For 2016, the market volume declined due to a reduction in demand from consumers, as well as an increase in the price of imported products. Import of machinery and equipment for the food and processing industry in 2015 compared with 2014 in value terms decreased by 9.0% (from 49.5 to 45.4 billion rubles).

Studies have shown that the share of domestic food equipment in the domestic market is about 20%, while the market for equipment for the food and processing industry in 2017 grew by 2% to 59.7 billion rubles, while import was to 48.4 billion rubles. A significant share of imports exists due to the fact that domestic manufacturers do not have competence in their production and there are no necessary technologies in the Russian market. This can be also explained by the weak level of cooperation between domestic companies [14].

As can be seen, a significant share of demand in the food and processing industry is offset by foreign high-tech equipment, which is an order of magnitude higher than Russian counterparts.

Currently, the major equipment suppliers in our country are large European producers such as Germany (20.4% of total imports in 2015, 26.7% in 2016) and Italy (19.2% in 2015 and 23.4% in 2016). Recently there has been an increase in imports from China (from 2.2% in 2014 to 4.1% in 2016), not only has it happened due to the price difference, but also to the presence of factories of European and American manufacturers in China [14].

At present, technological equipment in all branches of industry is modified, equipped with means of control, diagnostics and automation, which contributes to improving the quality and competitiveness of the products.

The main competitors of the developed self-learning software and hardware vision system on the Russian market are such producers of the equipment of the world's leading suppliers as: Baader, Marel, Carnitech, Cabinplant, Convenience Food Systems, etc.

The shortcomings of existing technologies include:

1. Presence of manual work in the process of visual inspection of products and management of technological equipment in the field of canning and testing their integrity.
2. There are issues on the existing differences in the weight dimensions of cans, requiring automation in the field of self-adjustment and the adaptation of the sealing machine to them.
3. Until now, self-learning software and hardware vision systems are not used to diagnose the air-tightness of food packaging under the conditions of in-line production.

Figure 1 shows the main competitors and the place of the company LLC "Intrlik" (the product being developed) in the Russian market.

Expert analysis showed that the main growth driver in the production of equipment for the fish processing industry is the general trend in Russia to increase the level of fish products processing, which will also be supported by the Ministry of Industry and Trade within the framework of the "Strategy for the Development of Mechanical Engineering for the Food and Processing Industries of the Russian Federation for the Period till 2030" [15].

According to this strategy, the Ministry of Industry and Trade expects to increase the production of food equipment by 4 times by 2025, and in particular, special attention is paid to the development of the following equipment, for [15]:

− fruit and vegetable industry;
− canning industry;
− fish processing industry.

In this regard, the main segments of consumers of the product being created are the food industry enterprises that use automated packaging lines, namely:

− the fishing industry;
− meat industry;
− fruit and vegetable canning industry.

Further in the Table 1 there is the information on production of the main types of food products in Russia for 2016–2017.

As can be seen from Table 1, there is a stable growth in the main types of canned products, so canned meat increased by 1% in comparison with 2016, canned fish natural 2.5%, canned fish in oil by 10.3%, and fish preserves in oil 1.5%.

This dynamics is caused by the modernization of lines and the growth of investments in food equipment. The total volume of investment for 2017 in equipment in Russia for the above-mentioned enterprises amounted to 180.7 billion rubles. Figure 2 shows the volume and capacity of the Russian market, to which the product is aimed.

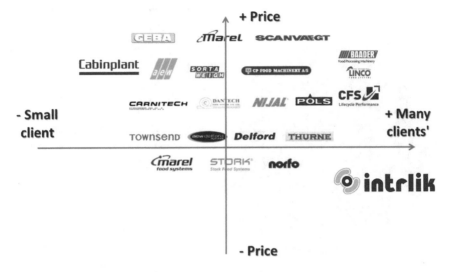

**Fig. 1.** The main competitors and the place of the company LLC Intrlik in the Russian market (Source: Compiled by the authors of the article)

**Table 1.** Production of main types of food products

| Indicator name | 2016 | 2017 |
| --- | --- | --- |
| Canned meat (meat-containing), including canned food for baby food, million cans | 643 | 649 |
| Fish, processed and preserved, crustaceans and molluscs, thousand tons | 4049 | 4152 |
| Including: | | |
| Canned fish natural, million cans | 198 | 203 |
| Canned fish in tomato sauce, million cans | 190 | 174 |
| Canned fish in oil, million conditional cans | 174 | 192 |
| Preserves fish, million conditional cans | 130 | 132 |
| Juices from fruit and vegetables, million conditional cans | 1262 | 1069 |
| Fruit and (or) vegetable nectars, million cans | 1180 | 1098 |
| Vegetables (except potatoes) and mushrooms, preserved without vinegar or acetic acid, other (except ready-made vegetable dishes), million cans | 1064 | 1203 |
| Vegetables (except potatoes), fruits, nuts and other edible parts of plants, processed or preserved with vinegar or acetic acid, million cans | 450 | 340 |

Source: Compiled by the authors according to Rosstat [16]

PAM (Potential Available Market) − the potential market volume is: 59.7 billion rubles. In Russia in 2015, there were 2908 enterprises engaged in the production of food products (the number of organizations/legal entities). On average, each enterprise had to pay 1,898 thousand rubles for installation of the complex with a technical vision system.

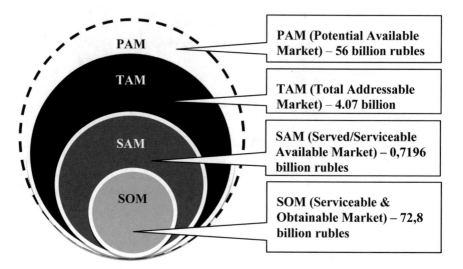

**Fig. 2.** The volume and capacity of the product market in Russia (Source: Compiled by the authors)

The preliminary cost of the developed model of the universal complex with the technical vision system for diagnosing the air-tightness of food packaging will amount to 1400 thousand rubles.

TAM (Total Addressable Market) − total volume of the target market − 2908 (potential number of clients in the Russian Federation) * 1400 thousand rubles (current average check) = 4.07 billion rubles;

SAM (Served/Serviceable Available Market) − available volume of the market − 514 (potential number of clients in the Russian Federation) * 1400 thousand rubles (current average check) = 0.7196 billion rubles;

SOM (Serviceable & Obtainable Market) − realistic achievable market volume within the current business model (in 3–4 years) − 52 (potential number of clients in the Russian Federation)) * 1400 thousand rubles (current average check) = 72.8 million. rub.

## 3.2   Risk Assessment of the Project

At the same time, it should be kept in mind that, like any market, the Russian machine-building market of the food industry is subject to a number of risks, the main of which are:

1. Inflationary processes
2. Scientific and technical risks
3. Prices for raw materials
4. Risks of erroneous demand forecasting
5. Risks of over-expenditures for production development

6. The investment environment
7. The state of the financial (banking) sector.

To assess the identified risks during the implementation of the R & D project, we will use the methodical approach developed by Polyakov [17].

The risk assessment for the project "Development, production and research of the prototype of a self-learning software and hardware vision system for the diagnosis of the air-tightness of food packaging under the conditions of in-line production" is presented in Table 2.

**Table 2.** Project risk assessment

| № | Risk | Loss probability | Risk probability | Level of effects | Risk level |
|---|------|------------------|------------------|------------------|------------|
| 1 | Inflationary processes | 65 | 0,7 | Substantial | **Inadmissible** |
| 2 | Scientific and technical risks | 62 | 0,8 | Substantial | **Inadmissible** |
| 3 | Prices for raw materials | 64 | 0,6 | Substantial | **Inadmissible** |
| 4 | Risks of erroneous demand forecasting | 45 | 0,5 | Mild | Justified |
| 5 | Risks of over expenditures for production development | 40 | 0,45 | Negligible | Justified |
| 6 | The investment environment | 35 | 0,3 | Disregarded | Acceptable |
| 7 | The state of the financial (banking) sector | 35 | 0,2 | Disregarded | Acceptable |

Source: Compiled by the authors

Figure 3 presents the matrix "Probability – losses" is presented before taking preventive measures to eliminate risks.

The above risks will be minimized:

1. Registration of intellectual property;
2. As the production base develops, it will be possible to reduce the cost of production and thereby reduce economic risks.
3. A clearer goal-setting and understanding of the project implementation team.
4. Determination of the necessary equipment and its specification, consideration of the experience of implementing IT systems in similar companies.
5. Adoption of a competently organized organizational structure, distribution of responsibility for business processes in accordance with the requirements of the BSC.
6. Planning a clear budget for the project, monitoring the implementation of the budget.
7. Timely and precise control over the execution of the project budget.
8. Appointment of a financial controller for the budget, creation of a reserve for unforeseen expenses.

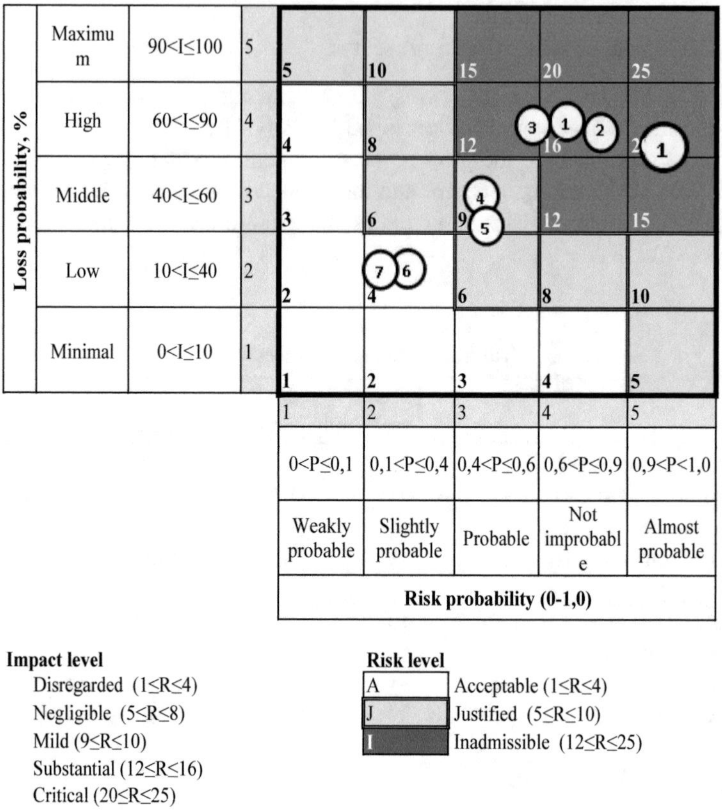

**Impact level**
Disregarded (1≤R≤4)
Negligible (5≤R≤8)
Mild (9≤R≤10)
Substantial (12≤R≤16)
Critical (20≤R≤25)

**Risk level**

| A | Acceptable (1≤R≤4) |
| J | Justified (5≤R≤10) |
| I | Inadmissible (12≤R≤25) |

**Fig. 3.** Matrix "Probability – losses" before taking preventive measures (Source: Compiled by the authors)

9. Personnel counseling and training, identification of risks using the matrix "Probability-loss" in the early stages of project planning
10. Clear delimitation of responsibilities and development of the project structure, introduction of regulations and workflow.
11. Appointment of those responsible for the order, adoption of regulations for all the procedures.

Risk assessment of the project after taking preventive measures (risks numbered) is presented in Table 3.

As can be seen from Fig. 4 after preventive measures adoption, only the risks associated with inflationary processes and the risks of raw material prices will be moderate in terms of impact, while at the risk level they are justified.

The matrix "Probability-losses" is presented below after taking preventive measures (risks numbered) (see Fig. 4).

**Table 3.** Risk assessment after taking preventive measures

| № | Risk | Loss probability | Risk probability | Impact level | Risk level |
|---|------|------------------|------------------|--------------|------------|
| 1 | Inflationary processes | 55 | 0,5 | Mild | Justified |
| 2 | Scientific and technical risks | 40 | 0,1 | Negligible | Acceptable |
| 3 | Prices for raw materials | 42 | 0,4 | Mild | Justified |
| 4 | Risks of erroneous demand forecasting | 45 | 0,4 | Mild | Justified |
| 5 | Risks of over expenditures for production development | 10 | 0,45 | Negligible | Justified |
| 6 | Investment Environment | 35 | 0,3 | Disregarded | Acceptable |
| 7 | The state of the financial (banking) sector | 35 | 0,2 | Disregarded | Acceptable |

Source: Compiled by the authors

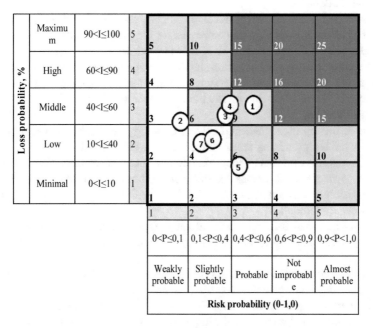

**Impact level**
Disregarded  (1≤R≤4)
Negligible  (5≤R≤8)
Mild (9≤R≤10)
Substantial (12≤R≤16)
Critical (20≤R≤25)

**Risk level**
A       Acceptable (1≤R≤4)
J       Justified  (5≤R≤10)
I       Inadmissible (12≤R≤25)

**Fig. 4.** Matrix "Probability – losses" after taking preventive measures (Source: Compiled by the authors)

### 3.3    Methodical Approaches to the Diagnostics Subsystem Creation for Food Packaging Air-Tightness in the Technological Process, Under the Conditions of Food Enterprise In-Line Production

A characteristic feature of increasing the efficiency of production and the quality of products is the automation of the technological process. At present, considerable attention has been paid to the automation of the canning lines at the food industry. However, on the most of the production lines, up-to-date systems of technical vision are not yet being applied, which could fill a gap in the implementation of continuous automatic control of the tightness of canned food.

At the same time, at the enterprises of the food industry, in particular in the fish processing and meat processing industry of the Kaliningrad region, there are no developed and approved methodological approaches to the introduction and use of such technical vision complexes to diagnose the air-tightness of food packaging under the conditions of in-line production. As shown earlier, the need for such methodological developments is very high and in demand by the leaders of food industry enterprises of the Russian Federation.

In this regard, the basis for the developed quality control system methodology is the real need of enterprises in modern complexes for the diagnostics of the air-tightness of food packaging under the conditions of in-line production.

At the same time, special attention should be given to the creation and introduction of continuous automated control systems for hermetic packaging of food products, under the conditions of in-line production, which is the main objective of the work under consideration.

To solve this problem microprocessor technology and technical vision should be used as one of the promising areas of application of new technical means.

Any industrial enterprise using production lines of conservation could easily develop its own practical actions for the introduction of a self-learning software and hardware vision system for the diagnosis of the air-tightness of food packaging using methodological materials.

It should be noted that the methodological approach being developed is comprehensive, systemic and covers all the existing interrelations in the technological process at work.

The use of computer vision systems allows you to find cans that have significant damage to the case automatically, focusing on their digital image banks. This will allow the rejection at all stages of production, including its early stages, when it is possible to unpack the product and return it to the beginning of the production line, which will lead to a significant economic effect.

To solve such a very important and actual technological task, the accumulated experience and modern scientific and applied potential was applied from the research and development theme *"Development of a method for diagnosing food packaging air-tightness under the conditions of in-line production for a prototype of a self-learning software and hardware vision complex for testing the food packaging air-tightness under the conditions of on-line production"*.

The subsystem to be created must be coordinated with other subsystems of the enterprise by regulating the diagnostics of the food packaging air-tightness: detection, segmentation and high-level processing, etc. The proposed subsystem is organically integrated into the technological process of on-line production, as a constantly performed function of continuous automated control of food packaging air-tightness, flow production. It is important that monitoring the control of the tightness of food packaging is continuous and automated, and not on time to time basis. In this case, the owner of the enterprise will have the confidence that the reject in production will be detected on a regular automatic basis and timely measures are taken to reduce the economic damage.

An illustration of the main stages of creating a subsystem for the air-tightness diagnostics of food packaging in the process, under the conditions of in-line production is presented in Fig. 5.

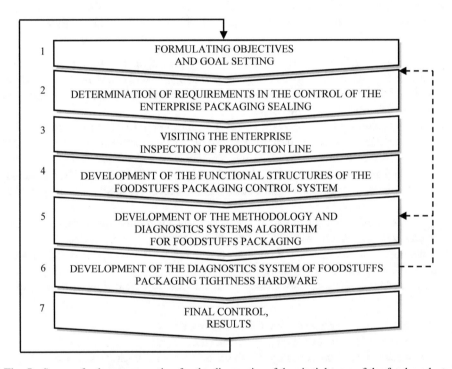

**Fig. 5.** Stages of subsystem creation for the diagnostics of the air-tightness of the food products packaging in the technological process, under the conditions of the food enterprise on-line production (Source: Compiled by the authors)

**First Step.** The goal is to "Introduce a self-learning software and hardware vision system to diagnose the tightness of food packaging in a production-in-process". The task to identify challenges for the enterprise is set. At the same time, the management of the enterprise makes a decision and the task is set in the framework of perfection.

**Second Phase.** At this stage, the study of economic and production information is received directly from the management of the enterprise, the magnitude of losses resulting from the reject, the cost of production, production capabilities, technical parameters of production, etc. is being studied. Next, an enterprise needs an assessment of the tightness of the package.

**Third Stage.**  Visiting the enterprise, inspection of the production line. Audit of production. Documenting. Obtaining all the necessary information about the techno-logical process in close interaction with the representatives of the enterprise in order to introduce into the technological process a self-learning software and hardware system of technical vision, based on the real needs and capabilities of the enterprise.

**Fourth Stage.**  The purpose of this stage is to carry out an examination of the pro-duction process, to identify the most problematic areas in the production line for the potential of creating a reject and document them. Then, using the information obtained by the method of synthesis and analysis of technological systems, design the operator model and develop the functional structure of the food packaging tightness control system.

**Fifth Stage.**  Development of the methodology and algorithm of the air-tightness system for food packaging. At this stage, digital images are obtained from one or more photosensitive cameras and its transformation into a control effect on the technological object and information display means.

Figure 6 below shows the main stages with the functions of the computer vision system that can be used in designing a prototype of a self-learning software and hardware vision system to diagnose the air-tightness of food packaging in a flow production environment using the classical computer vision scheme. In the neural network approach, detection, segmentation and high-level processing are combined in one step.

The problem itself can be represented as a classical two-class classification problem and, consequently, can be solved using machine learning algorithms. Since beforehand it is possible to select samples of hermetic and non-hermetic cans, this determines the choice of algorithms on the principle of "supervised learning".

The tasks of analysis and classification can be represented in the form of a conveyor:

1. Identification of the object under study on the image.
2. Separation of the investigated object from the image.
3. Transformation of the investigated object into a set of its characteristics.
4. Classification of the characteristics of the object under study.

The input from this conveyor is fed from the cameras of the technical vision subsystem, the output is a logical value of true/false.

In the specific task of analyzing and classifying images of tin cans on the basis of leak tightness under the conditions of in-line production, the camera angle at the time of the photograph is also fed to the input of the conveyor. This information is used when separating the object from the image.

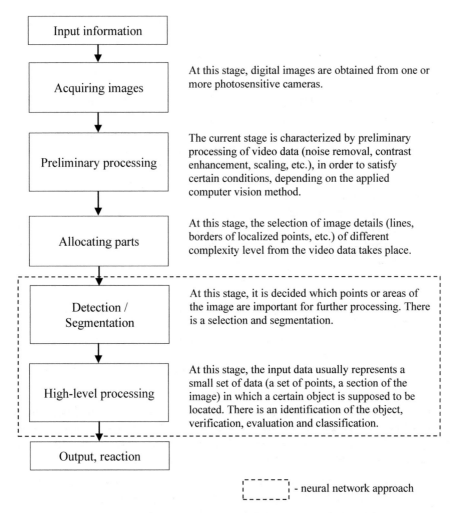

**Fig. 6.** Stages of the system for the diagnostics of hermetic sealing of food packaging under the conditions of in-line production (Source: Compiled by the authors)

General scheme of the algorithm for rejecting substandard packaging (at the detection stage):

1. Processing video images, splitting the image into separate frames for submission to the detection algorithm.
2. Then there are two possible options for working with the image, variant A and B:

Variant A: classical computer vision algorithm:
(a) Search for the object (can) on the general background using the pre-cascaded Haar cascades.
(b) Selecting the object (can) from the background based on the pre-assembled and processed set of images.

(c) Passage through the facility and its classification by the presence/absence of damage using a classification algorithm.
(d) Extraction of the coordinates of the damaged object for further processing.

Variant B: neural network approach:
(a) The only passage of the pre-accelerated convolutional network along the image and the search for any types of packaging damaged or not.
(b) Extraction of the coordinates of the damaged object for further processing.
3. The further operation of the algorithm does not depend on the chosen detection method, but it can also work in two versions:

Variant A: there is an operator:
(a) Drawing the coordinates of the object together with the object on the image, rendering the images as a video stream for visual control.
(b) Submission of an audiovisual signal to the operator when a damaged object appears.

Option B: automatic rejection:
(a) Drawing the coordinates of the object together with the object on the image, rendering the images as a video stream for visual control.
(b) Comparison of the received coordinates of the damaged object with the previously stored ones.
(c) When an object crosses a certain meta on the conveyor, signal is given to the tilting mechanism.

The implementation of a self-learning software and hardware vision system for the diagnosis of the air-tightness of food packaging in a production-line environment is highly dependent on the hardware platform and performance requirements. At this stage it is necessary to test and decide on the use of the most effective architecture variant.

**The Sixth Stage.** Development of the hardware part of the system for diagnosing the air-tightness of food packaging. Based on the requirements for the functions, functional scheme and the selected algorithm, a hardware selection satisfying the technical task is made (Application № C1-41610, project 41354, stage 1).

If the deviations in the methodology and algorithm of the air-tightness system for packaging food products are revealed as a result of developing a prototype a step back should be should be made down to step 5. Otherwise, if the goals and objectives of the project were not met, go to step 1.

**The Seventh Stage.** Final control, results. The produced and stable working self-learning software and hardware vision complex makes continuous diagnostics of the air-tightness of food products packaging under the conditions of in-line production, makes a conclusion about the tightness of the packaging, directs defective products to a special storage device, it determines the productivity of the packaging line and the percentage of defects, notifies the personnel in case of defective cans detection by signaling (color and sound). In the future, it is necessary to provide possibility of expanding the functionality of the system and to foresee the opportunity to specify the settings of the packaging machine, etc.

At the final stage of creating a self-learning software and hardware vision system, the final control and monitoring of diagnostic results is carried out. The results of the hardware-software system of technical vision are summed up. Then, a return to the 1st stage is made, with the purpose of considering another place that generates the reject in the technological process.

## 4 Discussion

One of the main trends in 2016 in the field of object detection was the transition to faster and more efficient detection systems. This can be seen in such approaches as Faster R-CNN [5], R-FCN [6], Multibox [10], SSD [7] and YOLO [9] as a step to joint calculations on the whole image. This is different from the resource-intensive subnets associated with the Fast/Faster R-CNN techniques. This technique is commonly called "end-to-end training/learning".

A trade-off between the accuracy and size of objects when objects are detected on different architectures is presented on Fig. 4. (*Source: Speed/accuracy trade-offs for modern convolutional object detectors Proceedings* [9]).

On the above Fig. 4. (*Source: Speed/accuracy trade-offs for modern convolutional object detectors Proceedings* [9]), the mAP indicator (mean Average Precision) is plotted along the vertical axis, and a variety of meta-architectures for each feature extraction unit (VGG, MobileNet... Inception Reset V2) are displayed on the horizontal axis.

On Fig. 4. (*Source: Speed/accuracy trade-offs for modern convolutional object detectors Proceedings* [9]) mAP reflects the average accuracy for small, medium and large objects. As can be seen from the diagram, the accuracy depends on a number of factors such as the size of the object, the meta-architecture and the feature extraction block. For this picture, the image size is fixed at 300 pixels.

From the above study it can be seen that the Faster R-CNN model stands out from among others, but it is important to note that this meta-architecture is much slower than more advanced approaches such as R-FCN. For a more detailed comparison of the performance of R-FCN, SSD and Faster R-CNN, see Huang et al. [9].

It should be noted that, because of the difficulties of an exact comparison of the techniques of machine learning, the authors of the article [9] consider these architectures to be "meta-architectures" and suggest that they be combined with different blocks of feature extraction and permissions, such as ResNet or Inception.

The comparison of the accuracy indicators of the YOLO and FASTER R-CNN algorithms is given in the Table 4. The results were obtained on training samples of VOC 2007 and VOC 2012.

From the data of Table 1, we can draw the following conclusions. The first version of YOLO runs much faster than Fast and Faster R-CNN, which is due to the use of the R-CNN approach in RPN, which requires significant computations, however, is slightly inferior in accuracy. The error in YOLO simultaneously includes localization errors and classification errors, unlike R-CNN, where they are accounted for independently, since the composite networks are trained independently. These features have been given special attention in the construction of the error function in the development of

**Table 4.** Comparison of the presented algorithms.

| Method | Learning samples | General precision | FPS (frames per second) |
|---|---|---|---|
| Fast R-CNN | VOC 2007 + 2012 | 69.7 | 0.6 |
| Faster R-CNN VGG-16 | VOC 2007 + 2012 | 72.9 | 7 |
| YOLO | VOC 2007 + 2012 | 63.7 | 44 |
| YOLOv2 | VOC 2007 + 2012 | 76.9 | 68 |

Source: Compiled by the authors

YOLO v2 and YOLOv3 [18]. Neural networks were developed using the Keras/tensorflow stack, or tensorflow.

So, the advantages of the YOLO approach: to find objects on the image and to classify them, one neural network is needed, in contrast to networks like R-CNN. Also, this algorithm can be used quite successfully in real-time systems, but this requires appropriate computing power. The highest performance is achieved when using the GPU.

In scientific articles [8, 11], it is also emphasized that it is necessary to find a compromise between the tendency of increasing the speed of the application, consumed by computing resources and preserving the quality of accuracy. The authors of the report also adhere to this opinion and develop their own algorithm, taking into account the above mentioned features.

## 5   Conclusions

Modern enterprises of the food and processing industry are inextricably linked with the production and packaging lines of food products. The wide application of automatic lines and automated production systems are inextricably linked with computer managers. At the same time, the efficiency of the production lines essentially depends on the degree of automation of technological operations in the production. The maintenance of a rational production structure within the technological process is now of primary importance.

At present, considerable attention has been paid to the automation of the canning lines at the food industry. However, the overwhelming majority of modern production lines still do not use automatic tightness control of finished products. Meanwhile, in most cases, the final control is carried out selectively, or manually, on the basis of a visual inspection [19].

Under air-tightness today we understand the ability of the shell (body), its individual elements and compounds, to prevent gas or liquid exchange between media separated by this shell [20]. According to GOST 26790-85 "tightness" - the property of the product or its elements, excluding penetration of gaseous and (or) liquid substances through them [21].

As can be seen from the above definitions, "tightness" is a mandatory indicator of the quality of canned products and an important property that, of course, must be taken into account in the manufacture and engineering of sealed objects.

In this regard, the main attention in the modernization of flow packaging lines should be given to the creation and implementation of self-learning systems of continuous automated control of the tightness of finished products, which is the main task of the prototype of the self-learning software and hardware vision system being developed.

The review of the results of research and operation of food industrial enterprises and canneries in terms of guaranteed fulfillment of this quality index confirm the advisability of further improvement of models, algorithms and devices of automated systems for monitoring and controlling the sealing of canned food under the conditions of in-line production. In particular, the sealing systems used do not contain reconfigurable control devices for the sealing machine and non-contact leak tightness control devices for finished products.

# References

1. Potapov AA, Gulyaev YuV, Nikitov SA, Pakhomov AA, German VA (2008) The latest image processing techniques. Fizmatlit, Moscow
2. Lukianitsa AA, Shishkin AG (2009) Digital processing of video images. Ay-Es-Es Press Moscow, Moscow
3. Veremeenko KK, Zheltov SY, Kim NV, Sebryakov GG, Krasil'shchikov MN (2009) Modern information technologies in problems of navigation and guidance of unmanned maneuverable flying vehicles. Fizmatlit, Moscow
4. Redmon J, Divvala S, Girshick R, Farhadi A, (2016) You only look once: unified, real-time object detection. In: Proceedings of the IEEE computer society conference on computer vision and pattern recognition, pp 779–788
5. Ren S, He K, Girshick R, Sun J (2015) Faster R-CNN: towards real-time object detection with region proposal networks. In: Advances in neural information processing systems, pp 91–99
6. Dai J, Li Y, He K, Sun J (2016) R-FCN: Object detection via region-based fully convolutional networks. arXiv preprint http://arxiv.org/1605.06409
7. Liu W, Anguelov D, Erhan D, Szegedy C, Reed S (2015) SSD: Single shot multibox detector. arXiv preprint http://arxiv.org/1512.02325
8. Hong S, Roh B, Kim K, Cheon Y, Park M. (2016) PVANet: lightweight deep neural networks for real-time object detection, 9 Dec 2016. http://arxiv.org/1611.08588v2 [cs.CV]
9. Huang J, Rathod V, Sun C, Zhu M, Korattikara A, Fathi A, Fischer I, Wojna Z, Song Y, Guadarrama S, Murphy K (2017) Speed/accuracy trade-offs for modern convolutional object detectors. In: Proceedings of 30th IEEE conference on computer vision and pattern recognition, CVPR 2017, pp 3296–3305
10. Szegedy C, Reed S, Erhan D, Anguelov D (2014) Scalable, high-quality object detection. arXiv preprint http://arxiv.org/1412.1441
11. Wu B, Iandola F, Jin PH, Keutzer K (2017) SqueezeDet: Unified, small, low power fully convolutional neural networks for real-time object detection for autonomous driving. In: IEEE Computer society conference on computer vision and pattern recognition workshops, 8014794, pp 446–454
12. Ng AY (2011) On optimization methods for deep learning. In: Proceedings of the 28th international conference on machine learning. Stanford University, p 8

13. Federal Law of 02.01.2000 N 29-FZ (Edited on 13.07.2015) On the quality and safety of food. "Rossiyskaya Gazeta", No. 5, January 10

14. Tass (2017) Minpromtorg counts on the growth of production of food equipment 4 times by 2025. https://tass.ru/ekonomika/4572642. Accessed 10 Oct 2018

15. Ministry of Industry and Trade of the Russian Federation (2018) Strategy of the development of mechanical engineering for the food and processing industry of the Russian Federation for the period until 2030. http://minpromtorg.gov.ru/docs/#!43220. Accessed 10 Oct 2018

16. Rosstat (2018) Russia in figures. Short. St. sub. Moscow

17. Polyakov RK, Khodzhaev RS (2007) Creation of a subsystem of risk management in the field of entrepreneurship: monograph. Baltic Inst Econ Finance, Kaliningrad

18. Redmon J (2018) YOLO: real-time object detection. University of Washington – CSE. https://pjreddie.com/darknet/yolo. Accessed 10 Oct 2018

19. Dolgin NA, Long NA (2012) Operator model of the production line for canned food from fried fish. Izvestia KSTU. Sci J Kaliningrad FGBOU HPE "KSTU" 27:69–73

20. The Great Soviet Encyclopedia (1969–1978) Soviet Encyclopedia, Moscow

21. GOST 26790-85 (1985) Technique of leakage. Terms and definitions

# Digital Ecosystem: Trends in the Retail Segment

N. S. Kisteneva[✉], D. V. Ralyk, E. V. Loginova, and T. E. Gorgodze

Samara State University of Economics, Samara, Russia
{kisteneva, dinarar}@inbox.ru,
loginovaevl1982@gmail.com, gorgodze_t@mail.ru

**Abstract.** The relevance of the contribution is determined by the growing trends in the development of flexible forms of interaction in the market, which can be implemented as part of ecosystems. The use of an ecosystem approach in economics is a natural response to changing needs of market participants and the development of information technology base. The purpose of the contribution is to study the trends in the development of retail trade as one of the basic segments of the digital economy, which is a driver of growth. When studying the development pathways of retail trade under deep penetration of information technologies in the activities of both individual enterprises and the economy as a whole, the system approach is mainly used. The contribution clarifies retail functions in the digital economy, describes the levels of penetration of information technology in the activities of trading companies. The materials of the contribution are of practical value for retail organizations that form a strategic vision for the development of the digital economy.

**Keywords:** Development trends · Digital economy · Digitalization · E-Commerce · Ecosystem · Retail

## 1 Introduction

The study of the development of economic ecosystems is a relatively new area of scientific knowledge, which is considered in the works of foreign and Russian scientists. An economic ecosystem can be represented as a community consisting of interacting and complementary enterprises and organizations (producers, intermediaries, suppliers, financial companies, trade associations, government institutions, etc.) and individuals (consumers) that are concentrated around a key product [21]. In practice, you can find an idea of the ecosystem as a platform for goods and services (marketplace), which offers integrated products and services that cover the widest possible range of consumer needs, or as a self-developing organization that uses organic management approaches and sees the company as a "living organism". From our point of view, these ideas about the ecosystem are complementary elements and allow considering the system from different points of view. The ecosystem is characterized by openness, dynamism, flexibility, self-organization, and evolution, a variety of forms and types, competition and cooperation relations [23]. Scientific studies of economic

S. Ashmarina et al. (Eds.): *Digital Transformation of the Economy: Challenges, Trends and New Opportunities*, pp. 609–621, 2019.
https://doi.org/10.1007/978-3-030-11367-4_59

ecosystems are devoted to modeling interactive relationships between enterprises [26], detailing and systematizing interaction [6], and identifying factors for changes in the company's life cycle [22].

The theory of ecosystems received a powerful impetus to the development due to the digital economy. In Russia, non-profit organizations (Association of Electronic Communications, Digital Economy (Autonomous Non-profit Organization) are active in studying the digital ecosystem. It is a link between the business community and government bodies. It assists in research and accumulates a database on experience of introducing digital technologies in the activities of individual companies. Russian scientists consider in some detail the issues of applying ecosystems in various areas of the economy [2, 11, 19]. Activities in the field of research and the formation of ecosystems are developing rapidly and, most importantly, they are supported by the scientific community, the largest market participants and state institutions.

As part of further theoretical and practical understanding of this trend, we will look at the problems of retail development as one of the basic segments in the digital ecosystem. In particular, we will focus on the transformation of traditional trading functions and the nature of relations with key market participants.

## 2  Materials and Methods

The basis of the study of the retail ecosystem in the digital economy is made up of such methods as general scientific, systemic, structural-analytical, allowing studying the main directions of retail trade in the digital economy. The focus of this contribution is on the online retail market, i.e. B2C market. The main sources of statistical information are reports of analytical agencies and non-profit organizations that explore the digital ecosystem, retail markets, online trading (Digital Economy, Russian Association of Electronic Communications, Association of Online Trading Companies, Consulting companies Admirad, AT Kearney, McKinsey).

## 3  Results

The digital ecosystem is a system of interconnected market segments that receive value added through the use of information technology. One of the most developed approaches describing the ecosystem is the research methodology of RAEC Company, which involves the allocation of 7 hubs: media and entertainment, marketing and advertising, finance and trade, infrastructure and communications, government and society, cyber security, education and personnel. Each hub is viewed from the point of view of 10 slices: analytics and data, development and design, AI and Big Data, regulation, business models, hard ware, mobile, platforms, Internet things, startups and investments. From the point of view of the level of penetration of technology, the authors single out: general (initial), professional (exchange of experience, master classes), specialized (making decisions, round tables). Based on data for 2017, Russia's digital economy is estimated at 4.35 trillion rubles, which is 5.06% of GDP. The results of the retail segment (included in the hub "Trade and Finance (Electronic Commerce)") reached 20% of the total ecosystem values and show an increase of 28% compared to 2016 [4, 14]. In terms of the structure of the digital economy and the growth dynamics of its individual parts, retail trade is the largest and fastest growing segment (Fig. 1).

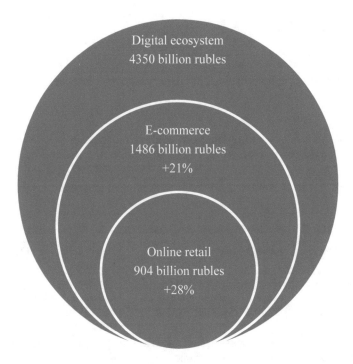

**Fig. 1.** The growth dynamics of the retail segment within the digital ecosystem ( Source: compiled by the authors)

The penetration rate of online sales in the market is estimated at 3% of retail trade turnover in Russia. However, the growth rate of online commerce is significantly higher than in the analog market: +28% compared to +1.2% in 2017. In general, we can say that the current situation reflects the global trend in the development of retail trade. It can be noted that the growth dynamics of online retail in Russia is slightly higher than the same indicator in the world: +28% compared to +23% [12, 24].

Assessing the attractiveness of Internet commerce in Russia, we turn to the Global Retail E-Commerce Index, which was developed by consulting company A.T. Kearney. The assessment is based on the following parameters: the size of the online market, consumer behavior, growth potential and infrastructure. In the rating of the market attractiveness, the retail e-commerce market in Russia ranks 8th. The market in Russia ranks 2nd in the ranking after China upon the growth rate, and this fact allows building optimistic forecasts. The market in Russia received the highest ratings upon sustainable consumer behavior, in which consumers demonstrate a movement towards Internet purchases, and the availability of the necessary infrastructure, which ensures the further development of electronic commerce (Table 1).

**Table 1.** The place of Russia in the global retail e-commerce market compared to leaders (according to The Global Retail E-Commerce Index).

| Criteria | Indicator | USA | China | Great Britain | Russia | Rank |
|---|---|---|---|---|---|---|
| Size of online market | 0,4 | 100 | 100 | 87,9 | 29,6 | 8 |
| Customer behavior | 0,2 | 83,2 | 59,4 | 98,6 | 66,4 | 9 |
| Growth potential | 0,2 | 22,0 | 86,1 | 11,3 | 51,8 | 2 |
| Infrastructure | 0,2 | 91,5 | 43,6 | 86,4 | 66,2 | 9 |
| Market attractiveness | 1,0 | 79,3 | 77,8 | 74,4 | 48,7 | 8 |

Source: compiled by the authors based on data of The Global Retail
E-Commerce Index [16, 18]

The study of features and problems of retailers' adaptation to new market conditions made it possible to identify the main directions of consistent optimization of methods and tools for commerce, marketing and logistics in the consumer market and to clarify the vision of key retail functions at the macro and micro levels (Table 2).

**Table 2.** Updating traditional retail functions in the digital economy

| Retail functions | Classical functions | Detailed content in the digital economy |
|---|---|---|
| Meeting the needs of the population in goods and services | Development of the optimal range of goods in accordance with the identified demand | Convenience of acquiring an unlimited number of types and varieties of goods and services in real time |
| Delivering goods to customers | Spatial movement of goods to the places of their sale | Spatial movement of goods to the places of their consumption |
| Maintaining a balance between supply and demand | Assortment and inventory management in accordance with the requirements of logistics | Anticipation of needs, high individualization of sales |
| Improving sales technology | Ensuring a high culture of service in the store | Continuous improvement of all forms of communication with the consumer |
| Promoting increased consumer personal time | Reducing time spent on service channels | Reducing the cost of finding information about a product and placing an order, omni channel marketing, creating an end-user ecosystem |

Source: compiled by the authors.

It should be noted that the comparison presented above is not aimed at identifying the fundamental contradiction between the classical and the new interpretation of retail functions. The goals are the same, only they use more relevant technologies and with higher requirements for retail service.

From the very beginning of its inception, the retail trade enterprise functioned to achieve internal and external competitive advantages. Today, their creation and maintenance is more difficult. A new level of competition in the new environment with a new consumer requires new approaches to integrate retail into the digital economy.

Effective planning of trade and technological processes in retail trade is already promoted not only by the competence of administrative and management personnel and automated business accounting. IT solutions allow for long-term cooperation, maximum loyalty of business partners and customers.

The influence of scientific and technical factors on the activity of the retail market was also reflected in the qualitative change in the distribution chain. There is a trend of multi-structural competition, integration of production and trade. The improvement of information technologies and the emergence of innovative developments in the field of mobile communications have led to the creation of own retail networks by manufacturers.

The transformation of the retail trade system in Russia and the definition of its contribution to the development of the digital economy are undoubtedly associated with the advent of the Internet, the technologies of which have led both to the emergence of new retail formats and intensification of innovative technologies to promote consumer goods and services.

The phased penetration of information technologies into the retail business of Russia can be represented as follows (Fig. 2).

Each level of digitalization of sales of goods and services to the final consumer in Russia has its own distinctive features and objective obstacles to the full and effective development.

## I. The development of hybrid retail formats

Due to the immediate need for a combination of the advantages of traditional (off line) and online commerce (on line), the notion of hybrid or combined retail network formats was formed. So, along with the traditional ones, the cybermarket and Outpost formats [25, 27, 28] became very popular.

According to experts, the insufficiently high level of the logistics component of the market is not the only obstacle to the development of new formats of the retail trade in Russia. It is necessary to take into account a low degree of public involvement in Internet commerce, the unwillingness to make an independent choice of complex technical products as described in the electronic catalog. The target audience of cybermarkets is youth and advanced Internet users. In addition, the mentality of Russian buyers, especially kinesthetics, is still largely affected, as they still prefer to see the product and evaluate it immediately before purchase.

At the same time, the terminals complement, but in no way replace, sales assistants, who, due to their professionalism and competence, are necessary for buyers when they have difficulties in the process of self-selection of goods and comparing their characteristics. And cybermarkets offer a solution to the problem of leveling deficiencies in the combination of two types of trade, covering consumer segments of both innovators and conservatives.

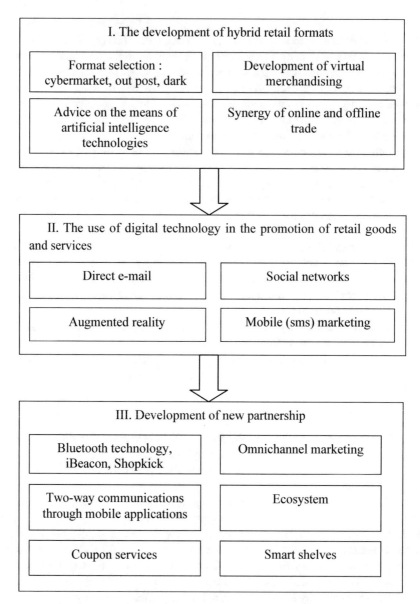

**Fig. 2.** The sequence of embedding retail in the digital economy of Russia (Source: compiled by the authors)

Cybermarkets in this combination of mutually complementary formats in the network are large hubs from which the goods arrive at Outpost pre-order points, which are a modern interpretation of the mini-market of consumer goods.

Cybermarkets successfully develop, mainly in large cities of Russia. In order to provide wider territorial coverage, including small cities, retailers are developing a network of Outpost minimarkets.

The Outpost format combines a pre-order point and an electronic catalog access point (terminal). The advantage of this format is reduced logistics costs associated with streaming processes in the micro logistics system of a retail trade enterprise, both in the sales area and in the warehouse, the total area of which is not more than $50^2$ m. In turn, the minimum cost of organizing and maintaining the format allows developing the network of pick-up points, ensuring their territorial availability for buyers.

The cybermarket, as a rule, fulfills the full potential, which gives the buyer a hybrid retail format along with modern technologies. A new cybermarket service has been created, allowing the buyer to quickly and easily determine the choice of goods on the site.

The virtual seller technology is an intelligent system that, in the "question-answer" mode, will help the buyer to choose the right product among the diverse assortment of different brands and prices.

In fact, the technology of intellectual service provides the necessary active interaction with the buyer, being a "virtual seller", which finds out the needs of the buyer and helps him make a choice, as well as the sales assistant in a real store.

Taking into account the European sales experience and the readiness of the FMCG market in Russia, local retailers are developing an experimental dark store format. This is a fundamentally new format for the country, which acquaints a company with the idea of reducing logistics costs. It happens not due to reduced sales and operational personnel, but due to reduced time spent by customer flows at points of sale. The high speed of order picking and delivery of goods contributes to the implementation of the above-mentioned social retail function - saving time for the buyer.

At the moment, the creation of hybrid format networks in Russia in various industries is a one-off project. It should be noted that the interest of investors in Internet commerce projects is growing. In recent years, e-commerce has been the main focus of venture capital investments in Russia. The potential of the e-commerce market in Russia is huge, and new formats are very flexible systems. With an increase in the youth audience and the involvement of the adult population in information technology, young online stores with multi-channel order networks and an established distribution system and product delivery will noticeably press the network retail giants.

Despite the development of digital forms of trade, merchandising, traditionally regarded as an off-line event, has also undergone a transformation and is described as "virtual". The consistency, visibility, structuring and visual appeal of the goods presented on the store's website, as well as in the case of stationary stores in the complex, ensure the ability of goods to be sold without the participation of the seller.

II. The use of digital technology in the promotion of retail goods and services

Against the background of the development of hybrid retail formats, companies are actively exploring technically advanced tools to increase sales efficiency. New conditions of competition, in which the majority of companies have reached approximately equal indicators of quality and customer service culture, force us to find ways to inform the buyer about updating the range of products and prices more convenient for the buyer.

Russia has a fairly large-scale mobile coverage: in 2017, 245 mobile phones per 100 households. Regardless of the complexity of the technical device of the consumer (ordinary phone or smartphone) allowed implementing sms-marketing, which demonstrated a high efficiency. Unlike direct email, messages are scanned by the addressee instantly and, if necessary, are stored until the client takes advantage of the received advertising offer. E-mail, which is actively used by 64.3%, is, as a rule, viewed in the presence of the Internet and technical means (smartphone, tablet, computer), but sms-message does not require any additional conditions for the recipient to familiarize [13].

The difficulties of distributing the recommended form of mobile promotion of retail goods and services include its relatively high cost and the possibility of negative emotions from the recipient as a reaction to annoying intrusive advertising. However, the distribution of sms-messages of the "pull" type (communication strategy of "pulling out") implies not only the initiation of potential customer buying activity, which is subsequently implemented directly in the store, where he voices his request for certain promoted products, but also adheres to the legality of advertising. The client gives prior consent to receive messages to his phone number, which eliminates a certain negative when reading them.

Unlike mobile marketing, namely, sending sms-messages to customers' phones, the distribution of advertising information via e-mail has less negative reaction from the audience, often perceiving the sender's perseverance as a certain obsession. In addition, the content of the advertising message is less limited, and you can send a larger amount of text with images of promoted products.

The cost of this communication event depends on the scale of the store. For non-networked retailers, direct mail costs are relatively small.

The disadvantages of traditional mailing of promotional materials by regular mail, such as changing the address of the client's residence, the high costs of printing and postal services are largely leveled in the promotion of goods in the electronic information space.

Foreign retailers also master the technology of "virtual reality", allowing customers to see 3D-visualization of future designs in a specially equipped room, for example, goods for repair, interior design, furniture, just before making a purchase decision [1, 8].

Currently, social networks are used by consumers not only for virtual communication, but also to search for information about their intended purchases.

More than 47% residents of Russia are actively using social networks, usually visiting via mobile devices [10].

Store management should take into account the distinctive feature of the current generation - the almost total presence of the target audience at various communication platforms can be regarded as an opportunity to replenish the customer base, informing it about updating the range of goods and price offers, expanding the technical possibilities to promote goods and services.

III.  Development of a new partnership

The development of mobile applications has become a new trend in the development of retail for many retailers. However, there are a number of shortcomings in their operation, especially in attracting those customers for whom visiting a real shopping outlet, rather than an e-commerce enterprise, is preferable. Through mobile marketing and direct mail, you can distribute information about events, for example, invitations to a closed (for owners of personal discount cards) sale.

Just as in the case of mobile marketing, the process of tracking the effectiveness of an advertising message is simplified. When the client visits the store, it is specified which means of communication contributed to his interest and made him visit the point of sale of goods.

One of the most advanced technical achievements, which are gaining immense popularity in the dialogue and cooperation "store-buyer", is the use of iBeacon beacons [20].

In the business world, there has been a long and active opportunity for energy-saving Bluetooth in the retail sector, contributing to the dissemination of relevant promotional and informational materials, improving the quality of customer service, as well as developing corporate intelligence and creating favorable working conditions for employees.

The iBeacon beacons allow you to establish an innovative communication between physical objects and gadgets, which creates a huge trading potential in retail.

Retailers can signal to the customer's smartphone about a passing or upcoming marketing event that could be the reason for a visit to the store.

A unique development, which is recommended to be tested by Russian retailers, is the beacon app Shopkick. This cooperation encourages bonuses to any action of the buyer, regardless of visiting the on-line or off-line store [1].

With moderate activity of using beacons (striving to eliminate the abuse of this opportunity to overload buyers with advertising information), the growth in retail sales may be due to direct sending messages, and to using the generated database and profiles for more personalized alerts to buyers, taking into account their consumer preferences.

Retailers in Russia are not fully and not so quickly able to integrate all the diversity of digital sales technologies into their business, partly because Bluetooth technology in Russia has not been rehabilitated yet, information has traditionally been sent to consumers via sms messages. Another reason is the rather high cost of technologies, inaccessibility for their implementation in small non-chain stores. However, there are practically no restrictions in attracting social networks to increase their own sales.

Another new trend in the development of retail sales was the promotion of goods and services involving coupon services, the most famous of which are Groupon, Kupibonus, Biglion, Vigoda, Kupikupon, etc.

The scheme of mutually beneficial cooperation through discount sites is quite simple. Customers, presenting a coupon at a point of sale, have the opportunity to purchase relatively expensive goods at a reasonable price. The store ensures large-scale popularity, the ability to attract new customers from related segments of the retail market, turning to coupon services for other goods and services. Advertising intermediaries in the market of collective purchases receive a corresponding commission [29].

With all the advantages of the presented method of goods promotion in retail, a number of limitations should be noted. The client needs the Internet, multifunctional devices for registration, purchase and printing of the coupon. In addition, the economic feasibility of attracting coupon services, taking into account the size of their remuneration for intermediary services, is achieved due to the initially inflated price for goods. The store deliberately creates the illusion of benefits for the buyer and, as a result, the problems of further maintaining customer loyalty. In this regard, we need a balanced approach to the selection and use of the advertising technology described above, careful calculation of the expected sales performance, high-quality, timely and guaranteed provision of promoted products at a retail outlet.

Another technology that significantly improves the quality of domestic logistics in retail is smart shelves. Due to various circumstances, the prices in the store are not timely updated or the working stock of goods in the sales area is not replenished. The situation of the lack of the right product on sale takes more than 4% of the annual income of large retailers, which prompted them to take into account the technology of "smart shelves", allowing monitoring prices and stocks in real time [3].

With the help of built-in sensors alert, store employees receive information about the items of the shelf, warning them about possible theft and the need to replenish stocks on the shelf.

Changes in prices are carried out automatically when they are corrected in the enterprise information system, providing buyers with extremely accurate information, and saving time on updating price tags by employees.

In addition, suppliers can be connected to the "smart shelves" system, which, in turn, thus maintains the intensity of the flow of goods and uninterrupted trade by automatically sending supplies to stores.

At present, it is difficult to assert the readiness of Russian retailers to implement the omnichannel marketing strategy. Unlike multichannel, where there is some disconnection of channels and devices, the guiding principle of omnichannel marketing is the organization of constant interaction with the buyer, regardless of his location and his chosen method of communication.

The creation of a consumer ecosystem is completely new and undeveloped in the retail in Russia. The development of a long-term partnership of stores and consumers is possible with the participation of service companies providing the sale of goods. In the future, it is necessary to have a single technological platform that can not only unite the desire of the subjects of the consumer market, but also realize the main goal of the retail trade - customer satisfaction with the maximum convenience of purchasing goods and services.

The development of the ecosystem allows satisfying a wide range of requirements of participants and customers within a single platform. Benefits for consumers are primarily in obtaining additional benefits while using several products and services through the "single window" system. Benefits for a retailer company are determined by increasing customer loyalty, sales growth through the use of additional channels and reducing risk. To form the ecosystem, a company needs to have a sufficient client base and necessary competences in the area of customer relationship management, including a high level of trust. A prerequisite is openness to change and a high level of adaptation to changing environmental conditions. The management of innovations based on the flexible interaction between key divisions of the company's organizational structure is of great importance in the ecosystem.

In the context of further research, it is necessary to pay attention to the integration of the ecosystem of the digital and real (analog) retail market, since the latter remains significant both in terms of the main economic results and is preferable for a number of products and consumers.

## 4 Discussion

One of the driving factors in the development of the digital ecosystem is retail trade, the features of its functioning under the conditions of the development of information technologies are considered in the works of Russian and foreign authors. V. P. Kupriyanovsky and co-authors consider in detail the technical aspects of retail digitalization [7]. V.B. Anfinogenov, analyzing innovative technologies in the retail trade, paid special attention to the problems of digital informatization of the retail trade in Russia [3]. In the global information space, the topic of retail digitalization is also revealed through technical devices to rationalize trade and technological processes. For example, C. Fuentes, K. Bäckström, K. Svingstedt describe the consequences of integrating smartphones into the trading and warehouse space of stores, as a result of which the system of relationships between market actors is transformed [15]. A new understanding of retail allows you to present it as socio-material assemblages. In the study of L. Bollweg, R. Lackes, M. Siepermann, A. Sutaj, P. Weber you can also get confirmation of the theses on new requirements for the formation of the external and internal competitive advantage of retailers [5]. Companies, adapting their services to the trends of the digital economy, ensure the loyalty of their customers. S. Claims, K. Quartier, J. Vanrie [9] considered the needs of retail professionals to acquire new competencies in the digital environment. The problem is highlighted by an insufficiently complete understanding of the importance of market changes when training retail professionals. The publication of M. Holmlund, T. Strandvik, I. Lähteenmäki has a practical value for creating under-studied ecosystems of retailers in Russia, which demonstrate the importance of multifaceted digital platforms for achieving maximum customer satisfaction [17]. The provisions on the prospects for the implementation of achievements of modern science and technology in the economy do not contradict each other, concentrating on certain aspects of its digitalization. The opportunities of the retail trade in Russia, which opened with the expansion of the Internet space, have been considered by domestic scientists and practitioners more than once. The study of the content of scientific publications and professional portals allowed systematizing the vision of key functions and confirm the logic of the level of retail immersion in the digital economy.

## 5 Conclusions

Economy digitization in Russia has set the goals for the development of modern information technologies in one of its most important sectors - retail. Closing the chain of economic ties in the process of movement of goods from producers to consumers, retail trade significantly affects the level of development of the digital ecosystem.

The introduction of innovative technologies for the retail sale of goods, such as attracting social media and mobile devices, ensures long-term loyalty of customers who value the convenience of choosing a product and paying for it. With these tools, the store creates brand loyalty both online and offline. In addition, modern technologies support the desired level of information exchange between the seller and the buyer, approaching the maximum customization of the sales offer.

The prospects and benefits of e-commerce become so indisputable that every year traditional federal retailers online are increasing. An interesting fact is that if earlier the electronic catalog was considered only as an attachment to a regular store, then today stores are becoming an attachment to the electronic catalog.

Regardless of how easy it is for a modern consumer to make purchases in the online store, in a number of cases, visiting the offline retail outlet is a more attractive and desirable event for most people. The reasons for this phenomenon include the continuous improvement of technologies for the retail sale of goods and customer service. Thus, elements of the digital and analog markets are integrated in retail, which may be a promising area of research.

# References

1. Akatkin YuM, Karpov OE, Konyavsky VA, Yasinovskaya ED (2017) Digital economy: conceptual architecture of the digital industry ecosystem. Bus Inform 4(42):17–28
2. Androsik Yu N (2016) Business ecosystems as a form of cluster development. In: Proceedings of BSTU. Series 5: economics and management 7(189). https://cyberleninka.ru/article/n/biznes-ekosistemy-kak-forma-razvitiya-klasterov/. Accessed 10 May 2018
3. Anfinogenov VB (2017) Innovative retail sector. Bull Saratov State Socio Econ Univ 4(68):99–102
4. Aptekman A, Kalabin V, Klintsov V, Kuznetsova E, Kulagin V, Yasenevets (2017) I Digital Russia: a new reality. https://www.mckinsey.com/ru/~/media/McKinsey/Locations/Europe%20and%20Middle%20East/Russia/Our%20Insights/Digital%20Russia/Digital-Russia-report.ashx/. Accessed 7 May 2018
5. Bollweg L, Lackes R, Siepermann M, Sutaj A, Weber, P (2016) Digitalization of local owner operated retail outlets: the role of the perception of competition and customer expectations. In: Pacific asia conference on information systems, PACIS: proceedings 458. https://aisel.aisnet.org/pacis2016/348/. Accessed 03 May 2018
6. Brown R, Mason C (2012) Raising the batting average: re-orientating regional industrial policy to generate more high growth firms. Local Econ 27(1):33–49
7. Business Insider (2015) 5 ways technology is revolutionizing the way we shop. http://www.businessinsider.com/sc/technology-changing-the-in-store-retail-experience-2015-10/. Accessed 14 Mai 201
8. Charlton G (2014) 12 more examples of digital technology in retail stores. https://econsultancy.com/blog/64408-12-more-examples-of-digital-technology-in-retail-stores/. Accessed 10 May 2018
9. Claes S, Quartier K, Vanrie J (2016) Reconsidering education in retail design: today's challenges and objectives proceedings - D and E: 10th international conference on design and emotion
10. Data Economy Russia 2024 (2018) Portal of non-profit Organization Digital Economy. https://data-economy.ru/2024/. Accessed 13 May 2018

11. Doroshenko SV, Shelomentsev AG (2017) Entrepreneurial ecosystem in modern socio-economic research. J Econ Theory 4:212–221
12. Ecommerce Admirad Report 2017/2018 (2018). https://blog.admitad.com/?p=12859/. Accessed 7 May 2018
13. Economy Runet 2017. Ecosystem of Russia's Digital Economy (2017). http://raec.ru/upload/files/de-itogi_booklet.pdf/. Accessed 10 May 2018
14. Ecosystem of the Digital Economy of Russia (2018) Research portal of the association of electronic communications. http://digitaleconomy.rf/. Accessed 10 May 2018
15. Fuentes C, Bäckström K, Svingstedt A (2017) Smartphones and the reconfiguration of retailscapes: stores, shopping, and digitalization. J Retail Consum Serv 39:270–278
16. Global Retail E-Commerce Keeps on Cleecking (2015). https://www.atkearney.com/documents/10192/5691153/Global+Retail+E-Commerce+Keeps+On+Clicking.pdf/abe38776-2669-47ba-9387-5d1653e40409/. Accessed 3 May 2018
17. Holmlund M, Strandvik T, Lähteenmäki I (2017) Digitalization challenging institutional logics: top executive sensemaking of service business change. J Serv Theory Pract 27 (1):219–236
18. Kearney AT (2017) The 2017 Global Retail Development Index. The age of focus. https://www.atkearney.com/documents/10192/12766530/The+Age+of+Focus%E2%80%93The+2017+Global+Retail+Development+Index.pdf/770c5a53-d656-4b14-bc6c-b0db5e48fdc1/. Accessed 13 May 2018
19. Kogdenko VG (2018) Specifics of analysis of companies operating in the digital economy. Econ Anal Theory Pract 1:424–428. https://doi.org/10.24891/ea.17.3.424/. Accessed 14 May 2018
20. Kylgin M (2016) iBeacon for retail - a detailed overview of the mechanisms of application. http://startup.today/article-mayachki-ibeacon-dlya-roznichnoj-torgovli-detalnyj-obzor-mexanizmov-ispolzovaniya/. Accessed 10 May 2018
21. Moore JF (2005) Business ecosystems and the view from the firm. Antitrust Bull Fall 58
22. Reeves M, Levin S, Ueda D (2016) Company as an ecosystem: survival biology. Harv Bus Rev Russ. https://hbr-russia.ru/biznes-i-obshchestvo/fenomeny/a17381/. Accessed 10 May 2018
23. Teece DJ (2007) Explicating dynamic capabilities, the nature and micro-foundations of (sustainable) enterprise performance. Strat Manag J 28(13):1319–1350
24. The Association of Internet Trade Companies (AITC) (2018). http://www.akit.ru/. Accessed 8 May 2018
25. The largest cybermarket of electronics Yulmart launches a new service - "Yulmart. Expert" (2013). http://www.dz.ru/news/press/2013-002/. Accessed 14 May 2018
26. Tian XH, Nie QK (2006) On model construction of enterprises' interactive relationship from the perspective of business ecosystem. South China J Econ 4:50–57
27. Vasfilov D (2013) Kibermarket - a new format of stores of household appliances and electronics. http://best-guide.ru/?p=2981/. Accessed 10 May 2018
28. Volkova O (2013) Through cybermarkets to the outpost format. http://www.retailer.ru/item/id/81034/. Accessed 10 May 2018
29. What are coupon services and is it possible to save with them? (2016). http://ecotonkosti.ru/chto-takoe-kuponnye-servisy.html/. Accessed 10 May 2018

# Integrated Reporting of Public Educational Institutions: Challenges of Modern Times

O. I. Averina[1], N. F. Kolesnik[1], and V. A. Manyaeva[2(✉)]

[1] National Research Ogarev Mordovia State University, Saransk, Russia
{oiaverina,kolesniknf}@mail.ru
[2] Samara State University of Economics, Samara, Russia
manyaeva58@mail.ru

**Abstract.** The research rationale is determined by the need to develop theoretical and methodological provisions of integrated reporting of public educational institutions based on the requirements of the digital economy.

The purpose of the study is to ground the conceptual approaches to the formation of integrated reporting of the public educational institution in the digital economy.

The leading approach to the study of this issue is the situational and systemic approaches in the framework of the theory and methodology formation of the integrated reporting of the public educational institution, allowing to monitor and analyze the key indicators of the responsibility centers.

According to the results of the study, they are as follows: a business model was developed and the responsibility centers of the public educational institution were identified in the conditions of the digital economy; the key indicators of the responsibility centers were determined, allowing to assess their results under the influence of the economy digitalization; the composition and the order of the integrated document formation, which meet the needs of all interested users as well as the requirements of the digital economy, were proposed.

The research materials can be useful for practitioners and academic specialists in the field of integrated reporting of public educational institutions in the process of the economy digitalization, teachers, postgraduate students, undergraduates and students who are studying economics and management in higher educational institutions.

**Keywords:** Integrated reporting · Business model · Responsibility centers · Public institutions of higher education · Digital economy · Key indicators

## 1  Introduction

Business development radically changes the key approaches to the implementation of social, environmental, production and management business processes, as well as the information space of economic entities. Globalization of the world economy and the development of economic theories make significant changes and complications in the Russian institutional environment. The importance of this process is confirmed by the necessity of Russia's transition to the digital economy. In 2017, the Edict of the President of the Russian Federation of 9 May 2017 № 203 "On the Information Society

© Springer Nature Switzerland AG 2020
S. Ashmarina et al. (Eds.): *Digital Transformation of the Economy: Challenges, Trends and New Opportunities*, pp. 622–634, 2019.
https://doi.org/10.1007/978-3-030-11367-4_60

development in the Russian Federation for 2017–2030" was issued. The programme "Digital economy in the Russian Federation" (hereinafter - the Programme) was approved by the order of the Russian Federation Government of 28 July 2017 № 1632 - p.

The main mode to ensure the efficiency of the digital economy is the implementation of data processing technology, which will reduce costs in the production of goods and services; development of an effective system of data collection, its processing, storage and submission to consumers, ensuring the needs of the state, business and citizens in the relevant and reliable information, based on the transparency and accountability of economic entities.

The section about statutory regulation is provided in the stages of development of digital economy directions in the "road-map" of the Programme. It specifies the task of creating the legal issues for the introduction of new rules for the collection of reporting, excluding the repetition of the collected information, providing ways to obtain it remotely, and is aimed at covering the society and the state requirements of the necessary data in real-time mode. This may fully relate to integrated reporting that is directed at stimulating economic activity of an economic entity.

In the documents of the International Integrated Reporting Council (IIRC), which was established in 2010, it is noted that integrated reporting is legitimately one of the main drivers of the corporate reporting system development.

Only a number of the largest domestic corporations representing the manufacturing and financial sectors of the economy are actively engaged in the preparation of integrated reporting. The weakest position in this process is occupied by the organizations of the public sector of the economy at the present time.

In different years, the following domestic and foreign researchers such as Arkhipenko [1], Bulyga [2], Chrisytensen et al. [3], Druzhilovskaya [5], Cooper and Owen [4], Gray et al. [9], Rudenko [16] Samuel et al. [17], Owen [14], Solomon and Maroun [18], Eccles et al. [19], Zaborovskaya and Kovyazina [20], Zyryanova and Tarnovskaya [21] etc., brought into focus of their works the matters of content and technique in formation of integrated reporting.

Commending the great contribution of scientists to the solution of such a complex problem, it should be noted that many issues were studied not during the development of the digital economy in the Russian Federation, in which data in the digital form are a key factor in all spheres of socio-economic activities.

Public institutions of higher education are full participants of economic market relations, therefore, the problem of formation of integrated reporting, which has to meet the digital economy requirements, for them, as well as for commercial establishments, is becoming increasingly vital. The integrated reporting involves the system of the interconnected financial and non-financial indicators designated for management of the performance of the organization as a whole, in the context of its business processes, types of activities, responsibility centers for achievement of the overarching aim.

The need in closing the gap between the private and public sectors organizations of the economy and the development of new approaches to the management information provision under the influence of digitalization have brought up the necessity and promptitude of this study preparation to date.

*The aim of the study* is to develop conceptual approaches to the of integrated reporting formation of public institutions of higher education in the digital economy.

The objectives of the study:

- to build a business model and identify responsibility centers of public institutions of higher education in the digital economy;
- define the key indicators for responsibility centers, allowing to assess their results in the digital economy;
- to propose the composition and procedure for the formation of an integrated document that meets the needs of all interested users on the basis of the information integration about the compliance with the digital economy requirements.

The scope of research is a set of theoretical, methodological and practical aspects of integrated reporting in the digital economy.

The research object is the integrated reporting in the digital economy of public institutions of higher education of the Russian Federation (hereinafter - universities).

The significance of the study lies in the development of theoretical aspects of the university integrated reporting role and place, the development of the procedure for establishment an integrated document that meets the needs of all interested users on the basis of information integration in the digital economy. The presented research has got the following structure: abstract, keywords, introduction, research methods, results, discussion, conclusions, references.

## 2    Materials and Methods

The study was conducted with the help of general scientific and special techniques and methods: observation and study of experience to date, methods of conceptual and socio-philosophical analysis, procedures of systemic, comparative and structural-functional analysis. Normative and legal acts in the field of reporting, scientific literature, current practice of accounting and reporting of public institutions of higher education were used as the materials of the research.

The experimental base of the research is the public institutions of higher education of the Russian Federation.

## 3    Results

### 3.1    Business Model and Responsibility Centers of Public Institutions of Higher Education in the Digital Economy

The Russian state university is the key link in the system of higher education, which is responsible to the state for the formation and development of the collective national intelligence. At the same time, the state university acts as a subject, and its source of financing is both budgetary and non-budgetary funds.

When creating a university business model, we took into account the specifics of its activities in the context of the digital economy, which is largely determined by the decision-making mechanism, by whom they are made, what they are focused on, and what the needs of interested users are. The business model of the university is presented in Fig. 1.

Since the system of higher professional education has not only got significant societal importance, but also is the foundation for the socio-economic development of the state, the formation of the integrated reporting should be based on the following principles: strategic direction and a vision for the future, the coherence of information, verifiability, accessibility, efficiency, consistency, publicity, transparency, complexity, productivity, effective strength, and meet the requirements of the digital economy.

When forming the integrated reporting, it is necessary to apply information technologies on a new technological basis, to promote projects for the implementation of electronic workflows, to create conditions for enhancing the credibility in electronic documents.

In this regard, in order to obtain information about the status and functioning of the business model of the university as a complex object it is proposed to form an integrated reporting, containing financial and non-financial information in the context of the following responsibility centers: costs, revenues, profit, investments, social responsibility, information security.

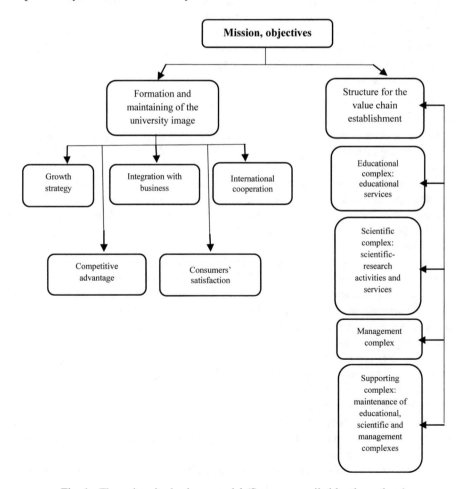

**Fig. 1.** The university business model (Source: compiled by the authors)

The suggested responsibility centers of the university are allocated based on the requirements of its strategic development and functioning in the digital economy.

*The cost centre* is a type of activity, the head of which controls and analyzes the costs arising in the process of its implementation and includes:

- the Vice-rector for academic studies, who is responsible for the educational process, providing: the formation of the basic competencies of the digital economy for all graduates and students; support for the advanced educational projects of the digital economy; the development or selection of an information system to support the student's individual competencies profile; participation in the postgraduate and master's degree programmes competition in each area of "cross-cutting" technologies, taking into account the results of monitoring personnel needs in the development of "cross-cutting" technologies of the digital economy; creation of organizational and methodological conditions, forms of appraisal, curricula, and methodological materials in the framework of the digital economy, etc.;
- the Vice-rector for science and research, who is responsible for scientific-research and innovation activities, including international scientific and technical cooperation in the areas of "cross-cutting" technologies regarding the digital economy, solution of research problems in the field of the digital economy, including in the framework of fund programmes that support scientific, scientific, and technical and innovative activities, etc.

*The revenue center-* is a type of activity, the head of which controls, analyses and influences the revenue collection from the provision of educational services, the performance of scientific-research works and services, activities ensuring the educational and research processes - it is the Vice-rector for academic studies and the Vice-rector for science and research.

*The profit centre* is a structural unit of the institution, the head of which controls and analyses the revenues and costs related to the sphere of its activity, that takes responsibilities for the financial results obtained in connection with the functioning of its subordinate structural unit – it is the Rector of the university, the Vice-rector for administrative and economic work.

*The investment centre* is a type of activity, the head of which participates in the formation of the investment policy of the university, determines the directions of capital expenditures and is responsible for the efficiency of their usage - it is the Vice-rector for investment activities. The investment centre is developing a program for the development of material and technical base, including broadband Internet access for educational institutions.

*The social responsibility centre* is a type activity, the head of which controls and analyses the expenditures for social policy of the university – it is a Vice-rector for academic studies, Vice-rector for social work. Their responsibilities include: development of a recommendation system for professional orientation of students used by the university on the basis of the map of new professions; providing conditions for the development of students' entrepreneurial competencies; development and implementation of professional development retraining programmes, continuous professional development of the professional and teaching staff, which ensures their readiness to implement modern educational process models, taking into account the requirements of the digital economy, etc.

*The information security centre* is a type of activity, the head of which ensures information security - it is the Vice-rector for informatisation. This manager's duties include: compliance with the requirements for the stability and security of communication networks, equipment and software, as well as ensuring control (supervision) over their compliance; compliance with the requirements for the use of domestic computer, server and telecommunications equipment at the facilities of the data processing infrastructure; implementation of analysis and assessment of information security risks and threats adequacy of existing standards of information security in systems, etc.

In order to carry out the functions assigned to the responsibility centers of the University, integrated reporting is needed, which will contain information about the indicators each centre is responsible for. The next section of the study considers this direction.

### 3.2 Key Indicators of the University Responsibility Centers that Allow to Assess their Results in the Digital Economy

The degree of the responsibility centre impact on the process of formation, distribution and use of resources depends on the power of its head.

It is expected to build an impact system on the basis of key indicators (financial and non-financial) for the responsibility centers and the motivation of their activities, including the conditions of the digital economy. In this matter, we should take into account the development of the indicators, the functioning area of a particular responsibility centre, and the possibility of its real impact on them.

Usually a kind of situation happens at the universities when the same indicators are influenced by several responsibility centers, that is why the responsibility for individual indicator is difficult to divide between them. The indicators composition of the university responsibility centers, the information about which will be reflected in the integrated reporting, is proposed below (Table 1).

The responsibility centres indicators should be carefully monitored, analyzed and assessed. For this purpose, it is necessary to form appropriate reporting, which should be detailed as much as possible and focused on the specifics of the certain responsibility center. This issue is being solved in the framework of determining the composition and procedure for the formation of integrated reporting, which defined the following research direction.

### 3.3 The Composition and Procedure of the University Integrated Reporting Formation in Respect of the Indicators' Analysis and Control of its Work in the Digital Economy

As a matter of the integrated reporting formation, it is necessary to develop a unified information accounting system, in such a way that it provides identified and the expected users' information requests. The integrated accounting system consists of interconnected target subsystems: accounting (financial), tax, management and

**Table 1.** The key indicators of the university responsibility centres

| Type of responsibility centre | Objectives | Key indicators | Responsibility centre |
|---|---|---|---|
| Cost centre | Control and analysis of costs | The percentage of numerical baselines completion of students' admission, including promising educational digital economy projects. | Vice-rector for academic studies |
| | Cost optimization | The amount of expenses for the implementation of educational activities, including promising educational digital economy projects. Share of costs (by type of costs) per one employee (per one academician), including promising educational digital economy projects | |
| | | The amount of spent on information systems support of a students' competencies individual profile. The amount of expenses spent on creation of organizational and methodical conditions, forms of appraisal, curricula, guidance materials within epy digital economy The amount of expenses spent on international scientific and technical cooperation in the areas of "cut-crossing" technologies in the field of the digital economy The amount of expenses spent on research works in the sphere of the digital economy | Vice-rector for science and research |
| Revenue centre | Control and analysis of revenues | The amount of revenue spent on educational activities, including promising educational digital economy projects | Vice-rector for academic studies |
| | Revenue optimization | The amount of grants spent on the implementation of postgraduate and master's programmes in each area of "cut-crossing" technologies The amount of grants that supports scientific, scientific and technical and innovation activities The amount of revenue that comes from scientific activities, including promising educational digital economy projects. Share of revenue (by type of income) per one employee (per one academician) | Vice-rector for science and research |
| Profit centre | Control and analysis of revenues, costs and financial results | Net operating result by type of an activity: educational, scientific and related one | Rector |
| | | Share of net operating result (by types) per one employee (per academician) The amount of costs, revenues and financial results for related university activities | Vice-rector for administrative and economic works |

(*continued*)

**Table 1.** (*continued*)

| Type of responsibility centre | Objectives | Key indicators | Responsibility centre |
|---|---|---|---|
| Investments centre | Control and analysis of investment cost and revenues. Assessment of cost-effectiveness of investment management | The amount of funds allocated for the implementation of investment activities<br>Effectiveness of investment management (standard for return on investment)<br>The amount of funds spent on proving broadband Internet access for educational institutions | Vice-rector for investment activities |
| Social responsibility centre | Control and analysis of costs on social development | The percentage of employed graduates<br>The ratio of the average salary to the industry average indicators and the road map benchmarks<br>The amount of costs spent on the development and maintenance of a recommendations system for students' professional orientation<br>The amount of costs spent to ensure conditions for the development of students' entrepreneurial competencies.<br>The amount of costs spent on the development and implementation of professional development and retraining programmes, continuous professional development of teaching staff, which ensures their readiness to implement modern educational process models, taking into account the requirements of the digital economy | Vice-rector for academic studies<br><br>Vice-rector for social work |
| Information security centre | Control and analysis of information security costs<br><br>Analysis and assessment of information security risks and threats adequacy of existing standards of information security in systems | A degree of information system security<br>The number of disclosure of confidential<br>The amount of costs for the implementation of the requirements for the stability and security of communication networks, equipment and software, as well as monitoring (supervision) of their compliance.<br>The amount of costs spent on the implementation of the requirements for the use of domestic computer, server and telecommunication equipment on the objects of data processing infrastructure | Vice-rector for informatization |

Source: compiled by the authors

statistical accounting. The ultimate goal of the integrated accounting system is to create an integrated document that meets the needs of all the interested users in information for making sound management decisions.

The composition of the integrated reporting should be responsive to the requests of responsibility centres to monitor, analyze and assess the indicators they are responsible for. The integrated reporting should contain information about the external environment, allowing to identify the impact of the global economy, the political and social factors actions, the potential of higher education, the competitors' composition and strength, identify their influence on the revenue and costs of the university. The university creates management and monitoring systems in all spheres of public life on the basis of information and communication technologies for these purposes in accordance with the Programme. The information considering the status of the external environment can be found in status reports about the mezo-and micro-economic environment. The composition of the indicators is determined by the university itself. The analysis tools may be SWOT-analysis, diagnostic analysis, comparative analysis, benchmarking etc.

The integrated reporting composition on the internal environment status is proposed as follows:

- a report on the costs of educational activities, including promising educational digital economy projects. This report allows us to calculate the following indicators: "the Share of costs (by type of costs) per employee (per academician), including promising educational digital economy projects"; the primecost of one student's tuition in educational programmes; the costs of institutions, faculties, branches implementing educational programmes;
- a report on the costs of information support systems of a student's competencies individual profile;
- a report on the costs of creating organizational and methodological conditions, forms of appraisal, cirricula, guidance materials in the framework of the digital economy;
- a report on the costs of international scientific and technical cooperation in the spheres of "cut-crossing" technologies in the field of digital economy;
- a report on the research work costs, including ones in the field of digital economy (by the expenditure type in the context of work);
- report on revenues from educational activities, including promising educational digital economy projects;
- a report on grants received (by types of grants and sources);
- a report on revenues from scientific-research activities, including ones in the field of digital economy (by the expenditure type in the context of work);
- a report on the allocated funds and expenses of investment activity;
- a report on expenditure on the university social policies (in the context of policies). Based on this report, the analysis and control of the expenses envisaged in the action plan are provided;
- report on the information security costs. Based on this report, the analysis and control of the expenses envisaged for accomplishing the information security requirements are provided;
- a report on financial results by type of activity.

The analysis of the indicators that are in the integrated reporting considering the internal environment is carried out by the responsibility centers. The analysis results are the starting point for assessing the likely future of the university status.

The information content to be presented in the integrated reporting demands a certain order of its formation, which comprises several procedures that meet the digital economy requirements:

- making adjustments to all elements of the accounting system, that is the ordering of data processing algorithms and the access to such data;
- development of internal regulations considering electronic document workflow in organizations, creation of conditions for enhancing the credibility in electronic documents;
- application of information technologies that ensure the implementation of increased labour productivity systems;
- coordination of responsibility centers activities through the development of accounting policy section in terms of management accounting in the order of integrated reporting;
- implementation of the social responsibility principles according to the international standard ISO 26000 "Guidance on social responsibility", developed by the International Organization for Standardization, the UN Global Compact, into the practice of managing the university.

In the context of the transition from the cost-based model of financing the university activities to the efficient model, integrated reporting should reflect all the essential facts that can affect the assessment of its performance and efficiency of activities, including in the digital economy.

## 4 Discussion

The problems of integrated reporting formation are in the focus of attention of both domestic and foreign researchers, but the issues of its formation in universities, especially in the period of digitalization of the economy, have not been properly reflected in scientific publications.

The shift in emphasis from the oversight function of accounting to communication one, especially in the framework of the digital economy, the development of new information requests of stakeholders, have changed approaches to reporting, have become a factor in the emergence of non-financial reporting, which was supported by many leading scholars in the field of economics, in particular Druzhilovskaya [5]. The decrease in users' confidence in financial data reporting, the insufficiently predictable character of the reporting data is addressed in the works of Grishchenko [10].

The research conducted by Bulyga [2], Gray et al. [9] partly touches upon the issues of the integrated reporting formation in public sector organizations and budget institutions. Selected issues of the integrated management reporting formation as the basis of the management analysis of the educational institution are being addressed within the work of Chrisytensen et al. [3].

In the study of the problems concerning the formation of integrated reporting, the issues of the integrated reporting forms content are often discussed, which for each educational institution has got a purely individual character. In this regard, the processes of building a business model of higher education institutions functioning are brought up to date. To that end, the authors have studied different approaches to building a university business model. In particular, T.V. Zyryanova and S. Tarnovskaya suggest using an approach that focuses on the internal environment of the organization from the position of the implemented business processes as the basis for building a business model [8]. According to N. V. Grishchenko, the approach focused on the consumers and the value creation for them should underpin the basis of building a business model [4].

Chrisytensen C.M., Johnson M.W., Kagermann H. understand the business model to be a set of interrelated elements: value suggestion; a profit formula; key resources; key processes [12]. In the view of M.J. Eyring, M.W. Johnson, and H. Nair, the main problem of a business model building often fails to meet the conditions and features of the external and internal environment [13]. According to O.V. Zaborov and M.G. Kovyazina, the university business model should be based on the unity of three forces in a society: universities, business and government [7].

Supporting the points of view of the above-mentioned researchers, we believe that, when building the university business model, it should be taken into account the specifics of its activities, determined largely by internal business processes aimed at creating a chain of values and focus on the formation and maintenance of the university image (reputation), as well as the requirements of the external environment- digitalization of the economy. The starting point for creating a business model is the personification of responsibility for decision-making and the specification of responsibility centres. The head of the responsibility centre is responsible only for the costs and revenues that they can control. Theoretically, this problem was developed several decades ago. In 1952 Higgins I. M. wrote about the accounting on the responsibility centres in his eponymous work on the accounting system, which is actualized by the organization so that costs are accumulated and reflected in reports at certain levels of management [14].

This accounting system is not new for commercial entities, it is adequately considered by both foreign and domestic specialists. However, according to N. A. Rudenko, the issues of combining responsibility centers and typical educational institutions have not been sufficiently developed, the peculiarity of universities is that most of them are state-owned, and therefore will not be able to fully adopt and implement the principles of market economy in their activities [11].

## 5    Conclusions

There is an objective need for research, comprehensively covering the theoretical and methodological issues of the integrated reporting formation that takes into account the challenges of the modern times to create the necessary conditions for the development of the digital economy in the Russian Federation.

Since the formation of integrated reporting is largely based on accounting information, the study proposes the composition and procedure for the integrated reporting formation, based on the following principles: strategic direction and focus on the digital economy, the coherence of information, verifiability, accessibility, efficiency, consistency, publicity, transparency, complexity, productivity, and effective strength.

The information basis for the integrated reporting should be based on the university business model building, the specification of responsibility centres, the ascertainment for their key financial and non-financial indicators that allow to assess their results in general and in the digital economy in particular.

**Acknowledgments.** The study was carried out within the framework of the state assignment of the Ministry of Education and Science of the Russian Federation, project No. 26.940.2017/4.6 "Managing Changes in the System of Higher Education Based on the Concept of Sustainable Development and Coordination of Interests".

# References

1. Arkhipenko NYu (2016) Global experience and perspectives of integrated reporting implementation in the Russian Federation. The journal Vestnik Professional'nyh Buhgalterov № 2:4–7
2. Bulyga RP (2015) The concept of presenting universities' public reporting. J Vestnik Financ Univ № 6:156–170
3. Chrisytensen CM, Johnson MW, Kagermann H (2008) Reinventing your business model. Harv Bus Rev, 51–59
4. Cooper SM, Owen DL (2007) Corporate social reporting and stakeholder accountability: the missing link accounting. Organ Soc 32(7–8):649–667. https://doi.org/10.1016/j.aos.2007.02.001
5. Druzhilovskaya TYu (2015) Problems of integrating reporting in modern scholars' publications. In: Druzhilovskaya TYu (ed) International accounting vol 11, pp 55–64
6. Eccles RG, Krzus MP (2010) One report: integrated reporting for a sustainable strategy. John Wiley & Sons, Hoboken, NJ
7. Eyring MJ, Johnson MW and Nair H (2011) New business models in emerging market. Harv Bus Rev, 85–95
8. Glushchenko AV, Egorova EM (2015) The integrated system of management reporting as a basis of managerial analysis at universities. J Account Anal Auditing № 4:64–75
9. Gray R, Dey C, Owen D, Evans R, Zadek S (1997) Struggling with the praxis of social accounting: Stakeholders, accountability, audits and procedures. Account Audit Account 10 (3):325–364. https://doi.org/10.1108/09513579710178106
10. Grishchenko AV (2017) The business model of an educational organization: problem statement. In: Trends in the development of science and education. Collection of scientific works, based on the materials of the XXVII international scientific and practical conference June 30, 2017 Part 2 Ed. SIC "L-Journal", pp 9–12. ljournal.ru/article/lj-30-06-2017-20.pdf
11. Higgins IM (1983) Organizational policy and strategic management. The Dryden Press, Chicago
12. Kogdenko VG (2014) Integrated reporting: problems of formation and analysis. In: Kogdenko VG, Melnik MV (eds) International accounting vol 10, pp 2–15
13. Kulikova LI, Gafieva GM (2014) Development of financial reporting principles. MediterrEan J Soc Sci 5(24):38–40. https://doi.org/10.5901/mjss.2014.v5n24p
14. Owen G (2013) Integrated reporting: a review of developments and their implications for the accounting curriculum. Account Educ 22(4):340–356. https://doi.org/10.1080/09639284.2013.817798
15. Rozhnova OV (2013) Relevant current issues oa financial reporting. In: Rozhnova OV (ed) International accounting vol 15, pp 2–8
16. Rudenko NA (2013) Financial responsibility centres as the basis of a special purpose fund costs and revenues in higher education institutes. Int Res J 7(14):50–54
17. Samuel A, Jr DiPiazza, Eccles Robert G (2003) The future of corporate reporting: building public trust. Alpina Publisher, Moscow

18. Solomon and W Maroun (2012) Integrated reporting: the new face of social, ethical and environmental reporting in South Africa. ACCA, London. http://www.accaglobal.co.uk/en/technical-activities/technical-library/integrated-reporting.html

19. The Landscape of integrated reporting: reflections and next steps (2010)/E-book/Eccles RG, Cheng B, Saltzman D (eds) Harvard Business School. http://www.hbs.edu

20. Zaborovskaya OV, Kovyazina MG (2015) Business model of higher education institutions иіт conditions of innovation development of the regional economics. Bus Educ Law Bull Volgogr Bus Inst 1(30):132–137

21. Zyryanova TV, Tarnovskaya SA (2012) Process approach modelling for management reporting purposes. Int Account 44(242):15–28

# Characteristics of Russian Government Financial Resources: Historical Overview and the Situation Under Digital Economy

T. M. Kovaleva[✉], E. N. Valieva, and E. V. Popova

Samara State Economic University, Samara, Russia
fikr@bk.ru, rad8063@yandex.ru, katirinna@mail.ru

**Abstract.** This paper presents the findings on government finances analysis. Income generation characteristics are studied. Territorial expense breakdown is considered throughout different historical periods of the country's social economic development. General differences and similarities in territorial budget sources of income and expenses powers are determined throughout the Russian Federation history.

**Keywords:** Budget income and expenses · Budget policy · E-budget Financial relations · State finances

## 1 Introduction

Characteristics of financial resources formation in the state are caused by a historical stage of its development. The study of the current stage of the RF's financial resources formation requires deep understanding of its historical perspective.

The study aims at examining specifics of financial resources formation on different historical stages of Russian state genesis.

Thus, the main tasks of the study were to analyze Russian financial resources formation throughout the period from 1913 to 2016 and to determine differences and similarities in the structure of the RF's financial resources throughout different historical periods. Part of the aim of this project was to determine advancement options for budget structuring and budget process in the age of digital economy.

## 2 Materials and Methods

The methodology applied in this study combined general scientific methods (synthesis, analysis, and generalization, analogy, historical and logical methods) and empirical knowledge methods (classification, observation, description and comparison).

The research data in this paper is drawn from official statistical sources on budgets of the Russian Empire, the Soviet Union and the Russian Federation. Previous research of financial relations in different economic periods by national and foreign scholars was analyzed.

© Springer Nature Switzerland AG 2020
S. Ashmarina et al. (Eds.): *Digital Transformation of the Economy: Challenges, Trends and New Opportunities*, pp. 635–645, 2019.
https://doi.org/10.1007/978-3-030-11367-4_61

## 3 Results

The structure of state authorities, their place in the management system, delegated powers and sources of funding are determined by the administrative-territorial division that has emerged as a result of historical development. The origin of budgetary relations in Russia is usually referred to the era of the reign of Ivan IV (the Terrible). The initial goal of Ivan IV's administrative and financial reform was to ensure the convenience of collecting taxes and *podat* (chiefery), as well as protecting the state from external enemies of the inner princely turmoil. Land, lip and tax reforms laid the foundations of local government. The village heads selected by the community were responsible for law and local order and collecting taxes to the Moscow Prince's treasury according to established rules.

The expenditure of treasury funds was carried out by the following orders: ambassadorial (foreign policy), *prikaz* (defense), local (land relations), *zemstvo* (protection of public law and order), etc. In contrast to Western Europe, the borders of medieval feudal principalities for the administrative-territorial division of Russia were not important, although taken into account. So, before the administrative reform of Peter I in Russia there were 166 *uezd* (counties), this is not counting many *volosts*, some of which were close to the counties. The Peter's reform of 1708 divided the country into 8 huge provinces, which were composed of cities and adjacent lands, as well as *razryad* and *prikaz*.

As the Russian empire was forming, another principle of administrative-territorial division was added - the national one, which made it possible to retain conquered territories within the state through providing the peoples of these lands with signs of their own statehood and certain privileges compared to other territorial units.

In the 19th century, the economy influenced significantly the country's administrative and territorial division. In addition to optimizing the collection of taxes, geographically and climatically homogeneous territories began to play a leading role. That allowed rationalizing the process of agricultural production, concentrating those territories around a single large urban center.

By 1917, there were more than 80 provinces, regions, and territories, 4 general governorates, 3 independent provinces with the rights of provinces in the Russian Empire. It should be noted that the economic and national components were the basis for the administrative and territorial division of the country. As the Russian empire was a unitary state, its financial management consisted in filling the treasury with revenues and making government expenses directly from the center. Researches of that time divided profitable sources into taxes, industrial incomes and duties. The predominant revenues in the budget of the Russian Empire were indirect taxes on consumer goods, including income from the wine monopoly. Direct taxes accounted for less than 10% of budget revenues [1].

Taxes at that time were considered "the most typical source of income on which the entire financial system of modern states is based". They were seen as "payment for such services provided, the usefulness of which is felt by all residents of the state, but which by their nature are not divisible." Direct taxes were divided into several types. First, there were personal taxes or *podushnaya podat* (capitation tax), which was levied in equal amounts from all citizens. By 1917, such taxes were preserved only in the "levying of nomadic peoples and in the kingdom of Poland." Secondly, there were property taxes, which included land tax, tax on buildings, and the main trade tax. Third, there were income taxes.

Industrial revenues included revenues from state-owned enterprises, which consisted of revenues of crown enterprises, including agricultural, and land leases, revenues from the state "wine monopoly", revenues from "state regalia" - minting coins and postal and telegraph, finally revenues received by the state from what is today called a public-private partnership [1].

Duties were levied for the provision of specific public services, which included services to ensure external security and internal security of individuals, the development of education, science and art, as well as to promote industry and commerce. Table 1 illustrates the income structure of the Russian Empire budget.

**Table 1.** Analysis of Russian budget revenues in 1913

| Revenues of the Russian budget | mln rub. | Percentage to total, % |
|---|---|---|
| Taxes (land, real estate, trade), levy on the collection of income of money capitals | 273 | 8 |
| Customs, sugar, liquor excise duty and collections | 708 | 20 |
| Stamp, court fees, charges on carry-over property | 231 | 7 |
| Revenues from the wine monopoly and postal institutions | 1025 | 30 |
| Revenues from crown railways | 1044 | 31 |
| Other revenues | 137 | 4 |
| Total revenue | 3421 | 100 |

Source: compiled by the authors based on the Annual report of Finance Ministry. Issue 1915. Pg., 1915 [2].

Half of the budget revenues were generated from indirect taxes - customs duties and fiscal monopoly taxes. Business taxes were less than ten percent, with the exception of profits from state-owned railways. The amount of income transferred to the provinces can be estimated on the basis of budget expenditures and the allocation of the territorial component analysis. This assessment though can be seen as indirect due to the limitations of budget reporting in the Russian empire.

Government spending was also divided into three large groups: real or actual expenses, fiscal expenses and industrial expenses.

Real expenses were targeted at:

- the provision of public goods and services, which included the financing of national defense and law enforcement, public education at lower levels, science and art;
- subsidies and support for higher education, religion, the implementation of the so-called public works necessary for society, but unattractive for private capital and industrial development;
- financing public administration;
- government loan service.

State spending was centralized. The provinces were financed from the state budget of the Russian Empire in order to provide cash for state expenditures in the territories of the provinces.

The results of our expert assessment of the provinces expenses in absolute terms, as well as the analysis of their departmental structure are presented in Table 2.

The findings presented in the table demonstrate that the Russian Empire was characterized by a significant centralization of public financial resources management. It should be noted that the provinces financed the education of the people, the state structures that carried out control activities. Some powers were in the areas of economics, finance and law enforcement.

**Table 2.** The spending's assigned to the Russian Empire provinces in 1913 in the context of departments

| Spending's | mln. rub. | Percentage to total, % | Proportion in ordinary expenses of the Russian Empire |
|---|---|---|---|
| Holy Synod | 39 | 12 | 73 |
| Ministry of Internal Affairs | 30 | 9,3 | 15 |
| Post Office | 2 | 0,6 | 2 |
| Ministry of Finance | 58 | 17,9 | 16 |
| Ministry of Foreign Affairs | 0,0 | 0,0 | 0,0 |
| Ministry of Education | 122 | 37,7 | 77 |
| Ministry of Railways | 3 | 0,9 | 0,0 |
| Ministry of Commerce and Industry, Main Directorate of Land Regulation, Main Directorate of Horse Breeding | 51 | 15,7 | 25 |
| Defense Ministry | 7 | 2,2 | 0,0 |
| Marine Ministry | 1 | 0,3 | 0,0 |
| State Control | 11 | 3,4 | 85 |
| Payments on government loans | – | – | – |
| Total spending | 324 | 100 | 10 |

Source: compiled by the authors based on the Annual report of Finance Ministry. Issue 1915. Pg., 1915 [2]

The basis of the administrative-territorial division of the RSFSR was laid in the late 20s - early 30s of the XX century. In the conditions of industrialization and collectivization, so-called "economic-economic regions" were formed with specific functions assigned to them within the framework of a national industrial production plan. These were:

– the presence of an industrial proletarian "core";
– placement of the population in relation to industrial and agricultural production;
– the number of people necessary for the organization of production;
– a certain combination of natural features;
– availability of transport areas linking this area with others in a single production chain;
– distribution system.

It was stated in the Constitution of the RSFSR of 1978 that the economy of the republics was an integral part of a single national economic complex, "encompassing all branches of social production, distribution and exchange in the territory of the USSR". At the same time, the management of the economy "was carried out on the basis of state plans for economic and social development, taking into account sectoral and territorial principles when combining centralized management with economic independence and initiative of enterprises, associations and other organizations". The jurisdiction of the republics included the conduct of a single socio-economic policy, economic management, as well as the development and approval of state plans for economic and social development, the budgeting and reporting on execution; management of state budgets for autonomous republics, budgets of territories, regions and cities of republican subordination.

Thus, in the USSR, a hierarchical structure of power was created, as well as a budgetary arrangement, in which the higher authorities determined the decisions of the lower bodies. The provinces and regions belonged to the local authorities and were endowed with financial resources from the budgets of the Union republics. There was no practice of independent management of state financial resources at the level of administrative-territorial entities in the Soviet Union. Since the 1980s, Western countries had carried out radical reforms of their financial systems [3]. In the USSR, economic reforms began in 1985. The structure of revenues and expenditures of the state budget of the USSR and the RSFSR in 1985 were analyzed to identify the budgetary powers of the authorities in the USSR (Table 3).

**Table 3.** Comparative analysis of the revenues of the Union budget and the budgets of the Union republics in the USSR in 1985

| № п/п | Type of revenue | The Union budget bln. rub. | (%) | The Union republics budgets bln. rub. | (%) | Proportion of the Union budget in the Union republics budgets (%) |
|---|---|---|---|---|---|---|
| | Total revenue | 114,91 | 100 | 137,3 | 100 | 84 |
| 1. | Revenue from the socialist economy in particular: | 101,23 | 88,1 | 119,3 | 86,9 | 85 |
| 1.1 | Turnover tax | 27 | 23,5 | 71 | 51,7 | 38 |
| 1.2 | Revenue from state enterprises profits | 74 | 64,4 | 46 | 33,5 | 161 |
| 1.3 | Income tax on cooperative and public enterprises | 0,23 | 0,2 | 2,3 | 1,7 | 10 |
| 2. | Taxes levied from population | 13 | 11,3 | 17 | 12,4 | 76 |
| 3. | Income from government domestic bonds | 0,68 | 0,6 | 1,0 | 0,7 | 68 |

Source: compiled by the authors based on The USSR State Budget 1981–1985 (Statistics Digest), M., Finance and Statistics, 1987–214 p. [4]

The data show that the revenues of the USSR state budget were divided almost equally between the center and the republics, which indicates the significant role of the latter in the implementation of the country's financial policy. The main tax income of the union budget was payments from profits, while the budgets of the republics got their major revenue from value added tax. Taxes from the population were divided equally.

It should be noted that the share of the main taxes in the income structure changed depending on the economic policy pursued by the central authorities of the USSR. So in the period of rapid industrial growth, the majority of income received accounted for deductions from the profits of enterprises. According to the Big Soviet Encyclopedia, payments from profits for the period from 1960 to 1973 increased more than 3 times (from 18.6 to 60 billion rubles). Their share in total revenues reached 32%. Closer to the 90s of the XX century, when the emphasis shifted from the development of industrial production to the maximum satisfaction of the needs of the population, turnover tax was put in the first place. Its burden was put on consumers. Therefore, the turnover tax had increased and amounted to almost a third of all revenues to the state budget. At the same time, the share of deductions from enterprises' profits for the period from 1981 to 1991 decreased from 28.8% to 23.4%.

The expenditures of the USSR state budget took into account the existing structure of income and were aimed at the growth of the main income sources, that is, the financing of industry and agriculture. The purpose of the USSR economic and financial policies was to ensure the growth of the economic power of the country and, as a consequence, the increase in the welfare of the population. Public consumption funds were also included in the tool set of welfare increase. The constant growth of incomes of the population was observed, which gave about 20% of budget revenues. Thus, public consumption funds became a noticeable source of replenishment of the state treasury. The distribution of expenses between the allied center and the republics in 1985 is analyzed in Table 4.

**Table 4.** Analysis of the expenditure structure of the USSR state budget for 1985

| Item of expenditure | The Union budget bln. rub. | % to the total | The Union republics budgets bln. rub. | % to the total | Proportion of the Union budget in the Union republics budgets (%) |
|---|---|---|---|---|---|
| Total expenditure, including | 203 | 100 | 183 | 100 | 111 |
| National economy | 119 | 59 | 101 | 55 | 118 |
| Social and cultural activities including science | 48 | 24 | 78 | 43 | 61 |
| Science | 13 | 6 | 1 | 0,5 | 1300 |
| Education | 6 | 3 | 30 | 16 | 20 |
| Print art television and radio broadcasting | 1 | 0,5 | 1 | 0,5 | 100 |
| Healthcare | 1 | 0,5 | 17 | 10 | 6 |
| Social security | 3 | 1 | 29 | 16 | 10 |
| Other social and cultural events | 24 | 12 | – | – | – |
| Defense | 19 | 9 | – | – | – |
| Providing for state government and administration | 1 | 0 | 2 | 1 | 50 |
| Other expenditures | 16 | 8 | 2 | 1 | 800 |

Source: compiled by the authors based on the USSR State Budget 1981–1985 (Statistics Digest), Moscow: Finance and Statistics (1987) [4]

As the table shows, the economy was financed statistically by the Union center and the republics. The powers to finance education, health and social support of the population were assigned to the regions. Science and defense were provided with financial resources exclusively from the Union budget. Comparison of the total amount of revenues and expenditures of the union budget and the budgets of the Union republics shows that the share of non-borrowed sources in republican budgets was about 75% and 60% in the Union budget. Thus, maintaining the financial sustainability of regional budgets was a priority goal.

Market reforms of the early 90s of the twentieth century radically changed the budget system in the Russian Federation [5]. The main distinguishing feature of the modern budget system of the Russian Federation is the independence of budgets, as well as the delimitation of revenue sources of budgets on various levels. Regional governments independently organize the budget process and are responsible for it at all stages. Consequently, as compared with the preceding historical periods, the role of administrative-territorial units in the achievement of federal strategic and tactical socio-economic goals and, in addition, responsibility for the results of state functions at the subfederal level is increasing. The findings presented in Tables 5 and 6 demonstrate how this affected the distribution of state revenues and expenditures between the federal budget and the consolidated budgets of the constituent entities of the Russian Federation.

The data in the table show that indirect taxes, namely VAT, excise taxes, customs duties prevail in the composition of federal budget revenues. As in many foreign countries, taxes are the main source of income [6]. Thus, the conclusion about the similarity in the structures of the Russian Federation and the Russian Empire budget revenues becomes evident enough. These taxes are a more reliable source of income compared with direct taxes, since they practically do not depend on the phase of the economic cycle.

The consolidated budget of the subjects of the Russian Federation is formed mainly at the expense of taxes on profits and income and taxes on property. The remaining sources are represented by shares of less than 7% (Tables 5 and 6).

Consequently, revenues to regional budgets are directly dependent on economic dynamics. Regional budgetary provision is more exposed to financial risks than federal level. The balance of the consolidated budget of the Russian Federation constituent entities is ensured by gratuitous receipts from the federal budget.

The types of revenues of the budget system are disproportionately distributed between the levels of state power: the volume of taxes on profits and income and gratuitous receipts at the regional level is 10 times more than the federal level, and the volume of indirect taxes is 20% of the federal budget. In general, the revenues of the federal budget are 26% higher than the revenues of the consolidated budget of the Russian Federation subjects.

Federal budget expenditures exceed expenditures of the subjects of the Russian Federation consolidated budget by 40%. An analysis of the structure of expenditures and their correlation between the levels of state power revealed relatively equal financing of national issues, the national economy, environmental protection, and the media. Financing of education, health care, housing and communal services is assigned primarily to the territorial authorities. Such entries as defense, national security, social

**Table 5.** Structure and ratio of revenues of the federal budget (FB) and the consolidated budget of the constituent entities of the Russian Federation (CBCERF) in 2016

| Revenues | FB (bln. rub.) | % to the total | CBCERF (bln.rub.) | % to the total | Ratio of CBCERF revenues to FB revenues (%) |
|---|---|---|---|---|---|
| Total receipts | 13460 | 100 | 9924 | 100 | 73,7 |
| Tax generated and non-tax revenue | 13308 | 98,9 | 8289 | 83,5 | 62,3 |
| Income taxes | 491 | 3,6 | 5299 | 53,4 | 1079,2 |
| VAT, excise duties | 3289 | 24,4 | 662 | 6,7 | 20,1 |
| Customs duties | 1976 | 14,7 | 0 | 0,0 | 0,0 |
| Total revenue tax | – | – | 388 | 3,9 | – |
| Property taxes | – | – | 1117 | 11,3 | – |
| Taxes on the use of natural resources | 2883 | 21,4 | 69 | 0,7 | 2,4 |
| Government duties | 94 | 0,7 | 39 | 0,4 | 41,5 |
| Tax debts | 1 | 0,0 | – | – | – |
| International economic activity | 2606 | 20,0 | – | – | – |
| The use of property | 1283 | 9,5 | 380 | 4,0 | 29,6 |
| Payments for the use of natural resources | 237 | 1,8 | 36 | 0,4 | 15,2 |
| Revenues from the provision of paid services | 142 | 1,1 | 50 | 0,5 | 35,2 |
| Income from the sale of tangible and intangible assets | 89 | 0,7 | 123 | 1,2 | 138,2 |
| Administrative fees and charges | 26 | 0,0 | 1 | 0,0 | 3,8 |
| Fines, sanctions, damages | 57 | 0,0 | 104 | 1,0 | 182,5 |
| Other non-tax revenues | 134 | 1,0 | 21 | 0,0 | 15,7 |
| Gratuitous receipts | 152 | 1,1 | 1635 | 16,5 | 1075,7 |

Source: compiled by the authors based on the data from: http://datamarts.roskazna.ru/

policy, financial sustainability supervision, servicing the public debt are all in the area of financial responsibility of federal authorities. Thus, it can be argued that the society represented by deputies placed the responsibility for maintaining internal and external socio-political stability on the federal government, while sub-federal authorities bear the responsibility for providing basic social services to the population.

In 2011, the Concept of creation and development of the state integrated information system for public finance management "Electronic budget" was adopted. The purpose of its implementation was to improve the efficiency of budget expenditures.

**Table 6.** Structure and ratio of expenditure of the federal budget (FB) and the consolidated budget of the constituent entities of the Russian Federation (CBCERF) in 2016

| Expenditures | FB (bln. rub.) | % to the total | CBCERF (bnl.rub.) | % to the total | Ratio of CBCERF expenditures to FB expenditures (%) |
|---|---|---|---|---|---|
| Total expenditures, including | 16416 | 100 | 9936 | 100 | 60,5 |
| Federal issues | 1095 | 6,7 | 625 | 6,3 | 57,1 |
| National defense | 3775 | 23,0 | 5 | 0,1 | 0,0 |
| National security and law enforcement | 1900 | 11,6 | 114 | 1,1 | 6,0 |
| National economy | 2302 | 14,0 | 2002 | 20,4 | 87,0 |
| Housing and utilities sector | 72 | 0,4 | 936 | 9,4 | 1300,0 |
| Environment protection | 63 | 0,4 | 22 | 0,2 | 34,9 |
| Education | 599 | 3,6 | 2547 | 25,6 | 425,2 |
| Culture and cinema | 87 | 0,5 | 340 | 3,4 | 390,8 |
| Health care | 506 | 3,1 | 1281 | 12,9 | 253,2 |
| Social policy | 4588 | 27,9 | 1654 | 16,6 | 36,1 |
| Sport | 59 | 0,4 | 212 | 2,1 | 359,3 |
| Mass media | 77 | 0,5 | 43 | 0,4 | 55,8 |
| Government debt service | 621 | 3,8 | 154 | 1,5 | 24,8 |
| Intergovernmental transfers to budgets of the budget system | 672 | 4,1 | 1 | 0,0 | 0,0 |

Source: compiled by the authors based on the data from: http://datamarts.roskazna.ru/

The electronic budget information system is intended to ensure openness, transparency and accountability of the activities of government bodies. It promotes improvements in the quality of financial management of public administration and allows the use of information technology in the management of public finances.

Initially, the creation of the system was planned for 2015. But due to the complexity of the implementation and the wide range of problems to be solved, the completion dates have been postponed to 2020. The "Electronic Budget" system allows forming a single information domain in which the integration of information flows in the field of public finance management is carried out. The electronic budget allows providing openness and accessibility for the public to the information on all-level budgets implementation, budgets of extra-budgetary funds and other financial relations.

## 4 Discussion

Studies on fiscal federalism have broken down into a fairly large number of independent areas. Foreign scholars consider these issues mainly in the framework of public financial management [7, 8]. An important area of research is the use of e-government [9]. Issues of fiscal federalism for the Russian Federation currently continue to be in the center of attention of such researchers as Davydov [10], Evstafieva [11], Lukin [12].

An independent domain of research has been presented by the study of the influence of fiscal federalism on the socio-economic development of the country. Opinions are divided here. Some scholars, including Goreglyad [13], Paikovich [14], Tatarkin [15], Marie-Laure Breuillé and Skerdilajda Zanaj [16] consider fiscal federalism to be the most important factor in stabilizing socio-economic development. Conversely, Strokov [17] and Shvetsov [18] believe that modern fiscal federalism is a source of problems for the state, including financial management.

Dashkeev and Turnovsky and Kutsuri has examined the sources of budget risks in the context of fiscal federalism [19, 20].

# 5    Conclusions

This paper has argued that the structure of federal budget expenditures in 2016 is more reminiscent of the distribution of powers in the Russian Empire. However, it is necessary to emphasize that modern Russia in the conditions of the digital economy, in contrast to the imperialist, is de jure a federal state delegating, in accordance with world practice, considerable powers to the sub-federal level of government.

In the framework of improving the budgetary system and the budget process, it is possible to use the experience of the USSR and to redistribute revenue sources and spending powers between the branches of state power in the Russian Federation. For example, it seems expedient to assign payments for the use of natural resources to the regions, to distribute parity between the center and the territories of VAT and excise taxes. In this case, it would be possible not only to reduce the volume of intergovernmental transfers, but also to transfer a significant share of social expenditures to the regional level.

The digital economy contributes to the growth of financial resources in the state, which means that a larger amount of funds will be used to maintain the authorities therefore the issue of optimal allocation of funds between levels of government. Thus, appropriate distribution tools (taxes, subsidies, subsidies, subventions, budget loans, etc.) have become extremely important.

Under the conditions of digital economy, the introduction and operation of the Electronic Budget system will allow:

(1)  making the results of financial and business activities and decision-making process transparent and open;
(2)  managing the financial resources of the authorities effectively, since the movement of funds through the electronic budget will allow the distribution of financial resources between levels of government;
(3)  creating a unified system for all participants in the budget process.

# References

1. Gorlov I (2014) Theory of finance: composition by Ivan Gorlov. Reprinted: Kazan. University Publishing House
2. Annual Report of Finance Ministry. Issue 1915. Sg., 1915
3. Lagoarde-Segot T, Paranque B (2018) Finance and sustainability: from ideology to utopia. Int Rev Financ Anal 44:80–92
4. The USSR State Budget CCCP 1981–1985 (Statistical Digest) (1987) Finance and Statistics, Moscow
5. Valieva EN, Milova LN, Dozhdeva EE, Lukin AG, Chapaev NK (2016) Development of scientific understanding of the essence of the fiscal control in Russia over the past 100 years. Int J Environ Sci Educ 11(15):7763–7781
6. Etro F (2016) Research in economics and public finance. Res Econ 70(1):1–6
7. Brigham E, Houston J (2012) Fundamentals of financial management, 12th edn. MG Reprograchics Inc., Philippines
8. Craing OB (2017) The politics of government financial management: evidence from state bonds. J Monet Econ 90:158–175
9. Puron-Cid G (2013) Interdisciplinary application of structuration theory for e-government: F case study of an IT-enable budget reform. Govern Inf Q 30(Supplement 1):46–58
10. Davydov MD (2014) Development of a financial mechanism for administering incomes of budgets of the budget system of the Russian Federation/Davydov MD, Valieva EN, Lukin AG. ZAO, Moscow «NESTO»
11. Evstafjeva AH (2015) The concept of fiscal tax federalism and its impact on the development of the tax administration system in Russia. Innov Dev Econ № 4 (28)
12. Lukin AG (2016) Financial Management and Federalism. Strategic guidelines for the development of economic systems in modern conditions: Sat. st. / ed Dubrovina NA (ed) Samara University Publishing House, Samara 4:129–136
13. Goreglyad VP (2016) Lessons of fiscal federalism. Federalism № 1(81):91–106
14. Paykovich PR (2016) Budgetary federalism as a factor of stabilization of the socio-economic development of the Russian Federation. Econ Entrepreneurship № 3-1 (68–71):172–174
15. Tatarkin AI Tatarkin DA (2016) Russian fiscal federalism in conditions of economic instability. Federalism № 3(83):9–26
16. Breuillé M-L, Zanaj S (2013) Mergers in fiscal federalism. J Public Econ 105:11–22
17. Strokov AI (2016) Actual theoretical and economic aspects of the development of fiscal federalism in the sphere of distribution of powers. Econ Entrepreneurship. № 12-1(77-1): 68–71
18. Shvetsov YuG (2017) The impasse of Russian fiscal federalism. Finance and credit 23. No. 19(739):1094–1107
19. Dashkeev VV, Turnovsky SJ (2018) Balanced-budget rules and risk-sharing in a fiscal union. J Macroecon 57:277–298
20. Kutsuri GN, Shanin SA, Frumina SV, Gardapkhadze T, Ivanova EV (2018) Russian practice of identifying and assessing budget risks. J Appl Econ Sci 13(3):711–719

# Current Problems of Enterprises' Digitalization

A. B. Vishnyakova[1](✉), I. S. Golovanova[1], A. A. Maruashvili[2],
P. Zhelev[3], and D. V. Aleshkova[1]

[1] Samara State University of Economics, Samara, Russia
{Angelina8105, irgolovanova}@yandex.ru,
dashajuly343@gmail.com
[2] New Higher Education Institute, Tbilisi, Georgia
annamaruashvili@mail.ru
[3] University of National and World Economy (UNWE), Sofia, Bulgaria
pzhelev@unwe.bg

**Abstract.** One of the key components of the economy of the Russian Federation at the present stage of economic development is industrial production. Such significant indicators as the country's GDP, the provision of the population with necessary products and services, labor productivity within the sectors of the national economy, the level of technological and environmental safety directly depend on the pace of development and the state of industrial sectors. The economic and defense potential of the country is largely determined by the dynamic development of enterprises of the machine-building complex. The state of the material and technical base and the level of innovation indicate the need to develop a number of measures aimed at improving the efficiency of large machine-building enterprises. Currently, due attention is not paid to the investment strategy development. Because of this, it is impossible to implement the most important programs of systemic modernization of the leading sectors of Russian economy and, above all, engineering. The relevance of deepening technological development within industrial production is assigned at the legislative level. In accordance with the decree of the President of the Russian Federation "On the Strategy of the Scientific and Technological Development of the Russian Federation", the main priority areas in the scientific and technological development of the country will be the development of relevant scientific and technological programs that will promote the innovative development of domestic products and services in the next 10–15 years [11]. The competitiveness of new developments will be largely ensured by digitalization and modernization of industrial enterprises.

**Keywords:** Digital economy · Digitalization · Economic system · Information management · Industrial production · Process automation

## 1 Introduction

Digitalization of Russian economy confidently moved to the implementation stage. The need for digitalization in current conditions is obvious. Modern digitalization is a completely new approach to the retrieval, storage and processing of information. The

S. Ashmarina et al. (Eds.): *Digital Transformation of the Economy: Challenges, Trends and New Opportunities*, pp. 646–654, 2019.
https://doi.org/10.1007/978-3-030-11367-4_62

role of information is huge in effective management decision making. In a market with a huge number of competitors, the speed and accuracy of decision making can significantly change the position of the enterprise on the market.

And, of course, information should be given a lot of attention. In many enterprises, pilot projects are being implemented to apply new approaches and digital technologies. However, it is not clear how much digitalization of machine-building industry should cost.

Digitalization in its general sense is understood as a socio-economic transformation initiated by mass introduction and assimilation of digital technologies, i.e. technologies of creation, processing, exchange and transfer of information [1]. Digitalization in a business context can be understood as digitization of information, generation of new understanding, and presentation that can help take more effective management decisions.

## 2 Materials and Methods

The results of the research of domestic and foreign authors have formed the theoretical and methodological basis of this research. The methods of the system, logical, structural and comparative analysis, statistical methods of information processing were used. The information basis of the research are also data of economic and analytical character from machine-building enterprises.

## 3 Results

The fourth industrial revolution ("Industry 4.0") involves the creation of digital enterprises based on digitization of all enterprise systems and their integration into the digital ecosystem together with partners involved in the value chain. The digitalization process is carried out through the operation of several interrelated processes, including digitization of value chains, introduction of digital products (using smart sensors), digital business models and a digital way to optimize costs, etc. Building digital capabilities takes time and effort. At the same time, it is important to move quickly in order not to lose the advantages of the "pioneer" over competitors [3].

Industry 4.0 will allow you to collect and analyze data on equipment, providing faster, more flexible and more efficient processes for the production of higher quality goods at reduced prices. This, in turn, will lead to an increase in productivity, stimulation of industrial growth and changes in the profile of the workforce, which ultimately will change the competitiveness of the enterprise. Currently nine technology trends are described that are building blocks of Industry 4.0, and explore their potential technical and economic advantages for manufacturers and suppliers of equipment for production.

Many of the nine advances in technology that form the basis for Industry 4.0 are already in production. However, with Industry 4.0, they will transform production: isolated, optimized cells will come together in the form of a fully integrated, automated and optimized product flow that can increase efficiency and change traditional production relations between suppliers, manufacturers and customers, as well as between man and machine [13].

The first fundamental technology of Industry 4.0 is analytics based on large data sets. As such, analytics has only recently appeared in the manufacturing world, where it optimizes product quality, saves energy and improves equipment maintenance. In the context of Industry 4.0, the collection and comprehensive assessment of data from various sources—production equipment and systems, as well as enterprise and customer management systems—will become the standard for real-time decision support.

For example, semiconductor manufacturer Infineon Technologies reduced product failures by matching data with a single chip, obtained at the testing stage at the end of the production process, with process data collected during the plate state phase earlier in the process. Thus, Infineon can identify patterns that help produce faulty chips early in the manufacturing process and improve product quality.

Next are autonomous robots. Manufacturers of many industries have long used robots to solve complex problems, but robots develop an even greater utility. They become more autonomous, flexible and collaborative. In the end, they will interact with each other and work safely alongside people and learn from them. These robots will cost less and have a larger range of capabilities than those used in production today.

For example, Kuka, a European manufacturer of robotic equipment, offers autonomous robots that interact with each other. These robots are interconnected, so that they can work together and automatically adjust their actions to fit the next unfinished product in the queue. High-quality sensors and control units ensure close collaboration with people. Similarly, an ABB industrial robot supplier launches a two-armed robot called YuMi, specifically designed to assemble products (such as consumer electronics) along with people. Two soft levers and digital vision ensure safe communication and recognition of parts.

The third technology is modeling. At the engineering stage, three-dimensional modeling of products, materials and production processes are already used, but in the future, modeling will be used more widely in factories. These simulations will use real-time data to mirror the physical world in a virtual model, which can include machines, products, and people. This allows operators to test and optimize equipment parameters for the next product in a linear mode in the virtual world before physical switching, thereby reducing equipment setup time and improving quality [2].

For example, Siemens and the German machine manufacturer have developed a virtual machine that can simulate machining parts using data from a physical machine. This reduces setup time for the actual processing by 80%.

The fourth technology is horizontal and vertical system integration. Most modern IT systems are not fully integrated. Production, suppliers and consumers are rarely closely related. Such departments as design, production and service are not integrated. Even the development itself - from products to installation - does not have full integration. But with Industry 4.0, enterprises, departments, functions, and capabilities will become much more cohesive as data networks evolve and activate truly automated value chains.

For example, Dassault Systèmes and boostaerospace launched a collaboration platform for European aerospace and defense industries. The airdesign platform serves as a common workspace for collaborative development and production and is available as a service in a private cloud. It manages the complex task of exchanging product and production data between several partners.

Fifth technology is the Industrial Internet of Things. Today, only some of the manufacturer's sensors and computers are connected to the network and use built-in computing. They are usually organized in a vertical automation pyramid, in which sensors and field devices with limited intelligent and automatic controllers are combined into a production control system. But with Industrial Internet of Things, more devices, sometimes including even unfinished products, will be enriched with built-in computing and connected using standard technologies. This allows field devices to interact with each other and with more centralized controllers, if necessary. It also decentralizes analytics and decision making, allowing it to respond in real time.

For example, Bosch Rexroth, a supplier of engine management and control systems, has equipped a production facility for valves with a semi-automated decentralized production process. Products are identified by radio frequency identification codes, and workstations "know" which production steps should be performed for each product and can be adapted to perform a specific operation.

The sixth technology is Information Security. Many enterprises still rely on management and production systems that are not connected or closed. By expanding connectivity and using standard communication protocols that come with Industry 4.0, the need to protect critical industrial systems and production lines from cybersecurity threats is increasing dramatically. The result is reliable communication, as well as complex identification and control of access to equipment and users.

The seventh technology is closely related to the previous one - Cloud Technologies. Businesses are already using cloud-based software for some enterprise applications and analytics, but with Industry 4.0, businesses will require a wider exchange of data between sites and company boundaries. At the same time, the performance of cloud technologies will improve, reaching a reaction time of just a few milliseconds. As a result, data and functionality will increasingly be introduced into the "cloud", which will provide more services based on data for production systems. Even systems that monitor and control processes can become cloudy.

Suppliers of production and execution systems are among the companies that have begun to offer cloud solutions.

To provide a quantitative understanding of the potential impact on industry by Industry 4.0 worldwide, production prospects in Germany were analyzed. And it was discovered that the fourth wave of technological progress would be beneficial in four areas.

Performance. Over the next 5–10 years, Industry 4.0 will cover most of enterprises, increasing productivity in all manufacturing sectors in Germany from 90 billion Euros to 150 billion Euros. Increased productivity due to the cost of transition, which excludes the cost of materials, will be from 15 to 25%. When material costs are taken into account, the productivity of between 5 and 8% is achieved. These improvements will vary by industry. For example, manufacturers of industrial components achieve some of the biggest performance improvements (from 20 to 30%), and car companies can expect an increase from 10 to 20%.

Revenue growth. Industry 4.0 will also contribute to revenue growth. Producer demand for improved equipment and new data applications, as well as consumer demand for a wider range of products are increasingly customized. All these factors will lead to an increase in revenues of about 30 billion Euros per year or about 1% of Germany's GDP [9].

Employment. In analyzing the impact of Industry 4.0 on German production, it was found out that its growth will result in a 6% increase in employment over the next ten years. In the short term, the tendency toward greater automation will lead to crowding out some of low-skilled workers who perform simple, repetitive tasks. At the same time, the increasing use of software, connectivity and analytics will increase the demand for employees with competencies in software development and IT technologies, such as mechatronics experts with software skills. Mechatronics is a field of technology that includes many engineering disciplines. This transformation of competencies is one of the key tasks for the future.

Investments. According to experts, the adaptation of production processes for Industry 4.0 will require German manufacturers to invest about 250 billion Euros over the next ten years (from 1 to 1.5% of producers' incomes). Estimated benefits in Germany illustrate the potential impact of Industry 4.0 on manufacturing worldwide.

The management of each company should pay attention to two tasks. First, it is necessary to get rid of the routine, which is overloaded with management, as quickly as possible. And today there are practically no technological limitations - everything that can be identified and formally described (standardized) can be automated, and it is a matter of price and potential value for the business. And secondly, the released resources should be directed to constant reflection - the search for ways to keep the consumer value of their product. This will require a lot of stress and time [7]. If earlier the increase in the competitiveness could be based on a gradual increase in efficiency and introduction of best industry practices, then the development of the digital economy today is largely based on cross-functional and cross-industry transfer of technologies and business models. Potential breakthrough ideas and technologies for your business today can be found in completely unexpected segments of the economy. Thus, digitalization allows you to create new business models as networks, digitizing routine procedures, those that can be standardized. Prescribing their algorithm will significantly improve the cost-effectiveness.

"Industry 4.0" assumes a powerful increase in investments in fixed assets and intangible assets that are accounted for within the enterprise. This can be seen in the example of foreign countries (for example, in Germany), if we present the statistics of investments in the described technologies 4.0 by year.

The relevance of digitalization in the world was launched in May 2010. The driver of this growth was the European Union (Digital Agenda for Europe, DAE), in order to support economic growth in Europe. Germany is not the only country that has realized the general trends and the real importance of the concept of "Industry 4.0". Currently, there are 15 approved programs in the field of digitalization of production in the world: Germany, China, Japan, Brazil, USA, UK, Estonia, Holland, Ireland, Sweden, Singapore, Philippines and Malaysia.

Industry 4.0 opens up new development prospects for all sectors of the economy, including industrial production. Over the past 10–15 years, global transformations of production in the direction of artificial intelligence have taken place in countries with the digital economy, and now they have quite digitalized, have high-tech and robotic production and they are connected to production networks [12].

The key digitalization technologies that emerged due to the concept of "Industry 4.0" include the following areas which will make the enterprise "digital":

- Big data analysis (data on products and production processes are integrated into large information systems, which facilitates the process of managing costs and resources);
- Autonomous work (introduction of robotics based on self-learning and self-optimization, facilitating the work of key employees);
- Simulation processes (simulations) (methods of virtual modeling of products, materials and processes used at the engineering stage are created, data is retrieved in real time to create a virtual copy of real production with the participation of machines, products and employees).
- Integration of IT systems (development of the common information space between functional units and individual companies)
- Cloud computing (use of software and analysis systems based on cloud platforms).

Thanks to such integrated processes, there are opportunities for modernization of most areas of management through the introduction of digital technologies within the industrial enterprise. So, thanks to the concept of "Industry 4.0", it became possible to obtain a number of advantages:

- Increase in production flexibility due to its quick changeover, dynamic changes in the production process;
- Efficiency of production management and the girth of a large number of credentials;
- Development of competitive advantages;
- Optimization of production, as well as product quality, environmental, technological and other types of safety;
- Development of new business opportunities, etc.

Another concept on the use of digital technology IIOT was created to standardize and unify the field of industrial digitalization due to the improvement of quality standards for digital technologies. The implementation of this technology involves computerization of all workplaces in the enterprise, development of a transmission system, automated information processing, integration of equipment and workstations into a single information network, etc.

- According to modern data, the IIOT concept allows reducing downtime (up to 10%), reducing maintenance costs, as well as improving the procedure for forecasting and preventing equipment failures (by 10%). Ultimately, the introduction of IIOT contributes to increased productivity and GDP growth. Further, there are some financial assessments of the IIOT development [8]:
- In the period from 2017 to 2023 this market will grow at an average annual rate of 14.36% and by 2023 its volume will be $700.38 billion;
- By 2025, the global IIOT market will reach 484 billion Euros;
- By 2030, the contribution of IIOT to the world economy in monetary terms will be more than $14 trillion, including: up to $6 trillion in the US and over $70 billion in Germany.

Available data indicate that there are significant positive operational, technological, managerial, environmental and other effects of digitization of production.

## 4  Discussion

Digitization of the industrial enterprise is significantly affected by concepts managing enterprise resources (ERP, MRP). The use of these concepts allows you to base the approach to cost automation, which comes into close contact with the digital technology. Almost all Western systems currently use the ERP system within their enterprise [8]. At the same time, there is no experience working with systems of this level in Russia yet.

Russia's digitalization experience is characterized by slow development and lagging behind other countries' advanced technologies. At the present stage of development, the process of digitalization in most Russian industrial enterprises is characterized by the following disadvantages:

- There is a slow increase in the level of mechanization and automation in the system of domestic production, while statistically the growth in the share of automatic equipment has slowed down over the last decade.
- There is a rather uncertain growth in the share of innovative products in certain industries of Russia from 2010 to 2016, as well as the uncertain proportion of organizations directly involved in the development of technological innovations.
- There is a lack of effective and high-quality cost management at the enterprise, poor quality cost accounting and a rare use of the innovative approach to management leads to low profitability of products.

According to the above data, it is clear that in Russia there is an extremely low base of automation and robotization in the industry, low awareness of technical management about digital technologies, including the training of specialists in this field. The lag of Russia from Germany, USA, Japan, China and South Korea averages 7–10 years.

## 5  Conclusions

Under these conditions, there are serious threats for the domestic industry both in the medium and long term, unless active and adequate measures are taken. In the medium term, they are connected with the fundamental impossibility of competing with leading industrial countries, and in the long term there will be an almost insurmountable technological barrier between Russia and the leading technologically advanced countries that rely on the concept of Industry 4.0. It will result in technological isolation.

The primary task of Russian industrial enterprises in the coming years is cardinal digitalization and modernization of production, the introduction of advanced world technologies that form the basis of Industry 4.0 that ensure the competitiveness of national products in markets.

These long-term goals will require government support for enterprises, including the development of state development programs and the provision of appropriate subsidies, consolidation of the work of the state, science and business to exchange ideas, experience, technology and knowledge for the joint implementation of projects on integrated digital ecosystems [15]. The cluster policy of Russia aimed at the development of production in the regions, actively implemented by the government, can be a good start to global transformations in industry and an increase in the rates of its annual development.

# References

1. Babkin AV, Chistyakova OV (2017) The digital economy and its impact on the competitiveness of entrepreneurship. Russian business. https://cyberleninka.ru/article/v/tsifrovaya-ekonomika-i-ee-vliyanie-na-konkurentosposobnost-predprinimatelskih-struktur. Accessed 12 Apr 2018
2. Decree of the President of the Russian Federation of December 1, 2016 No. 642 "On the strategy of the scientific and technological development of the russian federation"
3. Demyanova OV, Dimmieva AR (2018) Digitalization as a point of growth of an industrial enterprise. Socio-economic and legal foundations of innovative development. https://elibrary.ru/item.asp?id=35436326. Accessed 12 Apr 2018
4. Hoffmann R (2016) Investment opportunities in industry 4.0-industrial revolution "made in Germany". Ecovis. https://www.ecovis.com/focus-china/investmentopportunities-industry-4–0/. Accessed 12 Apr 2018
5. Khachirov AD, Khubulova VV (2018) Industry in the context of the digital economy. Bulletin of the academy of knowledge. https://elibrary.ru/item.asp?id=32758911. Accessed 12 Apr 2018
6. Korovin GV (2018) Problems of industrial digitalization in Russia. News of Ural State University of Economics. https://elibrary.ru/item.asp?id=35310777. Accessed 12 Apr 2018
7. Lamentova AYu (2018) Digitization of industry as a new strategy for economic development. Synergy of sciences. https://elibrary.ru/item.asp?id=35296212. Accessed 12 Apr 2018
8. Plotnikov VA (2018) Digitalization of production: theoretical nature and development prospects in the Russian economy. News of St. Petersburg State University of Economics. https://elibrary.ru/item.asp?id=35304372. Accessed 12 Apr 2018
9. Rosstat data [Electronic resource] http://www.gks.ru/free_doc/new_site//technol/3-04.xls/. Accessed 17 Apr 2018
10. Tarasov IV (2018) Technology Industry 4.0.: impact on increasing the productivity of industrial companies. Strategic decision and risk management. https://cyberleninka.ru/article/n/tehnologii-industrii-4-0-vliyanie-na-povyshenie-proizvoditelnosti-promyshlennyh-kompaniy. Accessed 17 Apr 2018
11. Tatarskikh BYa (2016) Issues of development of the material and technical base of Russian machine-building of Russia: the structural-dynamic aspect. Problems of enterprise development: theory and practice. https://cyberleninka.ru/article/n/strategicheskie-napravleniya-povysheniya-effektivnosti-mashinostroitelnogo-kompleksa-rossii. Accessed 17 Apr 2018
12. The data of Rosstat [Electronic resource] https://www.fedstat.ru/indicator/31429. Accessed 17 Apr 2018

13. Ustinova LN (2018) Industry 4.0-new challenges for the Russian production. Digital economy and industry 4.0.: new challenges. https://elibrary.ru/item.asp?id=32820238. Accessed 12 Apr 2018
14. Vaisman ED, Nikiforova NS (2018) Development of dynamic abilities of industrial enterprises in the digital economy. News of Ural State University of Economics. https://elibrary.ru/item.asp?id=35310780. Accessed 12 Apr 2018
15. Volkov VI (2004) Methods of expert evaluation of innovative projects. Bulletin of MGTU. Ser. "Engineering". https://cyberleninka.ru/article/n/metodika-ekspertnoy-otsenki-proektov-innovatsionnoy-napravlennosti. Accessed 12 Apr 2018

# Innovative Approaches to Quality Monitoring of Medical Services in the Digital Environment

A. D. Khairullina[1], A. V. Pavlova[1,2(✉)], N. V. Kalenskaya[1],
and G. R. Mukhametshina[1]

[1] Kazan Federal University, Kazan, Russia
{halbi,kalen7979,698817}@mail.ru, 930895@list.ru
[2] Volga Region State Academy of Physical Culture, Sport and Tourism,
Kazan, Russia

**Abstract.** The contribution presents a brief overview of companies providing private medical services in the Republic of Tatarstan. Based on the available statistical information and independently conducted field research, the participants of which were the heads of medical institutions, the authors have made conclusions regarding the main trends in improving quality monitoring of business modeling of the medical institution.

**Keywords:** Business process · Digital technologies · Hospital market ·
Market of modeling · Quality monitoring of medical services ·
Private medical services · Process approach · Reengineering

## 1 Introduction

For several years in a row, Russian companies have been living in the face of declining demand from consumers. But at the same time private medicine has grown, working exclusively for the domestic Russian consumer. According to the BusinesStat agency, the highest rates of growth in the market's value were observed in the private medical sector. Thus, the volume of private medical services in the Russian Federation increased from 425 billion in 2010 to 672 billion rubles in 2015 [8]. And although it is possible to be treated for money in the state or shadow sectors, the maximum increase is provided by the legal private segment. According to "BusinesStat", for 2015, taking into account the LCA, private medicine (along with shadow) occupies almost a third of the market of medical services in Russia.

Also, factors that stimulate the growth of the market of private medical services, according to the research by Romir Holding, include patient dissatisfaction with the quality of services in the public sector. In addition to the lack of necessary specialists or equipment, patients of state institutions complain about insufficient attention, the lack of an individual approach, and low qualifications of doctors in remote regions.

According to the news of Business Quarter DK.ru, the number of private medical institutions increased by 5.4% in Tatarstan in 2015. Bypassing the difficult economic situation, the turnover of all private medical institutions in the Republic increased by 4.5%, and reached 1.4 billion rubles in January-April 2016.

S. Ashmarina et al. (Eds.): *Digital Transformation of the Economy: Challenges,
Trends and New Opportunities*, pp. 655–668, 2019.
https://doi.org/10.1007/978-3-030-11367-4_63

Thus, there are a number of factors contributing to the growth of the market of private medical services in both Russia and the Republic of Tatarstan [3]. This dynamic has become possible against the background of growing public welfare and investment in medicine. And in recent years there has been a change in consumer behavior among Russians, who have begun to turn to private medical centers more often. Accordingly, there is an urgent need to model and reengineer business processes of medical institutions that ensure the high quality of the services provided.

The development of the theory of business modeling was considered in the work of such authors as Deming, Edwards [1], Osterwalder, Pigne I [6], Vader [10], Zeithaml [14], Ulumbekova [12], Uvarina Shushkin [9] and others.

Highly appreciating the contribution of these scientists and the results obtained by them, it must be noted that the level of development of the problem of modeling business processes in the medical industry remains clearly insufficient.

Currently, during the period of active medical reform in the Russian Federation, many authors in their scientific works raise the issue of management of medical institutions, quality management of medical car [4, 5]. At the same time, the existing literature on business process engineering and business modeling illustrates that there is no decision on such issues as the applicability of business modeling in medicine, the choice of key processes and resources, the parameters and methods for monitoring the performance of key business processes in the overall model of the company, the development and monitoring of specific algorithms to improve the company' performance. The lack of concrete, applicable in practice, unified methods causes difficulties in their implementation in the life of medical institutions.

In recent years, the government of the Russian Federation is actively seeking new mechanisms that could ensure the creation, implementation and further spread of innovative technologies in various fields of the national economy. It should be noted that the development of the Concept of Long-Term Socio-Economic Development for 2008–2020 was an extremely important step in this direction.

This concept defines the main directions for creating a national innovation policy. In this regard, the issue of developing and introducing new innovative technologies in medical care becomes particularly relevant. This is due to the fact that the increase in the availability and quality of medical care cannot be ensured without the use of new high-tech methods.

The increased interest in designing and optimizing business processes (BP) in medical institutions is associated with increased competition and consumer expectations in medicine, as well as opportunities to increase the performance of the medical institution through the use of computer technologies. Modern information technologies make it possible not only to optimize the process of designing business processes over time, but also to build models with the help of which it is possible to play various options and select the most effective ones from them.

The authors of this contribution analyzed the development trends of the market of medical services in the Republic of Tatarstan: they attempted to reengineer business processes of a medical institution.

The purpose of the study is to improve management tools for modeling business processes of medical institutions in order to improve the quality of the services provided.

## 2   Materials and Methods

As the analysis of statistical data of the Russian Federation showed, recent years have been characterized by a drop in demand from consumers. At the same time, private medicine shows steady growth. The volume of private medical services increased from 425 billion in 2010 to 672 billion rubles in 2015, according to the review "RBC. Market research" (Fig. 1).

As the statistics showed, this legal private segment gives the maximum growth. According to "BusinesStat", for 2015 and taking into account voluntary medical insurance (VMI), private medicine (along with shadow) occupies almost a third of the market of medical services in Russia (Fig. 2) [8].

The main factors behind the growth of the market of private medical services include the reduction in funding for the public sector and the number of public medical facilities. According to the forecast by Higher School of Healthcare Management, compared to 2013, public spending on medical care reduced by 8.6% in real prices in 2017. From 2000 to 2015, the number of hospital beds decreased by an average of 27.5% to 1.2 million.

**Fig. 1.** Volume of the market of private medical services in Russia, billion rubles (Source: compiled by the authors)

The natural volume of the public sector will continue to decline, including due to the transition to single-channel financing from the funds of mandatory medical insurance (MMI). In 2018, the availability of free medical care, which is directly determined by the expenditures of the budgets of the budget system, will decrease by 9% compared to 2013. The unavailability of medical public services will push consumers either to refuse treatment or to apply for full-time or online private consultation.

Also, factors that stimulate the growth of the private market of medical services, according to research by Romir Holding, include patient dissatisfaction with the quality of services in the public sector. In addition to the lack of necessary specialists or

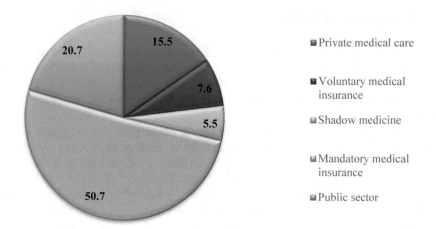

**Fig. 2.** Structure of the market of medical services in Russia (Source: compiled by the authors)

equipment, patients of state medical institutions complain about insufficient attention, the lack of an individual approach, and low qualifications of doctors in remote regions.

One of the most significant negative macroeconomic factors affecting the market of private medical services is the fall in real disposable incomes of the population, which continued in the period of 2014–2016. The impact of this factor is aggravated by the fact that the rise in prices for medical services outpaced the dynamics of the consumer price index. Throughout the period under review, with the exception of 2015, this advance was about 0.6–7.2% points (Fig. 3).

Thus, inflation remains the key driver for increasing the volume of the market of private medical services. According to the forecast of "BusinesStat", the rise in prices for medical services will slightly outpace the dynamics of the consumer price index in 2018 and 2019. This will restrain the increase in demand for private medicine even under the conditions of the Russian economy's recovery from the recession and the gradual resumption of income growth in 2017, because the rate of increase in income will be too low to significantly stimulate the demand for private medical services.

According to the KMPG forecast, the trend of new patients will continue to flow into private medicine, mainly in the lower price segment in the foreseeable future, due to further reduction in the availability of medical care in the MMI segment (especially in primary care, as a result of an increase in load on it), and due to the reduction of insurance coverage under corporate health insurance policies. The share of private medicine in the structure of private medical services will continue to grow and reach 57% by 2019 [2].

A promising factor in the growth of the private medicine segment is also the development of information technologies and the accompanying legislative framework, including the launch of applications to search for medical facilities or a doctor, for example, Yandex.Health, Health. Mail.ru or electronic medical records, for example, ONDOC, as well as the adoption of the law on telemedicine.

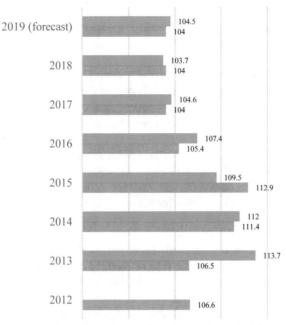

**Fig. 3.** Dynamics of consumer price index and price index of medical services, 2012–2019, % [7].

According to Business Quarter DK.ru (news portal), the number of private medical institutions increased by 5.4% in Tatarstan in 2015. Despite the difficult economic situation, the turnover of all private medical institutions in the Republic of Tatarstan added 4.5%, reaching 1.4 billion rubles in January-April 2016. Total from 2011 to 2015, the market of private medical services grew by 15.4%, while the turnover of private medical services for this period grew by 48.8%.

According to monitoring of the market of medical services published by experts in Kazan on dk.ru, dentists still provide up to 50% of the volume of the market of private medical services. At the same time, the share of private medicine in Kazan accounts for 22% of the market of all medical services provided by both private and municipal institutions of the republic [13]. Thus, the market of private medical services is still inferior in competition to public hospitals. A total of 494 companies operate in Kazan under the licenses "activities of hospital institutions", "activities in the field of medical care", and "general medical practice" as of October 2016. At the same time, 1,932 medical institutions belong to private medicine in the Republic of Tatarstan. Thus, the analysis showed that there are a number of factors contributing to the growth of the market of private medicine in both Russia and the Republic of Tatarstan.

In order to identify the features and patterns of growth in the market of private medical services in Tatarstan in 2012–2017 methods of conducting sociological research were used: a desk method of studying secondary data, which collects and analyzes the main key indicators for the pharmaceutical market of the Russian Federation [11]. More than 30 respondents from almost 10 heads of private medical institutions of the Republic of Tatarstan took part in the study.

An important element of business instruments for the development of private medical centers is the method for quality monitoring of their services. When using this method, all current and planned activities are monitored. In this case, the organization has the ability to quickly identify deviations from the plans, in a timely manner to monitor the implemented activities, to identify the company's performance, the results to achieve the planned indicators.

To monitor the quality of business processes, which the key value of any medical center, we can use a unified methodology for quality monitoring of business processes proposed by Uvarina [9].

From the standpoint of the proposed approach, quality monitoring (value of the offer) of services provided by a certain type of medical center can be viewed from the standpoint of customer satisfaction with the various business processes that they face when receiving services.

In accordance with this approach, quality monitoring (value of the offer) of services provided by a certain type of medical center is expediently understood as a comprehensive study, which allows characterizing the level of achievement of the organization's strategic goals in meeting customer needs.

The key point in quality monitoring of services received by clients of medical centers is their opinions on satisfaction and perceptions in four groups of factors - business processes of a certain type of medical center: treatment, service, infrastructure and marketing processes. To monitor the level of quality of services provided, it is proposed to use the so-called GAP model, which involves analyzing the gaps "customer expectations - customer perceptions". At the base of the model is a set of gaps. The level of quality of services received from the point of view of clients depends on the size and direction of the gap between their expectations and perceptions.

The proposed methodology for quality monitoring of services is based on two surveys of clients of medical centers in Kazan. Each of these surveys consists of 19 questions designed specifically for medical centers.

Secondary data of consulting companies RBC, BusinesStat, KPMG, regulatory and legal acts in the field of regulating the activities of medical institutions were used as an information base for the study. To substantiate the main provisions of the study, a systematic approach was used using a complex of general scientific methods:

– Comparative analysis - during critical analysis: approaches to the definition of the concepts of "business processes" and "business models", types of private medical centers, management approaches to the improvement of business processes;
– A mass survey of consumers in order to monitor the factors of attractiveness of private medical centers upon key business processes, as well as the quality of services provided in medical centers.

## 3  Results

The first survey, according to the methodology, refers to customer expectations, the second concerns perceptions of customers regarding the quality of services received in a private medical center. The rating was set on a five-point scale: 5 points means complete agreement with the statement, 1 point - complete disagreement. The remaining values (2, 3, and 4) reflected the degree of approximation to one or another extreme point of view.

For each of the 19 points, we calculated the indicator for quality monitoring of the service by subtracting the values of expectations from the obtained values of the level of perceptions (Fig. 4):

$$GAPi = Pi - Ei, \tag{1}$$

where
$Pi$ – is the level of the perceived quality of service upon the $i$-th factor ($i = 1...19$),
$Ei$ – is the level of the expected quality of service upon the $i$-th factor ($i = 1...19$).

The next step was the definition of four integrated indicators:

$$Isrb\ treatment = GAP1 + GAP2 + GAP3 + GAP4 + GAP5, \tag{2}$$

where *Isrb treatment* - is an integrated indicator of the gap in customer expectations upon profile indicators (therapeutic process).

$$Isrb\ infrastructure = GAP6 + GAP7 + GAP8 + GAP9 + GAP10 + GAP11, \tag{3}$$

where *Isrb infrastructure* - is an integrated indicator of the gap in customer expectations upon infrastructure indicators (supporting the process).

$$Isrb\ service = GAP12 + GAP13 + GAP14 + GAP15 + GAP16 + GAP17, \tag{4}$$

where *Isrb service* – is an integrated indicator of the gap in customer expectations upon indicators of contact personnel (service process).

$$Isrb\ marketing = GAP18 + GAP19 + GAP20 + GAP21, \tag{5}$$

where *Isrb marketing* - is an integrated indicator of the gap in customer expectations upon image indicators (the marketing process).

The zero values of indicators show that the levels of expectations and perceptions of the quality of services provided by the medical center coincide, that is, the expectations of customers are confirmed. Negative and positive values of the quality of services provided by the medical center indicate that the levels of customer expectations and perceptions do not match. A negative value means that customer expectations exceed perceptions (negative not confirmation). A positive value indicates that the level of perceptions exceed the level of customer expectations (positive not confirmation).

Approaching any value of indicators to zero or a positive value means a high level of the quality of services provided by the private medical center, and a negative value means a low level in this business process. The less negative values of integrated indicators, the higher the level of quality of services provided by the medical center as a whole and vice versa.

After the collection of information about customer expectations and perceptions was completed, it was analyzed, and the findings were applied to set standards and develop activities (Fig. 5) in order to improve the work of the medical center concerning 4 business processes (treatment, infrastructure, service and marketing).

Thus, the introduction of the proposed methodology will allow (Fig. 5):

- Monitor all business processes of the medical center;
- Promptly identify deviations in perception of the quality of medical services received by clients from their expectation;
- Adjust the directions and develop key measures aimed at improving the quality of services provided by the medical center.

Thus, with the help of a survey, a sociological survey of 50 patients of MC Genesis Group LLC was conducted. The existing gaps in the private medical center were calculated: MC Genesis Group LLC. The respondents were asked to rate 19 aspects of the clinic's work upon the expected and actual results on a 5-point scale. Then, for each question, the average score of the expected value (E) and its perceived value (P) was calculated, and the following indicator was also calculated:

$$GAPi = Pi - Ei, \tag{6}$$

where Pi – is the level of perceived quality of service upon the *i-th* factor (i = 1...21), Ei – is the level of the expected quality of service upon the *i-th* factor (i = 1...21).

The results of information processing led to the following conclusions.

High degree of customer satisfaction with MC Genesis Group LLC in the following areas of work:

- Treatment at the medical center should be effective (Gap4 = 0.2),
- Customers should be able to obtain additional services (taxi call, etc.) (Gap14 = 0.7),
- Communications should be available in the medical center (accessible phone, wi-fi, etc.) (Gap7 = 0.6),
- Customers are provided with up-to-date clinic information (Gap8 = 1.5).

These directions do not need to be adjusted.

The greatest deviation of the expected value from the perceived one was noted by the following factors:

- You can select the desired specialist from several similar ones in the medical center (Gap2 = −0.4),
- The medical center should be conveniently located (close to home/ work) (Gap6 = −0.3),
- There should be equipped parking close to the medical center (Gap9 = −0.4),
- The medical center has modern equipment and consumables (Ga11 = −1),
- Administrative staff is friendly during phone calls (Gap12 = −0.3),

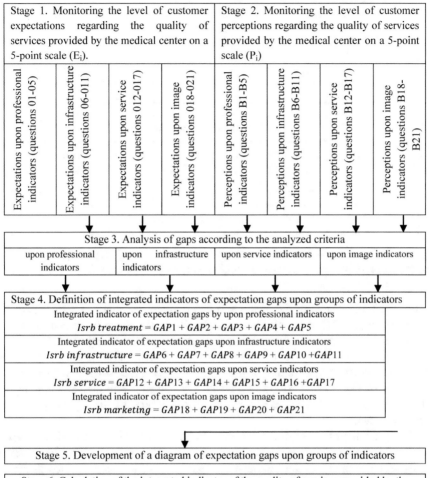

**Fig. 4.** Algorithm for quality monitoring of services provided by the medical center Genesis Group LLC (Source: compiled by the authors)

- Individual approach to each client (staff's ability to hear and solve problem situations) (Gap13 = −0.3),
- The medical center has a modern website (Gap18 = −0.4),
- Customers of the medical center are provided with high-quality printed materials (Gap19 = −1.3).

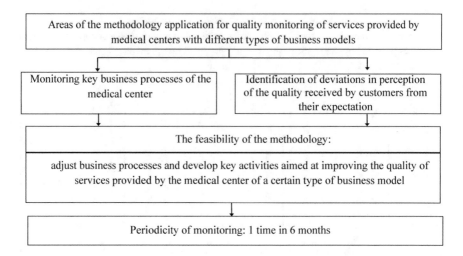

**Fig. 5.** The scope of the methodology for quality monitoring of services provided by key business processes of the medical center (Source: compiled by the authors)

Based on the results of the analysis, a gap diagram was also constructed for MC Genesis Group LLC (Fig. 6), which allows identifying the strengths and weaknesses of key business processes and determining the directions for their adjustment.

From the diagram presented in Fig. 6, it can be seen that in terms of the service process, customers' expectations of the model of the clinic being studied as part of monitoring (MC Genesis Group LLC) practically coincide with their perceptions, and in terms of the marketing, infrastructure and treatment process, the expectations are far from being met.

The possibility of modifying the method makes it possible to apply it to any type of medical institution with regard to unique values. The recommended frequency of quality monitoring of services provided by the medical center according to this method is 1 time in 6 months.

Thus, an important factor in project engineering is the determination of performance and efficiency indicators of the medical institution. In essence, performance refers to the achievement of organization's goals, i.e. it reflects the degree of implementation of the strategy, and efficiency rather refers to the assessment of the organization's resources used when implementing the strategy. Performance indicators for monitoring key business processes of the medical institution are complex, and all main components are in constant interaction and have a direct impact on each other.

Modification and testing of the methodology, developed by Yu. V. Uvarina, for quality monitoring of business processes of the private medical institution was carried out. This methodology allows monitoring the quality (value of the offer) of a certain type of medical service from the point of view of customer satisfaction with the various business processes that they face when receiving services: treatment, infrastructure, service and marketing.

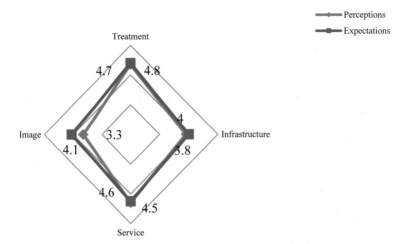

**Fig. 6.** Monitoring gaps between the expected and perceived level of the quality of services provided in the medical center of a highly specialized type - MC Genesis Group LLC (Source: compiled by the authors)

The methodology allowed determining the performance of business processes of the center. This analysis allowed us to identify areas for improving business processes in the medical center.

Business processes simultaneously play the role of standards for quality monitoring of medical care that each individual consumer of medical services receives, and the regulation of its main characteristics in accordance with the stages of the therapeutic and diagnostic process, prevention and rehabilitation. Without the introduction of business processes and medical and economic standards it is impossible to further develop the work of private medical centers. Based on this conceptual approach and taking into account the experience of leading world economies in the field of business process reengineering and the possibilities to use it in private medical centers, as well as personal practical experience, the methodology for monitoring business processes has been developed and adapted to the specifics of Russian medical institutions. In addition, a system of measures for the analysis, description, and optimization of existing business processes of the private medical center was compiled and substantiated.

A modified methodology for quality monitoring of services provided by key business processes in private medical centers, developed on the basis of business process reengineering methodology, was introduced in the process of reorganizing the management system of MC Genesis Group LLC. On the basis of the proposed recommendations, directions for improving business processes were identified, which will significantly improve the customer loyalty index of MC Genesis Group LLC, as well as the quality of medical care provided.

# 4 Discussion

The theoretical and methodological basis of the study was the work of A. Osterwalder and U.E. Deming dedicated to modeling high-quality business processes. However, business processes in the medical care system must be considered separately, without drawing any analogies with other industries. The study of modeling business processes in foreign medical care provides the basis for the following conclusions regarding its use in the Russian Federation. Firstly, it becomes expedient to introduce such process-oriented methodologies as reengineering of business processes in medical institutions of the Russian Federation, and then in the future it is necessary to switch to a system of continuous improvement. Secondly, innovations and radical transformations can only be implemented with the proper integration of appropriate combinations of modern information technologies and process-oriented management. Unfortunately, only a few medical institutions are fully familiar with them.

And finally, many medical institutions do not have resources necessary to develop qualitatively new models of functioning, as a result of which the process of introducing a process approach can be quite problematic.

It should be noted that the classical methodology of business process reengineering, as well as any other process management tool, is applicable to the medical care sector with great limitations. In this industry there are a large number of stakeholders at various levels of the hierarchy, as well as fairly stringent requirements, both from the internal consumer (staff) and external (patients). It follows that the right environment and a clear choice of institutions for the project implementation are crucial for such projects.

# 5 Conclusions

1. Thus, an important factor in project engineering project is the determination of performance and efficiency indicators of the medical institution. In essence, performance refers to the achievement of organization's goals, i.e. it reflects the degree of implementation of the strategy, and efficiency rather refers to the assessment of the use of the organization's resources during the implementation of the strategy. Performance indicators for monitoring key business processes of the medical institution are complex, and all main components are in constant interaction and have a direct impact on each other.
2. Modification and testing of the methodology, developed by Yu. V. Uvarina, for quality monitoring of business processes of the private medical company was carried out. This methodology allows monitoring the quality (value of the offer) of a certain type of medical service from the point of view of customer satisfaction with the various business processes that they encounter when receiving services: treatment, infrastructure, service and marketing.
3. The methodology allowed determining the performance of business processes of MC Genesis Group LLC. This monotoring allowed us to identify areas for improving business processes in the medical center.

4. Business processes simultaneously play the role of standards for quality monitoring of medical care that each individual consumer of medical services receives, and the regulation of its main characteristics in accordance with the stages of the therapeutic and diagnostic process, prevention and rehabilitation. Without the introduction of business processes and medical and economic standards today it is impossible to further develop the system of work of private medical centers. Based on this conceptual approach and taking into account the experience of leading world economies in the field of business process reengineering and the possibilities to use them in private medical centers, as well as personal practical experience, the methodology for monitoring business processes has been developed and adapted to the specifics of Russian medical institutions. In addition, a system of measures for the analysis, description, and optimization of existing business processes of a private medical center was compiled and substantiated.

5. A modified methodology for quality monitoring of services provided by key business processes in private medical centers, developed on the basis of business process reengineering methodology, was introduced in the process of reorganizing the management system of MC Genesis Group LLC. On the basis of the proposed recommendations, directions for improving business processes were identified, which will significantly improve the customer loyalty index of MC Genesis Group LLC, as well as the quality of medical care provided.

# References

1. Deming, WE (1993) The new economics for industry, government, and education. MIT Press, Boston, MA, p 132. ISBN: 0262541165
2. A look at the prospects for the development of the market of private medical services in the Russian Federation in 2017–2019. KPMG survey results. https://assets.kpmg.com/content/dam/kpmg/ru/pdf/2017/03/ru-en-research-on-development-of-the-private-medical-services-market.pdf. Accessed 3 July 2018
3. Appendix RBC + "Private Medical Services" to the magazine "RBC" (2016) № 9, p 95
4. Makova SE (2016) Formation and development of process-oriented management in medical institutions. Abstract https://refdb.ru/look/1391767-pall.html. Accessed 4 Mai 2018
5. Komrakov A (2016) State vices of private medicine. http://romir.ru/press/3279_07.11.2016/. Accessed 2 Jan 2017
6. Osterwalder A, Pigne I (2018) Construction of business models-Moscow, Skolkovo
7. Statistical compilation "Social and Economic Situation of Russia" (2014, 2015, 2016) Federal State Statistics Service
8. The market of medical services in Russia: structure, trends and prospects. BusinessStat. http://conference.apcmed.ru/upload/iblock/246/BusinesStat.pdf/. Accessed 5 Jan 2017
9. Uvarina YuA, Shushkin M (2016) Innovative business models of medical centers: marketing tools for analyzing the implementation of business processes. Innovations 1:99–107
10. Vader M (2015) Lean manufacturing tools: mini-guide to implement lean manufacturing techniques, 10th edn. Alpina Publisher, Moscow, p 151
11. Why the market of private medicine in Russia is growing and why it needs start-ups. https://vc.ru/p/starttrack-medicine/. Accessed 5 Jan 2017

12. Ulumbekova G It will not be available. Graduate School of Healthcare Management. http://www.
vshouz.ru/news/71.html?SSr=560133a26210fffffff27c__07e00b160a3817-7b05/. Accessed 5
Jan 2017
13. Experts predicted a reduction in the number of hospitals to the 1913 level (2017) RBC.
http://www.rbc.ru/society/07/04/2017/58e4feb59a794722462a85aa/. Accessed 2 Jan 2017
14. Zeithaml VA, Bitner MJ, Gremler DD (eds) (2017) Services marketing strategy, vol 1. Wiley
International Encyclopedia of Marketing, pp 31–38

# Cyber Risks for Insurance Company

G. N. Kaigorodova[1]([✉]), A. A. Mustafina[1], G. K. Pyrkova[1],
M. G. Vyukov[2], and L. M. Davletshina[1]

[1] Kazan Federal University, Kazan, Russia
{alfy2506,golsuorsil,guzel831,663499}@mail.ru
[2] State Budgetary Institution "Center for Advanced Economic Research
of the Academy of Sciences of the Republic of Tatarstan", Kazan, Russia
vjukovm@gmail.com

**Abstract.** The relevance of the contribution is due to the fact that financial intermediaries are among the most advanced companies using all modern IT-technologies. They are influenced by a high level of competition in this area, as well as by the requirements of the regulator (electronic reporting system, electronic sale of compulsory insurance policies, developed marketplace program). Digitalization, in addition to tangible benefits, comes laden with a number of risks, including the loss of business. The specific nature of the insurer increases the significance of these risks. In this regard, this contribution is aimed at identifying cyber risks for the insurance company as part of its specific business processes. The leading approach to the study of this problem was a systematic approach that allows considering business processes separately and insurance companies' activity in conjunction with cyber risks arising within each process. The authors systematize the factors causing an increased level of cyber risks for the insurer, carry out risk mapping and cyber risks for insurer's business processes, and identify business processes that are most at cyber risks. This allows getting the most adequate system for managing both risks in general and information security risks. The materials of the contribution are of practical value for insurance companies, since the consideration of the most dangerous risks for each business process will allow forming a relevant information security in the company.

**Keywords:** Cyber risk · Digitalization of business processes ·
Information security · Information technology · Insurance company ·
Insurer's business processes · Risk mapping

## 1  Introduction

In 2013, cyber risks ranked 15th place among business risks and only 6% of respondents named them among the three leading risks for their company. Currently, among the most significant risks for business are cyber risks - second in rank, after the risk of technology infusion in production. 40% of respondents noted significant cyber risks, such as cyber fraud, imperfection of IT-technologies and others. The value of cyber risks for business is constantly increasing. In 2017, only 30% of company executives recognized cyber risks; they considered market risks to be more important - volatility, competition, mergers and acquisitions, etc. [1].

The importance of cyber risks is growing. And since they are still poorly understood, they can be assessed mainly by indirect methods [17]. These risks are not isolated in any particular segment of the market, and they cover a wide variety of industries, different in scale of the company.

At the same time, the importance of digitization of the economy is growing. States and companies allocate significant resources for IT-technologies. This entails the growth of cyber risks both on a local and global scale. The budget of the national program "Digital Economy of the Russian Federation" is estimated at 3.5 trillion rubles for the period up to 2024. In this case, the costs of the section "Information Security" are estimated at $ 0.025 trillion rubles, that is, 0.7% of total costs on the program. These risks are to be recorded in economic development models of the state [13]. Obviously, cyber security issues should be addressed by resource capabilities of the business. Therefore, the issues of personnel reserve, investment in human capital are urgent [12].

This issue is of particular relevance for financial intermediaries. The specific nature of their activity is an active presence in the market, active digitalization of business, both in accounting aspects and marketing [2]. This increases the value of learning about cyber risks. Many researchers note the importance of cyber risks for the insurance company. Pooser et al. [10] note that the perception of cyber risks in insurance companies is growing.

The systematization of cyber risks within business processes of the insurance company is little studied. The problem is that each business process is associated with cyber risks and it is necessary to study their components and take into account the mutual influence of each of them.

The purpose of the contribution is to study cyber risks in conjunction with business processes of the insurance company. And achieving this purpose it is necessary to identify insurer's business processes and cyber risks, identify cyber risks for the insurance company and map cyber risks within the framework of the relevant business process.

## 2    Materials and Methods

The leading approach is a systematic approach, which considers insurance companies' activity from the point of view of a set of business processes that allow them to perform insurer functions. Cyber risks, both internal and external, are realized in diverse risks, differing both in probability and significance of the impact (potential damage) [7]. The systems approach allows linking insurer's business processes with known types of cyber risks.

The nature of risks is such that the same risk can lead to damages of various types and magnitudes. Therefore, it is necessary to identify the most significant and even critical risks. A risk mapping technique was used for these purposes. This technique allows identifying, prioritizing cyber risks for the insurance company, and assessing business processes that are most at corresponding risks. To compile a risk map, a survey of 87 managers of insurance companies was used.

Based on expert assessments, all results within certain limits were ranked as follows:

(1) Probability ranks: 5 - high; 4 - above average; 3 - average; 2 - below average; 1 - low;
(2) Significance ranks: 1 - boundary; 2 - significant; 3 - critical; 4 - catastrophic.

## 3  Results

According to research by NAFI Analytical Center, in 2017 half of Russian companies faced various cyber risks and 22% of them suffered significant financial losses. Large business is more at cyber risks compared to small and medium-sized businesses. It might seem that the difference is large, but 62% of the top management of large companies and 46–47% of small and medium-sized businessmen faced cyber attacks in 2017 [11]. At the same time, the majority of businessmen (60%) believe that cyber risks are minimal for their business. The average loss is about 300 thousand rubles, and the total loss of business amounted to 116 billion rubles in 2017 [11].

The companies in the financial sector most frequently face cyber risks. Therefore, the costs of financial companies to manage these risks are much higher than the costs of companies in other sectors. The factors causing a high level of risks in the financial sector are:

– Work with large data arrays and computerization of their processing systems;
– High level of competition, the possibility and necessity of promoting financial services through the Internet;
– Financial intermediation, respectively, concentration of huge financial resources, electronic payments increases the number of fraudsters, especially in the cyber sphere;
– Introduction of modern reporting standards, this leads to an increase in the business activity in the company's IT-system, as a rule, integrated with the systems of external users.

The study of cyber risks for insurance companies' activity is especially important. The functioning of the insurance business is a complex process. The insurance company should consider:

– Risks associated with doing business as an entrepreneurial activity;
– Risks associated with taking on customers' risks;
– Risks accompanying the insurer's investment activity;
– Technical risks associated with the creation of insurance funds.

**Table 1.** Identification of cyber risks in the relevant business environment of the insurer

| Business process | Types of activity | Potential damage from cyber risks |
|---|---|---|
| 1. Insurance marketing. Sales of insurance products. | - Market research;<br>- Promotion of insurance products;<br>- Retraining of agents and employees;<br>- Insurance premium | - Breaking-down business activity;<br>- Loss of customers;<br>- Loss of distinctive features of the insurance product;<br>- Loss of intellectual property;<br>- Loss of cash |
| 2. Pricing | - Ensuring the balance of tariffs;<br>- Ensuring the adequacy of tariffs | - Breach of contract;<br>- Loss of reputation;<br>- Product recall;<br>- Breaking-down business activity |
| 3. Insurance underwriting | - Selection of policyholders;<br>- Insurance rate for homogeneous groups;<br>- Individual characteristics of risks and regulation of coefficients | - Loss of the data array;<br>- Loss of commercial secrecy;<br>- Material damage;<br>- Regulatory actions and fines;<br>- Loss of customers;<br>- Loss of intellectual property |
| 4. Settlement of losses | - Claims from policyholders;<br>- Verification of documents confirming the fact of the insured event;<br>- Insurance payment | - Breach of contract;<br>- Loss of reputation;<br>- Loss of personal data;<br>- Loss of liability;<br>- Material damage |
| 5. Accounting for transactions | - Introduction of new insurance contracts into the system;<br>- Policy administration;<br>- Accounting of insurance premiums;<br>- Accounting of payments made | - Material damage;<br>- Loss of reputation;<br>- Loss of personal data;<br>- Regulatory actions and fines;<br>- Responsibility for network security;<br>- Losses payable due to loss of customer data |
| 6. Reinsurance | - Transfer of risks to reinsurance;<br>- Taking on risks in reinsurance | - Loss of commercial secrecy;<br>- Material damage;<br>- Loss of data;<br>- Loss of reputation |
| 7. Actuarial analysis | - Operations based on mathematical modeling | - Loss of the data array;<br>- Response costs;<br>- Material damage;<br>- Breaking-down business activity |

Source: compiled by the authors

In addition, the so-called cyber risks become relevant.

The insurer performs its functions in the form of specific business processes, each of which is currently subject to risks from the cyber environment. In Table 1, the authors carried out identification of possible cyber risks in the relevant business environment of the insurer.

Based on formulated risks to business processes of the insurance company, the authors have identified the risks of corresponding losses. A questionnaire survey of the heads of insurance companies' divisions allowed establishing the probability levels of risks and their significance for the relevant business process. Ranking of cyber risks is carried out. The results obtained are summarized in Table 2.

**Table 2.** Ranking of cyber risks

| Business process | Cyber risk | Probability rank | Relevance rank |
|---|---|---|---|
| 1. Insurance marketing. Sales of insurance products. | Risk of breaking-down business activity (R1-1) | 5 | 2 |
| | Risk of losing customers (R1-2) | 5 | 3 |
| | Risk of losing distinctive features of the insurance product (R1-3) | 1 | 1 |
| | Intellectual property risk (R1-4) | 1 | 1 |
| | Risk of losing money (R1-5) | 4 | 2 |
| 2. Pricing | Risk of breaching contract (R2-1) | 4 | 2 |
| | Reputation risk (R2-2) | 2 | 3 |
| | Risk of product recall (R2-3) | 1 | 1 |
| | Risk of breaking-down business activity (R2-4) | 3 | 2 |
| 3. Insurance underwriting | Data risk (R3-1) | 3 | 3 |
| | Risk of losing commercial secrecy (R3-2) | 3 | 3 |
| | Risk of material damage (R3-3) | 5 | 2 |
| | Risk of regulatory actions, fines (R3-4) | 4 | 2 |
| | Risk of losing customers (R3-5) | 5 | 2 |
| | Intellectual property risk (R3-6) | 4 | 1 |
| 4. Settlement of losses | Risk of breaching contract (R4-1) | 4 | 3 |
| | Reputation risk (R4-2) | 4 | 4 |
| | Personal data risk (R4-3) | 3 | 4 |
| | Liability risk (R4-4) | 3 | 2 |
| | Risk of breaching contract (R4-5) | 5 | 2 |
| 5. Accounting of operations | Risk of material damage (R5-1) | 4 | 3 |
| | Reputation risk (R5-2) | 4 | 3 |
| | Personal data risk (R5-3) | 2 | 4 |
| | Risk of regulatory actions, fines (R5-4) | 5 | 4 |
| | Risk of the person responsible for network security (R5-5) | 4 | 2 |
| | Risk of losses due to loss of customer data (R5-6) | 3 | 2 |
| 6. Reinsurance | Risk of losing commercial secrecy (R6-1) | 4 | 2 |
| | Risk of material damage (R6-2) | 4 | 3 |
| | Data risk (R6-3) | 2 | 1 |
| | Reputation risk (R6-4) | 1 | 4 |
| 7. Actuarial analysis | Risk of losing the data array (R7-1) | 4 | 2 |
| | Risk of response costs (R7-2) | 2 | 2 |
| | Risk of material damage (R7-3) | 4 | 3 |
| | Risk of breaking-down business activity (R7-4) | 2 | 2 |

Source: compiled by the authors

Based on the obtained results, the authors compiled a map of risks for insurer's business processes, as a result of cyber incidents (Fig. 1). The critical margin of tolerance for risk, as a rule, is the line delineating the upper right square. They require immediate attention from a management point of view. Those risks that are in the remaining three squares can be considered tolerant. But this does not mean that they cannot be managed.

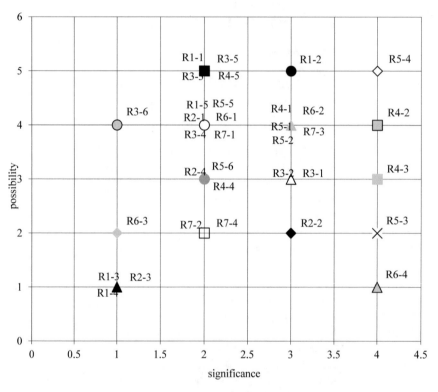

**Fig. 1.** Major cyber risks for insurer's business processes (Source: compiled by the authors)

Thus, the risk map allows the insurer to identify business processes that are most at cyber risks, identify responsibility centers for each division of the insurance company, define terms of reference to ensure an appropriate level of information security and ultimately create a cyber security culture in the company.

Mapping allows the insurer to:

– Identify business processes that are most at cyber risks, identify weaknesses ("human factor", excessive dependence on information systems);
– Make management decisions on cyber risk management and integrate them into a comprehensive risk management strategy in the insurance company;

- Identify "responsibility centers" for cyber security and assign this responsibility by the regulation of functions and access rights;
- Simulate a response plan for emerging cyber risks.

## 4   Discussion

Many researchers note the need for the active presence of the insurance company in the market of electronic services and mobile applications [9, 18].

What is cyber risk? It is necessary to distinguish between the possibility of damage as a result of the information system outrage due to imperfect equipment and software that cannot cope with a targeted attack on the company's IT system, datasets, and cash.

Mukhopadhyay et al. [8] defined cyber risks as risks associated with a malicious event in the cyber space that leads to loss of business reputation and money. Strupczewski [15] studied definitions of cyber risks and concluded that cyber risks should be considered as an operational risk in the organization's information and technological resources, the negative consequences of which can affect the confidentiality, availability and integrity of information or IT systems. The author also studied insurance defensibility [15].

In the study of CRO forum [4], the definition of cyber-risk covered any risks associated with the use and transmission of electronic data, including technological tools such as the Internet and telecommunications networks. In this case, the components of damage are:

- Damage from cyber attacks;
- Fraud committed as a result of the use of other people's data;
- Any liability arising from the use, storage and transmission of data;
- Availability, integrity and confidentiality of electronic information.

It is necessary to consider cyber risks not only as risks from unauthorized interventions, fraudulent actions of company employees. It is necessary to take into account the whole complex of cyber risks - including due to imperfect software, erroneous actions of employees, etc.

Pooser et al. [10] examined trends in the identification of cyber risks by insurance companies from 2006–2015, and concluded that the earlier insurance companies take cyber risk into their operations, the more successful they will be in their development. After 2013, all insurers under study identify cyber risk as a significant risk. The authors of the contribution analyze the need to account for cyber risks by business processes of the insurance company. Cyber risk analysis and insurance options are carried out by Biener et al. [3].

Shetty et al. [14] defined in their work that effective cyber risk management should include not only the use of insurance as a risk transfer mechanism, but also the provision of incentives for insurance companies to invest in cyber self-defense. Koch et al. [6] analyzed the formation of the insurance portfolio in modern conditions and noted the lack of insurance for cyber risks. Kaigorodova et al. [5] revealed directions for improving the information system of the insurance company.

The contribution complements the research in this field by systematizing risks for the insurance company from the cyber space and determining that it is necessary to analyze the issues of cyber risks in conjunction with the current business processes of the company.

## 5   Conclusions

The authors of the contribution considered cyber risk as a complex risk that is a risk to the development of the insurance business as a whole, since it affects insurer's business processes.

Business processes that are most at cyber risks:

- Settlement of losses, since they have three critical impacts on the possibility and significance of risks: risk of breaching contract, reputation risk, personal data risk;
- Accounting of transactions: risk of material damage and reputation risk, risk of regulatory actions and related fines;
- Insurance underwriting: data risk, risk of losing commercial secrecy.

At the same time, it is necessary to take into account losses incurred not only by one type of asset, but by all inter-related assets.

Understanding of business processes in terms of cyber risks, allows creating a complementary security in the IT system of the insurance company. At the same time, it makes possible to formulate requirements for each employee of the insurance company in terms of cyber security, taking into account the place and role in the relevant business process.

**Acknowledgments.** The contribution is performed according to the Russian Government Program of Competitive Growth of Kazan Federal University.

## References

1. Allianz Risk Barometer: Top Business Risks For 2018 (2018) Allianze. https://www.agcs.allianz.com/assets/PDFs/Reports/Allianz_Risk_Barometer_2018_EN.pdf. Accessed 15 Mai 2018
2. Alyakina DP, Khisamova GF (2014) Methodology for rating of insurance portfolio. Medit J Soc Sci 5(24):137–140. https://doi.org/10.5901/mjss.2014.v5n24p137
3. Biener Ch, Eling M, Wirfs JH (2015) Insurability of cyber risk: an empirical analysis. Geneva Pap Risk Insur Issues Pract 40(1):131–158. https://doi.org/10.1057/gpp.2014.19
4. CRO Forum Concept Paper on a Proposed Categorization Methodology for Cyber Risk (2016) https://www.thecroforum.org/2016/06/20/concept-proposal-categorisation-methodology-for-cyber-risk/. Accessed 15 Mai 2018
5. Kaigorodova GN, Mustafina AA, Alyakina DP (2018) Directions of improving information system of insurance company. J Phys (IOP Conference Series) 1015–1016. https://doi.org/10.1088/1742-6596/1015/4/042016

6. Kokh IA, Kaigorodova GN, Mustafina AA (2016) The research of conditions of insurance portfolio formation in the Russian practice. Int Bus Manag 10(23):5657–5662. https://doi.org/10.3923/ibm.2016.5657.5662

7. Krasnov IZ, Karelin OI (2014) Methods of analysis and risk assessment of the organization. Bull SibGAY 2(54): 37–44. https://elibrary.ru/download/elibrary_21938783_18185861.pdf

8. Mukhopadhyay A, Chatterjee S, Saha D, Mahanti A, Sadhukhan SK (2013) Cyber-risk decision models: to insure IT or not? Decis Supply Syst 56:11–26. https://doi.org/10.1016/j.dss.2013.04.004

9. Mustafina AA, Kaigorodova GN, Pyrkova GK, Alyakina DP, Syvorotkina KA (2017) Sanatorium and resort treatment as a factor of economic development in the republic of Tatarstan. Astra Salv 2:267–276. https://astrasalva.files.wordpress.com/2018/01/astra-salvensis-v-2017-supplement-no-2.pdf

10. Pooser DM, Browne MJ, Arkhangelska O (2018) Growth in the perception of cyber risk: evidence from U.S. P&C insurers. Geneva Pap Risk Insur Issues Pract 43(2):208–223. https://doi.org/10.1057/s41288-017-0077-9

11. Russian companies lost at least 116 billion rubles from Cyber attacks in 2017 (2017) NAFI. https://nafi.ru/upload/cybersecbusiness.pdf. Accessed 15 Mai 2018

12. Nabieva LG, Davletshina LM, Khairullina AD, Kulik EN (2016) Human capital as a factor in improving the competitiveness of the region. Int Bus Manag 10(23):5599–5602. https://doi.org/10.3923/ibm.2016.5599.5602

13. Sayfudinova NZ, Safiullin MR, Safiullin AR, Zaimullina MR (2016) Modeling of economic development system of the Russian Federation system. J Econ Econ Educ Res 17(2):334–346. https://www.abacademies.org/articles/jeeer-special-issue-2.pdf

14. Shetty S, McShane M et al (2018) Reducing informational disadvantages to improve cyber risk management. Geneva Pap Risk Insur Issues Pract 43(2):224–238. https://doi.org/10.1057/s41288-018-0078-3

15. Strupczewski G (2017) Ryzyko cybernetyczne jako wyzwanie dla branży ubezpieczeń w Polsce i na świecie. Finanse 10(1):251–271. http://www.knfpan.pan.pl/images/Fin._110-17_15-G.Strupczewski.pdf

16. Strupczewski G, Thlon M, Fijorek K (2016) Corporate insurance versus risk retention: an empirical analysis of medium and large companies in Poland. Geneva Pap Risk Insur Issues Pract 41(4):626–649. https://doi.org/10.1057/s41288-016-0005-4

17. Trynchuk VV (2017) Management of visual communications in insurance companies (on the example of using icons in logos). Probl Perspect Manag 15(2):319–331. http://dx.doi.org/10.21511/ppm.15(2–2).2017.02

18. Velichko NY, Kobersy IS, Radina OI, Khnikina TS, Ivashhenko SA (2017) Sales promotion in the marketing communications. Int J Appl Bus Econ Res 15(13):133–142. https://www.scopus.com/inward/record.uri?eid=2-s2.0-85025430730&partnerID=40&md5=c4525c478a24c2fd3e201ea7cde1d4f1

# Digital Technology in Insurance

A. A. Mustafina[1(✉)], G. N. Kaigorodova[1], P. D. Alyakina[1],
N. Y. Velichko[2], and M. R. Zainullina[1]

[1] Kazan Federal University, Kazan, Russia
{alfy2506,golsuorsil}@mail.ru, 345daria@gmail.com,
milyausha-zainul@list.ru
[2] Vesna Hotel & Spa, Sochi, Russia
velichkonu@mail.ru

**Abstract.** Under current conditions a company can rely on maintaining competitive positions only with proper use of information technology. This is important for companies operating in the field of financial intermediation, where it is important to use modern technologies to promote their products, ease the access to users, and accounting in accordance with IFRS and the regulator. The development of IT-technologies is at a high level in the banking and stock business. The insurance business is currently receiving opportunities for a technological breakthrough. The launch of mandatory sales of electronic liability insurance policies for motorists since the beginning of 2017 has accelerated digitization of the insurance market. The share of companies selling online policies increased to 69%. At the same time, an increase in sales of liability insurance electronic policies for motorists will contribute to further multiple increases in the volume of direct insurance. 85% of insurance companies use IT solutions in the sales process. Currently, 75% of insurance companies are engaged in the introduction of new IT products and solutions. Despite of the fact that the launch of electronic liability insurance for motorists has already demanded for significant investments in the development or improvement of IT systems from insurance companies, most of them continue to invest in new technologies. In this regard, this contribution is to study the impact of information technology on the development of direct insurance in Russia. The leading approach in the study was the method of covariance analysis (Ancova model), which allowed taking into account both quantitative and qualitative variables, as well as homogeneous and heterogeneous statistical data. The contribution reveals the relationship between the introduction of electronic policy systems in the company and the amount of insurance premiums. The study covers a group of the largest insurers in Russia. The research materials are of practical importance for insurers, because it gives the concept of a correlation between business digitalization and its level of profitability.

**Keywords:** Digital technologies · Direct insurance · Insurance company ·
Internet insurance products

© Springer Nature Switzerland AG 2020
S. Ashmarina et al. (Eds.): *Digital Transformation of the Economy: Challenges,
Trends and New Opportunities*, pp. 678–685, 2019.
https://doi.org/10.1007/978-3-030-11367-4_65

# 1   Introduction

The relevance of the research topic is that a higher sales ratio on the Internet means real money and high profits for insurance companies. McKinsey found that an auto insurer with an annual income of $10 billion could receive additional premiums of $400 million, increasing the online sales ratio by 20%. This value at stake gives insurers a powerful incentive to get rid of the gap in the digital potential that isolates it from the best performers.

The purpose of the contribution is to study the influence of digital technologies on the development of direct insurance. We set the following tasks: to study existing opinions about the relevance of introducing digital technologies in sales of insurance products; collect information on Russian insurance companies that have implemented digital technologies in direct insurance; analyze the influence of digital technology on the growth of insurance premiums.

Theoretical and methodological issues of using IT technologies in the insurance business are studied in the works of such authors as Huckstep [9], Adamova [2], Kaigorodova [11], Eling [7]. Direct insurance refers to the remote sale of policies (execution and calculation of their final cost) via the Internet or a call center without the participation of intermediaries - insurance agents. According to a survey conducted by RAEX [6] at the end of 2017, 85% of insurance companies in the process of selling insurance products use IT solutions related to the use of the Internet, against 76% a year earlier. Even more significant growth is observed in terms of the share of companies selling their policies through their website. Thus, according to data for 2016, 43% of insurers offered the possibility of online calculation and purchase of a policy. At the end of 2017, the share of such companies in the insurance market increased by 26 percentage points, to 69%. Currently, 71% of insurers selling through the site offer customers a wide range of insurance products: travel insurance, insurance against accidents and illnesses, voluntary medical insurance, boxed insurance of personal property, life insurance, liability insurance for motorists and others types of personal and property insurance. The remaining 29% of insurers, adhering exclusively to statutory requirements, limited themselves to selling electronic liability insurance policies to motorists.

At the same time, the level of penetration of mobile applications in insurance over the past year has not changed. Only 17% of insurers use mobile software when selling insurance services, including applications for interacting with agents. Positive dynamics is also noted in terms of the use of IT solutions by insurers associated with the use of the World Wide Web when settling insurance claims.

# 2   Materials and Methods

The authors used the method of covariance analysis of electronic policies in the development of direct insurance. The Ancova covariance analysis model is used to evaluate these measures - this is a regression model in which the indicators under study are both quantitative and qualitative. Qualitative analyzed variables (called dummy variables) are indicators that reflect the influence of a qualitative component containing qualitative signs of a second or more order in the model.

The specificity of the Ancova model is homogeneity and heterogeneity of the original statistical data. Statistical factors are usually homogeneous and they are observed in the same circumstances (date, period, education, gender). Heterogonous indicators are factors that can be fixed under different circumstances. In the case of heterogeneity, factors with more than two levels of quality can be introduced into the model. Dummy (dummy variables, artificial, binary, structural) indicators characterize, as a function, the effect of a quality indicator containing qualitative signs of a second or more order. The introduction of such indicators into the regression model necessitates the assignment of such indicators to the corresponding digital metrics, that is, it is necessary to convert qualitative indicators into quantitative ones.

Research using this type of regression necessitates following the rule of proper use of dummy indicators. If a qualitative variable has not two, but more than two values, then there is the possibility of a dummy indicator trap. It occurs if exactly k dummy indicators are used when modeling k values of a quality attribute. Then there is perfect multi-collinearity. In order to eliminate such a situation, the following rule should be applied: if a quality indicator has k alternative values, then only (k-1) dummy variables are used in the simulation.

To study the impact of introducing electronic policies into direct insurance with the parallel use of standard methods of interaction with customers, the authors used statistical information on 102 insurance companies in Russia, using or not using technologies of electronic sale of policies in a personalized offer of insurance products.

So, a dependent variable $(Y)$ is the amount of insurance premiums; an independent indicator $(x_1)$ is the cost of running a business in insurance companies when carrying out insurance activities related to attracting customers; a qualitative independent variable $(x_1z)$ is the introduction of electronic insurance policies.

At the same time $(x_1z)$ can take the following values:

$$x_1z = \begin{cases} 1, & \text{if the company uses texhnologies} \\ 0, & \text{if the company does not use technologies} \end{cases} \quad (1)$$

Then the Ancova model will look like this:

$$Y = a + bx_1 + cx_1z + e, \quad (2)$$

## 3  Results

The authors used the paired regression using the analysis package (regression) in MSExcel to assess the significance at the level of 95% of costs associated with attracting customers using standard methods. For this operation, the coefficient of determination $(R^2)$ is 0.9233. Thus, the model is justified by 92.3% of costs of doing business (attracting clients) using standard methods. So, a mathematical function explaining the econometric relationship was deducted (3).

$$Y = 161723,01 + 2,64x_1 + \varepsilon;$$ (3)

Table 1 presents the volume of insurance premiums and the introduction of electronic policies for 2017 for some insurance companies.

**Table 1.** The volume of insurance premiums and the introduction of electronic policies for 2017 for some insurance companies.

| Company/group of companies | Insurance premiums, thous. rubles | The cost of insurance operations, thous. rubles. | The share of electronic policies in premiums, % | The introduction of electronic policies (qualitative component - marker) |
|---|---|---|---|---|
| PECO-Garantia PJSC | 5738727,0 | 1624059,7 | 99,8 | 1 |
| Alfa Insurance Group | 4983180,0 | 1754079,4 | 100 | 1 |
| Ingosstrakh PJSC | 4761905,0 | 1642857,2 | 40,4 | 1 |
| SOGAZ Insurance Group | 3241079,0 | 547742,4 | 100 | 1 |
| Rosgosstrakh and Capital | 2919380,0 | 1234897,7 | 100 | 1 |
| Renaissance Insurance Group | 2298291,0 | 969878,8 | 61,1 | 1 |
| Tinkoff Insurance JSC | 1319359,0 | 456498,2 | 34,9 | 1 |
| VTB Insurance LLC | 265861,0 | 62743,2 | 66,6 | 1 |
| Russian Standard Insurance | 208850,0 | 20049,6 | 0 | 0 |
| ERGO Insurance Company | 156988,0 | 76924,1 | 0 | 0 |
| Astro-Volga JSC | 88444,0 | 27417,6 | 0 | 0 |
| Sberbank Life Insurance LLC | 79175,0 | 43546,3 | 100 | 1 |
| ERV Travel Insurance LLC | 70849,0 | 14878,3 | 0 | 0 |

Source: compiled by the authors.

The statistical significance of the econometric model is determined by the Fisher criterion. The observed $F_{actual}$ value was selected from ANOVA obtained during the regression. The critical $F_{critical}$ value was deducted using the MS Excel F function (probability; degree of freedom 1; degree of freedom 2).

$$F_{observ} = 340,9802, F_{Kp}(0,05; 1; 101) = 4,003982,$$ (4)

The inequality $F_{actual} > F_{critical}$ means that the econometric model is statistically significant. Thus, the relationship between the size of the volume of insurance premiums and the cost of doing business is justified.

Next, a quality metric was assigned. Multiple regression with qualitative variables was also carried out using the MS Excel analysis package (regression) to assess the significance of using digital technologies. As a result of the operation, the regression coefficient ($R^2$) is equal to 0.8614. Taking into account the qualitative variable, the model is reliable at 86.14%.

The mathematical function of the econometric model of covariance analysis is as follows:

$$Y = 163375,287 + 2,3548x_1 + 0,4657x_1z + \varepsilon, \tag{5}$$

where:

$$Y(x_1 = 0) = 163375,287 + 2,3548x_1 + \varepsilon;$$
$$Y(x_1z = 1) = 163375,287 + 2,3548x_1 + 0,4657x_1z + \varepsilon.$$

As can be seen from function (5), the introduction of electronic policies allows increasing the amount of insurance premiums by 0.4657 thousand rubles. If we average the observational data, the percentage increase is 14.07%.

$$F_{actual} = 180,3780; \quad F_{critical}(0,05;2;101) = 3,15593, \tag{6}$$

As can be seen from the presented data, $F_{actual} > F_{critical}$ econometric model of covariance analysis is statistically significant (reliability level 95%).

## 4  Discussion

The need to introduce information technology into business processes of insurance companies in Russia was emphasized in the works of such authors as Kaigorodova [10], Weingarth [20], Kokh [12]. According to research by individual scientists, the transformation of the insurance industry was quite late and it has to use the full potential of digital technology. However, digitalization will structurally change the creation of the value of insurance products in a variety of new ways to interact with customers, and will make the insurance product more perfect [7–9]. The insurance industry is indeed a bit behind the financial sector, in general, in terms of the introduction of digital technologies.

In general, the following research trends can be identified. Insurers and insurance intermediaries have a large database of their customers, while using social networks to promote their products and assess consumer behavior [1]. Therefore, it is necessary to pay more attention to the regulation of online sales [5, 19]. From a legal point of view, there are insurance sales when the insurer is required to properly inform the insured, but in online sales it is difficult to effectively provide such recommendations at the preliminary stage [5]. Moreover, insurance companies collect personalized information about their customers in order to make a flexible pricing policy [3]. On the one hand, under current competitive conditions, adaptation to the needs of customers is one of the main components of the competitive advantage and the possibility of turning an insurance product into a model product. This opinion was expressed in the works of many authors [11, 14, 17]. But on the other hand, it is difficult to apply it, since telematics and plans are difficult to apply when selling online [15]. The contribution complements research in this area, systematizing digital data on the practice of introducing digital technologies into direct insurance, as well as studying the impact of digital technologies on the growth of insurance premiums.

The forecasts for the development of digital insurance are given in the studies of McKinsey and Expert Ra [6]. The review of modern scientific literature on the problem under study showed a high degree of relevance of the issue and involvement of consulting companies in the research process.

## 5  Conclusions

According to insurance market experts, insurers cannot afford to suspend Digital Insurance Innovation. So, using the example of a call-center, you can see the costs of insurance companies when interacting with customers and how inefficient this work is. In call centers, insurance agents open several windows at once and manually reduce windows while the client is on the line. In this case, the windows are not integrated with each other, and the agent has to cut and insert data manually. At the same time, it has become the norm in the modern digital world that customers increasingly expect that they can buy/maintain/complain/compare/view the site and insurance products through their mobile phone or browser.

The advent of technology-based personalization technology is an important issue for the insurance industry. An interesting view exists on the transformation of an insurance product into a model product. Thus, advances in digital technology over the past decade have provided insurers with the means by which they can create "additional" insurance products. Once insurers have "won" a customer, they can keep them, using technologies such as telematics, mobile applications, and wearable devices to create ways to continuously connect and interact with customers.

As shown by the study on Russian insurance companies, the share of sales of electronic insurance policies affects the amount of insurance premiums.

**Acknowledgments.** The study is performed according to the Russian Government Program of Competitive Growth of Kazan Federal University.

## References

1. Abramovsky A, Kochenburger P (2016) Insurance online: regulation and consumer protection in a cyber world. In: Marano P et al (ed) The "dematerialized" insurance: distance selling and cyber risks from an international perspective, 5th edn. Springer, Cham. https://doi.org/10.1007/978-3-319-28410-1
2. Adamova M, Boudet J, Kalaoui H, Segev I (2018) How traditional insurance carriers can disrupt through personalized marketing, August 2018. McKinsey Company. https://www.mckinsey.com/industries/financial-services. Accessed 15 Mai 2018

3. Biener C, Eling M (2012) Insurance in the micro-insurance markets: analyzing problems and potential solutions. Geneva Risk Insurance Doc Quest Practice 37(1):77–107. https://doi.org/10.1007/978-1-137-57479-4_8

4. Cather AD (2018) Cream skimming: innovations in insurance risk classification and adverse selection. Risk Manag Insur Rev 21(2):335–366. https://doi.org/10.1111/rmir.12102

5. Chrissanthis ChS (2016) Online sales of insurance products in the EU. In: Marano P et al (ed) The "dematerialized" insurance: distance selling and cyber risks from an international perspective, 6th edn. Springer, Cham. https://doi.org/10.1007/978-3-319-28410-1

6. Electronic technologies in insurance: on the threshold of the epoch online (2018) Rating agency Expert Ra. https://raexpert.ru/researches/insurance. Accessed 15 Mai 2018

7. Eling M, Lehmann M (2018) The impact of digitalization on the insurance value chain and the insurability of risks. Geneva Pap Risk Insur Issues Pract 43(3):359–396. https://doi.org/10.1057/s41288-017-0073-0

8. Huckstep R (2017a) Digital transformation is the strategic imperative no insurer can ignore. Digital Insurance 66. https://www.the-digital-insurer.com/blog/insurtech-digital-transformation-strategic-imperative-no-insurer-can-ignore. Accessed 15 Mai 2018

9. Huckstep R (2017b) Metromile: the pioneers of digital engagement insurance. Digital Insurance 71. https://www.the-digital-insurer.com/blog/insurtech-metromile-pioneers-engagement-insurance. Accessed 15 Mai 2018

10. Kaigorodova GN, Mustafina AA, Alyakina DP (2018) Directions of improving information system of insurance company. IOP Conf Series: J Phys Conf Ser 1015 042016. https://doi.org/10.1088/1742-6596/1015/4/042016

11. Kaigorodova GN, Mustafina AA (2014) The influence of forms of insurance coverage organization on population's life quality. Mediterranean J Soc Sci 5(24):118–123. https://doi.org/10.5901/mjss.2014.v5n24p118

12. Kokh IA, Kaigorodova GN, Mustafina AA (2016) The research of conditions of insurance portfolio formation in the Russian practice. Int Bus Manag 10(23):5657–5662. https://doi.org/10.3923/ibm.2016.5657.5662

13. Mustafina AA, Kaigorodova GN, Pyrkova GK, Alyakina DP, Syvorotkina KA (2017) Sanatorium and resort treatment as a factor of economic development in the republic of Tatarstan. Astra Salvensis 2:267–276. https://astrasalva.files.wordpress.com/2018/01/astra-salvensis-v-2017-supplement-no-2.pdf. Accessed 2 Apr 2018

14. Nabieva LG, Davletshina LM, Khairullina AD, Kulik EN (2016) Human capital as a factor in improving the competitiveness of the region. Int Bus Manag 10(23):5599–5602. https://doi.org/10.3923/ibm.2016.5599.5602

15. Robertshou GS (2012) Online price comparisons sites: How technology has destabilised and transformed the UK insurance market. J Rev Pricing Manag 11(2):137–145. https://doi.org/10.1057/rpm.2011.5

16. Sayfudinova NZ, Safiullin MR, Safiullin AR, Zaimullina MR (2016) Modeling of economic development system of the Russian Federation system. J Econ Econ Educ Res 17(2):334–346. https://www.abacademies.org/articles/jeeer-special-issue-2.pdf. Accessed 6 Mai 2018

17. Trynchuk VV (2017) Management of visual communications in insurance companies (on the example of using icons in logos). Probl Perspect Manag 15(2):319–331. http://dx.doi.org/10.21511/ppm.15(2-2).2017.02

18. Velichko NY, Kobersy IS, Radina OI, Khnikina TS, Ivashhenko SA (2017) Sales promotion in the marketing communications. Int J Appl Bus Econ Res 15(13):133–142. https://www.scopus.com/inward/record.uri?eid=2-s2.0-85025430730&partnerID=40&md5=c4525c478a24c2fd3e201ea7cde1d4f1. Accessed 6 Mai 2018

19. Wang H-Ch (2016) E-commerce and distribution of insurance products: a few suggestions for an appropriate regulatory infrastructure. In: Marano P et al (ed) The "dematerialized" insurance: distance selling and cyber risks from an international perspective, 2nd edn. Springer, Cham. https://doi.org/10.1007/978-3-319-28410-1. Accessed 6 Mai 2018
20. Weingarth J, Hagenschulte J, Schmidt N, Balser M (2018) Building a digitally enabled future: an insurance industry case study on digitalization: How organizations rethink their business for the digital age. In: Urbach N (ed) Digitalization cases, 4th edn. Springer International Publishing, Cham. https://doi.org/10.1007/978-3-319-95273-4_13. Accessed 6 Mai 2018

# Crises and Digital Economy: The Territorial Aspect of the Problem of Networking of Stakeholders in the Food Markets

N. V. Sirotkina[1], O. G. Stukalo[2(✉)], N. V. Nikitina[3], and M. V. Filatova[2]

[1] Voronezh State University, Voronezh, Russia
docsnat@yandex.ru
[2] Voronezh State University of Engineering Technologies, Voronezh, Russia
stukalo_oksana@mail.ru.ru, fltvmrn@rambler.ru.ru
[3] Samara State University of Economics, Samara, Russia
nikitina_nv@mail.ru

**Abstract.** A research purpose is to analyze the impact of digital economy and inevitable crises on development of the territories. In the previous works the authors revealed the tendency to a setization of subjects of regional economy, including the local food markets. It became the basis for refining of prospects of network interaction of participants of the food markets in the conditions of the crises accompanying any evolutionary changes. Disaggregation of a research purpose allowed to allocate several specifying problems: typology of crises; forms of network interaction characteristic of transition to digital economy; features of the food sector as subsystems of regional social and economic system; development of the food markets. Lifting consistently each of the specified problems, it was succeeded to establish that the subjects of regional economy having knowledge, experience, resources and resource opportunities create network of competences which is a relevant evolutionary form of inter-action on economic space of the region. The network of competences applicable of the local food markets (the food sector of regional economy) differs in the list of participants and develops considering general and specific tendencies. The general background for development of network of competences creates tran-sition to digital economy. Any crises act as the internal and external factors influencing this process. Manifestation of an author's line item is that crises are considered as "creative destruction", and the fact of availability of crises doesn't change a general forward trajectory of development of regional economy.

**Keywords:** Competences network · Crisis · Digital economy · Digitalization · Food market · Regional economy

## 1 Introduction

The local food market is not just set of the sellers and buyers functioning in certain administrative-territorial borders today. Ideas of the local markets extend due to refining of interaction features of its participants. A relevant format of interaction is the

© Springer Nature Switzerland AG 2020
S. Ashmarina et al. (Eds.): *Digital Transformation of the Economy: Challenges,*
*Trends and New Opportunities*, pp. 686–692, 2019.
https://doi.org/10.1007/978-3-030-11367-4_66

network, and this is not about producers of the goods or services united by one brand. Forming and development of digital economy causes the necessity of expanded understanding of network and presentation of special requirements to its participants. In focus of attention of modern researchers there is a competences network. In relation to the food sector of economy of the region (territory) functioning by the principles of the local food market, competences are possibilities of participants to create the new way, culture and philosophy of production and consumption of food products promoting ensuring high level and quality of life of the population.

## 2  Materials and Methods

The concept of cluster and network development revealing the features of integration interaction in modern conditions which are that multi-branch, but the most competitive participants of the economic relations movable by selfish interests enter interaction on the terms of partnership, parity and voluntariness became reason for understanding competences network without forming at the same time legally arranged unions. Having applied symbiosis of a method of analogies (comparing subjects of competences network to subjects of social networks) and the morphological analysis, in the previous researches [9, 10] seven groups of subjects of the food sector of regional economy whose opportunities and participation form network of competences [8] were allocated. Differentiated depending on behavior type, subjects of the food sector of economy of the region are identified by means of the Russian term and its English translation so that the first letters of the English determinations formed the word "network" - network: "beginners" (newcomer), "apologists" (enthusiast), "trouble-makers" (troublemaker), "observers" (watcher), "creators" (originator), "moderators" (redactor), "consumers" (keen) [1, 2, 7]. Such formulation of the question, on the one hand, became possible, and, with another, - it was caused by need to conform to requirements of the digital economy which is meaning a new stage (revolution) in development of the economic relations and having the following characteristic features: redistribution of the most part of the created public wealth to the sphere of intellectual and organizational activities; development of remote forms of cooperation, crowd sourcing and outsourcing; a computerization and automation of an overwhelming part of transactions, including, connected with decision making; prompt dynamics of change of client (partner) preferences.

## 3  Results

Development of the food sector of economy of the region is under influence: world trend and nature of its communication with national economy; contextual history of development of a food subsystem of national social and economic system; the crises caused by technology, climatic, geopolitical and other factors [11]. As a rule, the recurrence is caused by internal (endogenous) structural disproportions, however the moment of approach of crisis and its depth essentially depend on external (exogenous) influences. The world trend of economic growth is determined by cycles of Kondratyev who managed with a fine precision to determine peak points of crisis and duration of cycles between them and also to predict reducing duration of cycles of economic development and to predict approach of "era" of permanent crisis [4]. Long-term cycles of Kondratyev (40–60 years) are followed by less long cycles of the Smith (20–22 years), Zhyuglyara (7–11 years), Kitchina (3–4 years). Generalizing approaches to the periodization of economic development which is carried out taking into account the arising crises and the observed cycles the following data on a trend of economic growth and development of the food sector of economy of the region (Table 1) were received:

**Table 1.** A periodization of development of the food sector of economy of the region (on the example of the Voronezh region) in the context of change of technological ways and technical scientific production cycles [9]

| Technological ways | Sequence of technical scientific production cycles | Stages of development of the food sector of economy of the region |
|---|---|---|
| The preindustrial ways which are based on muscular energy of the person and animals and the primitive technologies strengthening its | Primary technical cycle (till 14th century) which is characterized use of natural sources of energy (fire, wind, the muscular strength of the person and animals and use of primitive technologies (wheels, sails, a mail service) | Development of agriculture and cattle breeding for satisfaction of physiological need for food and later as satellite industries of a construction of the Petrovsky fleet |
| First industrial way (1785–1835) | The technical cycle of Renaissance, great geographical discoveries and publishing (the second half of the 13th century – 1750) which was marked by emergence of technologies of cultivation of new crops, enhancement of technologies in production of weapon, publishing, architecture and other spheres | Development of primitive forms of industrial conversion of agricultural raw materials, mainly a phytogenesis as the livestock production because of fixed robbery attacks of raids didn't develop |
| Second industrial way (1830–1890) | The technical cycle (1751–1890) which is characterized by energy use of water, steam, coal and mechanical production in the textile industry, a rail transport, shipbuilding | Cultivation of a set of crops and the organization of the relevant processing industries (flour-grinding, distilling, mead, fish processing, wax-refineries, barmy, acetic, etc.) |

*(continued)*

**Table 1.** (*continued*)

| Technological ways | Sequence of technical scientific production cycles | Stages of development of the food sector of economy of the region |
|---|---|---|
| The third industrial way (1880–1940) | A scientific and technical cycle (1880–1961) – application of the electric power, nuclear energy and energy of hydrocarbons, electrification and automobilization of all types of activity. The dominating science role during this period led to transformation of technical progress in scientific and technical | Lack of necessary means of communication constrained development of trade and directly production of food products. The organization of trade was performed by means of holding fairs. At the end of XIX – the beginning of the 20th century the specialization of the region which determined potential and the prospects of interregional interaction was created |
| The fourth industrial way (1930–1990) | A scientific production cycle (1957–2003) - preferable production of the knowledge-intensive products due to use of electrical, nuclear energy, chemical power sources and energy of hydrocarbons. Orientation to ensuring the sustainable balanced development | Development of production technologies, conversions and storages of agricultural raw materials allowed to create general structure of agro-industrial complex (the 70th of the 20th century) and to decomposirate it subsequently on the subsystems creating prerequisites for industry and regional interaction |
| Digital revolution | | |
| Fifth industrial way (1985–2035) | The first high-technology cycle (1983–2035) - use traditional and alternative energy sources for development of critical technologies. Total application of the information and communication technologies providing all types of the knowledge-intensive and high-technology activities | Forming of research and production clusters taking into account industry accessory of a kernel of a cluster; forming of technological platforms; development of network of competences |
| Post-industrial sixth way | | |
| Seventh socio-humanistic technological way | | |

Source: compiled by the authors

Crisis represents an indivisible element of social evolution and a necessary condition of development of an economic system. R. Nizhegorodtsev provides data that according to methodology of the theory of catastrophic crashes "economic systems are in principle nonequilibrium, and the processes proceeding in them are unstable so they small changes of some, at first sight, not considerable parameters, are capable to cause irreversible changes because of which the system passes into qualitatively new condition" [3].

For accumulation of positive manifestations and consequences of crisis it is necessary to take the preventive measures allowing to predict its origin and to use adequate tools, directed to elimination of a negative impact of crisis.

Observed as a result of imposing of various crises, the coincidence point bearish every time of a crisis cycle phase turns out at the level, above previous what corresponds to a general evolutionary forward positive economic trend. The modern situation in which there was a food sector of Russian regions is extremely difficult, but at the same time is so perspective. The coincidence of a bearish phase of various cycles against the background of total crisis observed until recently in the food sector allows to predict inevitable and essential growth. So, since the beginning of the 2000th the domestic economy entered a bearish wave of the fifth cycle of Kondratyev. However, according to the forecasts made by the Russian scientists by 2020 the bearish phase of a cycle will end, and subjects of regional economy will be able to function at qualitatively new level, "starting" with the line items which appeared above, than a bottom of the previous crisis and having saved up baggage of system innovations [6].

Imposing of various crises and coincidence in time of their bearish phase is the period of activization of managerial impacts from the state, i.e. time of development and implementation of special approaches to state regulation. So, origin of system of state regulation of agro-industrial production happened in the USA at the time of a great depression when there was an objective need of intervention of the state in affairs of agricultural industry as deep and destructive agricultural crisis matched deeper industrial crisis. The experience of state regulation got those years for the next years was enhanced also after World War II which strengthened processes of monopolization in the industry, influence of the state on development of agribusiness purchased the new forms which are beyond anti-recessionary actions.

# 4 Discussion

The analysis of developments of domestic and foreign authors (Polterovich (2017), Nizhegorodtsev (2003), etc.) demonstrates that on condition of effective use of state regulation machineries, in a condition of executive bodies of the government it appears to modify a trajectory of development of a cycle, i.e. to squeeze or extend its duration, to increase amplitude, etc. Especially it concerns medium-term cycles. The efficient instrument of crisis management are investments. As a result of investment support of investment projects, significant for separate segments of regional economy, medium-term cycles purchase the most acceptable V-shaped configuration differing in reducing duration of a bearish phase [3, 5]. The scheme of author's vision of development of the food sector of regional economy taking into account recurrence of its development (Fig. 1) is below included.

On a bearish phase of medium-term cycles discrete nature of the innovations purchasing a form system and basic is shown. This regularity has huge value for

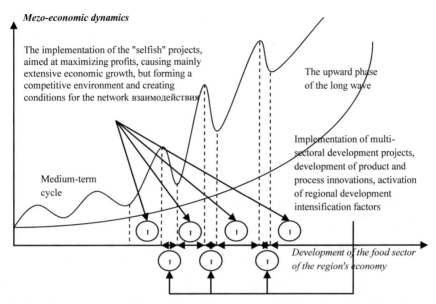

*Mezo-economic dynamics*

The implementation of the "selfish" projects, aimed at maximizing profits, causing mainly extensive economic growth, but forming a competitive environment and creating conditions for the network взаимодействия

The upward phase of the long wave

Implementation of multi-sectoral development projects, development of product and process innovations, activation of regional development intensification factors

Medium-term cycle

Development of the food sector of the region's economy

Designations: I – upward phase of a medium-term cycle; II - a bearish phase of a medium-term cycle

**Fig. 1.** Development of the food sector of economy of the region taking into account recurrence (Source: compiled by the authors)

development of the food sector of the region, in general regional economy and also national economy in the conditions of formation of digital way [10].

Here we will note that mass implementation of new technologies on a bearish phase of medium-term and also long-term cycles happens not in all economy but only in several industries but causes recovery from the crisis of all economy. Empirical confirmation of this phenomenon is experience of creation of the technological platforms causing the intensive growth of regional and national economy necessary for implementation of digital revolution.

## 5 Conclusions

Development of network of competences happens, mainly on a bearish phase of medium-term and long-term economic cycles, creating prerequisites for ensuring transition to qualitatively new level of development of social and economic system (the food sector of economy of the region). According to requirements of digital economy, participants of the local food markets aim at acquisition of new knowledge, experience, resources that is a driving incentive for their interaction as which format the network of competences acts. Functioning within network, its participants face internal and external threats which critical mass can cause crisis approach. However, crises in modern conditions, first, have property of "creative destruction", secondly, are permanent, thirdly, promote growth of economy as the cycle bottom every time appears at higher level. Considering these features, possible it is effective to manage network interaction of participants of the food sector of economy of the region.

# References

1. Doroshenko SV, Shelomentsev AG, Sirotkina NV, Khusainov BD (2014) Paradoxes of the "natural resource curse" regional development in the post-soviet space. Econ Reg 4(40):81–93. https://doi.org/10.17059/2014-4-6
2. Goncharov AY, Sirotkina NV (2015) The mechanism for man-aging a balanced development of regions with dominant economic activities. News High Educ Inst Technol Textile Ind 4(358):35–43
3. Nizhegorodtsev RM (2003) Nonlinear methods of prediction Cata-strophes in complex dynamic systems. In: Burkov VN, Novikov DA (eds) Theory of active systems: proceedings of scientific and practical international conference. Common, vol 1, pp 118–120. M.: IPU ran
4. Pantin VI, Aivazov EA (2012) Cycles of Kondratiev and the evolutionary cycles of the world system: rationale and potential. In: Kondratieff waves: dimensions and prospects, pp 136–155
5. Polterovich VM (2017) Strategies of socio-economic development development: science vs ideology. Probl Theor Econ 1:55–65
6. Sindyakin EM (2015) The impact of economic cycles on regional development modern problems of social-humanitarian Sciences. In: I am a collection of reports of the all-Russian scientific-practical conference (with international participation), pp 163–165
7. Sirotkina NV, Stukalo OG (2015) Clustering of the economic space of a region in the context of the formation of the food industry. Terra Economicus 13(3):99–109. https://doi.org/10.18522/2073-6606-2015-3-99-109
8. Stigler GJ (1961) The economics of information. J Polit Econ 69(3):213–225
9. Stukalo OG (2015) Conditions and prerequisites for the formation of the region's food industry. Bull Kursk State Agric Acad 135:31–38
10. Stukalo OG (2016) Cluster-network development of the model of food industry development in the Voronezh region. In: Collection of articles of scientific and practical international conference innovative processes in the scientific environment, pp 247–249
11. Wiliamson O (1996) Comparative economic organization: the analysis of discrete structural alternatives. Adm Sci Quartely 36(2):269–296

# Analysis and Assessment of Quality of Medical Services in Conditions of Digital Transformation

E. S. Rolbina, E. N. Novikova$^{(\boxtimes)}$, N. S. Sharafutdinova,
O. V. Martynova, and R. M. Akhmetshin

Institute of Management, Economics and Finance,
Kazan Federal University, Kazan, Russia
{rolbinaes,natabell22,olgavl982,renakhmet}@mail.ru,
novelena@list.ru

**Abstract.** The paper is devoted to use of quantitative methods of marketing analysis of the quality of medical services in conditions of implementation of «E-Healthcare» program in the Republic of Tatarstan. Informatization has direct impact on the progress in health care, both in the direction of development of the sphere itself, and monitoring the state of health of patients. Modern technologies change the form of work of medical organizations of various profiles and raise it to the qualitatively new level, including introduction of differentiated methods of diagnosis, treatment and prognosis of diseases into medical practice. The issue of developing electronic services for assessing patient satisfaction with the quality of service for further modeling the relationships of clients with a medical organization is becoming relevant. The authors have analyzed existing methods for assessing the quality of medical service in the service sector and substantiated quantitative studies using statistical methods for assessing the quality of medical services based on customer satisfaction. The measures on modernization of the Unified State Information System (USIS) "E-Healthcare", as well as measures to improve the quality of services provided by the medical organization are proposed.

**Keywords:** Consumer loyalty ·
Customer service satisfaction digitalization of healthcare · Electronic service ·
Service quality assessment

## 1 Introduction

The systematic informatization of healthcare sector of the Republic of Tatarstan has been carried out since 2011 in accordance with the Concept of USIS creation in the healthcare sector and methodological recommendations for equipping medical institutions with computer equipment and software for USIS regional level in the healthcare sector, as well as functional requirements for them. In the Republic of Tatarstan, within the Healthcare Modernization Program, 304.6 million rubles were allocated for informatization of healthcare. The funds were received from the federal budget (199.4 million rubles) and from the budget of the Republic of Tatarstan. These funds were

© Springer Nature Switzerland AG 2020
S. Ashmarina et al. (Eds.): *Digital Transformation of the Economy: Challenges,
Trends and New Opportunities*, pp. 693–702, 2019.
https://doi.org/10.1007/978-3-030-11367-4_67

used to finance the main stage of informatization- supply of equipment and performing the work on creation of local computer networks in the buildings of medical and preventive institutions of the Republic of Tatarstan. In total, the created infrastructure allowed connecting 500 buildings of medical organizations to the State Integrated Telecommunications System. The remaining buildings of medical organizations are scheduled to be connected to the Internet by the end of 2018.

In the framework of healthcare informatization, such projects are developed and operate as:

- The Unified State Information System "E-Healthcare of the Republic of Tatarstan" (hereinafter - USIS "EH RT"), which allows uniting all medical institutions in the single information environment. In USIS "EH RT", electronic medical record of the patient is maintained with display of the history of requests, services rendered to him, diagnoses and results of treatment;
- Medical information system in the State Autonomous Institution of Healthcare "Republican Clinical Oncological Dispensary of the Ministry of Healthcare of the Republic of Tatarstan";
- Information System "Central Archive of Medical Images".

Thanks to USIS "EH RT" during the first 12 months of 2017, the "Appointment to the doctor" service was used by Tatarstanians 14.2 million times (at the peak of incidence in November, 72% of patients made an appointment in electronic form).

Also in 2017, in the framework of the pilot project in Children's Republican Clinical Hospital, the voice input service for filling in medical examination protocols was tested. With the help of this service, 700 protocols of computed tomography were drawn up. Together with the Ministry of Healthcare of the Republic of Tatarstan, the strategy has been worked out to develop healthcare informatization until 2020. So in 2018, the task is to develop such services as:

1. Electronic line for provision of high-tech medical care (IVF, ophthalmology, orthopedics).
2. Electronic disability certificate, which will be put into industrial operation in April 2018.

The relevance of analyzing and researching the quality of medical services is determined by the state task in the field of modernization of the health care system, including the quality of medical services.

The works of Becker et al. [2], Ou et al. [5] reflected the issues of improving service quality and loyalty in the service sector. Issues of quality management of medical services were considered in the works of Sprauer et al. [8] and Yaghoubi et al. [9] were engaged in research in the field of evaluation of the quality of medical services. How artificial intelligence transforms population and personalized health was described in the work Shaban-Nejad et al. [7].

In this study, based on marketing research, the authors analyzed the quality of medical services and developed recommendations for improving the satisfaction and loyalty of clients of medical institution.

The purpose of the study was to evaluate and analyze the level of satisfaction with the quality of care and hospital conditions by patients of the Health Unit of Kazan Federal University и development of recommendations on use of data analysis methods in the «E-healthcare» system of RT.

The authors set research tasks aimed at assessing:

1. satisfaction with waiting conditions in the hospital reception room on the day of hospitalization;
2. satisfaction with the attitude of doctors and nurses;
3. satisfaction with the conditions of the hospital (nutrition, quality of cleaning of premises, lighting of rooms, temperature regime);
4. satisfaction with the organization and conducting the diagnostic studies;
5. patient loyalty and willingness to recommend the hospital to friends and relatives;
6. completeness of information on the official website.

## 2   Materials and Methods

To achieve the objectives in this study, quantitative methods of analyzing the degree of patient satisfaction with the quality of medical care were used. Based on the results of the assessment, it is possible to identify weaknesses and make recommendations for improving the performance of the medical institution. As a method of research, the personal interview was conducted with patients in hospital wards. The statistical sampling method was used. The sample size (n) is determined by the formula 1,

where

N - number of general population (according to data for 2016);

Δ - maximum sampling error (specified by the researcher);

$$n = \frac{t^2\sigma^2 N}{\Delta^2 N + t^2\sigma^2}$$

**Table 1.** Statistical characteristics of the study

| Address of the building | General population | Sample | Number of questionnaires, plan | Number of questionnaires, fact |
|---|---|---|---|---|
| RCH 2 1 a Chekhov | 7255 | 48,845 | 70 | 70 |
| HEMΠ -2, 2 Ershov | 11854 | 48,798 | 60 | 59 |
| RCH 2, 18 Volkov | 4620 | 48,485 | 50 | 48 |
| RCH 2, 51 B. Krasnaya | 3105 | 48,239 | 50 | 48 |
| Total: | 26834 | 194,37 | 230 | 225 |

Source: compiled by the authors.

$\sigma$ - dispersion (variance of estimates);

t - is the value at which F (t) takes the given value.

The study accepted the provision that the sample adequately reflects the population with the probability of 95%, the maximum sampling error is 5%. Thus, the computed sample was 48 respondents. The method of simple random sampling was used. The statistical characteristics of the study are presented in Table 1.

In order to determine the correlation dependence of satisfaction with the quality of patient care and the main characteristics of medical organizations, SPSS computer program was used for statistical processing of data.

# 3   Results

The results of the study are presented in Table 2. 117 people out of 225 respondents are loyal to the object studied (Table 3).

**Table 2.** Consolidated estimates of the main characteristics of hospitals

| Basic characteristics of hospitals | Evaluation of characteristics, % of satisfied[a] |
|---|---|
| Loyalty of patients (prefer RCH-2) | 52 |
| Satisfaction with waiting conditions in the reception room | 98 |
| Evaluations of doctor's courtesy and attentiveness | 99 |
| Evaluations of courtesy and attentiveness of nurse | 97 |
| Evaluation of explanation of investigation results by the doctor | 95 |
| Evaluation of diagnosing by the doctor | 98 |
| Evaluation of medical personnel competence of diagnostic room | 100 |
| Patient satisfaction with nutrition | 98 |
| Satisfaction with observance of silence at night | 95 |
| Satisfaction of patients with quality of cleaning | 100 |
| Satisfaction of patients with the conditions of medical care | 99 |
| Satisfaction of patients with sanitary-hygienic state of the hospital | 97 |

[a]The share of satisfied patients was calculated as the sum of positive ratings ("5" + "4" or "fully satisfied" + "partially satisfied", etc.)

Source: compiled by the authors.

The high degree of loyalty is characterized by the groups of patients under 25 (61%), 25–35 years and 35–45 years (57%). The loyalty of older age groups is somewhat lower: 45–60 years and over 60 years (43%). At the same time, analysis of readiness to recommend RCH-2 to friends and relatives showed that between 13% and 40% of hospital patients answered "I do not know yet", because the treatment was not over yet, and loyalty substantially depends on the result (Fig. 1).

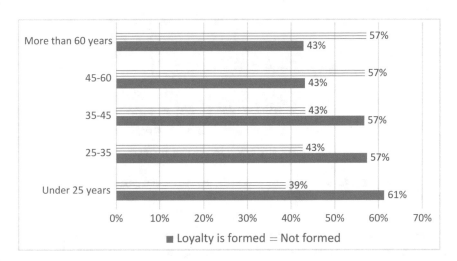

**Fig. 1.** Analysis of loyalty of hospital patients in terms of age groups. Source: compiled by the authors

The shares of loyal RCH-2 social groups of respondents range from 38 to 68%. The average assessment of the proportion of loyal patients is 50%. The analysis of readiness to recommend RCH-2 to friends and relatives showed that 74% were ready to recommend, 22% were not yet ready, probably because the treatment was not completed, and only 4% of patients were not going to recommend RCH-2 to anyone (Fig. 2).

Judging by consolidated estimates of the main characteristics of hospitals, satisfaction - the basis of loyalty is very high and ranges from 95 to 100%, so one should expect that after the end of treatment the proportion of loyal patients will be much higher.

Overall, patients' satisfaction with waiting conditions and the attitude of hospital admission personnel is high - more than 70%, but in categories older than 35 years, dissatisfaction is observed at the level of 2–3%.

The actions of doctors and nurses in the course of anaesthetizing procedures are assessed by all patients extremely highly - at the level of 96% for a set of excellent and good estimates, and only in the group of 60 or more years there is some dissatisfaction, not exceeding 4%.

Compared with previous the estimates of doctor's explanation of the assigned investigations and treatment in general are lower. Obviously, doctors should be advised to give more accessible and detailed explanations, which will help to increase satisfaction with investigations and appointments and increase loyalty.

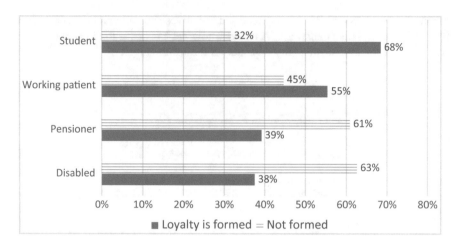

**Fig. 2.** Evaluation of loyalty of RCH-2 patients in terms of social groups. Source: compiled by the authors

Identification of changes in the state of health by the doctor, taking into account complaints and symptoms of the patient's malaise, received not the highest marks, taking into account the patient's complaints indicating that the doctor spends a lot of time to complete the documentation in the computer, it can be assumed that patients need to be paid more attention.

The conditions of waiting for hospitalization in the inpatient hospital completely or partially suit patients, dissatisfaction - 6% is observed in the group of disabled people, for whom special conditions are necessary, insignificant dissatisfaction - 1% is typical for working patients. In general, patient satisfaction with analgesic procedures management is high, dissatisfaction - 7% is observed in the group of people with disabilities and 1% in working patients. Besides students, all categories of patients have complaints about courtesy and attentiveness of nurses, maximum in disabled people (7%), 2% in pensioners and 4% in working patients.

The patients with disabilities, pensioners and working patients would like to have more detailed explanations; their dissatisfaction is in the range from 5% to 8%.

The highest level of communication with the doctor about changes in the health status and complaints of the patient was evaluated by the students. The remaining categories of patients highly evaluated, 3–4% of working and retired people, respectively, were not satisfied enough.

Complete dissatisfaction with food in the hospital is demonstrated at the level of 7% only by the disabled, 2% of the working patients are rather not satisfied, pensioners and students are fully or partially satisfied.

The competence of medical personnel in diagnostic rooms of hospitals is most highly estimated by pensioners: 92%, the minimum evaluation of students: 79%, the average evaluation is −84%.

The conditions for rendering medical care by hospital staff are highly evaluated, there are no "not satisfied" estimates, "more no than yes" estimates are within 2% for pensioners and 1% for working patients. Positive assessments of sanitary-hygienic state of hospitals fluctuate in the range from 94% to 100%, negative - from 6% to 1%. Within the framework of the study, the correlation of 15 basic characteristics of hospitals with patient loyalty was analyzed, of which only 2 - waiting conditions in the reception room and the conditions for rendering medical care correlate with loyalty.

The waiting conditions in the reception room and the conditions for rendering the medical care are characterized by significant but very weak relation with patient loyalty, since the correlation coefficients are significant and the degree of significance is very weak and P-value (probability of error in rejecting the null hypothesis) is less than 0.5 within 95% confidence interval.

**Table 3.** Analysis of correlation relationships

| Relation | Intervals of values | Correlation coefficient | P-value | Indicators |
|---|---|---|---|---|
| Strong or close | More than 0,70 | | | |
| Middle | 0,50 − 0,69 | | | |
| Moderate | 0,30 − 0,49 | | | |
| Weak | 0,20 − 0,29 | | | |
| Very weak | Less than 0,19 | 0,1762 | 0,0288 | Loyalty - waiting conditions in the reception room |
| | | 0,1584 | 0,0498 | Loyalty - the conditions for rendering the medical care |

Source: compiled by the authors.

# 4 Discussion

The quality of any service, including medical one, can be estimated at least by two components:

- as the process of providing the service and the beneficial effect obtained as the result of its provision. In this case, we are talking about quantitative factors that affect the quality of the service, they can be assessed using objective indicators.
- expectations of the client (patient) from the services rendered to him. In this case, it is about how the patient represents the process and the result of the service provided. This can include the whole complex of subjective representations of the client about how the service should be rendered, what service should be offered, what results should be obtained.

Based on the second approach to assessing the quality of services, the assessment of quality of medical services can be carried out on the basis of patient satisfaction surveys. Improving patient satisfaction is one of the long-term priorities of any medical organization, and therefore, continuous monitoring of the quality of medical services can serve as one of criteria for integrated assessment of activities of a medical organization.

Within actualization of the concept of relationship marketing, it should also be noted that the quality of the service is one of the elements of loyalty management system, including the sphere of medical services.

The most attention to quality of service in the service sector was given in the works of the following scientists: Alanazi et al. [1], Calnan and Sanford [3], Rolbina et al. [6]. Theoretical and methodological issues related to quality management of medical services are described in the works Ou et al. [5] and oth. The works of Calnan and Sanford [3] present the results of research in the field of assessing the quality of medical services.

However, the above works do not consider the possibilities of digitization of the healthcare sector and assessing the quality of medical services provided.

In conditions of digitalization of healthcare sphere in the Republic of Tatarstan, one of the elements of the Unified State Information System "Electronic Healthcare of the Republic of Tatarstan" is "Hospital for in-patients" electronic service. Its capabilities include:

- keeping records of hospitalization and stay of patient in medical institutions;
- accounting of all services rendered under the case;
- doctor's prescriptions;
- maintaining a sheet of diagnostic and medical prescriptions, diary entries and epicrises, preoperative concepts and protocols of operations;
- bed capacity monitoring;
- formation of accounts registers;
- generation of reports on the work of health care facilities;
- accounting of paid services.

However, there is no possibility of assessing the quality of medical services provided, which in our opinion is an omission on the part of service developers.

## 5   Conclusions

Proposals and wishes of patients to improve the quality of provided medical services in hospitals:

- to speed up the process of registration in the hospital;
- more of modern equipment;
- more conveniences for wheelchair users, more attention to the sick;
- more modern adhesive plasters should be available;
- nurses should be more responsive to doctors' instructions;
- to improve the supply of medicines, some drugs are not available;
- to tell the diagnosis more accurately;
- very poor attitude of junior medical staff towards the sick. Against the background of their rudeness, cruelty the merit of doctors fades away;

- to measure blood sugar at 18:00 and at 22:00;
- large queue for physical treatment and massage;
- to increase the salaries of staff;
- low number of outlets;
- to add shower cabins;
- well-maintained sanitary facilities should be in every ward;
- insufficient amount of qualitative drinking water;
- to purchase a microwave oven;
- more buckwheat, less cabbage;
- to offer oatmeal with fruit for breakfast;
- more vitamin salads;
- to provide 4 meals a day;
- to provide halal food;
- it is very hot in the wards, it is necessary to regulate the supply of heat;
- to install night lamps in the wards;
- It is overcrowded in the dining rooms for 3 departments;
- food is delicious, but the main dishes could be varied: to give rose hips, fruits, vinaigrette, vegetable salads;
- to increase the amount of protein foods in the diet;
- to post the menu.

The authors of the study developed recommendations for improving the quality of medical care, based on processing the patient survey results:

- The doctors should be recommended to provide more accessible and detailed explanations, especially for older patients and pensioners, which will increase investigation satisfaction and increase loyalty;
- To pay more attention to patients during the doctor's discovery of changes in health status, listening to complaints and identifying symptoms of malaise;
- Despite the fact that silence in the wards at night is almost always observed, patients of older age groups are most sensitive to this parameter, obviously, this requires placing in one ward the close as to the age patients;
- In the group of people with disabilities there is dissatisfaction with the waiting conditions for hospitalization in the hospital waiting room, which suggests that they need special conditions;
- Except for students in all categories of patients there are claims to courtesy and care of nurses, they should be more attentive, especially to older age categories.

There is complete dissatisfaction of the disabled with food in the hospital, most likely it is necessary to take into account the cause of disability in each case and prescribe the appropriate diet.

In order to increase the level of satisfaction and loyalty to the medical organization being studied, it is necessary:

to use the NPS indicator to analyze the level of loyalty and work on increasing it. The NPS indicator in the hospitals of the organization studied was 70 points. The systematic analysis of the level of satisfaction and loyalty will require the construction of marketing information system in the organization (Novikova [4]), which should become a part of the Unified State Information System "E-Healthcare".

It is necessary to automate the collection of information on the quality of medical services provided and accordingly develop a module to assess the level of customer satisfaction with the quality of services provided by medical institutions.

- to use the practice of collecting information from former patients of clinics;
- to introduce incentives and recognize the best employees on the basis of NPS.

# References

1. Alanazi HO, Abdullah AH, Qureshi KN, Ismail AS (2018) Accurate and dynamic predictive model for better prediction in medicine and healthcare. Irish J Med Sci 187(2):501–513
2. Becker E, Fishman EK, Horton KM, Raman SP (2016) Leading in the world of business and medicine: putting the needs of customers, employees and patients. First J Am Coll Radiol 13 (5):576–578
3. Calnan MW, Sanford E (2004) Public trust in health care: the system or the doctor? Quality Saf Health 13:92–97. https://doi.org/10.1136/qshc.2003.00900
4. Novikova E N (2015) Design of a marketing information system. Mediter J Soc Sci 6(1):141–145. https://doi.org/10.5901/mjss.2015.v6n1s3p141
5. Ou YC, Verhoef PC, Wiesel T (2017) The effects of customer equity drivers on loyalty across services industries and firms. J Acad Mark Sci 45(3):336–356. https://doi.org/10.1007/s11747-016-0477-6
6. Rolbina Elena S, Novikova Elena N, Sharafutdinova Natalya S, Martynova OV (2017) The study of consumer loyalty services. Ad Alta-J Interdiscip Res 2:248–253
7. Shaban-Nejad A, Michalowski M, Buckeridge DL (2018) Health intelligence: how artificial intelligence transforms population and personalized health. Npg Digital Med 1, Article number: 53. https://www.nature.com/articles/s41746-018-0058-9. Accessed 2 Mai 2018
8. Sprauer S, Döring D, Herb I, Küchenhoff H, Erhard MH (2012) Behaviour therapy in the practice management of a veterinary practice. Survey among practicing veterinarians and veterinarians specialized in behaviour therapy. Verhaltenstherapie im praxismanagement einer tierarztpraxis: Befragung praktischer und auf verhaltenstherapie spezialisierter tierärzte Tierarztliche Praxis Ausgabe K: Kleintiere - Heimtiere 40(2): 79–86
9. Yaghoubi M, Rafiei S, Alikhani M, Khosravizadeh O (2017) Modeling the brand loyalty of medical services in Iran's military hospitals. Ann Trop Med Public Health 10(4):841–846

# Electronic Interaction in the Sphere of Physical Culture and Sports Services in Russia

A. O. Aleksina, D. V. Chernova[(⊠)], and A. Y. Aleksin

Samara State University of Economics, Samara, Russia
{ms.anastasia1992, antonio407}@mail.ru,
danacher@rambler.ru

**Abstract.** Nowadays, all the fields of the economy are involved in the digital transformation processes, the field of physical culture and sports is not the exception. Thus, the development of a physical culture and sports services sector is impossible without electronic interaction of participants of the market. In different countries they use various ways of cooperation including the methods of electronic economy. The relevance of the topic also depends on systematization of the gained experience in Russia of the transmission from the traditional economy to the digitized one, and as a result digital transformation of the physical culture and sports field. With regard to it, this article is devoted to the definition of the main directions of electronic interaction between public management authorities, the sports organizations and consumers of services. Leading methods of this problem's research are the analysis of Russian literature dedicated to the field under consideration and the synthesis of experience gained by foreign authors, allowing to review the digital transformation forms of physical culture and sports services in an integrated way.

In this research study it is proved that information and communication technologies are effectively applied to coordination of government activity in the field of physical culture and sports development. The authors describe the highlights of functioning of the sports market services and reveal the definition of electronic interaction. They allocate general and particular moments of interaction between the participants of the physical culture and sports market in the electronic environment. The last are considered in three aspects: interaction of the services sector enterprises with consumers during the organization and holding sports activities; during the organization and holding sports and spectacular events; in educational and training process.

The materials of this article represent the practical significance for commercial and non-commercial organizations that provide physical culture and sports services, managing institutions ensuring physical culture and sports development as well as the consumers of physical culture and sports services.

**Keywords:** Digitalization · Digital transformation · Electronic economy ·
Electronic interaction · Information technologies ·
Information and communication technologies · Innovative technologies ·
Physical culture and sports services · Satisfaction of needs of the population

© Springer Nature Switzerland AG 2020
S. Ashmarina et al. (Eds.): *Digital Transformation of the Economy: Challenges,
Trends and New Opportunities*, pp. 703–713, 2019.
https://doi.org/10.1007/978-3-030-11367-4_68

# 1   Introduction

One of the tendencies of modern society development is its digitalization. We are witnesses of intensive penetration of information technologies into all areas of life and human activity [1]. Ensuring interaction of market participants in the electronic environment is resulting in this phenomenon.

Theoretical basis of research are scientific ideas: Bell, Masuda, Toffler [2] has formulated the theory of post-industrial society within the concept of information society. According to Y. Masoud: "Production of information product, but not product material will be the driving force of education and development of society" [3]. Toffler commented that with development and distribution of Worldwide network of ability of individuals and social groups to obtain information considerably have extended [4]. Thus, consumers, sitting in front of the screen, can obtain information practically on any question from any source worldwide.

In the course of the research the theoretical problems of services sector considered in researches of A.M. Babich have been also studied. He considered the conceptual provisions of the economic theory having the all-methodological importance for economy of the sphere of social services [5].

Also Andreff [6] works devoted to questions of economy, management and regulation of services sector of physical culture and sport and research of domestic scientists on this subject have been studied: Arestova L.V., Vasyukova V.A., Kuzmicheva E.V., Frolova O.Y. They claimed that to provide further effective use of sporting facilities, it is necessary to improve the system of sports management and attract of the population to systematic training [7–10].

However, complex consideration of the directions of electronic interaction of participants of the market of physical culture and sports services is not fully presented now, what explains the relevance of the subject of this research and its target orientation.

Objective of this research - to define the main tendencies of digital transformation of services sector of physical culture and sport.

For realization of goal the following problems were solved:

– to define the main forms of electronic interaction of managing institutions and the physical culture and sports organizations;
– to consider possible forms of electronic interaction of the organizations providing services of physical culture and sports with consumers: in the process of trainings and attraction to them; in the course of the organization and holding sports and spectacular actions; in educational training process.

Development of electronic interaction in the market of the considered services sector is connected with intensive solution of management problems, search of perfected forms and methods of training process management on the basis of introduction of scientific approaches, modern computer facilities and mathematical methods [11]. In this regard, increase in efficiency of educational and training process, transition to automated management which consists of development and deployment in all links of organizational structures of modern electronic innovations to become one of the most important conditions of development and improvement of control system in the field of physical culture and sport, improvement of activity of the sports organizations.

Most effectively information technologies (IT) help to coordinate activity of the executive authorities and subordinated organizations which are carrying out work in the field of physical culture and sport. These innovative, breakthrough technologies promote creation of necessary organizational, standard and legal economic and other conditions for satisfaction of needs of the population for physical culture and sport, including means of sport club, sport institutions, enterprises and other forms of work with the population in the community.

Without the use of information and communication technologies it is impossible to carry out and coordinate communications between the scientific organizations [12] and higher educational institutions and also introduce modern methods of management of sports branch.

IT is necessary for realization of personnel policy in the field of physical culture and sport; for the organization of training, professional development, certification of trainers and teachers and executives.

Digital transformation of services of physical culture and sport of participants of the sports services market provides automation of the following management processes: administrative, financial and economic, legislative, personnel and control activity of governing bodies; educational and training processes in the system of special training of athletes and experts; holding national and international competitions and other sporting events.

## 2  Materials and Methods

The complexity of the studied problem and complexity of the object of a research predetermined use of various methods at identification of the main directions of electronic interaction of participants of the market of services of physical culture and sport. Theoretical methods have served as the main methods of research.

At the first stage of studying of the question they used methods of scientific knowledge, the analysis and synthesis of the existing approaches of definition of services and the market of services of physical culture and sport.

At the second stage of the given research of studying of ways of electronic interaction in the market of services of physical culture and sport methods of induction and deduction and generalization were used.

In the third stage of research results of interaction of participants of the market of the considered services sector are processed and analyzed, the purposes for further phase of investigation are formulated.

## 3  Results

Considering possible forms of electronic interaction of participants of the market of sports services, we will allocate the most general moments of interaction of subjects with public authorities of management (see Fig. 1).

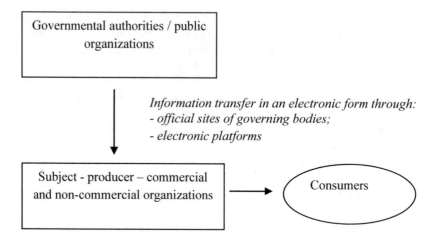

**Fig. 1.** Model of electronic interaction between the main subjects of the rendering process of physical culture and sports services. Source: compiled by authors

The main form of control over activity of the sports organizations and quality of rendering services of physical culture and sport is the system of reports on the basis of which the rating is formed. Now in Russia there is a set of ratings according to quality of rendering services among the commercial organizations. In public institutions since 2014 the active policy on carrying out independent assessment of quality of work of institutions and formation of public ratings of the sports organizations is also conducted. Data on institutions, information on assessment methods, public ratings and also opinions of consumers of social services on their quality are placed in open access on the uniform information www.bus.gov.ru portal and the official sites of subjects of governing body of physical culture and sport.

Thus, on the control system of services of physical culture and sport considerable influence is rendered by the information environment. So information on activity of the governing bodies, the subordinated organizations, market subjects rendering sports services are presented in Internet system. Interaction of society, the governing bodies and organizations providing services of physical culture and sport is also carried out through the information environment. It is shown in placement of information on the official sites of the executive authorities which are carrying out activity in the field of physical culture and sport about sporting events the current year ("the sports calendar"), existence of sports schools and their activity ("sports sections"), electronic forms of the appeal of citizens to governing bodies of the sphere of sports services; possibilities of the publication of reviews of the provided services on the official sites of the organizations; the administrative board of the considered services sector ("a sports personnel") [13].

One more aspect of interaction of public authorities and commercial, non-profit organizations of physical culture and sport consists in the organization of primary activity: issues of licenses, carrying out accreditation by executive authorities for the educational and sports organizations. This interaction originally happens by filling and attachment of a package of documents in an electronic form by the organization of the license applying for receiving for rendering the corresponding services or undergoing accreditation in appropriate authorities, or the organizations.

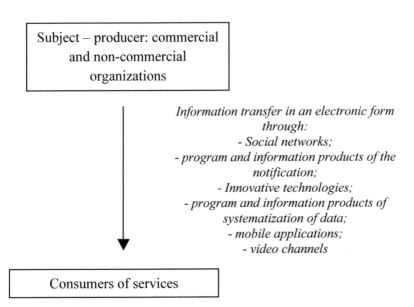

**Fig. 2.** Model of electronic interaction of the physical culture and sports services and consumers of services. Source: compiled by authors

The following form of electronic interaction between commercial and non-profit organizations of physical culture and sport can occur in the form of their cooperation on electronic trading floor (the official site - zakupki.gov.ru). This platform is created for formation of orders for purchase of goods, performance of work, rendering services.

Also we will note one more important form of electronic interaction of public authorities of the power, educational institutions, sports schools for development of physical culture and sport, promotion of a healthy lifestyle in Russia which is carried out due to realization of sports competitions and grants. All lists of necessary conditions, the list and examples of paperwork for participation of the organizations which are carrying out activity in the field of physical culture and sport is submitted on electronic platforms. Following electronic platforms, most popular in the Russian Federation: президентскиегранты.рф, mos.ru, etc. the official sites on realization of competitions and grants in territorial subjects of the Russian Federation, www.minsport.gov.ru, www.kias.rfbr.ru and some other. Submission of applications for participation in competitions and grants in the direction of physical culture and sport, also as well as in other directions, happens in an electronic form.

Having considered the general forms of electronic interaction of public authorities of management, commercial and non-profit organizations, we will pass to private (see Fig. 2).

## 3.1    Electronic Interaction with Consumers in the Course of Teaching Physical Culture and Sport and Attraction to Them

This interaction can be considered from three aspects.

Let's note that in Russia social networks are one of the main sources of informing the population on new tendencies of sports services and the advertising platform [14]. Let's consider percent of the users among the population of Russia registered on the most popular social networks (see Table 1) [15].

**Table 1.** Percentage of the population registered in social networks (2017)

| Social networks | Users (%) |
|---|---|
| Vkontakte | 61 |
| Odnoklassniki | 42 |
| Facebook | 35 |
| Instagram | 31 |

Source: compiled by authors

Use of social networks as way of electronic interaction of personnel of club with consumers is the most convenient form of communication and fast transfer of necessary and important information [16].

Let's consider electronic interaction between clients and administration by the sports organization (on the example of fitness club). So in fitness clubs found broad application in management - system "1C: Fitness club". It represents detailed information on clients of club – the statuses of clients, photos, full contact information, tags, the identity card, contracts, memberships and packages of services, mutual settlements and another. By means of this system the management and personnel can generate forms which allow them to do mailing of SMS notifications on clients [17]. Let's consider electronic interaction directly in the course of rendering sports services.

Control of process of trainings and, therefore, interrelation of the trainer and athlete can be carried out by means of information technologies. For example, there is a technology, the "clever" Adidas MiCoach ball, which is necessary for working off of the technology of blows and power, in improvement of accuracy of bends and transfers. In a ball sensors which determine all above-mentioned parameters are installed, and then on Bluetooth channel transfer them to the computer or the Smartphone. Such technology allows looking through and analyzing a trajectory, blow force, etc. [18].

## 3.2    Electronic Interaction with Consumers in the Course of the Organization and Holding Sports and Spectacular Actions

It's impossible to present the organization and holding sporting events without electronic ensuring refereeing, maintaining protocols of competitions, formation of the current information of each stage of a sporting event [19]. Creation of such information, ensuring fast systematization of protocols of large-scale competitions is impossible without information software products. Also electronic software products are used also at the initial stages of the organization of sports actions.

So, at the organization of the FIFA World Cup of FIFA 2018, the control system on sales of tickets by casual draw for an exception of unfair competition was created. Here by means of electronic interaction of participants of the market and football federation collection of information about fans their initial identification, replenishment of the budget was carried out.

Competitions of high level mean carrying out audio-and video reportings from sporting events, holding of conferences and briefings with TV and radio correspondents, reporters of media in audiences and on sport objects; the presentation of competitions in the Internet, creation of the Websites of competitions; radio - and television reports on sporting events. These ways of electronic impact on the population are carried out for promotion of physical culture and sport and ensuring leisure.

Let's consider a percentage ratio of obtaining information by the population from mass media (see Table 2) [20].

**Table 2.** Ways of obtaining sports information by the population of the Russian Federation (2017)

| Mass media | Percentage |
|---|---|
| TV | 64% |
| Internet | 57,50% |
| Video in internet | 38,80% |
| Radio | 18,40% |
| Newspaper | 15,10% |
| Social network | 14,80% |
| Website of TV channel | 5,30% |

Source: compiled by authors

Thus, the greatest percent of audience is tied to video content: 64% learn about sport from telecasts, 38,8% – from video on the Internet, 5,3% – on the website of TV channel.

Proceeding from the obtained information, we will consider dynamics of volumes of sports broadcastings per television channels in Russia (see Fig. 3).

According to Fig. 3 the tendency of decrease in volumes of sports broadcastings per television is observed since 2010. As the reason of it decrease in display of sports broadcasts on the most popular TV channels served: «Perviy», «Rossiya1», «Rossiya 24». It should be noted that since launch of TV channel «Match TV» in November, 2015, the share of TV viewing of sports broadcasts raised [21].

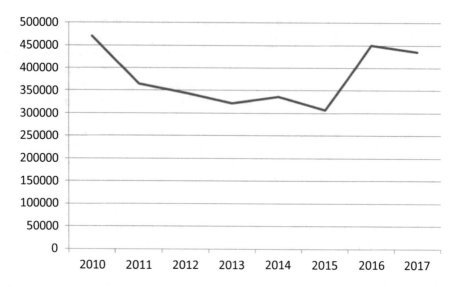

**Fig. 3.** Dynamics of volumes of sports broadcastings (min.) per television channels in Russia (2010–2017). Source: compiled by authors

### 3.3    Electronic Interaction with Consumers in Educational and Training Process

Such interaction is possible by means of various sports mobile applications which were created by IT specialists and the leading trainers for independent occupations of the population [22, 23].

The complex of trainings, the program of food and a day regimen which are formed under biological and physiological parameters of the person which performance will lead to a goal of the consumer are described in such appendices. Results of training process are fixed in the application [24, 25].

Other way of training in complexes of sports exercises and occupations physical culture and sport out of the sports organization, it is possible by means of remote rendering services. The most widespread information channel: www.youtube.com.

## 4    Discussion

Services of this branch include the sports and improving, educational and training, competitive activity and activities for their providing meaning the maintenance of network of sports constructions, the organization of service of their visitors during the occupations, the organization and providing sports competitions and spectacular actions, a professional training of shots, carrying out scientific research, trade, a hire and repair of the sports equipment and stock, service of sports insurance [26].

The sports service is the activity directed to satisfaction of need of the client for preservation and maintenance of health, spiritual and physical development of the personality, increase in sports professional skill.

Thus, "the market of services of physical culture and sport" - the difficult mechanism of regulation of set of the social and economic relations arising in an organized segment of the sports sphere for achievement of the certain result directed to preservation and maintenance of health, spiritual and physical development of the personality, increase in sports professional skill [27].

Development of the sphere of sports services is impossible without electronic interaction of participants of the market now.

Most of modern scientists define "electronic interaction" as process of influence of objects (subjects) at each other, for optimization of process of rendering of services by means of technical means, the Internet and modern mass media.

We believe that further studying of opportunities of electronic interaction of participants of the market of sports services is necessary and will be the main incitement of development of this branch.

# 5  Conclusions

Due to penetration and development of the Internet, electronic interaction of participants of the physical culture and sports market has become the main development of this branch. Information and communication technologies changed social life. They promote population involvement in physical culture and sports. Thus, development of sports applications and gadgets exert considerable impact on development of mass sport while development of sports innovative equipment and medical electronic technologies develop elite sport [28].

Websites are one of the most irreplaceable attributes of most companies nowadays because they play a role of the professional and qualified representative in the Internet which is available 24 h a day to all users at any place in the world.

Thus, electronic interaction in the considered services sector will allow us to establish internal and external connections of market participants by means of information and communication technologies and also provide opportunities [29]:

- to improve interaction between public authorities of government and sports organizations;
- to increase efficiency of information exchange;
- to improve planning and management in physical culture and sport organizations;
- to provide objective information regarding public opinion on the carried-out work of development of physical culture and sport in Russian Federation;
- to form public opinion which is favorable for the government.

High rates of penetration of information technologies into the area of physical culture and sport offer huge development prospects. Developed market and competitiveness of the sports organizations make services more available.

# References

1. Betz F (2011) Managing technological innovation—competitive advantage from change, 3rd edn. Wiley, Hoboken
2. Bell D (1973) The coming of post-industrial society, New York
3. Masuda Y (1980) The information society as post-industrial society, Tokyo
4. Toffler A (1980) The third wave. Morrow, New York
5. Babich AM (2014) Theoretical and methodological bases of the social sphere economy. Soc Policy Soc Partnersh № 12:5–17
6. Andreff V (2006) Television and sport. Domestic notes, № 6:168–178
7. Aristova LV (2014) Quality and competitiveness: strategy and priorities of increase in effective management of sporting facilities. Sport: economy, law, management. № 4:10–12
8. Vasyukova VA, Vorobyova IV, Kovalenko NP, Kushtova MH (2016) Innovative activity in physical culture and sports sphere as integral part of national economy element. The azimuth of scientific researches: economy and management T. 5. № 3(16):68–73
9. Kuzmicheva EV (2017) Global trends – the new ideas in sports management. The theory and practice of physical culture. № 10:34
10. Frolova OY, Horosheva TA (2015) Sports industry: commercial activity and policy of the Ministry of Sports of the Russian Federation. Vestnik NGIEI. № 9(52):P. 69–72
11. Chen JL (2012) The synergistic effects of IT-enabled resources on organizational capabilities and firm performance. Inf Manag 49(3/4):142–150
12. Luftman J, Zadeh HS, Derksen B, Santana M, Rigoni EH, Huang Z (2013) Key information technology and management issues 2012–2013: an international study. J Inf Technol 28 (4):354–366
13. Ministry of Sport of the Russian Federation. Development of physical culture and sports. http://www.minsport.gov.ru. Accessed 1 Oct 2018
14. Sariev SS, Chernova DV (2014) Social networks as an instrument of marketing communication in commercial activity. – Vestnik SSEU. № 11(121)
15. Chaffey D (2018) Global social media research summary. https://www.smartinsights.com. Accessed 10 Oct 2018
16. Aleksina AO, Chernova DV, Ivanova LA, Aleksin AY, Piskaykina MN (2018) The main directions in informatization of the sphere of physical culture and sports services. In: Perspectives on the use of New Information and Communication Technology (ICT) in the modern economy. Advances in Intelligent Systems and Computing. Springer,Cham, pp 473–479
17. https://solutions.1c.ru. Accessed 10 Oct 2018
18. Chen W (2013) Uwb technology application in football training load monitoring system development China sports science and technology, p 49
19. Chuang Yan (2014) Science, technology and development of competitive sports. Sport 2014:6
20. Mediascope: sports in Russia. http://mediascope.net. Accessed 10 Oct 2018
21. Ivan Ugrumov: the way Russia watches sports on TV. Research by Mediascope. http://sport-connect.ru. Accessed 5 Oct 2018
22. Middelweerd A, Mollee JS, van der Wal C, Brug J, Te Velde SJ (2014) Apps to promote physical activity among adults: a review and content analysis. Int J Behav Nutr Phys Act 11 (1):97
23. West JH, Hall PC, Hanson CL, Barnes MD, Giraud-Carrier C, Barrett J (2012) There's an app for that: content analysis of paid health and fitness apps. J Med Internet Res 14(3):e72

24. Dallinga JM, Mennes M, Alpay L, Bijwaard H, de la Faille-Deutekom, Marije Baart (2015) App use, physical activity and healthy lifestyle: a cross sectional study. BMC Public Health 15(1):833
25. Kranz M, Möllerb A, Hammerla N, Diewald S, Plötz T, Olivier P, Roalter L (2013) The mobile fitness coach: towards individualized skill assessment using personalized mobile devices. Pervasive Mob Comput 9(2):203–215
26. Kosogortsev VI (2016) Approaches to the classification of physical culture and sports services. Russian entrepreneurship, vol 17, № 4:573–584
27. Breedveld K, Scheerder J, Borgers J (2015) Running across Europe: the way forward. In: Scheerder J, Breedveld K, Borgers J (eds) Running across Europe: the rise and size of one of the largest sport markets. Palgrave Macmillan, Basingstoke, p 241–264
28. Grechishnikov AL, Levin AI (2016) Development of mass sports as the object management activity. Vestnik PAGS № 5(56)
29. Obozhina DA (2017) Management of physical culture and sports organization: tutorial. In: Obozhina DA (ed) and Sc. Min-y, Ural fed. uni. —Ekaterinburg : Ural uni. publ. 76 p

# Correction to: Gaps in the System of Higher Education in Russia in Terms of Digitalization

S. I. Ashmarina, E. A. Kandrashina, A. M. Izmailov,
and N. S. Mirzayev

**Correction to:**
**Chapter "Gaps in the System of Higher Education in Russia in Terms of Digitalization" in: S. Ashmarina et al. (Eds.):**
*Digital Transformation of the Economy: Challenges,*
*Trends and New Opportunities,*
**https://doi.org/10.1007/978-3-030-11367-4_43**

The original version of the book was inadvertently published with incorrect last author's name and e-mail address as "N. G. Mirzayev" and "mirzoev@mail.ru", respectively, which have been now corrected as "N. S. Mirzayev" and "mirzoev.n@mail.ru" in Chapter 43. The erratum chapter and the book have been updated with the changes.

The updated version of this chapter can be found at
https://doi.org/10.1007/978-3-030-11367-4_43

# Author Index

© Springer Nature Switzerland AG 2020
S. Ashmarina et al. (Eds.): *Digital Transformation of the Economy: Challenges, Trends and New Opportunities*, pp. 715–717, 2019.
https://doi.org/10.1007/978-3-030-11367-4

Printed in the United States
By Bookmasters